# INORGANIC BATTERY MATERIALS

# EIBC Books

*Application of Physical Methods to Inorganic and Bioinorganic Chemistry*
Edited by Robert A. Scott and Charles M. Lukehart
ISBN 978-0-470-03217-6

*Nanomaterials: Inorganic and Bioinorganic Perspectives*
Edited by Charles M. Lukehart and Robert A. Scott
ISBN 978-0-470-51644-7

*Computational Inorganic and Bioinorganic Chemistry*
Edited by Edward I. Solomon, R. Bruce King and Robert A. Scott
ISBN 978-0-470-69997-3

*Radionuclides in the Environment*
Edited by David A. Atwood
ISBN 978-0-470-71434-8

*Energy Production and Storage: Inorganic Chemical Strategies for a Warming World*
Edited by Robert H. Crabtree
ISBN 978-0-470-74986-9

*The Rare Earth Elements: Fundamentals and Applications*
Edited by David A. Atwood
ISBN 978-1-119-95097-4

*Metals in Cells*
Edited by Valeria Culotta and Robert A. Scott
ISBN 978-1-119-95323-4

*Metal-Organic Framework Materials*
Edited by Leonard R. MacGillivray and Charles M. Lukehart
ISBN 978-1-119-95289-3

*The Lightest Metals: Science and Technology from Lithium to Calcium*
Edited by Timothy P. Hanusa
ISBN 978-1-118-70328-1

*Sustainable Inorganic Chemistry*
Edited by David A. Atwood
ISBN 978-1-118-70342-7

*Metalloprotein Active Site Assembly*
Edited by Michael K. Johnson and Robert A. Scott
ISBN 978-1-11915983-4

*The Heaviest Metals: Science and Technology of the Actinides and Beyond*
Edited by William J. Evans and Timothy P. Hanusa
ISBN 978-1-11930409-8

*Inorganic Battery Materials*
Edited by Hailiang Wang and Boniface P.T. Fokwa
ISBN 978-1-11943199-2

# Encyclopedia of Inorganic and Bioinorganic Chemistry

The Encyclopedia of Inorganic and Bioinorganic Chemistry (EIBC) was created as an online reference in 2012 by merging the Encyclopedia of Inorganic Chemistry and the Handbook of Metalloproteins. The resulting combination proves to be the defining reference work in the field of inorganic and bioinorganic chemistry. The online edition is regularly updated and expanded. For information see:

**www.wileyonlinelibrary.com/ref/eibc**

# INORGANIC BATTERY MATERIALS

*Editors*

**Hailiang Wang**
*Yale University, New Haven, CT, USA*

**Boniface P.T. Fokwa**
*University of California, Riverside, CA, USA*

WILEY

This edition first published 2020


*Registered Office*
John Wiley & Sons Ltd, The Atrium, Southern Gate, Chichester, West Sussex, PO19 8SQ, UK

*Editorial Office*
The Atrium, Southern Gate, Chichester, West Sussex, PO19 8SQ, UK

For details of our global editorial offices, customer services, and more information about Wiley products visit us at www.wiley.com.

Wiley also publishes its books in a variety of electronic formats and by print-on-demand. Some content that appears in standard print versions of this book may not be available in other formats.



***Library of Congress Cataloging-in-Publication Data has been applied for.***

978-1-119-43199-2 (hardback)

Cover Design: Wiley
Cover Images: © Hailiang Wang, © 2019 International Union of Pure and Applied Chemistry

Set in 10/12pt TimesNewRomanMTStd by SPi-Global, Chennai, India

10 9 8 7 6 5 4 3 2 1

MIX
Paper from responsible sources
FSC www.fsc.org FSC® C013604

# Encyclopedia of Inorganic and Bioinorganic Chemistry

# Contents

# Contributors

**Hye Ryung Byon**
*Korea Advanced Institute of Science and Technology (KAIST), Daejeon, Republic of Korea*
- Lithium Oxygen Battery

**Shaowei Chen**
*South China University of Technology, Guangzhou, China*
*University of California, Santa Cruz, CA, USA*
- Structural Engineering of Cathode Materials for Lithium–Sulfur Batteries

**Jaephil Cho**
*Ulsan National Institute of Science and Technology (UNIST), Ulsan, Republic of Korea*
- How to Maximize the Potential of Zn–Air Battery: Toward Acceptable Rechargeable Technology with or without Electricity

**Francesco Ciucci**
*Hong Kong University of Science and Technology, Hong Kong, China*
- Modeling Solid State Batteries

**Camden DeBruler**
*Utah State University, Logan, UT, USA*
- Redox-Active Inorganic Materials for Redox Flow Batteries

**Kaoru Dokko**
*Yokohama National University, Yokohama, Japan*
- Lithium Oxygen Battery

**Elizabeth H. Driscoll**
*University of Birmingham, Birmingham, UK*
- Na-Ion Batteries: Positive Electrode Materials

**Laura L. Driscoll**
*University of Birmingham, Birmingham, UK*
- Na-Ion Batteries: Positive Electrode Materials

**Bin Fan**
*South China University of Technology, Guangzhou, China*
- Structural Engineering of Cathode Materials for Lithium–Sulfur Batteries

**Bo Hu**
*Utah State University, Logan, UT, USA*
- Redox-Active Inorganic Materials for Redox Flow Batteries

**Liangbing Hu**
*University of Maryland, College Park, MD, USA*
- Solid-State Electrolyte

**Maowei Hu**
*Utah State University, Logan, UT, USA*
- Redox-Active Inorganic Materials for Redox Flow Batteries

**Sooyeon Hwang**
*Center for Functional Nanomaterials, Brookhaven National Laboratory, Upton, NY, USA*
- TEM Studies on Electrode Materials for Secondary Ion Batteries

**Jang-Soo Lee**
*Ulsan National Institute of Science and Technology (UNIST), Ulsan, Republic of Korea*
- How to Maximize the Potential of Zn–Air Battery: Toward Acceptable Rechargeable Technology with or without Electricity

**Feng Li**
*University of Idaho, Moscow, ID, USA*
- S Electrode Materials

**Ligui Li**
*South China University of Technology, Guangzhou, China*
- Structural Engineering of Cathode Materials for Lithium–Sulfur Batteries

**Siyuan Li** *Zhejiang University, Hangzhou, China*
- Lithium Metal Anode

**Xiaolin Li** *Pacific Northwest National Laboratory, Richland, WA, USA*
- Silicon-Based Anodes for Advanced Lithium-Ion Batteries

**Feng Lin** *Virginia Tech, Blacksburg, VA, USA*
- Surface Chemistry of Alkali-Ion Battery Cathode Materials

**T. Leo Liu** *Utah State University, Logan, UT, USA*
- Redox-Active Inorganic Materials for Redox Flow Batteries

**Yingying Lu** *Zhejiang University, Hangzhou, China*
- Lithium Metal Anode

**Ziheng Lu** *Hong Kong University of Science and Technology, Hong Kong, China*
- Modeling Solid State Batteries

**Jian Luo** *Utah State University, Logan, UT, USA*
- Redox-Active Inorganic Materials for Redox Flow Batteries

**Wei Luo** *Tongji University, Shanghai, China*
- Solid-State Electrolyte

**Laura C. Merrill** *University of Notre Dame, Notre Dame, IN, USA*
- Multivalent Metallic Anodes for Rechargeable Batteries

**Joohyuk Park** *Ulsan National Institute of Science and Technology (UNIST), Ulsan, Republic of Korea*
- How to Maximize the Potential of Zn–Air Battery: Toward Acceptable Rechargeable Technology with or without Electricity

**Jelena Popovic** *Max Planck Institute for Solid State Research, Stuttgart, Germany*
- Chemistry of Soft Matter Battery Electrolytes

**Muhammad M. Rahman** *Virginia Tech, Blacksburg, VA, USA*
- Surface Chemistry of Alkali-Ion Battery Cathode Materials

**Philip N. Ross** *Lawrence Berkeley National Laboratory, Berkeley, CA, USA*
- Solid Electrolyte Interphase in Lithium-Based Batteries

**Jennifer L. Schaefer** *University of Notre Dame, Notre Dame, IN, USA*
- Multivalent Metallic Anodes for Rechargeable Batteries

**Divya Sehrawat** *University of New South Wales (UNSW Sydney), Sydney, Australia*
- Solid State and Materials Chemistry for Sodium-Ion Batteries

**Reza Shahbazian-Yassar** *University of Illinois at Chicago, Chicago, IL, USA*
- Application of In Situ Electrochemical-Cell Transmission Electron Microscopy for the Study of Rechargeable Batteries

**Neeraj Sharma** *University of New South Wales (UNSW Sydney), Sydney, Australia*
- Solid State and Materials Chemistry for Sodium-Ion Batteries

**Feifei Shi** *Stanford University, Stanford, CA, USA*
- Solid Electrolyte Interphase in Lithium-Based Batteries

**Peter R. Slater** *University of Birmingham, Birmingham, UK*
- Na-Ion Batteries: Positive Electrode Materials

**Junhua Song** *Pacific Northwest National Laboratory, Richland, WA, USA*
- Silicon-Based Anodes for Advanced Lithium-Ion Batteries

**Jennifer H. Stansby** *University of New South Wales (UNSW Sydney), Sydney, Australia; Australia Nuclear Science and Technology Organisation, Sydney, Australia*
- Solid State and Materials Chemistry for Sodium-Ion Batteries

**Dong Su** *Center for Functional Nanomaterials, Brookhaven National Laboratory, Upton, NY, USA*
- TEM Studies on Electrode Materials for Secondary Ion Batteries

**Xiaowei Teng** *University of New Hampshire, Durham, NH, USA*
- Electrode and Electrolyte Interaction in Aqueous Electrochemical Energy Storage

**Morgan L. Thomas** *Yokohama National University, Yokohama, Japan*
- Lithium Oxygen Battery

**Ting Hei Wan** *Hong Kong University of Science and Technology, Hong Kong, China*
- Modeling Solid State Batteries

**Nan Wang** *South China University of Technology, Guangzhou, China*
- Structural Engineering of Cathode Materials for Lithium–Sulfur Batteries

**Masayoshi Watanabe** *Yokohama National University, Yokohama, Japan*
- Lithium Oxygen Battery

**Raymond A. Wong** *Korea Advanced Institute of Science and Technology (KAIST), Daejeon, Republic of Korea*
- Lithium Oxygen Battery

**Wenda Wu** *Utah State University, Logan, UT, USA*
- Redox-Active Inorganic Materials for Redox Flow Batteries

**Zhaojun Xie** *Nankai University, Tianjin, China*
- $Li–CO_2$ Batteries

**Jixiang Yang** *Zhejiang University, Hangzhou, China*
- Lithium Metal Anode

**Wanli Yang** *Advanced Light Source, E O Lawrence Berkeley National Laboratory, Berkeley, CA, USA*
- Synchrotron-Based Soft X-Ray Spectroscopy for Battery Material Studies

**Wentao Yao** *Michigan Technological University, Houghton, MI, USA*
- Application of In Situ Electrochemical-Cell Transmission Electron Microscopy for the Study of Rechargeable Batteries

**Chuang Yu** *Delft University of Technology, Delft, The Netherlands*
- Solid-State Electrolyte

**Jingping Yu** *South China University of Technology, Guangzhou, China*
- Structural Engineering of Cathode Materials for Lithium–Sulfur Batteries

**Shuaibo Zeng** *South China University of Technology, Guangzhou, China*
- Structural Engineering of Cathode Materials for Lithium–Sulfur Batteries

**Ji-Guang Zhang** *Pacific Northwest National Laboratory, Richland, WA, USA*
- Silicon-Based Anodes for Advanced Lithium-Ion Batteries

**Jun Zhao** *South China University of Technology, Guangzhou, China*
- Structural Engineering of Cathode Materials for Lithium–Sulfur Batteries

**Zhen Zhou** *Nankai University, Tianjin, China*
- $Li–CO_2$ Batteries

# Series Preface

The success of the *Encyclopedia of Inorganic Chemistry* (EIC), pioneered by Bruce King, the founding Editor in Chief, led to the 2012 integration of articles from the *Handbook of Metalloproteins* to create the newly launched *Encyclopedia of Inorganic and Bioinorganic Chemistry* (EIBC). This has been accompanied by a significant expansion of our Editorial Advisory Board with international representation in all areas of inorganic chemistry. It was under Bruce's successor, Bob Crabtree, that it was recognized that not everyone would necessarily need access to the full extent of EIBC. All EIBC articles are online and are searchable, but we still recognized value in more concise thematic volumes targeted to a specific area of interest. This idea encouraged us to produce a series of EIC (now EIBC) Books, focusing on topics of current interest. These will continue to appear on an approximately annual basis and will feature the leading scholars in their fields, often being guest coedited by one of these leaders. Like the Encyclopedia, we hope that EIBC Books continue to provide both the starting research student and the confirmed research worker a critical distillation of the leading concepts and provide a structured entry into the fields covered.

The EIBC Books are referred to as *spin-on* books, recognizing that all the articles in these thematic volumes are destined to become part of the online content of EIBC, usually forming a new category of articles in the EIBC topical structure. We find that this provides multiple routes to find the latest summaries of current research.

I fully recognize that this latest transformation of EIBC is built on the efforts of my predecessors, Bruce King and Bob Crabtree, my fellow editors, as well as the Wiley personnel, and, most particularly, the numerous authors of EIBC articles. It is the dedication and commitment of all these people that are responsible for the creation and production of this series and the "parent" EIBC.

*Robert A. Scott*
University of Georgia
Department of Chemistry

*August 2019*

# Volume Preface

Rechargeable batteries, converting between chemical energy and electrical energy, are not only a currently dominant power source for portable electronics, cordless tools, electrochemical energy storage, and electric vehicles but are also closely related to the future of humanity's efficient and balanced energy consumption. The exponential growth of energy demand in our modern society has made rechargeable batteries one of the essential ingredients to satisfy our quest for more and more energy. Equally important is the fact that rechargeable batteries are playing a crucial role in climate change by helping to advance green energy technologies such as solar and wind. Consequently, it is urgent to build *better* batteries by making them more energy intensive, faster, more durable, safer, and cheaper. While incremental improvements of available batteries are being successfully pursued in the industry by design or architecture modifications, exploring and understanding new chemistries and materials will be necessary to address these challenges. This has stimulated extensive research on designing, synthesizing, and characterizing inorganic materials (broadly defined) for many battery technologies.

*Inorganic Battery Materials*, a newly added volume to *The Encyclopedia of Inorganic and Bioinorganic Chemistry*, is the product of excellent contributions from prominent experts from Asia, Australia, Europe, and North America. Selective overviews and critical perspectives are provided for battery technologies from the state-of-the-art lithium ion battery chemistry to more energy-aggressive chemistries involving lithium metal, sulfur, and oxygen; abundant metals including sodium, magnesium, calcium, and aluminum; aqueous battery chemistry; redox flow battery chemistry; solid state battery chemistry; and environmentally beneficial carbon dioxide battery chemistry. In addition to battery materials, extensive discussion is also dedicated to the development and employment of advanced structural characterization techniques, including synchrotron-based X-ray spectroscopy, vibrational spectroscopy, and transmission electron microscopy, to understand the structure–property correlations and working/failing mechanisms.

We would like to thank the editorial staff at Wiley for their expert guidance and assistance throughout the entire project. It could not have been completed without their steadfast help.

*Hailiang Wang*
Yale University,
New Haven, CT, USA

*Boniface P.T. Fokwa*
University of California,
Riverside, CA, USA

*August 2019*

# PART 1
# Chemistry of Li-Ion Battery Materials

# Silicon-Based Anodes for Advanced Lithium-Ion Batteries

**Junhua Song, Xiaolin Li and Ji-Guang Zhang**

*Pacific Northwest National Laboratory, Richland, WA, USA*



## 1 INTRODUCTION

Since the commercial introduction of lithium-ion batteries (LIBs) in 1990s by Sony Corporation, the quest for high energy density rechargeable batteries has become burgeoningly active.[1–3] In the past decades, the energy density has taken a threefold increment from the first-generation LIB (80 Wh kg$^{-1}$) to today's 240 Wh kg$^{-1}$.[4] The advancement of battery performance has been driven by several technical breakthroughs, such as the discovery of layered transition-metal oxide cathode material, $LiCoO_2$, in the 1980s by John Goodenough,[5,6] and using carbonaceous anode materials as Li-ion hosts. With the continuous efforts on the optimization of materials processing, improvements on the inactive components (including electrolyte, separator and binder, conductive additive, and current collector) and engineering progress in manufacturing, the ceiling of Li-ion's energy density is approaching its theoretical limits with the general concept of insertion anode and cathode pair. While technological achievements play a key role in boosting the LIBs' performance, the consumer demands, growing consciousness of sustainability, and pursue of cost-effectiveness in production have also driven the research for further improvements of the already powerful Li-ion chemistry. Nowadays, the surging number of the electrical vehicles (EVs) is gradually taking the dominating role in consuming high energy density LIBs, while their demands were marginally low in 2010 compared to those used in computers, consumer electronics, and cameras, also known as the 3C market.[7,8] However, the growing trend of electrifying the transportation systems requires a leap in the LIB technology to significantly extend the driving distance of EVs so that consumers can be freed from the range anxiety. Similar to the technical breakthroughs before the birth of Li-ion chemistry, finding a high-performance anode material holds a great promise meeting the increasing demand on LIBs' energy density, reliability, and low cost. The state-of-the-art graphite anode has been the top choice in today's LIBs with its low charge/discharge potential, moderate capacity of 372 mAh g$^{-1}$, and low manufacturing cost (Table 1). Among all the alternative anode materials, silicon (Si) is considered the most promising candidate for the next-generation high-energy LIBs for several reasons: (i) Si has a theoretical specific capacity of 4200 mAh g$^{-1}$ and volumetric capacity of 9786 mAh cm$^{-3}$ upon full lithiation ($Li_{4.4}Si$), which are ~10 times larger than graphite[9]; (ii) the lithiation/delithiation potential averages at ~0.4 V versus Li, which ensures a relatively high operation voltage in full cell configuration as well as prevents safety concerns caused by lithium plating at low anode voltage[10]; (iii) the potentially low toxicity, low cost, environmental friendliness, and natural abundance of Si render it the most suitable substitution to graphite anode. Nevertheless, the promises never come without a challenge. The efforts to make Si anodes practically adopted in LIBs

*Inorganic Battery Materials*. Edited by Hailiang Wang and Boniface P.T. Fokwa
 ISBN 978-1-119-43199-2

**Table 1** Comparison of the state-of-the-art anode materials for LIBs

| Anode | Graphite | Silicon | LTO | Lithium |
|---|---|---|---|---|
| Gravimetric capacity ($mAh\,g^{-1}$) | 372 | 4200 | 175 | 3862 |
| Volumetric capacity ($mAh\,cm^{-3}$) | 837 | 9786 | 613 | 2047 |
| Volume change (%) | 12 | 420 | 1 | ∞ |
| Average potential vs Li (V) | 0.05 | 0.4 | 1.6 | 0 |
| Cost | Low | Low | Medium | High |

have gone through a long research development process that is not completed yet.

Early trials in the 1970s using Li–Si alloy in electrochemical cells operated at elevated temperature >400 °C and suffered poor reversibility at room temperature owing to the large volume change.[11,12] Since then, tackling the structural damage and improving its room temperature performance have been the research focus of using Si-based anode in LIBs. In the early 1990s, Dahn *et al.* examined the $Si_xO_yC_z$ glass composite derived from Si-containing polymers. By tuning the Si content in the final product, his group was able to achieve a reversible capacity of 900 $mAh\,g^{-1}$ at room temperature in half-cell configuration.[13] Their results marked the possibility of replacing graphite with high-capacity Si as anode in LIBs. As materials science research entered the nano era, it also encouraged the exploration of using nano-sized Si as anode materials in the late 1990s. To fabricate nanoscale Si with controllable size, Si film was made with chemical vapor deposition (CVD) and able to achieve a capacity as high as 1000 $mAh\,g^{-1}$, despite the cyclability being quite limited due to the structural breakdown associated with huge volume change of the thin-film electrode.[14] Later at the beginning of the twenty-first century, several major material breakthroughs have taken place in the Si anode research. One is to construct composite Si alloys ($FeSi_2$, $NiSi_2$, and $BaSi_2$), which effectively extends the cyclic stability to 50 cycles.[15–17] The new compositing methods have also enabled nano Si particles to stably operate over 100 cycles.[18] More encouragingly, modification made on current collectors rendered the Si thin film (50 nm in thickness) anode cycle over 1000 cycles with reversible capacities as high as ~2000 $mAh\,g^{-1}$. Up to this point, the general challenges in applying Si anode in LIBs are gradually self-revealed.

Unlike the small volume change (~10%) during lithium uptake in graphite, Si expands up to 420% upon full lithiation.[19] The drastic volume swing stresses the integrity of the Si materials and brings destructive outcomes to the battery performance. The major problems associated with the Si anode during cycling are the following: (i) on the material level, the cracking and pulverization of Si particle sacrifice the electrical contact between particles and expose reactive new surface that consumes electrolyte in the following cycles; (ii) on the electrode level, delamination and electrical disconnections occur under huge interfacial stress generated; and (iii) on the cell level, the electrically segregated Si particles decrease the capacity utilization, facilitating thick insulating interphase build-up and hence leading to the rapid capacity fade (Figure 1). All of these issues accelerate electrolyte consumption rate, introduce safety hazard, and negatively affect the cell operation. Despite as challenging as it is, tremendous efforts have been made to improve the Si anode performance by designing beneficial nanostructures, improving interfacial properties, modifying inactive components, and optimizing testing conditions. A short summary of the challenges of Si materials and the corresponding solutions is listed in Table 2. Assisted by the knowledge gained in manipulating atoms and molecules at nanoscale, the Si anode research has entered an era with explosive progress on cell performance. To mitigate volume expansion, nanostructured materials with intentionally designed void space were designed and showed promising performance. The major consequence of volume expansion, that is, particle pulverization, was effectively tackled by the applying small size nano Si and constructing nanostructure with intentionally left void space. These nano strategies will be introduced and detailed in the Section 2 of this chapter. Leveraging the reduced particle swelling, the electrode-level engineering against swelling was achieved by selecting a right binder material, which was designed to tightly hold the active particles together under mechanical stress and, in some cases, to improve the conductivity of the particle. The progress on new genre of binder materials and their positive influence on electrode stability are covered in Section 3. Another negative outcome of Si volume expansion is the unstable solid-electrolyte-interphase (SEI). The cracked particle exposes new surface and initiates the growth of thick interfacial layer, which are detrimental to the cell impedance and electrolyte sustainability. The most effective way of addressing the interphase issue is through the electrolyte optimization and surface engineering. An appropriate choice of electrolyte additive is able to preferentially build a relatively stable SEI that can mitigate the electrolyte consumption during the long-term cell operation. The merits of electrolyte and interphase engineering beneficial for Si materials are highlighted in Section 3. In the end, the remaining challenges and practical aspects of Si-based anode materials are touched with the special focus on the full cell configuration. We hope the short walkthrough of the Si technology will help the readers have a grasp of the efforts made in improving the performance of electrochemical rechargeable batteries and appreciate the outcomes of the fundamental research brought to the development of practical Si-based LIBs.

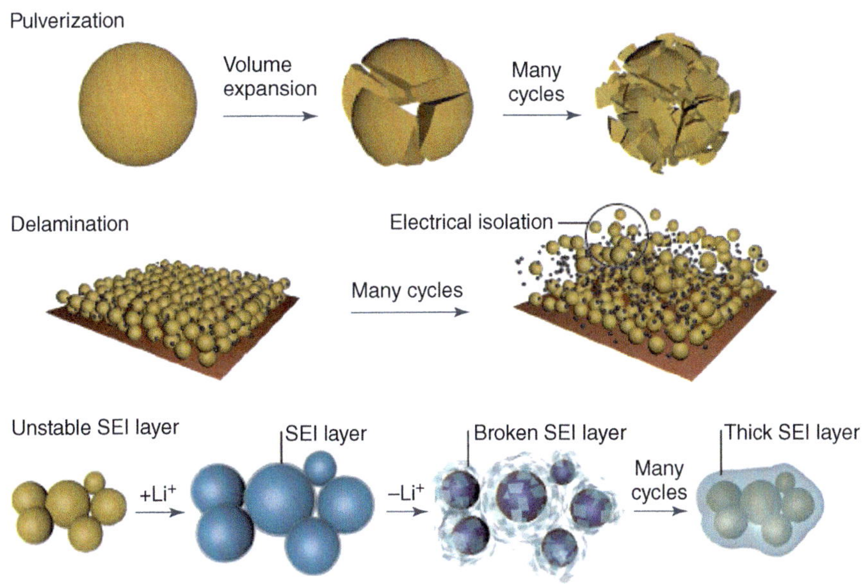


**Figure 1** Schematic illustration of the silicon anode degradation mechanism.[20] [Reproduced with permission from J. W. Choi, D. Aurbach, Nat. Rev. Mater. 2016, 1. © Springer Nature Limited, 2016]

**Table 2** Summary of the challenges and solutions to the Si anode in LIBs

| Challenge | Volume expansion | Particle pulverization | Electrode delamination | SEI formation |
|---|---|---|---|---|
| Solution | Nanostructured materials with predesigned void space | Reducing the primary particle size | Binder optimization | Artificial SEI |
| | | Surface coating | Reducing particle volume expansion | Electrolyte optimization |
| | | Nanostructure design | | Reduce surface area |

## 2 NANOSTRUCTURE DESIGN

Si anode stands out as the top candidate to replace graphite with its high specific capacity by forming $Li_{4.4}Si$ alloy. However, this large capacity is also accompanied by drastic volume expansion up to 420%. The change in material dimension puts serious threats to the materials integrity and interphase stability, which are supposed to be stable and intact during cell operation. In addition, the cracked Si particles also leave them out of reach from the conductive components in the electrode, and hence their capacitive contribution is lost. Nanoscience and technology bring hope for tackling the Si volume expansion issue, since downsizing the material dimension usually comes with unique mechanical, physiochemical properties and structural flexibility that are absent in the bulk counterpart.[21–23] Among the different nanostructures, nano Si particles with small diameters, Si nanowires (NWs), hollow Si nanospheres/tubes, yolk-shell structure, and porous silicon are the most promising designs in addressing the consequences associated with Si volume expansion.

As documented by early Si research, crystalline Si underwent a unique two-phase lithiation process, where the large tensile stress was built up to initiate the surface cracks. It was first reported by Liu *et al.*,[24] a strong size-dependent fracture of Si nanoparticle was observed in real time with in situ transmission electron microscopy (TEM). For particle size smaller than 150 nm, Si is free from cracking or facture upon full lithiation. This is due to an insufficient amount of strain energy release in driving crack propagation in Si particles smaller than the critical size. A more detailed study on the lithiation kinetics revealed that the changing mechanical stress at the reaction front altered the total driving force for lithiation reaction and hence results in slow lithiation process and structural breakdown (Figure 2a).[25] In fact, a critical size effect exits even in particles that are smaller than 50 nm. It was found that the 20 nm Si particles showed much improved cyclic stability than 5 and 10 nm.[31] Since the mechanical stress during lithiation/delithiation is the initiator of cracking, hollow structure with thin Si wall and void core was fabricated to solve the problem. Yao *et al.*[26] used 350-nm $SiO_2$ particles as hard templates to fabricate interconnected Si hollow

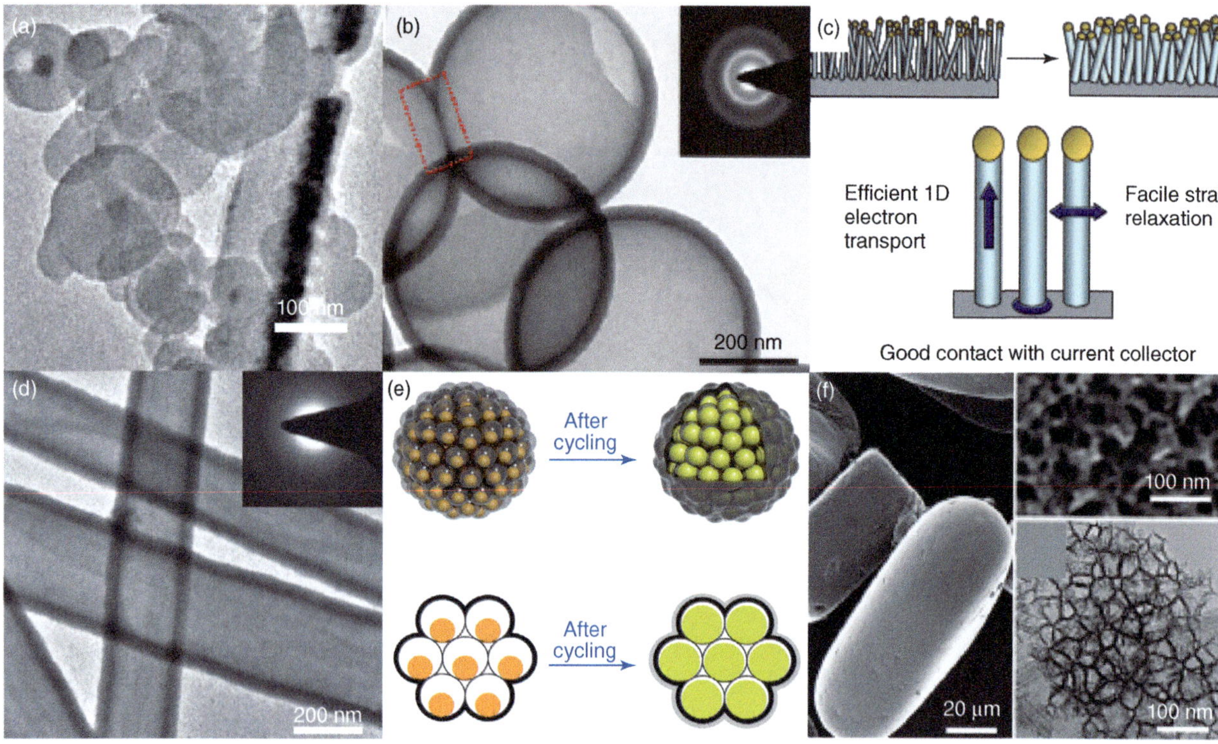


**Figure 2** Representative Si nanostructures. (a) Si nanoparticles at full lithiation state.[25] (b) Hollow Si nanospheres.[26] [Reprinted with permission from Y. Yao, M. T. McDowell, I. Ryu, H. Wu, N. A. Liu, L. B. Hu, W. D. Nix, Y. Cui, Nano Lett. 2011, 11, 2949–2954. © American Chemical Society 2011]; (c) Schematic illustration of Si nanowire.[27] [Reproduced with permission from C. K. Chan, H. L. Peng, G. Liu, K. McIlwrath, X. F. Zhang, R. A. Huggins, Y. Cui, Nat. Nanotechnol. 2008, 3, 31–35. © Springer Nature Limited, 2016]; (d) Hollow Si nanowire. Schematic illustration of Si nanowire.[28] [Reproduced with permission from H. Wu, G. Chan, J. W. Choi, I. Ryu, Y. Yao, M. T. McDowell, S. W. Lee, A. Jackson, Y. Yang, L. B. Hu, Y. Cui, Nat. Nanotechnol. 2012, 7, 309–314. © Springer Nature Limited, 2012]; (e) Graphical representation of pomegranate Si consisting of nano yolk-shell Si@C particles.[29] [Reproduced with permission from N. Liu, Z. D. Lu, J. Zhao, M. T. McDowell, H. W. Lee, W. T. Zhao, Y. Cui, Nat. Nanotechnol. 2014, 9, 187–192. © Springer Nature Limited, 2014]; (f) Images of mesoporous silicon sponge at different scale.[30] [Reproduced with permission from X. L. Li, M. Gu, S. Y. Hu, R. Kennard, P. F. Yan, X. L. Chen, C. M. Wang, M. J. Sailor, J. G. Zhang, J. Liu, Nat. Commun. 2014, 5, 4105. © Springer Nature Limited, 2014]

spheres (Figure 2b). It was simulated that the maximum tensile stress of solid sphere was about five times higher than that in hollow sphere at an equal volume of Si and hence make the hollow Si less susceptible to fracture. Owing to the structural advantages, the hollow spheres can deliver a capacity of 2725 mAh $g^{-1}$ at 0.1 C (1 C equals the current required to discharge the battery in 1 h) and retain over half of its capacity after 700 cycles at 0.5 C. The thin Si wall in the porous structure also gave the material a high rate capability, featured with 73% capacity retention at 5 C compared to 0.2 C (2417 mAh $g^{-1}$).

Besides downsizing the particle size and creating hollow structure, Si NWs also showed excellent properties in circumventing the particle fracture issue. Chan *et al.*[27] have demonstrated that with Si NWs grown directly on stainless steel current collectors. The unique wire structure showed high charge/discharge capacity close to theoretical value and retained most of the initial capacity with little fading during cycling. Although structural morphology and dimension changed after lithium uptake, the facile strain relaxation of Si NWs allowed it to be intact and undamaged during cycling (Figure 2c). Similar to the zero-dimensional hollow Si particles, the same concept was adopted in one-dimensional particles and demonstrated good electrochemical performance. Wu *et al.*[28] showed an anode material using silicon nanotube wrapped by an ion-permeable silicon oxide shell and were able cycle it for 6000 times in half-cell configuration with over 85% capacity retention. The double-walled tubular structure provided stable SEI layer, sufficiently accommodated the Si volume expansion, and exhibited eight times larger reversible capacity in large current test compared to conventional graphite anodes (Figure 2d).

Another effort to confront negative effects of Si volume expansion is to intentionally create void space to host the lithiated and/or cracked Si, that is, the yolk-shell structure. The signature of yolk-shell structure is the additional internal void space between the core particles and the outer shells that is purposely left during synthesis. The predesigned void space can accommodate Si volume expansion, which means the volume of yolk-shell particle as a whole stays the same during cycling. The internal room left is rather important since the integrity of the electrode depends on the dimensional stability of the active material. As volume changes are confined within the "shell", the safety issues related to the electrode/cell swelling can be largely evaded. The existence of the outer shell and void space also brings other advantages compare to solid Si: (i) inwardwise, the outer shell serves as a protective layer to prevent cracked Si "yolk" from losing electrical contact by scattering around upon volume change; (ii) outwardwise, the layer of carbon shell can help establish a robust and stable SEI layer, which would otherwise continuously grow on the broken Si surface during cycling; (iii) the conductive nature of the carbon also lifts the capacity utilization of the insulating Si, even if it is disintegrated inside the shell. Yi Cui 's group designed a yolk shell structure using nano Si particles encapsulated by a thin conformal and self-supporting carbon shell.[32] This structure demonstrated excellent capacity of 2833 mAh $g^{-1}$ at 400 mA $g^{-1}$ and showed impressive 74% capacity retention, after 1000 cycles with an average 99.84% Coulombic efficiency.

In the follow-up work, his group further fabricated a pomegranate structure consisting of myriads of nano yolk-shell Si particles.[29] This structure not only alleviated the mechanical stress induced by volume expansion of nano Si, but also successfully increased the tap density and volumetric capacity that are hard to achieve with nanomaterials (Figure 2e). The successful combination of the merits of both yolk-shell structure and hierarchical microsphere design showed excellent cyclic stability with 97% capacity retention after 1000 cycles with high Coulombic efficiency (99.87%) and volumetric capacity of 1270 mAh $cm^{-3}$. Conventionally, yolk shell structure has to go through multistep synthetic route, including $SiO_2$ coating, carbon coating, and acid etching. Although the structure is quite effective in lifting the Si performance, the chemical synthesis is usually time-consuming and hard to scale up in practical scenario. To address this problem, Li *et al.*[33] proposed a low-cost strategy by oxidizing the nano Si in air to create an $SiO_2$ surface layer followed by carbon coating and hydrofluoric acid etching of the $SiO_2$ layer. By controlling the oxidation and etching time, the void space of the structure can be easily tuned to buffer the Si expansion.

All the destructive outcomes of switching to high-capacity Si anode come from the changing volume of Si during cycling. Thus, preleft void space is important to the stability of active materials as well as the integrity of the electrode. Besides the yolk-shell structure, porous silicon also takes advantage of having predesigned pore volume to absorb the dimension change associated with lithiated Si. Although the benefits of using nano Si have brought hope of significantly boosting the energy density of LIBs, their unique properties also come with unavoidable drawbacks. Owing to the high surface area and complex synthesis conditions, anode with nano Si usually suffers from low tap/calendaring density, low Coulombic efficiency, and high cost. To address these issues, but at the same time preserve the merits of nano Si, are considered the paramount tasks in silicon structure design. Bulk porous Si seems to be the optimum solution, which has been found possessing the structural advantages lend from nanoporous materials without compromising the easy manufacturing ability and low cost of bulk Si. One exemplary design was the mesoporous silicon sponge structure by the research team at Pacific Northwestern National Laboratory (PNNL).[30] Using an electrochemical etching method, large silicon particles (20 μm) with rich mesopores demonstrated excellent performance as Li-ion anode. (Figure 2f) The micron-size silicon particle consisted of large pores up to 50 and 10 nm thin Si wall. In situ observation of the lithiation/delithiation process revealed that the volume change of the porous Si was merely ~30% compared to the 360% in the conventional Si. More importantly, there was no particle pulverization observed even after a thousand continuous cycles. This unique structure delivered an impressive capacity retention of 92% at high area-specific capacity of ~1.5 mAh $cm^{-2}$ over 300 cycles. It was an encouraging study that demonstrated the successful marriage between porous nanostructure design and manufacturing friendly bulk Si materials at practical testing conditions. While the antipulverization and volume accommodation structures are highly desirable features, simple and scalable syntheses are hard to meet using Si anode at nanoscale. On the basis of such concern, Yi *et al.*[34] have conceived a scalable method to synthesize micro-sized Si–C composite with interconnected nanopores as high-performance Li anode. They started with commercially available silicon monoxide (SiO, 20 μm). By thermal annealing, Si/$SiO_2$ composite was formed with intertwined Si and $SiO_2$ regions. After etching away the sacrificial $SiO_2$ matrix, a thin layer of carbon was coated on the porous Si using acetylene as carbon source. The resultant Si–C composite not only retained its bulk-size dimension and morphology, but also incorporated the three-dimensionally interconnected Si at nanoscale. Benefitting from the porous structure, the Si–C composite can reach a high capacity of 1459 mAh $g^{-1}$ after 200 cycles at 1 A $g^{-1}$ and retain 97.8% of its original capacity.

## 3 BINDER EFFECT

In electrode fabrication, binder plays an essential role gluing different parts in electrode slurry casted on the current collector. Ideally, the binder should stay intact and responsive to the stress generated during electrochemical environment. The application of Si anode, however, poses a great challenge to the existing binder selection, poly vinylidene fluoride (PVDF), that is commonly used in graphite anode. This is mainly caused by the weak Van der Waals forces established between Si and PVDF binder, which is not sustainable to hold the adjacent particles together or the electrodes to the current collector during cycling. Clearly, binders with high elastic modulus and strength are particularly desirable for fabricating stable Si anode. In addition to the volume-expandable issue, the poor electrical conductivity of Si can also be addressed using binders that are highly conductive and soluble in nontoxic solvents.

To overcome the drawbacks of the PVDF binder, polymers with strong interaction with Si particles are under the radar. Carboxymethyl cellulose (CMC) was once considered the standard binder for Si anode, as it offered much improved cycling stability compared to PVDF. The bonding between carboxyl groups in CMC and hydroxyl groups on Si surface provides enhanced strength and minimized the stress and volume change in lithiated Si.[35] Polyacrylic acid (PAA) was also tested as an advanced binder for Si anode and shows improved electrochemical stability.[36] It is believed that the high concentration of COOH groups plays the determining role for its superior performance by forming strong bonding with OH groups on the Si surface. The remaining carboxyl groups of the PAA binder also help forming ionically conductive COOLi groups that are beneficial of the developing flexible SEI layer during cycling. On the basis of the findings in CMC and PAA study, alginate, a natural polysaccharide rich in carboxylic groups, was applied as binder for high-capacity Si anode.[37] By choosing alginate extracted from grown algae, the binder possessed the strongest interaction with Si surface while maintaining mechanical rigidity even at high current density and prolonged cycling. Polymer has the flexible nature, which is friendly to size-changeable Si anode. Compared to the relative brittle PVDF binder, a polymer-coated layer can stretch and contract during Si charge/discharge, while maintaining an intimate contact on the Si surface is of great interest. Inspired by the action of a pulley, Choi *et al.*[38] incorporated a small portion of ring-slide polyrotaxane (PR), consisting of polyethylene glycol (PEG) threads and $\alpha$-cyclodextrin rings functionalized with 2-hydroxypropyl moieties, into conventional linear binder, PAA. The PR–PAA binder exhibited extremely high elasticity and stress dissipation ability, which enabled micron-size Si particle free from pulverization during cycling. Even at commercial-level testing scenario, the Si anode with PR–PAA binder shows stable cycling over 400 times with minimum capacity fade.

Alongside the good mechanical properties, binders with electrical conductivity are also desirable as Si particle might detach from the conductive additive in the electrode after repeated expansion/contraction. Polyaniline (PANI) is one of the conductive polymers that show positive effect on Si anode. Unlike regular binders, where their interactions with active material are made by mechanical mixing and usually result in nonuniform surface coating, Wu *et al.*[39] managed to in situ polymerize PANI and form conformal coating layer on the Si surface. The whole coating process took only 2 min, thus rendering it a quite scalable approach (Figure 3a). Moreover, the high viscosity of the hydrogel can be used directly as electrode paste and casted onto a current collector. The results of having PANI coating layer are significant; the Si anode under optimized synthetic condition showed a record-breaking cyclic stability of 5000 cycles and retained 90% of its capacity at current density of $6\,A\,g^{-1}$. Notably, the voltage plateau remained in the range 0.01–0.3 V at high current density of $3\,A\,g^{-1}$, indicating an enhanced alloying kinetics benefited from the conductive polymer network. Another class of conductive binder is polypyrrole (PPy). The effectiveness of PPy-based binder, however, highly depends on the uniformity and thickness applied in Si anode. To obtain a uniform and thickness controllable conductive coating, polypyrrole-ferric oxide (PPy-Fe) coordination complex was prepared so that it can chemically interact with the surface functional groups on Si particles. The binder provided a continuous electronic conduction and served as a protective layer for stable SEI formation. Under the optimized condition, the PPy-Fe-coated anode can stably operate over 800 cycles with 95% capacity retention at current density of $1.2\,A\,g^{-1}$. Incredibly, the discharge capacity reaches $4000\,mAh\,g^{-1}$ at $300\,mA\,g^{-1}$, representing a 95% of the theoretical capacity utilization of the Si anode. The PPy binder also possess good elasticity to accommodate the volume expansion of Si during charge/discharge. To experimentally examine this idea, Luo *et al.* applied a conductive PPy-coated nano Si. According to the in situ TEM study, the coating layer acted as mechanical confinement that buffers the expanded Si during cycling, providing a more sustainable surface layer than the naturally grown SEI layer.[41]

The low-temperature synthesis and scalability are important aspects for using conductive polymer as binder material. These attributes were demonstrated using poly(3,4-ethylenedioxythiophene) polystyrene sulfonate (PEDOT) as a bifunctional additive for nano Si particles.[42] Different from the conventional conductive network formed by binder and carbon additive, the PEDOT polymer gave much continuous and uniform electron pathway in the electrode fabrication and can be produced in large quantity. They were able to achieve a high lithiation capacity of $3685\,mAh\,g^{-1}$, corresponding to ~88% of capacity utilization of the theoretical value of Si. Another type of conductive polymeric binder that showed good electrochemical performance was the poly(9,9-dioctylfluorene-co-fluorenone-co-methylbenzoic ester (PFFOMB).[43] The polymer has high electric conductivity and robust mechanical binding force, which stay

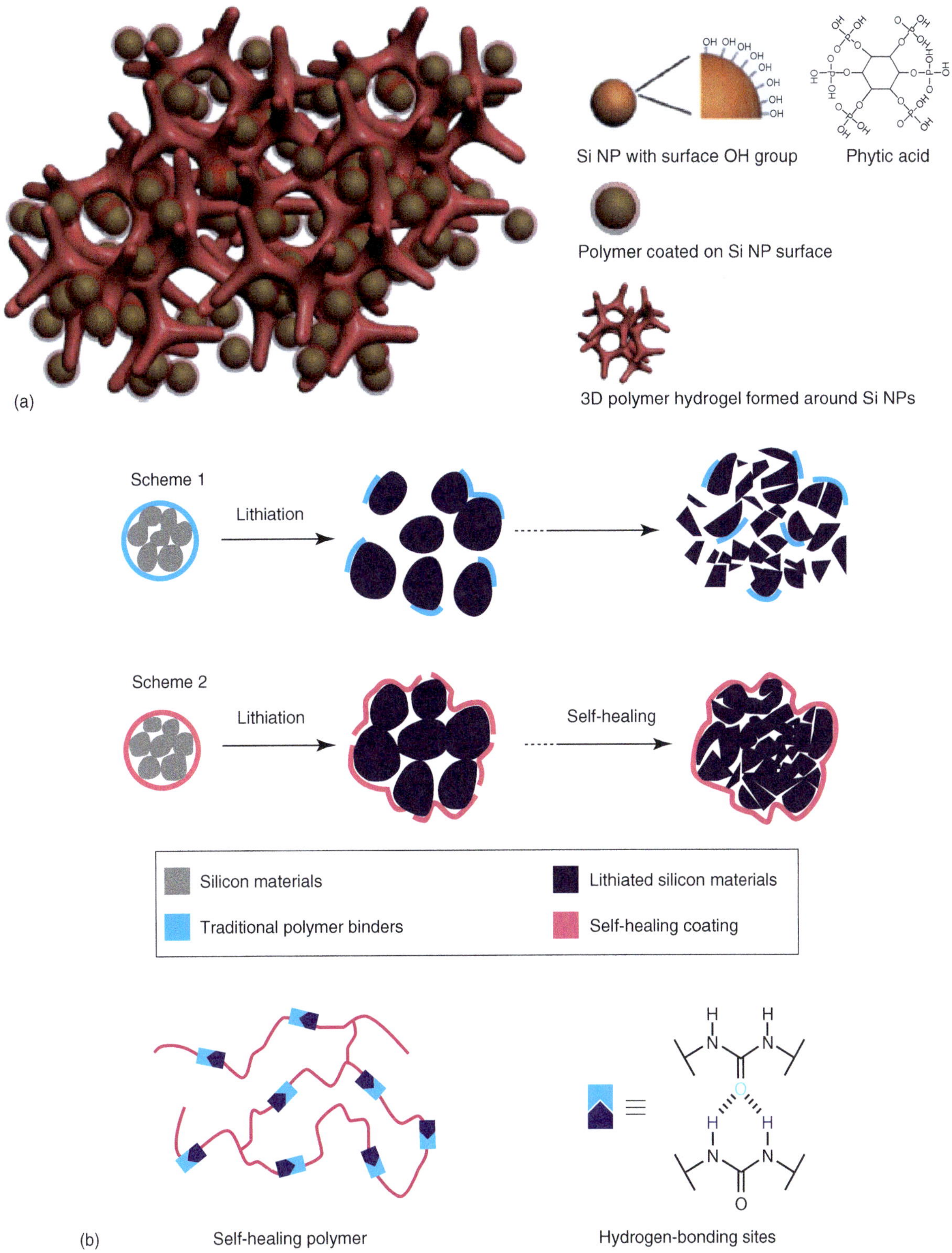


**Figure 3** Different kinds of polymer binders. (a) Graphical representation of the in situ-grown PANI on the nano Si particles.[39] [Reproduced with permission from H. Wu, G. Yu, L. Pan, N. Liu, M. T. McDowell, Z. Bao, Y. Cui, Nat. Commun. 2013, 4, 1943. © Springer Nature Limited, 2013]; (b) Schematic comparison of the structural behavior of Si particle with conventional and self-healing polymer binder.[40] [Reproduced with permission from C. Wang, H. Wu, Z. Chen, M. T. McDowell, Y. Cui, Z. A. Bao, Nat. Chem. 2013, 5, 1042–1048. © Springer Nature Limited, 2013]

intimately on the Si surface and accommodated volume change during lithiation/delithiation. As a result, the PFFOMB coated electrode delivered 2100 mAh $g^{-1}$ after 650 cycles, far exceeding the performance of pristine Si without any conductive coating. The motivation of using elastic polymer coating on Si anode originates from the nature of the volume changeable Si during cycling, where brittle coating might experience structural breakdown. Nevertheless, even polymer coating after cycling can suffer gradual degradation under prolonged mechanical stress and harsh electrochemical environment. To address this problem, polymeric binder that can heal itself seems like a good approach to address the concern. Wang *et al.*[40] developed a self-healing polymer with dynamic hydrogen bonds. The polymer has the ability of repeatable healing even after destructive structural change, which is particularly desirable for Si anode (Figure 3b). Meanwhile, the polymer shows extraordinary strechability, even after being stretched to 300% of its original length; no structural breaking was observed. The high elasticity and self-healing features enabled the coated Si show 10 times longer cycle life compared to conventional binders.

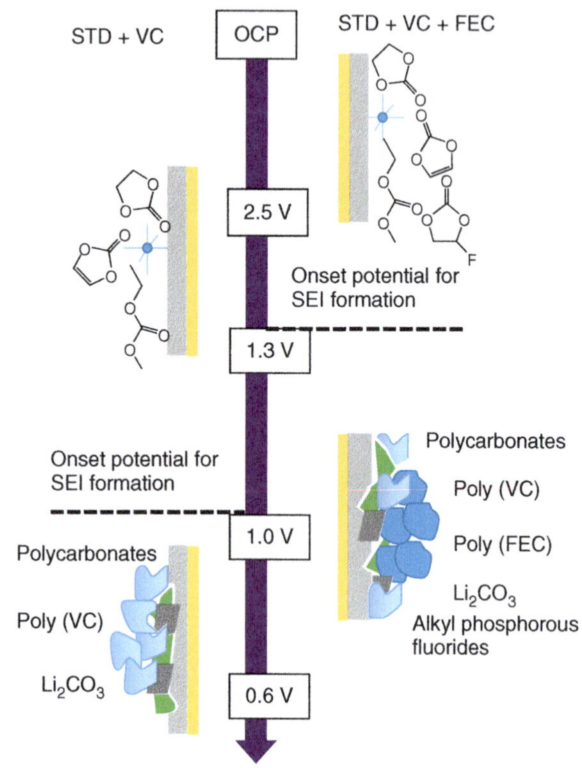


**Figure 4** Graphical illustration of electrolyte additives and the SEI forming mechanism with FEC and VC electrolyte additives on Si anode.[49] [Reprinted with permission from Y. B. Yohannes, S. D. Lin, N. L. Wu, J. Electrochem. Soc. 2017, 164, A3641–A3648. Copyright 2017, The Electrochemical Society]

## 4 ELECTROLYTE OPTIMIZATION AND INTERPHASE ENGINEERING

As mentioned in the previous sections, SEI layer is the key enabler for the long-term cyclic stability of the Si anode. In an ideal scenario, the SEI should form once during the first charge/discharge cycle and remain stable on the Si surface for the rest of the cell life. However, if the volume of Si particle constantly changes during cycling, the destruction and reforming of SEI layer become inevitable, which deplete the electrolyte, increase the cell resistance, and eventually cause capacity fade.[44] Thus, constructing a robust and flexible SEI layer without sacrificing its ion permeability is of vital importance for Si performance. The composition and structure of the SEI are largely dependent on the electrolyte formula. A good example is the addition of fluoroethylene carbonate (FEC) in conventional electrolyte of ethylene carbonate (EC) and diethyl carbonate (DEC) or dimethyl carbonate (DMC). The main reduction product of FEC consists of LiF and $-CHF-OCO_2-$ compounds, which are uniformly deposited on the surface of Si surface and hence alleviate further lithium salt and solvent decomposition during cycling.[45] Microscopically, the Si electrode cycled in FEC containing electrolyte also exhibited much smoother surface with minimum cracking and delamination. Since vinylene carbonate (VC) is a major part of the reduction products of FEC, it was also tested as the electrolyte additive for Si anode.[46] After the initial development of SEI layer upon VC reduction, the interfacial film limited the contact between Si and electrolyte and maintained the mechanical integrity of the electrode. Detailed comparisons of the additive effect between FEC and VC revealed that both of them can mitigate electrode cracking during cycling and provide improved thermal resistance at high-temperature operation.[47,48] Despite sharing influences on the Si SEI layer, FEC and VC are reduced in different electrochemical routes, where FEC has slight higher reductive potential and generates more alkyl phosphorous fluorides species (Figure 4).[49] Other popular electrolyte additives in carbon-based LIBs are also examined in Si anodes, including lithium bis(oxalate)borate (LiBOB)[50] and lithium difluoro(oxalate)borate (LiDFOB).[51] Their protective mechanism, advantages, and drawback are listed in Table 3.

Another class of electrolyte that shows unique properties in forming stable SEI layer on Si are ionic liquids (ILs), which are molten salts consisting of organic cation and inorganic/organic anions at room temperature and free of molecular compounds.[44] ILs offer high ionic conductivity and wide electrochemical window, negligible vapor pressure, and low flammability. Pioneering work on IL-based electrolyte used 1-methyl-1-propylpiperidinium bis(trifluoromethylsulphonil)imide

**Table 3** Summary of electrolyte additive found effective on improving Si anode performance

| Additive | Mechanism | Advantages | Drawbacks |
|---|---|---|---|
| FEC | Stable SEI layer with LiF | Improved cycling stability and reversible capacity | Gas generation |
| VC | Reduced organic and LiF content and stable SEI film | Improved thermal resistance, Coulombic efficiency, and cycling stability | Gas generation |
| LiDFOB | Improved flowability and stable SEI | Improved temperature tolerance | Pricy and limited effect |
| LiBOB | Protected counter electrode | Improved thermal stability | Gas generation and high impedance |

with 1 M Li bis(trifluoromethylsulphonil)imide (LiTFSI) and demonstrated good cathodic stability, which are less shared in IL family.[52] Practical feasibility of IL was demonstrated by applying bis(fluorosulfonyl)imide (FSI)-based IL in Si/lithium nickel manganese cobalt oxide (NMC) cylindrical full cell and showed remarkable cyclic stability up to 1000 cycles.[53] The highly reversible electrochemical cycles are supported by the stable SEI layer formed by the IL electrolyte. Overall, ILs offer tunable SEI properties, including high-temperature resistance and strong mechanical confinement of the expandable Si and hence open a new research direction for constructing high-energy Si-based LIBs.

In addition to controlling SEI formation through electrolyte optimization, the surface coating is an effective approach to improve interfacial stability. Moreover, the coating layer on Si has an additional role of being physical protector under the constant volume change during cycling. The coating layer on the Si anode has to sustain the mechanical stress generated with Si expansion, but also promotes forming stable SEI layer. It will alleviate the continuous growth of SEI layer and the Si particle breakdown that accelerates the process in one shot.

Generally, metal oxide coatings used in Si are also active toward Li by forming Li-containing alloy or intercalated compound. The interaction between $Li^+$ and metal oxides usually happens around 1 V versus $Li/Li^+$, at which electrolyte decomposition occurs and SEI formation starts. One good example was reported with titanium oxide ($TiO_2$) as outer shell for nano silicon particles.[54] The $TiO_2$ has five times greater strength than an amorphous carbon shell and hence can survive over 20 MPa calendaring process typically used in industrial electrode manufacturing. The great compressibility of the $TiO_2$ lead to a threefold increase of Si tap density from 0.4 to 1.4 g $cm^{-3}$ and provided stable SEI layer under sever Si volume change, and consequently leads to a high-loading full cell with two times larger gravimetric and volumetric capacity than the same cell with a graphite anode.

Besides metal oxides, another type of metal-based coating found beneficial to Si anode is titanium nitride (TiN).[55] A 5-nm coating achieved by atomic layer deposition (ALD) on the Si nanowire was able to double its capacity retention and high rate performance compared to the uncoated counterpart. The existence of the coating layer also prevents Si from disintegration with more reversible volume expansion and provide stable SEI layer. The irreversible bonding between Li and TiN, however, would consume the limited lithium source in practical full cell set-up. Thus, coating layer with built-in lithium source seems more feasible. Lee *et al.*[56] applied a multicomponent coating formula, consisting of lithium silicate ($Li_2SiO_3$ and $Li_2Si_2O_5$) and lithium titanate ($Li_4Ti_5O_{12}$) on the nano Si surface. There are several advantages of the synergistically coupling of the multifunctional shell layers: (i) $Li_2SiO_3$ and $Li_2Si_2O_5$ have good mechanical strength capable of preventing fracture damage of Si particles; (ii) $Li_4Ti_5O_{12}$ is active toward $Li^+$ intercalation and hence provides an effective Li-ion diffusion path; (iii) both oxide coatings assist a stable SEI formation on the Si particle surface; (iv) multifunctional coating layers significantly reduce the heat generation and confront the volume fluctuation during the battery operation; and (v) tap density of Si particles can be relatively increased by coating Si surface with $Li_4Ti_5O_{12}$. With the above benefits, the coated Si anodes demonstrated highly stable long-term cycling performance with 65% capacity retention after 1000 cycles at 1 C and showed excellent rate capability with 89% capacity at 10 C compared to 0.2 C (Figure 5).

## 5 PRACTICAL PERSPECTIVES OF Si-BASED ANODES IN FULL CELLS

At the beginning stage of the Si anode research, thin electrodes with low active materials loading (<1 mg $cm^{-2}$) are generally adopted in half-cell configuration, where an Li metal chip serves as the reference and counter electrode. In most half-cell testing, impressive performance with high capacity utilization and long cycle life are reported. However, the use of lithium metal and the low areal capacity can mask some critical issues associated with Si anodes and may result in misleading information

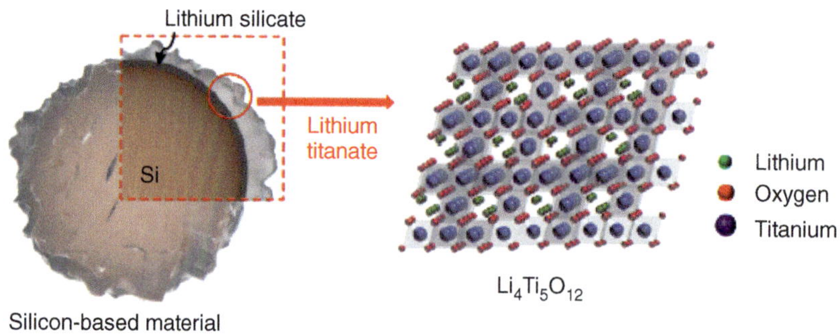


**Figure 5** Schematic illustration of lithium silicate ($Li_2SiO_3$ and $Li_2Si_2O_5$) and lithium titanate ($Li_4Ti_5O_{12}$) coating on Si surface.[56] [Reproduced with permission from J. I. Lee, Y. Ko, M. Shin, H. K. Song, N. S. Choi, M. G. Kim, S. Park, Energ. Environ. Sci. 2015, 8, 2075–2084. © Wiley-VCH 2015]

for designing practical full cells. In half-cells with thin Si electrodes, (i) the infinite Li supply could compensate lithium loss from the low Coulombic efficiency of Si; (ii) the influence of electrode expansion might be mitigated and limited to a small level; and (iii) scalable and cost-effective synthesis might not be necessary owing to the small amount of Si required for low areal loading. In practical full-cell configuration, however, low Coulombic efficiency would lead to quick depletion of the limited Li-ions and cost the cycle life of the battery. Not to mention the electrode expansion post severe safety concern. Thus, properly analyzing and hence selecting an appropriate Si material and structure require electrochemical examination of Si anodes with practical loading in full-cell configuration.

One good example was made in mesoporous silicon–graphite composite electrodes.[57] A practical electrode with the loading of ~3 mAh cm$^{-2}$ was prelithiated and used in a full cell against NMC cathode and achieved stable cycling over 300 cycles. A notable observation was that the porous silicon structure can effectively accommodate the electrode swelling during cycling. The results demonstrated that engineered nanostructure was capable of providing competitive performance at practical level. Si-modified graphite composite was recently developed by depositing Si on graphite particles using CVD technology. Minseong Ko *et al.* controlled the percentage of the Si nanolayer embedded in graphite carbon hybrids and demonstrated large-scale production of ~5 kg per batch. Their material marked an impressively small electrode swelling of 38% at areal capacity over 4 mAh cm$^{-1}$ (~6.5 mg cm$^{-2}$), only 23% larger than pure graphite and half of the physically blended graphite/silicon composite electrode. The full cell using this hybrid material was able to achieve high volumetric capacity of 1042 Wh l$^{-1}$ and comparable cyclic stability to pure graphite anode. In an effort of demonstrating the high volumetric capacity of Si anode and its practical compatibility with existing battery format, Son *et al.*[59] assembled a commercial format 18 650 cylindrical cell with specific areal capacity of 3.8 mAh cm$^{-2}$. The

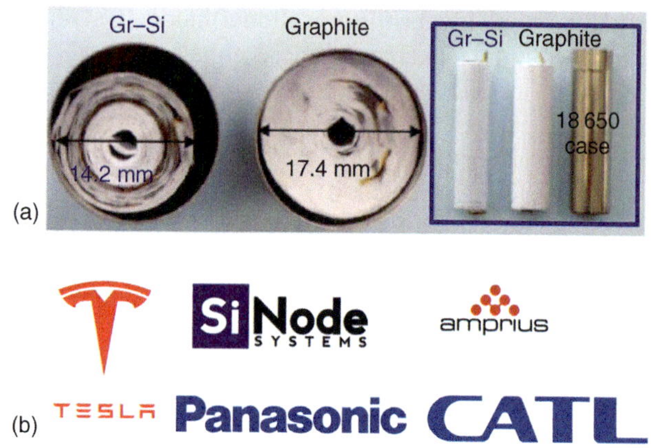


**Figure 6** (a) Digital image of 18 650 cylindrical cell made with Gr–Si anode.[59] [Reproduced with permission from I. H. Son, J. H. Park, S. Kwon, S. Park, M. H. Rümmeli, A. Bachmatiuk, H. J. Song, J. Ku, J. W. Choi, J.-m. Choi, Nat. Commun. 2015, 6, 8393. © Springer Nature Limited, 2015]; (b) Companies involved in Si-based LIBs research and development

anode used graphene-wrapped Si nanoparticles (Gr–Si) and was able to achieve volumetric energy densities of 972 and 700 Wh l$^{-1}$ at initial and the 200th cycles, which are 1.8 times higher than the graphite-based LIBs (Figure 6a).

## 6 CHALLENGES AND OUTLOOK

Transferring the technology from lab to practical applications has always been challenging. Since the first high-temperature electrochemical Li–Si cell, 40 years have passed to reach the current state of energy density and cycle life with Si-based Li-ion batteries. With the continuous progress in materials research and engineering, efforts across both academia and industries, the practical implementation Si anode are on the horizon.

Companies pioneering in Si anode development and commercialization (Figure 6b) are making solid progress delivering the promise of replacing graphite in high-energy EV and aviation batteries.[60] It is anticipated that the imminent application of Si in practical LIBs would be compositing graphite with small amount of Si particles. According to the full cell energy density increment with respect to anode capacity, the gain would level out only if the specific capacity of anode passes $1000\,mAh\,g^{-1}$.[61] Until then, there are still tasks to complete with increasing Si content in the electrode. Firstly, the low Columbic efficiency (compared to graphite) of Si and volume expansion upon lithiation might require more electrolyte to sustain a long-term cycling, thus compromising the cell-level energy density.[62] Secondly, the issues originated from electrode expansion remain as noticeable concerns against battery safety, despite the use of nanostructured Si showing encouraging potential absorbing a major portion of volume fluctuation during cycling. Lastly, the cost of using Si materials might not be comparably optimistic as its natural abundance, since novel nano Si processing and nanostructure fabrication need considerable investment on the infrastructure. As promising as it may look, these intrinsic drawbacks need to be taken care of in the future development of anode with increasing Si content. Nevertheless, hope must prevail. With the years of research accumulation, advancements in existing battery engineering, and the rapidly growing electrical transportation market, advanced Si materials will soon be an indispensable part of next-generation high energy density LIBs.

## 7 ABBREVIATIONS AND ACRONYMS

ALD = atomic layer deposition; CMC = carboxymethyl cellulose; CVD = chemical vapor deposition; DEC = diethyl carbonate; DMC = dimethyl carbonate; EC = ethylene carbonate; EV = electrical vehicle; FEC = fluoroethylene carbonate; FSI = bis(fluorosulfonyl)imide; IL = ionic liquids; LiBOB = lithium bis(oxalate)borate; LIBs = lithium-ion batteries; LiDFOB = lithium difluoro(oxalate)borate; LiTFSI = Li bis(trifluoromethylsulphonil)imide; NMC = lithium nickel manganese cobalt oxide; NWs = nanowires; PAA = polyacrylic acid; PANI = polyaniline; PEDOT = poly(3,4-ethylenedioxythiophene) polystyrene sulfonate; PEG = polyethylene glycol; PFFOMB = poly(9,9-dioctylfluorene-co-fluorenone-co-methylbenzoic ester; PNNL = Pacific Northwestern National Laboratory; PPy-Fe = polypyrrole-ferric oxide; PPy = polypyrrole; PR = polyrotaxane; PVDF = poly vinylidene fluoride; SEI = solid-electrolyte-interphase; TEM = transmission electron microscopy; TiN = titanium nitride; $TiO_2$ = titanium oxide; VC = vinylene carbonate

## 8 REFERENCES


1. S. Megahed and B. Scrosati, *J. Power Sources*, 1994, **51**, 79.
2. K. Ozawa, *Solid State Ion.*, 1994, **69**, 212.
3. T. N. K. Tazawa, *Prog. Batter. Sol. Cells*, 1991, **9**, 209.
4. M. Hagen, D. Hanselmann, K. Ahlbrecht, R. Maca, D. Gerber and J. Tubke, *Adv. Energy Mater.*, 2015, **5**, 1401986.
5. K. Mizushima, P. C. Jones, P. J. Wiseman and J. B. Goodenough, *Mater. Res. Bull.*, 1980, **15**, 783.
6. M. Winter, J. O. Besenhard, M. E. Spahr and P. Novak, *Adv. Mater.*, 1998, **10**, 725.
7. M. Li, J. Lu, Z. Chen and K. Amine, *Adv. Mater.*, 2018, **30**, e1800561.
8. J. Tarascon, *Electrochem. Soc. Interface*, 2016, **25**, 79.
9. J. Graetz, C. C. Ahn, R. Yazami and B. Fultz, *Electrochem Solid State*, 2003, **6**, A194.
10. W. J. Zhang, *J. Power Sources*, 2011, **196**, 13.
11. V. P. Nikolaev, A. G. Morachevskii, A. I. Demidov and E. V. Bairachnyi, *J. Appl. Chem. USSR*, 1980, **53**, 1549.
12. B. A. Boukamp, G. C. Lesh and R. A. Huggins, *J. Electrochem. Soc.*, 1981, **128**, 725.
13. A. M. Wilson, G. Zank, K. Eguchi, W. Xing and J. R. Dahn, *J. Power Sources*, 1997, **68**, 195.
14. S. Bourderau, T. Brousse and D. M. Schleich, *J. Power Sources*, 1999, **81**, 233.
15. H. Dong, X. P. Ai and H. X. Yang, *Electrochem. Commun.*, 2003, **5**, 952.
16. H. Dong, R. X. Feng, X. P. Ai, Y. L. Cao and H. X. Yang, *Electrochim. Acta*, 2004, **49**, 5217.
17. M. S. Park, Y. J. Lee, S. Rajendran, M. S. Song, H. S. Kim and J. Y. Lee, *Electrochim. Acta*, 2005, **50**, 5561.
18. M. Holzapfel, H. Buqa, W. Scheifele, P. Novak and F. M. Petrat, *Chem. Commun.*, 2005, **28**, 1566.
19. L. Y. Beaulieu, K. W. Eberman, R. L. Turner, L. J. Krause and J. R. Dahn, *Electrochem. Solid State*, 2001, **4**, A137.
20. J. W. Choi and D. Aurbach, *Nat. Rev. Mater.*, 2016, **1**, 1.
21. C. Lee, X. D. Wei, J. W. Kysar and J. Hone, *Science*, 2008, **321**, 385.
22. J. T. Hu, T. W. Odom and C. M. Lieber, *Acc. Chem. Res.*, 1999, **32**, 435.
23. X. L. Li, X. R. Wang, L. Zhang, S. W. Lee and H. J. Dai, *Science*, 2008, **319**, 1229.
24. X. H. Liu, L. Zhong, S. Huang, S. X. Mao, T. Zhu and J. Y. Huang, *ACS Nano*, 2012, **6**, 1522.
25. M. T. McDowell, I. Ryu, S. W. Lee, C. M. Wang, W. D. Nix and Y. Cui, *Adv. Mater.*, 2012, **24**, 6034.
26. Y. Yao, M. T. McDowell, I. Ryu, H. Wu, N. A. Liu, L. B. Hu, W. D. Nix and Y. Cui, *Nano Lett.*, 2011, **11**, 2949.
27. C. K. Chan, H. L. Peng, G. Liu, K. McIlwrath, X. F. Zhang, R. A. Huggins and Y. Cui, *Nat. Nanotechnol.*, 2008, **3**, 31.

28. H. Wu, G. Chan, J. W. Choi, I. Ryu, Y. Yao, M. T. McDowell, S. W. Lee, A. Jackson, Y. Yang, L. B. Hu and Y. Cui, *Nat. Nanotechnol.*, 2012, **7**, 309.
29. N. Liu, Z. D. Lu, J. Zhao, M. T. McDowell, H. W. Lee, W. T. Zhao and Y. Cui, *Nat. Nanotechnol.*, 2014, **9**, 187.
30. X. L. Li, M. Gu, S. Y. Hu, R. Kennard, P. F. Yan, X. L. Chen, C. M. Wang, M. J. Sailor, J. G. Zhang and J. Liu, *Nat. Commun.*, 2014, **5**, 4105.
31. H. Kim, M. Seo, M. H. Park and J. Cho, *Angew. Chem. Int. Ed.*, 2010, **49**, 2146.
32. N. Liu, H. Wu, M. T. McDowell, Y. Yao, C. M. Wang and Y. Cui, *Nano Lett.*, 2012, **12**, 3315.
33. X. L. Li, P. Meduri, X. L. Chen, W. Qi, M. H. Engelhard, W. Xu, F. Ding, J. Xiao, W. Wang, C. M. Wang, J. G. Zhang and J. Liu, *J. Mater. Chem.*, 2012, **22**, 11014.
34. R. Yi, F. Dai, M. L. Gordin, S. R. Chen and D. H. Wang, *Adv. Energy Mater.*, 2013, **3**, 295.
35. J. S. Bridel, T. Azais, M. Morcrette, J. M. Tarascon and D. Larcher, *Chem. Mater.*, 2010, **22**, 1229.
36. A. Magasinski, B. Zdyrko, I. Kovalenko, B. Hertzberg, R. Burtovyy, C. F. Huebner, T. F. Fuller, I. Luzinov and G. Yushin, *ACS Appl. Mater. Interfaces*, 2010, **2**, 3004.
37. I. Kovalenko, B. Zdyrko, A. Magasinski, B. Hertzberg, Z. Milicev, R. Burtovyy, I. Luzinov and G. Yushin, *Science*, 2011, **334**, 75.
38. S. Choi, T. W. Kwon, A. Coskun and J. W. Choi, *Science*, 2017, **357**, 279.
39. H. Wu, G. Yu, L. Pan, N. Liu, M. T. McDowell, Z. Bao and Y. Cui, *Nat. Commun.*, 2013, **4**, 1943.
40. C. Wang, H. Wu, Z. Chen, M. T. McDowell, Y. Cui and Z. A. Bao, *Nat. Chem.*, 2013, **5**, 1042.
41. L. L. Luo, P. Zhao, H. Yang, B. R. Liu, J. G. Zhang, Y. Cui, G. H. Yu, S. L. Zhang and C. M. Wang, *Nano Lett.*, 2015, **15**, 7016.
42. T. M. Higgins, S. H. Park, P. J. King, C. Zhang, N. MoEvoy, N. C. Berner, D. Daly, A. Shmeliov, U. Khan, G. Duesberg, V. Nicolosi and J. N. Coleman, *ACS Nano*, 2016, **10**, 3702.
43. G. Liu, S. D. Xun, N. Vukmirovic, X. Y. Song, P. Olalde-Velasco, H. H. Zheng, V. S. Battaglia, L. W. Wang and W. L. Yang, *Adv. Mater.*, 2011, **23**, 4679.
44. J. H. Song, B. W. Xiao, Y. H. Lin, K. Xu and X. L. Li, *Adv. Energy Mater.*, 2018, **8**.
45. C. Xu, F. Lindgren, B. Philippe, M. Gorgoi, F. Bjorefors, K. Edstrom and T. Gustafsson, *Chem. Mater.*, 2015, **27**, 2591.
46. M. Ulldemolins, F. Le Cras, B. Pecquenard, V. P. Phan, L. Martin and H. Martinez, *J. Power Sources*, 2012, **206**, 245.
47. I. A. Profatilova, C. Stock, A. Schmitz, S. Passerini and M. Winter, *J. Power Sources*, 2013, **222**, 140.
48. C. C. Nguyen and B. L. Lucht, *J. Electrochem. Soc.*, 2014, **161**, A1933.
49. Y. B. Yohannes, S. D. Lin and N. L. Wu, *J. Electrochem. Soc.*, 2017, **164**, A3641.
50. A. Rezqita, M. Sauer, A. Foelske, H. Kronberger and A. Trifonova, *Electrochim. Acta*, 2017, **247**, 600.
51. S. Dalavi, P. Guduru and B. L. Lucht, *J. Electrochem. Soc.*, 2012, **159**, A642.
52. V. Baranchugov, E. Markevich, E. Pollak, G. Salitra and D. Aurbach, *Electrochem. Commun.*, 2007, **9**, 796.
53. D. M. Piper, T. Evans, K. Leung, T. Watkins, J. Olson, S. C. Kim, S. S. Han, V. Bhat, K. H. Oh, D. A. Buttry and S. H. Lee, *Nat. Commun.*, 2015, **6**, 6230.
54. Y. Jin, S. Li, A. Kushima, X. Q. Zheng, Y. M. Sun, J. Xie, J. Sun, W. J. Xue, G. M. Zhou, J. Wu, F. F. Shi, R. F. Zhang, Z. Zhu, K. P. So, Y. Cui and J. Li, *Energy Environ. Sci.*, 2017, **10**, 580.
55. A. Kohandehghan, P. Kalisvaart, K. Cui, M. Kupsta, E. Memarzadeh and D. Mitlin, *J. Mater. Chem. A*, 2013, **1**(41), 12850.
56. J. I. Lee, Y. Ko, M. Shin, H. K. Song, N. S. Choi, M. G. Kim and S. Park, *Energy Environ. Sci.*, 2015, **8**, 2075.
57. X. L. Li, P. F. Yan, X. C. Xiao, J. H. Woo, C. M. Wang, J. Liua and J. G. Zhang, *Energy Environ. Sci.*, 2017, **10**, 1427.
58. M. Ko, S. Chae, J. Ma, N. Kim, H. W. Lee, Y. Cui and J. Cho, *Nat. Energy*, 2016, **1**(9), 16113.
59. I. H. Son, J. H. Park, S. Kwon, S. Park, M. H. Rümmeli, A. Bachmatiuk, H. J. Song, J. Ku, J. W. Choi and J.-m. Choi, *Nat. Commun.*, 2015, **6**, 8393.
60. I. C. Stefan, Amprius. Inc, 2018 DOE Vehicle Technologies Office Annual Merit Review, 2018.
61. T. Placke, R. Kloepsch, S. Duhnen and M. Winter, *J. Solid. State. Electr.*, 2017, **21**, 1939.
62. S. A. Freunberger, *Nat. Energy*, 2017, **2**, 27091.

# Surface Chemistry of Alkali-Ion Battery Cathode Materials

**Muhammad M. Rahman and Feng Lin**

*Virginia Tech, Blacksburg, VA, USA*



## 1 INTRODUCTION

Alkali metal ion batteries have revolutionized the modern world through the widespread implementation of consumer electronic devices.[1,2] Moreover, they hold the potential to reduce environmental pollution and help create a sustainable future by commercializing electric vehicles[3] and storing renewable energy in stationary electrochemical energy storage devices (i.e., batteries).[4,5] Even though these energy storage devices provide reasonable energy and power density, further performance improvement is necessary for their widespread applications in many other energy and powder demanding fields. In order to achieve such goals, careful performance tuning of electrode materials is necessary. Cathode materials are one of the most vital components in alkali metal ion batteries. They most often determine the energy, power, and cycle life of a battery.[6–9] To date, different families of cathode materials have been investigated, including layered oxide materials,[10–14] spinel-structured materials,[15–20] conversion and alloy electrodes.[21–23] Among these, layered and spinel transition metal oxide electrodes have garnered significant attention because of their improved electrochemical performance, low volume expansion, and reasonably long cycle life.[24,25] However, understanding the phenomena that create instability challenges is a prerequisite for further tuning these layered and spinel oxide cathode materials. It has been widely reported that the surface of a material behaves differently from the bulk.[13,26] Performance degradation phenomena mostly originate in the surface.[27–29] Some of the key issues that can affect the long-term stability of layered and spinel oxide cathode materials include, but are not limited to, (i) transition metal dissolution, migration, and segregation;[30–37] (ii) electrolyte decomposition and gas evolution;[38–44] (iii) surface reconstruction;[45–47] and (iv) chemomechanical degradation of cathode materials.[48–53] Transition metal dissolution from layered and spinel oxide cathodes has been extensively observed, which not only results in the loss of active materials from the cathode but also leads to extensive impedance development.[54] Metal dissolution is usually triggered by a complex interaction between the electrode and the electrolyte. The electrolyte, most commonly $LiPF_6$ dissolved in an organic carbonate solvent, is inherently unstable when operating the battery at a voltage higher than 4.5 V versus $Li^+/Li$ because of crossing the electrolyte decomposition potential.[55,56] Moreover, the electrode surface may act as a catalyst to accelerate the degradation process.[57] Thus, electrolyte decomposition is very often observed while cycling a battery, which may result in the formation of different degradation products as well as gas evolution.[58] The degradation may reduce ionic conductivity of the electrolyte and, hence, even curtail the cycle life and performance of a battery. Furthermore, the surface of layered cathode materials may be unstable and can reconstruct from the layered to the spinel/rock salt structure.[46] Such reconstructions may hinder ion diffusion into and out of the active particles and result in impedance development. Lin *et al.* demonstrated that such surface reconstructions are further induced by the electrode–electrolyte interaction.[47] Moreover,

*Inorganic Battery Materials*. Edited by Hailiang Wang and Boniface P.T. Fokwa
 ISBN 978-1-119-43199-2

surface reconstructions may also result in transition metal dissolution from the cathode surface. Lastly, these materials may undergo heterogeneous phase transformation and anisotropic volume changes during electrochemical cycling, which can exert mechanical stress in the cathode particles.[59,60] In order to relieve stress, cracks can form in the particles that can expose fresh surfaces. These exposed fresh surfaces will inevitably accelerate all the degradation processes mentioned earlier through electrode–electrolyte reactions. Moreover, surface reconstruction and gas evolution such as lattice oxygen evolution may also result in crack formation.[50] Hence, it is clear that all these surface phenomena are interconnected. Triggering one process may initiate the other, which may accelerate the overall degradation of cathode materials. Therefore, it is important to understand the origin and propagation of these processes. Many strategies have been attempted to stop, reduce, or delay these processes, which include doping and coating of cathodes to reduce transition metal dissolution, surface reconstruction, and chemomechanical breakdown.[61–65] In addition, purposefully designing the transition metal distribution in a cathode particle can result in stable cathode materials with high electrochemical performance.[66,67] Variations in transition metal distribution on the surface than the bulk can enable a stable cathode electrolyte interphase (CEI), which may enhance the cycle life of a battery. Moreover, adding electrolyte additives is a common strategy to reduce the extent of electrolyte decomposition.[68] Alternatively, additives can also decompose to form a protection/passivation layer during the earlier stages of cycling, thus preventing the electrolyte from extensive decomposition.[69] In this chapter, we present the origin of these key surface-related issues that can undermine the cycle life of a battery. We discuss the performance outcome and formation mechanism of these processes. Moreover, stabilization techniques such as elemental substitution, coating, electrolyte additives, and engineering transition metal distribution are also introduced along with the mechanistic insight into stabilization. We believe that understanding these surface phenomena of cathode materials along with the state-of-the-art stabilization techniques will help rationalize the design principles for the improvement of cathode materials in alkali metal ion batteries.

## 2 SURFACE CHEMISTRY OF CATHODE MATERIALS

### 2.1 Transition Metal Dissolution, Migration, and Segregation

Transition metals have been used as the key elements for most cathode materials because of their ability to function as rapid redox centers for electrochemical reactions and to form frameworks with oxygen anions to host alkali metal ions. They also play a crucial role in determining the overall performance and cycling stability of batteries. One of the key issues with long-term stability of cathode materials is transition metal dissolution and segregation from the surface of cathode particles. The dissolved transition metal cations can subsequently migrate and deposit onto anodes. Such dissolution is one of the most exigent reasons for capacity fading because it results in both loss of active materials from cathodes and impedance development on the anode side because of the ensuing deposition of the dissolved transition metal ions as metals through reduction. This, in turn, increases the electronic conductivity of the SEI on the anode side, which leads to aggravated electrolyte decomposition. Thus, the deposited transition metals on the anode can act as an active reaction center for accelerated electrolyte decomposition.[34,37,70,71] Similar cases were widely observed and comprehensively documented in Mn-based cathodes, most prominently in spinel $LiMn_2O_4$ cathodes.[37] However, Mn dissolution is not uncommon in other Mn-based cathodes including layered oxide cathode materials.[30] The aqueous acidic treatment of $LiMn_2O_4$ causes delithiation of the material and subsequent formation of the delithiated $\lambda$-$MnO_2$ phase.[72] This happens along with the disproportionation reaction of $Mn^{3+}$ ions into $Mn^{2+}$ and $Mn^{4+}$ ions. $Mn^{2+}$ ions then dissolve in the reaction medium because of the increased solubility. Such a dissolution process begins from the surface. With the removal of $Li^+$ from the surface of the material, more and more $Li^+$ can be removed from the bulk because of the diffusion of the $Li^+$ from the bulk to the surface due to the concentration gradient. During delithiation, electrons can hop between $Mn^{3+}$ that can lose/accept electrons to produce $Mn^{2+}$ that is susceptible to dissolution. This process can continue until the material is fully delithiated. Figure 1 summarizes electron hopping in $Mn^{3+}$ cations and their subsequent dissolution through the aforementioned disproportionation reaction.

For a long time, the disproportionation reaction was the accepted mechanism to explain the transition metal dissolution from cathode materials. However, in an electrochemical cell, this can only be true at low operating voltages where Mn cations can stay in the $Mn^{3+}$ oxidation state. However, operating at a higher voltage, $Mn^{3+}$ will be oxidized to $Mn^{4+}$ due to electrochemical oxidation, and such a mechanism is not applicable to explain transition metal dissolution. Therefore, an improved understanding of the mechanism is necessary to explain the dissolution. At higher voltages, the phase transformation mechanism can explain Mn dissolution from the spinel-structured $LiMn_2O_4$.[73] Operating at a high voltage, there is a two-phase reaction with the soluble $MnO_2$ as a phase. The

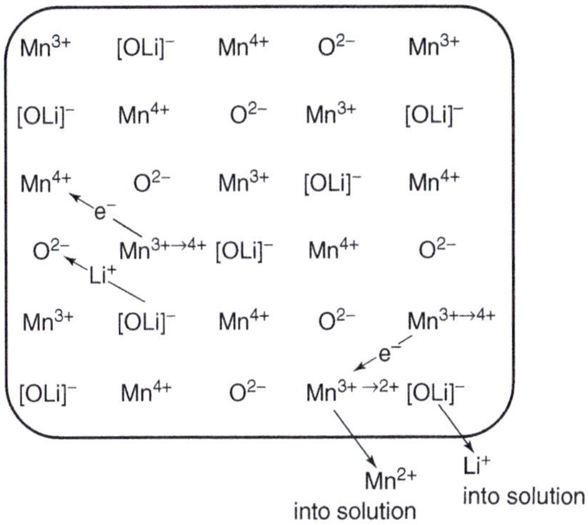


Figure 1 Electron hopping mechanism to produce soluble $Mn^{2+}$ ions from $LiMn_2O_4$ material when treated with aqueous acidic solution. $Mn^{3+}$ can accept/lose electron to produce $Mn^{2+}$, while the concentration of the surface $Mn^{4+}$ increases.[72] [Reproduced with permission from J. C. Hunter, J. Solid State Chem., 1981, 39, 142–147 © Elsevier, 1981]

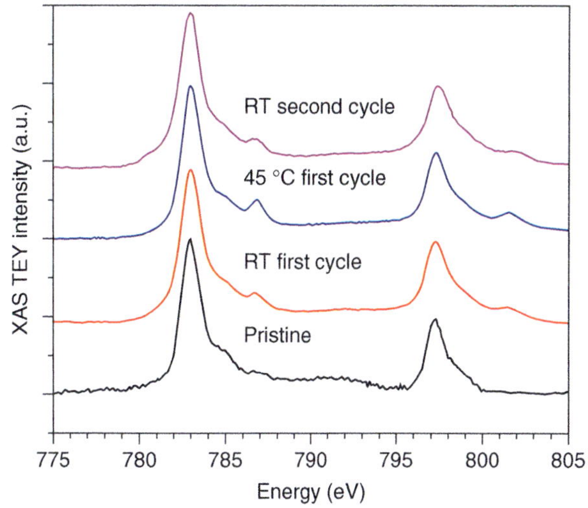


Figure 2 Soft X-ray absorption spectroscopy (XAS) spectra of Co L-edge after first cycle and second cycle at room temperature and after first cycle at 45 °C cycling.[75] [Reprinted with permission from J. D. Steiner, L. Mu, J. Walsh, M. M. Rahman, B. Zydlewski, F. M. Michel, H. L. Xin, D. Nordlund and F. Lin, ACS Appl. Mater. Interfaces, 2018, 10, 23842–23850 © 2018 American Chemical Society]

reaction can be shown as follows:

$$LiMn_2O_4 \rightarrow LiMn_{2-y}O_{4-2y} + yMnO_2 \text{ (soluble)}$$

Moreover, at a higher degree of delithiation, the spinel $Li_{1-x}Mn_2O_4$ becomes more reactive and acts to decompose the organic electrolyte. The decomposition of the electrolyte may lead to the formation of acidic species, which, in turn, can accelerate the transition metal dissolution.[41] As the degree of delithiation is higher at higher charging voltages, this may explain the dissolution mechanism at higher states of charge.

The dissolution of transition metals beyond Mn cations was also observed in many cathode materials. Co dissolution from the layered $LiCoO_2$ cathode was reported, which can be reasoned as the migration of the $Co^{3+}$ into the $Li^+$ layer. An irreversible phase transformation of rhombohedral phase to monoclinic phase was also observed.[74] Similar Co dissolution was also reported in other layered oxide cathode materials including $LiNi_{1-x-y}Mn_xCo_yO_2$ (NMC) material for lithium-ion batteries. Recently, our research has shown the Co dissolution from $LiNi_{0.8}Co_{0.1}Mn_{0.1}O_2$ through soft X-ray absorption spectroscopy (soft XAS).[75] The soft XAS spectra (Figure 2) clearly show a peak at around 786 eV that can be attributed to the Co cation complexes with organic ions. The complex could have been generated by the reaction between dissolved Co cations and electrolyte components.

Normally, transition metal dissolution starts from the very beginning of electrochemical cycling and keeps on aggravating in subsequent cycles. Recent studies on the layered NMC materials for lithium-ion batteries have shown evidence of both Co and Mn dissolution.[30,32] However, Ni dissolution from such materials is much harder due to the formation of NiO-like rock salt phase at high states of charge. Pieczonka *et al.* demonstrated the presence of Mn nanoparticles on the graphite anode after cycling with NMC in a full cell for 100 cycles (Figure 3).[30] Zhang *et al.* also observed similar transition metal dissolutions in a full cell with the NMC cathode and graphite anode.[35] There was deposition of nanoparticles on the surface of the anode at higher charging voltages. Energy-dispersive X-ray spectroscopy (EDX) on those particular nanoparticles also proved the presence of Mn, Co, and Ni, with the amount of Mn being higher than the other two. Thus, it is proved that transition metal dissolution is not only observed in spinel Mn-based materials but also in layered oxide materials.

Layered oxide cathodes for sodium-ion batteries also undergo similar transition metal dissolution. The dissolution problem is more severe in Mn-based cathodes. In layered oxide cathodes for sodium-ion batteries, Mn dissolution can take place in the early stage of cycling.[76] In a recently reported compositionally heterogeneous $Na_{0.9}Cu_{0.2}Fe_{0.28}Mn_{0.52}O_2$ material, we observed no significant changes in the Fe and Cu oxidation states at the surface (Figure 4a,b). In contrast, Mn underwent extensive reduction at the beginning of first charge (Figure 4c), which is most likely related to the CEI formation. Such reduction is most often irreversible, and Mn did not return to its pristine oxidation state ($Mn^{4+}$) even at higher states of charge (4.0 V in Figure 4c). Therefore, even though starting with predominantly $Mn^{4+}$ in the pristine material,

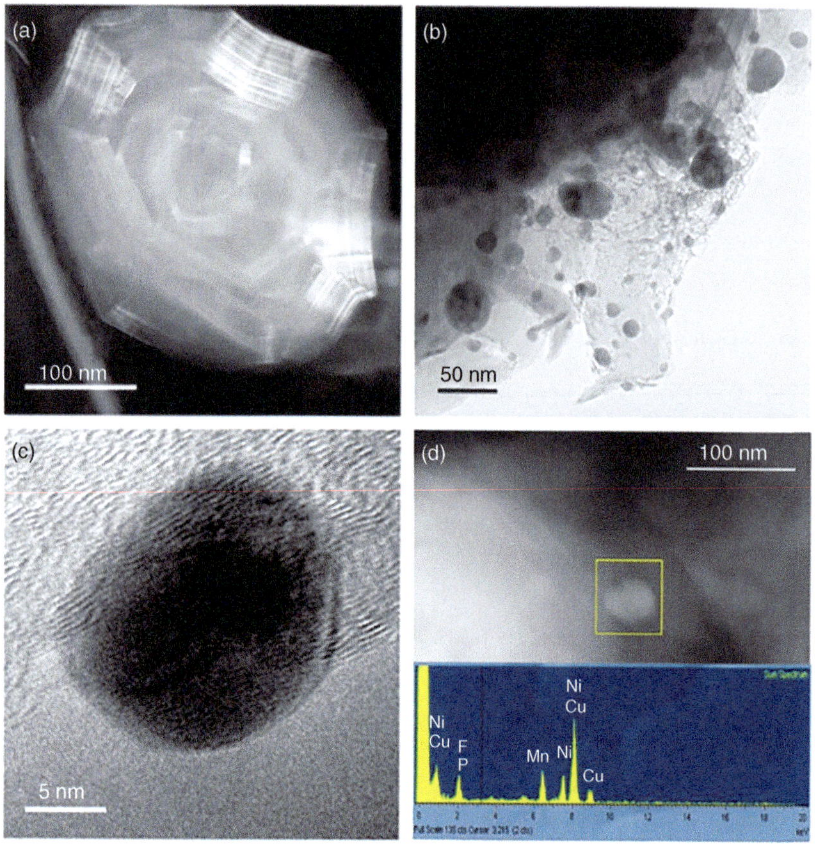


**Figure 3** (a) Microstructure of a pristine graphite particle as visualized through scanning transmission electron microscopy–annular dark-field imaging (STEM–ADF). (b) TEM image of nanoparticles deposited on the graphite anode after cycling. (c) High-resolution TEM image of an Mn nanoparticle deposited on the graphite anode after cycling. (d) Energy-dispersive X-ray spectroscopy (EDX) on the highlighted region on the graphite anode showing the presence of Ni and Mn. The Cu signal was from the Cu TEM grid.[30] [Reprinted with permission from N. P. W. Pieczonka, Z. Liu, P. Lu, K. L. Olson, J. Moote, B. R. Powell and J.-H. Kim, J. Phys. Chem. C, 2013, 117, 15947–15957 © 2013 American Chemical Society]

Mn can be rapidly reduced to $Mn^{3+}$ and $Mn^{2+}$ upon electrochemical cycling thus leading to Mn dissolution.

This result signifies the crucial role of Mn in the interfacial reaction between the cathode and the electrolyte, which, in turn, also determines the overall stability of the cathode material. This irreversible change is expected to aggravate at progressive electrochemical cycling. Indeed, through transmission X-ray tomography, we have shown that $Na_{0.9}Cu_{0.2}Fe_{0.28}Mn_{0.52}O_2$ suffered an extensive surface elemental rearrangement upon long-term cycling. We observed that the surface of the material after 400 cycles was more Mn populated than the pristine material (Figure 5). We conjectured that Mn segregation to the surface resulted from Mn dissolution and migration from the primary particles in the bulk. A similar observation was reported in Li/Mn-rich layered oxide cathodes.[36]

Even though it is widely accepted that Mn dissolves as $Mn^{2+}$ in the electrolyte, the fate of such dissolved $Mn^{2+}$ ions is highly debated. One of the rational explanations would be the dissolved $Mn^{2+}$ migrates onto the anode side due to the concentration gradient and gets deposited there. $Mn^{2+}$ is expected to be reduced to metallic Mn because of the lower operating potential of anodes such as lithium metal and graphite anodes relative to the redox potential of $Mn^{2+}/Mn^0$.[34] This process can increase the irreversible capacity derived from a cell. However, $Mn^{2+}$ ions may not undergo reduction on the anodes that are operated at voltages higher than that of the $Mn^{2+}$ reduction reaction. Zhan *et al.* studied the relationship between capacity fading and transition metal dissolution in the $LiMn_2O_4$ spinel cathodes and a series of anode materials including delithiated $LiFePO_4$, $Li_4Ti_5O_{12}$, graphite, and Li.[34] In contrast to the traditional belief, this study showed that the oxidation state of the deposited Mn was $Mn^{2+}$ rather than $Mn^0$. Thus, $Mn^{2+}$ can be deposited on the surface of the anodes. In order to explain such phenomena, the authors proposed an ion exchange metathesis reaction, where $Mn^{2+}$ exchanges position with mobile lithium alkyl carbonate from the SEI layer. The mechanism in Figure 6 hypothesizes the aforementioned metathesis reaction.

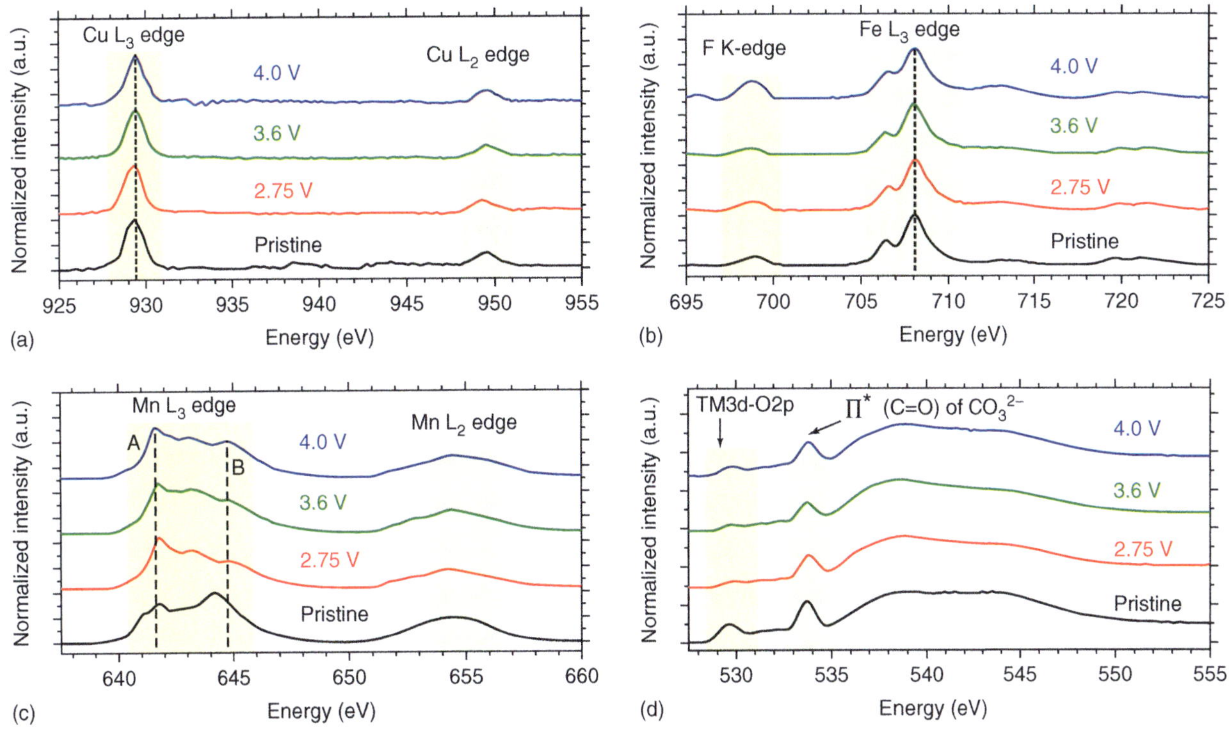


**Figure 4** Soft XAS spectra in the total electron yield (TEY) mode of (a) Cu L-edge, (b) Fe L-edge, and (c) Mn L-edge. Peaks labeled as "A" and "B" vary in intensity as a function of charge. (d) O K-edge for $Na_{0.9}Cu_{0.2}Fe_{0.28}Mn_{0.52}O_2$ electrodes at different states of charge. The variation in the intensity of O K-edge is similar to that of the Mn L-edge, signifying that the Mn3d–O2p hybridization may be dominant on the surface during cycling.[76] [Reproduced with permission from M. M. Rahman, Y. Xu, H. Cheng, Q. Shi, R. Kou, L. Mu, Q. Liu, S. Xia, X. Xiao, C.-J. Sun, D. Sokaras, D. Nordlund, J.-C. Zheng, Y. Liu and F. Lin, Energy Environ. Sci., DOI:10.1039/C8EE00309B © Royal Society of Chemistry, 2018]

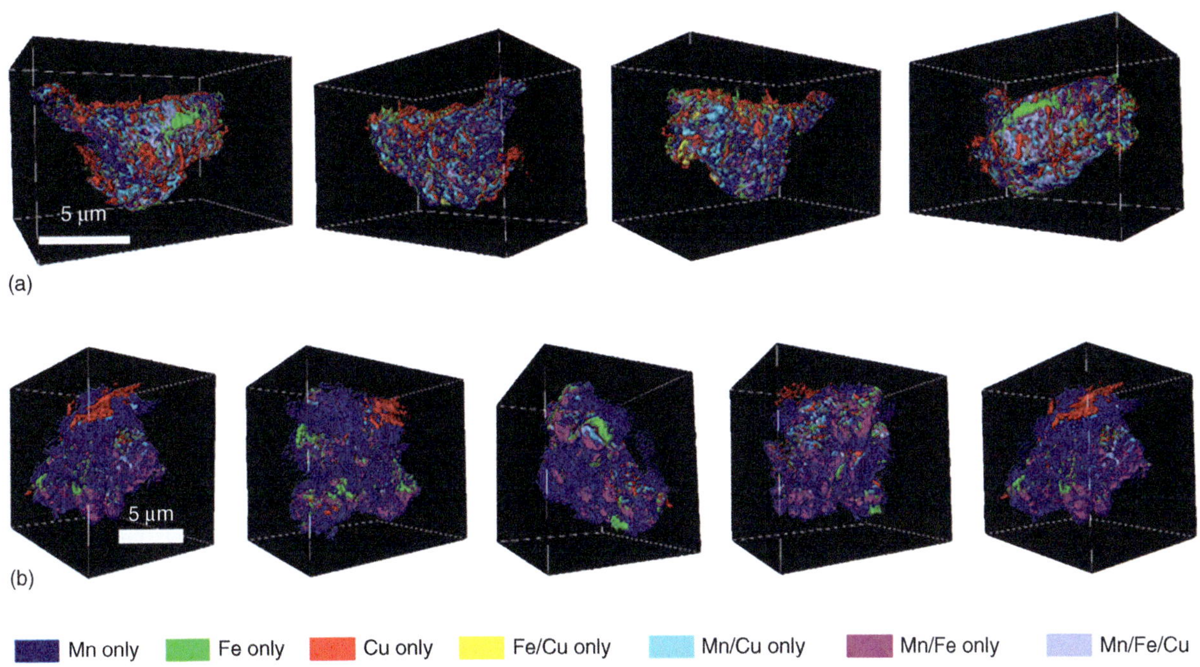


**Figure 5** Transmission X-ray tomography of (a) pristine $Na_{0.9}Cu_{0.2}Fe_{0.28}Mn_{0.52}O_2$ and (b) cycled $Na_{0.9}Cu_{0.2}Fe_{0.28}Mn_{0.52}O_2$ at 1C for 400 cycles (2–4 V vs $Na^+/Na$). The color bars below show different elemental association on the particle. Note that the Mn only association significantly increased after cycling, indicating Mn segregation to the surface of the particle upon long-term cycling

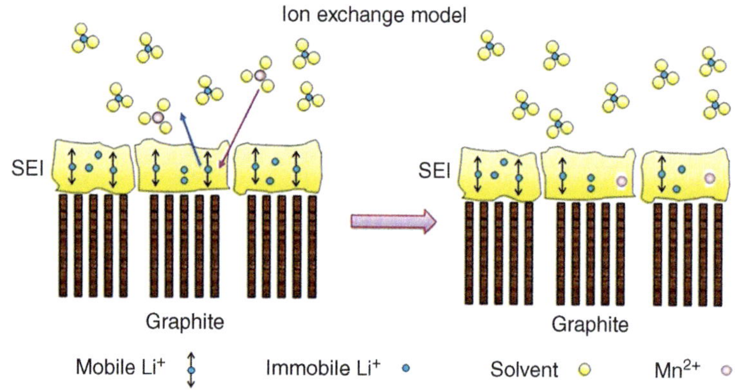


**Figure 6** Ion exchange mechanism proposed by Zhan *et al.* in order to explain the $Mn^{2+}$ deposition on the anode surface.[34] [Reproduced with permission from C. Zhan, J. Lu, A. Jeremy Kropf, T. Wu, A. N. Jansen, Y.-K. Sun, X. Qiu and K. Amine, Nat. Commun., 2013, 4, 2437 © Springer, 2013]

## 2.2 Electrolyte Decomposition and Gas Evolution

In an electrochemical cell, the electrolyte allows for ion conduction between the electrodes, thus maintaining the reversibility of the electrochemical reaction. The stability of the electrolyte plays an important role in determining the cycle life of a battery. Nearly all commercial alkali metal ion batteries use organic electrolytes. Some of the most popular organic electrolyte solvents include ethylene carbonate (EC), dimethyl carbonate (DMC), diethyl carbonate (DEC), ethyl methyl carbonate (EMC), and propylene carbonate (PC). The salts used in lithium- and sodium-ion batteries are mostly lithium hexafluoro phosphate ($LiPF_6$), lithium perchlorate ($LiClO_4$), lithium bis(trifluoromethanesulfonyl)imide (LiTFSI), sodium hexafluoro phosphate ($NaPF_6$), and sodium perchlorate ($NaClO_4$). A typical electrolyte would be a solution of one or more of the aforementioned solvents and salts.

The stability of the electrolyte and the electrodes are interconnected. The electrolyte may undergo side reactions with both electrodes.[40–42,77] The anode of a battery may act as a reducing agent and undergo redox reaction with the aprotic solvent of the electrolyte. The $Li^+$ and $Na^+$ in the battery along with the aprotic solvent can form insoluble Li and Na compounds, which can deposit on the interfacial region of the electrode materials.[58] Reduction reactions of the electrolyte can take place on both metal anodes and insertion-type carbon anodes. On the contrary, the cathode can act as an oxidizing agent and oxidizes the aprotic solvent, which forms insoluble reaction products on the cathode surface. The deposited products on the electrodes are usually ion conductive and protect the electrode–electrolyte interface from extensive degradation. In general, there is an urge to operate batteries at high voltages to achieve energy density closer to the theoretical values. However, operating at a higher voltage in lithium-ion batteries (>4.5 V) will increase electrolyte oxidation.[55] In such cases, the stability of the interphase can be suppressed by the degradation rate of the electrolyte. Through these reactions, electrolyte solvents form various degradation products including $Li_2CO_3$, $CO_2$, and other soluble/insoluble products. Among many chemical entities found in these interphases, the most predominant are aldehydes, hydrogen fluoride, alcohols, lithium ethoxide, and products of transesterification reactions.[42] Along with the interphase formation, transition metals from the cathode surface also undergo gradual dissolution due to the reaction with the electrolyte solvent. Thus, it is necessary to prevent extensive electrolyte decomposition to ensure long-term cycling stability.

Even though the electrolyte decomposition is a widely observed phenomenon, the mechanisms of the decomposition processes are still not clearly understood. In order to have a clear understanding of the decomposition mechanism, it is necessary to identify the reaction products. Takeda *et al.* identified various degradation products from the electrolyte (PC) using liquid chromatography (LC) and mass spectrometry (MS) at different temperatures (30 and 60 °C) and on long-term cycling (up to 40 cycles).[44] The analysis indicated various organophosphates and other organic compounds dissolved in the electrolyte. Four compounds of mainly carbonates and oligomers were identified at both 30 °C and 60 °C. Besides these compounds, eight organophosphates were identified at 60 °C only. Therefore, the study indicated that at higher temperatures degradation is much more severe and complex. The authors hypothesized that the presence of $POF_3$ in the electrolyte triggers the formation of these organophosphates by acting as an intermediate. $POF_3$ was formed by the decomposition of electrolyte solvents in the presence of P-containing salts such as $LiPF_6$.

A more mechanistic understanding of the electrolyte decomposition process can be found from the work of Jarry *et al.*[42] They employed Raman and fluorescence spectroscopic measurements on $Li_xNi_{0.5}Mn_{1.5}O_{4-\delta}$ after cycling and on graphite electrodes (Figure 7). From the fluorescent background profile (Figure 7a,b), the presence of Mn(III) tris(acetylacetonate) on the surface of the cathode and Mn(II) bis(acetylacetonate) on the surface of the anode were identified. Moreover, normalized X-ray absorption near edge structure (XANES) spectra supported the presence of these species (Figure 7c–f) by comparing with the XANES spectra of various standards.

Thus, it can be inferred that with the decomposition of the electrolyte, the dissolved Mn(III) ion formed a complex with acetylacetonate and subsequently migrated to the anode. The migrated Mn(III) complex was then reduced to Mn(II) complex upon reaction with the graphite electrode. Figure 8 shows the postulated formation mechanism of the Mn/Ni complex with such a diketonate ligand.

According to the proposed mechanism, a proton-coupled electron transfer (PCET) mechanism can be accounted for the formation of these metal complexes. The whole process is initiated upon delithiating the material at a higher voltage. On charging to above 4.2 V, Mn(III) ions of $Li_xNi_{0.5}Mn_{1.5}O_{4-\delta}$ near the oxygen vacancies are oxidized to Mn(IV). Mn(IV) can form coordination bonds with the DEC molecules due to the orbital energy overlap between the Mn(IV) d orbitals and the carbonyl group of DEC. The lone pair electron of oxygen of the carbonyl group can then reduce Mn(IV) to Mn(III) along with the formation of the DEC cation radical. The surface oxygens of $Li_xNi_{0.5}Mn_{1.5}O_{4-\delta}$ can then attack the DEC cation radical, which results in hydroxyl species and acetaldehyde formation. The acetaldehyde then further reacts with the hydroxyl group, producing acetaldehyde radical. Two such radicals can combine to make a bidentate ligand, which then weaken the Mn—O bond of the $Li_xNi_{0.5}Mn_{1.5}O_{4-\delta}$ material, resulting in the formation of the Mn(III) diketonate complex. The electrolyte decomposition is continued by the repetition of such reactions.

Electrolyte decomposition reactions can be moderated by the nature of cathode materials. Li/Mn-rich layered cathodes with local $Li_2MnO_3$ domains are one such example. These cells deliver a large first charge capacity, which is much higher than that of conventional layered oxide cathodes, partially due to the Li extraction from $Li_2MnO_3$ domains and oxygen extraction.[78,79] The extracted oxygen during the first charge can be evolved as molecular oxygen or react with the electrolyte in the electrode–electrolyte interphase to form various degradation products including gases. Freunberger *et al.* proposed that the reactive oxygen radical upon reaction with the electrolyte can form various organic and inorganic carbonates, acetates, $CO_2$ and $H_2O$.[80] Berg and coworkers used online electrochemical mass spectrometry (OEMS) to identify numerous gases formed from the decomposition of the electrolyte including $POF_3$, $CO_2$, $H_2O$, $H_2$, and $O_2$.[56] However, there is still much debate on the actual origin of the evolved $CO_2$ and CO during cycling of an electrochemical cell. In addition to the electrolyte decomposition, inorganic carbonate impurities on the cathode surface may also account for gas evolution such as $CO_2$ and/or CO. Renfrew *et al.* demonstrated that the surface $Li_2CO_3$ impurities on both Ni-rich and Li/Mn-rich cathodes account for most of the evolved $CO_2$ and CO while the lattice oxide accounts for the evolved oxygen.[81] In order to clearly identify the source of these gases, the authors did isotope labeling of the lattice oxygen of the cathode materials as well as the surface carbonate impurities with $^{18}O$ to decouple the electrode from the electrolyte. The authors galvanostatically charged and discharged the material at a 0.1 Li $h^{-1}$ rate followed by a period of rest. $CO_2$ and CO evolution was detected at low voltage charging as can be seen in the blue-shaded region of Figure 9(a). However, at a higher voltage of charging (at about 4.8 V), the amount of $CO_2$ and CO increased significantly (the red-shaded region) including the observed evolution of molecular oxygen (the red-shaded region in Figure 9a). The authors argued that the evolved $CO_2$ and CO came mostly from the $Li_2CO_3$ impurities on the surface of the cathode while the source of the molecular oxygen was the crystal lattice. Their calculation showed that the total $^{18}O$ enrichment was 23%, which is significantly larger than the evolved $CO_2$/CO and the total $Li_2CO_3$ impurities. Moreover, they argued that based on the binomial distribution of the $^{18}O$, the evolved oxygen had to come from the lattice oxygen of the cathode material. Besides, the measured isotope ratio in $CO_2$ followed closely to that of the predicted ratio based on the decomposition from the $Li_2CO_3$ alone (Figure 9c). Thus, $Li_2CO_3$ was pinpointed as the predominant source of $CO_2$/CO, even though minor contribution from the electrolyte decomposition cannot be ruled out completely.

Even though the source of evolved gas is highly debatable, these gases can be detrimental to the cell performance because they can create excess pressure on the cell housing and can compromise the physical integrity of the cell. Moreover, the electrolyte decomposition can reduce the lifetime of a battery by disrupting the reversibility of the electrochemical reaction. Hence, in order to ensure safer and long-term cycling of the cell, it is important to stabilize the electrolyte against undesirable side reactions. Various approaches including electrolyte additives have been gaining popularity to fulfill such objectives. These approaches are discussed in detail later.

### 2.3 Surface Reconstruction

Surface structural transformation of layered oxide materials can directly impact the electrochemical performance of alkali metal ion batteries. Such surface reconstruction can adversely affect the implementation

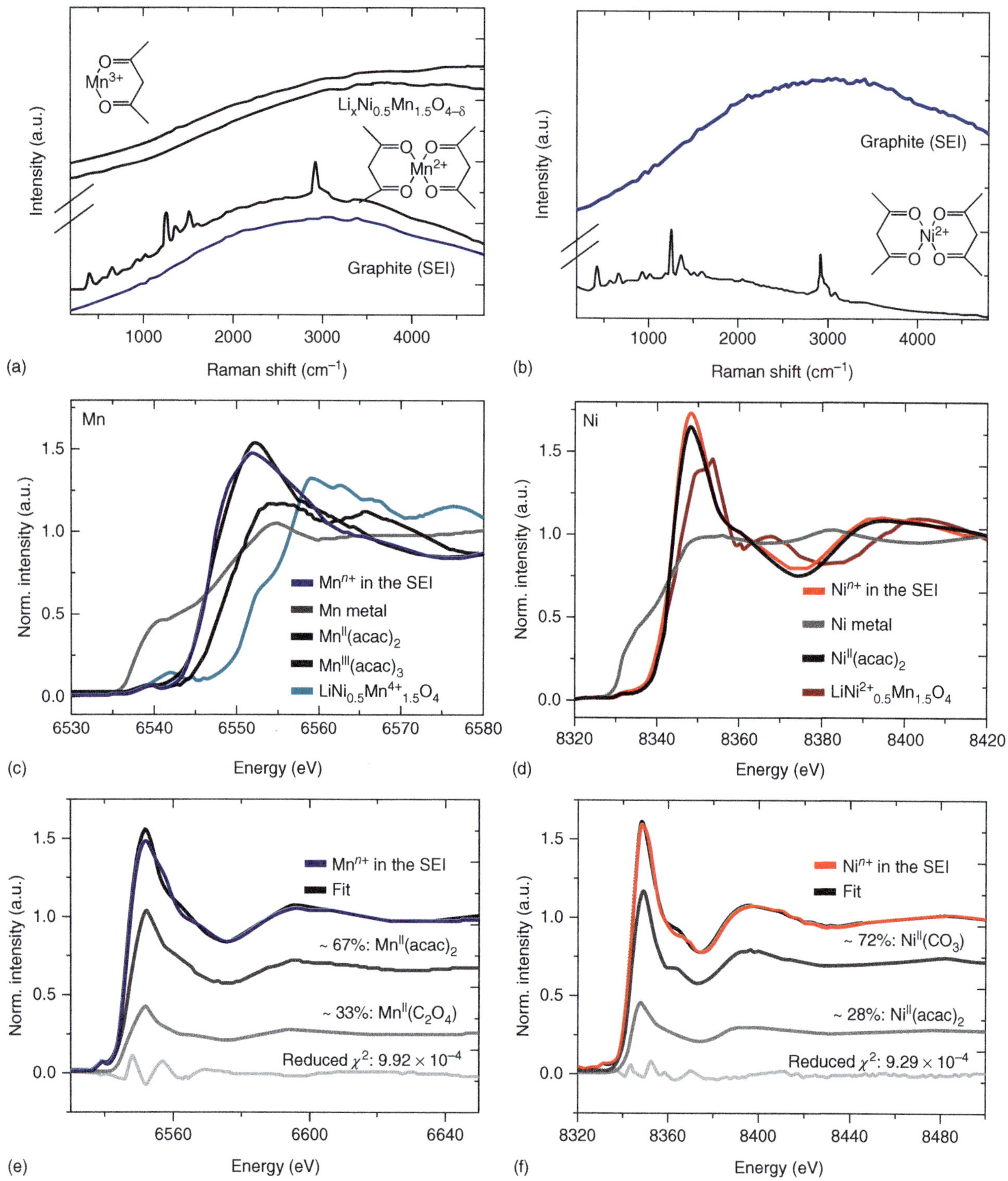


**Figure 7** Identifying the Mn and Ni species on $LiNi_{0.5}Mn_{1.5}O_{4-\delta}$ electrode and graphite anode through (a) and (b) ex situ Raman spectra indicating the presence of Mn(III), Mn(II) acetylacetonates and Ni(II) acetylacetonates, (c) and (d) normalized X-ray absorption near edge spectra (XANES) of Mn and Ni K-edges at the cycled graphite anode, $LiNi_{0.5}Mn_{1.5}O_{4-\delta}$, and of different standards. The peaks of the standard samples verified the presence of various Mn(II/III) acetylacetonate species, and (e) and (f) XANES spectra with linear combination fitting of the graphite electrodes after cycling with Mn(III), Mn(II), and Ni(II) acetylacetonate.[42] [Reprinted with permission from A. Jarry, S. Gottis, Y.-S. Yu, J. Roque-Rosell, C. Kim, J. Cabana, J. Kerr and R. Kostecki, J. Am. Chem. Soc., 2015, 137, 3533–3539 © 2015 American Chemical Society]

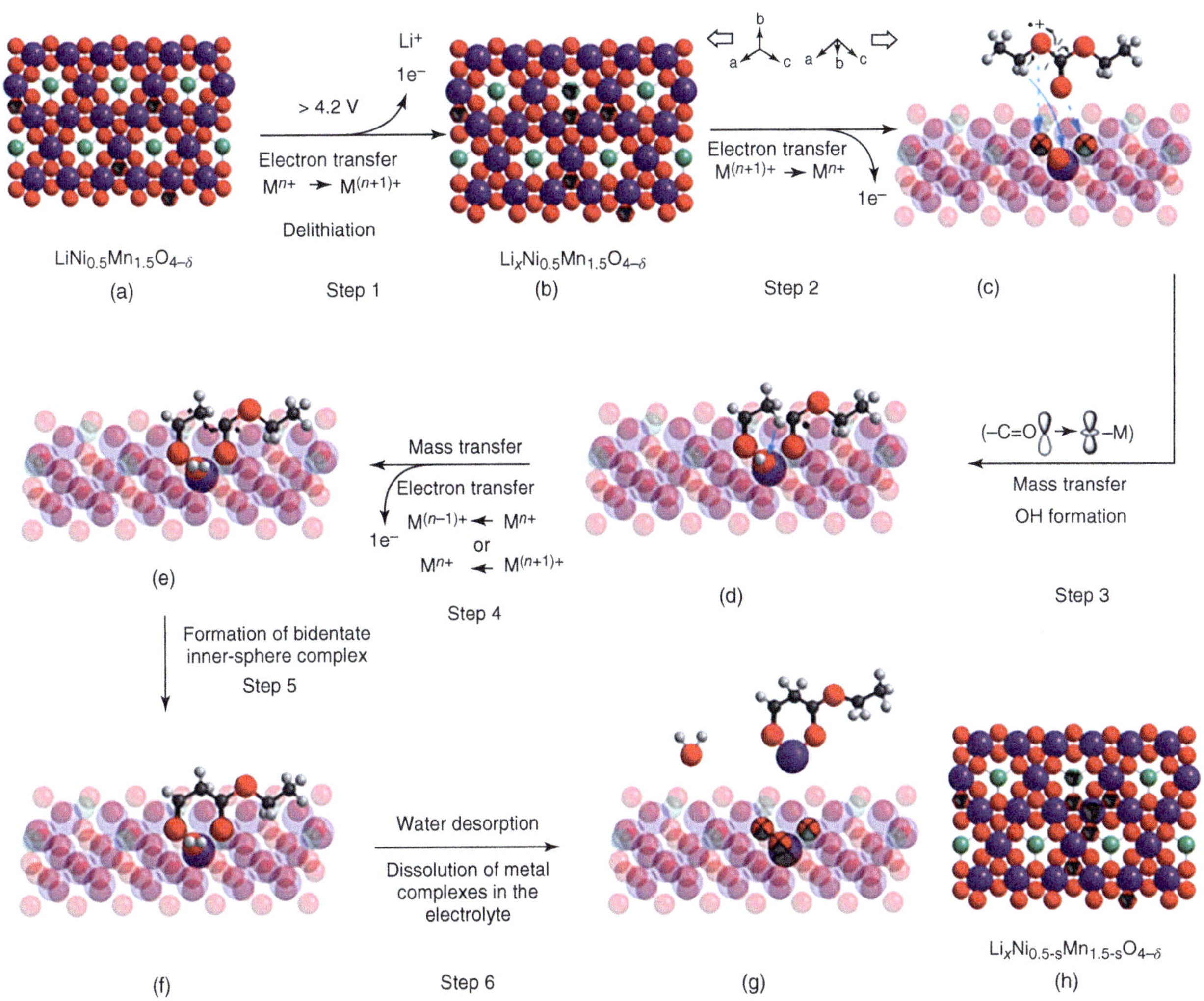


**Figure 8** Proposed formation mechanism of the metal acetylacetonate complexes through oxidation of diethyl carbonate (DEC). (a) Pristine $LiNi_{0.5}Mn_{1.5}O_{4-\delta}$ surface along the (111) crystal plane. Step 1: delithiation of the material and oxidation of the transition metals along with the formation of oxygen vacancies (b). Step 2: metal reduction and initiation of the cation radical through electron transfer (c). Step 3: cation radical adsorption on the surface of the cathode along with C—H bond cleavage to form $OH^-$ and acetaldehyde adsorption on the nearby oxygen site (d). Step 4: formation of $H_2O$ with the interaction of C—H bond and the OH along with surface metal reduction (e). Step 5: radical termination with C—C bond and complex formation (g). Step 6: surface product desorption including water and the complex. (h) The surface of the electrode after the reaction.[42] [Reprinted with permission from A. Jarry, S. Gottis, Y.-S. Yu, J. Roque-Rosell, C. Kim, J. Cabana, J. Kerr and R. Kostecki, J. Am. Chem. Soc., 2015, 137, 3533–3539 © 2015 American Chemical Society]

of high-voltage batteries because the reconstruction is accelerated by high-voltage operation as discussed later. Moreover, these transformations can be directly related to the observed high first cycle Coulombic inefficiency, impedance development, poor rate capability, and capacity fading.[47] Thus, it is important to understand such degradation mechanisms so that preventive measures can be taken to increase the cycle life of batteries.

During cycling, a structural transformation from the layered phase to the rock salt phase and/or spinel phase is widely observed on the surface region of the particle.[45–47] These transformations can result from electrochemical activation by the removal of alkali metal ions from the material as well as a complex interaction between the electrode and the electrolyte. Recent studies on both layered NMC cathodes and Li/Mn-rich cathodes have demonstrated the presence of these reconstruction layers.

Zheng *et al.* reported surface degradation of $Li[Li_{0.2}Ni_{0.2}Mn_{0.6}]O_2$ particles.[82] This Li/Mn-rich material shows a large irreversible voltage plateau on the first charge and the charge capacity achieved is beyond what can be delivered by oxidizing $Ni^{2+}$ alone. The extensive $Li^+$ removal from the cathode in the high voltage range

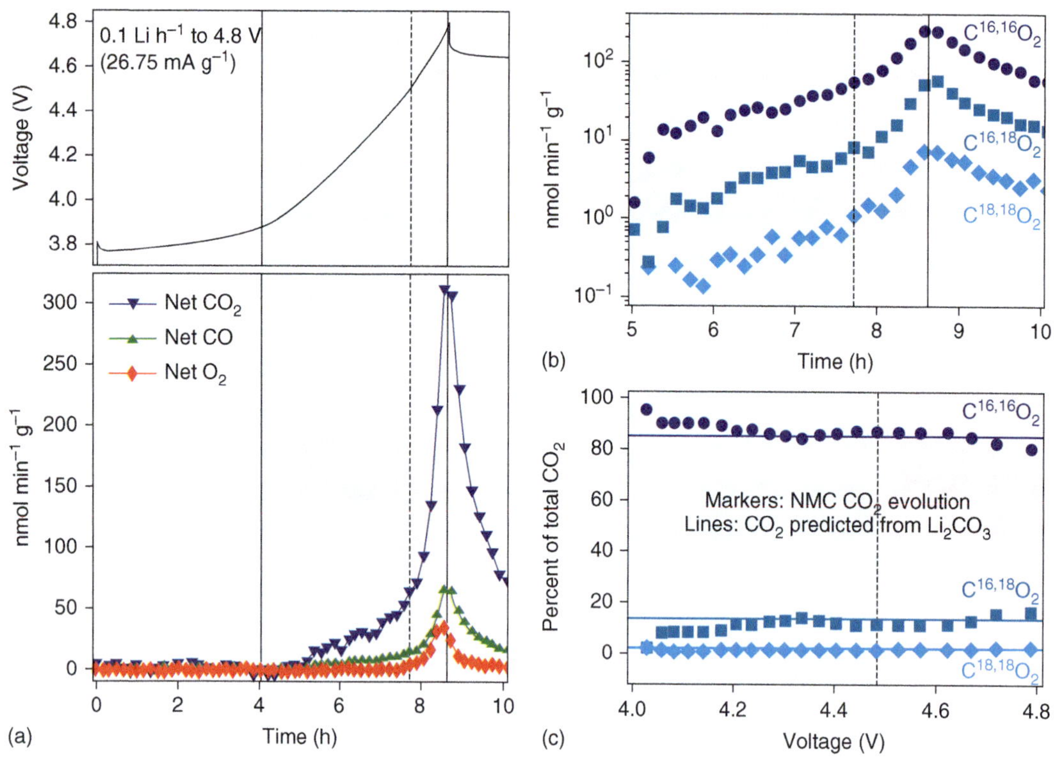


**Figure 9** Oxygen evolution study of $LiNi_{0.6}Co_{0.2}Mn_{0.2}O_2$ during galvanostatic cycling. The oxygen is isotopically labeled by $^{18}O$ for decoupling the effect of the electrolyte. (a) Studying different gas evolution during cycling at 0.1 Li h$^{-1}$ rate. $CO_2$/CO evolution was detected in the blue-shaded region and the red-shaded region indicates where $O_2$ evolution was detected along with these gases. (b) Evolution rate of different isotopic $CO_2$ species ($^{16}O$ and $^{18}O$) and (c) assessment of various isotopic $CO_2$ evolution against voltage. The lines represent the predicted $CO_2$ evolution from $Li_2CO_3$ degradation.[81] [Reprinted with permission from S. E. Renfrew and B. D. McCloskey, J. Am. Chem. Soc., 2017, 139, 17853–17860 © 2017 American Chemical Society]

along with a large lattice oxygen loss from the material may account for this high first charge capacity. Lattice oxygen loss may also result in structural changes from the layered to the spinel structures at the surface of the particles. Such transformation is an outcome of the migration of the transition metal into the Li site along with distortion of the lattice oxygen. The migration of transition metals and subsequent surface structural reconstruction contribute to the observed capacity fading and voltage instability. To visualize these surface changes, the authors used aberration-corrected scanning transmission electron microscopy (STEM) coupled with electron energy loss spectroscopy (EELS) (Figure 10). They observed degraded surface features along with microcracks and fragmented regions on the surface of the cathode particles (Figure 10a–d). The fragmented particles still contained all the transition metals but in a reduced states compared to the bulk (Figure 10j–l).

Moreover, they found that the surface layered structure was completely destroyed. In addition, transition metal migration to the Li site was also evident and, thus, the formation of the spinel phase from the layered structure. Similar surface reconstruction from the layered to spinel was also observed by Gu *et al.* in layered composite oxide cathodes, for example, $Li_{1.2}Ni_{0.2}Mn_{0.6}O_2$ and $Li_{1.2}Ni_{0.1}Mn_{0.525}Co_{0.175}O_2$ on operating at higher voltages (>4.5 V) after long-term cycling (300 cycles).[83]

Analogous transition metal migration onto the Li site was also reported in $Li_{1.2}Mn_{0.61}Ni_{0.18}Mg_{0.0.1}O_2$ by Boulineau *et al.*[84] In the cycled material, a progressively increasing Ni content was found when going from the bulk to the surface with a large quantity of Ni in the interslab region where the Li sits. This again gives an indication of surface reconstruction from layered to spinel/rock salt structure (Figure 11a,b) on electrochemical cycling.

The concomitant densification of the material was also observed upon transition metal migration. The authors contemplated that transition metals, mostly Mn, can migrate to the bulk Li vacancies created by the removal of $Li^+$ during cycling. However, as each Mn is surrounded by 3Li/Ni atoms, removing Li can enable bulk Mn diffusion while Ni is locked in its place by Mn. Consequently, with the progressive diffusion of Mn into the Li vacancies in the bulk, a concentration gradient was observed with Ni-rich surface and Mn-rich bulk. Moreover, this migration would be aided by the removal of oxygen, which is initiated by the

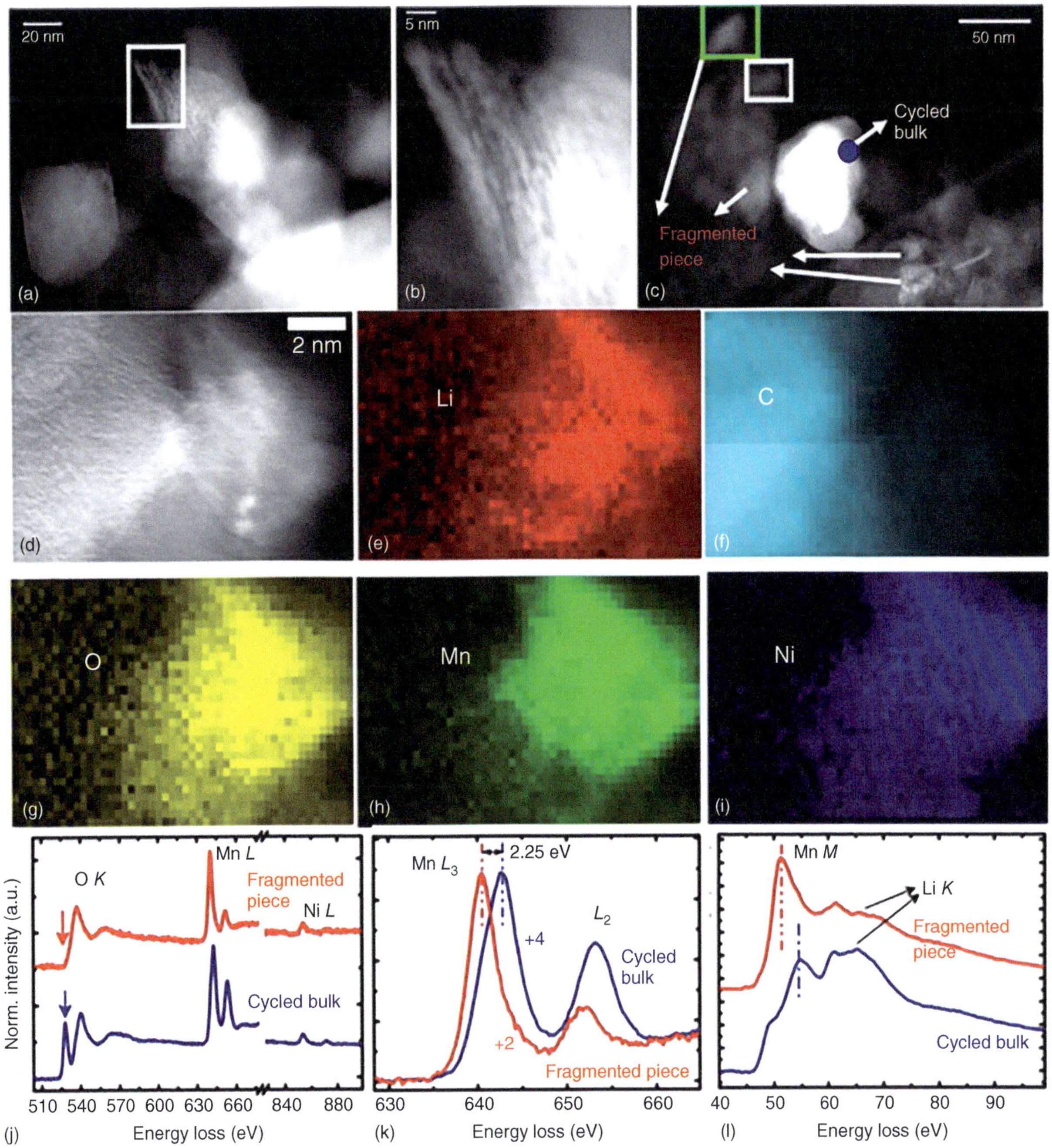


**Figure 10** (a) and (c) STEM images of the material after cycling. (b) and (d) Zoomed-in images of the boxed portion of (a) and (c), respectively. Elemental mapping of (e) Li, (f) C, (g) O, (h) Mn, and (i) Ni of the cycled cathode nanoparticles. (j) O K-, Mn L-, and Ni L-edges in fragmented pieces and cycled bulk of the particles. (k) Mn L-edge comparison and (l) comparison of Mn M-edge to that of the Li K-edge.[82] [Reprinted with permission from J. Zheng, M. Gu, J. Xiao, P. Zuo, C. Wang and J.-G. Zhang, Nano Lett., 2013, 13, 3824–3830. © 2013 American Chemical Society]

high degree of Li removal on operating at high voltages. These migrations result in shrinkage of the particles due to densification. Thus, surface reconstruction in Li/Mn-rich layered oxide materials from the layered to spinel structure also results in volume shrinkage, which may accelerate the degradation.

Xu *et al.* used the first principle calculation to explain the surface reconstruction phenomenon in Li/Mn-rich layered oxide cathode materials.[85] In these materials, Li sitting in the transition metal layer are edge shared with three octahedral Li in the lithium layer. During charging, when these octahedral Li are removed, they migrate from

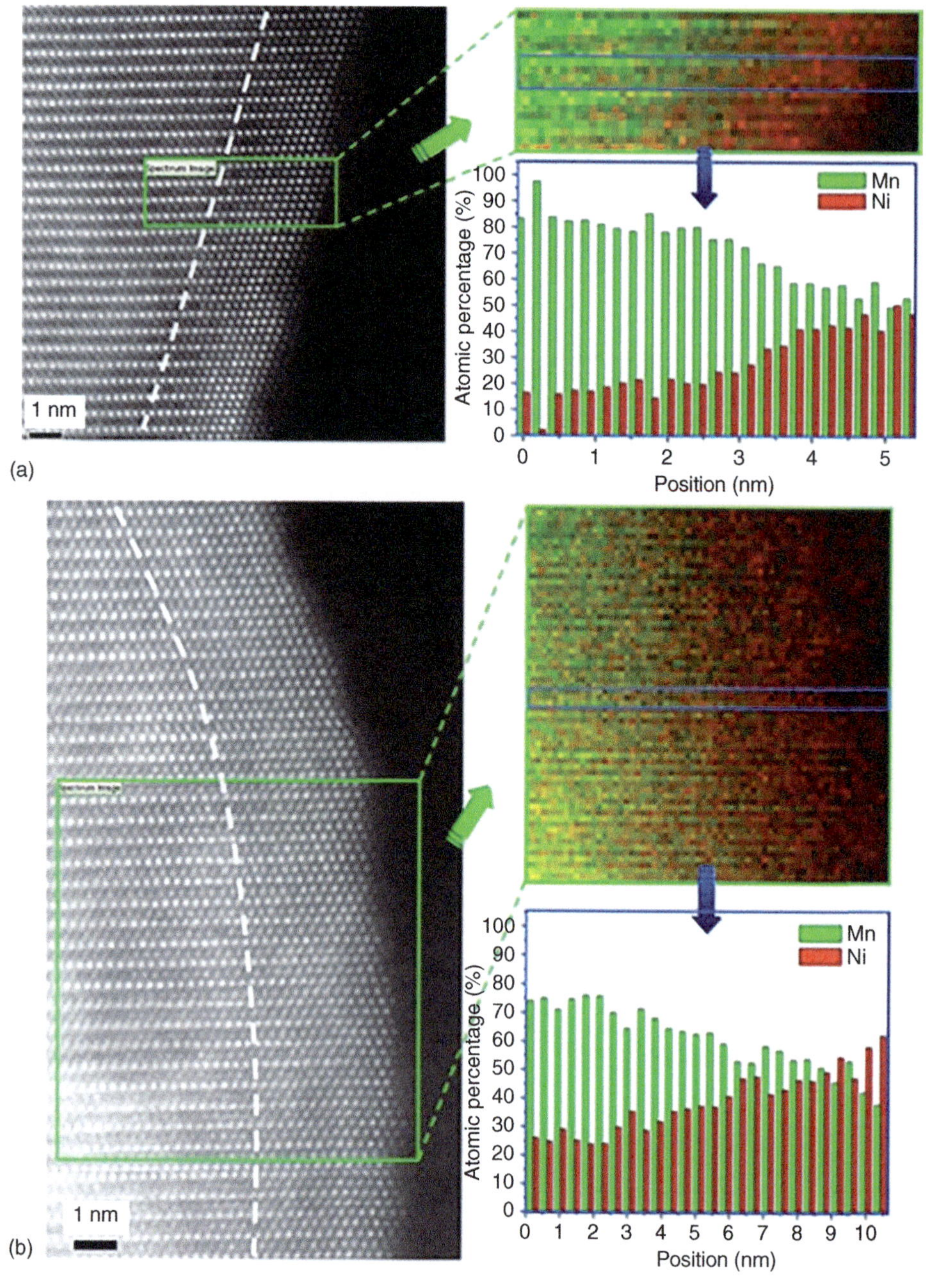


**Figure 11** High-resolution STEM–HAADF images, chemical maps, and evolution of Mn and Ni concentration at (a) C/2 rate and (b) C/10 rate after 50 cycles. The white dotted lines show the boundary between the bulk layered structure and the surface spinel/rock salt structures.[84] [Reprinted with permission from A. Boulineau, L. Simonin, J.-F. Colin, C. Bourbon and S. Patoux, Nano Lett., 2013, 13, 3857–3863. © 2013 American Chemical Society]

the octahedral site to the tetrahedral site of the crystal lattice. The Li in the transition metal will also migrate to the tetrahedral site in opposite site forming an Li–Li dumbbell (Figure 12c). Similar to the above-described process, transition metals can also migrate to the tetrahedral site. The author's calculations showed that $Ni^{2+}$ required the least energy for this form of migration, thus increasing the viability of Li/Ni cation mixing in materials containing Ni. However, the proposed Li–Li dumbbells are very stable and will not be affected by charging to voltages as high as 4.8 V. These dumbbells would block the vacant sites of the transition metal layer, which might cause low Coulombic efficiency. The similarity in the Li and transition metal migration would imply that the Li/Ni intermixing will be continued on subsequent cycles. Keeping along with the previous studies and the experimental

**Figure 12** Schematic representation of Li–Li dumbbell formation. (a)–(c) represent the progressive formation of these dumbbells with increasing Li removal. Color scheme: red: Li, green: Mn, blue: Ni, yellow: vacancy.[85] [Reproduced with permission from B. Xu, C. R. Fell, M. Chi and Y. S. Meng, Energy Environ. Sci., 2011, 4, 2223–2233. © Royal Society of Chemistry, 2011]

data gathered by the authors, they contemplated that such intermixing would be more severe in the surface region of the particle.

Most of these studies discussed thus far focused mostly on electrochemical activation of surface reconstruction, that is, the reconstruction process initiated by the removal of Li through electrochemical cycling. Furthermore, the evidence gathered in these studies mainly used techniques that were spatially resolved and thus could not take into account the heterogeneous nature of cathode materials.

A study by Lin *et al.* on stoichiometric NMC (Ni, Co, and Mn) materials ($LiNi_{0.4}Mn_{0.4}Co_{0.18}Ti_{0.02}O_2$) for lithium-ion batteries attempted to present a more complete picture of surface reconstruction.[47] This study provided evidence that surface reconstruction is also present in stoichiometric layered oxide materials in addition to the Li/Mn-rich layered oxide materials. In this study, the authors analyzed both the chemical evolution of the material and the surface structural transformation after long-term cycling. The authors used ensemble-averaged soft XAS in total electron yield (TEY) and fluorescence yield (FY) to study the dynamic evolution of the oxidation states of transition metals after a number of cycles in an electrochemical cell. The TEY mode mostly gives information on the surface region of the particle, whereas the FY mode is a more subsurface-sensitive technique. Thus, this information allowed the authors to correlate the surface chemical changes to the bulk changes of a cathode particle to get a more complete picture. The soft XAS spectra gathered on the NMC material after cycling showed gradually decreased oxidation states of both Co and Mn, whereas Ni mostly maintained its +2 oxidation state. Mn and Co were more reduced as determined from the TEY mode than that from the FY mode (Figure 13a,b). Hence, the authors argued that the surface of the cathode particles were mostly populated with the reduced Mn and Co species. Moreover, a calculation based on the linear combination of $Mn^{2+}$ and $Mn^{4+}$ showed that the surface layer was about 2 nm thick, quantified by the EELS. Reduction of these transition metals was also evident by just merely exposing the cathode particles to the electrolyte (Figure 13c,d).

Such a reduced transition metal layer on the surface of the particle is influenced by a high-voltage (upper cutoff voltage to 4.7 V) charging condition. The authors showed that the surface reduced layer is more prominent when charging to 4.7 V compared to charging to 4.3 V. The authors argued that the surface reduced layer may act as a passivation layer that can reduce the performance of a cell by thwarting ion diffusion. In addition to electrochemical activation, the reduced layer can be a direct result of the electrode–electrolyte interaction. The authors showed that a similar reduced layer was formed upon exposing the cathode particles in an electrolyte solution (Figure 13c,d).

Furthermore, the authors gathered direct visualization of the reduced layer through STEM (Figure 14). The STEM images clearly showed the presence of the rock salt phase on the particle surface. Moreover, the formation of this phase was orientation dependent with a larger degree of formation on the direction of Li diffusion, that is, across the <001> plane. This also relates the electrochemical activation aspect of the surface reduced layer formation as the formation is preferred in the direction of Li removal. Furthermore, some loosely bound transition metal atoms were also observed on the surface region (Figure 14f). These loose particles are most susceptible to dissolution. The loose particles contained transition metals in the reduced state which is consistent with our previous discussion.

### 2.4 Chemomechanical Degradation of Cathode Materials

Chemomechanical breakdown of the layered oxide cathode materials, that is, crack formations, plays a

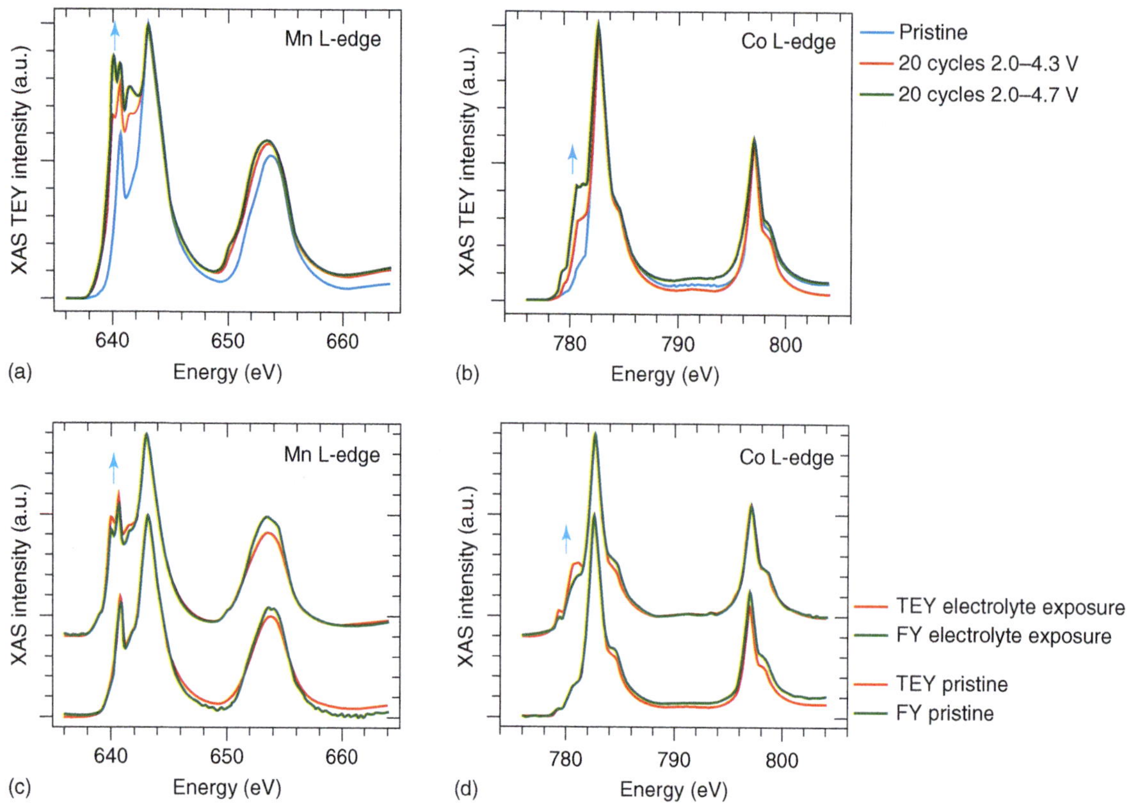


**Figure 13** Soft XAS spectra of $LiNi_{0.4}Mn_{0.4}Co_{0.18}Ti_{0.02}O_2$ in both total electron yield (TEY) mode and fluorescence yield (FY) mode. (a) Mn L-edge and (b) Co L-edge of pristine and cycled samples for 20 cycles. (c) Mn L-edge and (d) Co L-edge of pristine powder and after exposure to electrolyte. Both electrochemical cycling and electrolyte exposure reduced the transition metals as indicated by the rising intensity of the low energy shoulder. The extent of reduction was greater in samples cycled to 4.7 V.[47] [Reproduced with permission from F. Lin, I. M. Markus, D. Nordlund, T.-C. Weng, M. D. Asta, H. L. Xin and M. M. Doeff, Nat. Commun., 2014, 5, 3529. © Springer, 2014]

major role in deciding the cycle life of a battery. During electrochemical removal of alkali metal ions, all layered oxide materials suffer from heterogeneous phase transformation, lattice oxygen release, and anisotropic volume changes, which are compounded by the random orientation of primary particles.[59,60] All these changes produce mechanical stress in the materials, which can create cracks in the particles. These cracks open up fresh surfaces from the bulk of the material, increasing the effective surface area of the material. The fresh surfaces can inevitably further increase degradation of the material, discussed earlier including surface reconstruction, transition metal dissolution, electrolyte decomposition, and gas evolution, because of electrolyte infiltration into the cracks. Moreover, extensive crack formation can also result in detachment of cathode particles from the bulk, which may render them electrochemically inactive. Besides, cracks can cause poor grain-to-grain connection, resulting in poor electrical conductivity.[49,52] Hence, all these processes combined can severely affect the electrochemical performance of a cell. Thus, it is important to understand the initiation and propagation of these cracks in order to design stable high-performance cathode materials.

Most layered oxide materials are a collection of small-sized single-crystal particles with random orientations called primary particles. Agglomerations of these primary particles are known as secondary particles. Cracks can originate at both the primary particles and in between the primary particles along the grain boundary. Cracks in the primary particles are known as intragranular cracks, whereas cracks in between the primary particles are called intergranular cracks.[49,52] Intergranular cracks are usually larger than intragranular cracks, but intragranular cracks can be much larger in number than their counterparts. Based on the origin of these cracks, the reason behind crack formation can be divided into two main categories: intrinsic origins and extrinsic origins. The intrinsic origin of crack formation is related to the cathode material itself. Factors such as transition metal composition, particle size, heterogeneous phase transformations, and anisotropic volume changes are intrinsic origins of cracks.[53] Extrinsic origins of cracks can be attributed to

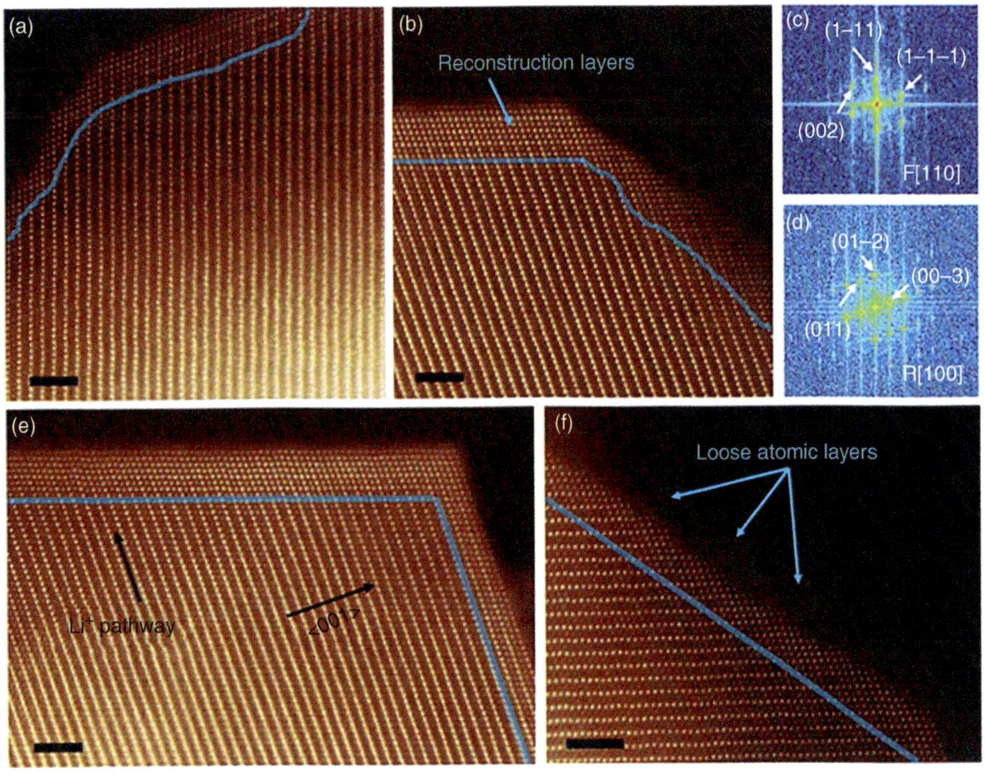


**Figure 14** ADF STEM images of the NMC particles (a) after exposing in electrolyte for 30 h, (b) after 1 cycle from 2.0 to 4.7 V, (c) and (d) FFT result showing the surface reconstruction layer and the layered bulk structure of the material, (e) orientation dependence on the surface reconstruction layer thickness. The thicker reconstruction layer was observed along the Li diffusion pathway, and (f) image of a loose atomic layer on the surface of the cathode particles. This atomic layer is most susceptible to dissolution during cycling.[47] [Reproduced with permission from F. Lin, I. M. Markus, D. Nordlund, T.-C. Weng, M. D. Asta, H. L. Xin and M. M. Doeff, Nat. Commun., 2014, 5, 3529. © Springer, 2014]

the electrochemical cycling condition of cathode materials such as cycle number, cycling rate, and cutoff voltage of charging.

Crack formation is also related to surface reconstruction of cathode particles. In Ni-rich layered NMC cathode materials ($LiNi_xCo_yMn_zO_2$), extensive surface reconstruction is observed when $x > 0.8$ in the materials.[47] This reconstruction is associated with particle volume shrinkage, which results in crack formation (see Section 2.3 for more details). In addition, Ni-rich NMC materials suffer from extensive anisotropic volume changes, which add up to the overall crack formation processes. Such crack formation exemplifies one of the intrinsic reasons of crack formation. Lim *et al.* reported such anisotropic volume changes in the NMC811 material.[86] They showed that the *a* lattice parameter of the material gradually decreased with increasing Li removal. However, the *c* lattice parameter increased up to 75% Li removal and decreased thereafter. Volume changes similar to this would obviously create strain in the material. Figure 15 shows the variation of strain in the material at different states of charge and Li removal, derived from a phase field model analysis. From Figure 15, the authors proved that distinct phase separation along with severe structural distortions can be observed at 85% Li removal from the material. Moreover, these distortions would be heterogeneously generated because of hindrance in diffusion along the *c* direction. Such changes in the material can cause 3.9% contraction in the nanodomain, which can shrink the size of the primary particles. Gaps will be generated between the primary particles undergoing these changes, which can translate to formation of intergranular cracks. Moreover, the authors argued that these anisotropic volume changes can cause severe mechanical instability and can result in the formation of intragranular cracks within the primary particles of the material.

Yan *et al.* demonstrated the relation between charging voltage and the intragranular crack formation, which can be exemplified as an extrinsic origin of crack formation.[52] They chose to study the commercial NMC333 ($LiNi_{1/3}Mn_{1/3}Co_{1/3}O_2$) material to understand the formation of these cracks. Their study showed that at lower charging voltages, for example, 4.2 and 4.5 V, the degree of crack formation was significantly lower than that at

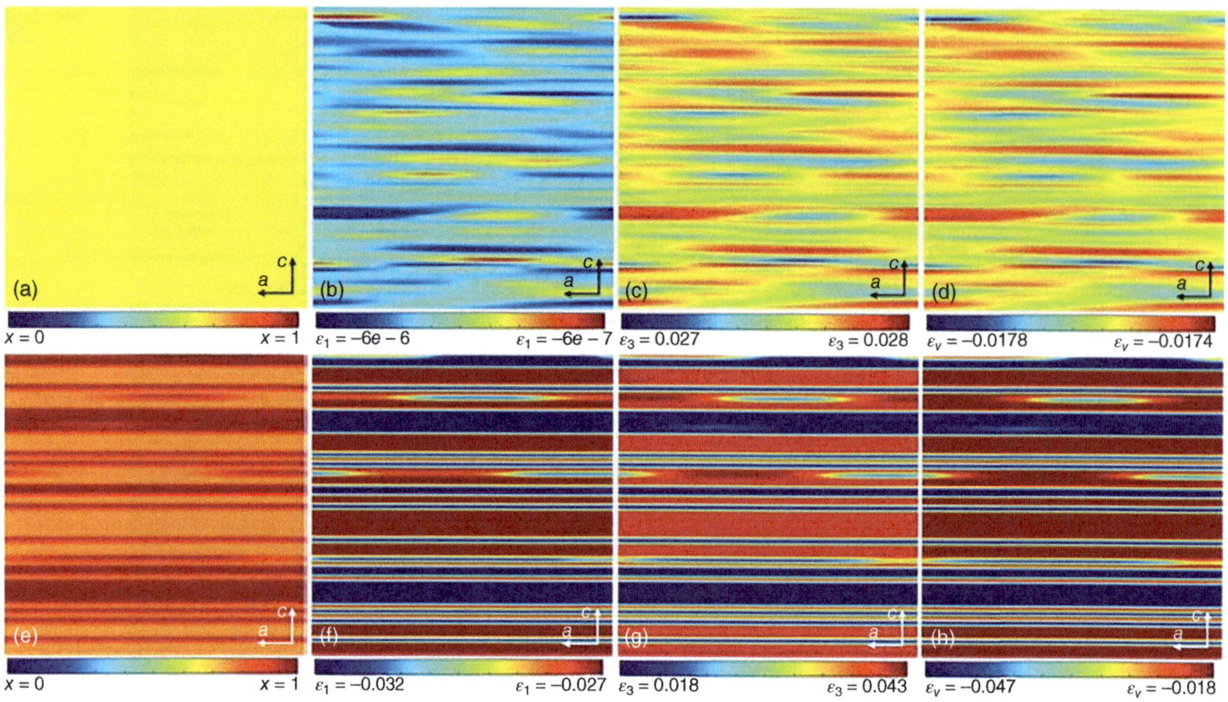


**Figure 15** Mesoscale phase transformation in $Li_{1-x}Ni_{0.8}Co_{0.1}Mn_{0.1}O_2$ at room temperature and at different Li removal. (a–d) are for 65% Li extraction and (e–h) are for 85% Li removal during charging of the material. The figures show the distributions in the *ac* plane at different Li removal. $\varepsilon_1$, $\varepsilon_3$, and $\varepsilon_V$ are strain in *a* direction, strain in *c* direction, and volumetric strain, respectively.[86] [Reproduced with permission from J.-M. Lim, T. Hwang, D. Kim, M.-S. Park, K. Cho and M. Cho, Sci. Rep., 2017, 7, 39669. © Springer 2017]

4.7 V. Through high-angle annular dark-field imaging (HAADF), the authors identified two different categories of cracks. One is a relatively larger crack featured by two free surfaces appearing to be parallel to each other as indicated by the yellow arrows in Figure 16. These large cracks are parallel to the (003) plane. The other relatively smaller cracks appear as dark strips, which are also parallel to the (003) plane. Hence, it signifies that all these cracks propagate predominantly along the (003) direction. The *c* spacing of these dark contrasted strips is larger than the usual *c* spacing between the transition metal layers. Thus, the authors conjectured that these cracks might have been generated by the splitting of the two adjacent transition metal layers. Moreover, these dark contrast strips are connected to the bigger cracks. This indicates that these strips can develop onto larger cracks with continued cycling of the battery. Hence, the authors named these strips as premature cracks.

Liu *et al.* reported intergranular crack formation in $LiNi_{0.8}Co_{0.15}Al_{0.05}O_2$ (NCA) cathode materials and attributed it to be one of the main reasons for capacity fading of the material on long-term cycling.[49] Extensive intergranular cracking was observed on cycled secondary particles as evident by the dark lines in the secondary particles from the transmission X-ray tomography (Figure 17). As the primary particle size did not increase after cycling (from XRD peak width analysis), these cracks had to originate along the grain boundary, which, in other words, are intergranular cracks. They reasoned that the anisotropic volume changes during Li extraction might be accounted for such crack formation. Such volume change generation was explained earlier in this section.

Oxygen release has been widely observed in Li/Mn-rich materials, which is considered as a key reason for voltage fading. However, in stoichiometric layered oxide materials, how lattice oxygen functions in charge compensation and leads to crack formation are not clearly understood. Recently, our group has demonstrated that lattice oxygen release from stoichiometric NMC materials may also result in crack formation under thermal abuse conditions. A delithiated NMC442 material (about 50% Li removal) was heated to 230 °C during the *in situ* TEM study to exponentially accelerate the oxygen release and phase transformation. This allowed us to observe the chemomechanical breakdown of the cathode material in a reasonable timeframe. Upon heating the material, several microcracks appeared and grew to about 70 nm within 15 min (Figure 18). Upon further heating, the cracks continued to grow and eventually led to the separation

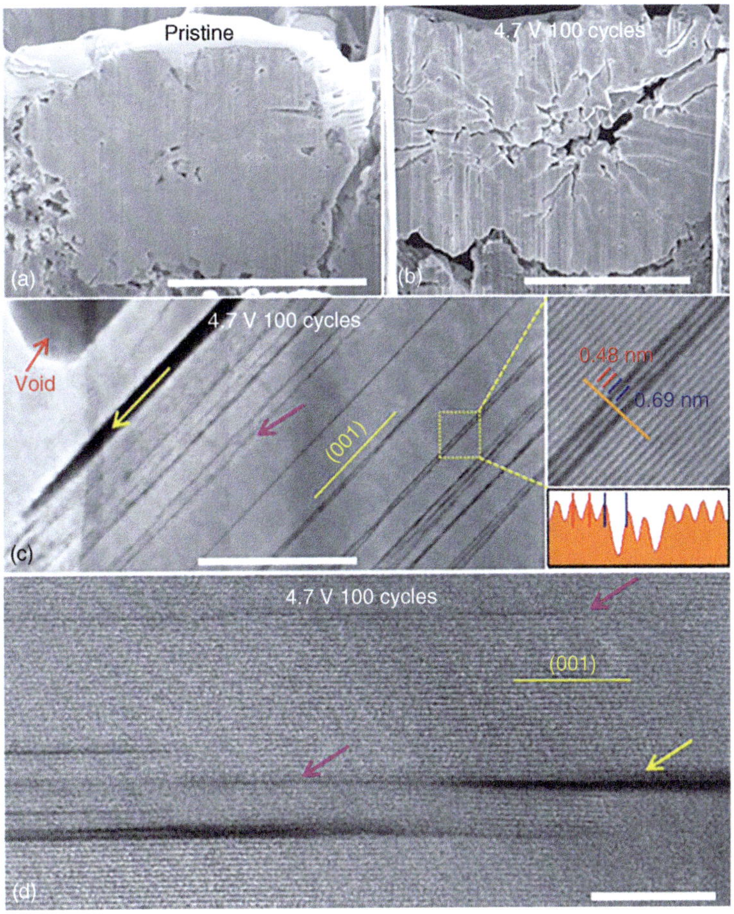


**Figure 16** Cross-sectional SEM images of pristine secondary particles (a) and particles after cycling for 100 cycles (b) of $LiNi_{1/3}Co_{1/3}Mn_{1/3}O_2$. The high cutoff voltage is 4.7 V. (c) and (d) STEM-HAADF images of the cycled particles with a high cutoff voltage of 4.7 V. The yellow arrows indicate the real cracks with two separate and parallel surfaces. The pink arrows indicate the premature cracks that can grow into bigger cracks (indicated by the yellow arrows).[52] [Reproduced with permission from P. Yan, J. Zheng, M. Gu, J. Xiao, J.-G. Zhang and C.-M. Wang, Nat. Commun., 2017, 8, 14101. © Springer 2017]

between primary particles and the exfoliation of primary particles (i.e., intragranular cracks).

Our spectroscopic and finite element analysis proved oxygen release and extensive phase transformation contributed to the formation of intergranular and intragranular cracks that were observed. The surface transition metals were more reduced than the bulk, likely resulting from the oxygen release from the surface of the material. Due to the surface oxygen loss, surface reconstruction took place, which built up the stress through the grain boundaries. The intergranular cracks eventually formed to release the stress (Figure 19a,b). Furthermore, we modeled the impact of surface oxygen vacancies on the mechanical properties of individual primary particles (Figure 19c,d). Upon oxygen release due to heating, the surface, with enriched oxygen vacancies, underwent a larger lattice expansion relative to the bulk part of the particle, creating a compressive stress field near the surface of the material and tensile stress at the bulk. As a result, these stresses led to the chemomechanical breakdown of NMC grains (Figure 19).

## 2.5 Surface Stabilization of Cathode Materials

Coating electrode materials has been a common strategy to mitigate severe cathode–electrolyte side reactions. Through mitigating the side reactions, the coating layer can not only inhibit electrolyte degradation but also prevent transition metal reduction. As transition metal reduction, for example, $Mn^{3+}$ reduction to $Mn^{2+}$, has been marked as a key pathway for transition metal dissolution, preventing reduction can mitigate the dissolution problem. Moreover, in many cases electrolyte decomposition can yield acidic products such as HF.[40] The harsh nature of this species can further accelerate transition metal dissolution. The surface coating can also act as a scavenger for this acidic species. Meng and

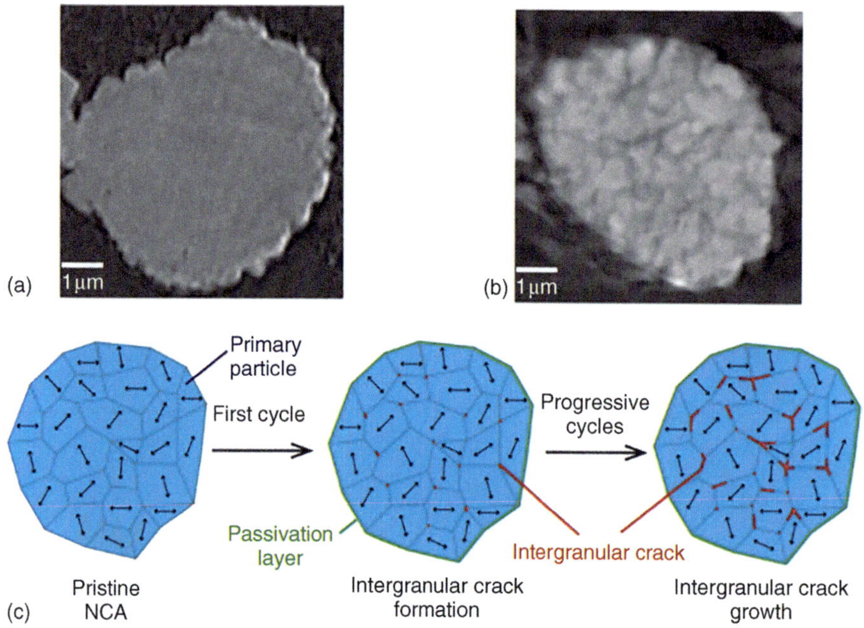


**Figure 17** Cross-sectional transmission X-ray tomograms of (a) pristine $LiNi_{0.8}Co_{0.15}Al_{0.05}O_2$ and (b) cycled $LiNi_{0.8}Co_{0.15}Al_{0.05}O_2$. (c) Schematic illustration of intergranular crack formation in the material. Arrows pointing at different directions represent the orientation of that particular grain.[49] [Reprinted with permission from H. Liu, M. Wolf, K. Karki, Y.-S. Yu, E. A. Stach, J. Cabana, K. W. Chapman and P. J. Chupas, Nano Lett., 2017, 17, 3452–3457. © 2017 American Chemical Society]

coworkers demonstrated that $Al_2O_3$ coating on the surface of the cathode particles can yield stable cycle life with high Coulombic efficiency.[87] Qin *et al.* reported that ultrathin $TiO_2$ coating through atomic layer deposition (ALD) on the surface of $LiNi_{0.6}Co_{0.2}Mn_{0.2}O_2$ cathode materials can improve the cycle life of batteries.[88] They hypothesized that the improved cycle life may have originated from the protection of $TiO_2$ from HF attack, electrolyte decomposition, and transition metal dissolution. Wang and coworkers reported $Li_2TiO_3$ coating on the surface of $LiNi_{0.8}Co_{0.1}Mn_{0.1}O_2$ for improving electrochemical performance of the cathode material.[61] The material had 98% capacity retention at 1C rate after 170 cycles. They reasoned that the coating suppressed the direct contact between the cathode and the electrolyte, enhancing $Li^+$ diffusion, and prevention of pulverization of the active materials during cycling. Recently, our group has demonstrated that the cathode–electrolyte side reaction can be put into good use to stabilize cathode materials.[62] Pretreatment of layered oxide cathode materials by exposing them to electrolyte solvents was demonstrated to produce a conformal coating layer on the surface of the cathode particles. The presence of such coating layers was proved through TEM imaging (Figure 20). In addition, the X-ray spectroscopic analysis showed that the surface of these cathode particles was populated with reduced transition metals, similar to the surface reconstruction layer discussed earlier. As a result, the transformed surface layer acted as a barrier between the cathode and the electrolyte and when cycled in a battery can reduce the extensive electrolyte decomposition. This can in turn, to a great extent, increase the cycle life of the battery.

Elemental doping/substitution can reduce the degree of heterogeneous phase transformation and anisotropic volume change, which may prevent the chemo-mechanical breakdown of cathode particles. Moreover, surface doping/substitution can effectively curtail the extent of surface reconstruction, which can reduce the crack formation by limiting the interfacial stress buildup along grain boundaries. Wolf-Goodrich *et al.* reported a Ti-substituted NMC 422 material ($LiNi_{0.4}Co_{0.2}Mn_{0.2}O_2$) with improved long-term cycling performance.[89] Soft XAS study revealed that Co and Mn in the material without Ti were more reduced state compared with the substituted material. Hence, the Ti substitution strategy can effectively prevent rock-salt-like phase formation on the surface of the cathode particle. Furthermore, Mg doping in $LiNi_{0.80}Co_{0.15}Al_{0.05}O_2$ can provide a short-range ordering of Ni cations around $Mg^{2+}$. This provides stronger bonding in the structure and can inhibit NiO-like phase formations on the surface.[65] On the other hand, phase transformation can be inhibited by Al and La codoping in $LiCoO_2$ materials.[64] $LiCoO_2$ used in commercial batteries cannot be operated at high voltages because of structural instability, thus limiting the amount of energy from

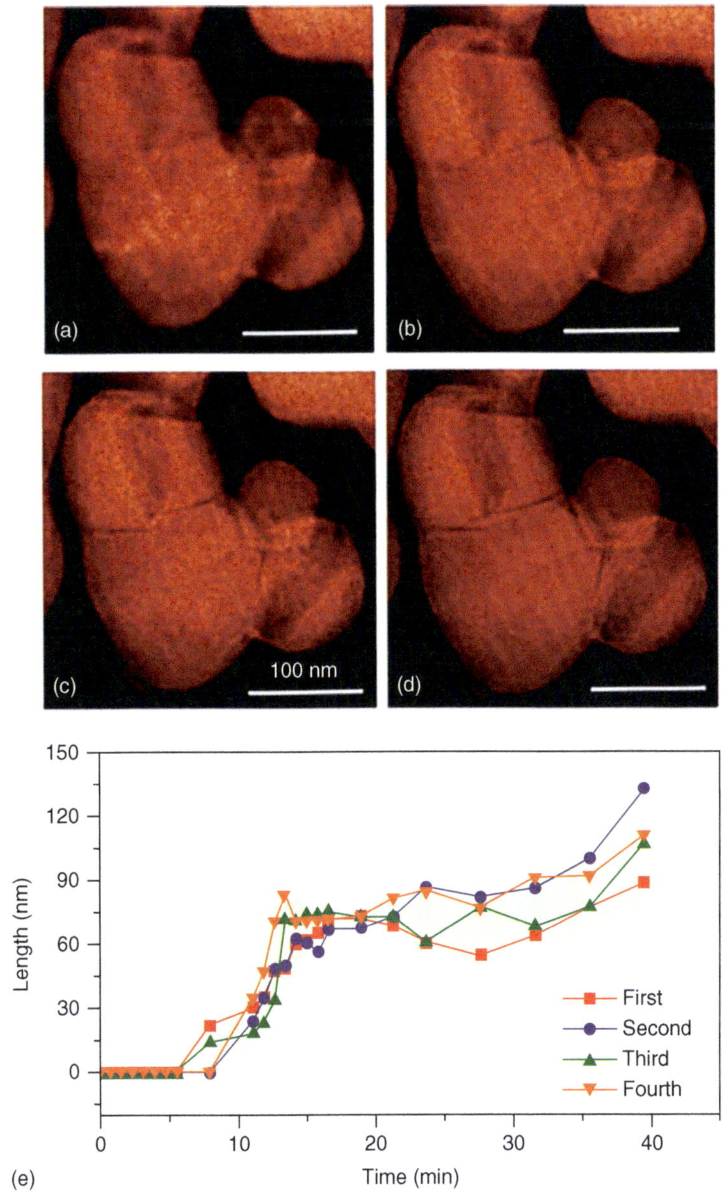


**Figure 18** Visualizing the intergranular cracks in delithiated $Li_{1-x}Ni_{0.4}Mn_{0.4}Co_{0.2}O_2$ under thermal abuse conditions after (a) 0 min, (b) 12 min, (c) 26 min, and (d) 40 min. (e) Evolution of crack length as a function of time.[50] [Reprinted with permission from L. Mu, R. Lin, R. Xu, L. Han, S. Xia, D. Sokaras, J. D. Steiner, T.-C. Weng, D. Nordlund, M. M. Doeff, Y. Liu, K. Zhao, H. L. Xin and F. Lin, Nano Lett., 2018, 18, 3241–3249. © 2018 American Chemical Society]

the material. However, La and Al codoping can induce structural stability at high voltages, thus increasing the amount of energy this material can deliver. La doping can increase the *c* lattice parameter. On the other hand, Al can act as a positively charged centre and suppress phase transformation to increase the structural stability at high voltage.

Designing cathode materials with the preferential distribution of elemental composition has been reported to yield stable cycle life for electrochemical cells. Using such a design strategy, Sun *et al.* reported on stabilization of Ni-rich NMC materials for lithium-ion batteries.[90] They designed the core-and-shell NMC material with the core of the material being Ni rich and the shell being Mn rich (Figure 21). Thus, the material successfully used the high energy of the Ni rich bulk while effectively using the stable nature of the Mn-rich shell to reduce surface degradation phenomena. However, it is not clear how such a design would inhibit Mn dissolution that is commonly observed in Mn-containing cathodes.

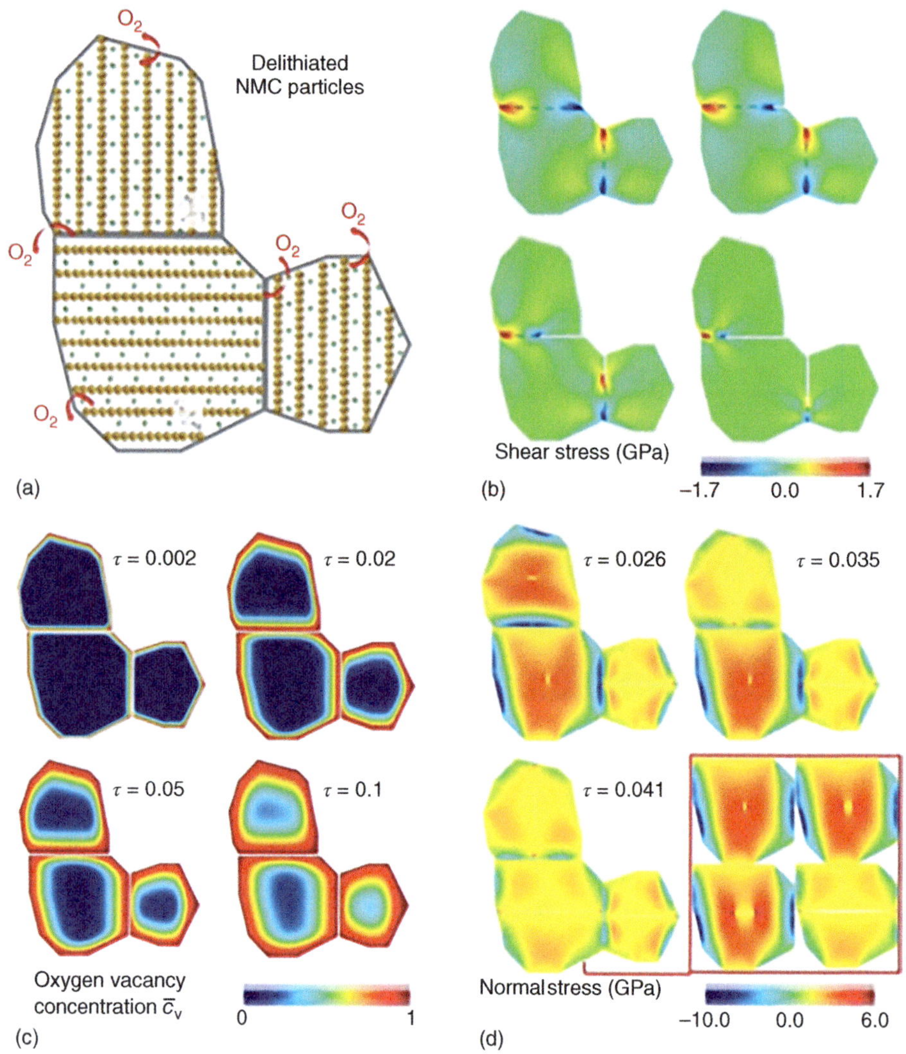


**Figure 19** (a) Finite element modeling of the randomly oriented NMC crystals, (b) shear stress progression along the grain boundary and phase-transformation-induced intergranular crack formation in the NMC material. (c) Oxygen vacancy profiling and (d) nucleation and propagation of intragranular cracks initiated through oxygen release.[50] [Reprinted with permission from L. Mu, R. Lin, R. Xu, L. Han, S. Xia, D. Sokaras, J. D. Steiner, T.-C. Weng, D. Nordlund, M. M. Doeff, Y. Liu, K. Zhao, H. L. Xin and F. Lin, Nano Lett., 2018, 18, 3241–3249. © 2018 American Chemical Society]

Sun *et al.* later adopted the core-and-shell-type distribution of transition metals to full-gradient transition metal distribution where the Mn content of the NMC particles decreased as a function of depth while the Ni content increased linearly with depth.[91] Such materials can also utilize the advantages provided by the core-and-shell NMC materials.

Lastly, adding additives to electrolytic solution has been reported as an effective method to reduce the extent of electrolyte decomposition and also to prevent some other observed cathode-surface-related phenomena. These additives can form a film on the surface of the electrodes to implement their protective effect.[68] Ma *et al.* reported a boron-based anion receptor, tris(2,2,2-trifluoroethyl)borate (TTFEB), capable of improving the interface between Li/Mn-rich cathode and the electrolyte.[69] TTFEB was demonstrated to reduce the inorganic Li species in the cathode–electrolyte interphase and curtail the reduction of $Mn^{4+}$. The last effect is especially important because it can reduce the extent of transition metal dissolution. Similar performance improvement was reported in $LiNi_{1/3}Co_{1/3}Mn_{1/3}O_2$ cathode upon the addition of 0.5 wt% TTFEB in the electrolyte.[92] Lithium benzimidazole has also shown promise as a novel electrolyte additive to $LiPF_6$ salt for operating at high voltages and temperatures.[93] A polyionic liquid–solid interphase on the cathode surface responsible for performance

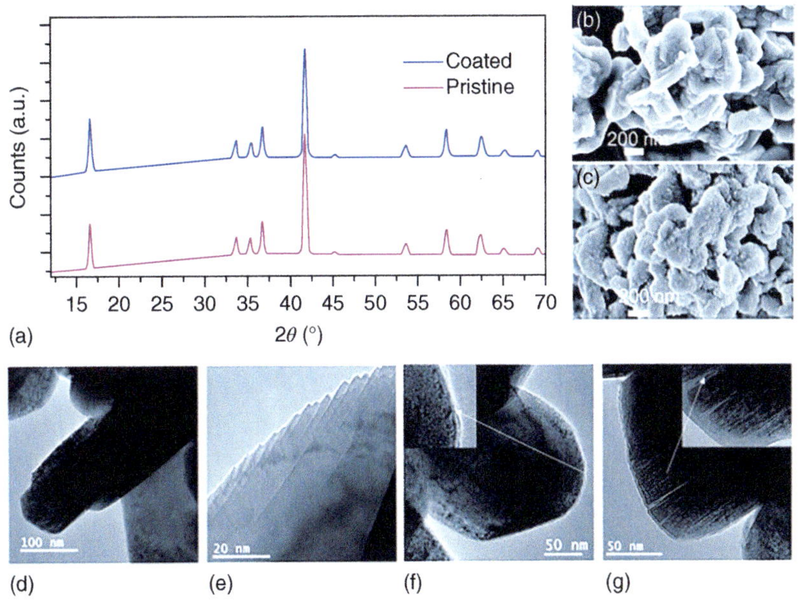


**Figure 20** (a) XRD pattern of the coated and pristine $NaNi_{1/3}Fe_{1/3}Mn_{1/3}O_2$. (b) and (c) SEM images of pristine and coated material, respectively. TEM images of (d) and (e) pristine powder at different magnification, and (f) and (g) coated powder at different magnification.[62] [Reprinted with permission from L. Mu, M. M. Rahman, Y. Zhang, X. Feng, X.-W. Du, D. Nordlund and F. Lin, J. Mater. Chem. A, 2018, 6, 2758–2766. © 2018 American Chemical Society]

improvement was formed due to the reaction between the $CF_3$-substituted benzimidazole ring and the lattice $O^{2-}$ of Li-rich materials and ethylene carbonate. Similar performance improvements were also observed for sodium-ion batteries. Komaba *et al.* reported performance improvement of $NaNi_{1/2}Mn_{1/2}O_2$ cathode and hard carbon anode in $NaClO_4$/PC electrolyte with FEC as additive.[94]

## 3 CONCLUSION

Alkali metal ion batteries hold the potential to change how we think and perceive about energy consumption through implementing energy storage devices of various scales. However, the instability issues related to the surface of cathode materials curtail their long-term usability. The interconnected nature of these surface issues can accelerate degradation of cathode materials. Thus, understanding these issues is of vital importance for long-term applications of the alkali metal ion batteries. In this chapter, we summarized and explained some important surface-related issues such as transition metal dissolution, electrolyte decomposition, surface reconstruction, and chemo-mechanical breakdown of cathode particles. The chapter also provided some insights into the methods for improving the long-term cycling performance of these batteries. Finally, we believe that understanding these surface processes will accelerate the development of a knowledge base to promote further development of stable cathode materials for alkali metal ion batteries.

## 4 ACKNOWLEDGMENT

The authors would like to acknowledge Virginia Tech Startup Fund for the support in their research.

## 5 RELATED ARTICLE

Solid State and Materials Chemistry for Sodium-Ion Batteries

## 6 ABBREVIATIONS AND ACRONYMS

ALD = atomic layer deposition; CEI = cathode electrolyte interphase; DEC = diethyl carbonate; DMC = dimethyl carbonate; EC = ethylene carbonate; EDX = energy-dispersive X-ray spectroscopy; EELS = electron energy loss spectroscopy; EMC = ethyl methyl carbonate;

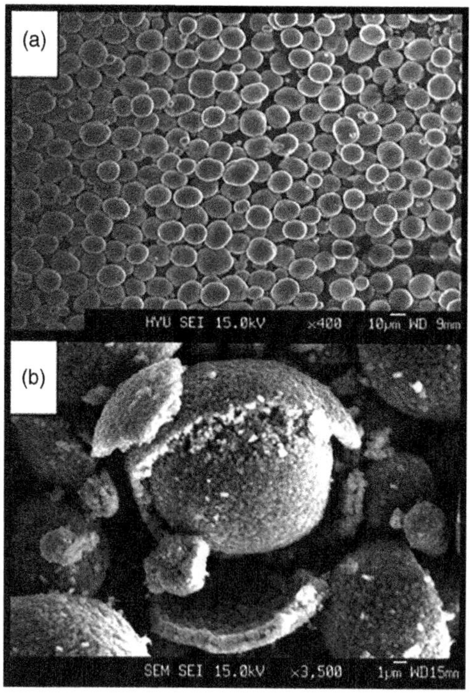

Figure 21 SEM images of (a) $Li[(Ni_{0.8}Co_{0.1}Mn_{0.1}O_2)_{0.8}(Ni_{0.5}Mn_{0.5})_{0.2}]O_2$ and (b) after partially crushed in agate mortar. The core–shell structure is evident from the material morphology.[90] [Reprinted with permission from Y.-K. Sun, S.-T. Myung, M.-H. Kim, J. Prakash and A. Khalil, J. Am. Chem. Soc., 2005, 127, 13411–13418. © 2005 American Chemical Society]

FY = fluorescence yield; HAADF = high-angle annular dark field imaging; LC = liquid chromatography; LiTFSI = lithium bis(trifluoromethanesulfonyl)imide; NCA = $LiNi_{0.8}Co_{0.15}Al_{0.05}O_2$; NMC = $LiNi_xMn_yCo_zO_2$; OEMS = online electrochemical mass spectroscopy; PC = propylene carbonate; PCET = proton-coupled electron transfer; SEI = solid–electrolyte interphase; SEM = scanning electron microscopy; soft XAS = soft X-ray absorption spectroscopy; STEM–ADF = scanning transmission electron microscopy–annular dark-field imaging; TEY = total electron yield; TTFEB = tris(2,2,2-trifluoroethyl)borate; XANES = X-ray absorption near edge structure; XRD = X-ray diffraction.

## 7 REFERENCES


1. J.-M. Tarascon and M. Armand, *Nature*, 2001, **414**, 359.
2. M. Armand and J.-M. Tarascon, *Nature*, 2008, **451**, 652.
3. J. Tollefson, *Nature*, 2008, **456**, 436.
4. D. H. Doughty, P. C. Butler, A. A. Akhil, N. H. Clark and J. D. Boyes, *Electrochem. Soc. Interf.*, 2010, **19**, 49.
5. Y. Li, Y. Lu, C. Zhao, Y.-S. Hu, M.-M. Titirici, H. Li, X. Huang and L. Chen, *Energy Storage Mater.*, 2017, **7**, 130.
6. N. Nitta, F. Wu, J. T. Lee and G. Yushin, *Mater. Today*, 2015, **18**, 252.
7. J. Xu, S. Dou, H. Liu and L. Dai, *Nano Energy*, 2013, **2**, 439.
8. L. Li, Y. Zheng, S. Zhang, J. Yang, Z. Shao and Z. Guo, *Energy Environ. Sci.*, DOI:10.1039/C8EE01023D.
9. L. P. Wang, L. Yu, X. Wang, M. Srinivasan and Z. J. Xu, *J. Mater. Chem. A*, 2015, **3**, 9353.
10. M. H. Han, E. Gonzalo, G. Singh and T. Rojo, *Energy Environ. Sci.*, 2015, **8**, 81.
11. N. Ortiz-Vitoriano, N. E. Drewett, E. Gonzalo and T. Rojo, *Energy Environ. Sci.*, 2017, **10**, 1051.
12. R. Jung, M. Metzger, F. Maglia, C. Stinner and H. A. Gasteiger, *J. Electrochem. Soc.*, 2017, **164**, A1361.
13. C. Tian, F. Lin and M. M. Doeff, *Acc. Chem. Res.*, 2018, **51**, 89.
14. J. Yan, X. Liu and B. Li, *RSC Adv.*, 2014, **4**, 63268.
15. D. Liu, W. Zhu, J. Trottier, C. Gagnon, F. Barray, A. Guerfi, A. Mauger, H. Groult, C. M. Julien, J. B. Goodenough and K. Zaghib, *RSC Adv.*, 2014, **4**, 154.
16. Y. Xia and M. Yoshio, 'Lithium Batteries', Springer US, Boston, MA, 2009, p.361.
17. N. Hung Vu, P. Arunkumar and W. Bin Im, *Sci. Rep.*, 2017, **7**, 45579.
18. J. Lu and K. S. Lee, *Mater. Technol.*, 2016, **31**, 628.
19. X. Liu, X. Wang, A. Iyo, H. Yu, D. Li and H. Zhou, *J. Mater. Chem. A*, 2014, **2**, 14822.
20. J. R. Kim and G. G. Amatucci, *J. Electrochem. Soc.*, 2016, **163**, A696.
21. S.-H. Yu, X. Feng, N. Zhang, J. Seok and H. D. Abruña, *Acc. Chem. Res.*, 2018, **51**, 273.
22. F. Wu and G. Yushin, *Energy Environ. Sci.*, 2017, **10**, 435.
23. C. Liu, Z. G. Neale and G. Cao, *Mater. Today*, 2016, **19**, 109.
24. P. K. Nayak, L. Yang, W. Brehm and P. Adelhelm, *Angew. Chem. Int. Ed.*, 2018, **57**, 102.
25. S. Guo, Q. Li, P. Liu, M. Chen and H. Zhou, *Nat. Commun.*, 2017, **8**, 135.
26. F. Lin, Y. Liu, X. Yu, L. Cheng, A. Singer, O. G. Shpyrko, H. L. Xin, N. Tamura, C. Tian, T.-C. Weng, X.-Q. Yang, Y. S. Meng, D. Nordlund, W. Yang and M. M. Doeff, *Chem. Rev.*, 2017, **117**, 13123.
27. H.-H. Sun and A. Manthiram, *Chem. Mater.*, 2017, **29**, 8486.
28. J. Cabana, B. J. Kwon and L. Hu, *Acc. Chem. Res.*, 2018, **51**, 299.
29. R. Hausbrand, G. Cherkashinin, H. Ehrenberg, M. Gröting, K. Albe, C. Hess and W. Jaegermann, *Mater. Sci. Eng. B*, 2015, **192**, 3.

30. N. P. W. Pieczonka, Z. Liu, P. Lu, K. L. Olson, J. Moote, B. R. Powell and J.-H. Kim, *J. Phys. Chem. C*, 2013, **117**, 15947.
31. N. V. Faenza, Z. W. Lebens-Higgins, P. Mukherjee, S. Sallis, N. Pereira, F. Badway, A. Halajko, G. Ceder, F. Cosandey, L. F. J. Piper and G. G. Amatucci, *Langmuir*, 2017, **33**, 9333.
32. M. Evertz, F. Horsthemke, J. Kasnatscheew, M. Börner, M. Winter and S. Nowak, *J. Power Sources*, 2016, **329**, 364.
33. B. Vortmann-Westhoven, M. Winter and S. Nowak, *J. Power Sources*, 2017, **346**, 63.
34. C. Zhan, J. Lu, A. Jeremy Kropf, T. Wu, A. N. Jansen, Y.-K. Sun, X. Qiu and K. Amine, *Nat. Commun.*, 2013, **4**, 2437.
35. H. Zheng, Q. Sun, G. Liu, X. Song and V. S. Battaglia, *J. Power Sources*, 2012, **207**, 134.
36. F. Yang, Y. Liu, S. K. Martha, Z. Wu, J. C. Andrews, G. E. Ice, P. Pianetta and J. Nanda, *Nano Lett.*, 2014, **14**, 4334.
37. C. Zhan, T. Wu, J. Lu and K. Amine, *Energy Environ. Sci.*, 2018, **11**, 243.
38. E. Rosenberg, C. Kanakaki, A. Amon, I. Gocheva and A. Trifonova, *Bulg. Chem. Commun.*, 2017, **49**, 242.
39. Z. Zeng, W.-I. Liang, H.-G. Liao, H. L. Xin, Y.-H. Chu and H. Zheng, *Nano Lett.*, 2014, **14**, 1745.
40. M. Broussely, P. Biensan, F. Bonhomme, P. Blanchard, S. Herreyre, K. Nechev and R. J. Staniewicz, *J. Power Sources*, 2005, **146**, 90.
41. J. Vetter, P. Novák, M. R. Wagner, C. Veit, K.-C. Möller, J. O. Besenhard, M. Winter, M. Wohlfahrt-Mehrens, C. Vogler and A. Hammouche, *J. Power Sources*, 2005, **147**, 269.
42. A. Jarry, S. Gottis, Y.-S. Yu, J. Roque-Rosell, C. Kim, J. Cabana, J. Kerr and R. Kostecki, *J. Am. Chem. Soc.*, 2015, **137**, 3533.
43. H. Yang, G. V. Zhuang and P. N. Ross, *J. Power Sources*, 2006, **161**, 573.
44. S. Takeda, W. Morimura, Y.-H. Liu, T. Sakai and Y. Saito, *Rapid Commun. Mass Spectrom.*, 2016, **30**, 1754.
45. P. Yan, A. Nie, J. Zheng, Y. Zhou, D. Lu, X. Zhang, R. Xu, I. Belharouak, X. Zu, J. Xiao, K. Amine, J. Liu, F. Gao, R. Shahbazian-Yassar, J.-G. Zhang and C.-M. Wang, *Nano Lett.*, 2015, **15**, 514.
46. K. Jarvis, C.-C. Wang, M. Varela, R. R. Unocic, A. Manthiram and P. J. Ferreira, *Chem. Mater.*, 2017, **29**, 7668.
47. F. Lin, I. M. Markus, D. Nordlund, T.-C. Weng, M. D. Asta, H. L. Xin and M. M. Doeff, *Nat. Commun.*, 2014, **5**, 3529.
48. Y. Xu, E. Hu, K. Zhang, X. Wang, V. Borzenets, Z. Sun, P. Pianetta, X. Yu, Y. Liu, X.-Q. Yang and H. Li, *ACS Energy Lett.*, 2017, **2**, 1240.
49. H. Liu, M. Wolf, K. Karki, Y.-S. Yu, E. A. Stach, J. Cabana, K. W. Chapman and P. J. Chupas, *Nano Lett.*, 2017, **17**, 3452.
50. L. Mu, R. Lin, R. Xu, L. Han, S. Xia, D. Sokaras, J. D. Steiner, T.-C. Weng, D. Nordlund, M. M. Doeff, Y. Liu, K. Zhao, H. L. Xin and F. Lin, *Nano Lett.*, 2018, **18**, 3241.
51. R. Panat, *Thin Solid Films*, 2015, **596**, 174.
52. P. Yan, J. Zheng, M. Gu, J. Xiao, J.-G. Zhang and C.-M. Wang, *Nat. Commun.*, 2017, **8**, 14101.
53. K. Min and E. Cho, *Phys. Chem. Chem. Phys.*, 2018, **20**, 9045.
54. J. A. Gilbert, I. A. Shkrob and D. P. Abraham, *J. Electrochem. Soc.*, 2017, **164**, A389.
55. L. Xia, B. Tang, L. Yao, K. Wang, A. Cheris, Y. Pan, S. Lee, Y. Xia, G. Z. Chen and Z. Liu, *Chem. Select*, 2017, **2**, 7353.
56. A. Guéguen, D. Streich, M. He, M. Mendez, F. F. Chesneau, P. Novák and E. J. Berg, *J. Electrochem. Soc.*, 2016, **163**, A1095.
57. P. N. Ross, *Catal. Lett.*, 2014, **144**, 1370.
58. P. Arora, R. E. White and M. Doyle, *J. Electrochem. Soc.*, 1998, **145**, 3647.
59. L. Mu, Q. Yuan, C. Tian, C. Wei, K. Zhang, J. Liu, P. Pianetta, M. M. Doeff, Y. Liu and F. Lin, *Nat. Commun.*, 2018, **9**, 2810.
60. A. O. Kondrakov, A. Schmidt, J. Xu, H. Geßwein, R. Mönig, P. Hartmann, H. Sommer, T. Brezesinski and J. Janek, *J. Phys. Chem. C*, 2017, **121**, 3286.
61. K. Meng, Z. Wang, H. Guo, X. Li and D. Wang, *Electrochim. Acta*, 2016, **211**, 822.
62. L. Mu, M. M. Rahman, Y. Zhang, X. Feng, X.-W. Du, D. Nordlund and F. Lin, *J. Mater. Chem. A*, 2018, **6**, 2758.
63. G. Li, Z. Zhang, R. Wang, Z. Huang, Z. Zuo and H. Zhou, *Electrochim. Acta*, 2016, **212**, 399.
64. Q. Liu, X. Su, D. Lei, Y. Qin, J. Wen, F. Guo, Y. A. Wu, Y. Rong, R. Kou, X. Xiao, F. Aguesse, J. Bareño, Y. Ren, W. Lu and Y. Li, *Nat. Energy*, DOI:10.1038/s41560-018-0180-6.
65. A. H. Tavakoli, H. Kondo, Y. Ukyo and A. Navrotsky, *J. Electrochem. Soc.*, 2012, **160**, A302.
66. S.-T. Myung, H.-J. Noh, S.-J. Yoon, E.-J. Lee and Y.-K. Sun, *J. Phys. Chem. Lett.*, 2014, **5**, 671.
67. F. Lin, D. Nordlund, Y. Li, M. K. Quan, L. Cheng, T.-C. Weng, Y. Liu, H. L. Xin and M. M. Doeff, *Nat. Energy*, 2016, **1**, 15004.
68. S. S. Zhang, *J. Power Sources*, 2006, **162**, 1379.
69. Y. Ma, Y. Zhou, C. Du, P. Zuo, X. Cheng, L. Han, D. Nordlund, Y. Gao, G. Yin, H. L. Xin, M. M. Doeff, F. Lin and G. Chen, *Chem. Mater.*, 2017, **29**, 2141.
70. J. Wandt, A. Freiberg, R. Thomas, Y. Gorlin, A. Siebel, R. Jung, H. A. Gasteiger and M. Tromp, *J. Mater. Chem. A*, 2016, **4**, 18300.
71. D. H. Jang, Y. J. Shin and S. M. Oh, *J. Electrochem. Soc.*, 1996, **143**, 2204.
72. J. C. Hunter, *J. Solid State Chem.*, 1981, **39**, 142.
73. Y. Xia, Y. Zhou and M. Yoshio, *J. Electrochem. Soc.*, 1997, **144**, 2593.
74. G. G. Amatucci, J. M. Tarascon and L. C. Klein, *Solid State Ion.*, 1996, **83**, 167.

75. J. D. Steiner, L. Mu, J. Walsh, M. M. Rahman, B. Zydlewski, F. M. Michel, H. L. Xin, D. Nordlund and F. Lin, *ACS Appl. Mater. Interf.*, 2018, **10**, 23842.

76. M. M. Rahman, Y. Xu, H. Cheng, Q. Shi, R. Kou, L. Mu, Q. Liu, S. Xia, X. Xiao, C.-J. Sun, D. Sokaras, D. Nordlund, J.-C. Zheng, Y. Liu and F. Lin, *Energy Environ. Sci.*, DOI:10.1039/C8EE00309B.

77. J. B. Goodenough and Y. Kim, *Chem. Mater.*, 2010, **22**, 587.

78. P. Lanz, C. Villevieille and P. Novák, *Electrochim. Acta*, 2014, **130**, 206.

79. F. Amalraj, M. Talianker, B. Markovsky, D. Sharon, L. Burlaka, G. Shafir, E. Zinigrad, O. Haik, D. Aurbach, J. Lampert, M. Schulz-Dobrick and A. Garsuch, *J. Electrochem. Soc.*, 2013, **160**, A324.

80. S. A. Freunberger, Y. Chen, Z. Peng, J. M. Griffin, L. J. Hardwick, F. Bardé, P. Novák and P. G. Bruce, *J. Am. Chem. Soc.*, 2011, **133**, 8040.

81. S. E. Renfrew and B. D. McCloskey, *J. Am. Chem. Soc.*, 2017, **139**, 17853.

82. J. Zheng, M. Gu, J. Xiao, P. Zuo, C. Wang and J.-G. Zhang, *Nano Lett.*, 2013, **13**, 3824.

83. M. Gu, I. Belharouak, J. Zheng, H. Wu, J. Xiao, A. Genc, K. Amine, S. Thevuthasan, D. R. Baer, J.-G. Zhang, N. D. Browning, J. Liu and C. Wang, *ACS Nano*, 2013, **7**, 760.

84. A. Boulineau, L. Simonin, J.-F. Colin, C. Bourbon and S. Patoux, *Nano Lett.*, 2013, **13**, 3857.

85. B. Xu, C. R. Fell, M. Chi and Y. S. Meng, *Energy Environ. Sci.*, 2011, **4**, 2223.

86. J.-M. Lim, T. Hwang, D. Kim, M.-S. Park, K. Cho and M. Cho, *Sci. Rep.*, 2017, **7**, 39669.

87. J. Alvarado, C. Ma, S. Wang, K. Nguyen, M. Kodur and Y. S. Meng, *ACS Appl. Mater. Interf.*, 2017, **9**, 26518.

88. C. Qin, J. Cao, J. Chen, G. Dai, T. Wu, Y. Chen, Y. Tang, A. Li and Y. Chen, *Dalton Trans.*, 2016, **45**, 9669.

89. S. Wolff-Goodrich, F. Lin, I. M. Markus, D. Nordlund, H. L. Xin, M. Asta and M. M. Doeff, *Phys. Chem. Chem. Phys.*, 2015, **17**, 21778.

90. Y.-K. Sun, S.-T. Myung, M.-H. Kim, J. Prakash and A. Khalil, *J. Am. Chem. Soc.*, 2005, **127**, 13411.

91. Y.-K. Sun, Z. Chen, H.-J. Noh, D.-J. Lee, H.-G. Jung, Y. Ren, S. Wang, C. S. Yoon, S.-T. Myung and K. Amine, *Nat. Mater.*, 2012, **11**, 942.

92. L. Wang, Y. Ma, P. Wang, S. Lou, X. Cheng, P. Zuo, C. Du, Y. Gao and G. Yin, *J. Electrochem. Soc.*, 2017, **164**, A1924.

93. F.-M. Wang, S. A. Pradanawati, N.-H. Yeh, S.-C. Chang, Y.-T. Yang, S.-H. Huang, P.-L. Lin, J.-F. Lee, H.-S. Sheu, M.-L. Lu, C.-K. Chang, A. Ramar and C.-H. Su, *Chem. Mater.*, 2017, **29**, 5537.

94. S. Komaba, T. Ishikawa, N. Yabuuchi, W. Murata, A. Ito and Y. Ohsawa, *ACS Appl. Mater. Interf.*, 2011, **3**, 4165.

# PART 2
# Lithium Metal Battery Materials

# $Li-CO_2$ Batteries

**Zhaojun Xie and Zhen Zhou**

*Nankai University, Tianjin, China*



## 1 INTRODUCTION

Nowadays, $CO_2$ is receiving much attention because it is one of the leading greenhouse gases, especially due to its steadily increasing concentration in the atmosphere. Balancing the release of $CO_2$ into the environment with the increasing world energy demand is one of the most defining challenges, for billions of tons of fossil fuels (coal, natural gas, and petroleum products) are consumed each year. A lot of research has been expended to solve this issue, and some developments have been achieved. Among the widespread studies on the recycling and utilization of $CO_2$, the exploitation of $CO_2$ in energy storage devices, such as $Li-CO_2$ batteries, represents a novel approach and attracts much attention.[1] The system mainly consists of a Li metal anode, a separator, an electrolyte, and an air cathode (Figure 1). Obviously, rechargeable $Li-CO_2$ batteries can reduce fossil fuel consumption and mitigate the "greenhouse effect" due to the reversible reaction between lithium and $CO_2$. Besides, they are potential energy sources for scientific exploration and future immigration on other planets, such as Venus and Mars, where more than 95% of the atmosphere is $CO_2$.

Many efforts have been devoted since the prototype of $Li-CO_2$ batteries was reported. Many encouraging developments in terms of cathode matcrials, reaction mechanism, and battery configuration have been achieved. However, there are still many challenges to be overcome, including the high overpotential during the charging process, poor cyclability, and inferior rate capability. Except for the operation temperature, reactant gas and electrolytes, all these problems more or less result from the sluggish dynamic formation and decomposition of $Li_2CO_3$, especially for the latter.[2] The electrical insulation and thermodynamic stability of $Li_2CO_3$ make it extremely difficult to decompose, while the high charge voltage would lead to the oxidative decomposition of many battery components. As a result, the development of $Li-CO_2$ batteries mainly focuses on the decomposition of $Li_2CO_3$, optimizing the electrode structure, and introducing high active catalysts and electrolyte additives. Besides, the chemistry of $Li_2CO_3$ in $Li-O_2$ batteries is crucial, because $CO_2$ is highly soluble in organic solvents and can react with the intermediate species or the discharge products to form $Li_2CO_3$.

## 2 REACTION MECHANISM

The reaction mechanism for $CO_2$-based batteries can be roughly classified into two categories according to the working gas, $CO_2/O_2$ mixture, or pure $CO_2$. Except for the catalytic effect of cathodes, electrolyte composition also plays an important role in the discharge and charge processes. In the following sections, the reaction mechanisms of $O_2/CO_2$ batteries and pure $CO_2$ batteries are discussed separately.

### 2.1 $O_2/CO_2$-Based Batteries

For $O_2/CO_2$-based batteries, the preferential reaction is the electrochemical formation of superoxide anion ($O_2^{\bullet-}$).[3] Indeed, the general discharging plateau of

*Inorganic Battery Materials*. Edited by Hailiang Wang and Boniface P.T. Fokwa
 ISBN 978-1-119-43199-2

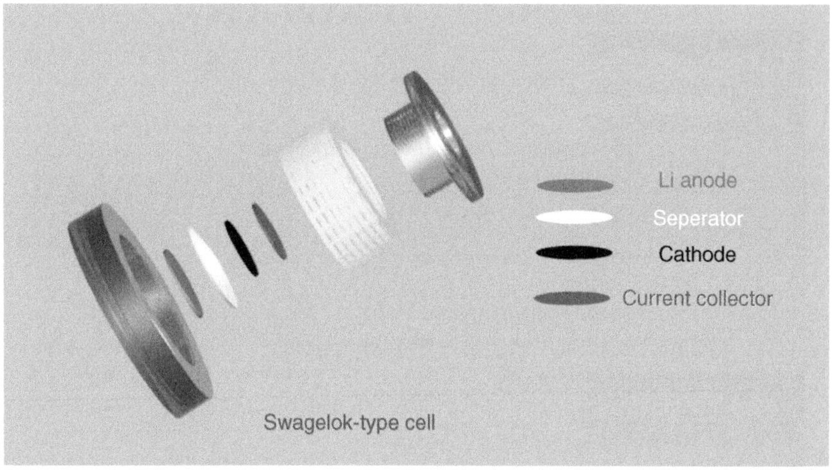


**Figure 1** Schematic configuration of Swagelok-type Li–$CO_2$ batteries

metal–$O_2/CO_2$ batteries is identical to that of metal–$O_2$ batteries. In view of the fast reaction between $O_2^{\bullet-}$ and $CO_2$, which has been successfully applied to $CO_2$ sensors or molten-carbonate fuel cells, the subsequent reaction is the complex transformation of superoxocarbonate ($CO_4^{\bullet-}$). Therefore, the initial reaction of $O_2/CO_2$-based batteries during discharge is proposed as follows:

$$O_2 + e^- \rightarrow O_2^{\bullet-} \quad (1)$$

$$O_2^{\bullet-} + CO_2 \rightarrow CO_4^{\bullet-} \quad (2)$$

In addition, the discharge process is strongly affected by the electrolyte system, including solvent species,[4] electrolyte concentration,[5] and the metal ions.[6,7]

### 2.1.1 *Li-$O_2/CO_2$ Batteries*

$Li_2CO_3$ is the most common discharge product for Li–$O_2/CO_2$ batteries, and its formation route has been studied intensively. First, $Li_2CO_3$ could form through the decomposition of electrolytes or carbon attacked by $O_2^{\bullet-}$. For example, carbonate electrolytes in Li–$O_2$ batteries have been known to be attacked by $O_2^{\bullet-}$ and generate $Li_2CO_3$. Secondly, the chemical reaction between $CO_2$ and $Li_2O_2$ will generate $Li_2CO_3$ as a solid product. As reported by Gowda *et al.*,[8] the discharge product of $Li_2CO_3$ in Li–$O_2/CO_2$ batteries with 1,2-dimethoxyethane (DME) electrolytes is formed simply due to the chemical reaction between $CO_2$ and $Li_2O_2$, while the higher capacity results from changes in the deposit morphology. Mekonnen *et al.*[9] further studied the reaction by density functional theory (DFT) computations and galvanostatic charge–discharge measurements. They observed that the dominant electrochemistry for ether-based Li–$O_2/CO_2$ batteries is the formation of $Li_2O_2$ via a $2e^-/O_2$ process. Thirdly, $Li_2CO_3$ could also form through the further conversion of $CO_4^{\bullet-}$ with $CO_2$ and $Li^+$. By using quantum mechanical simulations combined with experimental verification, Lim *et al.*[4] found that the electrolyte solvation effect can leverage the reaction pathway at the initial complex formation step by altering the potential energy surface (Figure 2). For low dielectric electrolytes such as DME, $Li_2O_2$ is first generated and spontaneously reacts with $CO_2$ to form $Li_2CO_3$. On the other side, a high dielectric electrolyte such as dimethyl sulfoxide (DMSO) favors the electrochemical activation of $CO_2$ to $CO_4^{\bullet-}$ or $C_2O_6^{2-}$, yielding only $Li_2CO_3$. On the contrary, the decomposition of $Li_2CO_3$ is complex and cannot be a simple reaction as

$$Li_2CO_3 \rightarrow 2Li^+ + 2e + 1/2O_2 + CO_2 \quad (3)$$

Through the use of isotope labeling and gas chromatography–mass spectrometry (GC–MS) method, Yang *et al.*[10] suggested that $Li_2CO_3$ decomposes into $CO_2$ and superoxide radicals, while the latter are finally consumed by the electrolyte solvent. Even though the stability of electrolytes was ensured, Li–$O_2/CO_2$ batteries based on $Li_2CO_3$ cycle exhibit low voltaic efficiency because of the high charging potentials (>4 V).

Notably, both the solvation state of $Li^+$ and the structure of entire electrolyte systems can be essentially adjusted and controlled by the concentration of salts with specific anions, especially in the electrolyte with high dielectric solvents such as DMSO. Enlightened by the unique features of super-concentrated electrolytes, Qiao *et al.*[5] introduced an aggregated electrolyte system composed of DMSO-sheathed contact-ion-pair. Compared with diluted electrolytes, the polymeric fluid network composed in their specific-introduced electrolyte would effectively stabilize

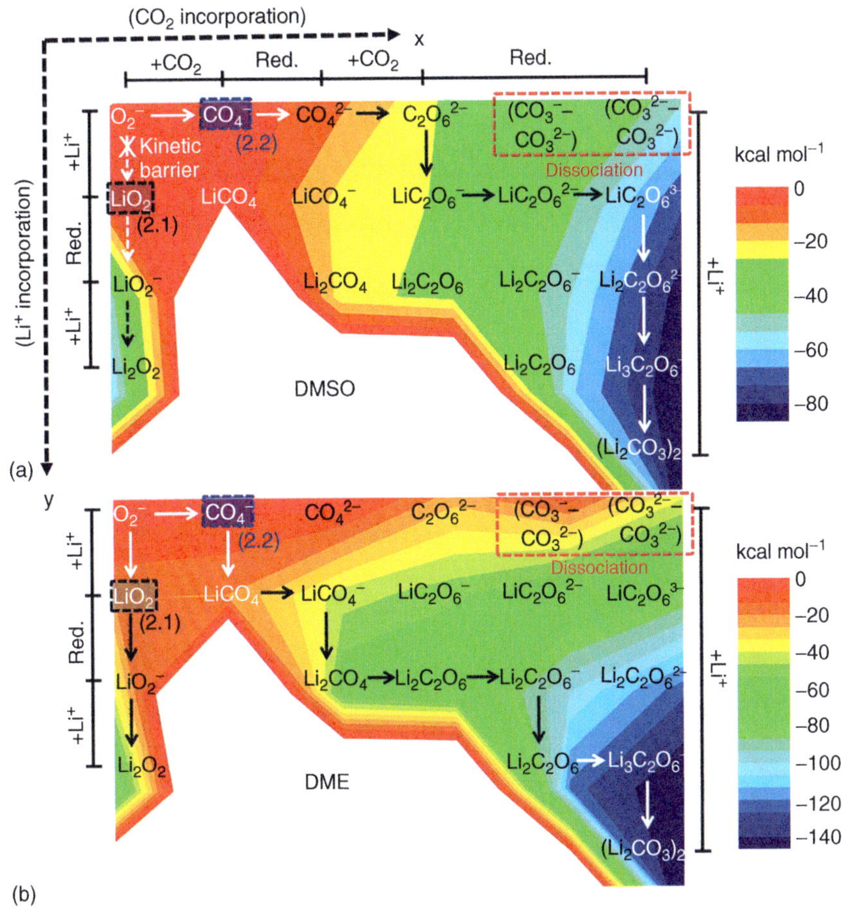


**Figure 2** Two-dimensional energy surface map including every possible intermediate species from the dioxygen anion radical of $O_2^-$ to $Li_2O_2$ and $Li_2CO_3$ with (b) DME and (a) DMSO electrolytes. [Reprinted with permission from H. Lim, H. Lim, K. Park, D. Seo, H. Gwon, J. Hong, W. Goddard, H. Kim, K. Kang, *J. Am. Chem. Soc.*, 2013, **135**, 9733–9742. Copyright 2013 American Chemical Society]

the intermediate of peroxodicarbonate ($C_2O_6^{2-}$) generated from

$$CO_4^{\bullet -} + O_2^{\bullet -} \rightarrow CO_4^{2-} \quad (4)$$

and

$$CO_4^{2-} + CO_2 \rightarrow C_2O_6^{2-} \quad (5)$$

By limiting the further reduction of $C_2O_6^{2-}$ into $Li_2CO_3$, the Li–$O_2$/$CO_2$ battery cycled successfully with a obvious decrease in the charge potential (from 4.2 to 3.5 V) and a remarkable boost of the energy efficiency (Figure 3a). Different reaction routes of $CO_2$ were well elucidated by in situ Raman spectra and differential electrochemical mass spectrometry (DEMS) methods (Figure 3b–f). However, $CO_2$ still does not participate in the redox conversion yet.

### 2.1.2 *Other Metal–$O_2$/$CO_2$ Batteries*

In contrast to Li–$O_2$/$CO_2$ batteries, Archer group focused on other attractive anode materials such as Na[6] and Al[7] for $O_2$/$CO_2$-based batteries. These metal–$O_2$/$CO_2$ batteries can be used as platforms for $CO_2$ capture from an effluent gas mixture and electricity supply in the form of primary batteries. Compared with lithium anodes, this design strategy is appropriate because of the high theoretical capacity and the abundance of Na and Al. Both Na–$O_2$/$CO_2$ and Al–$O_2$/$CO_2$ batteries exhibited higher discharge capacity than pure $O_2$-based batteries, while the discharging plateaus were close to that of the formation of $O_2^{\bullet -}$. Ex situ X-ray diffraction (XRD) and Fourier transform infrared spectroscopy (FTIR) results showed that $Na_2C_2O_4$ was the dominant discharge product for Na–$O_2$/$CO_2$ batteries with ionic liquid electrolytes, while $Na_2CO_3$ and $Na_2C_2O_4$ coexisted for batteries with tetraethylene glycol dimethyl ether (TEGDME) electrolytes. The formation of $Na_2CO_3$ in the TEGDME-based batteries was similar to $Li_2CO_3$ resulting from the

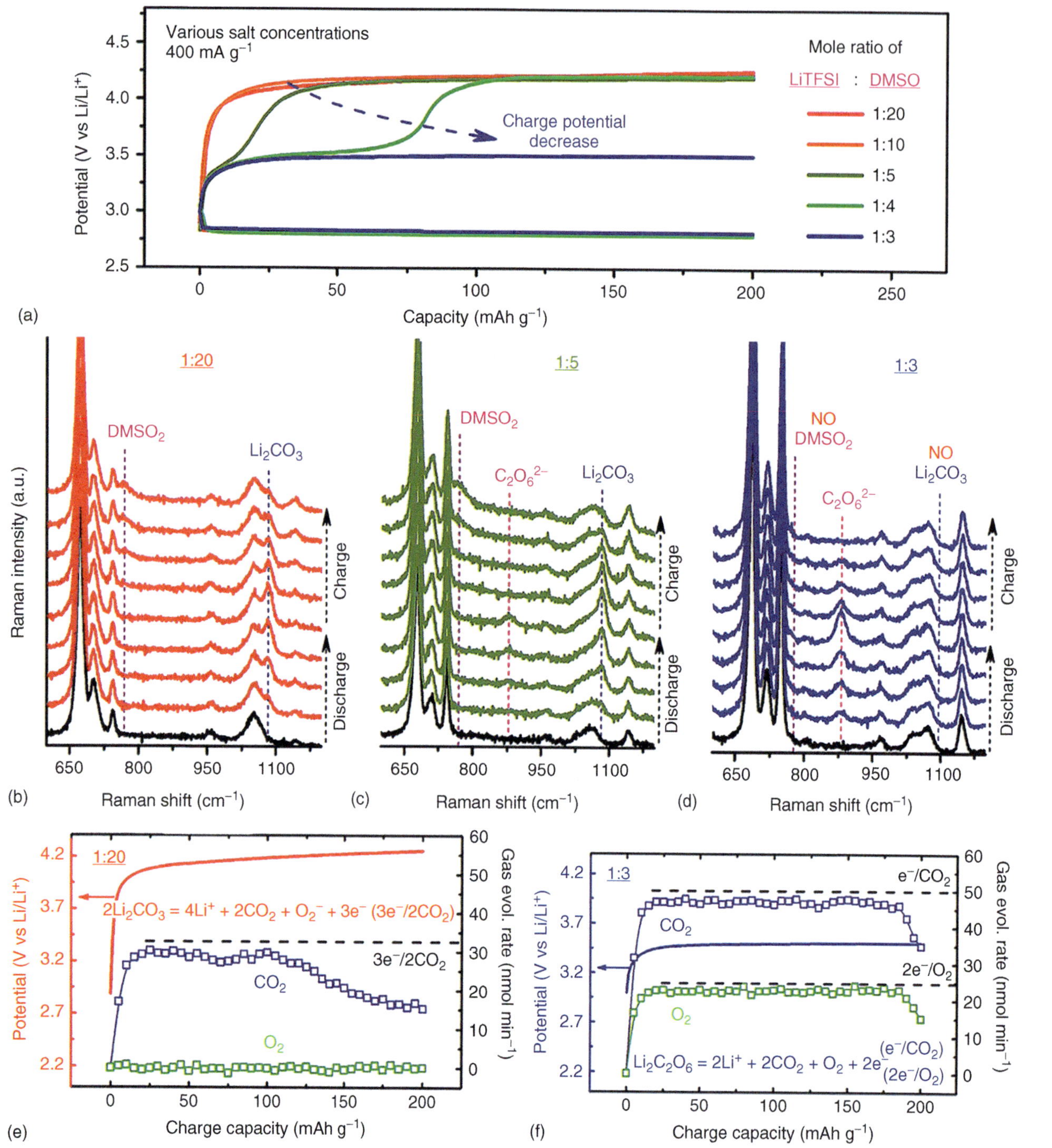


**Figure 3** Electrochemistry and spectroscopy characterizations. (a) Voltage profiles of Li–$O_2$/$CO_2$ (1:1, volume ratio) batteries operate in lithium bis(trifluoromethane sulfonimide) (LiTFSI)/DMSO electrolyte with different mole ratios (current density: 400 mA g$^{-1}$). (b–d) In situ Raman spectra recorded during the first cycle from batteries with various LiTFSI:DMSO mole ratios: (b) 1:20; (c) 1:5; (d) 1:3. (e,f) DEMS results of gas evolution rates for $CO_2$ and $O_2$ during charging in batteries with different LiTFSI:DMSO mole ratios: (e) 1:20 and (f) 1:3. [Reproduced with permission from Ref. 5. Copyright 2018 Royal Society of Chemistry]

conversion of $CO_4^{\bullet-}$ or $C_2O_6^{2-}$ with $Na^+$. In the case of $Na_2C_2O_4$, $CO_2$ is the active material and involved in the electrochemical reactions. Interestingly, $Al_2(C_2O_4)_3$ was found as the discharge product for the Al–$O_2$/$CO_2$ battery based on the 1-ethyl-3-methylimidazolium chloride/aluminum chloride electrolyte. The formation of oxalates involves the further conversion of $CO_4^{2-}$ (from reaction 4) with $CO_2$, which can be expressed as

$$CO_4^{2-} + CO_2 \rightarrow C_2O_4^{2-} \qquad (6)$$

Obviously, metal ions and solvents play an important role in the discharge process of $O_2/CO_2$-based batteries. The above metal–$O_2/CO_2$ batteries prove that $CO_2$ can work as an active material in the electrochemical conversions.

## 2.2 Pure $CO_2$-Based Batteries

Because carbonates are not the electrochemical reductive product of $CO_2$, the formation of low valence carbon compounds is the critical factor for pure $CO_2$-based batteries. Therefore, the reaction mechanisms for the discharge and charge processes play an important role in all pure $CO_2$-based batteries.

### *2.2.1 Li–$CO_2$ Batteries*

The first primary Li–$CO_2$ battery at moderate temperatures was reported by Archer group.[11] The theoretical equilibrium potential of an electrochemical reaction can be estimated from the formula $E = -\Delta G/zF$, where $\Delta G$ is the change in Gibbs free energy, $z$ is the number of electrons transferred per mole of products, and $F$ is the Faraday constant. Therefore, the battery potential can be estimated based on the thermodynamic data, which could provide indirect evidence for the reaction mechanism. The measured battery discharge potentials at different temperatures are listed in Figure 4. The theoretical equilibrium potential 1 (the dashed line) was calculated based on the reaction

$$2Li + 2CO_2 \rightarrow Li_2CO_3 + CO \qquad (7)$$

and the theoretical equilibrium potential 2 (the dotted line) was estimated based on the reaction

$$4Li + 3CO_2 \rightarrow 2Li_2CO_3 + C \qquad (8)$$

The measured battery potential initially approaches the dashed line, but then surpasses it. Based on Tafel theory, the actual discharge potential cannot exceed the theoretical equilibrium potential, indicating the proposed reaction (7) is incorrect. In addition, CO was not detected from the gas phase composition by DEMS. Thus, the discharge process was considered as reaction (8), where lithium carbonate was also a principal discharge product. However, the formation of carbon was not further investigated due to the application of carbon cathodes.

In order to confirm the existence of carbon in the discharge products, Li[12] and Zhou group[13] used a porous gold and a platinum net cathode for their room-temperature Li–$CO_2$ batteries, respectively. Although amorphous carbon was confirmed by surface enhanced Raman spectroscopy and electron energy loss spectroscopy (EELS), it is not clear whether the reaction mechanism is the same between precious metal and carbon cathodes. In the following studies, both Wang[14] and Zhou groups[15] separately found that transmission electron microscope (TEM) and high-resolution transmission electron microscope (HRTEM) could allow screening the thin-layered carbon along with agglomerated $Li_2CO_3$ particles. By using in situ DEMS technique, the two groups also confirmed the consumption and evolution of $CO_2$ during the discharge–charge cycle, but not in the equivalent amount. Compared with the formation of $Li_2CO_3$ and amorphous carbon, the reverse reaction remains much more ambiguous.

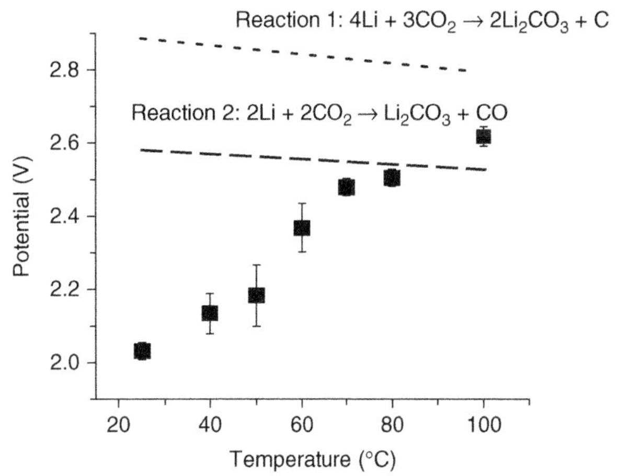


**Figure 4** The comparison of theoretical equilibrium potentials with actual discharge potentials at different temperatures. [Reproduced with permission from Ref. 11. Copyright 2013 Royal Society of Chemistry]

Surprisingly, after a typical discharge platform of the Li–$CO_2$ battery about 2.5 V, Qiao *et al.*[16] discovered a new plateau around 1.8 V, which can display an extended discharge capacity. Based on the corresponding in situ Raman spectra, a new discharge product of $Li_2O$ emerged. According to the reported first-principles computations[13] and the theoretical thermodynamic calculations, the newly observed discharge plateau was assigned as

$$4Li + CO_2 \rightarrow 2Li_2O + C \qquad (9)$$

Most notably, they further provided a flexible strategy for the practical design of both $CO_2$ fixation techniques and renewable energy storage devices (Figure 5). The discharge processes of both strategies are the electrochemical reduction of $CO_2$ species, resulting in the synchronous formation of $Li_2CO_3$ and carbon via reaction (8). However, the subsequent charge process differs starkly between these two strategies. By adopting a specific catalyst (ruthenium), the reversible Li–$CO_2$ battery can be

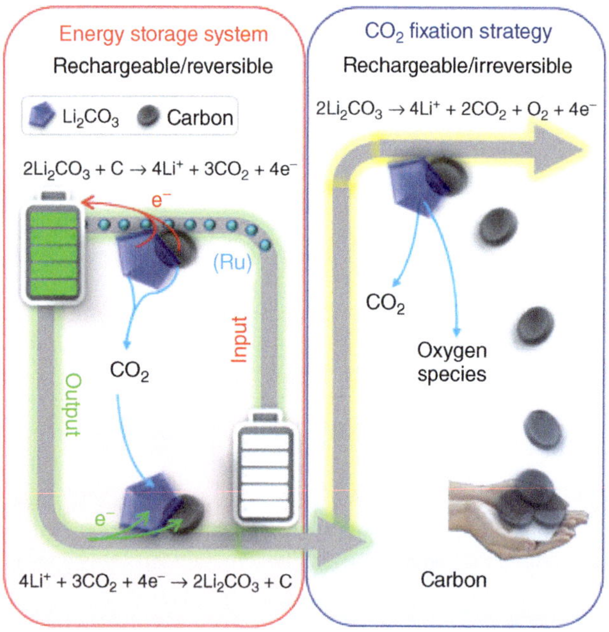


**Figure 5** Schematic for the achievement of an energy storage system (reversible process) and a $CO_2$ fixation strategy (irreversible process) via Li–$CO_2$ electrochemistry technology. [Reproduced with permission from Ref. 16. © Elsevier, 2017]

realized. On the contrary, the charging process is based on the decomposition of $Li_2CO_3$ alone, and carbon remains fixed. All the abovementioned conclusions were validated through the in situ Raman technology combined with DEMS results. Obviously, by the rational design of specific cathode materials, both $Li_2CO_3$ and carbon species can be oxidized during charging. In another word, reversible room-temperature Li–$CO_2$ batteries can be obtained.

### 2.2.2 *Other Metal–$CO_2$ Batteries*

Along with the fast development of Li–$CO_2$ batteries, efforts have also been devoted on other anode materials such as Na and Al for pure $CO_2$-based batteries. By using a Na anode, an ether-based electrolyte, and a designed cathode with electrolyte-treated multiwall carbon nanotubes, Hu *et al.*[17] realized the rechargeable room-temperature Na–$CO_2$ battery. The reaction in the Na–$CO_2$ battery was regarded as

$$4Na + 3CO_2 \leftrightarrow 2Na_2CO_3 + C \tag{10}$$

The reversible formation and decomposition of $Na_2CO_3$ were verified by in situ Raman, ex situ XRD, and X-ray photoelectron spectroscopy (XPS). By measuring the evolved gas in the charging process, they also confirmed that $CO_2$ can form reversibly. Similar to Li–$CO_2$ batteries, a silver nanowire cathode was used to illustrate carbon involved in the battery reaction. The superior reversibility of the carbon product was indeed detected by EELS. More recently, Ma *et al.*[18] have reported a rechargeable room-temperature Al–$CO_2$ battery consisting of an Al foil anode, a Pd-coated nanoporous gold pieces cathode, and an ionic liquid-based electrolyte. The discharge products of $Al_2(CO_3)_3$ and C were characterized directly by confocal micro-Raman spectroscopy, XPS, FTIR, and EELS. In consequence, the rechargeable Al–$CO_2$ batteries should follow the equation

$$4Al + 9CO_2 \leftrightarrow 2Al_2(CO_3)_3 + 3C \tag{11}$$

Therefore, without special catalysts, the common reaction of metal–$CO_2$ batteries can be summarized as

$$xM + 3CO_2 \rightarrow 2M_{x/2}CO_3 + C \tag{12}$$

## 3 INFLUENCE FACTORS OF Li–$CO_2$ BATTERIES

The performance of a Li–$CO_2$ battery is affected by many factors such as reactant gases, operation temperature, electrolytes, and cathode materials. As discussed in Section 2.1.1, $CO_2$ does not work as a redox material in Li–$O_2$/$CO_2$ batteries. Therefore, in this section, only pure $CO_2$-based Li–$CO_2$ batteries are discussed.

### 3.1 Operation Temperature

It is well known that the performance of a lithium ion battery is greatly affected by temperature, and the development of low-temperature lithium ion batteries with excellent performance is significant for military and aerospace industry. The performance of a Li–$CO_2$ battery is also influenced by temperature, and the effect was first studied by Archer group.[11] Figure 6 shows the galvanostatic discharge curves of Li–$CO_2$ batteries in the range of 60–100 °C at a fixed current density of 0.05 mA cm$^{-2}$ and a final potential of 2.2 V. Obviously, both the discharge potential and capacity were enhanced with increasing temperature. A high operation temperature could offer at least two key advantages. Along with the reduction of the transport barrier at the cathode–electrolyte interface, the dynamic performances will be improved. In addition, the solubility of $Li_2CO_3$ in the electrolyte will be increased; thus, its deposition on the electrode will be prolonged.

### 3.2 Cathode Structure

Besides investigations on the operation temperature of Li–$CO_2$ batteries, Archer group investigated the effect of cathode surface area (SA) and pore volume

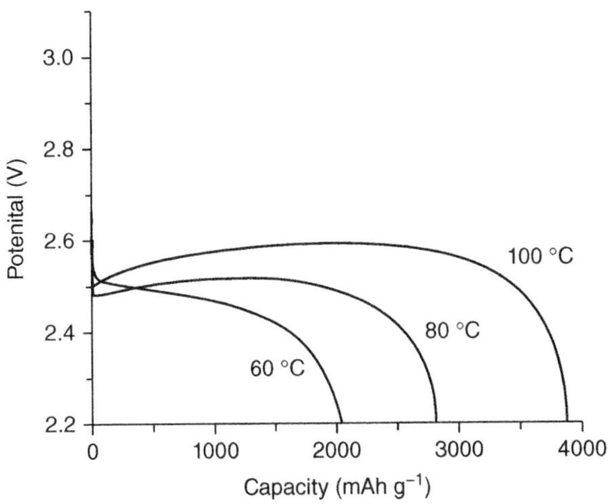


**Figure 6** Galvanostatic discharge curves of Li–$CO_2$ batteries operated in the range of 60–100 °C at a current density of 0.05 mA cm$^{-2}$ and final potential of 2.2 V. [Reproduced with permission from Ref. 11. Copyright 2013 Royal Society of Chemistry]

(PV).[11] Carbon materials with different porosities, pore size distributions, and SAs were used to construct primary Li–$CO_2$ batteries working in the range of 20–100 °C. They found that the carbon cathode, super P (SA 60 m$^2$ g$^{-1}$, PV 0.3 cm$^3$ g$^{-1}$), with the lowest specific surface area exhibited a thermal triggering effect on its performance. On the other side, the temperature dependence was considerably weaker for carbon cathodes with the specific surface area as high as 1020 m$^2$ g$^{-1}$ (PV: 4.45 cm$^3$ g$^{-1}$) and 1300 m$^2$ g$^{-1}$ (PV: 2.2 cm$^3$ g$^{-1}$). This work provides a possible alternative solution to realizing the room-temperature Li–$CO_2$ by utilizing cathode materials with high surface area and appropriate structure. Particularly, with Ketjen black (KB) cathodes and TEGDME-based electrolytes, Li group[12] revealed that Li–$CO_2$ batteries can discharge and charge reversibly at room temperature. The full discharge–charge performance of their Li–$CO_2$ batteries at a current density of 30 mA g$^{-1}$ delivered seven reversible cycles with a decent capacity of around 1000 mAh g$^{-1}$.

Since the electrochemical reaction occurs on the solid–electrolyte–gas triple phase, the structure of the air cathode is critical for the battery performance. Generally, materials with large specific surface, high electronic conductivity, hierarchical porous configurations, and uniformly distributed nanostructured catalysts are expected to exert a prominent effect on the performance of Li–$CO_2$ batteries, by providing substantial active sites, effective transfer pathways for $CO_2$ and lithium ions, lower charge transfer resistance, and considerable storage space for discharge products. After all, the accumulation of insoluble discharge products such as solid $Li_2CO_3$ precipitates can block the transfer pathway of $CO_2$ and lithium ions, resulting in an increase in battery resistance. During the development of Li–$CO_2$ batteries, various strategies of cathode configurations, including some successfully applied ones in fuel cells and metal–air batteries, have been explored. Different kinds of materials used as Li–$CO_2$ air cathodes are discussed at length in Section 4.

### 3.3 Electrolytes

Due to the high decomposition potential of $Li_2CO_3$ and the resultant superoxide radicals, the stability of electrolytes can greatly affect the performance of Li–$CO_2$ batteries. Compared with solid–solid reactions during the charge process, reactions in the solid–liquid interface could occur much more easily. The addition of a redox mediator has been proved to be a good strategy for metal–air batteries. On the other side, flexible liquid-free Li–$CO_2$ batteries were also introduced to meet the requirement of wearable electronics.

#### *3.3.1 Redox Mediators*

Exploring through the theoretical calculation study, Zhou group[19] found that the following equation could be utilized:

$$2Li_2CO_3 + C + 2Br_2 = 3CO_2 + 4LiBr \qquad (13)$$

the electromotive force value of this reaction is −0.39 V. The oxidation–reduction reaction between $Br_2$ and $Li_2CO_3$ was confirmed by ultraviolet (UV)–vis spectroscopy and XRD analysis. Commercial KB was chosen as the cathode material of Li–$CO_2$ batteries, and TEGDME/LiTFSI (lithium bis(trifluoromethane sulfonimide)) electrolyte with or without 50 mM LiBr was used. The Li–$CO_2$ batteries with LiBr could deliver a discharge capacity of about 11 500 mAh g$^{-1}$ at a current density of 50 mA g$^{-1}$ and cycle more than 38 times stably with a cut-off capacity of 500 mAh g$^{-1}$ at a current density of 100 mA g$^{-1}$ (Figure 7). Even at a current density of 200 mA g$^{-1}$, a low charge voltage around 4.0 V was observed, and the battery could work for over 16 cycles. They found that LiBr could be beneficial for the formation of desired morphology of discharge products, and the newly formed $Br_2$ during charge process could promote the decomposition of discharge products chemically. Compared with developing solid catalysts and novel cathode structures, their work proved that the utilization of redox mediators in Li–$CO_2$ batteries is a promising new approach to improve the electrochemical performances.

#### *3.3.2 Polymer Electrolytes*

In order to adapt the flexible or bendable power sources, polymer electrolytes are required for the development of Li–$CO_2$ batteries. Hu *et al.*[20] first introduced a liquid-free composite polymer electrolyte (CPE) of

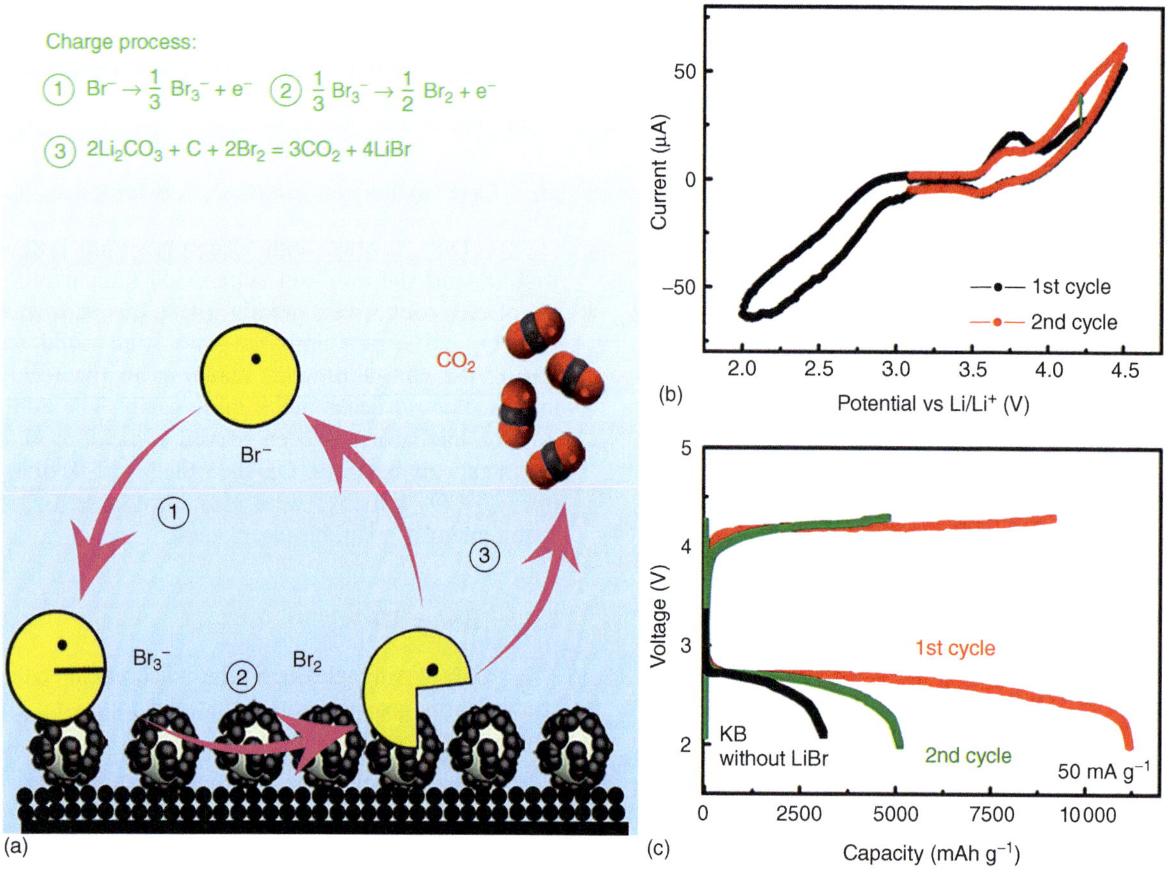


**Figure 7** (a) Proposed mechanism for the charging process with LiBr as the redox mediator in Li–$CO_2$ batteries. (b) Cyclic voltammetry curves of Li–$CO_2$ batteries with LiBr at a scan rate of 0.1 mV $s^{-1}$. (c) Discharge–charge curves at 50 mA $g^{-1}$. [Reproduced with permission from Ref. 19. Copyright 2017 Wiley-VCH Verlag GmbH & Co. KGaA, Weinheim]

poly(methacrylate) (PMA)/poly(ethylene glycol) (PEG)-$LiClO_4$-3 wt% $SiO_2$ into a Li–$CO_2$ battery. In the gel state, this organic–inorganic hybrid electrolyte has a high ionic conductivity of ~0.07 mS $cm^{-1}$ at 55 °C and could directly coat carbon nanotubes (CNTs) on the current collector to form an integrated CPE@CNT cathode (Figure 8). Specifically, the incorporated CPE@CNT cathode owns a large active reaction zone on the CNT surface, and the assembled Li–$CO_2$ batteries could run for 100 cycles with a controlled capacity of 1000 mAh $g^{-1}$ at 55 °C. Benefiting from the robust configuration of the Li/CPE@CNT cathode with reduced interface resistance, their lab-scale pouch-type batteries (5–16 $cm^2$) exhibited a large reversible capacity of 993.3 mAh, high energy density of 521 Wh $kg^{-1}$, and long operation time of 220 h at different bending degrees (0–360°).

Considering the fact that the low ionic conductivity of liquid-free electrolytes limits the cycling capability and rate performance at room temperature, Li *et al.*[21] further introduced a gel polymer electrolyte (GPE) with high ionic conductivity to

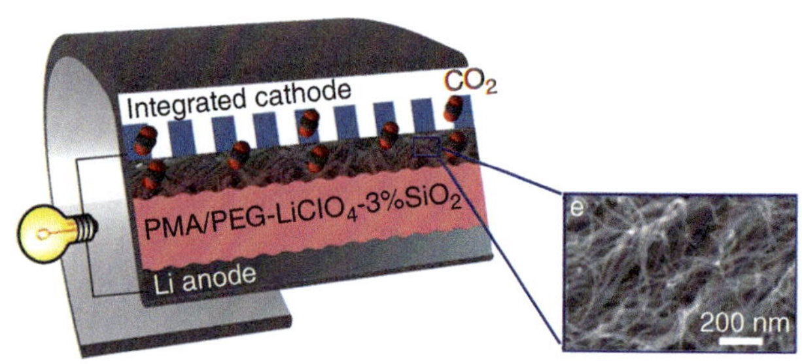


**Figure 8** Structure of all-solid-state Li–$CO_2$ batteries with CNT cathodes. [Reproduced with permission from Ref. 20. Copyright 2017 Wiley-VCH Verlag GmbH & Co. KGaA, Weinheim]

rechargeable Li–$CO_2$ batteries. The compact and thermally stable GPE was prepared through the irradiation of a mixture composed of LiTFSI/TEGDME solution, poly(vinylidene fluoride-co-hexafluoropropylene)/*N*-methyl-2-pyrrolidinone solution and 2-hydroxy-2-methyl-1-phenyl-1-propanone/trimethylolpropane ethoxylate triacrylate solution. CNT-based cathodes were used.

Compared with the contiguous polymerlike and crystalline discharge product in conventional $Li-CO_2$ batteries with liquid electrolytes, $Li_2CO_3$ formed in the GPE-based $Li-CO_2$ batteries exhibits a particle-shaped morphology with poor crystallinity. Thus, the GPE-based battery showed much enhanced electrochemical performance. Their $Li-CO_2$ batteries could run for 60 cycles with a cut-off capacity of 1000 mAh $g^{-1}$ at 100 mA $g^{-1}$, and the maximum applied current density could be increased to 500 mA $g^{-1}$.

## 4 CATHODE MATERIALS

Air cathodes play a key role in $Li-CO_2$ batteries and directly affect the electrochemical performances, including the specific capacity, the round-trip efficiency, the rate capability, and the cycling stability. As discussed in Section 3.2, the air cathode should possess at least three characteristics: a durable porous structure for the diffusion of gas and deposition of discharge product, good electrical conductivity to facilitate electron transportation, and highly catalytic activity for the reversible conversion of $CO_2$. In the following sections, different materials used as air cathodes for $Li-CO_2$ batteries are discussed in detail. These materials can be approximately divided into four main groups: carbon materials, composite materials, metal organic frameworks (MOFs), and binder-free cathode materials.

### 4.1 Carbon Materials

Until now, carbon materials have been widely adopted as air cathodes for $Li-CO_2$ batteries, which can be classified into three groups: commercial carbon materials, functional carbon materials, and heteroatom-doped carbon materials. As shown in Sections 3.1 and 3.2, commercial carbon materials were the initial cathode materials used in $Li-CO_2$ batteries.

When using carbonates in energy storage systems, Zhou group noticed that $CO_2$ could be used by electrocatalytic conversion reactions. They successfully introduced functional carbon materials graphene (Figure 9a)[13] and CNTs[22] as air cathodes for the first time, and the capacities of room-temperature $Li-CO_2$ batteries were dramatically

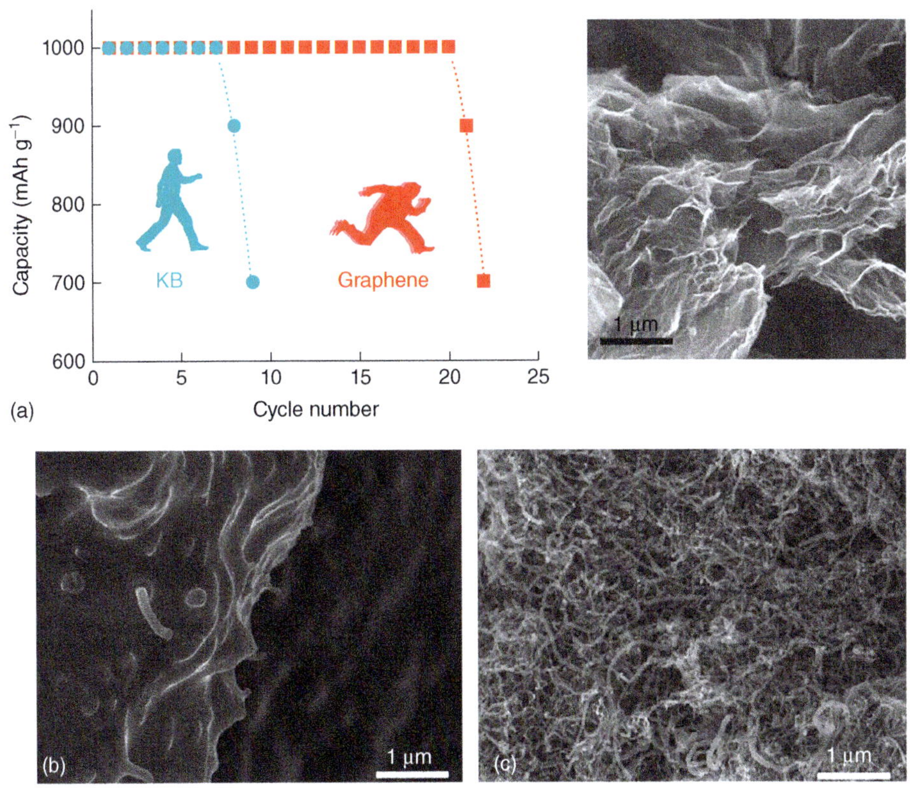


**Figure 9** (a) Schematic illustration of $Li-CO_2$ batteries based on graphene cathodes. [Reproduced with permission from Ref. 13. Copyright 2015, Wiley-VCH Verlag GmbH & Co. KGaA, Weinheim] (b) SEM images of the discharged. [Reproduced with permission from Ref. 22. Copyright 2015 Royal Society of Chemistry] and (c) recharged CNT cathodes. [Reproduced with permission from Ref. 22. Copyright 2015 Royal Society of Chemistry]

enhanced to 14 722 and 8379 mAh $g^{-1}$ at a current density of 50 mA $g^{-1}$, respectively. Furthermore, with a controlled capacity of 1000 mAh $g^{-1}$ at 50 mA $g^{-1}$, the batteries with graphene or CNT cathodes can be stably cycled for 20 or 29 cycles. The remarkable performances of their batteries can be attributed to excellent electrical conductivity, large surface area, and high electrochemical stability of the graphene and CNT cathodes. Interestingly, Li–$CO_2$ batteries with CNT cathodes demonstrate better rate capability and cycling performance. The design of active materials with connected 3D spaces is a promising strategy, for 3D networks, not only benefit the transport of electrons, $CO_2$ and lithium ions but also have more space for the deposition of discharged products without clogging the cathode surface (Figure 9b and c). Thus, CNT cathodes were further applied in polymer electrolyte-based Li–$CO_2$ batteries (Section 3.3.2). Besides, graphene and CNTs have been vastly used as substrate materials for the dispersion of metal or metal–oxide nanoparticles.

The active sites of pure carbon materials are mainly from the surface defects, while carbon structures can be modified by heteroatom doping, which, in turn, would endow them with enhanced catalytic activity or special electronic properties. Qie *et al.*[23] reported that B,N-codoped holey graphene (BN-hG) can be used as efficient bifunctional cathode catalysts for rechargeable Li–$CO_2$ batteries (Figure 10). A reversible charge capacity of 14 996 mAh $g^{-1}$ was obtained at a current density of 0.3 A $g^{-1}$ and the initial coulombic efficiency was as high as 93.5%. Moreover, their Li–$CO_2$ batteries exhibited superior long-term cycling stability up to 200 cycles with a controlled capacity of 1000 mAh $g^{-1}$ at 1.0 A $g^{-1}$. The authors considered that the unique porous holey nanostructure and highly active bifunctional effect are critical to low polarization, excellent rate performance, and superior cycling stability. This work demonstrated the possibility of using low-cost, metal-free, bifunctional carbon cathode catalysts to improve the kinetics of electrochemical formation and decomposition of $Li_2CO_3$ in Li–$CO_2$ batteries.

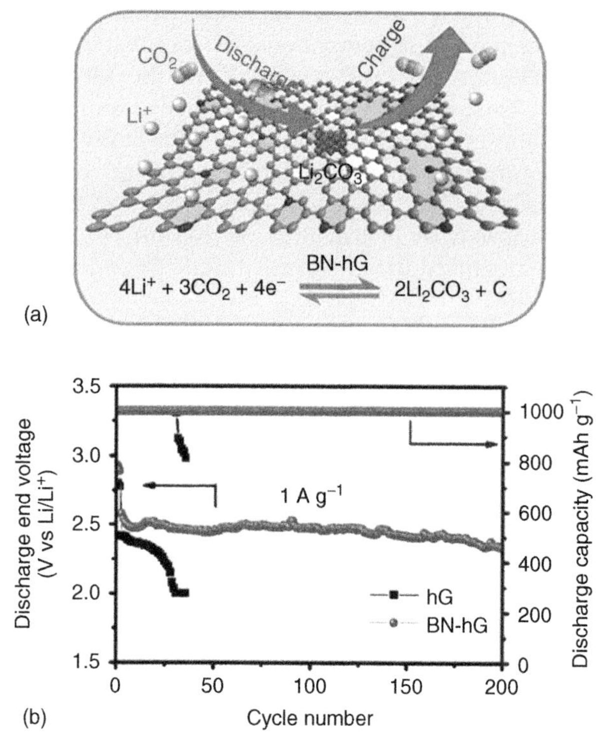


**Figure 10** (a) Schematic representation of the corresponding $CO_2$ reduction and evolution reactions on the BN-hG cathode. (b) Cycling performance of Li–$CO_2$ batteries with hG and BN-hG cathodes. [Reproduced with permission from Ref. 23. Copyright 2017 Wiley-VCH Verlag GmbH & Co. KGaA, Weinheim]

## 4.2 Composite Materials

As discussed, the cathode materials should have not only a large surface area and appropriate porous structure to store the insoluble discharge products but also enough active sites to catalyze the decomposition of $Li_2CO_3$ during charge process. Yang *et al.*[10] found that the self-decomposition of $Li_2CO_3$ at high overpotential would cause the degradation of electrolytes. In order to promote the charge process of Li–$CO_2$ batteries, they synthesized a Ru@Super P cathode catalyst where Ru nanoparticles are evenly deposited on Super P carbon through a solvothermal method.[24] The Li–$CO_2$ battery delivered a discharge capacity of 8 229 mAh $g^{-1}$ at 100 mA $g^{-1}$, and the initial coulombic efficiency of the full discharge–recharge test was 86.2%. The main discharge products were identified as $Li_2CO_3$ and carbon by XRD and Raman spectra. Compared with Super P carbon cathode, the charge potential can be dramatically reduced to below 4.4 V. With a cut-off capacity of 1000 mAh $g^{-1}$, the Li–$CO_2$ battery could be cycled for over 70 times at current densities of 100, 200, and 300 mA $g^{-1}$, indicating high electrocatalytic activity and stability of the Ru@Super P cathodes. In situ surface-enhanced Raman scattering (SERS) and GC–MS tests confirmed that Ru can effectively promote the decomposition of $Li_2CO_3$ and carbon during the charge process, and electrolyte degradation can be largely avoided in the operating potential range (Figure 11). This study shows that the exploration of highly efficient catalysts for effectively reducing the charge overpotential is crucial in achieving reversible Li–$CO_2$ batteries.

Hou *et al.*[25] reported rechargeable Li–$CO_2$ batteries with molybdenum carbide/carbon nanotube ($Mo_2C$/CNT) cathodes, which were prepared by carbothermal reduction of the ball-milled mixture of molybdenum trioxide and CNTs. Owing to the excellent electrical conductivity of 3D CNT networks and substantial catalytic sites of $Mo_2C$ nanoparticles, this cathode material

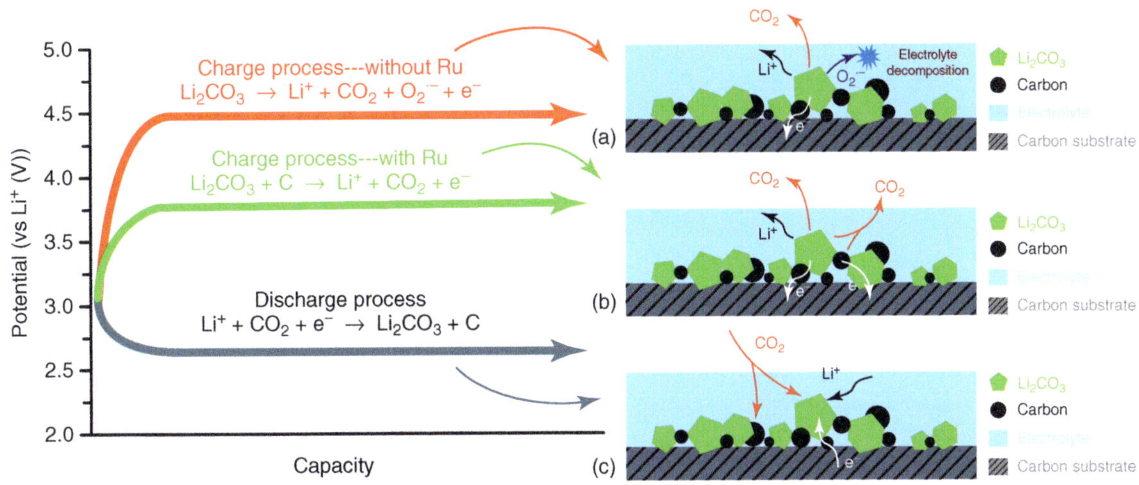


Figure 11 Schematic diagram of the reaction mechanism for the charging process of Li–$CO_2$ batteries with or without the Ru catalyst. [Reproduced with permission from Ref. 24. Copyright 2017 Royal Society of Chemistry]

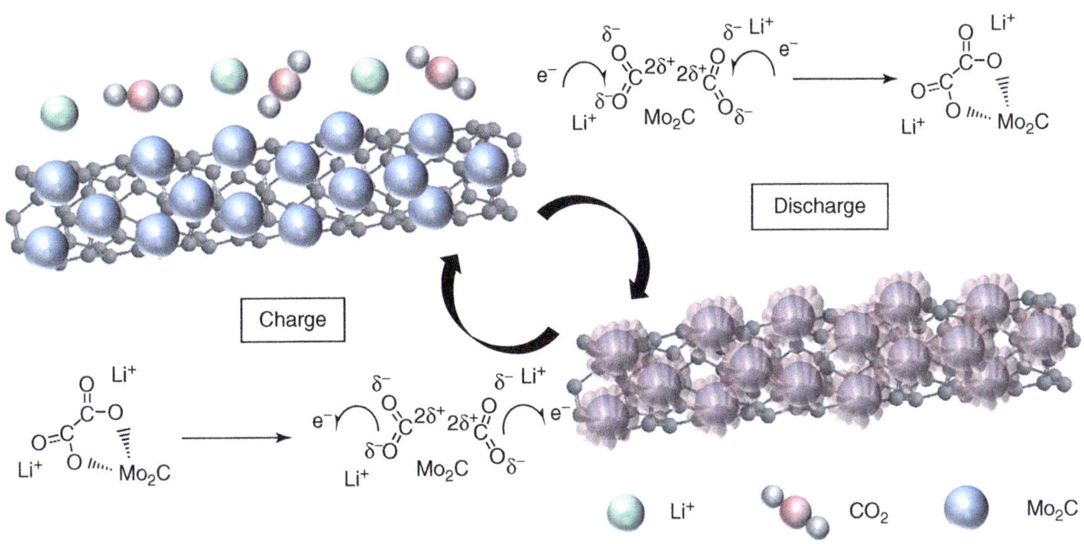


Figure 12 Schematic illustration of reactions during discharge and charge of $Mo_2C$/CNT in the Li–$CO_2$ battery. [Reproduced with permission from Ref. 25. Copyright 2017 Wiley-VCH Verlag GmbH & Co. KGaA, Weinheim]

exhibited a high round-trip efficiency of 77% and improved cycling performance of 40 cycles with a fixed capacity of 100 μA h at the current of 20 μA. In contrast to the formation of $Li_2CO_3$ and carbon, the amorphous discharge product $Li_2C_2O_4$–$Mo_2C$ was detected by a series of ex situ characterizations. As can be seen from Figure 12, $CO_2$ was reduced at the $Mo_2C$/CNT electrode surface during discharge, forming $Li_2C_2O_4$, and then this intermediate product was stabilized by $Mo_2C$, forming an amorphous discharge product that can be easily decomposed. Hence, the charge plateau was dramatically reduced below 3.5 V. Although $Mo_2C$ was consumed in the discharge process, this work provides fresh ideas for catalyst design to improve the energy efficiency of Li–$CO_2$ batteries.

By using a hydrothermal treatment and postannealing method, Zhou group[15] prepared a composite of Ni nanoparticles highly dispersed on N-doped graphene (Ni-NG) with three-dimensional porous structure (Figure 13). Their Li–$CO_2$ batteries with Ni-NG cathodes exhibited a superior discharge capacity of 17 625 mAh $g^{-1}$ at a current density of 100 mA $g^{-1}$, which is 2.7 times that of their batteries with graphene cathodes. On the basis of this highly efficient cathode, the batteries could sustain more than 100 cycles at a cut-off capacity of 1000 mAh $g^{-1}$ with low

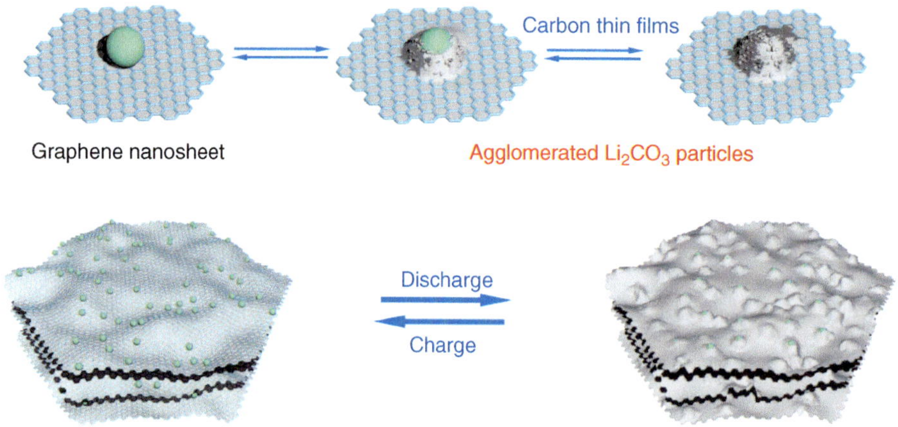


**Figure 13** Schematic illustration of Li–$CO_2$ batteries with Ni-NG cathodes. [Reproduced with permission from Ref. 15. Copyright 2018 Wiley-VCH Verlag GmbH & Co. KGaA, Weinheim]

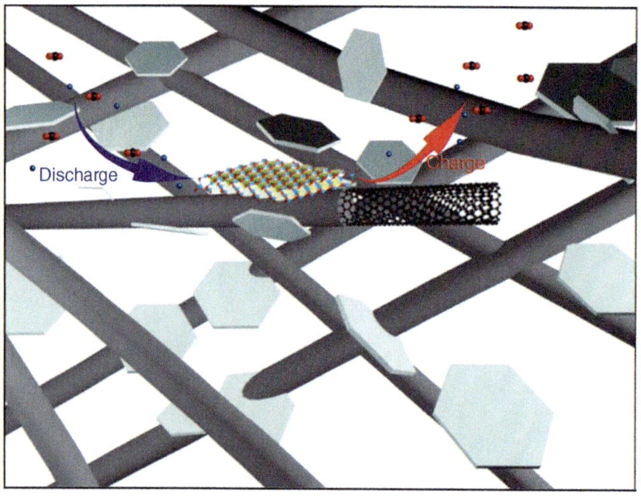


**Figure 14** Schematic diagram of NiO–CNT composite. [Reproduced with permission from Ref. 26. Copyright 2018 Royal Society of Chemistry]

overpotential at the current density of $100\,mA\,g^{-1}$. Particularly, Ni-NG cathodes allowed screening the appearance/disappearance of agglomerated $Li_2CO_3$ particles and carbon thin films upon the discharge/charge process. This may represent a critical step toward better understanding of the electrochemical processes of Li–$CO_2$ batteries. In addition, the reversible electrochemical reaction of $CO_2$ was verified by in situ DEMS. Moreover, first-principles computations further proved that Ni nanoparticles are active sites for the reaction of Li and $CO_2$, which is considered to be advantageous for the discharge process.

Considering the catalytic activity and cycling stability of the transition metal oxides, Zhou group[26] further prepared a NiO–CNT composite (Figure 14) via hydrothermal reaction and calcination method and applied it in Li–$CO_2$ batteries. NiO nanoplates are evenly dispersed on CNTs, which ensures both uniform distribution of active sites and high electrical conductivity. The composite shows high catalytic activity and great stability in the operation of the batteries. Their batteries displayed an outstanding coulombic efficiency of 97.8% during the first cycle of full discharge–charge tests and could sustain coulombic efficiency of 91.7% even at the fifth cycle. With a cut-off capacity of $1000\,mAh\,g^{-1}$ at the current density of $50\,mA\,g^{-1}$, the charge potential sustained around 4.0 V for the initial 20 cycles. XPS results show that the high catalytic activity of NiO could fully realize the decomposition of $Li_2CO_3$. This work provides new choices

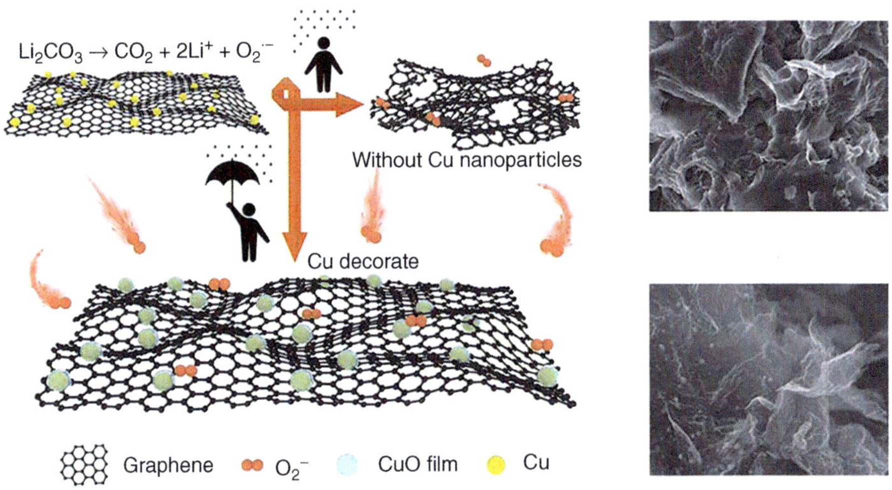


**Figure 15** Schematic illustration of Cu-NG as a cathode in Li–$CO_2$ batteries. [Reproduced with permission from Ref. 27. Copyright 2018 Royal Society of Chemistry]

and strategies in pursuing alternative cathode materials for high-performance rechargeable Li–$CO_2$ batteries.

Zhang *et al.*[27] reported that a composite of Cu nanoparticles highly dispersed on N-doped graphene (Cu-NG) can be used as an efficient cathode for Li–$CO_2$ batteries (Figure 15). Their batteries delivered a charge capacity of up to 13 676 mAh $g^{-1}$ with a coulombic efficiency of 92.0% at 200 mA $g^{-1}$. When the Li–$CO_2$ batteries were tested with a cut-off capacity of 1000 mAh $g^{-1}$ at a current density of 200 mA $g^{-1}$, excellent round-trip efficiencies during cycling were observed. The voltage gaps between discharge and charge were as low as 0.77, 0.99, and 1.26 V for the 1st, 20th, and 50th cycle. The excellent catalytic activity was in agreement with the earlier studies about electrochemical reduction of $CO_2$ in aqueous solutions. Moreover, the authors suggested that the generated 3- to 5-nm-thick CuO film would effectively protect the cathode during additional cycling from the corrosion of superoxide radicals, which are generated from the self-decomposition of $Li_2CO_3$. This work could open up new strategies to develop highly efficient and stable cathodes for Li–$CO_2$ batteries.

### 4.3 Metal–Organic Frameworks

With the rapid utilization of MOFs in $CO_2$ capture, separation, and catalytic conversion into useful compounds, the potential of MOFs as porous catalysts in Li–$CO_2$ batteries was certified by Li *et al.*[28] for the first time (Figure 16). In view of structural versatility and functional tunability, eight porous MOFs as well as two nonporous materials, $MnCO_3$ and MnO, were investigated for the capability of $CO_2$ capture and $Li_2CO_3$ decomposition. Among them, Mn(II)-MOFs showed enabled efficient and reversible performances as cathode catalysts for Li–$CO_2$ batteries. Specifically, $Mn_2$(2,5-dioxido-1,4-benzenedicarboxylate) delivered a remarkable discharge capacity of 18 022 mAh $g^{-1}$ at 50 mA $g^{-1}$, while $Mn(HCOO)_2$ retained low charge potential of ~4.0 V even for over 50 cycles at 200 mA $g^{-1}$. The authors found that the deposition of $Li_2CO_3$ during discharge is sensitively affected by MOFs, where nanostructured morphology with a uniform dispersion tends to grow into the whole electrode. The distribution of the discharge products is critical in mitigating the growth of charge transfer resistance and the conversion of $Li_2CO_3$ back to $CO_2$ during the charge process. Although the reported system is preliminary, it can provide useful design principles for enhancing the reversibility and energy efficiency of Li–$CO_2$ batteries.

### 4.4 Binder-Free Cathodes

Although the nanoparticles of metal or metal oxide in composite materials are evenly distributed, the use of conductive additives and polymer binders to assemble cathodes could not ensure the uniform distribution of catalysts. Meanwhile, the insulating polymer binders may degenerate during the discharge/charge process and severely impede the efficient diffusion and transport of $Li^+$, gas, and electrons. Therefore, a binder-free cathode is a good choice for fabricating a Li–$CO_2$ battery. Binder-free cathodes can be prepared in two methods: By modifying a conducting substrate material with various functional precursors and by creating a conductive fiber network with active sites directly.

Zhu *et al.*[14] used nickel foam as the substrate material for in situ formation of N-doped wormlike carbon with embedded MoFeNi and MoC nanoparticles (MFCN) through a floating catalyst chemical vapor

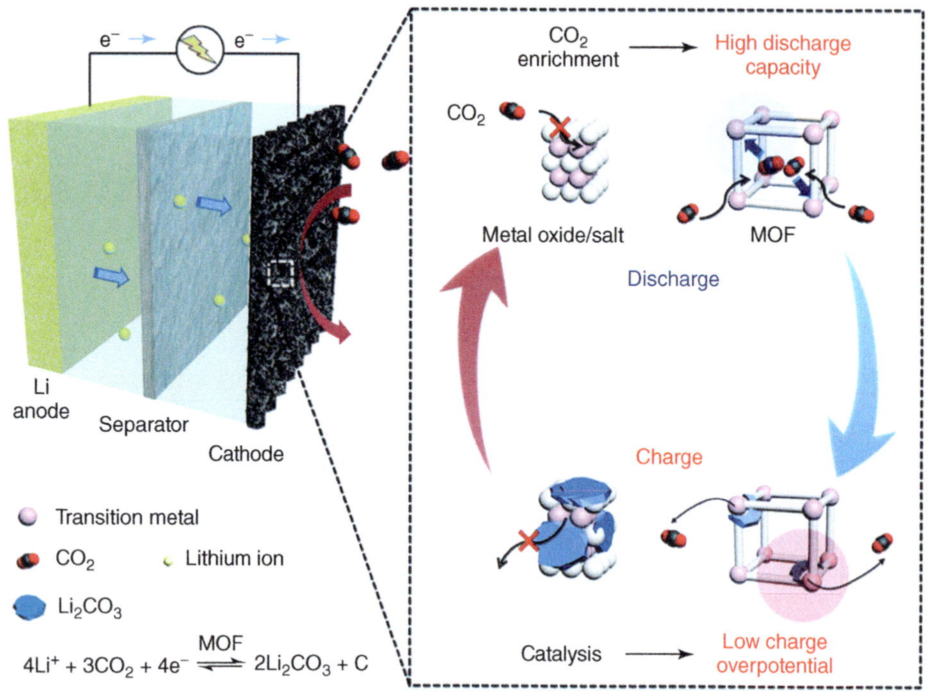


**Figure 16** Scheme of the advantages of a Li–$CO_2$ battery equipped with a MOF-based $CO_2$ electrode. [Reproduced with permission from Ref. 28. Copyright 2018 Royal Society of Chemistry]

deposition method. The MFCN composite can be used directly as a binder-free cathode to fabricate Li–$CO_2$ batteries, because the ductility of the nickel foam is well maintained. Their batteries delivered discharge capacities of 8827, 5695, and 1461 mAh $g^{-1}$ at current densities of 100, 200, and 400 mA $g^{-1}$, respectively. Furthermore, with a controlled capacity of 1000 mAh $g^{-1}$ at 100 mA $g^{-1}$, the initial discharge plateau is about 2.75 V, while the following charge plateau is only approximately 3.79 V. In addition, the charging potential is still below 4.0 V after 50 cycles and stable discharge/charge plateaus can be observed even for over 90 cycles. The authors believed that the remarkable electrochemical performance can be attributed to the high $CO_2$ capture ability of N-doped carbon and the excellent electrocatalytic activity toward the $CO_2$ reduction and evolution reactions (Figure 17).

Different from the synthesis mentioned earlier, electrospinning was a direct method to produce free-standing materials with tunable porous structures. Electrospinning can also avoid the use of substrate material (such as Ni foam) by forming conducting carbon nanofiber networks directly in the anneal process. Zhou group[29] prepared iridium/carbon nanofibers (Ir/CNFs) as free-standing cathodes for Li–$CO_2$ batteries via the electrospinning method, which successfully combined the high catalytic activity of Ir nanoparticles and the unique structure superiority of carbon nanofiber networks (Figure 18). The Ir/CNFs achieved an extremely high energy density of 11.54 mAh $cm^{-2}$ and ultra-long operation time of 1200 h, demonstrating the vast potential for practical applications. Interestingly, they found that Ir/CNFs can change the morphology of discharge products from polymerlike films into particlelike aggregations, which further contributes to the discharge and charge process. This work offers a new insight into the design of catalysts and unique architecture structure of cathodes for high-performance Li–$CO_2$ batteries.

### 4.5 Summary

The electrochemical performances of Li–$CO_2$ batteries with various cathode materials are summarized in Table 1. Commercial carbon blacks were first used as active materials in Li–$CO_2$ batteries, while functional carbon materials such as graphene and CNTs exhibit much better electrochemical properties. In addition, the catalytic activity of carbon materials can be further enhanced by heteroatom doping. Particularly, composite materials with nanoparticles of metals or metal oxides highly dispersed on the functional carbon matrix show superior catalytic activities in Li–$CO_2$ batteries, which, in turn, reduce the overpotential during the charging process and improve the cyclability. Some special materials, such as MOFs, were also explored. Considering the weakness of conductive additives and polymer binders, binder-free cathodes were investigated to meet the demand of practical applications.

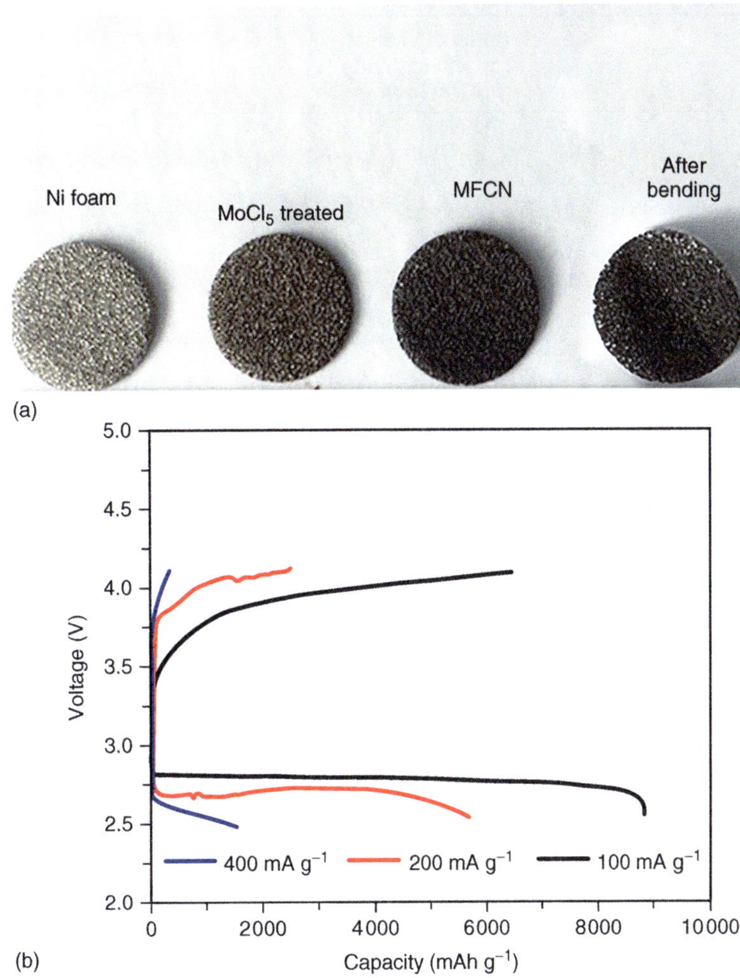


**Figure 17** (a) Optical photo of Ni foam, $MoCl_5$-treated Ni foam, MFCN, and MFCN after bending; (b) full-range rate performances of Li–$CO_2$ batteries with MFCN cathodes at different current densities. [Reproduced with permission from Ref. 14. © Elsevier 2017]

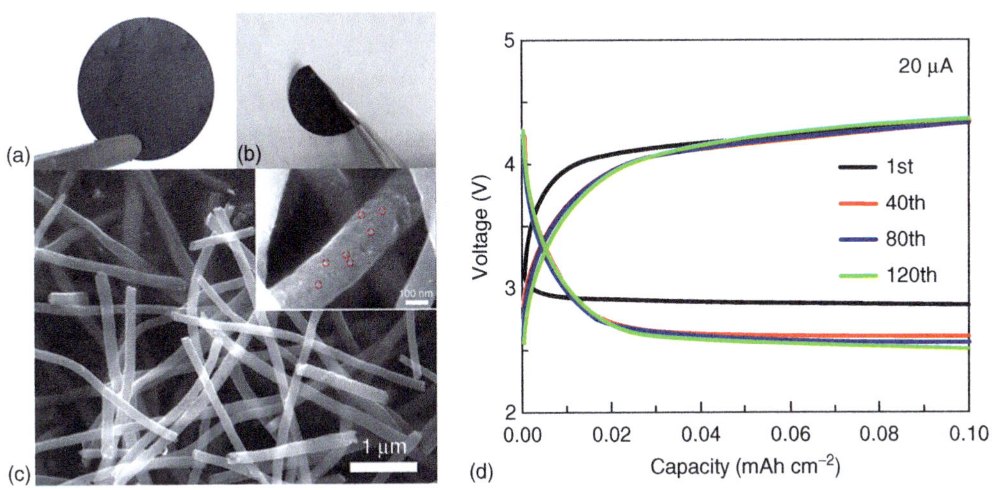


**Figure 18** Optical photo of (a) the free-standing Ir/CNF film cathode and (b) after bending; (c) SEM images of Ir/CNF; (d) cycling stability of Li–$CO_2$ batteries with free-standing Ir/CNF films as air cathodes. [Reproduced with permission from Ref. 29. Copyright 2018 Wiley-VCH Verlag GmbH & Co. KGaA, Weinheim]

**Table 1** Summary of the electrochemical performances of state-of-the-art Li–$CO_2$ batteries

| Cathode catalyst | Electrolyte | Initial discharge capacity/current (mAh $g^{-1}$/mA $g^{-1}$) | Coulombic efficiency | Cycle capacity/current (mAh $g^{-1}$/mA $g^{-1}$) | Cycle number | Refs |
|---|---|---|---|---|---|---|
| Super P | LiTFSI in [bmim][Tf2N] | ~2500 (80 °C) | — | — | — | 11 |
| Super P | $LiCF_3SO_3$ in TEGDME | 6062/100 | 10.9% | 1000/100 | 20 | 24 |
| KB | $LiCF_3SO_3$ in TEGDME | 1032/30 | — | ≈1000/30 | 7 | 12 |
| KB | LiBr and LiTFSI in TEGDME | 11 500/50 | 81% | 500/100 | 38 | 19 |
| | | | | 500/200 | 16 | |
| Graphene | LiTFSI in TEGDME | 14 722/50 | — | 1000/50 | 20 | 13 |
| | | | | 1000/100 | 10 | |
| CNTs | LiTFSI in TEGDME | 8378/50 | — | 1000/50 | 29 | 22 |
| | | | | 1000/100 | 22 | |
| CNTs | CPE | 950/100 (55 °C) | — | 1000/100 | 16 | 20 |
| CNTs | GPE | 8536/50 | — | 1000/100 | 60 | 21 |
| | | 5139/100 | | | | |
| BN-hG | LiTFSI in TEGDME | 16 033/300 | 93.5% | 1000/100 | 200 | 21 |
| Ru@super P | $LiCF_3SO_3$ in TEGDME | 8229/100 | 86.2% | 1000/100 | 70 | 24 |
| | | | | 1000/200 | 70 | |
| | | | | 1000/300 | 70 | |
| $Mo_2C$ | $LiCF_3SO_3$ in TEGDME | 1150 μAh/20 μA | — | 100 μAh/20 μA | 40 | 25 |
| Ni-NG | LiTFSI in TEGDME | 17 625/100 | 55.6% | 1000/100 | 101 | 15 |
| NiO-CNT | LiTFSI in TEGDME | 9000/50 | 97.8% | 1000/50 | 42 | 26 |
| | | | | 1000/100 | 25 | |
| Cu-NG | LiTFSI in TEGDME | 13 676/200 | 92.0% | 1000/200 | 50 | 27 |
| $Mn_2$(dobdc) | LiTFSI in TEGDME | 18 022/50 | — | 1000/200 | 50 | 28 |
| $Mn(HCOO)_2$ | LiTFSI in TEGDME | 15 510/50 | — | 1000/200 | 50 | 28 |
| MFCN | LiTFSI in TEGDME | 8827/100 (free-standing) | 73.4% | 1000/100 | 50 | 14 |
| Ir/CNFs | LiTFSI in TEGDME | 21 528/50 | 93.1% | 1000/50 | 45 | 29 |
| | | 18 813/100 | 65.1% | 1000/100 | 27 | |
| | | 11.54 mAh/40 μA (free-standing) | — | 100 μAh/20 μA | 120 | |

[bmim][Tf2N], 1-Butyl-3-methylimidazolium bis(trifluoromethanesulfonyl) imide.

With the development of air cathodes as well as other components, the electrochemical performances of Li–$CO_2$ batteries will be improved gradually.

## 5 CONCLUSION

The rechargeable Li–$CO_2$ battery offers a novel approach not only for clean recycling and utilization of $CO_2$ but also for advanced energy storage devices. Along the process of study, it is found that the performance of a Li–$CO_2$ battery is affected by many factors, including the reactant gas, the operation temperature, electrolytes, and cathode materials. Many encouraging developments in terms of cathode materials, reaction mechanism, and battery compositions have been achieved. However, there are still considerable obstacles including low discharge capacity, weak rate capability, high charge overpotential, and poor cyclability still need to be overcome. All these problems more or less result from the sluggish dynamic formation and decomposition of $Li_2CO_3$, while air cathodes play a key role and directly influence the battery performance. Up to now, various strategies of cathode configurations have been used in Li–$CO_2$ batteries, which can be roughly divided into four main groups: carbon materials, composite materials, other special materials like MOFs, and binder-free cathode materials. Particularly, binder-free cathode materials will contribute to the development of practical applications and could further meet the requirement of the flexible or bendable Li-$CO_2$ battery when combined with polymer electrolytes. With continued and extensive development of innovative cathode materials as well as other components, the electrochemical performances of Li-$CO_2$ batteries can be dramatically improved. Considering that $CO_2$ can be supplied as supercritical fluid or dry ice, Li–$CO_2$ batteries will play an important role in next-generation energy storage systems.

## 6 ABBREVIATIONS AND ACRONYMS

[bmim][Tf2N] = 1-butyl-3-methylimidazolium bis(trifluoromethanesulfonyl) imide; BN-hG = B,N-codoped holey graphene; CNTs = carbon nanotubes; CPE

= composite polymer electrolyte; Cu-NG = Cu nanoparticles highly dispersed on N-doped graphene; DEMS = differential electrochemical mass spectrometry; DFT = density functional theory; DME = 1,2-dimethoxyethane; DMSO = dimethyl sulfoxide; EELS = electron energy loss spectroscopy; FTIR = Fourier transform infrared spectroscopy; GC-MS = gas chromatography–mass spectrometry; GPE = gel polymer electrolyte; HRTEM = high-resolution transmission electron microscope; Ir/CNFs = iridium/carbon nanofibers; KB = Ketjen black; LiTFSI = lithium bis(trifluoromethane sulfonimide); MFCN = N-doped wormlike carbon with embedded MoFeNi and MoC nanoparticles; MOFs = metal organic frameworks; $Mo_2C$/CNT = molybdenum carbide/carbon nanotube; Ni-NG = Ni nanoparticles highly dispersed on N-doped graphene; PV = pore volume; SA = surface area; SERS = surface-enhanced Raman scattering; TEGDME = tetraethylene glycol dimethyl ether; TEM = transmission electron microscope; UV = ultraviolet; XPS = X-ray photoelectron spectroscopy; XRD = X-ray diffraction

## 7 REFERENCES


1. Z. Xie, X. Zhang, Z. Zhang and Z. Zhou, *Adv. Mater.*, 2017, **29**, 1605891.
2. X. Li, S. Yang, N. Feng, P. He and H. Zhou, *Chin. J. Catal.*, 2016, **37**, 1016.
3. K. Takechi, T. Shiga and T. Asaoka, *Chem. Commun.*, 2011, **47**, 3463.
4. H. Lim, H. Lim, K. Park, D. Seo, H. Gwon, J. Hong, W. Goddard, H. Kim and K. Kang, *J. Am. Chem. Soc.*, 2013, **135**, 9733.
5. Y. Qiao, J. Yi, S. Guo, Y. Sun, S. Wu, X. Liu, S. Yang, P. He and H. Zhou, *Energy Environ. Sci.*, 2018, **11**, 1211.
6. S. K. Das, S. Xu and L. A. Archer, *Electrochem. Commun.*, 2013, **27**, 59.
7. W. I. Al Sadat and L. A. Archer, *Sci. Adv.*, 2016, **2**, e1600968.
8. S. R. Gowda, A. Brunet, G. M. Wallraff and B. D. McCloskey, *J. Phys. Chem. Lett.*, 2013, **4**, 276.
9. Y. S. Mekonnen, K. B. Knudsen, J. S. G. Mýrdal, R. Younesi, J. Højberg, J. Hjelm, P. Norby and T. Vegge, *J. Chem. Phys.*, 2014, **140**, 121101.
10. S. Yang, P. He and H. Zhou, *Energy Environ. Sci.*, 2016, **9**, 1650.
11. S. Xu, S. K. Das and L. A. Archer, *RSC Adv.*, 2013, **3**, 6656.
12. Y. Liu, R. Wang, Y. Lyu, H. Li and L. Chen, *Energy Environ. Sci.*, 2014, **7**, 677.
13. Z. Zhang, Q. Zhang, Y. Chen, J. Bao, X. Zhou, Z. Xie, J. Wei and Z. Zhou, *Angew. Chem. Int. Ed.*, 2015, **54**, 6550.
14. Q.-C. Zhu, S.-M. Xu, Z.-P. Cai, M. M. Harris, K.-X. Wang and J.-S. Chen, *Energy Storage Mater.*, 2017, **7**, 209.
15. Z. Zhang, X.-G. Wang, X. Zhang, Z. Xie, Y.-N. Chen, L. Ma, Z. Peng and Z. Zhou, *Adv. Sci.*, 2018, **5**, 1700567.
16. Y. Qiao, J. Yi, S. Wu, Y. Liu, S. Yang, P. He and H. Zhou, *Joule*, 2017, **1**, 359.
17. X. Hu, J. Sun, Z. Li, Q. Zhao, C. Chen and J. Chen, *Angew. Chem. Int. Ed.*, 2016, **55**, 6482.
18. W. Ma, X. Liu, C. Li, H. Yin, W. Xi, R. Liu, G. He, X. Zhao, J. Luo and Y. Ding, *Adv. Mater.*, 2018, 1801152.
19. X.-G. Wang, C. Wang, Z. Xie, X. Zhang, Y. Chen, D. Wu and Z. Zhou, *ChemElectroChem*, 2017, **4**, 2145.
20. X. Hu, Z. Li and J. Chen, *Angew. Chem. Int. Ed.*, 2017, **56**, 5785.
21. C. Li, Z. Guo, B. Yang, Y. Liu, Y. Wang and Y. Xia, *Angew. Chem. Int. Ed.*, 2017, **56**, 9126.
22. X. Zhang, Q. Zhang, Z. Zhang, Y. Chen, Z. Xie, J. Wei and Z. Zhou, *Chem. Commun.*, 2015, **51**, 14636.
23. L. Qie, Y. Lin, J. W. Connell, J. Xu and L. Dai, *Angew. Chem. Int. Ed.*, 2017, **56**, 6970.
24. S. Yang, Y. Qiao, P. He, Y. Liu, Z. Cheng, J.-j. Zhu and H. Zhou, *Energy Environ. Sci.*, 2017, **10**, 972.
25. Y. Hou, J. Wang, L. Liu, Y. Liu, S. Chou, D. Shi, H. Liu, Y. Wu, W. Zhang and J. Chen, *Adv. Funct. Mater.*, 2017, **27**, 1700564.
26. X. Zhang, C. Wang, H. Li, X.-G. Wang, Y. Chen, Z. Xie and Z. Zhou, *J. Mater. Chem. A*, 2018, **6**, 2792.
27. Z. Zhang, Z. Zhang, K. Cao, P. Liu and Z. Zhou, *J. Mater. Chem. A*, 2018, **6**, 3218.
28. S. Li, Y. Dong, J. Zhou, Y. Liu, J. Wang, X. Gao, Y. Han, P. Qi and B. Wang, *Energy Environ. Sci.*, 2018, **11**, 1318.
29. C. Wang, Q. Zhang, X. Zhang, X.-G. Wang, Z. Xie and Z. Zhou, *Small*, 2018, **14**, 1800641.

# S Electrode Materials

**Feng Li**

*University of Idaho, Moscow, ID, USA*



## 1 INTRODUCTION

With the invention and commercialization of the rechargeable lithium-ion (Li-ion) batteries in the late twentieth century, they have become a dominant source of energy storage for portable, mobile applications ranging from mobile phones to laptops to electric cars. With the ever-increasing need of even higher energy density of rechargeable batteries, especially for electric vehicles and power microgrid, Li-ion battery is no longer meeting the requirement because its energy density has reached its theoretical limit. In the past couple of decades, researchers have been trying to find a replacement with large energy density and specific capacity. To this end, a lot of research has been done on lithium–sulfur (Li–S) battery as a most promising candidate.

Li–S battery, with metallic lithium as anode and elemental sulfur as cathode, is considered as one of the most promising next-generation energy-storage system after Li-ion battery. This is primarily because of its high theoretical specific capacity of 1675 Ah $kg^{-1}$ and energy density of 2600 Wh $kg^{-1}$ delivered during the chemical reaction of sulfur with lithium to form $Li_2S$. In contrast, in a traditional Li-ion battery system, the capacities of the insertion-oxide cathodes and the graphite anode are limited to ~250 and ~370 Ah $kg^{-1}$, respectively, and the theoretical energy density is limited to ~420 Wh $kg^{-1}$.[1] On the other hand, sulfur is cheap due to its wide availability, which makes it a cost-effective cathode material. However, there are several challenges associated with Li–S battery that must be overcome in order to make it practical. First, sulfur and lithium sulfide are both insulators with a low conductivity of $5 \times 10^{-30}$ S $cm^{-1}$ at 25 °C, and conductive materials such as carbon need to be incorporated to the cathode electrodes. Second, the intermediate lithium polysulfides (PSs) formed during the chemical reaction are soluble in the liquid electrolyte, which results in the so-called shuttle effect. Thirdly, the electrochemical lithiation and delithiation processes experience serious volume change of around 80%. These challenges limit the actual specific capacity and cycle stability of Li–S battery. The cycle life and efficiency of Li–S battery must be improved in order to make practical Li–S batteries.

This chapter first discusses the fundamental principle of Li–S battery, the challenges it faces, and some possible solutions. Then, different types of composite materials for sulfur cathode are described in detail.

## 2 ELECTROCHEMISTRY OF LITHIUM–SULFUR BATTERIES

### 2.1 Basic Cell Structure

A basic Li–S cell consists of a lithium metal anode, an organic electrolyte, and a sulfur composite or compound cathode, as shown in Figure 1. Sulfur cannot be used as cathode independently because sulfur and lithium sulfide are both insulators and they have no stable crystal structures to form a robust solid electrode when the temperature increases to about 100 °C. Therefore, conductive host materials must be used to accommodate sulfur to chemically react with lithium and form $Li_2S$. Furthermore,

*Inorganic Battery Materials*. Edited by Hailiang Wang and Boniface P.T. Fokwa
 ISBN 978-1-119-43199-2

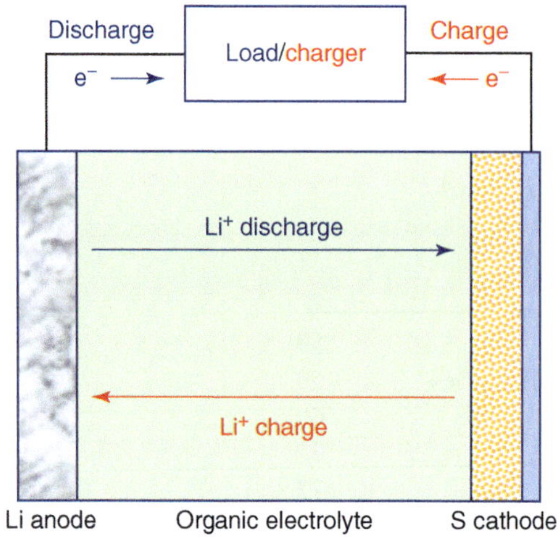


**Figure 1** Schematic diagram of an Li–S cell and its operations

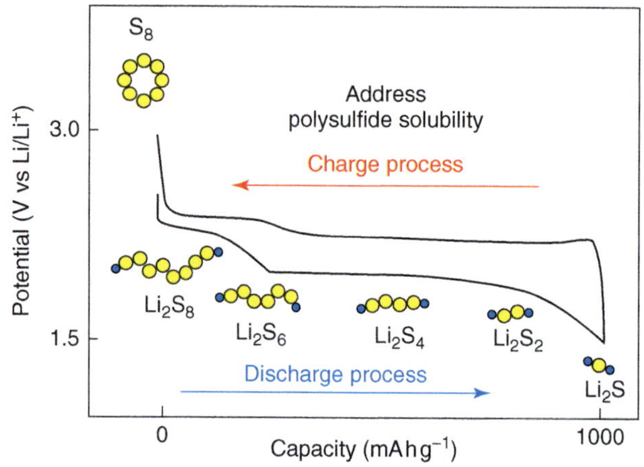


**Figure 2** Typical discharge/charge voltage profiles of an Li–S cell. [Reproduced with permission from P. G. Bruce, S. A. Freunberger, L. J. Hardwick, and J. Tarascon, *Nat. Mater.*, 2012, **11**, 19–29. © Springer Nature, 2012]

the host material should have a matrix structure to account for the volume expansion resulting from conversion reaction of the active materials and accommodate lithium ions.[1,2]

## 2.2 Chemical and Electrochemical Reactions

During the discharge process, the lithium metal of the anode is oxidized to form lithium ions and electrons. The Li ions travel through the electrolyte and move to the sulfur cathode, while the electrons pass through the external circuit and reach the sulfur cathode. At the cathode, sulfur is reduced to form lithium sulfide by accepting lithium ions and electrons. During the charge process, the process works backward, as summarized below:

- Discharge process:
  Anodic reaction at negative electrode: $2Li \rightarrow 2Li^+ + 2e^-$
  Cathodic reaction at positive electrode: $S + 2Li^+ + 2e^- \rightarrow Li_2S$
  Overall: $2Li + S \rightarrow Li_2S$
- Charge process:
  Cathodic reaction at positive electrode: $Li_2S \rightarrow S + 2Li^+ + 2e^-$
  Anodic reaction at negative electrode: $2Li^+ + 2e^- \rightarrow 2Li$
  Overall: $Li_2S \rightarrow 2Li + S$

All in all, the discharge process is based on the reaction of Li ions from the Li anode with the sulfur cathode to form solid $Li_2S$, but the detailed process is much more complicated because of the several intermediates of PS. Under normal conditions, sulfur atoms form cyclic octatonic molecules $S_8$ ring. As illustrated in Figure 2, the octasulfur (cyclo-$S_8$) at the cathode is first reduced to $Li_2S_8$, and then $Li_2S_6$ recognized as the upper voltage plateau of around 2.3 V. As the discharge continues, lower order lithium polysulfide (LiPS), that is, $Li_2S_4$, and then $Li_2S_2$ are formed, represented as the lower voltage plateau of about 2.1 V. After the intermediate PS, the insoluble $Li_2S$ is formed in the end. During the charging process, $Li_2S$ is converted back to $S_8$ with some intermediate LiPSs.

## 2.3 Challenges

The intermediate LiPSs are soluble to the organic electrolyte solution. The high-order PSs may dissolve to the electrolyte and diffuse to the Li anode to be randomly precipitated onto the anode and or chemically reduced to lower order PSs. These soluble lower order PSs can then diffuse back to sulfur cathode to be oxidized. This PS shuttle effect is essentially an internal short, leading to self-discharge and low Coulombic efficiency. Some PSs may remain in the electrolyte and do not diffuse back to the sulfur cathode, leading to capacity fading. PS shuttle effect is one of the most challenging problems in the design of Li–S batteries.

The sulfur electrode experiences a significant volume expansion of about 80% upon forming lithium sulfide due to the density difference between the two materials (2.03 vs 1.66 g cm$^{-3}$). On the other hand, both sulfur and lithium sulfides are nonconductive either to electrons or ions. Furthermore, unlike other cathode materials, octasulfur is soft and cannot form standalone electrodes. All these challenges necessitate special conductive host materials to accommodate sulfur at the cathode.

# 3 SULFUR CATHODE HOST MATERIALS

To overcome the PS dissolution and shuttle effect and other aforementioned challenges, conductive matrix host materials have been utilized to form sulfur composites to accommodate and encapsulate sulfur and withhold long-chain LiPS from dissolving to the electrolyte. These include carbonaceous nanomaterials or nanocarbons, metal compounds, conducting polymers, and their combinations.

### 3.1 Carbonaceous Nanomaterials

Most of the host cathode materials used for Li–S batteries are carbon materials, including graphene, porous carbon, carbon nanotube (CNT), carbon nanofiber (CNF), and carbon spheres. In general, the porous structure of the carbon materials can be classified into three types based on the pore size: micropore, mesopore, and macropore. Micropore has a diameter of less than 2 nm and is ideal for accommodating and immobilizing the active materials. The mesopore has a diameter between those of micropore and macropore, that is, 2 and 50 nm. This type of pore can help enhance sulfur encapsulation and improve the transportation of lithium ion and electrolyte. Macropore has a diameter greater than 50 nm and is mainly derived from an interwoven network of CNTs or CNFs. It can facilitate electrolyte immersion and thus suppress PS migration.

#### 3.1.1 *Graphene*

Graphene is a thin sheet of single-layer carbon atoms with a honeycomb (hexagon) lattice. It is the parent form of all graphitic structures of carbon including graphite, CNTs, and fullerenes. Graphene is one million times thinner than paper and is considered as a two-dimensional (2-D) material. Graphene can be used as a host material to anchor active materials for electrochemical storage applications due to its high electronic conductivity, large surface area, high mechanical strength, and flexibility.

**Figure 3** (a) SEM image of bare S. (b) Field-emission scanning electron microscope (FE-SEM) images of GNS and (c) S-GNS composites. [Reproduced with permission from J. Wang, L. Lu, M. Choucair, J. A.Stride, X. Xu, H. Liu, *J. Power Sources*, 2011, **16**, 7030–7034. © Elsevier, 2011]

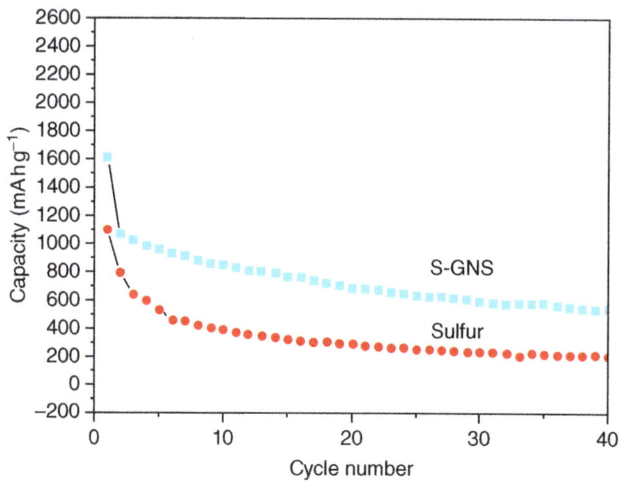


**Figure 4** Discharge capacities versus cycle number for the S-GNS composite and pure S electrodes. [Reproduced with permission from J. Wang, L. Lu, M. Choucair, J. A.Stride, X. Xu, H. Liu, *J. Power Sources*, 2011, **16**, 7030–7034. © Elsevier, 2011]

Many different approaches have been proposed and tested to utilize graphene as the host material for sulfur electrode. The two most basic ones are graphene nanosheet (GNS) and graphene oxide (GO) sheet.

In Wang's research,[4] highly porous GNSs were mixed with elemental sulfur forming a mixture designated S-GNS. The melted sulfur was infiltrated into the layer of graphene at 200 °C for 6 h, and then the vaporized sulfur was coated onto the graphene with the temperature increased to and held at 300 °C for 3 h. Scanning electron microscope (SEM) images in Figure 3 show that the sulfur particles melted and evenly distributed throughout the GNS. The capacity and cycle life of Li–S battery are significantly improved using the S-GNS composite in comparison to the bare S electrode, as shown in Figure 4.

Ji and coworkers immobilized S on GOs by a simple chemical reaction deposition to form graphene oxide–sulfur (GO–S) nanocomposite cathodes for Li–S battery followed by a low-temperature (155 °C) heat-treatment process. This heat treatment melted and diffused the bulk S on the surface into the pores of the GO. It partially removed and/or chemically modified some of the functional groups on the GO surface and improved the electronic conductivity of the GO–S nanocomposites. Some of the bulk S that was not directly attached to the GO layers was also removed. Figure 5(a) shows the SEM image of the as-synthesized GO–S composite, and Figure 5(b) shows a schematic of a representative pattern of GO immobilizing S. The GO–S nanocomposite cathodes show excellent capacity stability of about 1000 mA h g$^{-1}$, as shown in Figure 5(c).

One problem with the abovediscussed method is that the PSs formed during the discharge process can still diffuse out of the graphene structure and lead to the shuttle effect; thus, the cycling performance still needs to be improved. Wang *et al.* synthesized a graphene–sulfur composite material by wrapping poly(ethylene glycol) (PEG)-coated submicrometer sulfur particles with mildly oxidized graphene oxide (mGO) sheets decorated with carbon black nanoparticles. Figure 6 shows a schematic of the synthesis steps, and Figure 7 shows the SEM image of the graphene–sulfur composite. The PEG serves as a "cushion" in the hybrid structure. The PEG and graphene-coating layers can prevent the PSs from diffusing out of

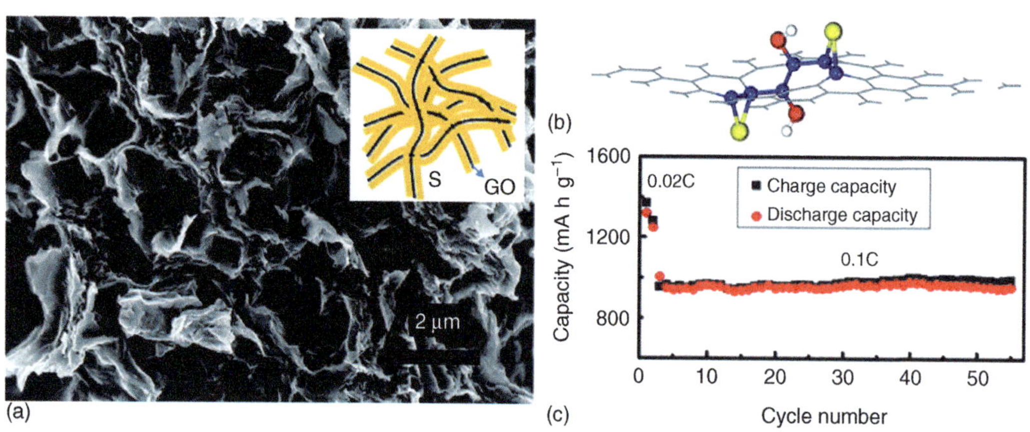


**Figure 5** (a) SEM image of the GO–S nanocomposite after heat treatment in Ar at 155 °C for 12 h. (b) Representative pattern of GO immobilizing S. The hydroxyl enhances the binding of S to the C—C bond due to the induced ripples by epoxy or hydroxyl group. Yellow, red, and white balls denote S, O, and H atoms, respectively, while the others are C atoms. Note that the C atoms bonding to S or O are highlighted as blue balls. (c) Cycling performance at a constant current rate of 0.1C after initial activation processes at 0.02C for two cycles. [Reprinted with permission from L. Ji, M. Rao, H. Zheng, L. Zhang, Y. Li, W. Duan, J. Guo, E. J. Cairns, Y. Zhang, *J. Am. Chem. Soc.* 2011, **133**, 18522–18525. © 2011 American Chemical Society]

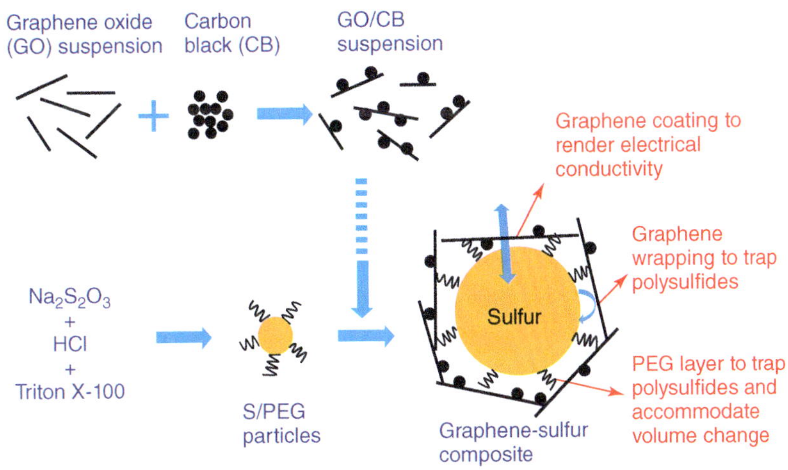


**Figure 6** Schematic of the synthesis steps for a graphene–sulfur composite, with a proposed schematic structure of the composite. [Reprinted with permission from H. Wang, Y. Yang, Y. Liang, J. T. Robinson, Y. Li, A. Jackson, Y. Cui, and H. Dai, *Nano Lett*. 2011, **11**, 2644–2647. © American Chemical Society]

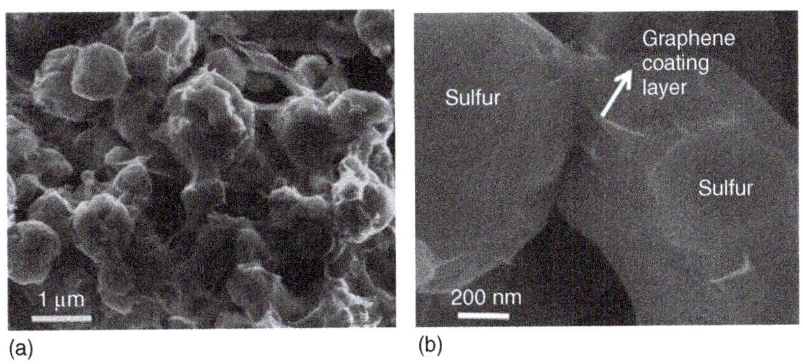


**Figure 7** · SEM characterization of graphene–sulfur composite at low (a) and high (b) magnifications. [Reprinted with permission from H. Wang, Y. Yang, Y. Liang, J. T. Robinson, Y. Li, A. Jackson, Y. Cui, and H. Dai, *Nano Lett*. 2011, **11**, 2644–2647. © American Chemical Society]

the cathode and accommodate volume expansion of the active material during the discharge process. The carbon black decoration improves the electrical conductivity of the sulfur cathode. The Li–S battery based on this graphene–sulfur composite electrode exhibits high and stable specific capacity of ~600 mA h $g^{-1}$ over more than 100 cycles.

### 3.1.2 *Meso/Microporous Carbon Sphere*

Microporous carbon spheres with pore size less than 2 nm can be used to constrain the electrochemical active material and reaction inside the narrow micropores by strong adsorption. Zhang and coworkers prepared a sulfur–carbon sphere composite by encapsulating sulfur into micropores of carbon sphere by thermal treatment of a mixture of sublimed sulfur and carbon spheres.[7] A schematic of the constrained electrochemical reaction process inside the micropores of the sulfur–carbon sphere composite cathode is shown in Figure 8. This type of sulfur–microporous carbon cathode showed a long cycle life for 500 cycles with the remaining capacity of about 650 mA h $g^{-1}$.

Li *et al*. studied mesoporous carbon (MC) with tunable pore sizes (22, 12, 7, and 3 nm) and pore volumes (from 1.3 to 4.8 $cm^3$ $g^{-1}$) with sulfur in the pores as a MC–sulfur composite electrode for Li–S batteries. It is found that MC materials with a larger pore volume provide higher maximum sulfur loadings, but the battery performance showed little difference for different MC–S composites when the pores are fully filled with sulfur. On the other hand, partial sulfur filling and surface modification can promote the electrical contact between sulfur and the MC, reducing the dissolution and diffusion

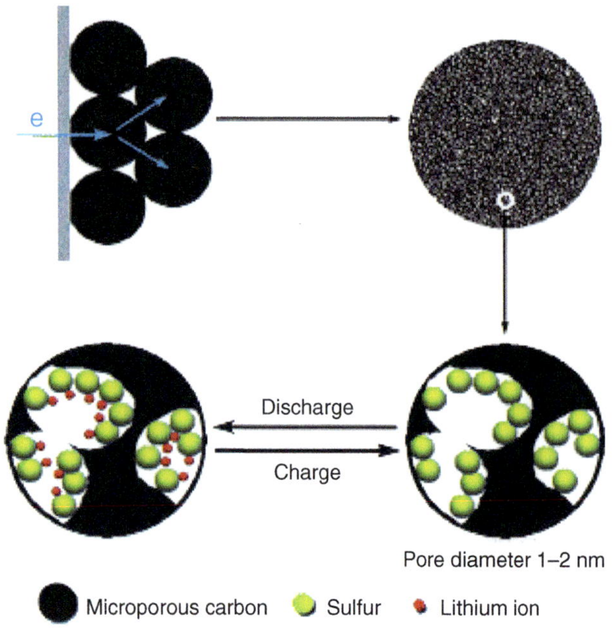


**Figure 8** A schematic of the constrained electrochemical reaction process inside the micropores of the sulfur–carbon sphere composite cathode. [Reproduced with permission from B. Zhang, X. Qin, G. R. Li, and X. P. Gao, *Energy Environ. Sci.*, 2010, **3**, 1531–1537. © Royal Society of Chemistry, 2011]

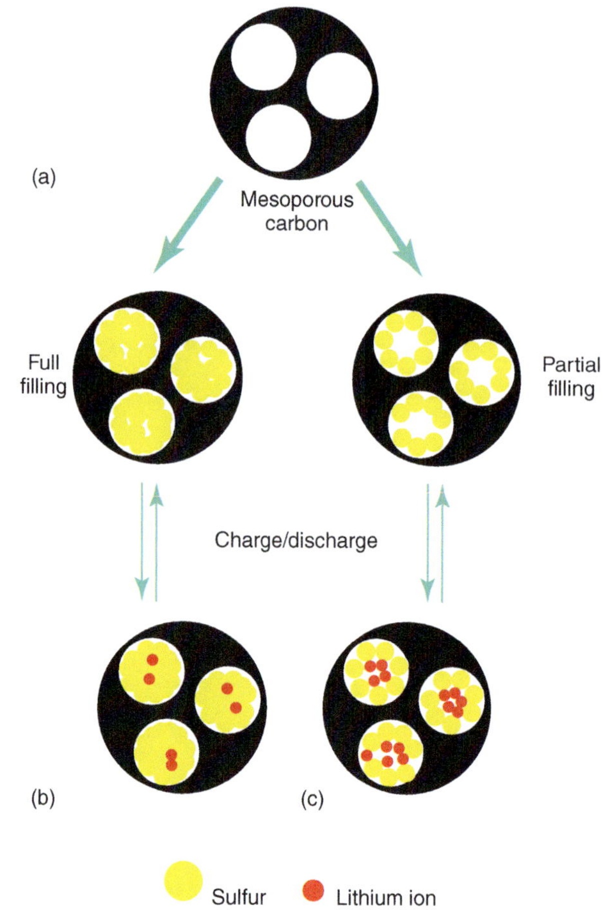


**Figure 9** Schematic diagram of MC–S composite electrodes for Li–S batteries: (a) MC. (b) MC with pores filled with sulfur. (c) MC with pores partially filled with sulfur. [Reproduced with permission from X. Li, Y. Cao, W. Qi, L. V. Saraf, J. Xiao, Z. Nie, J. Mietek, J. Zhang, and J. Liu, *J. Mater. Chem.*, 2011, **21**, 16603–16610. © Royal Society of Chemistry, 2011]

of PS anions and providing a steady supply of lithium ions, thus significantly improving the battery performance. For example, an MC with a pore size of 22 nm and 50% sulfur filling in the mesopores provides S–MPC composites with an initial capacity as high as 1390 mA h g$^{-1}$ and a capacity of 840 mA h g$^{-1}$ after 100 cycles. A schematic of MC spheres with full and partial filling is shown in Figure 9.

### 3.1.3 *Carbon Nanotubes*

CNTs take the form of cylindrical carbon molecules, which can be treated as wrapped single-layer graphene, that is, single-walled carbon nanotubes (SWCNTs), or coaxial multilayer graphene, or multiwalled carbon nanotubes (MWCNTs). It is a natural semi one-dimensional (1-D) nanostructure. With the extremely high aspect ratio, that is, the length-to-diameter ratio, which can be as high as 132 000 000:1, excellent mechanical properties, and electrical conductivities, CNTs are attractive to form sulfur cathode composite materials. Typical CNTs have diameters ranging from 0.4 to 40 nm. CNTs can form interwoven conductive network in the sulfur-based composites. Han and coworkers demonstrated that a simple addition of MWCNTs to sulfur cathodes can increase the specific capacity from ~400 mA h g s$^{-1}$ (without additive) to 498 mA h g s$^{-1}$ with additive. The capacity retention can be improved from 25% to 62% after 50 cycles.[9] To further increase the contact between MWCNTs and sulfur, thereby to improve the electrical conductivity, utilizing the capillary effect, sulfur can be coated onto the outer surface of the CNTs by mixing the two species together with low sulfur viscosity at 155 °C, as shown in Figure 10.[10]

### 3.1.4 *Carbon Nanofibers*

CNFs are cylindric nanostructures with graphene layers arranged as stacked cones, cups, or plates. CNTs can be treated as a special type of CNFs with graphene layers wrapped into perfect cylinders. While CNTs are hollow with small diameters, CNFs are typically solid with larger diameter from 50 nm to submicrons. CNFs, like CNTs, also have excellent electrical conductivity and mechanical strength, suitable for sulfur-containing composite materials.

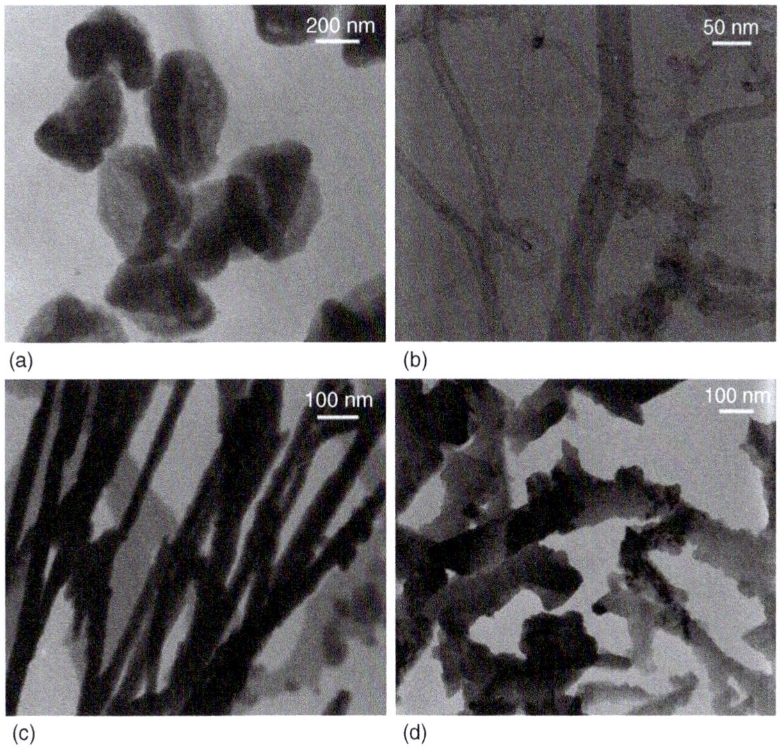


**Figure 10** TEM images of (a) elemental sulfur, (b) MWCNTs, (c) S-coated-MWCNTs (50% sulfur), and (d) S-coated-MWCNTs (80% sulfur). [Reproduced with permission from Ref. 10. © Elsevier, 2009]

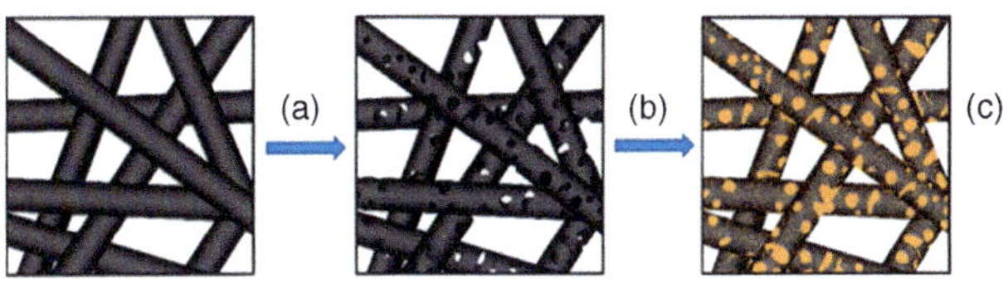


**Figure 11** Schematic illustration of the experimental procedures: (a) Carbonizing electrospun PAN/PMMA bicomponent nanofibers to obtain porous CNFs. (b) Incorporating S into the porous CNF nanostructure via solution-based chemical reactions to obtain porous CNF–S nanocomposites. (c) Thermal treatment of the as-formed porous CNF–S nanocomposites to remove S outside of the pores. [Reproduced with permission from L. Ji, M. Rao, S. Aloni, L. Wang, E. J. Cairns, and Y. Zhang, *Energy Environ. Sci.*, 2011, **4**, 5053–5059 © Royal Society of Chemistry, 2011]

Ji *et al.* developed porous CNF-S composites as cathode electrode for rechargeable Li–S cells, as shown in Figure 11. It is shown that the high electroconductive porous CNFs provide a high surface area to adsorb and disperse S and suppress the solubility of PS intermediates in organic solvent-based electrolytes, thus improving the electrochemical performance of the sulfur electrode. Nearly 1400 mA h $g^{-1}$ initial discharge capacity at 0.05C rate was achieved, with 85% capacity retention after 30 cycles.

In the abovediscussed CNT/CNF-S composites, sulfur is only coated onto the outer surface. Sulfur and PS active materials are directly exposed to the electrolytes. Zheng *et al.* developed hollow CNF to encapsulate sulfur only into the inner surface of the tube, thus effectively trapping the PSs inside the tube.[12] With this hollow CNF-encapsulated sulfur cathodes based on anodic aluminum oxide (AAO) template, a high specific capacity of 1400 mA h g $s^{-1}$ was achieved at 0.2C. However, a capacity decay of 400–500 mA h $g^{-1}$ appeared in 50 cycles.

It is worth noting that Guo *et al.* developed hollow CNTs using an AAO template to impregnate sulfur to the inner surface of the tube, as shown in Figure 12.[13] A high specific capacity of 1500 mA h g $s^{-1}$ was demonstrated, with 47% capacity retention after 100 cycles at 0.1C.

### 3.1.5 *Doping of Heteroatoms*

Although carbon-based host materials, including graphene, porous carbon, CNTs, and CNFs, show superior electrical conductivity, they absorb polar PSs only via physical interaction based on structure modification due to their nonpolarity leading to insufficiency in retarding the shuttle effect of PSs during the cycling processes; thereby, the obvious capacity fading occurs after long cycles. To improve the PS chemisorption and further suppress the shuttle effect,

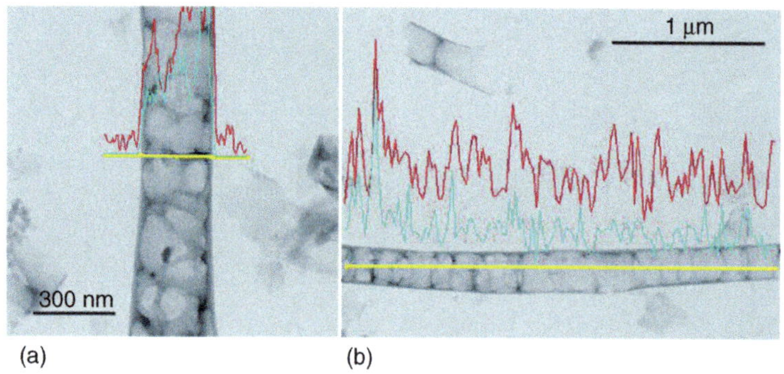


**Figure 12** TEM images and EDS (energy-dispersive X-ray spectroscopy) line scanning of the sulfur-impregnated disordered carbon nanotubes at 300 °C (a) along the cross section and (b) along the axis direction. The carbon signal is red and the sulfur signal is blue. [Reprinted with permission from Ref. 13. © American Chemical Society, 2011]

heteroatoms, such as nitrogen (N) and oxygen (O), can be doped to the aforediscussed carbon host materials, especially graphene. Heteroatom doping, as a chemical modulation, can regulate the chemical and physical properties of the carbon hosts significantly. The heteroatom-doped carbon conductive matrix materials as sulfur hosts can improve the absorption of PSs because of the enhanced polarity. Heteroatoms such as N and O can significantly enhance the interaction between the carbon host and the PSs to prevent the shuttle of PSs, leading to high capacity and high Coulombic efficiency.

It has been shown that the codoping of carbon host material such as porous carbon and graphene with nitrogen and sulfur or phosphorous can significantly improve the Li–S battery performance such as specific capacity, cycling stability, and Coulomb efficiency.[14] Zhu *et al.* studied the mechanism of enhanced carbon cathode performance by nitrogen doping and observed that the strong interaction between the N-doped C and S is through O functional groups, while N doping makes O functional groups more reactive toward S.[15]

On the basis of the density function theory (DFT), Hou *et al.* studied the heteroatom doping of different atoms to several carbon hosts such as graphene and CNTs.[16] The authors demonstrated that N and O heteroatoms with extra pair of electrons as electron-rich donors act as Lewis base sites to interact with strong Lewis

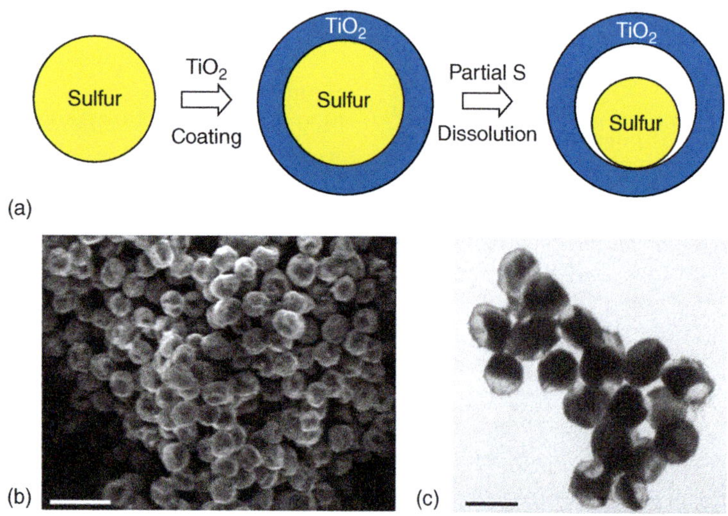


**Figure 13** (a) Schematic of the synthetic process that involves coating of sulfur nanoparticles with $TiO_2$ to form sulfur–$TiO_2$ core–shell nanostructures, followed by partial dissolution of sulfur in toluene to achieve the yolk–shell morphology. (b) SEM image and (c) TEM image of as-synthesized sulfur–$TiO_2$ yolk–shell nanostructures. (b) Scale bar, 2 μm. (c) Scale bar, 1 μm. Through large-ensemble measurements, the average nanoparticle size and $TiO_2$ shell thickness were determined to be 800 and 15 nm, respectively. [Reproduced with permission from Ref. 18. © Springer Nature, 2013]

acid of terminal Li atoms in lithium (poly)sulfides via dipole–dipole electrostatic interaction.

## 3.2 Metal Compounds

### *3.2.1 Metal Oxides*

While the carbon materials restrain PSs mainly by physical absorption, metal oxides with a strong polar surface trap PSs by the strong chemical bonding between the metal ions and PSs, restraining the shuttle effect. Furthermore, the discharge products ($Li_2S$ and $Li_2S_2$) are favorably adsorbed onto the surface of the metal oxides, increasing the cycle life. The representative metal oxides as sulfur cathode materials include magnesium nickel oxide ($Mg_{0.6}Ni_{0.4}O$), alumina ($Al_2O_3$), titania ($TiO_2$), Magnéli phase titanium suboxide ($Ti_4O_7$), and manganese dioxide ($MnO_2$).

Metal oxides were first studied as additives to sulfur cathode to suppress the PS migration and diffusion. For example, large surface area alumina nanoparticles were added to sulfur cathode to help in the adsorption of LiPSs. The specific capacity was observed to increase by around 50% at 0.06C. The cycling performance was also found to be relatively stable, but the cycle life was very short, only 25 cycles, which needed to be significantly improved.[17]

In order to improve the cycling stability, a yolk–shell morphology of titanium dioxide ($TiO_2$)–sulfur composite was designed, as shown in Figure 13.[18] The $TiO_2$ shell physically confined PS within the shell, suppressing the shuttle effect. Meanwhile, the internal void space between the $TiO_2$ shell and the sulfur yolk accommodated the sulfur volume expansion upon discharge. The $TiO_2$–sulfur yolk–shell nanoarchitecture showed much improved cycling stability in comparison to $TiO_2$–sulfur core–shell and uncoated sulfur structure, with a capacity decay rate of only 0.033% per cycle after 1000 cycles.

In general, metal oxides, such as $Al_2O_3$ and $TiO_2$, are usually semiconducting, which limits the overall cathode conductivity. To solve this problem, highly conductive $Ti_4O_7$ was investigated as a host material for sulfur cathode.[19,20] $Ti_4O_7$ has a high electronic conductivity of $\sim 10^3\ S\ cm^{-1}$. Besides that, oxygen-deficient $Ti_4O_7$ possesses undercoordinated titanium atoms at the step sites, which bond strongly with LiPSs and sulfides through titanium–sulfur interaction. The $Ti_4O_7$–S cathodes showed higher reversible capacity and significantly improved cycling stability than its $TiO_2$–S counterpart. $Ti_4O_7$–S cathode exhibited an intimal capacity of $1069\ mA\ h\ g^{-1}$ with 88% capacity retention over 100 cycles at a rate of C/5. Pang *et al.* proposed a surface-enhanced redox chemistry of PSs to explain the significant decrease in PS concentration and a slow and controlled deposition of lithium sulfide during discharge, as shown in Figure 14.[20]

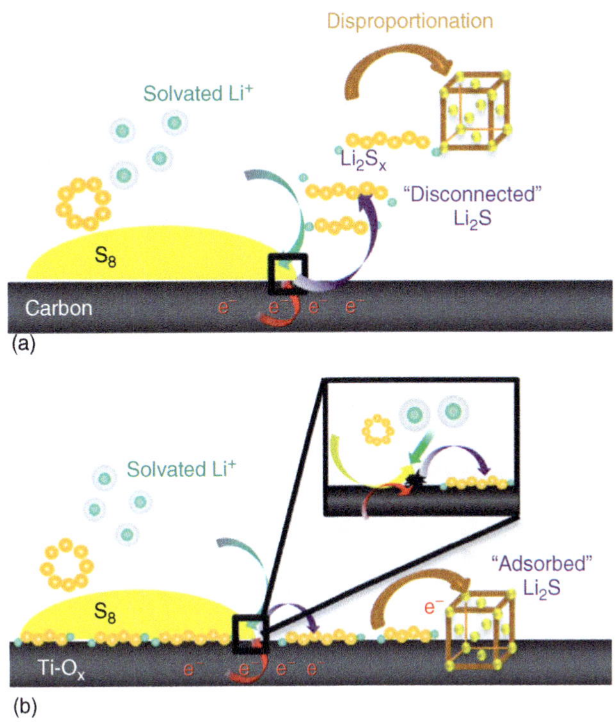


**Figure 14** (a) On reduction of $S_8$ on a carbon host, LiPSs ($Li_2S_X$) desorb from the surface and undergo solution-mediated reactions leading to broadly distributed precipitation of $Li_2S$. (b) On reduction of $S_8$ on the metallic polar $Ti_4O_7$, LiPSs adsorb on the surface and are reduced to $Li_2S$ via surface-mediated reduction at the interface. [Reproduced with permission from Ref. 20. © Springer Nature, 2014]

Manganese dioxide ($MnO_2$) was also reported to entrap PSs in the sulfur cathode, but with a different mechanism.[21] The host $MnO_2$ nanosheet, as a host material, reacts with LiPSs to form thiosulfate species on the surface of the $MnO_2$ nanosheet. These surface-bound thiosulfate groups then catenate the newly formed soluble "higher" PSs to form polythionates and convert them to insoluble "lower" PSs. The polythionates formed on the $MnO_2$ nanosheet surface function as a redox shuttle to anchor the PSs and thus suppress the shuttle effect, resulting in high-performance sulfur cathode. A high specific capacity of $1120\ mA\ h\ g^{-1}$ with 92% capacity retention after 200 cycles at C/5 was observed for this $MnO_2$–sulfur composite cathode-based Li–S cell.

### *3.2.2 Metal Sulfides*

Dichalcogenides such as $TiS_2$ were once explored as cathode materials for secondary lithium batteries. The 3D $S_8/TiS_2$ hybrid form was fabricated as a cathode for lithium battery by thermal reaction with elemental sulfur filled into the porous $TiS_2$ form.[22] A schematic of the process is shown in Figure 15. Different from the carbon and

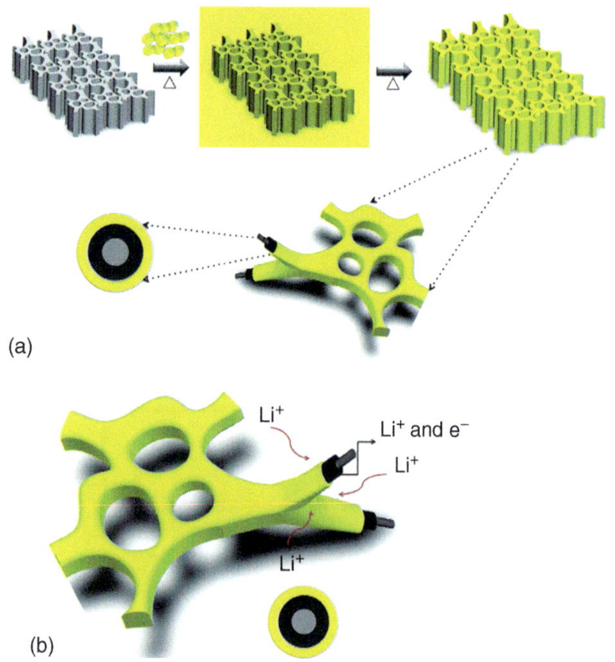


**Figure 15** (a) One-step method to prepare sulfur-infused $TiS_2$ foam. (b) Diagram of the cross section of the foam during discharge. Yellow: sulfur; dark gray: $TiS_2$; gray: Ti. [Reproduced with permission from Ref. 22. © Royal Society of Chemistry, 2015]

metal oxides solely as additive or coating host materials for sulfur cathode, metal sulfides function as secondary active materials in the hybrid sulfur–metal sulfide hybrid cathode, besides providing the supporting matrix for sulfur. $TiS_2$ has an operating voltage window (1.5–3.0 V) compatible with that of sulfur. $TiS_2$ has low specific capacity, but it is well known for its high rate and cycling stability, which sulfur lacks. $TiS_2$ is also semiconducting, which can enhance both the electronic and ionic conductivities of the hybrid system. $TiS_2$ anchors strongly to PSs and thereby limits the soluble LiPS loss to the electrolyte. Despite the very different electrochemical reactions with Li, $TiS_2$ and sulfur were found to work synergistically. The porous 3D metal foam provides good access of electrolyte to the active materials and provides free space to accommodate the volume expansion of sulfur during the discharge process.

### 3.3 Conductive Polymers

As polymers can be synthesized in many ways, while sulfur can be synthesized at very low temperatures, various sulfur-conducting polymer composites have been developed to increase the electric conductivity and minimize the release and shuttle effect of PSs. The elastic and flexible nature of the polymers can partially accommodate the volumetric expansion of the sulfur during the cycling process. On the other hand, the spacious architecture of the polymers can allow the Li to diffuse to the internal sulfur. The chemical interaction between PSs and the polymer chains can restrain PSs from dissolving to the electrolyte and suppress the shuttle effect. Moreover, some conductive polymers are electrochemically active, and thus can be used as a partial active material to improve the capacity of the sulfur–polymer composite cathode. A variety of conducting polymers, such as polypyrrole (PPy), polyaniline (PANI), poly(3,4-(ethylenedioxy)thiophene) (PEDOT), and polyacrylonitrile (PAN), have been studied as a conductive matrix and/or barrier for blocking PSs in the Li–S cells.

Wang *et al.* studied the composite of sulfur powder coated with conductive PPy for Li–S cell by a simple chemical polymerization method.[23] The SEM images of bare S powder and S-PPy composite are shown in Figure 16. The conductive PPy served in the composite as a conducting additive, an active material, and an adsorbing agent for PS and sulfur. Compared with bare sulfur, the initial discharge capacity was increased from 1100 to 1280 mA h $g^{-1}$ after

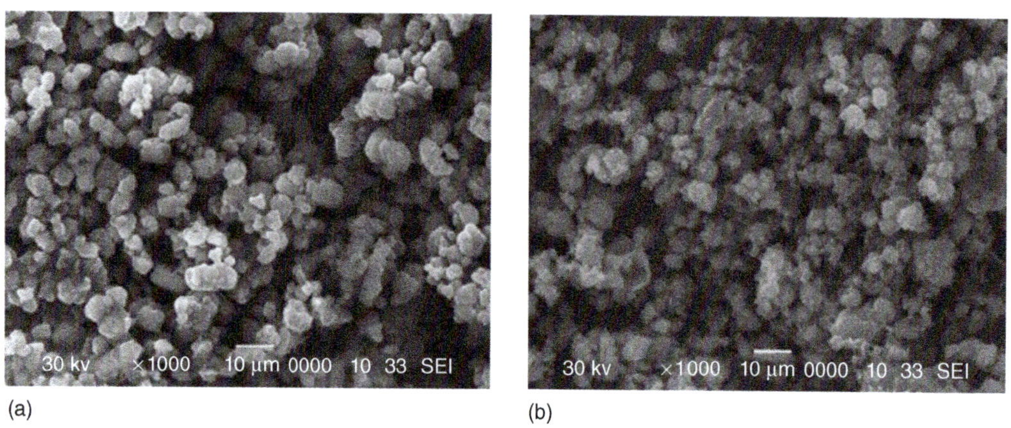


**Figure 16** SEM images of (a) bare S powder and (b) S-PPy composite. [Reproduced with permission from Ref. 23. © Elsevier, 2006]

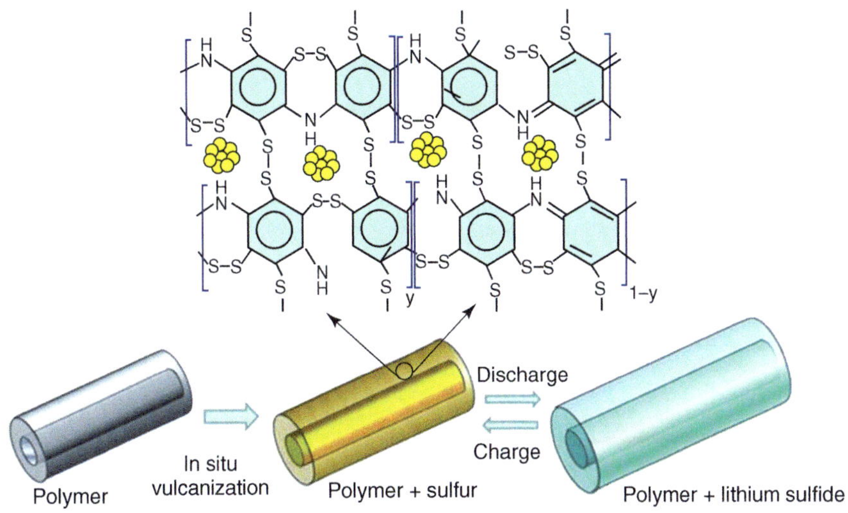


**Figure 17** Schematic illustration of the construction and discharge/charge process of the SPANI-NT/S composite. [Reproduced with permission from Ref. 24. © WILEY-VCH Verlag GmbH & Co., 2012]

sulfur powder was coated with PPy. The cycling durability also showed improvement. However, the capacity decay was observed over 20 cycles, indicating that the PS was only partially restrained to the polymer.

To improve the cycle life, Xiao *et al.* synthesized self-assembled polyaniline nanotubes (PANI-NT) for sulfur encapsulation.[24] The formed three-dimensional (3-D), cross-linked structurally stable sulfur-polyaniline-nanotube (SPANI-NT) polymer backbone has both inter- and intrachain disulfide bonds interconnected through in situ vulcanization. A schematic of the construction and discharge/charge process of the composite is shown in Figure 17. The polymer provides strong physical and chemical confinement to the sulfur and PS. This SPANI-NT/S composite exhibited a specific capacity of 568 mA h $g^{-1}$ with 76% capacity retention over 500 cycles at 1C, representing a superior long-term cycling stability and rate capability. Furthermore, the flexible polymeric nanotube structure was shown to be able to partially accommodate the volumetric change of sulfur during cycling.

Li *et al.* compared the performance of the cells made from PANI-, PPy-, and PEDOT-encapsulated hollow sulfur nanospheres and found that PEDOT-S showed the best cycling stability in long-term charge/discharge process.[25] A high discharge capacity of 780 mA h $g^{-1}$ was achieved at a current rate of C/2 after 500 cycles. Theoretical simulation showed that PEDOT possessed a much stronger binding energy with lithium atom in Li–S than PANI and PPy, and thus effectively inhibits PS dissolution. PEDOT also has the best electric conductivity among the three. The long-term cycling stability and high rate performance of the sulfur composite cathode were found to decrease in the order of PEDOT > PPy > PANI. The fabrication process of the conductive polymer-coated hollow sulfur nanospheres and the cycling performance of the cells are illustrated in Figure 18.

### 3.4 Multicomponent Host Materials

To serve as a host material for the sulfur cathode, a matrix material should have high specific surface area to achieve a high sulfur loading, good electric conductivity and good mechanical properties, and be able to confine the PSs to suppress their diffusion into electrolyte and the shuttle effect. Each type of host material has its advantages and also limitations. For example, while graphene has good electric conductivity and mechanical robustness and flexibility, it cannot confine PSs well. In order to overcome the limitations of certain single host material, the combination of different host materials has been investigated to form multicomponent host materials to improve the overall performance of the sulfur cathode, especially toward multidimensional structures. The multicomponent host materials can be binary, ternary, and quaternary.

Yang *et al.* developed a graphene-based layered structure with porous carbon uniformly covered on both sides of the graphene sheets, as illustrated in Figure 19.[26] The graphene acts as the electronic conductive channel to ensure rapid charge transfer, and the pores of the porous carbon act as PS reservoirs to confine the PSs in the cathode and alleviate the shuttle effect. The graphene–porous carbon host material synthesized showed much better electrochemical performance and cycle stability in comparison to those of graphene–sulfur composites.

Besides the electric conductivity of the composite material, PSs shuttle effect, and volume expansion, sulfur loading in the composite is another problem that needs to be addressed for Li–S batteries. The single S-C composites

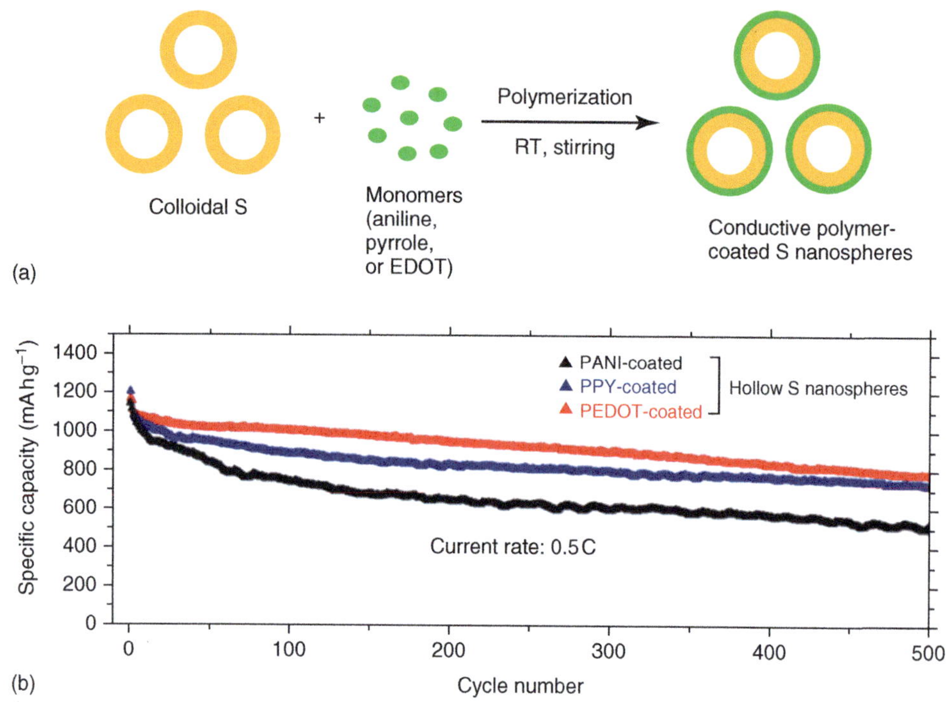


**Figure 18** Schematic illustration of the fabrication process of conductive polymer-coated hollow sulfur nanospheres (a), and cycling performance of the cells made from hollow sulfur nanospheres with PANI, PPY, and PEDOT coatings of ~20 nm at C/2 rate for 500 cycles (b). [Reprinted with permission from W. Li, Q. Zhang, G. Zheng, Z. W. She, H. Yao, and Y. Cui, *Nano Lett.*, 2013, **13** (11), 5534–5540. © American Chemical Society, 2013]

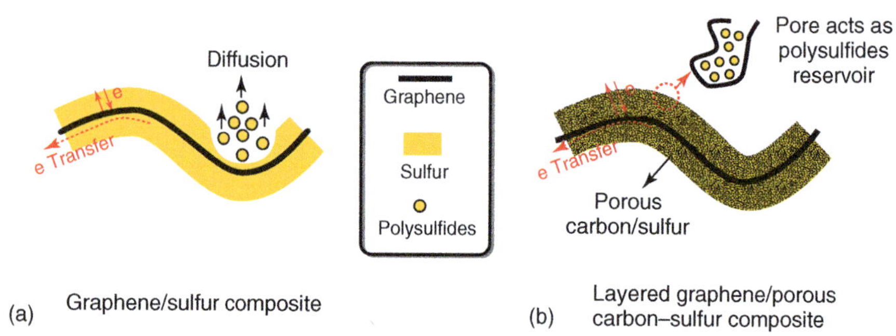


**Figure 19** (a) The schematic architecture of the graphene/sulfur composite. Graphene acts both as a substrate to disperse the sulfur and an electronic conductive channel. However, polysulfides can readily diffuse outside the cathode. (b) The schematic architecture of the layered graphene/porous carbon–sulfur composite, proposed in this work. An extra thin layer of porous carbon uniformly covers both the surfaces of the graphene, and sulfur is highly dispersed in its pores. The porous carbon acts as a polysulfide reservoir to alleviate the shuttle effect. [Reprinted with permission from X. Yang, L. Zhang, F. Zhang, Y. Huang, and Y. Chen, *ACS Nano*, 2014, **8** (5), 5208–5215. © American Chemical Society, 2014]

mostly have a sulfur loading between 40 and 70 wt%, which is relatively low as a higher sulfur loading can exacerbate electric conductivity problem of the cathode composite. This low sulfur loading results in low overall specific capacity and low energy of the Li–S cells. Du *et al.* developed a wet chemical synthesis route of preparing S-C nanocomposite.[27] Specifically, three different types of nanocarbons—zero-dimensional (0-D) Ketjen black (KB), 1-D CNT, and 2-D pristine graphene (PG)—were integrated into highly efficient 3D conducting networks in certain organic solvents such as *N*-methyl-2-pyrrolidone (NMP), as illustrated in Figure 20. The transmission electron microscopy (TEM) images of starting nanocarbons and as-obtained composite are shown in Figure 21.

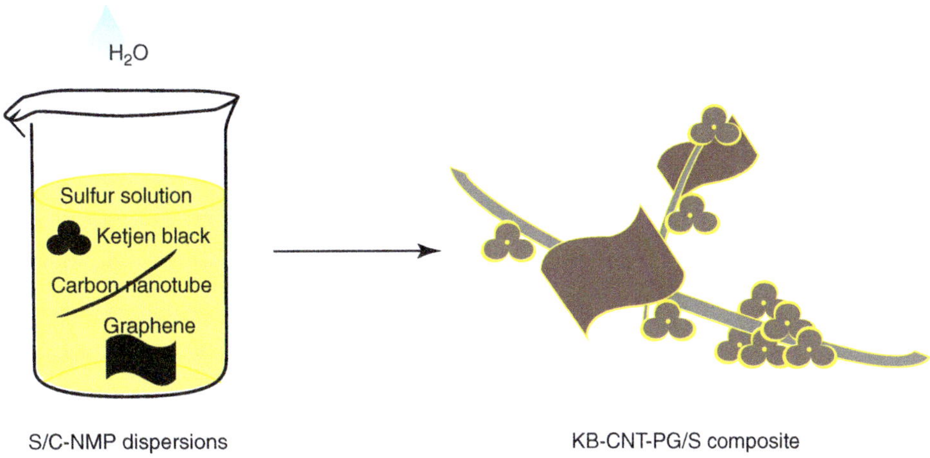


**Figure 20** Schematic illustration of the wet chemical synthesis procedure of the nano KB-CNT-PG/S composite. [Reprinted with permission from W. Du, Y. Yin, X. Zeng, J. Shi, S. Zhang, L. Wan, and Y. Guo, *ACS Appl. Mater. Interfaces*, 2016, **8**, 3584–3590. © American Chemical Society, 2015]

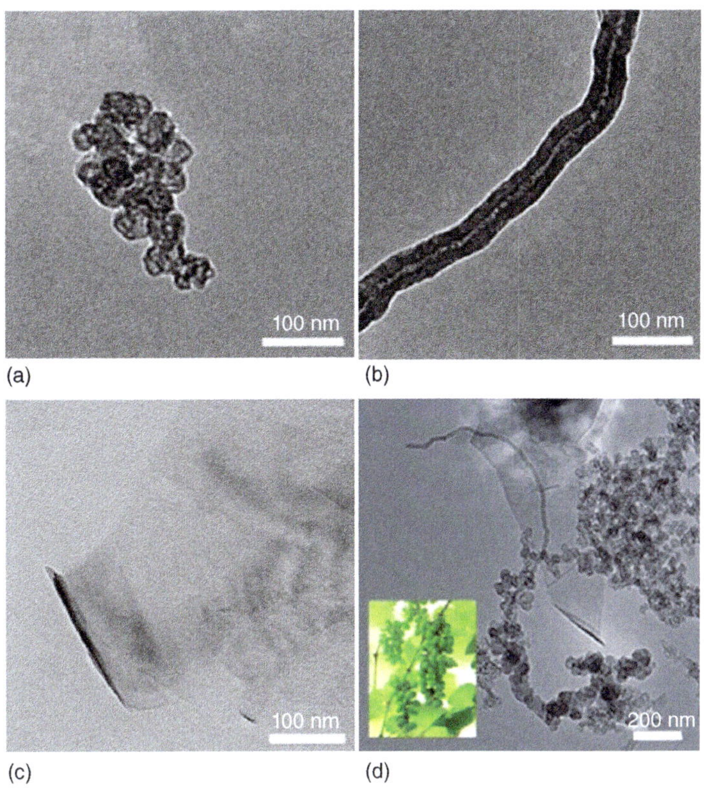


**Figure 21** TEM images of starting nanocarbons including (a) KB, (b) CNT, and (c) PG and (d) as-obtained KB-CNT-PG/S composite with grapevine-like structure. The inset image in d shows an optical image of the grapevine. [Reprinted with permission from W. Du, Y. Yin, X. Zeng, J. Shi, S. Zhang, L. Wan, and Y. Guo, *ACS Appl. Mater. Interfaces*, 2016, **8**, 3584–3590. © American Chemical Society, 2015]

The coassembly of these different types of nanocarbons effectively restrains aggregation of single carbon nanomaterial. The intrinsic high electric conductivity and large surface area of the nanocarbons are retained, and sulfur is uniformly distributed across the carbon hybrids in hierarchical carbon networks with very high loading of 90 wt%. High specific capacities up to 1115 mA h $g^{-1}$ for composite and 1239 mA h $g^{-1}$ for sulfur at 0.02C are presented. Excellent electrochemical performances were attained.

## 4 SUMMARY

This chapter addresses sulfur cathode materials for Li–S batteries. The advantages and limitations of sulfur electrode material are analyzed by discussing the basic electrochemical principles of the Li–S battery. Different types of single host materials are then discussed, including nanocarbons, metal compounds, and conductive polymers. Multicomponent host material to combine the advantages and overcome the limitations of each single host material are then discussed. The organization of the chapter is not trying to be inclusive of all research findings in the past, which is impossible limited by the space, but more weighted on the basic principles with representative examples.

## 5 RELATED ARTICLES

Li–$CO_2$ Batteries; Lithium Metal Anode; Lithium Oxygen Battery; Electrode and Electrolyte Interaction in Aqueous Electrochemical Energy Storage; Silicon-Based Anodes for Advanced Lithium-Ion Batteries; Na-Ion Batteries: Positive Electrode Materials; Structural Engineering of Cathode Materials for Lithium–Sulfur Batteries; Surface Chemistry of Alkali-Ion Battery Cathode Materials

## 6 ABBREVIATIONS AND ACRONYMS

AAO = anodic aluminum oxide; CNF = carbon nanofiber; CNT = carbon nanotube; DFT = density function theory; FE-SEM = field-emission scanning electron microscope; GNS = graphene nanosheet; GO = graphene oxide; KB = Ketjen black; Li-ion = lithium ion; LiPS = lithium polysulfide; Li–S = lithium–sulfur; MC = microporous carbon; mGO = mildly oxidized graphene oxide; MWCNT = multiwalled carbon nanotube; NMP = *N*-methyl-2-pyrrolidone; PAN = polyacrylonitrile; PANI = polyaniline; PEDOT = poly(3,4-(ethylenedioxy)thiophene); PEG = poly(ethylene glycol); PG = pristine graphene; PPy = polypyrrole; PS = polysulfide; SEM = scanning electron microscope; SWCNT = single-walled carbon nanotube; TEM = transmission electron microscopy; 1-D = one-dimensional; 2-D = two-dimensional; 3-D = three-dimensional

## 7 REFERENCES

1. A. Manthiram, Y. Fu, S. Chung, C. Zu and Y. Su, *Chem. Rev.*, 2014, **114**(23), 11751.
2. A. Eftekhari and D. Kim, *J. Mater. Chem. A*, 2017, **5**, 17734.
3. P. G. Bruce, S. A. Freunberger, L. J. Hardwick and J. Tarascon, *Nat. Mater.*, 2012, **11**, 19.
4. J. Wang, L. Lu, M. Choucair, J. A. Stride, X. Xu and H. Liu, *J. Power Sources*, 2011, **16**, 7030.
5. L. Ji, M. Rao, H. Zheng, L. Zhang, Y. Li, W. Duan, J. Guo, E. J. Cairns and Y. Zhang, *J. Am. Chem. Soc.*, 2011, **133**, 18522.
6. H. Wang, Y. Yang, Y. Liang, J. T. Robinson, Y. Li, A. Jackson, Y. Cui and H. Dai, *Nano Lett.*, 2011, **11**, 2644.
7. B. Zhang, X. Qin, G. R. Li and X. P. Gao, *Energy Environ. Sci.*, 2010, **3**, 1531.
8. X. Li, Y. Cao, W. Qi, L. V. Saraf, J. Xiao, Z. Nie, J. Mietek, J. Zhang and J. Liu, *J. Mater. Chem.*, 2011, **21**, 16603.
9. S. C. Han, M. S. Song, H. Lee, H. S. Kim, H. J. Ahn and J. Y. Lee, *J. Electrochem. Soc.*, 2003, **150**, A889.
10. L. Yuan, H. Yuan, X. Qiu, L. Chen and W. Zhu, *J. Power Sources*, 2009, **2**(15), 1141.
11. L. Ji, M. Rao, S. Aloni, L. Wang, E. J. Cairns and Y. Zhang, *Energy Environ. Sci.*, 2011, **4**, 5053.
12. G. Zheng, Y. Yang, J. Cha, S. S. Hong and Y. Cui, *Nano Lett.*, 2011, **11**, 4462.
13. J. Guo, Y. Xu and C. Wang, *Nano Lett.*, 2011, **11**, 4288.
14. Q. Pang, J. Tang, H. Huang, X. Liang, C. Hart, K. C. Tam and L. F. Nazar, *Adv. Mater.*, 2015, **27**, 6021.
15. P. Zhu, J. Song, D. Lv, D. Wang, C. Jaye, D. Fischer, T. Wu and Y. Chen, *Phys. Chem. C*, 2014, **118**(15), 7765.
16. T.-Z. Hou, X. Chen, H.-J. Peng, J.-Q. Huang, B.-Q. Li, Q. Zhang and B. Li, *Small*, 2016, **12**, 3283.
17. Y. J. Choi, B. S. Jung, D. J. Lee, J. H. Jeong, K. W. Kim, H. J. Ahn, K. K. Cho and H. B. Gu, *Phys. Scr.*, 2007, **T129**, 62.
18. Z. W. Seh, W. Li, J. J. Cha, G. Zheng, Y. Yang, M. T. McDowell, P.-C. Hsu and Y. Cui, *Nat. Commun.*, 2013, **4**, 1331.
19. X. Tao, J. Wang, Z. Ying, Q. Cai, G. Zheng, Y. Gan, H. Huang, Y. Xia, C. Liang, W. Zhang and Y. Cui, *Nano Lett.*, 2014, **14**, 5288.

20. Q. Pang, D. Kundu, M. Cuisinier and L. F. Nazar, *Nat. Commun.*, 2014, **5**, 4759.
21. X. Liang, A. Garsuch and L. F. Nazar, *Angew. Chem. Int. Ed.*, 2015, **54**, 3907.
22. L. Ma, S. Wei, H. L. Zhuang, K. E. Hendrickson, R. G. Hennig and L. A. Archer, *J. Mater. Chem. A*, 2015, **3**, 19857.
23. J. Wang, J. Chen, K. Konstantinov, L. Zhao, S. H. Ng, G. X. Wang, Z. P. Guo and H. K. Liu, *Electrochim. Acta*, 2006, **51**(22), 4634.
24. L. Xiao, Y. Cao, J. Xiao, B. Schwenzer, M. H. Engelhard, L. V. Saraf, Z. Nie, G. J. Exarhos and J. Liu, *Adv. Mater.*, 2012, **24**, 1176.
25. W. Li, Q. Zhang, G. Zheng, Z. W. She, H. Yao and Y. Cui, *Nano Lett.*, 2013, **13**(11), 5534.
26. X. Yang, L. Zhang, F. Zhang, Y. Huang and Y. Chen, *ACS Nano*, 2014, **8**(5), 5208.
27. W. Du, Y. Yin, X. Zeng, J. Shi, S. Zhang, L. Wan and Y. Guo, *ACS Appl. Mater. Interfaces*, 2016, **8**, 3584.

# Lithium Metal Anode

**Siyuan Li, Jixiang Yang and Yingying Lu**

*Zhejiang University, Hangzhou, China*



## 1 INTRODUCTION

Lithium (Li) metal is a promising anode material for next-generation secondary batteries due to its relatively high theoretical capacity (3860 mA h $g^{-1}$), low redox potential (−3.04 V vs the standard hydrogen electrode), and low density (0.534 g $cm^{-3}$), thus attracting the intensive researches to develop lithium metal-based batteries (LMBs) since the 1960s (Figure 1a).[1–3] In the 1980s, Moli *et al.*[4] firstly tried to commercialize the metal lithium secondary battery of Li/$MoS_2$ in mobile phones but failed by severe battery explosion, warning the safety hazards induced by lithium metal chemistry (Figure 1b). Specifically, there are two main obstacles when directly using metallic lithium as the anode. One is that Li metal problematically forms dendrite-like lithium electrodeposition during the repeated plating/stripping process.[5] The uneven ion flux caused by heterogeneous surface morphologies and chemistries tends to trigger dendrite nucleation. The sharp and dendritic lithium, the real culprit for cell short-circuit and explosion, can spontaneously grow, amplify the surface roughness, and pierce the separator, making LMBs a potential bomb for commercial use. The other limit is that the hyperactive chemistry of lithium metal causes low Coulombic efficiency (CE) and short lifespan of cells.[6] The intrinsic low redox potential of lithium metal determines that it can easily react with any organic electrolytes, cathode materials, and the products shuttled from the cathode (e.g., polysulfides in the case of lithium–sulfur batteries). These irreversible reaction process increases lithium consumption at the anode, rendering the increasing cell impedance and low capacity retention. In addition to the above problems, the related issues such as large volume expansion, inactive lithium deposition (dead lithium), electrolyte consumption, and unstable solid electrolyte interphase (SEI) also cause the electrode pulverization and tremendous internal resistance, impeding the practical application of Li metal anodes.[7,8] In the 1990s, with the emergence of "rocking-chair" Li-ion batteries, researches moved their attentions to the safer technology.[9]

The state-of-the-art Li-ion batteries using graphite (372 mA h $g^{-1}$) as anode material is a single-ion-intercalation reaction upon each cycling.[10] Li ions can intercalate into the interlayer of graphite, therefore greatly decrease the possibility of lithium growth. However, the supporting materials do not provide electric energy, which greatly limits the theoretical energy density of LIB cells (Figure 1c).[11] When pairing graphite with commercial transition metal oxide (TMO) cathodes such as $LiCoO_2$ (LCO), $LiMn_2O_4$ (LMO), $LiFePO_4$ (LFP) or $LiNi_xMn_yCo_zO_2$ (NMC), fewer of them are capable to obtain specific energy density higher than 250 Wh $kg^{-1}$.[12] With the increasing demand on energy density of batteries, the research interest turned back to LMBs because of their step-change energy improvement over LIBs. When replacing the common

*Inorganic Battery Materials*. Edited by Hailiang Wang and Boniface P.T. Fokwa
 ISBN 978-1-119-43199-2

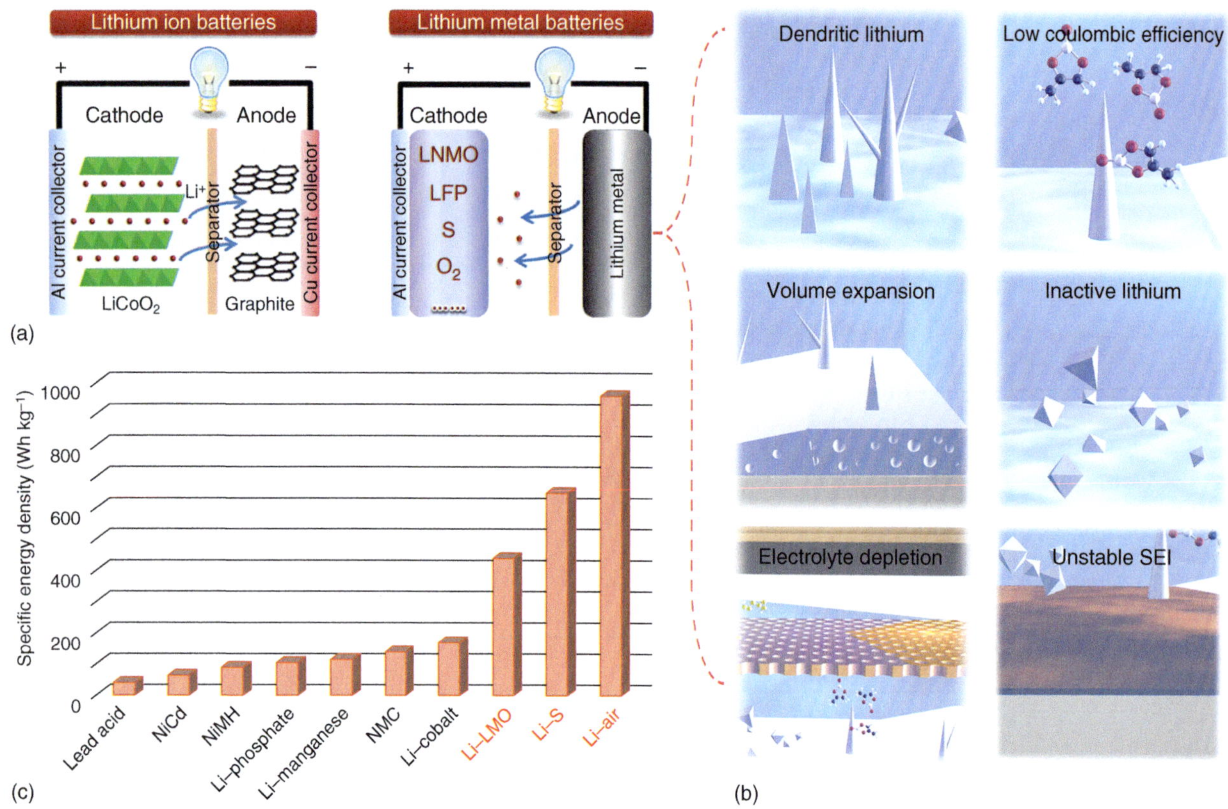


**Figure 1** Schematic diagrams of: (a) lithium-ion batteries and lithium-metal batteries [Reproduced with permission from Ref. Y. Guo, H. Li, T. Zhai, Adv. Mater., 2017, 29, 1700007. © John Wiley and Sons, 2017]. (b) The main problems that hinders the utilization of lithium-metal anodes: dendritic lithium, low Coulombic efficiency, volume expansion, inactive lithium, electrolyte depletion and unstable SEI. (c) Typical energy densities of lead-, nickel-, lithium-ion and lithium metal based batteries

graphite in LIBs with lithium metal anode, the battery can offer a higher specific energy density of 440 Wh $kg^{-1}$ when paired with TMO and even 650 Wh $kg^{-1}$ when paired with conversion-type sulfur cathode (based on the total weight of cell).[8] These lithium-metal-based batteries provide excellent potential to be utilized in high-energy storage systems such as large electric grids and electrical vehicles.

In order to understand the failure mechanisms of lithium metal anodes, various celebrated progress has been made to simulate the growth behavior of lithium dendrites and propose effective methods to address the "Gordian knot" of dendrite and other related issues. In terms of theoretical simulations, quantities of modeling strategies have been developed, such as Electromigration-Limited Model, Surface-Tension model, Brownian Statistical Simulation Model, and so on, which are focused on electromigration, interface mechanical modulus, and ions diffusion, respectively. These models simulate and predict the nucleation process, growth mechanism, and electroplating behaviors during lithium ion deposition, providing a theoretical basis for solving the harsh issue of lithium dendrite growth. On the experimental basis, many approaches focused on the surface chemistry, dimensional structure, and physical barriers have been put forward to improve the stability of the lithium electrodeposition. For instance, enhancing the mechanical modulus and uniformity of the SEI layer is a type of approach to suppress dendrite growth and improve the CEs, such as constructing robust ex situ formed artificial SEI layers and adding additives into electrolytes.[13] Another method category to prolong the cell lifetime is to build a stable framework to host the deposited Li and lower the local current densities, which has been proved to effectively decrease the dendrite nucleation. Moreover, various factors such as charging type and state of electrolytes can also regulate the Li ion diffusion and its diffusion coefficient in the bulk electrolyte, thus achieving the uniform lithium ion flux approaching the electrolyte/electrode interface.

In Section 2 of this chapter, we first introduce the characterization instruments to observe the surface of lithium anode and the morphologies of metallic dendrites. In Section 3, the intrinsic property of pristine SEI such as its components and structure as well as its effect on dead lithium formation is thoroughly illustrated. Section 4 gives the basic models to simulate the dendrite growth with an emphasis on the factors influencing the growth of dendrite such as current density and interfacial mechanical modulus, together with the numerous effective methods to suppress dendritic lithium proliferation and improve CEs in Section 5. Finally, a perspective that is expected to drive progress in the development of LMBs is given in Section 6.

## 2 CHARACTERIZATION OF LITHIUM DENDRITE AND ANODE SURFACE CHEMISTRY

Characterization of dendritic lithium under microscope is an effective tool to understand the mechanism of $Li^+$ electrodeposition behavior. In the last four decades, morphology characterization methods such as optical microscopy, scanning electronic microscopy (SEM), transmission electron microscope (TEM), and component identification strategies such as X-ray photoelectron spectroscopy (XPS), Fourier transform infrared spectroscopy (FTIR) have been utilized to study the lithium anode surface and dendrite formation.

SEM is a common and useful characterization method to analyze the lithium surface morphology during repeated deposition/dissolution cycles (Figure 2a).[14] Dey *et al.*[18] firstly used SEM to probe the surface change in lithium–thionyl chloride inorganic electrolyte system, after which different structures of lithium dendrites were observed by ex situ SEM, including needle-like, mossy-like, and tree-like metallic structures. However, due to the high vacuum condition of SEM, in situ observation of lithium metal growth becomes extremely difficult for cells using liquid-like electrolytes. Therefore, in situ observation techniques such as macroscopic optical microscopy and in situ TEM were employed. Optical microscopy not only provides images of lithium dendrites but also can record its growth process as a video, which is helpful to understand the continuous process of dendrite germination (Figure 2b).[15] Brissot *et al.*,[19] Stark *et al.*,[20] and Howlett *et al.*[21] used optical microscopy to observe the dendrite growth from the sectional side. The growth behavior and short-circuit mechanism in Li/Li cells can be completely recorded. Different from optical microscopy, in situ TEM can offer more detailed dendrite images at nanoscale (Figure 2c). Ghassemi *et al.*[16] and Huang *et al.*[22] observed the clear growth behavior of Li fiber from the lithium metal surface by using in situ TEM. In situ TEM can even offer the direct observation of $Li^+$ nucleation on anode surface, which provides the basis for future modeling of lithium nucleation. Besides these in situ techniques, AFM is another powerful tool to identify the interface morphology in both in situ and ex situ forms (Figure 2d).[17] The results of AFM observation are more intuitive than those of TEM and SEM, because AFM images not only provide a 3D topography of the surface but also quantify the surface roughness. Until now, numerous models of AFM detection have been developed to examine the surface roughness and interface fluctuation during in situ cycling and to measure the mechanical strength of the SEI layer.[23,24]

In the other aspect, lithium chemistry component has been identified by methods such as XPS, Raman, and FTIR. These in situ or ex situ detection methods can investigate the influence of electrolyte, lithium salt, and additive on the chemical compositions of lithium metal surface. The components of the lithium surface are generally distinguished by the analysis of the peak positions in the spectrum. For example, Aurbach *et al.*[25] used FTIR to identify surface chemical changes of lithium

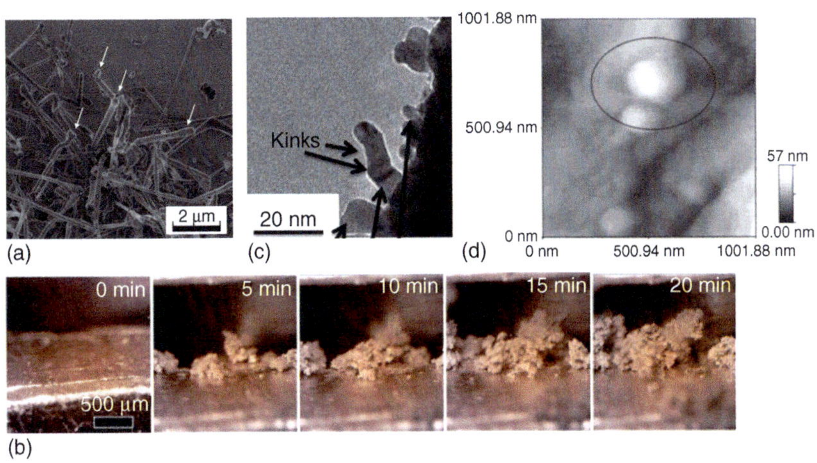


**Figure 2** Characterization of lithium dendrite growth by using different techniques of (a) SEM. [Reproduced with permission from Ref. 14. © Elsevier, 2014]; (b) In situ optical microscopy. [Reproduced with permission from Ref. 15. © John Wiley and Sons, 2017]; (c) TEM. [Reproduced with permission from Ref. 16. © American Institute of Physics, 2011]; (d) AFM. [Reproduced with permission from Ref. 17. © American Chemical Society, 2000]

electrodes in different lithium salt solutions. Wenzel *et al.*[26] used in situ XPS to study the stability of solid-state electrolytes when in contact with lithium metal surfaces. However, these methods have some shortcomings, such as the overlap of peak positions, missing some characteristic peaks (particularly in the infrared spectrum), and so on.

In short, these advanced technologies provide visible insights into the complex surface evolution for LMB-based energy storage systems. SEM and XPS/FTIR are the most common methods to characterize morphologies and surface chemistry, while other methods also provide effective identification for surface physical and chemical properties.

## 3 INTRINSIC PROPERTY OF SEI LAYER ON LITHIUM METAL SURFACE

The negative redox potential of Li metal not only offers superior high energy density but also brings the by-product named solid-electrolyte interface (SEI) covered on the surface of Li metal. Due to the high reactivity of metallic lithium, Li is thermodynamically unstable with any organic solvent and can continuously induce the reduction reactions with electrolyte to generate a passivation layer upon fresh Li. This thin SEI allows the $Li^+$ to pass while exhibiting electronic isolation, which effectively avoids the further side reactions between active Li and electrolyte.[27,28] However, this fragile SEI layer cannot completely endure the repeated Li plating/stripping during cells operation, resulting in the repeated break down and regrowth of SEI, which further prompts the tendency of dendrite lithium growth on surface.[29] In this section, the physical structures and chemical components of the SEI are extensively illustrated to understand the internal relationship between dendrite growth and heterogeneous SEI layer.

### 3.1 Structure and Composition of SEI

The passivation layer on Li metal was first found by Dey in 1970 and Peled named it as SEI in 1979.[30,31] With a decade of efforts, researchers discovered that the SEI layer is not a single-component film and consists of various organic ($LiCO_2$-R, alkoxides, polyolefins, etc.) and inorganic (LiF, $Li_2O$, $Li_2CO_3$, etc.) components. In 1997, Peled *et al.*[32] proposed a generally accepted "mosaic" model to describe the heteronomous property of the SEI layer (Figure 3a), after which many following works demonstrated that the SEI layer also exhibits a dual-layer structure in thickness direction. The dense layer near the anode consists of different inorganic products, which enables the fast lithium-ion transport. Meanwhile, the upper porous layer comprising polymeric content offers enough flexibility to accommodate lithium deposition. Shi *et al.* used density functional theory (DFT) to calculate the conductive type of $Li^+$ in this binary film. They found that in a $Li_2CO_3$-dominated inorganic layer, lithium ions diffuse

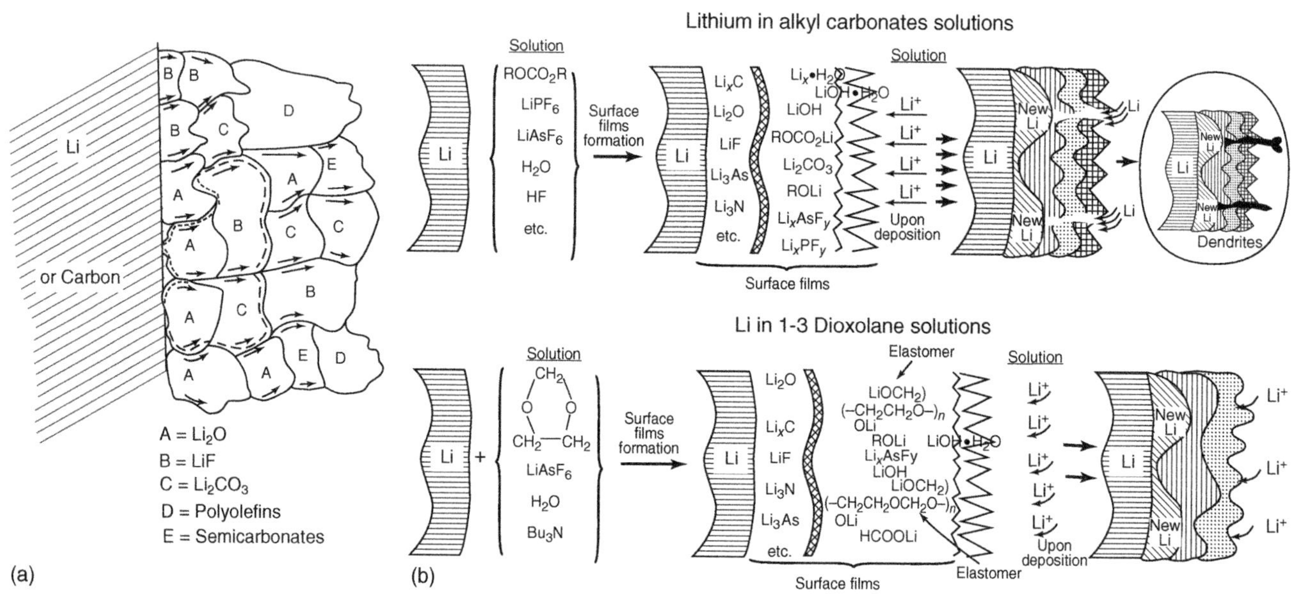


**Figure 3** Scheme of SEI structure. (a) Mosaic structure of original SEI. [Reproduced with permission from Ref. 32. © Electrochemical Society, Inc., 1997]; (b) Comparison of typical components in SEI layer of alkyl carbonate-based and ether-based electrolyte. The substitution of DOL for conventional alkyl carbonates significantly improves the flexibility due to the partial polymerization of DOL, and thus maintain the integrity of SEI layer. [Reproduced with permission from Ref. 33. © Elsevier, 2000]

through the lattices by a knock-off mechanism with low energy barriers, and in the outer organic layer, they diffuse through the ordinary pore diffusion mechanism.[34] This model fundamentally explains the lithium ion conduction in the SEI film, but the diversity of SEI materials still needs to be explored.

The components of the SEI are not constant, but closely related to the composition of the electrolyte. The solvents, lithium salts, and other reaction intermediates can react with lithium to construct the SEI film synergistically. For example, solvents such as ethylene carbonate (EC), diethyl carbonate (DEC), 1,3-dioxolane (DOL), 1,2-dimethoxyethane (DME), polycarbonate (PC), and lithium salts such as lithium hexafluorophosphate ($LiPF_6$), lithium perchlorate ($LiClO_4$), and lithium triflate ($LiSO_3CF_3$) have been deeply investigated.[33,35] In general, $Li_2CO_3$ can be found in all carbonate-based electrolytes and some ether-based electrolytes due to the decomposition of solvents.[36,37] LiF can be found in F-containing additives or lithium salts such as fluoroethylene carbonate (FEC) and $LiPF_6$. Moreover, some organic species such as $ROCO_2Li$ and ROLi can also be formed in EC or DEC electrolyte. Different from the carbonate-based solvents, the ether-based solvents such as DOL can produce—$(CH_2CH_2O)_nOLi$ elastomer by the ring-opening reaction and polymerization of DOL molecules. As-formed SEI shows more flexibility compared to those of EC and DEC to endure the surface fluctuation from lithium deposition. A more detailed comparison of typical components in alkyl carbonate-based and ether-based electrolyte is presented in Figure 3(b).[33]

The components of SEI still cannot be clearly identified and completely described due to intrinsic limits of detection methods such as peak masking or peak offset in XPS and infrared spectroscopy. Therefore, exploring the composition of the SEI film is still an important issue for further understanding of lithium deposition behavior.

### 3.2 Effect of SEI Layer on Dendritic Lithium Formation

SEI layer with heterogeneous and mosaic-like structures will directly affect the ionic concentration gradient near the lithium metal surface. During deposition, the area with a high concentration of metal cations will preferentially form a small protrusion of lithium metal, which is considered the onset of lithium dendrite growth.[38] However, most simulation models do not take into account the effect of SEI due to the complexity of the components and unintelligible diffusion mechanism of Li ions. In order to reveal the veil of the SEI, its effect on lithium dendrite formation has been enthusiastically pursued in recent years.

Cohen and Aurbach[39] firstly used the in situ atomic force microscopy to investigate the lithium surface morphology and found that SEI is intrinsically nonuniform on nanometer or even on micrometer scale, which leads to localized deposition and stripping the parts of the SEI that possesses higher ionic conductivity. Therefore, the uneven lithium deposition can be attributed to the unstable ion transport through the SEI layer (Figure 4a). The SEI components with high $Li^+$ conductivity serve as the ionic pathway for fast lithium electrodeposition. This uneven deposition behavior results in an internal stress between the shape-changing lithium layer and the SEI layer. When cells operate under high current densities or repeated charging/discharging, the SEI is not able to endure the stress and lithium nuclei can easily pierce the original SEI, thus attracting more lithium ions to deposit on the tips of the surface bump, which amplifies the surface roughness.[40] On the other hand, because of the nonreplenishable counterions, the anions will deplete near the negative electrode upon charging, cause the large local space charge and the consequent internal stress, and generate more fresh hotspots for lithium deposition. The newly exposed dendritic lithium reacts with electrolytes and forms new SEI at the electrolyte/electrode interface, gradually consuming the electrolytes irreversibly. Such pernicious redeposition of lithium metal can eventually cause LMBs failing quickly with rapid capacity fading or even hazardous cell short-circuit.[3,41]

Diverging significantly from previous models that emphasize on the transport of ions during dendrite formation, Steiger[14] model gives a more detailed dendrite growth process through in situ light microscopy observation. The results demonstrated that lithium dendrites can grow from the defects of lithium wires, such as substrate lithium interface, kinks, or the region close to the tips, which is called the defect-based insertion growth process (Figure 4b). The lithium nuclei firstly appear under an inert inhomogeneity within the SEI by Li insertion at the substrate, after which Li insertion takes place at the tip, kink, or base part, resulting in kinked whisker lithium needles. This kinked structure is also consistent with the observation of Cui and coworkers[42] by using cryo-electron microscopy.

Other characterization models have also been put forward. Jäckle and Groß[43] pointed out that the morphology of needle/nodular lithium is related to the $Li^+$ cation mobility on the surface components, especially the $Li^0$ atom self-diffusion in the SEI, thus proposed a surface nucleating and diffusion model. Ely and García believed that the deposition of lithium is a typical heterogeneous nucleation process. The morphology of the deposit is closely related to the substrate. By performing numerical simulations in both thermodynamics and kinetics, five regimes of heterogeneous nucleation behavior are identified, including a nucleation suppression regime, a long incubation time regime, a short incubation time regime, an early growth regime, and a late growth regime.[44] These models obtained by direct observations or indirect theoretical simulations provide a theoretical basis for the subsequent material design.

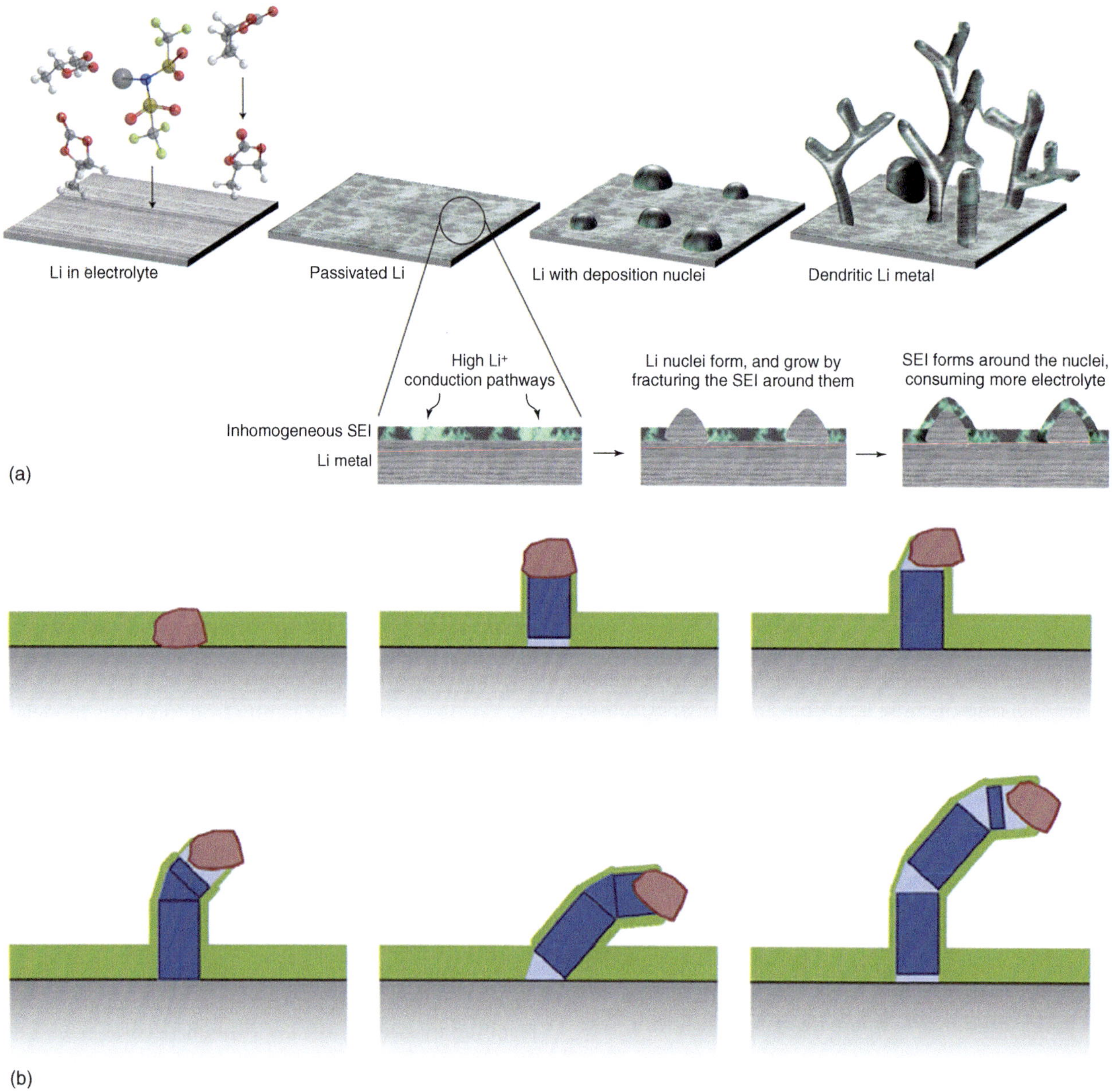


**Figure 4** Scheme of dendrite growth due to the heterogeneity of SEI layer. (a) The growth of deposited nuclei progresses by preferential transport of $Li^+$ on high ionic conduction pathway in SEI, thus inducing the crack of the fragile SEI layer, which further leads to rapid growth of dendritic lithium. [Reproduced with permission from Ref. 13. © Nature publishing group, 2016]; (b) Schematic describing growth of a Li needle. The lithium prefers to deposit under an inert inhomogeneity within the SEI, after which Li insertion can be taken place at tip, kink or base part, resulting in a kinked whisker lithium needle. [Reproduced with permission from Ref. 14. © Elsevier, 2014]

### 3.3 Dead Lithium Layer

The performance of LMBs is strongly related to the growth of lithium dendrites and the formation of dead lithium. As early as 1974, Selim and Bro[45] found that the CE and capacity dropped very quickly when testing the LMBs. After disassembling the battery for ex situ observation, they found that the original smooth lithium surface was covered with thick mossy lithium. The same phenomenon was also observed by Yoshimatsu[46] and Arakawa *et al.*[47] that needle-like lithium led to the formation of mossy lithium that cannot contact the electrode surface electrically. This electrochemically unrecoverable Li is known as "dead" Li that cannot obtain electrons and redeposit to the bulk lithium in the cycling process later. Steiger *et al.*[48] deeply investigated the formation mechanism of dead Li and proposed an insertion/dissolution process based on their own Steiger model (see Section 3.2). When lithium deposits, $Li^+$ can insert in

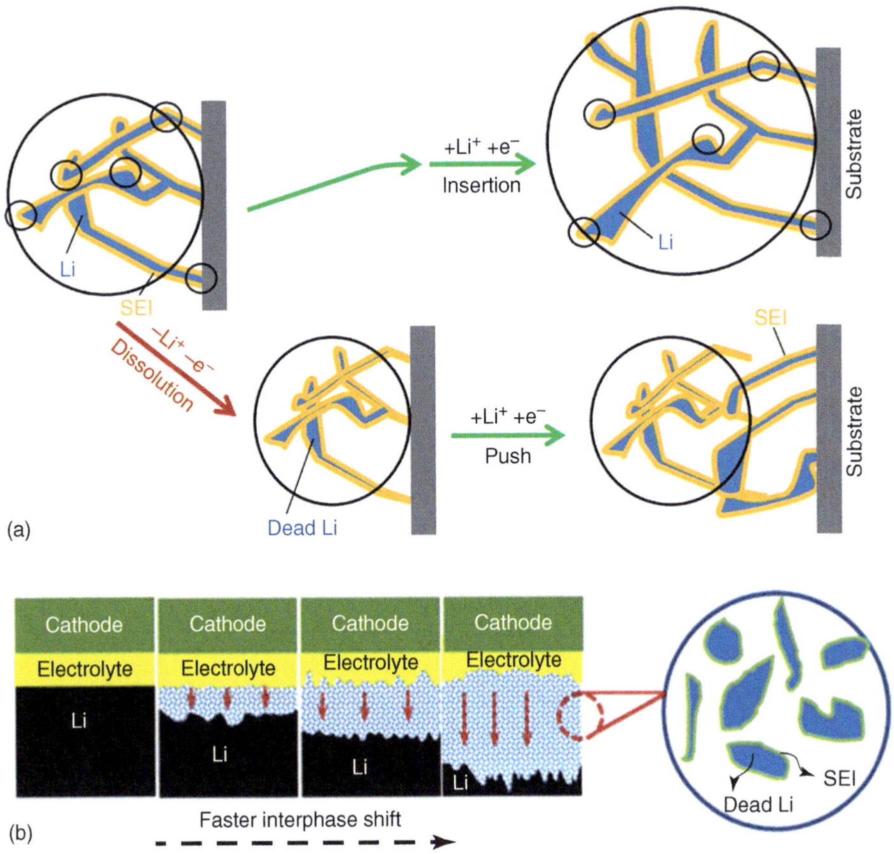


**Figure 5** Schematic of formation mechanism for dead lithium. (a) Dead lithium formation due to the dissolution of thinner part in lithium "stem". [Reproduced with permission from Ref. 48. © Elsevier, 2014]; (b) Electrode pulverization due to the fast interface shift between active lithium to dead lithium. As-formed mossy lithium layer will greatly increase the internal resistance, causing the failure of lithium metal batteries. [Reproduced with permission from Ref. 49. © John Wiley and Sons, 2015]

the tip or kink of the lithium "stem", making the lithium dendrites more uneven in the radial direction (Figure 5a). Conceivably, if the $Li^+$ dissolves in discharging process, the lithium in the fine stem structures is consumed before that of the branches. This behavior further leads to the breakage of lithium "stems", and the detached part will be passivated by the electrolyte to form a complete SEI layer on the surface. Such isolated lithium will not get or lose the electrons from the bulk lithium due to the electron-isolated property of the SEI, just physically attaching to the top of needle lithium and accelerating the formation of the mossy lithium layer.

The as-formed dead lithium layer gradually aggravates the cycling performance of batteries. For example, the existence of dead lithium reduces the active surface area for lithium deposition, thus increasing the effective current density on lithium surface, which will speed up the growth of dendritic lithium. The distorted ionic pathway in dead lithium layer will increase the effective diffusion distance of lithium ions, resulting in low diffusion kinetics and high internal resistance.[49] As a result, high polarization can be always found in dendritic LMBs due to the high resistance of mossy lithium layer. Meanwhile, the fresh lithium and the nonreplenishable electrolytes will be gradually depleted and the battery will fail due to the dendrite penetration and huge internal resistance (Figure 5b). Therefore, reducing the generation of mossy lithium has a significant impact on the development of LMBs.

## 4 MODELING OF LITHIUM DENDRITE GROWTH

Dendrite growth and "dead" metal are common in metallic lithium (Li), zinc (Zn), copper (Cu), silver (Ag) during electrodeposition process. Especially for lithium metal, the dendrite morphology becomes more severe and unpredictable due to the complex surface chemistry of the SEI, which cannot be found in Zn or Cu electrode in most cases.[50] In the case of lithium, general macroscopic observation methods such as SEM and AFM become much weaker on predicting the dendrite growth behavior and provide less information on dendrite formation. In order to deeply excavate the mechanism of dendrite formation, a great number of theoretical models have been developed to theoretically analyze the growth process and the pattern of lithium dendrites, thus fundamentally clarifying the growth principle of lithium dendrites.

General understanding of dendrite formation can be attributed to the uneven localized lithium ion distribution in the cell. When the cell is fully polarized, lithium ions in the bulk electrolyte tend to migrate from the positive to negative electrode and be reduced to metallic lithium on the surface. However, the heterogeneous surface results in a varied distance between two electrodes, leading to different distances of the diffusion pathway for lithium ion propagation, which is considered a reason for uneven lithium flow. The protrusion and mosaic-like SEI layer also enlarge this uneven $Li^+$ flux. The tip effect of surface protrusion attracts more $Li^+$ to aggregate around the tip, while heterogeneity of the SEI layer allows $Li^+$ to diffuse through high-lithium-conductivity pathway. Both behaviors induce high localized $Li^+$ flux at the electrode/electrolyte interface and further lead to dendritic lithium deposition. Based on these understandings, some widely acknowledged models that emphasize on the electromigration and interface properties of Li ions have been proposed, such as electromigration model and surface tension model.[51] In addition, several influencing factors such as current density and mechanical strength on lithium nucleation and growth will also be included in this section.

## 4.1 Onset Time for Dendrite Growth Based on Ion Transport Property

Different from the one-time electrodeposition of primary cells, lithium metal secondary batteries with repeated plating and stripping process greatly increase the difficulty of simulation. To simplify the modeling procedure, most models are simulated in a binary lithium/polymer electrolyte system consisting of lithium salt (lithium bis(trifluoromethylsulphonyl)imide (LiTFSI), $LiClO_4$, etc.) and polymer matrix (typically poly (ethylene oxide)). Moreover, the cell is designed in one or two dimensions without considering the effect of surface chemistry like SEI film. In the open-circuit condition, the electrolyte remains in a steady state without convective mass transfer or any ionic concentration gradient. Although the assumptions in these models are inconsistent with the actual situation, they still serve as a fundamental model to guide the anode protection in practice.

In a polarized cell, the lithium ions transport to the negative electrode to obtain electrons while anions migrate toward the positive electrode to reach equilibrium. Although the consumption of lithium ions on the negative electrode can be gradually compensated by the lithium source on the metal electrode, the anions are nonreplenishable thus the distribution gradient of ions appears to be uneven under microscopic conditions. From this point, Brissot *et al.*[52] and Chazalviel *et al.*[53] proposed an electromigration model to study ion concentration gradients during cycling.

Chazalviel firstly calculated the distribution of electrostatic potential and ion concentrations in a dilute electrolyte in a thin, rectangular symmetrical cell. In region I of bulk electrolyte phase, the ion transport is governed by diffusion. While in region II near the interface, the ion transport is governed by electric-field migration. During Li ion deposition process, the concentration of anions depletes quickly in the electrode surface, creating a large electric field near the electrode/electrolyte interface, which is called "space charge". Brissot divided these two parts in the interelectrode space with the boundary condition of

$$\frac{\partial C}{\partial x}(x = 0) = \frac{-J}{eD\left(1 + \frac{\mu_c}{\mu_a}\right)}$$

where $D$ refers to the ambipolar diffusion constant, $\mu$ is the mobility of anions or cations, $J$ is the effective electrode current density, and $e$ is the electronic charge.

If $d_C/d_x < 2C_0/L$, the ionic concentration profile evolves to a steady state, where the concentration gradient is constant. While $d_C/d_x > 2C_0/L$, the concentration will become zero at the negative electrode at a time $\tau$, called the "Sand's time", which is considered the onset of dendrite growth.

$$\tau = \pi D\left(\frac{C_0 e}{2Jt_a}\right)^2$$

where $t_a$ is the anionic transport number: $t_a = \frac{\mu_a}{\mu_a+\mu_c}$.

At Sand's time, a large "space charge" is formed on the surface region. Lithium dendrites tend to grow with a velocity of anion migration rate, that is:

$$v = -\mu_a E$$

where $E$ is the electric field strength. This prediction was validated in LiTFSI/poly(ethylene oxide) (PEO) systems by Brissot that the growth speed of the dendrites seems to depend only on the anionic mobility and the electric field.

Chazalviel electromigration model gives important guidelines for future LMB design, that is, enhancing the ionic conductivity to delay "Sand's time" or fixing the anions to decrease the dendrite growth velocity. These methods have been further proven by various experimental works.

## 4.2 Effect of Discharge Conditions and Mechanical Properties

It is widely known that high current density and large charge capacity induce rapid dendrite growth and thus accelerate the deterioration of LMBs. Based on the Chazalviel electromigration model, the onset time for dendrite growth is inversely proportional to current density ($\tau \sim J^{-2}$). This result is also consistent with the finding by Monroe and Newman[54] that the propagation velocity of the dendrite tip is proportional to the effective current density ($v_{\text{tip}} \sim i_n$), which means reducing large, local current density can be useful in delaying the growth velocity of the dendrite, thus prolonging "Sand's time". Chazalviel model also proposed a more detailed equation to describe the effect of current density on dendrite growth:

$$J = \frac{2eC_0D}{t_aL}$$

where $L$ represents the interelectrode distance. When $L$ approaches infinity, $J$ tends to zero, at which "Sand's time" tends to infinity. Thus, there will be no dendrite growth in principle.[19,52] However, it is contrary to Rosso's[55] observations that dendrites appeared even in a small current density, and they attributed this phenomenon to intrinsic heterogeneity of the SEI. Seong *et al.* studied the relationship between charge capacity, current density, and dendrite growth in lithium powder electrodes and proposed an empirical formula to describe the dendrite growth behavior. The area under the line of empirical formula can effectively suppress dendrite growth, which was fitted through experimental tests.[56]

Surface mechanical strength is another important factor influencing dendrite formation. Monroe and Newman[57,58] proposed a theory based on the linear stability analyses on factors such as compressive stress, deformation stress, and surface tension in lithium/polymer cell. When the current density at the deformed surface was higher than that at the nondeformed surface, surface roughness will be increased, which led to a stability parameter of $\Delta\mu$:

$$\frac{\Delta\mu_{e^-}}{A\cos(\omega x)} \equiv \Delta\mu \geq 0$$

where $\Delta\mu_{e^-}$ is the potential change within electrodes caused by local strain or interfacial shape change, $A\cos(\omega x)$ represents the periodic displacement of the surface, and $\Delta\mu$ is the stability parameter. When $\Delta\mu < 0$, the electrode roughness reduces and results in a stable condition with no dendrite formation. Calculations showed that only when the shear modulus of the electrolyte is three times ($\sim 10^9$ Pa) higher than that of the studied PEO, lithium dendrites cannot grow, thus preventing the penetration of sharp lithium dendrites. This surface tension model sheds light on cell modification by using high shear modulus separator, SEI or solid-state electrolyte to suppress dendrite proliferation.

In addition to current density, charge capacity, and surface tension, many other factors such as charging type and temperature affect the lifespan of LMBs. For example, by changing the charging type from galvanostat to pulse can effectively prolong the lifetime of LMBs;[59] Aryanfar *et al.*[60] established a coarse-grain Monte Carlo (CG-MC) model to simulate the diffusion behavior of lithium ions in the electrolyte under various temperatures and found that high temperatures contributed to the inhibition of lithium dendrites. These models provide a theoretical basis for subsequent material design. In the future, a large unified model containing all of the abovementioned factors needs to be proposed to more accurately simulate the capacity decay behavior and dendritic growth behavior of LMBs.

## 5 STRATEGIES TO PROTECT METALLIC LITHIUM ANODE

To successfully prevent metallic lithium anodes from dendrite growth and improve the stability of the anode/electrolyte interface, researchers have proposed many effective strategies based on the insights from theoretical works. They can be generally divided into five different aspects, including adding additives to electrolytes, using safer electrolytes such as solid-state electrolytes, changing battery operation mode, making artificial SEI, and modifying the anode structure. This section of the chapter will extend these five aspects in detail.

### 5.1 Additives

Doping additives to electrolytes in LMBs is a simple strategy to suppress lithium dendrite growth. However, only a small number of additives are effective in protecting lithium metal anodes. As additives, the specific materials added to electrolytes are expected to be of small amount.[61] They do not affect the important properties of electrolytes obviously, such as ionic conductivity and viscosity, but lead to stable interface formation and uniform Li deposition. A novel additive can improve performance not only in the electrochemical aspect but also in other aspects, such as the safe use of the battery. For example, NuLi and coworkers[62] used tris(2,2,2-trifluoroethyl) phosphite (TTFP) as additives, which stabilize the lithium metal anode and inhibit the flammable nature of organic electrolytes, leading to a much safer and more efficient system. Additives can mainly be divided into two types—one contributes a more uniform lithium deposition, called electrodeposition-regulation additives, and the other induces favorable reactions to form a stable interface and suppress dendrite growth, called reaction-forming additives.

#### *5.1.1 Electrodeposition-Regulation Additives*

One of the most typical electrodeposition-regulation additives was reported by Zhang and coworkers

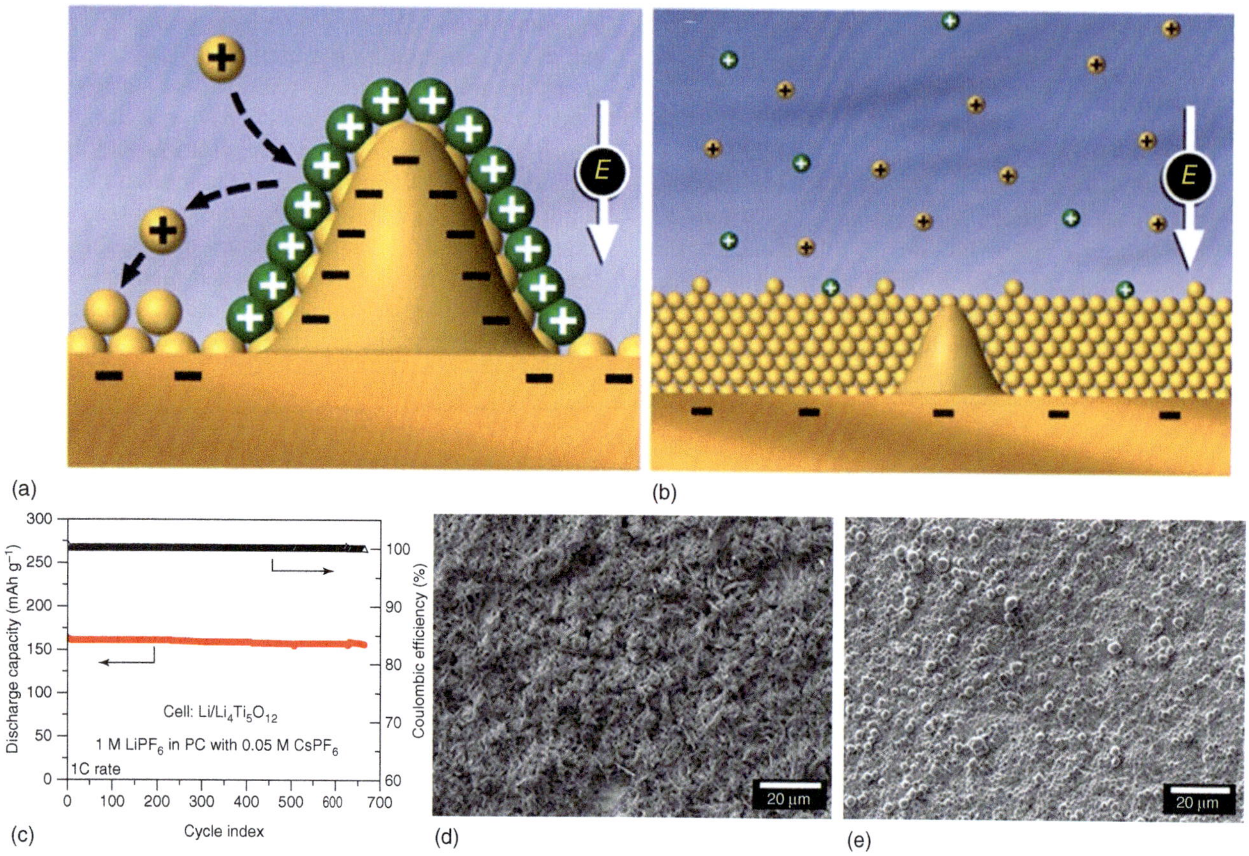


**Figure 6** (a,b) Illustration of Li deposition process based on the self-healing electrical shielding mechanism. (c) Long cycling performance of Li|$Li_4Ti_5O_{12}$ cell at 1C rate. (d) Dendritic Li film deposited on a copper substrate in a control electrolyte of 1 M $LiPF_6$ in PC for 1 h. (e) The same film after another 14 h deposition in the electrolyte with 0.05 M $Cs^+$. [Reprinted with permission from Ref. 63. © American Chemical Society, 2013]

in 2013.[63] They used a small concentration of alkali cations as additives. Based on a self-healing electrostatic shield mechanism, alkali cations, such as $Cs^+$ and $Rb^+$, form an electrostatic shield at the anode surface, which is helpful for uniform deposition. As illustrated in Figure 6(a), because of their lower reduction potentials than that of $Li^+$, $Cs^+$ and $Rb^+$ cannot be reduced on the anode surface. When an Li tip appear on the surface, $Cs^+$ and $Rb^+$ cations will be attracted to the tip surface and accumulate around the Li tip, due to the corona effect of the electrostatic shield. The preaccumulated cations form a positive electrostatic field, counteracting the effect of the original negative electrostatic shield of anode and impelling $Li^+$ cations to deposit on the root of the tip rather than the peak. Thus, as shown in Figure 6(b), $Li^+$ can deposit more adjacently and dendrite growth is successfully depressed. Two experiments were done to demonstrate the theoretical assumption. Using an electrolyte of 1 M $LiPF_6$/PC with 0.05 M $CsPF_6$ additive, the Li|$Li_4Ti_5O_{12}$ cell showed excellent cycling stability and rendered only 4% capacity fade after 660 charge and discharge cycling at 1C (Figure 6c), illustrating that $Cs^+$ containing additives is effective to stabilize LMBs. Once lithium deposits on a copper substrate in a control electrolyte of 1 M $LiPF_6$ in PC for 1 h, the dendritic structure can be seen under SEM analysis (Figure 6d and e). The copper substrate was then transferred to the electrolyte with 0.05 M $Cs^+$ for deposition for another 14 h, as shown in Figure 6(e), the surface became much smoother, proving the adjacent deposit effect of $Cs^+$.

Other than $Cs^+$ and $Rb^+$, many other additives were also found to be effective, such as $Na^+$, $K^+$, alkaline earth metal ions (including $Ca^{2+}$, $Sr^{2+}$, $Ba^{2+}$), N-methyl-N-butylpiperidinium.[20,64–66] For example, Archer and coworkers reported another noticeable work, inspired by DFT calculation results.[67,68] Different kinds and amounts of Li halide salts were applied as additives in conventional PC electrolytes and significantly improved the cell lifetime, with the most pronounced effect found in 30 mol% LiF/PC electrolytes, contributing to nearly 25-fold cell lifetime enhancement. Owing to low diffusing energy and high surface energy created by LiF additives, lithium ions migrate much faster in the LiF-containing SEI and thus stable deposition is realized.[69]

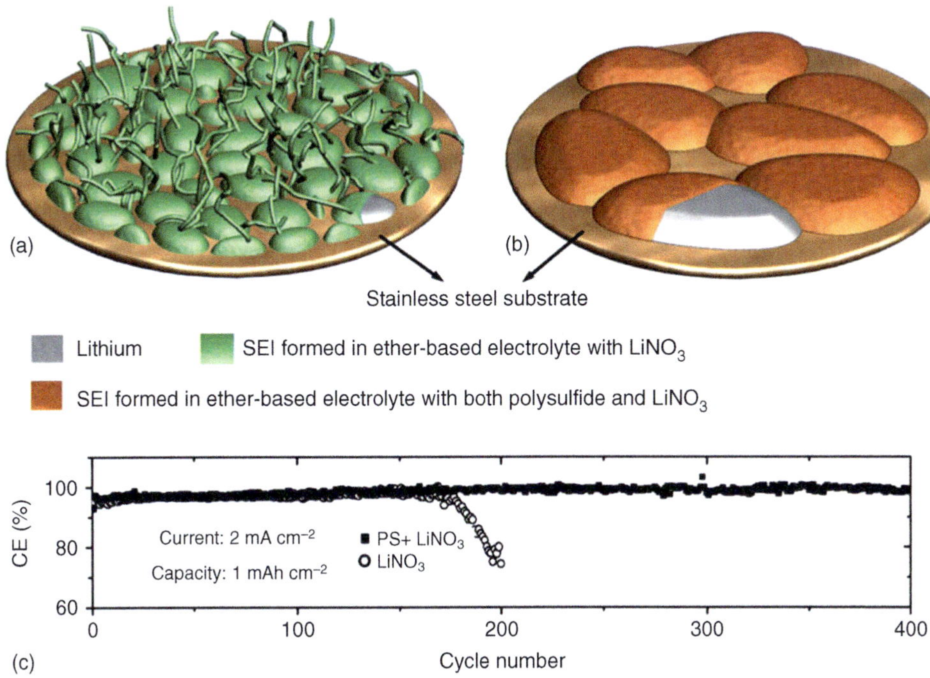


**Figure 7** Schematic illustration showing the morphology difference of lithium deposited on the stainless steel substrate in the two electrolytes (both contain lithium nitrate), but (a) without lithium polysulfide (b) containing lithium polysulfide. (c) Cycling performance of Li symmetric cell using the electrolyte containing 0.18 M $Li_2S_8$ and 5 wt% $LiNO_3$ in DOL/DME electrolyte at a current density of 2.0 mA cm$^{-2}$ with a deposition capacity of 1 mAh cm$^{-2}$. [Reproduced with permission from Ref. 71. © Nature Publishing Group, 2015]

### 5.1.2 *Reaction-Forming Additives*

Reaction-forming additives can react with the lithium metal anode, form an artificial SEI film at the electrolyte/electrode interface, and prevent further irreversible reactions since most organic electrolytes can react with the lithium anode because of its lowest negative potential. An efficient reaction-forming additive should meet three basic criteria.

1. Compared to lithium salt in electrolytes, the additive should preferentially react with the lithium anode. In other words, it should have higher highest occupied molecular orbital (HOMO) and lower lowest unoccupied molecular orbital (LUMO).
2. The reaction products, as the components of SEI, should have small ion transfer impedance and allow ions to pass through.
3. The formed SEI film should be dense and continuous and should not be brittle, tolerating the volume change during the cycling process.

One of the most typical additives is $LiNO_3$, which has been widely used in Li–S batteries. $LiNO_3$ was initially found to participate in the formation of stable interface between the lithium metal anode and liquid electrolyte. Zhang[70] reported that $LiNO_3$ plays a key role in effectively suppressing the reaction between the dissolved lithium polysulfide and lithium metal anode. Recently, Cui and coworkers[71] studied the synergetic effect of lithium polysulfide and $LiNO_3$ to prevent lithium dendrite growth. As illustrated in Figure 7(a), when using electrolytes containing $LiNO_3$, the SEI layer composed of $LiN_xO_y$ can cover the whole anode surface, but it cannot accommodate the big volume change during cycling, due to its weak mechanical strength. Thus, cracks and pits appear on SEI, whose impedances are lower than in other places, leading to lithium dendrite growth. By contrast, when using electrolytes containing both $LiNO_3$ and lithium polysulfide, as shown in Figure 7(b), lithium polysulfide will react with cracks and pits in the initial stage of the deposition. Thus, lithium ions can be well distributed on the whole surface, forming pancake-like morphology without dendrites. The stable SEI results in excellent cycling stability in Li–Cu cells, rendering CE of 99% even after 300 cycles at 2 mA cm$^{-2}$, 1 mA h cm$^{-2}$ (Figure 7c).

Other than $LiNO_3$, many other additives have been proved to be beneficial to form stable SEI and protect lithium metal anode, including $O_2$,[72] $N_2$,[73] $CO_2$,[74] $SO_2$,[75] nitrous oxides,[76] trialkylamines,[77] aromatic compounds,[78] vinylene carbonate (VC),[79] FEC,[80] 2-methylfuran,[81] succinic anhydride,[82] 2-(triphenylphosphoranylidene) succinic anhydride,[83] $Li[N(SO_2F)_2]$ (LiFSI),[84] lithium bis(oxalato)borate (LiBOB),[85] lithium oxalyldifluroborate

(LiODFB),[86] Py13TFSI ionic liquid (IL),[87] $Li[(FSO_2)(n\text{-}C_4F_9SO_2)N]$,[88] 1,4-dioxane,[89] bromide ionomers,[90] methyl viologen,[91] poly(ethyl α-cyanoacrylate),[92] $AlCl_3$,[93] $LaNO_3$,[94] $KNO_3$,[64] HF,[95] and even trace of water.[96]

## 5.2 Safer Electrolytes

Electrolyte is a necessary part of a cell system, generally made of nonaqueous liquid solvent and lithium salt. The composition of SEI and the surface condition of the lithium metal anode largely depend on the electrolyte, including its solvent, lithium salt type, and concentration, because it is the electrolyte that directly contacts and reacts with the anode. The most popular electrolytes of current LIBs are 1 M $LiPF_6$ in carbonate solvents, such as propylene carbonate (PC), EC, dimethyl carbonate (DMC), DEC, and ethyl methyl carbonate (EMC). These types of electrolyte solutions show excellent performance in graphite-based LIBs, both in Coulombic efficiency and cell lifetime. However, the carbonate-based electrolytes do not make enough contributions to dendrite growth depression and long cell lifetime in LMBs. When applied in LMBs such as Li–air or Li–S batteries, the carbonate-based electrolytes decompose because of the nucleophilic attack by reduced oxygen species or polysulfides. Thus, new electrolyte systems should be designed for LMBs.[97]

### *5.2.1 Superconcentrated Electrolytes*

Superconcentrated electrolytes with high concentration of lithium salts can be effective in LMBs because there remain adequate anions ($TFSI^-$) near metallic lithium surface, which can keep balance of cations ($Li^+$) and avoid the depletion of ionic strength, thus delaying the onset of Sand's time. There are less free solvent molecules in superconcentrated electrolytes with more fixed solvent molecules by lithium salts, resulting in IL-like bulk properties of electrolytes.[98] The first research about superconcentrated electrolytes was reported by Ogumy and coworkers[99] in 2008, with PC electrolytes containing different concentrations of $LiN(SO_2C_2F_5)_2$. By increasing the concentration, the cycling stability largely improved, and TEM results suggested that a thinner and more favorable SEI can be obtained in higher-concentration electrolyte solutions. For example, as shown in Figure 8(a), the SEI is about 35 nm in 1.28 mol $kg^{-1}$ and about 20 nm in 3.27 mol $kg^{-1}$.

Hu and coworkers[100] studied the influence of LiTFSI concentrations in DOL:DME (1:1 by volume) electrolytes in Li–S batteries. As presented in Figure 8(b), 7 M LiTFSI electrolyte can not only effectively suppress lithium dendrite growth and shape change in the metallic lithium anode but also inhibit lithium polysulphide dissolution. Apart from LiTFSI-DOL:DME (V:V = 1:1), other superconcentrated electrolyte systems, such as LiFSI–AN, acetonitrile,[102] LiFSI–DME,[103] and $LiNO_3$–DMSO, dimethyl sulfoxide,[104] were demonstrated to be effective on LMBs. Nonetheless, the biggest problem of superconcentrated electrolytes is their relatively higher cost for lithium salts.

### *5.2.2 Ionic-Liquid-Based and Solid-State Electrolytes*

ILs have many favorable properties as electrolytes for LMBs, including incombustibility, low melting point, solubility with almost all organics, and acceptable ionic conductivity.[105]

Archer and coworkers have attached short IL chains on silica nanoparticles and used them as electrolyte additives.[101] The employment of silica undermines the weakness of ILs of limited lithium transference number and low mechanical strength. 1-Methy-3-propylimidazolium (IM) and 1-methy-3-propylpiperidinium (PP)-based IL were grafted on 10 nm silica nanoparticles (Figure 8c) and added in 1 M LiTFSI/PC solution. The hybrid electrolytes significantly inhibit the migration of anions and extend Sand's time ($\tau_s$), inducing a much safer lithium anode.[106,107] These IL-$SiO_2$ electrolytes, especially PP-based electrolytes, facilitate more than a 10-fold increasing lifetime of LMBs and display far more uniform Li deposition. Similar to $SiO_2$-IL, polymethyl methacrylate (PMMA)-IL system can provide significant stability in LMBs as well.[108]

The currently used liquid electrolytes have some disadvantages such as leakage, flammability, low mechanical modulus. Changing the form of the electrolyte from a liquid state to a solid state can effectively prevent the above-mentioned disadvantages (Figure 8d). For instance, the high mechanical modulus of the solid electrolyte inhibits the growth of lithium dendrites, while solid-state and non-flammable properties improve the safety when using LMBs. This type of electrolytes can be widely used in large-scale production of metal-based batteries.[7]

Solid-state electrolytes can be divided into ceramic electrolytes, solid polymer electrolytes, and ceramic/polymer composite electrolytes. More detailed, ceramic electrolytes include lithium phosphorous oxynitride (LiPON), $Li_3N$, perovskite-type ($Li_{0.34}Nd_{0.55}TiO_3$), $Na^+$ superionic conductor (NASICON)-type ($LiTi_2(PO_4)_3$), argyrodite-type ($Li_6PS_5Cl$), garnet-type ($Li_7La_3Nb_2O_{12}$) ceramic, sulfide system ($Li_2S$–$P_2S_5$), etc. Solid polymer electrolyte systems are mostly based on Li-conduction polymers such as PEO, poly(propylene oxide) (PPO), poly(ethylene glycol) (PEG).[109] However, solid electrolytes are also facing a difficult problem, that is, lower ionic conductivity ($10^{-3}$–$10^{-5}$ S $cm^{-1}$) and poor interfacial contact, which is considered to be the Achilles heel of solid electrolytes. In the future, the research hot point should be focused on increasing the ionic conductivity of solid electrolytes and reducing the interfacial resistance.

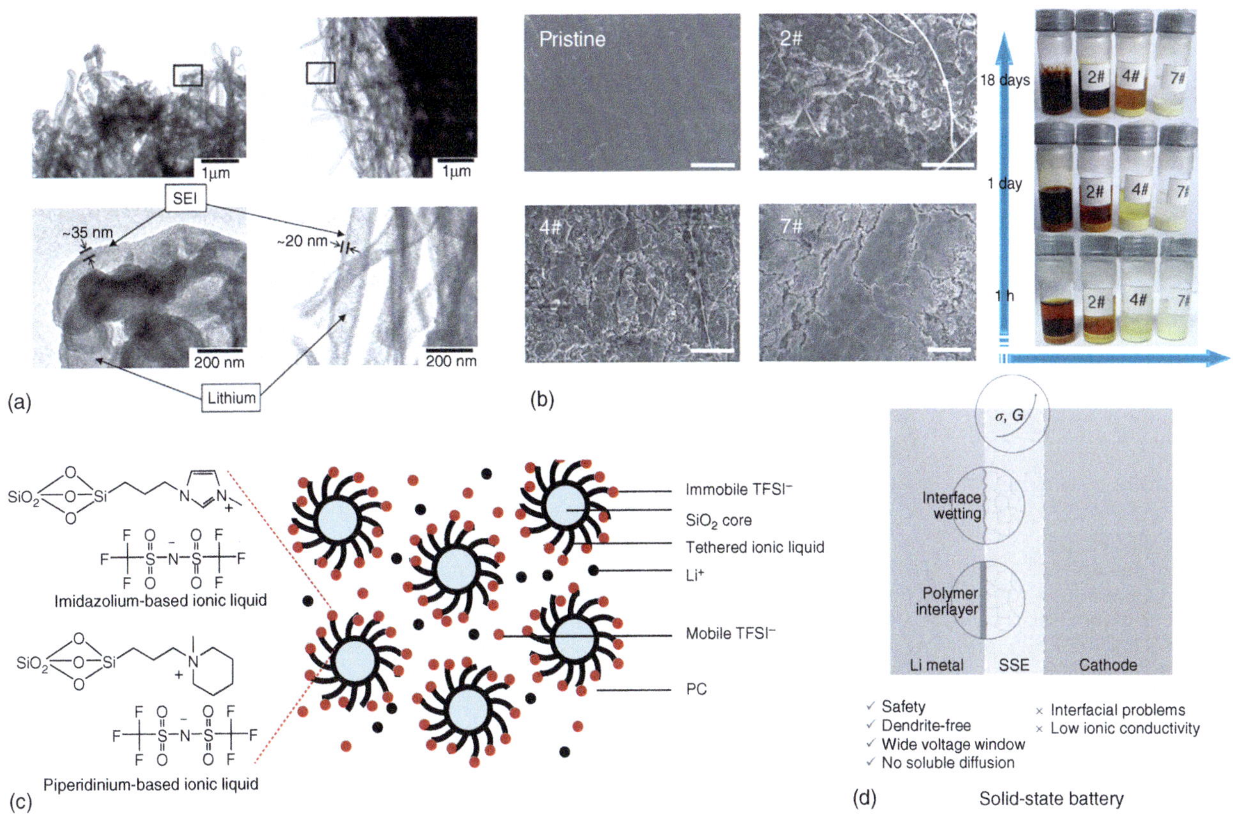


**Figure 8** (a) TEM images of the anode after the 30th cycle in 1.28 mol kg$^{-1}$ LiN($SO_2C_2F_5$)$_2$/PC (left) and 3.27 mol kg$^{-1}$ LiN($SO_2C_2F_5$)$_2$/PC (right). [Reproduced with permission from Ref. 99. © Elsevier, 2008]; (b) The morphologies of lithium metal using different 2 M, 4 M and 7 M LiTFSI electrolyte and the color changes of four samples with different salt concentrations containing the same amount of $Li_2S_8$. [Reproduced with permission from Ref. 100. © Nature Publishing Group, 2013]; (c) Schematic representation of the $SiO_2$-IL in 1 M LiTFSI/PC hybrid electrolyte system. [Reproduced with permission from Ref. 101. © John Wiley and Sons, 2012]; (d) Schematic of advantages, problems, and current strategies to address the problems of Li metal anode using solid-state electrolytes. [Reproduced with permission from Ref. 7. © John Wiley and Sons, 2017]

## 5.3 Battery Operation Mode

Dendrite growth can be curbed when using different battery operation modes, as migration and diffusion behavior of lithium ions varying with or without external voltage applied.[59] Different from common constant current charging mode, pulse current charging mode, which was firstly applied in lead acid batteries to remove lead sulfate and prolong battery life, has been approved to be effective in inhabiting dendrites and decreasing side reactions (Figure 9a).[111,112]

To understand the function of pulse current charging mode, we first introduce the mechanism of dendrites growth under pulsed current. Lithium dendrites appear when the anions are depleted near the negative electrode. Then these dendrites grow as lithium cations diffuse and migrate to tips under applied electric field. Pulsed charging mode can inhibit dendrite growth as it has two periods with different functions. One is charging at high current density ($T_{on}$), which triggers nucleation spots. Another following period is rest period ($T_{off}$), which encourage lithium ions diffusion and eliminate the depletion of cations near the electrode surface. Such charging mode has also been applied to uniform electrodeposits of other metals, such as Ni, Cu, and Au.[113]

Whittingham and coworkers[114] adopted pulse current charging mode into lithium electrodeposition system to modify the morphology, and observed the big difference of morphology results under the two different charging mode. Needle-like dendrites appeared under constant current charging mode, while larger lithium particles emerged under pulse current charging mode. To explain the interesting phenomenon, coarse-grained (CG) model was employed, with lithium dendrite growth process considered as a random process.[59] Based on the CG model and Monte Carlo calculation, a post-optimized model named CG-MC model was proposed, which speculated the optimal condition when $T_{on}$ = 1 ms, $T_{on}/T_{off}$ = 1:3. And this conjecture was later demonstrated by experiment (Figure 9b).[60] Systematic research about pulse current

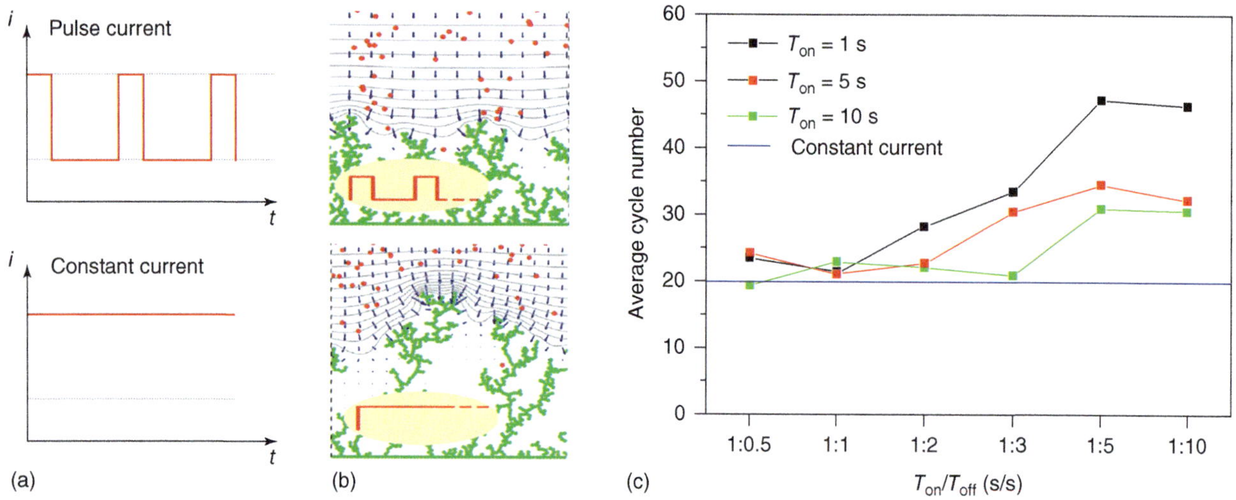


**Figure 9** (a) Basic waveforms for pulse current and constant current. (b) Calculation results of $Li^+$ transport using a CG-MC method. Gray lines: equipotential contours. Blue vectors: the electric field. Blue dots: Li in dendrites. Red dots: $Li^+$ ions in solution. [Reprinted with permission from Ref. 60. © American Chemical Society, 2014]; (c) Average cell lifetime as a function of $T_{on}/T_{off}$ ratios. [Reproduced with permission from Ref. 110. © AAAS, 2017]

charging mode at the molecular level was done by Lu and coworkers.[110] They used molecular dynamics (MD) simulation to calculate the diffusivity of lithium ions in LiTFSI-PC electrolyte, and ascribed the changes of diffusivity to the different solvation structure of lithium ions under pulse current charging mode. When the $T_{on}/T_{off}$ ratio is optimal (1:3 and 1:5), the solvation structure have more PC molecules, leading to a loose association between cations and anions. As a result, a higher diffusivity of lithium ions was achieved and dendrites growth was inhibited successfully (Figure 9c). The cell lifetime under pulse current charging mode ($T_{on}/T_{off} = 1{:}5$) was 2.4 times higher than normal constant current charging mode.

Similar to the effect of using high current in $T_{on}$ during pulsed current charging, researchers have found that using large current density to increase the nucleation site in the early stages can effectively extend the life of LMBs. Cui *et al.*[115] studied the effects of different current densities on nucleation by depositing lithium metal on Cu foil. They found that at high current densities, the number of lithium nuclei increased and the radius became smaller. Based upon this understanding, they proposed that high current density and instantaneous nucleation of Li during galvanostatic electrodeposition will be helpful to achieve the uniform Li metal anode. Koratkar and coworkers[116] also proved the high current density (>15 mA cm$^{-2}$) can result in a dendrite-free anode morphology. They found that higher currents cause the dendrites to produce a thermal effect, accelerating the diffusion of Li atoms over the tip of dendrites, thereby smoothing the surface and forming a dense and stable layer structure. When applied to Li–S batteries, the batteries with self-healing exhibit higher capacity.

### 5.4 Artificial Solid Electrolyte Interface

SEI layer formed on the lithium anode comes from the chemical reactions between electrolytes and lithium metal. It usually contains lithium salts and polymeric lithium products. Its compositions highly depend on the electrolyte solvent, lithium salt, electrolyte additive, operation temperature and operation mode.[117] SEI on lithium is different from graphite anode, as lithium anode owns the lowest negative electrochemical potential that it can react with all electrolytes thermodynamically. Different from relatively stable SEI on graphite, the SEI layer on lithium anode should be able to tolerate large volume change and avoid dendrite growth. The spontaneously formed SEI are usually not uniform in more physical morphology and chemical composition, leading to repeated break-rebuild and continuous consumption of electrolytes. An ideal SEI for lithium metal anode should have at least two extra characteristics. First, the outer part of the SEI should not be brittle in order to accommodate the volume changes during lithium plating and striping while the inner part of the SEI should have good mechanical property in order to suppress the dendritic shoot-out. Second, the physical morphology and the chemical composition of the SEI should be uniform along the lithium anode in order to provide uniform ion diffusion properties and stabilize the lithium deposition.

As most of the spontaneously formed SEI cannot meet the above requirements, many artificial SEIs have emerged, which can be classified as chemical and physical pretreatments before battery cycling.

### 5.4.1 *Chemical Pretreatment*

Chemical pretreatment is a method, making use of controllable chemical reactions between lithium metal and doped initiator to form artificial SEI film, which contains specific components to protect anode. Chemical pretreatment to form SEI includes gas processing and liquid processing. Gas processing is a facile method to build artificial SEI. Using DFT calculation, Koch *et al.*[118] studied influence of different gases, including $N_2$, $CO_2$, $O_2$, and $SO_2$. They found that different gases favor different lithium surface termination. For example, $O_2$, $N_2$ and $CO_2$ favor Li(110), $F_2$ favors Li(100) surface, while $SO_2$ favors Li(111). $SO_2$ pretreatment contributes to a thin insulating passivation layer while the others lead to metallic adlayers. $N_2$ dissociation into N-adatoms will yield the most elastically compliant adlayer. On the contrary, all the other molecularly dissociated adlayers have elastic stiffness, or equivalently, reduced elastic compliance.

Other than gas processing, liquid processing have also been done to construct defensive thin films on lithium metal.[119] The most typical work was done by Zhang and coworkers,[120] who precycled lithium metal in a LiTFSI (1.0 M)–$LiNO_3$ (5 wt%)–$Li_2S_5$ (0.02 M)–DOL/DME ternary salt electrolyte, forming SEI on lithium anode ex situ. After that, this anode with ex situ formed SEI paired with both sulfur and $LiNi_{0.5}Co_{0.2}Mn_{0.3}O_2$ cathodes and effectively enhanced the cycling performance (Figure 10a). It also demonstrated that the ex situ formed SEI can exist and maintain its function in both LiTFSI-DOL/DME and $LiPF_6$-EC/DEC electrolytes.

Cui and coworkers[121] designed an artificial SEI by using $Cu_3N$ nanoparticles and styrene-butadiene rubber (SBR) (Figure 10b). Once in contact with metallic lithium, $Cu_3N$ nanoparticles react and form $Li_3N$, and this reaction is partial reversible during cycling process. The special SEI shows high mechanical modulus, favorable flexibility and low lithium ion migration impendence, which combines the strength of $Li_3N$ ($Li_3N$ has a $10^{-3}$–$10^{-4}$ S cm$^{-1}$ ionic conductivity at room temperature) and SBR (SBR has good mechanical flexibility). Owing to the high ionic conductivity and mechanical flexibility, there are less cracks forming on the surface, successfully suppressing the further decomposition of electrolyte and the growth of lithium dendrites.

### 5.4.2 *Physical Pretreatment*

Different from chemical pretreatment whose SEI is usually hard to control (both in thickness and uniformity), physical pretreatment improves the reliability of SEI formation. It includes atomic layer deposition (ALD), flash evaporation, magnetron sputtering and spin-coating. There are many materials demonstrated to be useful, such as $Al_2O_3$,[123] carbon film,[122] Nafion,[124] polyvinylidene fluoride (PVDF)–$Li_2CO_3$,[125] poly(vinylene carbonate-co-acrylonitrile) (P(VC-co-AN)) layer,[126] poly vinylidenefluoride-hexafluoro propylene (PVDF-HFP)[127] and PVDF-HFP-$Al_2O_3$.[128] Some favorable components of spontaneously formed SEI have also been adopted to ex situ formed SEI, such as LiF[129] and $Li_3PO_4$,[130] by physical pretreatment.

Cui and coworkers successfully applied thin amorphous carbon as SEI with special nano-scale hollow sphere structure, through hard template method (Figure 10c).[122] The artificial SEI made of carbon which is chemically stable with lithium, owns a high mechanical module about

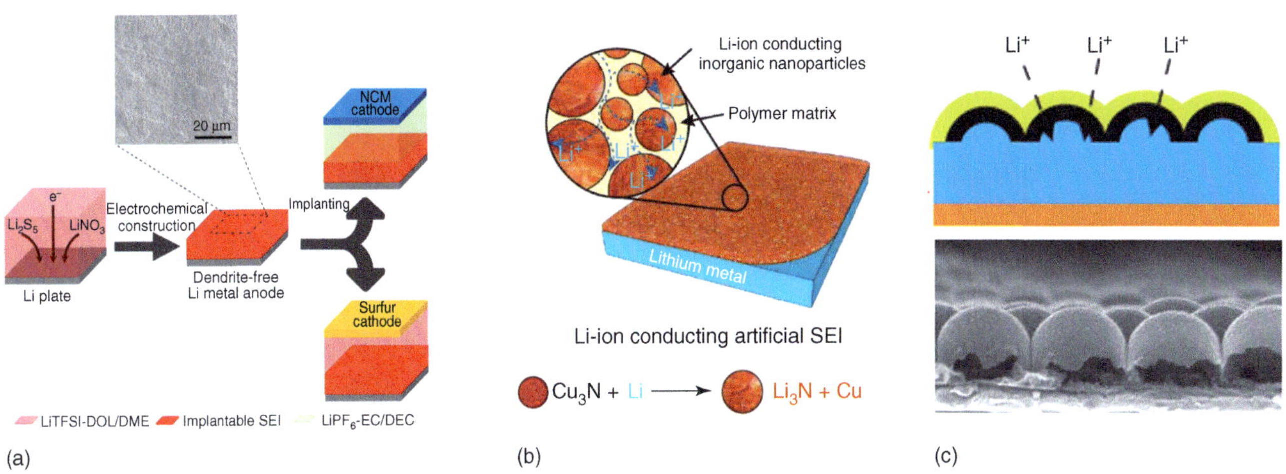


**Figure 10** (a) Scheme of implantable SEI on Li metal anode produced by in situ chemical pretreatment. [Reproduced with permission from Ref. 120. © Cell Press, 2017]; (b) Schematic illustration of the protection mechanism of the $Cu_3N$ + SBR composite artifcial SEI. [Reproduced with permission from Ref. 121. © John Wiley and Sons, 2017]; (c) Scheme and cross-sectional SEM of hollow carbon nanosphere-coated electrode. [Reproduced with permission from Ref. 122. © Nature publishing group, 2014]

200 GPa, which is thought to be enough to suppress lithium dendrites growth. Its thin thickness of about 20 nm leads to almost unchanged internal impedance, and its weak connection with the metal current collector makes it easy for the film to move up and down to adjust the volume change during cycling.

As for the polymer materials listed earlier, they have large lithium migration impedance because of intrinsic nature of polymers. In this perspective, they are not suitable to construct artificial SEI, although they have favorable mechanical properties. To conquer this shortage of polymer materials, nano-pores are introduced into SEI, acting as the pathways of lithium ions transport.[131] As a result, a uniform distribution of lithium ions flux is be obtained in the polymer-based SEI layer.

Physical pretreatment is an effective method to control the components and structure of SEI. Choosing correct physical method, almost all materials can be homogeneously introduced on the anode surface. The artificial SEI created by this method needs to be aged with electrolytes before cycling in order to obtain good contact between electrolytes and the anode.

To sum up, both pretreatment methods are effective to form an artificial SEI, protecting lithium metal anode. Different method has its own characteristics and scope of applications. Favorable effect will be obtained when appropriate pretreatment is adopted, and much is expected as integrating these methods together properly.

## 5.5 Modified Anodes

### *5.5.1 Anode with Alloy Property*

Lithium can form alloy with many metals, including Cu,[132] Al,[133] Ni,[134] Au,[122] Ag,[135] Sn,[136] In,[137] Zn[138], and so on. Using Li-based alloys as anode is an effective strategy to suppress dendrite growth and stabilize the active Li. However, alloy anode faces with the severe problem of large volume expansion, which will pulverize the anode and result in quick capacity fading. In order to solve this problem, Cui and coworkers[139] reacted Li metal with silicon and confined Li–Si nanoparticles into graphene framework to yield high stability and dendrite-free alloy anode (Figure 11a). The Li–Si alloy is well confined in the porous structure of graphene framework.

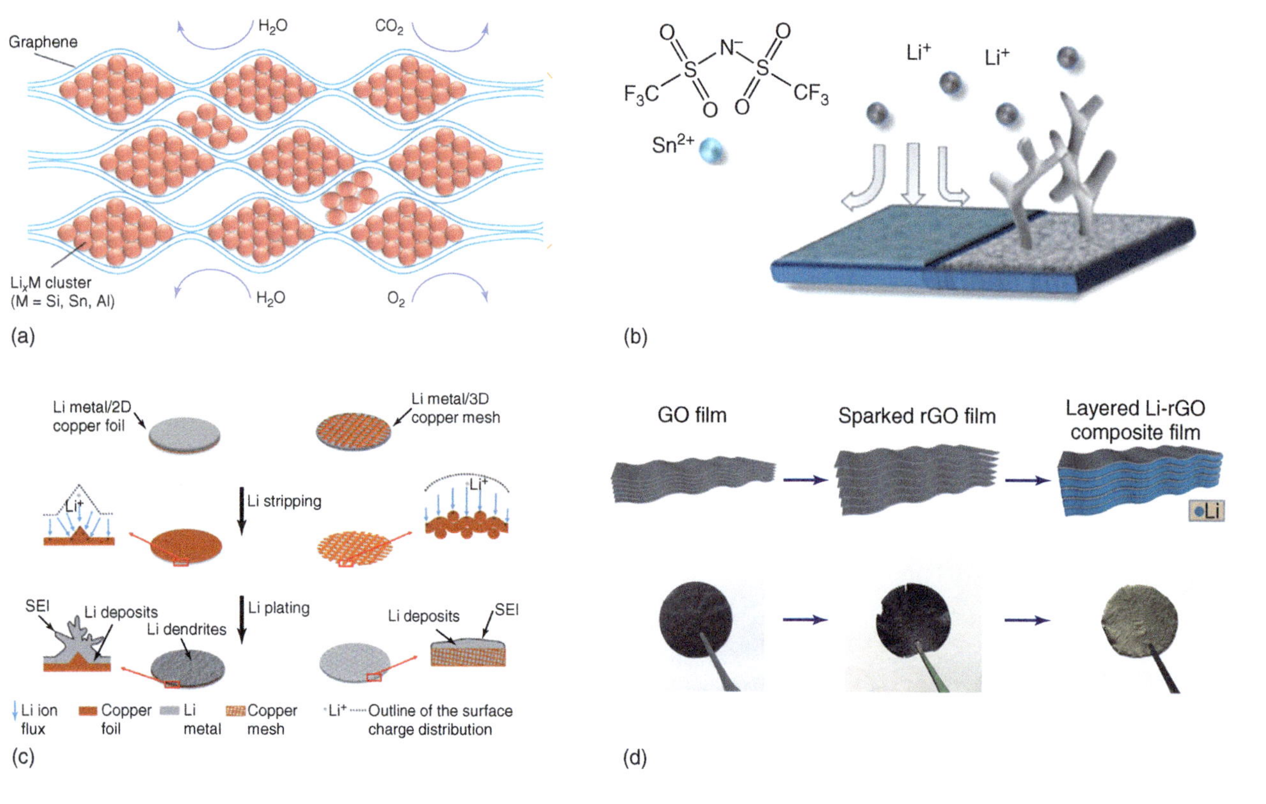


**Figure 11** (a) Scheme of the microstructure of the Li–Si alloy/graphene anode. [Reproduced with permission from Ref. 139. © Nature publishing group, 2017]; (b) Lithium anodes with alloy surface based on facile and fast Sn deposition on reactive metals produced by ion exchange. [Reproduced with permission from Ref. 40. © Nature publishing group, 2018]; (c) Scheme of lithium stripping–plating behaviors on 2D or 3D Cu current collectors. [Reproduced with permission from Ref. 140. © John Wiley and Sons, 2017]; (d) Scheme of 3D layered Li–rGO composite film. [Reproduced with permission from Ref. 141. © Nature publishing group, 2016]

As-prepared anode showed higher capacity and longer lifetime in Li/$LiFePO_4$ full cells and exhibited stability against oxygen, moisture and polysulfides.

Recently, some works also found that Li anode with alloy surface will largely facilitate the diffusion of $Li^+$ at electrolyte/electrode interface, leading to a fast and uniform ion transport behavior. Nazar *et al.*[142] constructed an in situ lithium-based alloy interface ($Li_{13}In_3$, LiZn, $Li_3Bi$ or $Li_3As$) on the metallic Li to suppress the growth of lithium dendrites. They found that the lithium-rich alloy layer exhibited a fast ionic diffusion rate, and lithium metal was deposited under the protective layer, indicating that this alloy interface can effectively improve the uniformity of lithium. Archer and coworkers[40] constructed the Li alloy interface by the reaction between metallic Li and MTFSI (M = Sn, In or Si), which also proved that fast ion transport at the interface can reduce the growth of lithium dendrites (Figure 11b).

### 5.5.2 *Three-Dimensional Anode*

The key factor of lithium dendrite nucleation and growth is the inhomogeneous lithium ion distribution, which leads to asymmetric nucleation. The electrostatic field on lithium metal plate of heterogeneous nucleation further induces lithium dendrite growth. Three-dimensional anode can suppress dendrite nucleation and growth. As it has higher specific surface area, consequently reducing the local current density and inducing uniform lithium ions flux (Figure 11c).[140]

One straightforward method to fabricate three-dimensional anodes is to build three-dimensional frameworks by lithium metal itself. For example, lithium powder anode and the lithium foil with microneedle surface pretreatment have been adopted as anode, as alternatives of the planar lithium foil.[143,144] However, the 3D lithium framework can be disappeared after several cycles, as the newly deposited lithium is not 3D any more.

The mechanism of conductive framework is to maintain a homogeneous electric field during lithium deposition, consequently suppressing dendrite growth. Guo and coworkers[145] studied the effectiveness of using 3D Cu framework with different pore diameters (by volume). When the pore diameter is 170 μm, the conductive framework cannot suppress lithium dendrite growth. When the diameter reaches as low as 2.1 μm, the electrochemical performance is improved significantly. Many other conductive frameworks have been developed, including 3D Cu collector,[71] Cu nanowires,[146] carbon nanotubes (CNTs),[147] graphene matrix,[148] carbon nanofibers,[149] porous carbon film,[150] 3D graphene@Ni scaffold,[151] $Li_7B_6$[152] and fibrous metal felt.[153] Although there have been many conductive frameworks showing dendrite-free characteristic, careful design in cycling current density and the 3D pore size is highly important since lithium can deposit on the surface layer of the 3D conductive framework instead of forming compact lithium electrodeposits.

Lithiophilic framework owns nonconductive polar surface, inducing a higher binder energy which can attract lithium ions to balance the electrostatic tip effect of roughness of Cu foil or previously formed lithium dendrites. The consequently homogeneous distribution of lithium ions can lead to dendrite-free morphology and stable cycling of metallic lithium metal anode. Many materials have been employed as lithiophilic framework, such as polar Kimwipe paper,[154] lithiophilic oxidized polyacrylonitrile nanofiber,[155] metal–organic framework (MOF),[156] and N-doped graphene.[157] After placed with polar functional groups, even glass fibers can act as framework, which has a binding energy gap of 1 eV. Different from conductive framework, the nonconductive polar framework gets out of the risk that lithium metal plates on the surface of the framework, thus building up a compact newly deposited lithium.

Cui and coworkers[141] provided a melting-infusion strategy, successfully fabricating lithium metal anode with lithiophilic graphene oxide (GO) as framework (Figure 11d). When the nonconductive GO contacts with molten lithium metal, a spark reaction induces molten lithium to enter the pore of the GO and a change from GO to reduced graphene oxide (rGO) happened consequently. This Li-rGO anode presented dendrite-free morphology and favorable performance during cycling at high C rates. Another interesting work was proposed by Hu and coworkers, which adopted carbonized wood (C-wood) as framework that possesses many low-tortuosity channels.[158] In order to improve the affinity between molten lithium and C-wood, lithiophilic ZnO coating was applied. The Li/C-wood anode shows enhanced cycling performance with only 90 mV in overpotential after 225 cycles at 3 mA $cm^{-2}$. It can be attributed to the channels that can regulate the lithium ions flux and accommodate the anode volume change.

## 6 SUMMARY AND OUTLOOK

Lithium metal anode, with extremely high capacity and lowest redox potential, has long been considered as a "holy grail" for electrical energy storage. Unfortunately, the commercial application of LMBs is severely hindered by its low CE and safety hazards, which is more or less due to the dendrite formation. In the past 40 years, researchers have proposed several methods to observe and simulate the dendrite formation, further developing strategies to suppress the lithium dendrite growth. The microscopic morphology and process of proliferation of lithium dendrite are identified by analytical measurements including SEM, AFM, TEM, and optical microscopy. The study on the surface SEI layer brought about the theory of SEI-induced lithium dendrite proliferation and formation of electrochemically isolated "dead" Li. In light of the simulation on the diffusion of lithium ion in the cell, the multiple models, including well-established Chazalviel electromigration model, surface tension model, and so on, were proposed and provide important insights of the material modification of LMBs. Based on the understanding of the lithium dendrite formation in LMBs, methods concerning on suppressing lithium dendrite and improving the cell CE are categorized as the electrolyte additives, structural electrolyte development, new battery operation mode, composite electrode design, and so on. Integrated solutions involving above methods applying to a battery system remain to be developed.

In order to widely apply the lithium metal anode, the theoretical simulation and the practical experiment are of equal importance and combined effectively. Initially, progress needs to be made in the in situ observation during battery cycling, such as using cryo-electron microscopy or in situ AFM, to observe transport behavior in sensitive metals. Due to the insufficient research toward the SEI formation mechanism, composition, and the dynamic behavior of the lithium ion within, the research of the characteristics of the SEI layer and relative simulation should maintain continuous attention in order to provide guidance for designing artificial SEI in the future. The innovative design and application of electrolyte additives, solid-state electrolyte and structured anode in the soft packing battery significantly facilitate the commercialization of the LMBs. As new understanding and innovative strategies springing up, LMBs will no doubt have an ever greater impact on our lives in the years to come.

## 7 RELATED ARTICLE

Li–$CO_2$ Batteries

## 8 ABBREVIATIONS AND ACRONYMS

ALD = atomic layer deposition; AN = acetonitrile; CE = Coulombic efficiency; CG model = coarse-grain model; CG-MC model = coarse-grain Monte Carlo model; CNTs = carbon nanotubes; DEC = diethyl carbonate; DFT = density functional theory; DMC = dimethyl carbonate; DME = 1,2-dimethoxyethane; DMSO = dimethyl sulfoxide; DOL = 1,3-dioxolane; EC = ethylene carbonate; EMC = ethyl methyl carbonate; FEC = fluoroethylene carbonate; FTIR = Fourier transform infrared spectroscopy; GO = graphene oxide; HOMO = highest occupied molecular orbital; IL = Ionic liquid; IM = 1-methy-3-propylimidazolium; LCO = lithium cobalt oxide; LFP = lithium iron phosphate; LiBOB = lithium bis(oxalato)borate; LiFSI = lithium di(fluorosulfonyl)imide; LiODFB = lithium oxalyldifluroborate; LiPON = lithium phosphorous oxynitride; LiTFSI = lithium bis(trifluoromethylsulphonyl)imide; LMBs = lithium metal batteries; LMO = lithium manganese oxide; LMO = lithium transition-metal oxide; LUMO = lowest unoccupied molecular orbital; MD = molecular dynamics; MOF = metal-organic framework; NASICON = $Na^+$ superionic conductor; NMC = lithium nickel manganese cobalt; PC = polycarbonate; PEG = poly(ethylene glycol); PEO = poly(ethylene oxide); PMMA = polymethyl methacrylate; PP = 1-methy-3-propylpiperidinium; PPO = poly(propylene oxide); PVDF = polyvinylidene fluoride; PVDF-HFP = poly vinylidenefluoride-hexafluoro propylene; rGO = reduced graphene oxide; SBR = styrene–butadiene rubber; SEI = solid electrolyte interphase; SEM = scanning electronic microscopy; TEM = transmission electron microscopy; TMO = transition metal oxide; TTFP = tris(2,2,2-trifluoroethyl) phosphite; VC = vinylene carbonate; XPS = X-ray photoelectron spectroscopy

## 9 REFERENCES


1. Y. Guo, H. Li and T. Zhai, *Adv. Mater.*, 2017, **29**, 1700007.
2. H. Kim, G. Jeong, Y.-U. Kim, J.-H. Kim, C.-M. Park and H.-J. Sohn, *Chem. Soc. Rev.*, 2013, **42**, 9011.
3. X. B. Cheng, R. Zhang, C. Z. Zhao and Q. Zhang, *Chem. Rev.*, 2017, **117**, 10403.
4. M. S. Whittingham, *Chem. Rev.*, 2004, **104**, 4271.
5. L. Li, S. Li and Y. Lu, *Chem. Commun.*, 2018, **54**, 6648.
6. S. Chu, Y. Cui and N. Liu, *Nat. Mater.*, 2016, **16**, 16.
7. C. Yang, K. Fu, Y. Zhang, E. Hitz and L. Hu, *Adv. Mater.*, 2017, **29**, 1701169.
8. D. Lin, Y. Liu and Y. Cui, *Nat. Nanotechnol.*, 2017, **12**, 194.
9. M. Armand and J. M. Tarascon, *Nature*, 2008, **451**, 652.
10. M. Li, J. Lu, Z. Chen and K. Amine, *Adv. Mater.*, 2018, **30**, e1800561.
11. I. Buchmann, 'Encyclopedia of Inorganic and Bioinorganic Chemistry', John Wiley & Sons, Inc., New York, 2015.
12. S. Li, M. Jiang, Y. Xie, H. Xu, J. Jia and J. Li, *Adv. Mater.*, 2018, **30**, 1706375.
13. M. D. Tikekar, S. Choudhury, Z. Tu and L. A. Archer, *Nat. Energy*, 2016, **1**, 16114.

14. J. Steiger, D. Kramer and R. Mönig, *J. Power Sources*, 2014, **261**, 112.
15. Q. Zhao, Z. Tu, S. Wei, K. Zhang, S. Choudhury, X. Liu and A. Archer Lynden, *Angew. Chem. Int. Ed.*, 2017, **57**, 992.
16. H. Ghassemi, M. Au, N. Chen, P. A. Heiden and R. S. Yassar, *Appl. Phys. Lett.*, 2011, **99**, 123113.
17. Y. S. Cohen, Y. Cohen and D. Aurbach, *J. Phys. Chem. B*, 2000, **104**, 12282.
18. A. N. Dey, *Electrochim. Acta*, 1976, **21**, 377.
19. C. Brissot, M. Rosso, J. N. Chazalviel, P. Baudry and S. Lascaud, *Electrochim. Acta*, 1998, **43**, 1569.
20. J. K. Stark, Y. Ding and P. A. Kohl, *J. Electrochem. Soc.*, 2013, **160**, D337.
21. P. C. Howlett, D. R. MacFarlane and A. F. Hollenkamp, *J. Power Sources*, 2003, **114**, 277.
22. J. Y. Huang, L. Zhong, C. M. Wang, J. P. Sullivan, W. Xu, L. Q. Zhang, S. X. Mao, N. S. Hudak, X. H. Liu, A. Subramanian, H. Fan, L. Qi, A. Kushima and J. Li, *Science*, 2010, **330**, 1515.
23. S. Shiraishi and K. Kanamura, *Langmuir*, 1998, **14**, 7082.
24. Y. Zhang, J. Qian, W. Xu, S. M. Russell, X. Chen, E. Nasybulin, P. Bhattacharya, M. H. Engelhard, D. Mei, R. Cao, F. Ding, A. V. Cresce, K. Xu and J.-G. Zhang, *Nano Lett.*, 2014, **14**, 6889.
25. D. Aurbach, I. Weissman, A. Zaban and O. Chusid, *Electrochim. Acta*, 1994, **39**, 51.
26. S. Wenzel, T. Leichtweiss, D. Krüger, J. Sann and J. Janek, *Solid State Ion.*, 2015, **278**, 98.
27. S.-P. Kim, A. C. T. v. Duin and V. B. Shenoy, *J. Power Sources*, 2011, **196**, 8590.
28. M. R. Palacin and A. de Guibert, *Science*, 2016, **351**, 1253292.
29. D. Aurbach, E. Zinigrad, Y. Cohen and H. Teller, *Solid State Ion.*, 2002, **148**, 405.
30. A. N. Dey, *J. Electrochem. Soc.*, 1970, **117**, C248.
31. E. Peled, *J. Electrochem. Soc.*, 1979, **126**, 2047.
32. E. Peled, D. Golodnitsky and G. Ardel, *J. Electrochem. Soc.*, 1997, **144**, L208.
33. D. Aurbach, *J. Power Sources*, 2000, **89**, 206.
34. S. Shi, P. Lu, Z. Liu, Y. Qi, L. G. Hector Jr, H. Li and S. J. Harris, *J. Am. Chem. Soc.*, 2012, **134**, 15476.
35. D. Aurbach, A. Zaban, A. Schechter, Y. Ein-Eli, E. Zinigrad and B. Markovsky, *J. Electrochem. Soc.*, 1995, **142**, 2873.
36. N. Nitta, F. Wu, J. T. Lee and G. Yushin, *Mater. Today*, 2015, **18**, 252.
37. K. Ushirogata, K. Sodeyama, Z. Futera, Y. Tateyama and Y. Okuno, *J. Electrochem. Soc.*, 2015, **162**, A2670.
38. L. Fan, L. Zhuang Houlong, W. Zhang, Y. Fu, Z. Liao and Y. Lu, *Adv. Energy Mater.*, 2018, **8**, 1703360.
39. D. Aurbach and Y. Cohen, *J. Electrochem. Soc.*, 1996, **143**, 3525.
40. Z. Tu, S. Choudhury, M. J. Zachman, S. Wei, K. Zhang, L. F. Kourkoutis and L. A. Archer, *Nat. Energy*, 2018, **3**, 310.
41. W. Zhang, H. L. Zhuang, L. Fan, L. Gao and Y. Lu, *Sci. Adv.*, 2018, **4**, eaar4410.
42. Y. Li, Y. Li, A. Pei, K. Yan, Y. Sun, C.-L. Wu, L.-M. Joubert, R. Chin, A. L. Koh, Y. Yu, J. Perrino, B. Butz, S. Chu and Y. Cui, *Science*, 2017, **358**, 506.
43. M. Jäckle and A. Groß, *J. Chem. Phys.*, 2014, **141**, 174710.
44. D. R. Ely and R. E. García, *J. Electrochem. Soc.*, 2013, **160**, A662.
45. R. Selim and P. Bro, *J. Electrochem. Soc.*, 1974, **121**, 1457.
46. I. Yoshimatsu, T. Hirai and J. i. Yamaki, *J. Electrochem. Soc.*, 1988, **135**, 2422.
47. M. Arakawa, S.-i. Tobishima, Y. Nemoto, M. Ichimura and J.-i. Yamaki, *J. Power Sources*, 1993, **43**, 27.
48. J. Steiger, D. Kramer and R. Mönig, *Electrochim. Acta*, 2014, **136**, 529.
49. D. Lu, Y. Shao, T. Lozano, W. D. Bennett, G. L. Graff, B. Polzin, J. Zhang, M. H. Engelhard, N. T. Saenz, W. A. Henderson, P. Bhattacharya, J. Liu and J. Xiao, *Adv. Energy Mater.*, 2015, **5**, 1400993.
50. W. Xu, J. Wang, F. Ding, X. Chen, E. Nasybulin, Y. Zhang and J.-G. Zhang, *Energy Environ. Sci.*, 2014, **7**, 513.
51. Z. Li, J. Huang, B. Yann Liaw, V. Metzler and J. Zhang, *J. Power Sources*, 2014, **254**, 168.
52. C. Brissot, M. Rosso, J. N. Chazalviel and S. Lascaud, *J. Power Sources*, 1999, **81–82**, 925.
53. J. N. Chazalviel, *Phys. Rev. A*, 1990, **42**, 7355.
54. C. Monroe and J. Newman, *J. Electrochem. Soc.*, 2003, **150**, A1377.
55. M. Rosso, C. Brissot, A. Teyssot, M. Dollé, L. Sannier, J.-M. Tarascon, R. Bouchet and S. Lascaud, *Electrochim. Acta*, 2006, **51**, 5334.
56. I. W. Seong, C. H. Hong, B. K. Kim and W. Y. Yoon, *J. Power Sources*, 2008, **178**, 769.
57. C. Monroe and J. Newman, *J. Electrochem. Soc.*, 2005, **152**, A396.
58. C. Monroe and J. Newman, *J. Electrochem. Soc.*, 2004, **151**, A880.
59. M. Z. Mayers, J. W. Kaminski and T. F. Miller, *J. Phys. Chem. C*, 2012, **116**, 26214.
60. A. Aryanfar, D. Brooks, B. V. Merinov, W. A. Goddard, A. J. Colussi and M. R. Hoffmann, *J. Phys. Chem. Lett.*, 2014, **5**, 1721.
61. R. Tao, X. X. Bi, S. Li, Y. Yao, F. Wu, Q. Wang, C. Z. Zhang and J. Lu, *ACS Appl. Mater. Interfaces*, 2017, **9**, 7003.

62. J. L. Wang, F. J. Lin, H. Jia, J. Yang, C. W. Monroe and Y. N. NuLi, *Angew. Chem. Int. Ed.*, 2014, **53**, 10099.

63. F. Ding, W. Xu, G. L. Graff, J. Zhang, M. L. Sushko, X. Chen, Y. Shao, M. H. Engelhard, Z. Nie, J. Xiao, X. Liu, P. V. Sushko, J. Liu and J. G. Zhang, *J. Am. Chem. Soc.*, 2013, **135**, 4450.

64. W. S. Jia, C. Fan, L. P. Wang, Q. J. Wang, M. J. Zhao, A. J. Zhou and J. Z. Li, *ACS Appl. Mater. Interfaces*, 2016, **8**, 15399.

65. J. K. S. Goodman and P. A. Kohl, *J. Electrochem. Soc.*, 2014, **161**, D418.

66. B. R. Wu, Q. Liu, D. B. Mu, H. L. Xu, L. Wang, L. L. Shi, L. Gai and F. Wu, *RSC Adv.*, 2016, **6**, 51738.

67. Y. Y. Lu, Z. Y. Tu and L. A. Archer, *Nat. Mater.*, 2014, **13**, 961.

68. Y. Y. Lu, Z. Y. Tu, J. Shu and L. A. Archer, *J. Power Sources*, 2015, **279**, 413.

69. Y. Ozhabes, D. Gunceler and T. A. Arias, *Physics*, 2015, arXiv:1504.05799.

70. S. S. Zhang, *Electrochim. Acta*, 2012, **70**, 344.

71. W. Li, H. Yao, K. Yan, G. Zheng, Z. Liang, Y. M. Chiang and Y. Cui, *Nat. Commun.*, 2015, **6**, 7436.

72. Y. Malik, D. Aurbach, P. Dan and A. Meitav, *J. Electroanal. Chem. Interfac. Electrochem.*, 1991, **282**, 73.

73. T. Momma, H. Nara, S. Yamagami, C. Tatsumi and T. Osaka, *J. Power Sources*, 2011, **196**, 6483.

74. Y. Ein-Eli and D. Aurbach, *J. Power Sources*, 1995, **54**, 281.

75. R. D. Rauh and S. B. Brummer, *Electrochim. Acta*, 1977, **22**, 75.

76. M. Uchiyama, S. Slane, E. Plichta and M. Salomon, *J. Power Sources*, 1987, **20**, 279.

77. P. Dan, E. Mengeritsky, D. Aurbach, I. Weissman and E. Zinigrad, *J. Power Sources*, 1997, **68**, 443.

78. K. M. Abraham, *J. Power Sources*, 1985, **14**, 179.

79. J. K. Stark, Y. Ding and P. A. Kohl, *J. Electrochem. Soc.*, 2011, **158**, A1100.

80. X. Q. Zhang, X. B. Cheng, X. Chen, C. Yan and Q. Zhang, *Adv. Funct. Mater.*, 2017, **27**, 1605989.

81. M. Morita, S. Aoki and Y. Matsuda, *Electrochim. Acta*, 1992, **37**, 119.

82. G. B. Han, J. N. Lee, J. L. Dong, H. Lee, J. Song, H. Lee, M. H. Ryou, J. K. Park and M. L. Yong, *Electrochim. Acta*, 2014, **115**, 525.

83. J. Jeong, J. N. Lee, J. K. Park, M. H. Ryou and M. L. Yong, *Electrochim. Acta*, 2015, **170**, 353.

84. R. Miao, J. Yang, X. Feng, H. Jia, J. Wang and Y. Nuli, *J. Power Sources*, 2014, **271**, 291.

85. H. Xiang, P. Shi, P. Bhattacharya, X. Chen, D. Mei, M. E. Bowden, J. Zheng, J. G. Zhang and W. Xu, *J. Power Sources*, 2016, **318**, 170.

86. F. Wu, J. Qian, R. Chen, J. Lu, L. Li, H. Wu, J. Chen, T. Zhao, Y. Ye and K. Amine, *ACS Appl. Mater. Interfaces*, 2014, **6**, 15542.

87. N. W. Li, Y. X. Yin, J. Y. Li, C. H. Zhang and Y. G. Guo, *Adv. Sci.*, 2016, **4**, 1600400.

88. F. Zheng, M. Qiang, P. Liu, M. Jie, Y. S. Hu, Z. Zhou, L. Hong, X. Huang and L. Chen, *ACS Appl. Mater. Interfaces*, 2016, **9**, 4282.

89. R. Miao, J. Yang, Z. Xu, J. Wang, Y. Nuli and L. Sun, *Sci. Rep.*, 2016, **6**, 21771.

90. S. Choudhury, C. T. Wan, W. I. Al Sadat, Z. Tu, S. Lau, M. J. Zachman, L. F. Kourkoutis and L. A. Archer, *Sci. Adv.*, 2017, **3**, e1602809.

91. H. Wu, Y. Cao, L. Geng and C. Wang, *Chem. Mater.*, 2017, **29**, 3572.

92. Z. Hu, S. Zhang, S. Dong, W. Li, H. Li, G. Cui and L. Chen, *Chem. Mater.*, 2017, **29**, 4682.

93. H. Ye, Y. X. Yin, S. F. Zhang, Y. Shi, L. Liu, X. X. Zeng, R. Wen, Y. G. Guo and L. J. Wan, *Nano Energy*, 2017, **36**, 411.

94. S. Liu, G. R. Li and X. P. Gao, *ACS Appl. Mater. Interfaces*, 2016, **8**, 7783.

95. S. Shiraishi, K. Kanamura and Z. Takehara, *J. Electrochem. Soc.*, 1999, **146**, 1633.

96. J. Qian, W. Xu, P. Bhattacharya, M. Engelhard, W. A. Henderson, Y. Zhang and J. G. Zhang, *Nano Energy*, 2015, **15**, 135.

97. S. A. Freunberger, Y. Chen, Z. Peng, J. M. Griffin, L. J. Hardwick, F. Bardé, P. Novák and P. G. Bruce, *J. Am. Chem. Soc.*, 2011, **133**, 8040.

98. Q. Ma, Z. Fang, P. Liu, J. Ma, X. Qi, W. Feng, J. Nie, Y. S. Hu, H. Li and X. Huang, *ChemElectroChem*, 2016, **3**, 531.

99. S.-K. Jeong, H.-Y. Seo, D.-H. Kim, H.-K. Han, J.-G. Kim, Y. B. Lee, Y. Iriyama, T. Abe and Z. Ogumi, *Electrochem. Commun.*, 2008, **10**, 635.

100. L. Suo, Y. S. Hu, H. Li, M. Armand and L. Chen, *Nat. Commun.*, 2013, **4**, 1481.

101. Y. Lu, S. K. Das, S. S. Moganty and L. A. Archer, *Adv. Mater.*, 2012, **24**, 4430.

102. Y. Yamada, K. Furukawa, K. Sodeyama, K. Kikuchi, M. Yaegashi, Y. Tateyama and A. Yamada, *J. Am. Chem. Soc.*, 2014, **136**, 5039.

103. J. Qian, W. A. Henderson, W. Xu, P. Bhattacharya, M. Engelhard, O. Borodin and J. G. Zhang, *Nat. Commun.*, 2015, **6**, 6362.

104. N. Togasaki, T. Momma and T. Osaka, *J. Power Sources*, 2016, **307**, 98.

105. H. Zheng, J. Qin, Y. Zhao, T. Abe and Z. Ogumi, *Solid State Ion.*, 2005, **176**, 2219.

106. M. D. Tikekar, L. A. Archer and D. L. Koch, *Sci. Adv.*, 2016, **2**, e1600320.

107. M. D. Tikekar, L. A. Archer and D. L. Koch, *J. Electrochem. Soc.*, 2014, **161**, A847.

108. Y. Li, K. W. Wong, Q. Dou and K. M. Ng, *J. Mater. Chem. A*, 2016, **4**, 18543.

109. L. Fan, S. Wei, S. Li, Q. Li and Y. Lu, *Adv. Energy Mater.*, 2018, **8**, 1702657.

110. Q. Li, S. Tan, L. Li, Y. Lu and Y. He, *Sci. Adv.*, 2017, **3**, e1701246.

111. A. Kirchev, A. Delaille, M. Perrin, E. Lemaire and F. Mattera, *J. Power Sources*, 2007, **170**, 495.

112. M. A. Monem, K. Trad, N. Omar, O. Hegazy, B. Mantels, G. Mulder, P. V. D. Bossche and J. V. Mierlo, *Appl. Energy*, 2015, **152**, 143.

113. M. S. Chandrasekar and M. Pushpavanam, *Electrochim. Acta*, 2008, **53**, 3313.

114. H. Yang, E. O. Fey, B. D. Trimm, N. Dimitrov and M. S. Whittingham, *J. Power Sources*, 2014, **272**, 900.

115. A. Pei, G. Zheng, F. Shi, Y. Li and Y. Cui, *Nano Lett.*, 2017, **17**, 1132.

116. L. Li, S. Basu, Y. Wang, Z. Chen, P. Hundekar, B. Wang, J. Shi, Y. Shi, S. Narayanan and N. Koratkar, *Science*, 2018, **359**, 1513.

117. M. Gauthier, T. J. Carney, A. Grimaud, L. Giordano, N. Pour, H. H. Chang, D. P. Fenning, S. F. Lux, O. Paschos and C. Bauer, *J. Phys. Chem. Lett.*, 2016, **6**, 4653.

118. S. L. Koch, B. J. Morgan, S. Passerini and G. Teobaldi, *J. Power Sources*, 2015, **296**, 150.

119. N. W. Li, Y. X. Yin, C. P. Yang and Y. G. Guo, *Adv. Mater.*, 2016, **28**, 1853.

120. X. B. Cheng, C. Yan, X. Chen, C. Guan, J. Q. Huang, H. J. Peng, R. Zhang, S. T. Yang and Q. Zhang, *Chemistry*, 2017, **2**, 258.

121. Y. Liu, D. Lin, P. Y. Yuen, K. Liu, J. Xie, R. H. Dauskardt and Y. Cui, *Adv. Mater.*, 2017, **29**, 1605531.

122. G. Zheng, S. W. Lee, Z. Liang, H. W. Lee, K. Yan, H. Yao, H. Wang, W. Li, S. Chu and Y. Cui, *Nat. Nanotechnol.*, 2014, **9**, 618.

123. A. C. Kozen, C. F. Lin, A. J. Pearse, M. A. Schroeder, X. Han, L. Hu, S. B. Lee, G. W. Rubloff and M. Noked, *ACS Nano*, 2015, **9**, 5884.

124. J. Song, H. Lee, M. J. Choo, J. K. Park and H. T. Kim, *Sci. Rep.*, 2015, **5**, 14458.

125. K. I. Chung, J. D. Lee, E. J. Kim, W. S. Kim, J. H. Cho and Y. K. Choi, *Microchem. J.*, 2003, **75**, 71.

126. S. M. Choi, I. S. Kang, Y. K. Sun, J. H. Song, S. M. Chung and D. W. Kim, *J. Power Sources*, 2013, **244**, 363.

127. D. J. Lee, H. Lee, Y. J. Kim, J. K. Park and H. T. Kim, *Adv. Mater.*, 2016, **28**, 857.

128. H. Lee, J. L. Dong, Y. J. Kim, J. K. Park and H. T. Kim, *J. Power Sources*, 2015, **284**, 103.

129. L. Fan, H. L. Zhuang, L. Gao, Y. Lu and L. A. Archer, *J. Mater. Chem. A*, 2017, **5**, 3483.

130. L. Wang, Q. Wang, W. Jia, S. Chen, P. Gao and J. Li, *J. Power Sources*, 2017, **342**, 175.

131. B. Zhu, Y. Jin, X. Hu, Q. Zheng, S. Zhang, Q. Wang and J. Zhu, *Adv. Mater.*, 2017, **29**, 1603755.

132. B. Baretzky, W. Eckstein and R. P. Schorn, *J. Nucl. Mater.*, 1995, **224**, 50.

133. R. D. Armstrong, O. R. Brown, R. P. Ram and C. D. Tuck, *J. Power Sources*, 1989, **28**, 259.

134. B. Predel, 'Li–Ni (Lithium–Nickel)', Springer, Berlin Heidelberg, 1997.

135. A. D. Pelton, *Bull. Alloy Phase Diagr.*, 1986, **7**, 223.

136. C. J. Wen and R. A. Huggins, *J. Solid State Chem.*, 1980, **35**, 376.

137. W. A. Alexander, L. D. Calvert, R. H. Gamble and K. Schinzel, *Can. J. Chem.*, 1976, **54**, 1052.

138. J. Wang, P. King and R. A. Huggins, *Solid State Ion.*, 1986, **20**, 185.

139. J. Zhao, G. Zhou, K. Yan, J. Xie, Y. Li, L. Liao, Y. Jin, K. Liu, P. C. Hsu, J. Wang, H. M. Cheng and Y. Cui, *Nat. Nanotechnol.*, 2017, **12**, 993.

140. Q. Li, S. Zhu and Y. Lu, *Adv. Funct. Mater.*, 2017, **27**, 1606422.

141. D. Lin, Y. Liu, Z. Liang, H. W. Lee, J. Sun, H. Wang, K. Yan, J. Xie and Y. Cui, *Nat. Nanotechnol.*, 2016, **11**, 626.

142. X. Liang, Q. Pang, I. R. Kochetkov, M. S. Sempere, H. Huang, X. Sun and L. F. Nazar, *Nat. Energy*, 2017, **2**, 17119.

143. J. Heine, S. Krüger, C. Hartnig, U. Wietelmann, M. Winter and P. Bieker, *Adv. Energy Mater.*, 2014, **4**, 288.

144. M. H. Ryou, Y. M. Lee, Y. Lee, M. Winter and P. Bieker, *Adv. Funct. Mater.*, 2015, **25**, 834.

145. C. P. Yang, Y. X. Yin, S. F. Zhang, N. W. Li and Y. G. Guo, *Nat. Commun.*, 2015, **6**, 8058.

146. L. L. Lu, J. Ge, J. N. Yang, S. M. Chen, H. B. Yao, F. Zhou and S. H. Yu, *Nano Lett.*, 2016, **16**, 4431.

147. S. Jin, S. Xin, L. Wang, Z. Du, L. Cao, J. Chen, X. Kong, M. Gong, J. Lu and Y. Zhu, *Adv. Mater.*, 2016, **28**, 9016.

148. R. Mukherjee, A. V. Thomas, D. Datta, E. Singh, J. Li, O. Eksik, V. B. Shenoy and N. Koratkar, *Nat. Commun.*, 2014, **5**, 3710.

149. T. T. Zuo, X. W. Wu, C. P. Yang, Y. X. Yin, H. Ye, N. W. Li and Y. G. Guo, *Adv. Mater.*, 2017, **29**, 1700389.

150. H. K. Kang, S. G. Woo, J. H. Kim, S. R. Lee and Y. J. Kim, *Electrochim. Acta*, 2015, **176**, 172.

151. K. Xie, W. Wei, K. Yuan, W. Lu, M. Guo, Z. Li, Q. Song, X. R. Liu, J. G. Wang and C. Shen, *ACS Appl. Mater. Interfaces*, 2016, **8**, 26091.

152. X. Zhang, W. Wang, A. Wang, Y. Huang, K. Yuan, Z. Yu, J. Qiu and Y. Yang, *J. Mater. Chem. A*, 2014, **2**, 11660.

153. H. Lee, J. Song, Y. J. Kim, J. K. Park and H. T. Kim, *Sci. Rep.*, 2016, **6**, 30830.

154. C. H. Chang, S. H. Chung and A. Manthiram, *Adv. Sustain. Syst.*, 2017, **1**, 1600034.

155. Z. Liang, G. Zheng, C. Liu, N. Liu, W. Li, K. Yan, H. Yao, P. C. Hsu, S. Chu and Y. Cui, *Nano Lett.*, 2015, **15**, 2910.

156. L. Wen, Y. Mi, W. Zhe, Y. Zhong, Z. Wu and H. Wang, *Chem. Sci.*, 2017, **8**, 4285.

157. R. Zhang, X. R. Chen, X. Chen, X. B. Cheng, X. Q. Zhang, C. Yan and Q. Zhang, *Angew. Chem. Int. Ed.*, 2017, **56**, 7764.

158. Y. Zhang, W. Luo, C. Wang, Y. Li, C. Chen, J. Song, J. Dai, E. M. Hitz, S. Xu, C. Yang, Y. Wang and L. Hu, *Proc. Natl. Acad. Sci. USA*, 2017, **114**, 3584.

# Lithium Oxygen Battery

**Raymond A. Wong and Hye Ryung Byon**

*Korea Advanced Institute of Science and Technology (KAIST), Daejeon, Republic of Korea*

and

**Morgan L. Thomas, Kaoru Dokko and Masayoshi Watanabe**

*Yokohama National University, Yokohama, Japan*



## 1 INTRODUCTION

The current trend toward electrification (i.e., including hybrid electric vehicles (EVs), and drones) can be considered a paradigm shift. The improved understanding of intercalation chemistries initiated the development of LIBs since the 1970s, with commercialization occurring in 1991, driven in part by the downsizing of electronics. However, due to the urgency in supplanting fossil fuel technologies, there is a need for batteries with even higher energy densities in order to realize advanced EVs and to effectively harness energy from intermittent sources such as solar and wind. In effect, this has provided the impetus for research into "beyond LIB," in the form of conversion–reaction type systems capable of multi-electron transfer such as metal–air and metal–sulfur batteries.

Amongst these, the lithium–oxygen (Li–$O_2$) or lithium–air system is noteworthy for the exceptionally high theoretical specific energy density of ~3460 Wh $kg^{-1}$ (up to six times that of LIBs, Table 1). The practical specific energy is projected to be 500–900 Wh $kg^{-1}$, giving the possibility of providing >550 km in driving range for EVs compared to the ~160 km driving range for the current LIBs with a specific energy of ~140 Wh $kg^{-1}$.[1,2] Such a high gravimetric theoretical energy density arises due to low mass of active materials at the positive electrode, that is, effective zero mass of $O_2$ from ambient air or the solid discharge product (lithium peroxide, $Li_2O_2$) that is still lower weight than typical LIB positive electrodes. Another attractive aspect of Li–$O_2$ batteries is the potential of reduced costs if the positive electrode is comprised of carbonaceous materials in contrast to LIBs which often require more expensive elements, such as cobalt, leading to a negative effect on the critical metric of cost per kilowatt hour (\$ kW $h^{-1}$). It is noted that with regard to nomenclature, there has been some intermingling of the terms "Li–$O_2$ batteries" and "Li–air batteries" in published research. Here-on, in this chapter, we refer to "Li–$O_2$ batteries" as those using a pure $O_2$ source, and "Li–air batteries" as those using ambient air (i.e., presence of other gases/humidity, covered in Section 4). Regarding the major categories of Li–$O_2$ batteries, this chapter will focus specifically on the nonaqueous variant of Li–$O_2$ batteries, while we note that aqueous-based systems have been previously reviewed.[3]

The first report of nonaqueous Li–$O_2$ batteries is attributed to Abraham and Jiang in 1996,[4] while

*Inorganic Battery Materials*. Edited by Hailiang Wang and Boniface P.T. Fokwa
 ISBN 978-1-119-43199-2

**Table 1** Comparison of the lithium–oxygen (Li–$O_2$) battery with alternative and "beyond LIB" chemistries

| | Discharge composition | Theoretical specific capacity ($mAh\,g^{-1}$) | Standard cell potential (V) | Theoretical specific energy density ($Wh\,kg^{-1}$) |
|---|---|---|---|---|
| Li–ion (LIB) | $LiCoO_2$ | 140[(a)] | 3.8 | 520[(a)] |
| Li–S | $Li_2S$ | 1168 | 2.24 | 2610 |
| Zn–$O_2$ | ZnO | 1318 | 1.65 | 2170 |
| Mg–$O_2$ | $MgO_2$ | 952 | 2.91 | 2770 |
| Na–$O_2$ | $NaO_2$ | 488 | 2.27 | 1110 |
| Li–$O_2$ | $Li_2O_2$ | 1168 | 2.96 | 3460 |

[(a)]Assumption is that only half of the lithium ($0.5e^-$) with $LiCoO_2$ can be reversibly intercalated.

research interest increased significantly beginning in the past decade (from the late 2000s). In terms of development, nonaqueous Li–$O_2$ batteries draw on existing knowledge from LIBs and proton exchange membrane fuel cells (PEMFCs). The utilization of a metallic Li negative electrode with $Li^+$ as the main charge carrier in an aprotic electrolyte solution are adapted from LIBs while the gas diffusion electrode and ingress of $O_2$ gas at the positive electrode are analogous to PEMFCs. It indicates that the overall reactions are based on $O_2$ reduction and evolution at the positive electrode, and Li/$Li^+$ redox at the negative electrode. A schematic of a Li–$O_2$ battery and the overall discharge–recharge reactions are shown in Figure 1. Empirically, the main discharge product from the oxygen reduction reaction (ORR) is known to be solid-state $Li_2O_2$ at room temperature, which indicates a 2 electron ($e^-$) transfer occurs in nonaqueous Li–$O_2$ batteries. During recharge, the solid $Li_2O_2$ decomposes to evolve $O_2$ gas. This electrochemical reaction is intriguing on a fundamental level, as no cleavage of molecular $O_2$ gas is observed, in contrast to the typical $4e^-$ electrochemical pathway in an aqueous electrolyte solution (as in PEMFCs), and three phases are involved in discharge and recharge process, that is, $O_2$ gas, $Li^+$ in liquid electrolyte and solid-state $Li_2O_2$. However, the three-phase reaction may be complicated with regard to the gas flow dynamics (and partial pressure

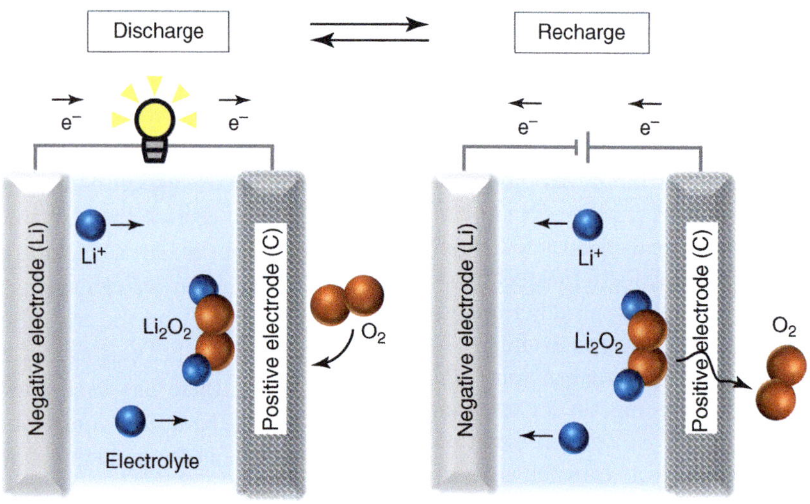


$$\text{Negative electrode: } 2Li \underset{\text{Recharge}}{\overset{\text{Discharge}}{\rightleftharpoons}} 2Li^+ + 2e^-$$

$$\text{Positive electrode: } O_2 + 2Li^+ + 2e^- \underset{\text{Recharge}}{\overset{\text{Discharge}}{\rightleftharpoons}} Li_2O_2$$

$$\text{Overall: } 2Li + O_2 \underset{\text{Recharge}}{\overset{\text{Discharge}}{\rightleftharpoons}} Li_2O_2 \quad E^o = 2.96\text{ V vs Li/Li}^+$$

**Figure 1** Schematic of a nonaqueous Li–$O_2$ cell with the discharge (oxygen reduction reaction, ORR) and recharge reactions (oxygen evolution reaction, OER) with chemical equations and thermodynamic reversible potential, $E^o = 2.96$ V vs Li/$Li^+$). The cell configuration shows the negative and positive electrodes comprised of Li metal and porous carbon, respectively of $O_2$ when air is used), dissolution of $O_2$ species in aprotic medium, and physical and electrical properties of $Li_2O_2$, which lead to crucial challenges in the characterization of performance in $Li-O_2$ batteries.

At the current stage, there are scientific and engineering hurdles to be overcome in $Li-O_2$ batteries originating from the insulating nature of $Li_2O_2$, contaminants from air and the instability of both aprotic solvent and carbonaceous electrode in the presence of reactive $O_2$ species (i.e., superoxide radical, peroxide and singlet oxygen), which are formed during electrochemical reactions. These challenges manifest themselves as key battery cell-level issues highlighted by (i) limitations in discharge capacities, (ii) high recharge overpotential, and (iii) poor cycling performance. Of note, the development of a metallic Li negative electrode to mitigate volume expansion, formation of dendritic and dead Li are the other challenges common in the development of beyond LIBs, which are mainly elaborated upon in the ***Lithium Metal Anode*** chapter.

In this chapter, we elaborate upon the current state-of-the-art understanding of the various chemical and electrochemical phenomena observed in research-level $Li-O_2$ batteries, including where appropriate a discussion on the solid-state chemistry of electro-generated $Li-O_2$ compounds, and the coordination environment of the $Li^+$ (and superoxide radical anion) in the electrolytes commonly studied. We also discuss the key challenges underpinning $Li-O_2$ batteries along with the notable chemical, electrochemical, and materials science approaches taken to enhance battery performance.

## 2 $O_2$ REDOX IN $Li^+$-CONTAINING APROTIC SOLUTIONS

The fundamentals of $O_2$ electrochemistry in aprotic electrolyte solutions have been widely explored since the 1980s while its application to $Li-O_2$ cells has been extensively investigated since around 2010, after the first demonstration in 1996. A variety of studies to understand the discharge and recharge processes of $Li-O_2$ cells have been conducted using delicate ex situ and in situ imaging, chemical identification and electrochemical analysis tools. As a result, it has been revealed that beyond the simple $Li-O_2$ electrochemical equations (Figure 1), the overall electrochemical and chemical processes are strongly influenced by the materials composition of the electrolyte solution and positive electrode, additives (or impurities) and structure of $Li_2O_2$. These factors also determine the nucleation, growth and decomposition processes of the $Li_2O_2$ occurring in $Li-O_2$ cells and critically affect the cell performance. In this section, we introduce an overview of the current understanding gained through these studies.

### 2.1 $Li-O_2$ Electrochemistry from a Molecular Point of View

In $O_2$ gas-saturated nonaqueous (aprotic) media, it has been well established that ORR/oxygen evolution reactions (OER) occur via one-electron transfer. Sawyer *et al.*[5] observed the reversible $O_2/O_2^-$ redox couple by cyclic voltammetry (CV) and noted that the reduction of $O_2$ to $O_2^-$ Equation (1) is accompanied by ion-pair formation with bulky molecular tetraethylammonium ($TEA^+$) cations.

$$O_2 + e^- \rightleftharpoons O_2^- \quad (1)$$

However, in contrast to the weak interaction of tetraalkylammonium cations with superoxide (the superoxide radical anion is generally written as $O_2^-$ or $O_2^{\bullet -}$, and hereafter we use the $O_2^-$ notation), the highly Lewis acidic $Li^+$ cation interacts strongly with $O_2^-$. Incrementally increasing the Li salt concentration in a tetrabutylammonium ($TBA^+$)-based solution results in a potential shift for both the cathodic (ORR) and anodic (OER) peaks and reduction in the current response (Figure 2a). The products formed at the gold (Au) working electrode surface have been identified from in situ surface-enhanced Raman spectroscopy (SERS).[6] A distinct peak ascribed to the O–O stretch of lithium superoxide ($LiO_2$) appears at 1137 cm$^{-1}$ ("2" in Figure 2b) after maintaining the potential at 2.2 versus $Li/Li^+$ (hereafter all potentials are with respect to $Li/Li^+$) for 2 min, which corresponds to the cathodic peak in the CV curve (Figure 2a). Prolonging the period at 2.2 V reveals a pronounced new peak at 808 cm$^{-1}$, assigned to the O–O stretch of lithium peroxide ($Li_2O_2$, "3" in Figure 2b) while concomitantly, the signal of $LiO_2$ diminishes with time. This in situ analysis reveals that the initial 1e$^-$ transfer produces $LiO_2$ (Equation 2) followed by formation of $Li_2O_2$ via chemical disproportion of $LiO_2$ (Equation 3).

$$O_2 + Li^+ + e^- \rightarrow LiO_2 \quad (2)$$

$$2LiO_2 \rightarrow Li_2O_2(s) + O_2 \quad (3)$$

Along with this chemical disproportion route to $LiO_2$, subsequent electrochemical reaction of $LiO_2$ to form $Li_2O_2$ can also occur (Equation 4).

$$LiO_2 + Li^+ + e^- \rightarrow Li_2O_2(s) \quad (4)$$

The preferred pathway for $Li_2O_2$ formation (i.e., Equations (3) and (4) is determined from the cathodic potential, degree of $LiO_2$ solvation, and reaction kinetics (see Section 2.2.2)). The direct 2e$^-$ transfer to form $Li_2O_2$ has also been mentioned in the literature, although step-wise 1e$^-$ transfer reactions are generally expected to be favored. We note that further reduction toward lithium

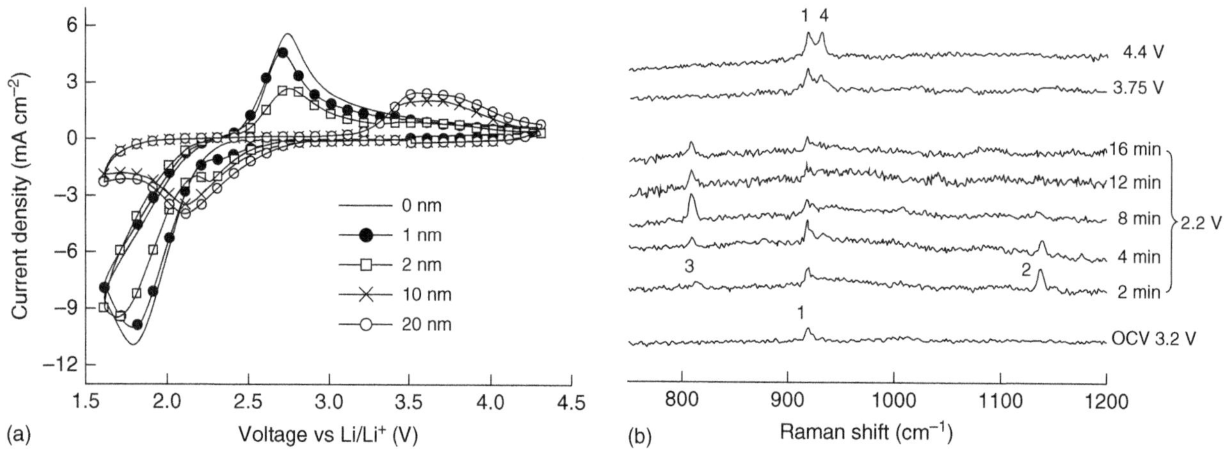


**Figure 2** Cyclic voltammetry (CV) and surface-enhanced Raman spectroscopy (SERS) during $O_2$ reduction and re-oxidation on Au. (a) CV voltammograms of Au electrode in $O_2$-saturated 0.1 M $Bu_4NClO_4/CH_3CN$ containing various concentrations of $LiClO_4$ as indicated. The scan rate was $1.0\ V\ s^{-1}$. (b) SERS spectra performed in $O_2$-saturated 0.1 M $LiClO_4/CH_3CN$, collected at a series of times and at the reduction potential of 2.2 V versus Li/Li$^+$ followed by other spectra at the oxidation potentials shown. The peaks are assigned as follows: (1) C–C stretch of $CH_3CN$ at 918 cm$^{-1}$, (2) O–O stretch of $LiO_2$ at 1137 cm$^{-1}$, (3) O–O stretch of $Li_2O_2$ at 808 cm$^{-1}$, (4) Cl–O stretch of $ClO_4^-$ at 931 cm$^{-1}$. [(b) Reproduced with permission from Peng, Z.; Freunberger, S. A.; Hardwick, L. J.; Chen, Y.; Giordani, V.; Bardé, F.; Novák, P.; Graham, D.; Tarascon, J. M.; Bruce, P. G., Angew. Chem. Int. Ed 2011, 50, 6351–6355. © Wiley-VCH, 2011]

oxide ($Li_2O$, Equation 5) is not kinetically favored within the typical cutoff cathodic potentials of >2 V, which is supported by the absence of any evidence pointing to overall $O_2$ (O=O) bond cleavage at room temperature.[7]

$$O_2 + 4Li^+ + 4e^- \rightleftharpoons 2Li_2O, \quad E^o = 2.91\ V\ \text{vs}\ Li/Li^+ \qquad (5)$$

The anodic CV response in the presence of sufficient $Li^+$ exhibits a broad feature spanning a range of oxidative potentials (Figure 2a). This indicates a complicated and multi-step OER process in decomposing $Li_2O_2$. In addition, $LiO_2$ intermediate was not detected by in situ SERS during the anodic reaction, which indicates a preference for the direct 2e$^-$ transfer pathway (Equation 6) over sequential 1e$^-$ transfer steps.

$$Li_2O_2(s) \rightarrow 2Li^+ + O_2(g) + 2e^- \qquad (6)$$

Most significantly, the considerable shift in anodic potential in comparison to the reversible $O_2/O_2^-$ potential reveals a large overpotential (defined as the difference between the measured potential and the thermodynamic potential ($E^o$= 2.96 V vs Li/Li$^+$)), indicating that $Li_2O_2$ decomposition is more demanding.

### 2.2 Application of Li–$O_2$ Electrochemistry to Batteries

Li–$O_2$ electrochemistry was first applied in nonaqueous battery cells by Abraham and Jiang in 1996 and has since been widely studied from the standpoints of science and engineering.[4] A typical Li–$O_2$ cell configuration utilized at research laboratory scale is shown in Figure 3(a). The cell is composed of a Li negative electrode||separator with electrolyte solution||porous positive electrode, all sandwiched between two current collectors. The $O_2$ gas must have facile access to the positive electrode where ORR and OER occur, which is typically supplied from a $O_2$ tank attached to the cell or continuously flows in/out of the cell by means of an external supply. For the positive electrode, generally carbonaceous materials have been employed owing to their attractive characteristics relating to cost, weight, abundance, porosity, conductivity, ease of electrode fabrication, and surface property tunability. The nonaqueous electrolyte medium should possess chemical and electrochemical robustness along with sufficient ionic conductivity. In this regard, ether-based glymes containing 0.5–1 M Li salt, such as LiTFSI ($LiN(SO_2CF_3)_2$), have been widely employed as the electrolyte of choice owing to their comparatively higher tolerance to the presence of nucleophilic $O_2$ species (i.e., $O_2$, $O_2^-$, $O_2^{2-}$) in comparison to carbonate-based electrolytes.

Typical battery cell tests are carried out at a constant current (i.e., galvanostatic mode), resulting in prolonged ORR and OER processes. Discharge entails the continuous deposition and formation of bulk $Li_2O_2$, which is expected to fully decompose during the subsequent recharge. A representative potential profile corresponding to the first discharge–recharge cycle is shown in Figure 3(b) revealing a constant potential plateau throughout the discharge process, while a substantial potential rise emerges upon recharge. The notable polarization observed during recharge, in conjunction with the cyclability challenges,

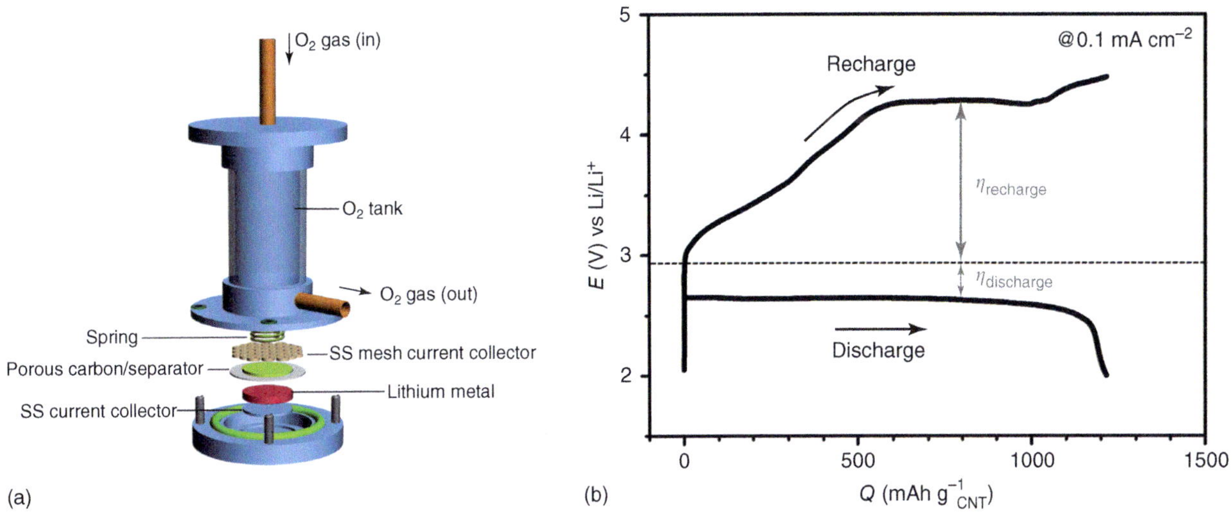


**Figure 3** Li–$O_2$ cell configuration and galvanostatic discharge–recharge profile. (a) Typical Li–$O_2$ cell configuration containing metallic Li negative electrode, porous carbon positive electrode and other components (current collectors, separators) and ether-based electrolyte. The top portion hermetically seals the cell and contains the $O_2$ supply. (b) Representative galvanostatic (constant current) discharge–recharge profile of a Li–$O_2$ cell for the first cycle. The dashed line indicates the thermodynamically reversible potential of 2.96 V versus $Li/Li^+$, and $\eta_{discharge}$ and $\eta_{recharge}$ denote the respective overpotentials. The cell contained a carbon nanotube positive electrode with 0.5 M LiTFSI/tetraglyme as the electrolyte and was discharge–recharged at 0.1 mA $cm^{-2}$

attests that at the present stage, Li–$O_2$ batteries are still far from practical application. The pivotal reason for such poor performance is sluggish decomposition of the insulating $Li_2O_2$ discharge product and occurrence of parasitic side reactions, which to a certain degree can be mitigated through the structural engineering of $Li_2O_2$ and use of promising catalysts (Section 3). In this section, we elaborate on the intrinsic physical properties of $Li_2O_2$ and the process of nucleation, growth, and decomposition during discharge–recharge.

### 2.2.1 *Structure and Electronic Properties of Crystalline $Li_2O_2$*

Crystalline $Li_2O_2$ at ambient conditions belongs to the $P6_3/mmc$ space group. The O–O bond length is approximately 1.56 Å, and the two Li are inequivalent, occupying sites with trigonal and octahedral coordination, respectively.[8,9] $Li_2O_2$ has wide-bandgap and p-type semiconducting characteristics. In assessing the possible mechanisms for charge transport in $Li_2O_2$, Radin *et al.* identified negative lithium vacancies ($V_{Li}^-$, i.e., absence of $Li^+$) and hole polarons ($p^+$, self-trapped at the oxygen dimer) as the dominant charge carriers for ionic and electrical conduction, respectively.[10] These two charge carriers were predicted to exhibit different directional dependencies. Although the $V_{Li}^-$ migration (i.e., ionic conductivity) was assumed to be largely isotropic, the $p^+$ hopping (i.e., electrical conductivity) was predicted to be anisotropic. Nevertheless, the magnitude of the conductivity for these charge carriers in estimated in the range $10^{-19}$–$10^{-20}$ S $cm^{-1}$ (with concentrations in the order of $10^7$ $cm^{-3}$), indicating that the pure bulk material is essentially insulating (in a similar range as experimental values of conductivity for Teflon). Further work[11] considered the reasonable suggestion that electrochemically generated, micron order $Li_2O_2$ contains defects and possible doping, which enhance ionic and electrical conductivity. Moreover, noncrystalline (i.e., amorphous) $Li_2O_2$ was estimated to exhibit a substantially increased ionic conductivity of 12 orders of magnitude and relatively modest increase in electrical conductivity of up to four orders of magnitude.[12] These findings provide a possible mechanism to suppress the large polarization, and the related experimental efforts and progress is introduced in Section 3. Here, we further take the opportunity to note that experimental data to complement and confirm theoretical calculations on the structure and charge transport properties of $Li_2O_2$ have been limited by experimental difficulty in preparing this material with rigorous purity.

### 2.2.2 *$Li_2O_2$ Nucleation and Growth during Discharge*

The nucleation and growth process largely determines the properties and characteristics of the generated $Li_2O_2$ and thus has a significant impact on battery cell performance. It has been well established that the principle discharge product is $Li_2O_2$, which forms at a constant rate during discharge. Representatively, in situ X-ray diffraction (XRD) analysis reveals a linear increase in peak intensity

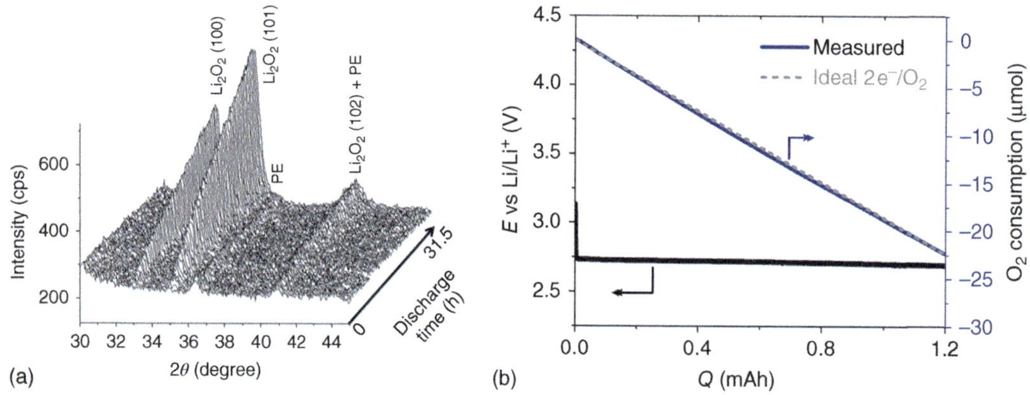


**Figure 4** In situ X-ray XRD and gas analysis during discharge. (a) In situ XRD during discharge in the 2$\theta$ region showing the $Li_2O_2$ (101), (100), and (102) reflections. The cell consisted of a carbon (Super P) electrode and 0.5 M LiTFSI/triglyme at a current rate 0.1 mA $cm^{-2}$. [Reprinted with permission from Lim, H.; Yilmaz, E.; Byon, H. R., J. Phys. Chem. Lett. 2012, 3, 3210–3215. © American Chemical Society, 2012]; (b) In situ pressure monitoring during discharge. The black and blue solid lines indicate discharge potential and the corresponding gas consumption, respectively. The dashed line denotes the ideal $2e^-/O_2$. The positive electrode consists of a binder-free multi-walled carbon nanotube electrode. The cell test was performed with 0.5 M LiTFSI in tetraglyme at a current rate of 50 mA $g_{carbon}^{-1}$ and fixed capacity of 1.2 mAh

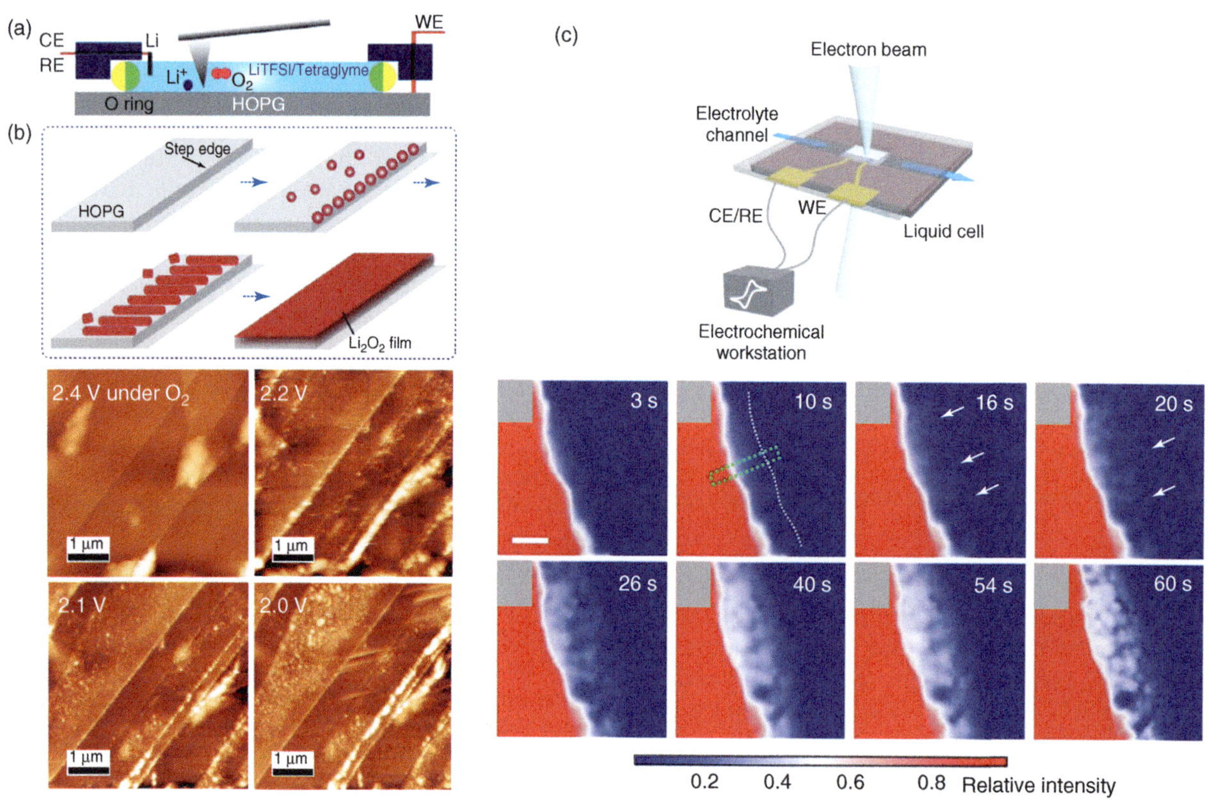


**Figure 5** In situ imaging of the evolution of $Li_2O_2$ morphology during formation using EC-AFM and HAADF-STEM. (a) EC-AFM setup and cell configuration utilizing 0.5 M LiTFSI/tetraglyme with HOPG as the working electrode. (b) Schematic view and AFM images showing the progression of $Li_2O_2$ formation during reduction. [Reprinted with permission from Wen, R.; Hong, M.; Byon, H. R., J. Am. Chem. Soc. 2013, 135, 10870–10876. © American Chemical Society 2013]; (c) Sequential HAADF-STM imaging during discharge of a liquid-cell containing 1.0 M $LiClO_4$/DMSO at a current rate of 1.3 μA $cm^{-2}$. The enhanced contrast from the electrode surface to the dotted line in the top row, suggests the presence of charged or polarized species, while further discharge in the bottom row shows solid nanoparticles constituting a $Li_2O_2$ film-like structure. [Source: Liu [16], https://www.ncbi.nlm.nih.gov/pubmed/29453422. Licensed under CC BY 4.0]

with the $Li_2O_2$ (101) and (100) reflections (Figure 4a).[13] In situ pressure monitoring also consistently identifies that the cumulative $O_2$ gas consumption equates to a $2e^-$ transfer process throughout discharge (Figure 4b).[7,14]

The in situ visualization of the $Li_2O_2$ nucleation and growth processes has been carried out with various microscopy techniques. The electrochemical atomic force microscopy (EC-AFM) technique was employed for in situ imaging with a liquid-based model cell at ambient conditions, which is analogous to the cell environment of typical Li–$O_2$ cells[15] (Figure 5a). Upon cathodic reduction, nucleation takes the form of nanoparticulate seeds (3–10 nm in size) that align along the edge plane of a highly orientated pyrolytic graphite (HOPG) working electrode (Figure 5b). The particles then migrate throughout the HOPG surface followed by lateral growth and film formation at 2.0–2.1 V. Aside from EC-AFM, which provides a viewpoint which is perpendicular to the electrode surface, high-angle annular dark field scanning transmission electron microscopy (HAADF-STEM) analysis has revealed characteristics of the same process across the electrode/electrolyte interface.[16] Sequential snapshots in a liquid cell show the $Li_2O_2$ nanoparticles forming via the electrolyte solution (Figure 5c). These in situ monitoring results suggest the pivotal roles of the electrode surface and electrolyte solution in the $Li_2O_2$ nucleation and growth processes, which results in different structural characteristics of the $Li_2O_2$ formed (vide infra).

In general, two morphologies of $Li_2O_2$ can form with different structural characteristics upon prolonged discharge: (i) toroidal particles of micrometer order with high crystallinity and, (ii) film-like with lower crystallinity and nanometer order thickness (Figure 6a and b). The factors that determine the structure are highly related to the discharge conditions including solvation ability of the electrolyte solution, surface properties of the electrode and discharge current density.[18] A prevalent solvation ability of the electrolyte solution can lead to the favored dissolution of the $LiO_2$ intermediate that is formed after the first electron transfer of $O_2$ (Equation 2, Figure 6c). The dissolved $LiO_2$ supersaturates in the electrolyte solution and nucleates, and thus seeds the $Li_2O_2$ via disproportionation (Equation 3).[19] After the $Li_2O_2$ is seeded on the electrode, growth continues from the supply of dissolved $LiO_2$. In this aptly-named, solution-mediated pathway, the overall rate of nucleation and growth is determined

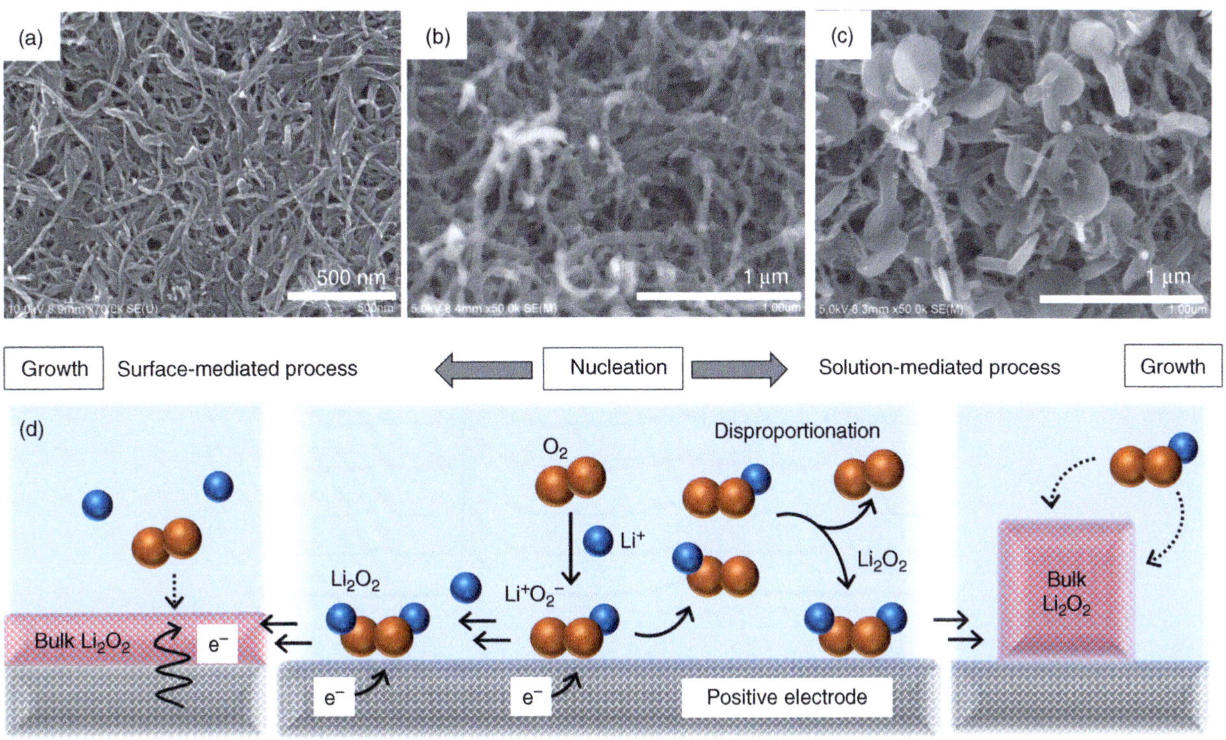


**Figure 6** $Li_2O_2$ formation pathways. (a–c) SEM images of (a) pristine CNT, and after discharge in 0.5 M $LiClO_4$/tetraglyme (b) containing <20 ppm $H_2O$ yielding film-shaped $Li_2O_2$ and (c) containing 2000 ppm $H_2O$ yielding toroidal $Li_2O_2$ (d) sketch illustrating the solution-based and surface-based pathways of $Li_2O_2$ nucleation and growth during discharge. [(d) Reprinted with permission from Yang, C.; Wong, R. A.; Hong, M.; Yamanaka, K.; Ohta, T.; Byon, H. R., Nano Lett. 2016, 16, 2969–2974. © American Chemical Society 2016]

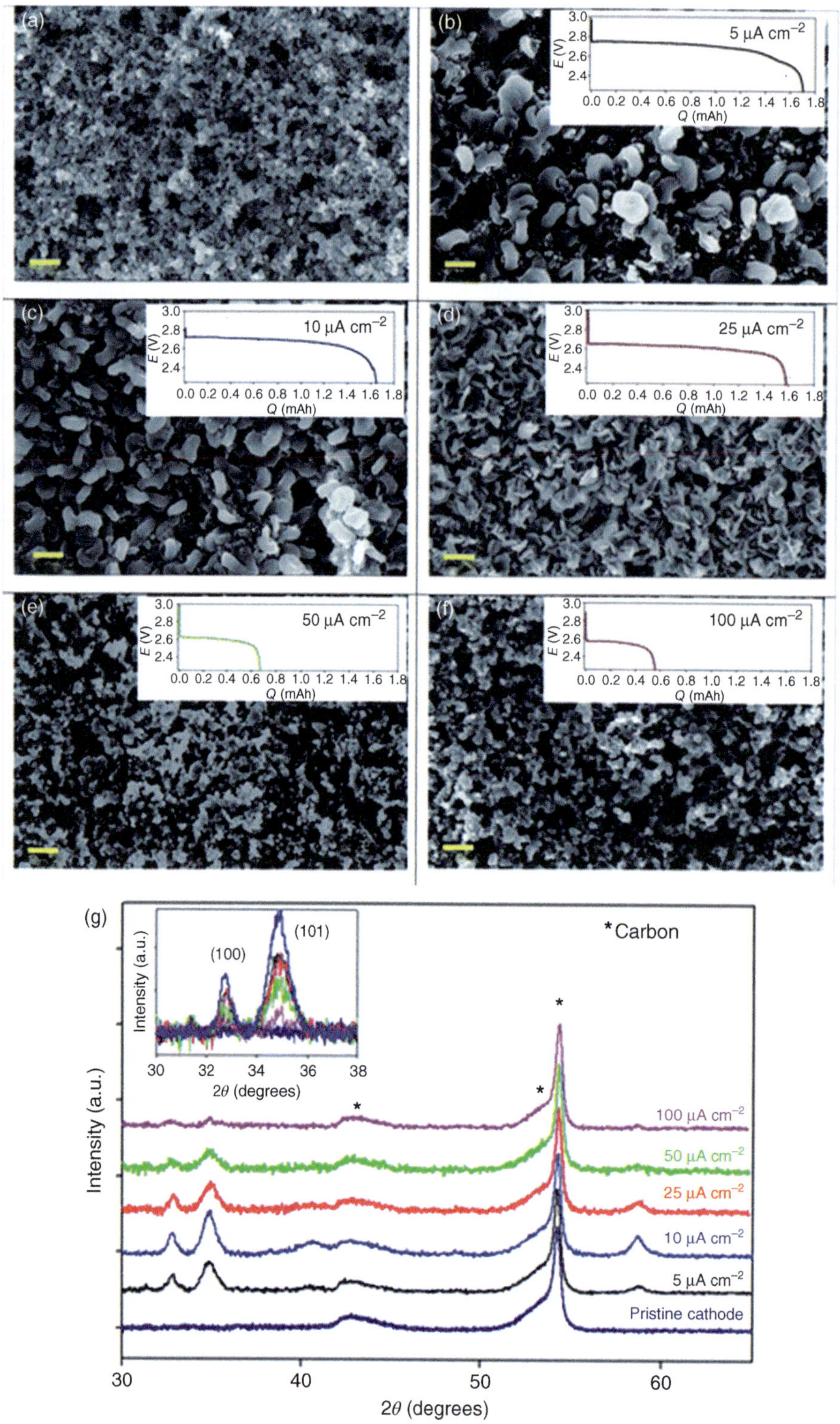


**Figure 7** Effect of current density on $Li_2O_2$ structural characteristics. SEM images of (a) the pristine positive electrode and after full discharge at (b) 5 μA cm$^{-2}$, (c) 10 μA cm$^{-2}$, (d) 25 μA cm$^{-2}$, (e) 50 μA cm$^{-2}$, and (f) 100 μA cm$^{-2}$, with the corresponding discharge curves (inset). Scale bar = 400 nm. (g) XRD patterns after full discharge at the corresponding current rates. The asterisk symbol indicates peaks attributed to the carbon electrode. The inset displays a close-up of the (100) and (101) reflections of $Li_2O_2$. [Reproduced with permission from Adams, B. D.; Radtke, C.; Black, R.; Trudeau, M. L.; Zaghib, K.; Nazar, L. F., Energy Environ. Sci. 2013, 6, 1772–1778. © Royal Society of Chemistry]

from the mass transport of $LiO_2$, thus is expected to be a slow process to allow for the formation of toroidal $Li_2O_2$ particles with large dimensions (up to the micrometer regime) and with high crystallinity. In contrast, the $LiO_2$ intermediate can be confined to the electrode if the electrode surface has a prevalent adsorption affinity for the $LiO_2$ and thereby nullifies solvation. This is termed the surface-mediated pathway and leads to $Li_2O_2$ seeding via a second electron-transfer step (Equation 4), Figure 6d) and growth continues as a thin film that is deposited along the electrode surface with lower crystallinity.

The two abovementioned pathways are also dependent on the discharge current density and overpotential. Low current density or low overpotentials favor $LiO_2$ dissolution and result in large toroidal $Li_2O_2$ with higher capacity (Figure 7a–d). On the contrary, a rapid discharge process at high overpotential prompts the formation of a thin $Li_2O_2$ film leading to lower capacity (Figure 7e and g).[18,20,21]

It is important to note that both the solution- and surface-mediated pathways generally coexist and compete with each other to varying degrees at a given discharge condition. Even when toroidal $Li_2O_2$ particles are predominantly formed, a thin-film product is also observed together on the electrode surface. On the other hand, in situations where the surface-mediated pathway is dominant, the thickness of $Li_2O_2$ film is typically tens of nanometer which is thicker than the electron-tunneling distance of <10 nm.[22] This indicates the contribution of dissolved $LiO_2$ to some degree contributes to the continuous deposition of $Li_2O_2$ attained via disproportionation.

### 2.2.3 *$Li_2O_2$ Decomposition during Recharge*

The decomposition of $Li_2O_2$ exclusively occurs via charge transport in the hypothetically insulating $Li_2O_2$ since there is no solution-mediated pathway. This leads to sluggish decomposition and observations of high recharge potential reaching over 4.0 V. The decomposition behavior of $Li_2O_2$ is contingent on its structural characteristics. In situ XRD conclusively shows a nonlinear decomposition rate of the $Li_2O_2$ crystalline reflections (Figure 8a).[13,23] The initial 50% state of recharge (SOC) exhibits a modest decrease in the $Li_2O_2$ reflection peaks, while the decrease rate in the latter 50% process is more rapid as sufficient overpotential and recharge time have passed. At the end of recharge (to the equivalent discharge capacity), the complete disappearance of $Li_2O_2$ reflections is observed. On the other hand, different insights can be gained from in situ gas analysis. The $O_2$ gas evolution as characterized by on-line electrochemical mass spectrometry (OEMS), or with a comparable technique referred to as differential electrochemical mass spectrometry (DEMS), shows a broadly comparable quantity of $O_2$ gas in both the initial and latter 50% SOC (Figure 8b).[14] In particular, considerable $O_2$ evolution occurs from <3.7 V, indicating the occurrence of significant $Li_2O_2$ decomposition during the initial stages. Taken together, these results infer that facile decomposition initiates from the defect-dominant and noncrystalline $Li_2O_2$ at lower overpotential, which cannot be detected in the XRD measurement, followed by the slower elimination of crystalline $Li_2O_2$ at higher overpotential. At over 70% SOC, unintended $CO_2$ gas evolution is observed as the $O_2$ gas evolution rate diminishes. The origin of $CO_2$ gas

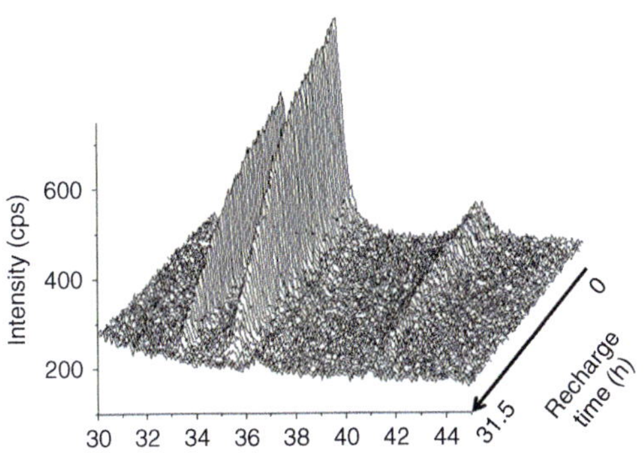


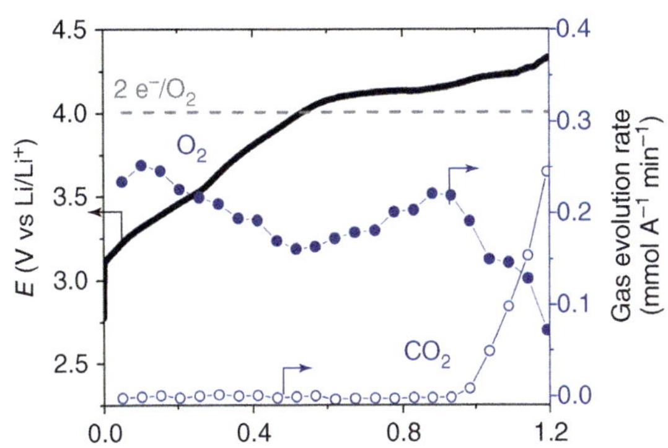


**Figure 8** In situ XRD and gas analysis during recharge. (a) In situ XRD during recharge in the $2\theta$ region showing the $Li_2O_2$ (101), (100), and (102) reflections. The cell consisted of a carbon (Super P) electrode and 0.5 M LiTFSI/triglyme at a current rate of 0.1 mA $cm^{-2}$. Recharge was performed to the equivalent capacity of discharge. [Reprinted with permission from Lim, H.; Yilmaz, E.; Byon, H. R., J. Phys. Chem. Lett. 2012, 3, 3210–3215. © American Chemical Society, 2012]; (b) In situ OEMS during recharge. The black solid line indicates the recharge potential, and the blue colored closed and open circles denote $O_2$ and $CO_2$ gas evolution, respectively. The dashed line denotes the ideal $2e^-/O_2$ gas evolution rate. The positive electrode consists of a binder-free multi-walled carbon nanotube electrode. The cell test was performed with 0.5 M LiTFSI/tetraglyme at a current rate of 50 mA $g_{carbon}^{-1}$ following discharge to the fixed capacity of 1.2 mAh. [Reprinted with permission from Wong, R. A.; Yang, C.; Dutta, A.; O, M.; Hong, M.; Thomas, M. L.; Yamanaka, K.; Ohta, T.; Waki, K.; Byon, H. R., ACS Energy Lett. 2018, 3, 592–597. © American Chemical Society 2018]

production is the oxidation of lithium carbonate (e.g., $Li_2CO_3$) side products, which form during recharge from the degradation of electrolyte solution and carbonaceous electrode (Section 3). The electrons per oxygen molecule ($e^-/O_2$) metric exceeds the ideal $2e^-/O_2$ (dashed line in Figure 8b) throughout recharge and in particular, a notable increase in the $e^-/O_2$ is observed at >3.7 V, demonstrating the contribution of unintended oxidation reactions. Overall, the total $O_2$ gas recovered from recharge is only ~60% relative to the $O_2$ consumption from discharge.[14,24,25]

With an aim of alleviating the high polarization, the initial $Li_2O_2$ decomposition process occurring at low overpotential has been studied with in situ imaging techniques, which can give insights into identifying the

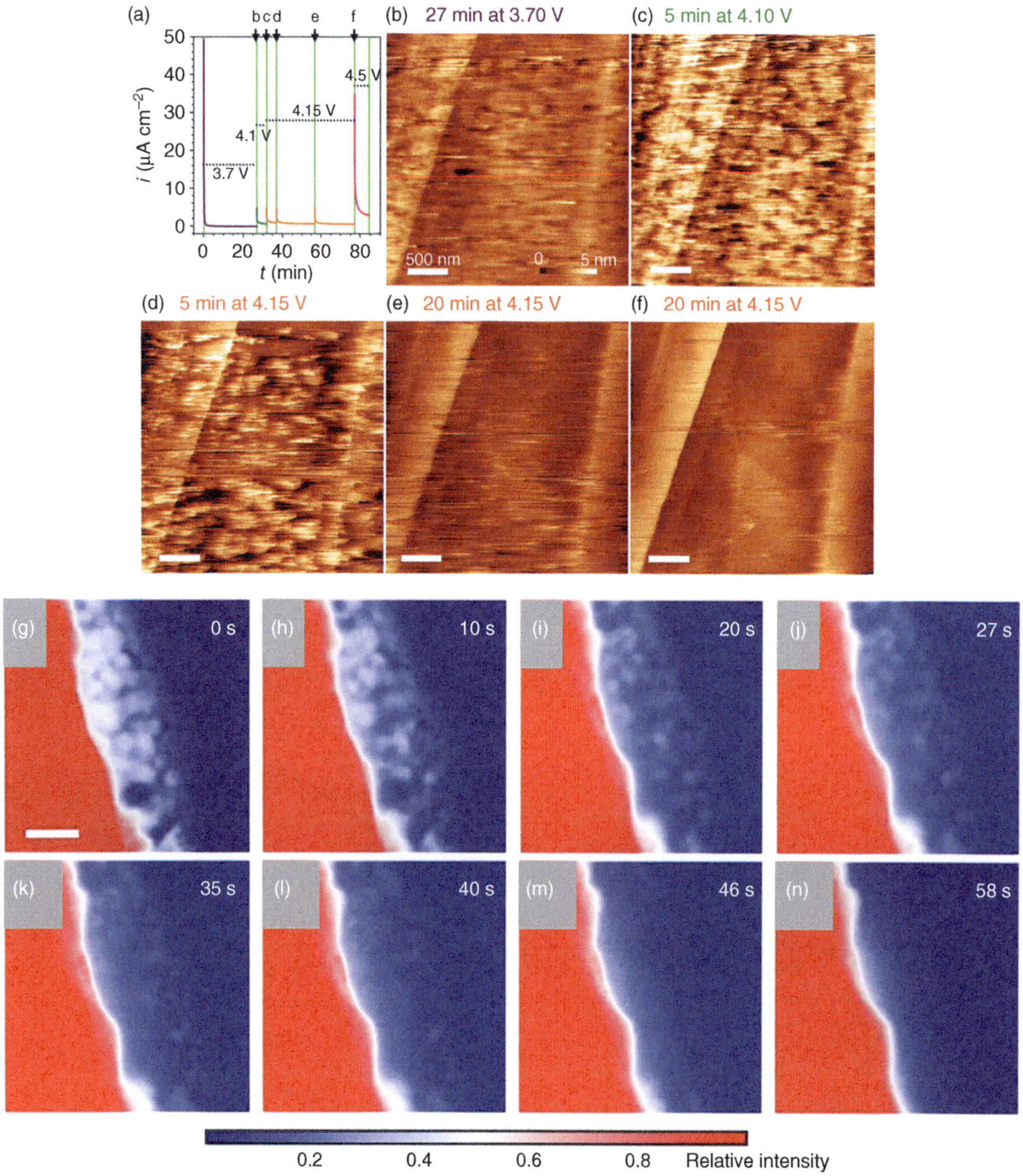


**Figure 9** In situ probing of morphology changes during $Li_2O_2$ oxidation. (a) Potentiostatic profile corresponding to the EC-AFM images in (b–f) taken after holding at the time and anodic potentials indicated utilizing 0.5 M LiTFSI/tetraglyme. Prior to the application of anodic potentials, cathodic reduction was performed under potentiostatic conditions at 2.2 V. (g–n) Sequential in situ HAADF-TEM imaging during recharge in a liquid-cell containing 1.0 M $LiClO_4$/DMSO at a current rate of 1.3 μA $cm^{-2}$. [(g–n) Source: Hong [26], https://www.nature.com/articles/s41598-018-21503-w#rightslink. Licensed under CC BY 4.0]

facile oxidation route in $Li_2O_2$. EC-AFM reveals that the decomposition of $Li_2O_2$ film begins at the thinner parts (ca. ~2 nm thickness), which forms small pits at <3.7 V (Figure 9a and b).[26] When the potential is over 3.7 V, the pits develop into cracks in the $Li_2O_2$ film, and the lateral-direction shrinkage of $Li_2O_2$ occurs before its complete elimination (Figure 9c–f). In addition, in situ HAADF-STEM shows the initial decomposition of $Li_2O_2$ occurring in the vicinity of electrode surface (Figure 9g–n).[16] These findings suggest accessibility to a facile electron-transfer route in the initial stage of recharge.

In addition, a plausible explanation for the facile and inhomogeneous decomposition is the elimination of $Li^+$, termed delithiation, from the $Li_2O_2$ surface yielding $Li_{2-x}O_2$, and subsequent $O_2$ evolution (Equations 7 and 8):

$$Li_2O_2 \rightarrow Li_{2-x}O_2 + xLi^+ + xe^- \quad (7)$$

$$Li_{2-x}O_2 \rightarrow O_2 + (2-x)(Li^+ + e^-) \quad (8)$$

Computational simulations envision fast kinetics for delithiation that is only accessible at low overpotentials (<400 mV), while the typical $2e^-$ transfer (Equation 6) with negligible contribution of delithiation proceeds at higher potential.[27,28] Experimentally, the quasi-equilibrium state of $Li_{2-x}O_2$ has been suggested from electrochemical intermittent titration techniques and Rietveld refinement.[23] In addition, according to Equation (7) with $x = 1$, $LiO_2$ can also be formed at the $Li_2O_2$ surface. The dissolved $LiO_2$ is indeed detected exclusively at a low potential (~3.55 V) using the rotating ring-disk electrode (RRDE) technique, albeit at low amounts.[26] In all, such facile decomposition is likely to occur from smaller-sized and highly defective (less crystalline) $Li_2O_2$, which is essential to enhance the proportion of evolving $O_2$ gas, decrease the large polarization and suppress side reactions.

## 3 MAIN CHALLENGES AND RESEARCH EFFORTS TO IMPROVE PERFORMANCE IN Li–$O_2$ BATTERIES

The key challenges relating to Li–$O_2$ cell performance manifest themselves in the limitation of discharge capacity, high recharge overpotential (>1 V), and poor cyclability (typically <100 cycles even under optimized conditions). These issues originate from the insulating properties of $Li_2O_2$ and presence of highly reactive $O_2$ species ($O_2^-$, $O_2^{2-}$, and singlet $O_2$) formed during the discharge and recharge processes. The insulating $Li_2O_2$ hinders electron transfer from the electrode, which provokes the sudden drop in potential and termination of discharge. The sluggish charge transfer with $Li_2O_2$ is the primary cause for an increase in recharge overpotential. Moreover, the reactive $O_2$ species and high recharge potential induce the degradation of the electrolyte solution and carbonaceous electrode. To understand and mitigate these issues, significant research progress has been made and we discuss the origin of these challenges and the research approaches taken to improve Li–$O_2$ performance.

### 3.1 Limitation of Discharge Capacity

It was initially assumed that the discharge capacity in Li–$O_2$ cells was governed by the available pore space of the electrode. The limiting factor in such a scenario is the slow $O_2$ mass transport within the electrolyte solution, leading to pore blockage by $Li_2O_2$ resulting in the termination of discharge.[21,29] However, it is now understood that the most critical factor that governs capacity is slow charge transport in the wide-bandgap $Li_2O_2$.[21,22] The termination of the galvanostatic discharge profile is exemplified by a sudden drop in potential just prior to the end of discharge, in what is commonly referred to as "sudden death" (Figure 3b). Through experiments and modeling, Lau and Archer showed that sudden death is caused by the elimination of accessible surface area of electrode due to the coverage of insulating $Li_2O_2$.[21] The fact that the capacity decreases with increasing current density signifies the prominence of the surface-mediated $Li_2O_2$ growth pathway that rapidly passivates the entire electrode surface, thus blocking the electron-transfer route.

#### *3.1.1 Electrolyte Solution Engineering*

Owing to adverse effects of $Li_2O_2$ passivation through the surface-mediated pathway, a viable approach to extend discharge capacity is to promote solution-mediated $Li_2O_2$ nucleation and growth.[2] A good solvation ability toward $LiO_2$ is a pivotal prerequisite for the solution-mediated pathway and is dependent on the electron withdrawing and donating properties of the electrolyte solution. Systematic studies with various aprotic solvents and electrolyte salts have established that the solvents' Gutmann acceptor numbers (ANs) and donor number (DNs) can be used as descriptors. High DN number solvents are generally more able to stabilize $Li^+O_2^-$ due to the favorable interaction of $Li^+$ associated with $Li^+O_2^-$.[30] As a representative example, at moderate current densities, dimethyl sulfoxide (DMSO) with a DN of 30 yields a higher discharge capacity in comparison to ether-based glymes and acetonitrile with DNs of 20 and 14, respectively. The evidence for formation of large toroidal $Li_2O_2$ with higher DN solvents also corroborates the predominant solution-mediated pathway.[2,18,30] In a similar manner, the donor ability of the anion of the electrolyte salt also determines the solubility of $LiO_2$. Kwon *et al.* noted that weakly Lewis basic anions such as $TFSI^-$ ($[N(SO_2CF_3)_2]^-$)

and BETI$^-$ ([N(SO$_2$CF$_2$CF$_3$)$_2$]$^-$) may limit the solubility/lifetime of $O_2^-$ from $Li^+O_2^-$, as determined by the assessment of $Li_2O_2$ quantity in the separator (vide infra) and relative rate of chemical disproportionation.[31] On the contrary, anions with higher Lewis basicity, such as TfO$^-$ ($CF_3SO_3^-$), TFA$^-$ ($CF_3CO_2^-$) and $NO_3^-$ tend to enhance the solubility/lifetime.[31] This then leads to higher discharge capacities, with larger toroidal $Li_2O_2$ crystals for the more associative anions, as reported by Aurbach and coworkers[32] which is particularly effective when employed with low DN solvents. It has been speculated that the structure–property relationships for lithium coordination also plays a role, with formation of solvated, long-lived triple ions (i.e., [anion$^-$–$Li^+$–$O_2^-$]$^-$) for a highly associative anion (TfO$^-$).[33]

An alternative method to enhance discharge capacity is the introduction of soluble additives. A protic additive in ppm quantities, such as water or acids having a high AN (AN of 54.8 for water) can sufficiently stabilize $O_2^-$ from $Li^+O_2^-$ in the electrolyte solution, forming large toroidal $Li_2O_2$ particles (Figure 6b and c).[19,34] However, possible protonation of the electrolyte solvent by protic additives and the production of lithium hydroxide (LiOH) at both the positive and negative electrodes should be considered before applying this strategy.[17,34]

Working with a different mechanism, the addition of a redox mediator (RM) (e.g., ethyl viologen or quinone derivatives) promotes solution-phase discharge. For example, the 2,5-di-*tert*-butyl-1,4-benzoquinone (DBBQ) RM is first reduced to DBBQ$^-$ and the [$Li^+$–DBBQ$^-$] complex binds and reduces $O_2$. Analogous to a typical ORR process, both disproportionation and direct electron-transfer (electrochemical) processes form $Li_2O_2$ when DBBQ is regenerated.[35] In effect, all two-electron-transfer processes take place away from the electrode, which considerably enhances the capacity as well as rate capability during discharge.

It is worth noting that despite the enlarged capacity, the solution-mediated pathway can lead to the loss of $Li^+O_2^-$ away from positive electrode and invariably causes an irreversible loss of capacity. The dissolved $Li^+O_2^-$ can be transported to the separator and Li negative electrode, followed by disproportionation, and indeed a significant amount of toroidal $Li_2O_2$ is observed on the separator.[20,31] Nonetheless, exploiting electrolyte engineering to elongate the discharge capacity can potentially be an attractive approach when paired with additional methods such as the use of soluble RMs (see Section 3.2.3) to decompose the large crystalline $Li_2O_2$ during recharge.

### 3.2 High Recharge Overpotential

The large potential polarization is particularly evident during the recharge process and is a critical reason for substandard cell performance with low round-trip efficiency (typically ~70%) and poor cyclability in Li–$O_2$ cells.[36] The central reason for the potential rise is sluggish decomposition of insulating $Li_2O_2$, while this process is accompanied by severe side reactions as the electrolyte solution and carbonaceous electrode are not stable at >3.5 V. To mitigate these possible detrimental scenarios, suppression of recharge potential is vital, and thus studies have focused on increasing the rate of $Li_2O_2$ decomposition. There have been various attempts to improve $Li_2O_2$ decomposition rate and here three overall strategies, structural engineering of $Li_2O_2$, addition of heterogeneous catalysts and RMs are introduced.

#### *3.2.1 Structural Engineering of $Li_2O_2$*

It has now been well established that the morphological and crystalline characteristics of $Li_2O_2$ have a major influence on the overpotential and mechanism for recharge.[18] Typically, the toroidal $Li_2O_2$ shows higher crystallinity and larger dimensions, while the film-like $Li_2O_2$ is paired with lower crystallinity and a smaller dimension (i.e. thin). They exhibit different decomposition characteristics. The film-like $Li_2O_2$ decomposes at lower recharge overpotential than the toroidal particles. Morphologically, the smaller size (tens of nanometer thickness) and larger contact area with the electrode are beneficial for electron transfer from electrode to film-like $Li_2O_2$. In terms of crystallinity, higher ionic and electrical conductivities are present with a more defective, that is, less crystalline structure, thus the limitation of charge transport inside $Li_2O_2$ can to a certain degree be mitigated. The ionic and electrical conductivities of noncrystalline (amorphous) $Li_2O_2$ were computationally calculated to be $10^{-8}$ and $10^{-16}$ S cm$^{-1}$, respectively, which are 12 and 4 orders of magnitude higher than crystalline $Li_2O_2$ which are both in the range of $10^{-19}$–$10^{-20}$ S cm$^{-1}$.[12] In comparison, the experimentally measured ionic and electrical conductivities of pelletized powders of chemically synthesized amorphous $Li_2O_2$ are around $7 \times 10^{-8}$ and $5 \times 10^{-9}$ S cm$^{-1}$, respectively. These conductivities equate to an enhancement of two to three orders of magnitude over pelletized powders of commercial crystalline $Li_2O_2$. The significantly larger values measured from $Li_2O_2$ than theoretically predicted values (see Section 2.2.1) are presumably due to poor purity.[37]

$Li_2O_2$ that is less-crystalline and film-shaped can be formed in situations where the surface-mediated pathway is favored and can be manipulated based on the properties of the electrolyte solution and electrode surface. The combination of a weakly Lewis basic anion and low DN solvent can suppress the dissolution of $LiO_2$. In addition, increasing the concentration of $Li^+$ salt can limit the solubility/lifetime of the $LiO_2$ as the proportion of free solvent that is available for solvation decreases.[25,38] However, even with similar electrolytes and similar galvanostatic test conditions (i.e., moderate current density), there have been conflicting reports of either toroidal or film-like $Li_2O_2$. One possibility for this discrepancy is the ingress of $H_2O$ impurities arising from the cell components and the possibility of cell leakage. To ensure that the surface-mediated pathway is the predominant route, an attractive approach is *anchoring* or confining $O_2$ species ($O_2$, $LiO_2$, etc.) to the electrode surface.

The *anchor* should have a stronger affinity for $O_2$ and $LiO_2$ adsorption to counter the propensity of the solvation of $LiO_2$. $RuO_2$ and $CeO_2$ have been shown to be appropriate candidates owing to the presence of oxygen vacancies. During discharge, initial $Li_2O_2$ nucleation occurs within the vicinity of the $RuO_2$ and $CeO_2$ nanoparticles, followed by the lateral growth along the carbon nanotube (CNT) support, which results in the formation of film-like, defective $Li_2O_2$.[17,39] Even upon increasing the $H_2O$ concentration (up to ~2000 ppm) causing a switch to the solution-mediated pathway, the film shape of $Li_2O_2$ is sustained with the $CeO_2$/CNT electrode while this is met with insignificant changes in crystallinity and a moderate increase in discharge capacity. This is in stark contrast to the appearance of toroidal and crystalline $Li_2O_2$ with the bare CNT electrode (absence of $CeO_2$), which is accompanied with a substantial increase in capacity. On the subsequent recharge, the film-shaped $Li_2O_2$ evolves a greater amount of $O_2$ gas at low recharge potential (<3.7 V) in comparison to toroidal $Li_2O_2$, indicating the facile decomposition of film-shaped $Li_2O_2$.

Further studies have shown that it is possible to achieve a similar effect with carbonaceous electrodes without heavy and expensive transition metal oxides, by introducing anchoring moieties to the electrode surface. For instance, the provision of oxygen functional groups such as carbonyl, epoxy, ester, and carboxyl moieties onto the CNT electrode leads to the formation of entirely amorphous $Li_2O_2$ films, in contrast to highly crystalline $Li_2O_2$ particles produced from highly uniform and graphitized CNT.[40] It is apparent that the lower recharge overpotentials from amorphous $Li_2O_2$ film is accompanied by a greater amount of $O_2$ evolution during the initial stage of recharge. However, the presence of oxygen functional groups on the CNTs may initiate unintended side reactions at >3.5 V, manifesting into the low observed cyclability.

It has been shown that forming ultrathin and one-dimensional (1-D) nanostructures of amorphous $Li_2O_2$ notably enhances the decomposition rate (Figure 10).[41] The low volume of this $Li_2O_2$ nanostructure and amorphousness improve charge transport while the enlarged surface area of 1-D $Li_2O_2$ in contact with the electrolyte solution in all directions critically contributes to enable

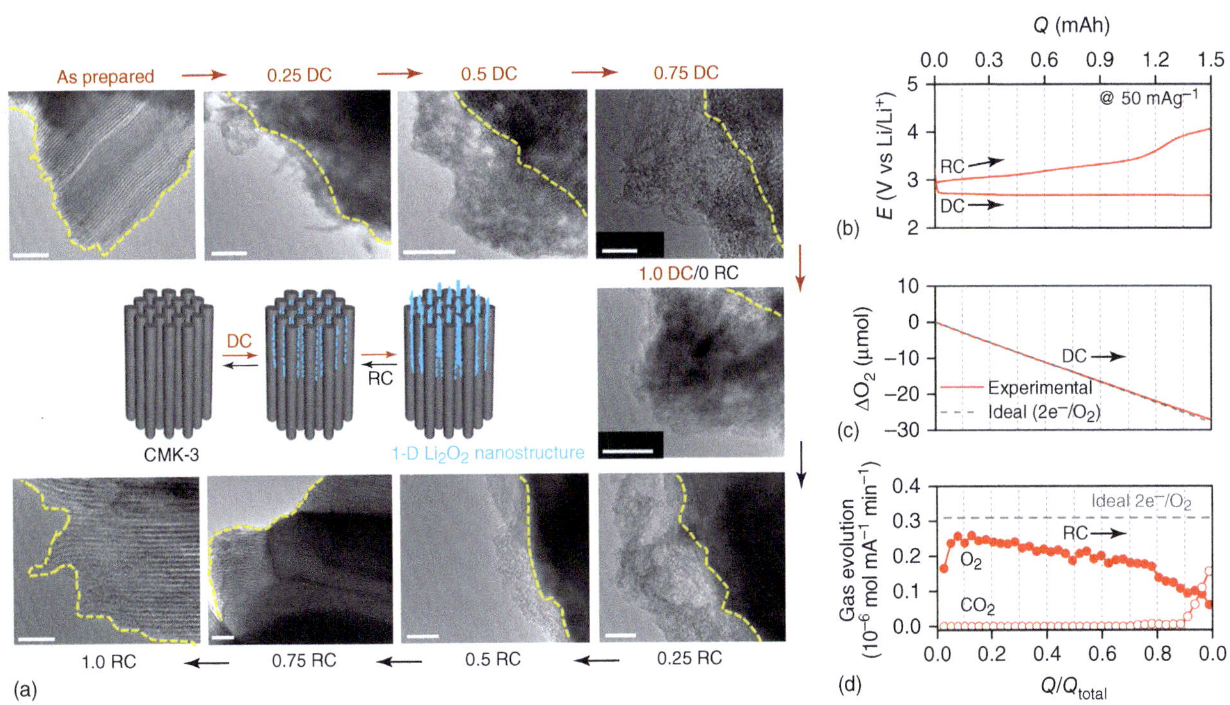


**Figure 10** 1-D nanostructured $Li_2O_2$ formed from a mesoporous CMK-3 electrode. (a) TEM images of $Li_2O_2$ at sequential depth of states for discharge (DC) and recharge (RC). The dashed yellow lines are guides to indicate the CMK-3 surface. The schematic illustration depicts formation and decomposition of 1-D $Li_2O_2$. The scale bars indicate 50 nm. (b) Discharge–recharge profile in 0.5 M LiTFSI/tetraglyme at 50 $mAg^{-1}$. (c) In situ monitoring of $O_2$ consumption during discharge. (d) In situ OEMS during recharge with the solid and open dots denoting $O_2$ and $CO_2$ evolution, respectively. [(d) Reproduced with permission from Dutta, A.; Wong, R. A.; Park, W.; Yamanaka, K.; Ohta, T.; Jung, Y.; Byon, H. R., Nat. Comm. 2018, 9, 680. © Nature publishing group 2018]

facile delithiation. Such a unique structure of $Li_2O_2$ can be formed when a mesoporous carbon electrode of CMK-3 is utilized. The $Li_2O_2$ is locked in place with its dimensions and shape defined by the hexagonal and mesoporous channels of CMK-3 where $O_2$ and $Li^+$ are confined. The resulting 1-D $Li_2O_2$ forms over the CMK-3 surface presumably due to the fast clogging of mesopores by $Li_2O_2$ (Figure 10a). The $Li_2O_2$ smoothly decomposes below 3.7 V contributing to a round-trip efficiency of ~80% (Figure 10b–d). This achievement is a noteworthy as being able to reduce the polarization in the absence of catalysts that can often provoke further side reactions (see Sections 3.2.2 and 3.2.3). In all, these approaches have demonstrated a profound suppression of recharge overpotential in $Li-O_2$ cells through the structural engineering of $Li_2O_2$.

### 3.2.2 *Heterogeneous Electrocatalysis to Reduce Recharge Overpotential*

Driven by the necessity to suppress the large recharge overpotentials, researchers have explored the use of heterogenous electrocatalysts such as nanostructured precious metals (Pt, Pd, Au, and Ru) or metal oxides ($RuO_2$, $MnO_2$, $Co_3O_4$, etc.) that are typically decorated onto a carbon support with loadings of 20–40 wt%.[36,42] The inspiration behind catalysts is largely based on vast amount of existing work on aqueous OER catalysis.[43] However, the fundamental difference from aqueous OER is that as discharge proceeds, the catalyst surface is passivated by insoluble $Li_2O_2$. This can result in a different working mechanism toward an increase in OER kinetics from typical heterogeneous electrocatalysts, and thus the term of "promoter" has been often used to reflect this difference.[44] While the precise enhancement mechanism remains unresolved and is an active area of research, the vast majority of reports on catalysts have shown suppressed recharge overpotentials, often coupled with enhanced cyclability.[42]

Experimental and computation studies have shown that catalysts can alter the $Li_2O_2$ morphology and crystallinity since these catalysts affect the binding affinity with the $O_2$ species in comparison to the unaltered carbonaceous electrode (Section 3.2.1).[18,39] Yilmaz *et al.* reported that the use of $RuO_2$ nanoparticles on CNT promotes the formation of film-like and defective $Li_2O_2$ resulting in a low recharge potential of ~3.5 V.[39] However, the $Li_2O_2$ structure is not the sole reason for the low potential which is maintained throughout the recharge process because when a similar $Li_2O_2$ structure is formed with $CeO_2$ nanoparticles on CNT, the potential rises to >4.0 V after the initial stage of recharge.[17] There have also been other proposed mechanisms involving metal ion dissolution from the nanocatalysts and the interplay with $Li_2O_2$. Schwenke *et al.* suggested that dissolution of metal ions from nanocatalysts occurs, with the dissolved metal ions possibly acting as a RM,[45] although the existence of soluble metal ions is still being debated. By computational simulations, Radin *et al.* predicted that the doping of metal ions from catalysts that are in close proximity to $Li_2O_2$ can improve the electrical/ionic conductivity providing a means for facile decomposition.[11] Furthermore, Black *et al.* and Lu and Shao-Horn proposed that catalysts promotion the formation and transport of lithium deficient $Li_2O_2$ ($Li_{2-x}O_2$).[44,46] It has also been reported that iridium (Ir) nanocatalysts on graphene electrode can stabilize $LiO_2$ resulting in $LiO_2$-based batteries that operate with an overall $1e^-$ transfer.[47] Although the evidence of stable, long-lived $LiO_2$ has been debated,[48] this battery has shown surprisingly low recharge potential of ~3.3 V for over 40 cycles.

To elucidate the role of nanocatalysts and to find a correlation between recharge potential and cyclability, systematic studies utilizing in situ gas analysis have been conducted. There is significant emphasis on characterizing the predominant recharge processes because of the possibility of enhanced parasitic side reactions arising from catalysts. For instance, McCloskey and Addison initially showed that Pt nanoparticles on XC72 carbon leads to low amounts of $O_2$ evolution upon recharge, thus strongly suggests the exacerbation of parasitic side reactions.[43] Further, by examining an assortment of nanocatalysts, Wong *et al.* showed that despite the presence of nanocatalysts resulting in a lowered recharge overpotential (reflecting facile $Li_2O_2$ decomposition) for the first cycle (Figure 11a and b), there is a lack of correlation between the recharge potential and cyclability (Figure 11d and e).[14] Instead, the reduced recharge overpotential is reflected in the reduced potential for $CO_2$ gas evolution (Figure 11c) indicating that nanocatalysts promote the decomposition of side products. Moreover, catalysts that are highly active in suppressing the recharge overpotential (i.e., Ru and Pt) exhibit greater overall instability at higher potentials and nonselectivity toward side reactions, which leads to side product accumulation and ultimately poor capacity retention upon cycling despite maintaining low recharge potential. These results suggest that recharge potential itself is not representative of the true electrochemical reactions occurring in $Li-O_2$ cells. Moreover, the development of stable heterogeneous catalysts to mitigate the high potential needs proper analysis to ensure that side reactions are not exacerbated.

### 3.2.3 *Redox Mediators*

The use of RMs has also been widely studied, primarily as a means to lower the high overpotentials typically observed during recharge. Ideally, the soluble RM should decompose $Li_2O_2$ at its surface via chemical reaction, thus the central issue of the sluggish charge transport within the insulating $Li_2O_2$ during the electrochemical reaction is bypassed. The conceptual model shown in Figure 12(a)

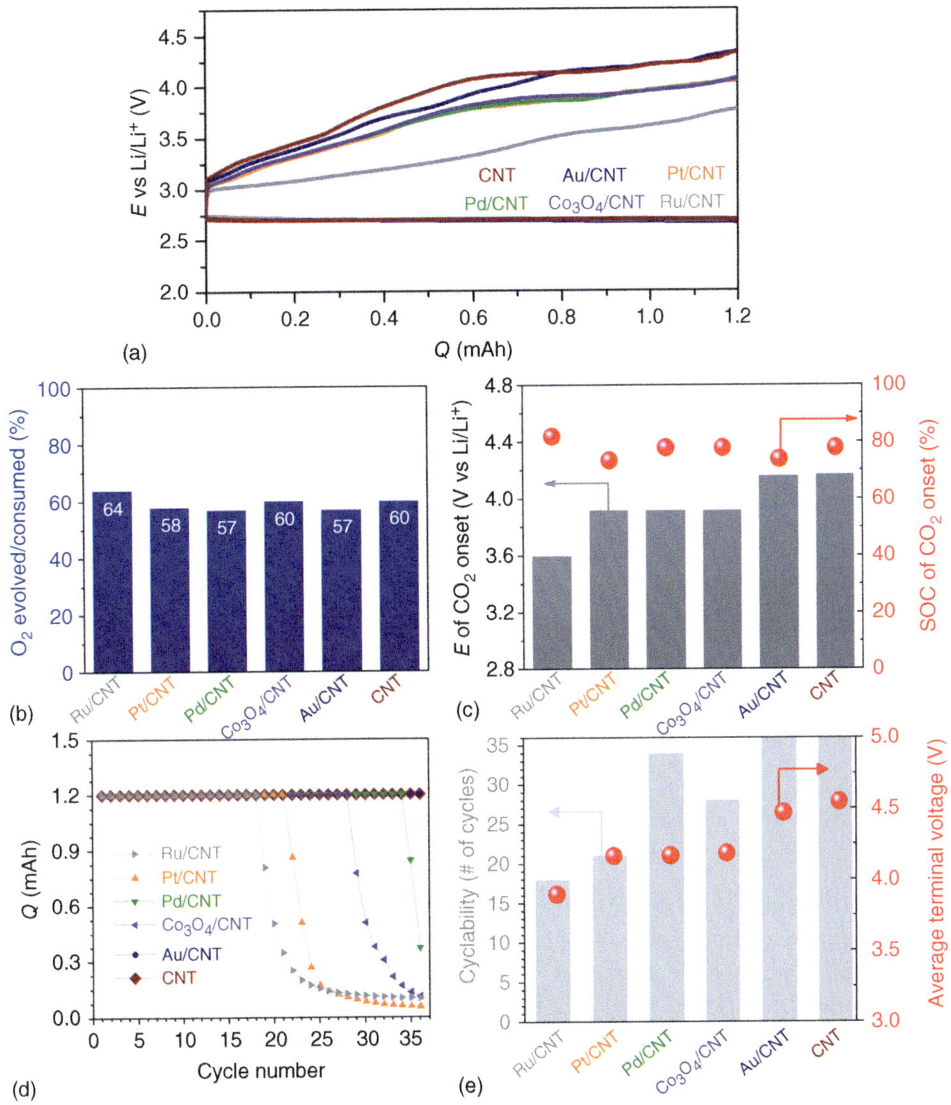


**Figure 11** In situ gas analysis of the first cycle of catalyst-containing cells and long-term cyclability. (a) Discharge–recharge profiles of Ru/CNT, Pt/CNT, Pd/CNT, $Co_3O_4$/CNT, Au/CNT, and CNT electrodes utilizing 0.5 M LiTFSI/tetraglyme at 50 mA $g^{-1}$ and fixed capacity of 1.2 mAh. (b) Corresponding $O_2$ evolved/consumed ratios obtained from in situ gas analysis. (c) Corresponding onset potential for $CO_2$ evolution and SOC of $CO_2$ evolution. (d) Capacity retention plot of cells performed at 50 mA $g^{-1}$ and a fixed capacity of 1.2 mAh. (e) Summary of cycling tests and average terminal voltage throughout cycling. [Reprinted with permission from Wong, R. A.; Yang, C.; Dutta, A.; O, M.; Hong, M.; Thomas, M. L.; Yamanaka, K.; Ohta, T.; Waki, K.; Byon, H. R., ACS Energy Lett. 2018, 3, 592–597. © American Chemical Society 2018]

highlights the stages involved in the redox mediation during recharge. Firstly, the RM is electrochemically oxidized on the positive electrode (Equation 9). The oxidized mediator species, which may be neutral or charged, then diffuses to the $Li_2O_2$ discharge product. Subsequently, chemical oxidation of the $Li_2O_2$ occurs, and the reduced form of the RM is regenerated (Equation 10).

$$RM - e^- \rightarrow RM^+ \quad (9)$$

$$2RM^+ + Li_2O_2 \rightarrow 2RM + 2Li^+ + O_2(g) \quad (10)$$

This process then repeats, with the cell maintaining the oxidation potential of the RM until all $Li_2O_2$ is depleted. To accomplish RM-mediated $Li_2O_2$ decomposition, the thermodynamic redox potential of the RM should be higher than that of $Li_2O_2$ (2.96 V vs $Li/Li^+$) prompting electron transfer from $Li_2O_2$ surface to $RM^+$ according to Equation (10), while it should also be lower than the potential where side reactions exacerbate (typically <3.7 V). Therefore, the redox potential of the RM is an important parameter in determining the recharge potential in $Li–O_2$

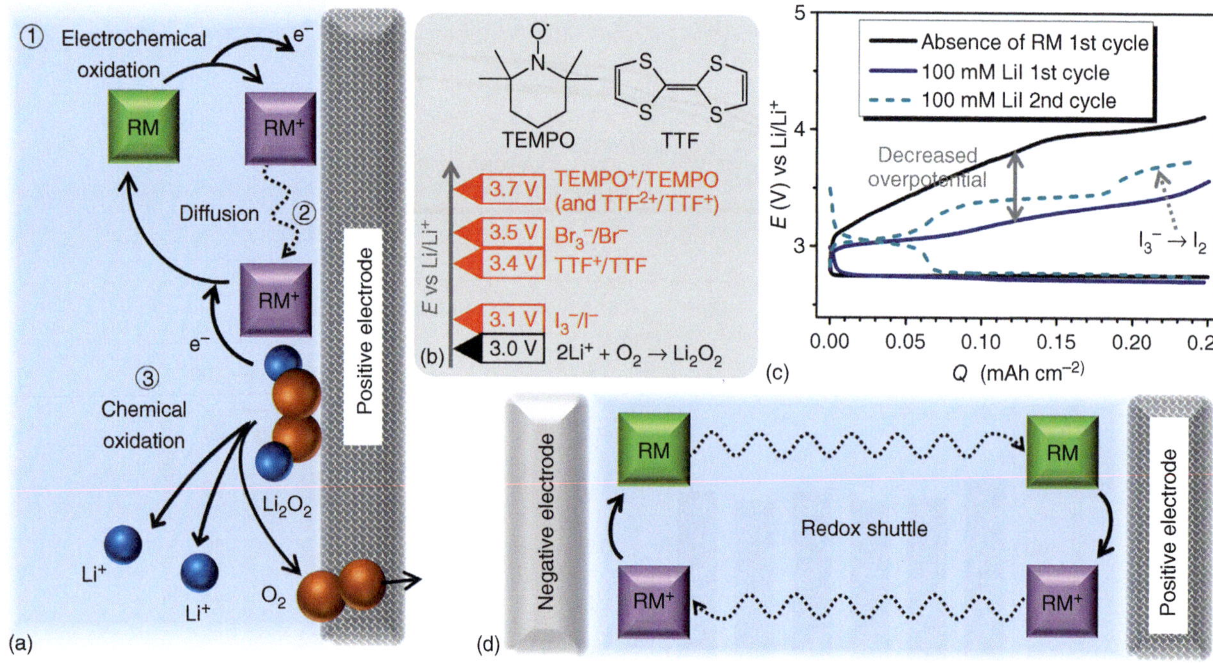


**Figure 12** The role of redox mediator in Li–$O_2$ cells. (a) Sketch illustrating the concept of redox mediation as applied to the recharge process in a Li–$O_2$ cell. (b) Sketch illustrating reversible redox potential of several commonly studied RMs (from reported CV-derived values in diglyme or tetraglyme-based electrolyte solutions with 0.2–1.0 M LiTFSI.[49–52] (c) Experimental data demonstrating the decrease in overpotential observed during charging upon addition of 100 mM LiI to 1 M LiTFSI in tetraglyme, and possible presence of $I_2$; (d) Sketch illustrating the redox shuttle effect

cells. In addition, the stability, electron-transfer kinetics, and diffusivity of the RM are other pivotal parameters.

From this starting point, various candidates for RMs have been investigated. Some of the more noteworthy are shown in Figure 12(b), including organic RMs such as tetrathiafulvalene (TTF), 2,2,6,6-tetramethylpiperidin-1-yl)oxyl (TEMPO) and lithium salts of simple halides ($I^-$, $Br^-$).[49,50,52,53] In particular, the triiodide/iodide ($I_3^-/I^-$) couple has been thoroughly investigated, with a redox potential of ~3.1 V in a typical 0.5–1.0 M LiTFSI/glyme solution.[52] The decrease in overpotential upon employing a LiI RM is shown in Figure 12(c). Although seemingly a well-defined and straightforward redox reaction, complex side-reactions, particularly in the presence of protic species (including $H_2O$), have been demonstrated, leading to the formation of lithium hydroxide (LiOH) as a side product.[54] In addition, there is the possibility for the further oxidation of the $I_3^-$ to iodine ($I_2$), that is, the possible involvement of a second RM based on the $I_2/I_3^-$ couple (with a potential of ~3.6 V).

Regarding organic RMs, TEMPO stands out as a stable organic radical species. As shown in Figure 12(b), the TEMPO+/TEMPO couple has a suitable redox potential of ~3.7 V in a typical glyme-based electrolyte.[49] Considering the galvanostatic recharge profile, this RM enables a constant potential of ~3.7 V throughout the recharge process,[49] and fast decomposition of $Li_2O_2$ has been also verified by in situ AFM studies.[26] A major benefit of such organic RMs is the possibility for tuning the redox potential by the addition of substituents with electron-withdrawing or electron-donating properties. This approach offers promise to further develop a truly task-specific RM for Li–$O_2$ batteries, providing optimum redox potential, kinetics, solubility and suppression of redox shuttle (vide infra) can be achieved.

Aside from the criterion for a suitable redox potential, the stability of the RM in the operational discharge and recharge potential range and also in the presence of the various $O_2$ species should be considered. For example, the reduction of TEMPO at a potential of ≤~2.2 V is poorly reversible since it yields the unstable TEMPO−.[49] As a result, sustained exposure of a TEMPO-containing electrolyte to low discharge potentials must be avoided. Indeed, decomposition of RMs during cell operation has been frequently reported, which leads to a gradual loss of efficiency for the RM, and hence potential rise upon cycling since cell charging gradually reverts to that observed in the absence of RM.[49] The aforementioned LiI RM offers another specific example of this, as the further oxidation of the mediator (i.e., the $I_2/I_3^-$ couple, vide supra, observed most clearly as an assigned $I_3^-$ oxidation plateau during the second charge in Figure 12(c)) may generate volatile $I_2$, which in turn may be lost from the electrolyte, as identified by in situ gas analysis, although there is scope for control of these by tuning the electrolyte composition.[52]

Indeed, for RM species in solution, the physicochemical properties of the electrolyte including choice of solvent, salt, and concentration/composition, have a profound influence on the electrochemical and solution properties. The electrolyte composition (including $Li^+$ activity) determines not only the redox potential (of RM, and indeed both the cathodic ($Li_2O_2$) and anodic ($Li/Li^+$) reactions), solubility, and diffusivity of the RM, but also the kinetics of electron transfer between the $RM^+$ and $Li_2O_2$.[52,55] Moreover, for RMs that are present in anionic form, the influence of contact ion pair formation between $Li^+$ and the anion, and hence promotion of different $Li^+$ coordination environments should be considered (see also Section 3.1.1).

One of the most critical issues remaining for RMs is the redox shuttle. During recharge, the oxidized form of the RM may diffuse to the Li negative electrode, and subsequently be reduced, then the reduced form itself is once again oxidized at the positive electrode, and so on (Figure 12d). This process will inevitably contribute to the observed Faradaic capacity of a cell during recharge. More seriously, the degradation of RM due to its crossover from the negative electrode of Li causes the loss of active RM, and hence recharge potential increase. Various efforts to modify the separator or Li electrode surface offer potential to restrict the redox shuttle effect (and are also relevant to the lithium–sulfur battery). These include the use of a separator modified with a polyanion to hinder crossover of the cationic form of RM,[56] although further work to ensure specificity of the interaction for the RM cation may be required. Moreover, for suppression of reaction at the Li electrode, efforts to modify the Li surface have included incorporating an $Al_2O_3$/PVdF-HFP coating,[57] or a thin (~30 μm) layer of conductive carbon resulting in a reaction-suppressing SEI.[58]

Overall, RMs continue to be an attractive option for lowering the recharge potential of $Li-O_2$ batteries. Recent findings, reviewed here, have highlighted key metrics and issues that must be addressed in order to enable their implementation with stability and longevity.

### 3.3 Poor Cyclability

Systematic in situ and ex situ studies have typically been limited to the first cycle or initial few cycles due to the cumulative effect of side reactions upon cycling. The side reactions drive the degradation of electrolyte (solvent and salt) and carbonaceous electrode in the presence of reactive $O_2$ species and are exacerbated at high recharge overpotential. This results in the deposition of lithium carbonates and lithium carboxylates on the electrode. Moreover, a buildup of impurities such as $CO_2$ and $H_2O$ along with the formation of mossy and dead Li at the negative electrode is detrimental toward cycling.[59] The potential profile in Figure 13(a) shows a more sudden rise in recharge potential to >4.25 V from the second cycle, reflecting the enhanced proportion of side reactions. However, despite such high potential, the elimination of side products is incomplete, resulting in the electrode becoming passivated (Figure 13b and c) leading to capacity fading and leads to inevitable cell failure. Typically, $Li-O_2$ cells achieve only ~50 cycles at 250 mA $g^{-1}{}_{electrode}$ and a limited capacity of 1000 mAh $g^{-1}{}_{electrode}$ when the carbonaceous electrode is 1–2.5 mg $cm^{-2}{}_{geometric\ electrode\ area}$,[60] while the cyclability is

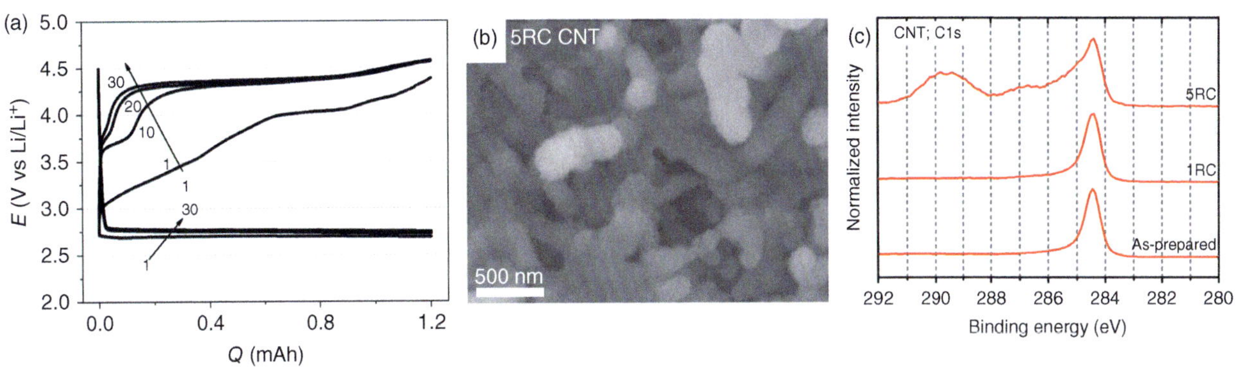


**Figure 13** Cyclability of $Li-O_2$ cell containing CNT using 0.5 M LiTFSI/tetraglyme. (a) Potential profile at a current rate of 50 mA $g^{-1}$ and fixed capacity of 1.2 mAh. [Reprinted with permision from Wong, R. A.; Yang, C.; Dutta, A.; O, M.; Hong, M.; Thomas, M. L.; Yamanaka, K.; Ohta, T.; Waki, K.; Byon, H. R., ACS Energy Lett. 2018, 3, 592–597. © American Chemical Society 2018]; (b) SEM image of 5th recharged (RC) CNT electrode. The test was carried out at 0.1 mA $cm^{-2}$ and a cut-off discharge potential of 2.0 V with equivalent capacities for recharge. The pristine electrode is similar to Figure 6(a). (c) Corresponding C 1 s XPS spectrum with pristine and 1st recharged CNT electrode. The intense and broad peak appearing at 289–290 eV indicates $CO_3{}^{2-}$ arising from deposited lithium carbonates. [Reprinted with permission from Hong, M.; Choi, H. C.; Byon, H. R., Chem. Mater. 2015, 27, 2234–2241. © American Chemical Society 2015]

also strongly dependent on the extent of capacity, current density, mass of electrode, and other cell conditions.

The main cell failure mechanism is ascribed to the accumulation of side products leading to electrode passivation. Therefore, approaches to enable the facile decomposition of lithium carbonates and lithium carboxylates have been explored using nanocatalysts such as porous nickel oxide (NiO) nanoplates resulting in increased capacity retention upon cycling.[60] Other approaches include the development of more stable electrolyte and electrode materials to effectively eliminate the origins of the side reactions. In this section, efforts on the development of advanced electrolytes and noncarbon electrodes are introduced.

### *3.3.1 Development of Stable Electrolytes*

Electrolyte development remains one of the most challenging topics in the Li–$O_2$ battery field, due to the susceptibility of electrolytes employed to-date toward side reactions. Apart from the basic characteristics of the electrolyte solution such as electrochemical potential range, vapor pressure, dielectric constant, viscosity, DN, AN, etc., the chemical stability of the aprotic solvent with the active $O_2$ species crucially determines the stability of Li–$O_2$ cells. Early on, conventional carbonate-based electrolyte solvents (propylene carbonate in particular) were used in Li–$O_2$ cells, which led to predominant formation of lithium carbonates after discharge.[61] This is due to the susceptibility of carbonate solvents to nucleophilic attack by $O_2^-$ species. Soon afterward, the improved stability against $O_2^-$ was confirmed with DMSO, *N,N*-dimethylacetamide (DMAc) and ether-based glymes, although they are still not ideal solvents. DMSO can be deprotonated by the nucleophilic $O_2^-$, and there have been reports of continuous reaction of DMSO with metallic Li.[62] DMSO also has a relatively low oxidation potential and thus cannot be used above a limited recharge potential of ~4 V. In comparison, DMAc has exceptional nucleophilic stability supported by computational prediction[63] and a wide electrochemical window of ~5 V.[64] Walker *et al.* demonstrated that the chemical reactivity of DMAc with Li electrode could be significantly hindered by using $LiNO_3$ as the salt, forming a stable $Li_2O$ layer on Li, and leading to high reversibility for >80 cycles.[65] Glymes (i.e., glycol methyl ethers) have been extensively used within the Li–$O_2$ battery field owing to the formation of a relatively stable SEI on the Li electrode and a reasonable recharge potential in the range of ~4.2 V.[66] Nevertheless, ether-based solvent systems (particularly glymes), similar to most aprotic solvents, are not completely inert to the nucleophile attack of $O_2^-$ and singlet $O_2$ during discharge and initial recharge, and to $Li_2O_2$ during recharge with high potential (>3.5 V).[67] Such degradation causes the formation of lithium carbonates and lithium carboxylates on the positive electrode throughout the discharge–recharge process for the first cycle. It has been reported that lithium carbonates are derived from degradation of both aprotic electrolyte solvent and carbonaceous electrode while lithium carboxylates arose from the electrolyte solvent.[68]

Apart from such aprotic solvents, various other types of electrolytes have been explored. Polymer-based solid electrolytes consisting of poly(ethylene oxide) (PEO) combined with a lithium salt revealed the production of $Li_2O_2$ and LiOH as the discharge products.[69] One of the limitations of solid-polymer electrolytes (SPEs) applied for Li–$O_2$ cells is the low discharge capacity, because of restricted $O_2$ supply by SPEs. This issue is mitigated by establishment of a new electrode structure: the thin (5–40 nm) coating of CNT with PEO and its integration into a 3-D structure incorporating sufficient void space, enabling concomitant supply of $Li^+$, electrons and $O_2$ gas.[70] With increased active surface area, however, significant side products of LiOH, methyl ether, and $H_2$ were detected by in situ gas analysis, suggesting severe hydrogen abstraction by reactive $O_2$ species, for example, $O_2^-$, and water impurities.

A recently rapidly developing field is related to the use of solvent-in-salt systems. The molar ratio of the Li salt to aprotic solvent is substantially increased compared to conventional electrolytes, such that the proportion of noncoordinating solvent molecules substantially decreases. In a work considering solutions of LiTFSI in DMSO from conventional concentration (0.1–1.0 M, DMSO/$Li^+$ 137:1–12:1) to high concentration (up to 3.4 M, DMSO/$Li^+$ 2:1), Tatara *et al.* demonstrated that the highly concentrated electrolytes provide a means to limit the dissolution of the $Li^+O_2^-$ intermediate (as mentioned in Section 3.1.1) due to the reduction in the activity of *free*, that is, uncoordinating solvent, and speculated that this may provide a means to improve electrolyte stability by preventing reaction with reactive $O_2$ species.[38]

Included within this category of highly concentrated electrolytes are the solvate ionic liquids (solvate ILs), formed from the equimolar combination of a lithium salt with a weakly Lewis basic anion (such as LiTFSI) and a chelating solvent (such as triglyme or tetraglyme). The solvent coordinates the $Li^+$, forming a relatively long-lived solvate cation of $[Li(glyme)]^+$, with negligible free glyme in solution.[71] The resulting room-temperature liquid is, therefore, considered as being composed of only solvate cations and anions, and it is termed a solvate IL.[72] Similar to other concentrated electrolytes (vide supra), the solvate ILs exhibit lower solubility of $Li^+O_2^-$. Kwon *et al.* demonstrated the lower solubility by employing a RRDE method. In comparing LiTFSI/triglyme 1:4 (conventional concentration of Li salt ~1.2 M) with 1:1 (i.e., solvate IL, ~3.1 M), negligible ring current was observed for the solvate IL (see Figure 14(a) and (b)).[25] Although the decreased solubility of $Li^+O_2^-$ inevitably leads to lower capacities, the lack of dispersion of discharge product

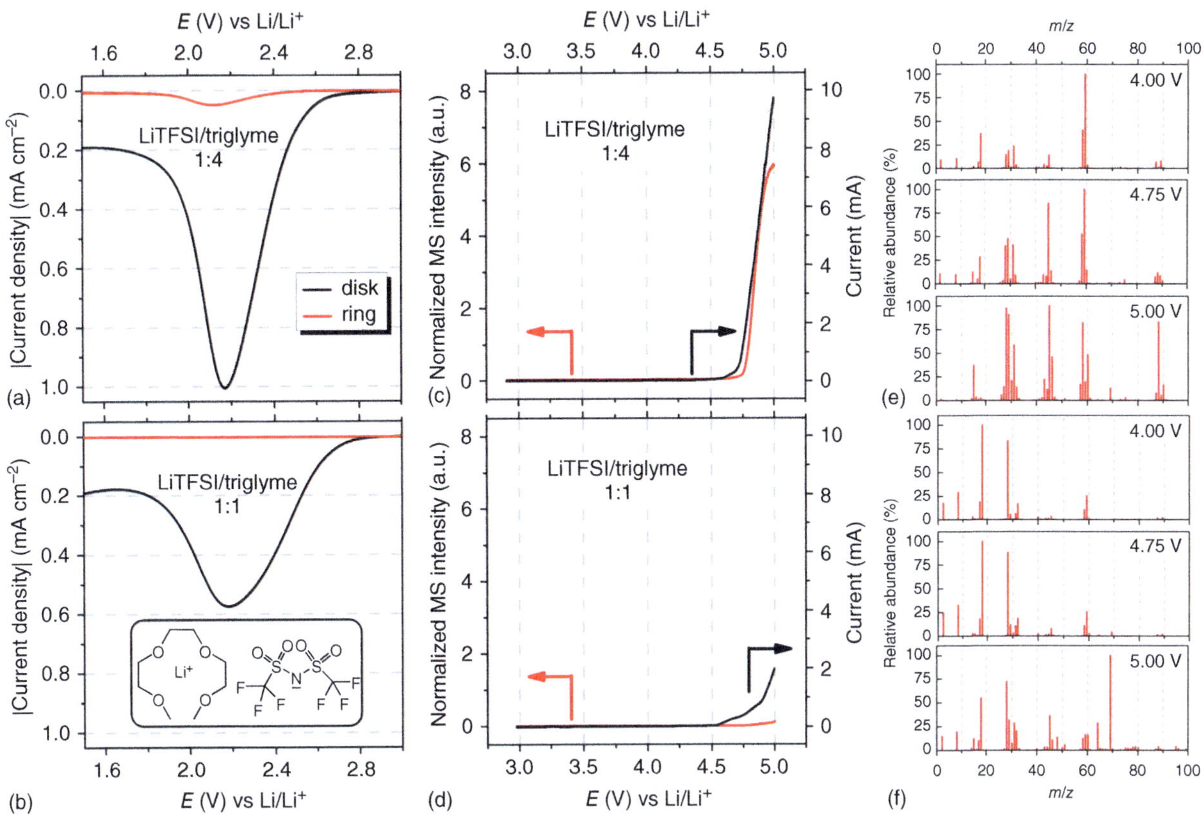


**Figure 14** Analysis of solvate ionic liquids (solvate ILs) as $L-O_2$ battery electrolytes (a, b) Rotating ring disk electrode (RRDE) responses for LiTFSI/triglyme of 1:4 and 1:1 respectively on GC disk/GC ring electrodes (1000 rpm, $100\,mV\,s^{-1}$, ring biased to 3.3 V). Inset in (b) shows the structure of the solvate IL. (c, d) Current and intensity of $m/z = 15$ mass fragment during LSV of as-prepared cells under helium at $0.1\,mV\,s^{-1}$ with LiTFSI/triglyme of 1:4 and 1:1 electrolytes respectively, and (e, f) corresponding overall mass spectra at 4.00, 4.75, and 5.00 V. [Reprinted with permission from Kwon, H.-M.; Thomas, M. L.; Tatara, R.; Oda, Y.; Kobayashi, Y.; Nakanishi, A.; Ueno, K.; Dokko, K.; Watanabe, M., ACS Appl. Mater. Interfaces 2017, 9, 6014–6021. © American Chemical Society 2017]

throughout the cell (i.e., remote from the electrode) may conversely aid in enabling longevity by preventing loss of active material (as mentioned in Section 3.1.1). Thus, these highly concentrated electrolytes also offer the possibility for promoting the surface-mediated pathway rather than the solution-mediated mechanism for discharge. A further advantage of such systems is the increased oxidative stability, due to a decrease in the highest occupied molecular orbital (HOMO) level for the solvent when complexed to $Li^+$.[73] This should enable the electrolyte to be used at higher recharge potentials, and thus may alleviate some of the decomposition problems faced with conventional concentration electrolyte systems. Indeed, when employing such electrolytes in $Li-O_2$ batteries, Kwon *et al.* noted a reduced proportion of glyme-based side products as determined by in situ gas analysis for the solvate IL as compared to the lower concentration counter-parts,[25] see Figure 14(c)–(f), although it should be noted that side reactions have not been completely eliminated. Another benefit of such systems is in the low volatility compared to conventional electrolytes, which provides an opportunity for developing a low flammability and nonevaporating electrolyte for use in an open-type battery system, where conventional electrolytes would slowly evaporate over time and may require a costly solvent capture and separation system.[74] This advantageous low volatility also extends to the application of conventional ILs as a component of the electrolyte.[75]

Indeed, a variety of conventional ILs in combination with lithium salt have been studied as electrolytes for $Li-O_2$ batteries. One of the key merits of ILs is the ability to easily tune the electrochemical window and physicochemical/solution properties based on selection of appropriate cation and anion combinations, for ILs based on commonly used cations. Therefore, some notable examples with PP13-TFSI (*N*-methyl-*N*-propylpiperidinum TFSI), DEME-TFSI (*N*,*N*-diethyl-*N*-methyl-*N*-(2-methoxyethyl) ammonium-TFSI), and PYR14-TFSI (*N*-butyl-*N*-methylpyrrolidinium bis(trifluoromethanesulfonyl)imide) showed low recharge potentials (<3.5 V) in the early stages.[76,77] In particular, DEME-TFSI has emerged as a promising candidate providing lower viscosities (80 mP s at 30 °C for 0.2 mol kg$^{-1}$ LiTFSI in DEME-TFSI) than most conventional ILs, and continues to generate interest.[78] However, stability issues have hindered IL development. Cations of many ILs are reactive with various $O_2$ species and the resulting hydrogen abstraction produces side products of LiOH or $H_2$ gas evolution during recharge.[79,80] In addition, some specific cations such as imidazolium and pyrrolidinium are incompatibility with metallic Li electrode.[81,82]

Widely reported issues with stability for aprotic solvents based on organic structures have driven interest in systems where the organic components are no longer present, possibly preventing side reactions. In this regard, inorganic molten salts have become promising candidates. Appropriate mixtures of alkali metal salts form a eutectic with a reasonably low melting point. Nevertheless, the eutectic point in many cases may still exceed the melting point of metallic lithium (~180 °C). However, Giordani *et al.* employed a binary mixture of $LiNO_3$–$KNO_3$ (melting point ~125 °C), or ternary mixture of $LiNO_3$–$KNO_2$–$CsNO_3$ (melting point ~90 °C).[83] For the former, a low recharge potential of ~2.83 V for more than 40 cycles at 150 °C provides the starting point for a promising future direction in development of stable electrolytes. It should be noted that lithium carbonates production is also observed from the first cycle, which may be attributed to the unstable carbonaceous electrode at operating conditions. This indicates that efforts to hinder side reactions should be achieved by consideration of both electrolyte solution and electrode concomitantly (vide infra). More recent developments have provided promise of a battery based on $Li_2O$ formation through a 4 $e^-$ transfer process, based on this nitrate eutectic at high temperature.[84]

### 3.3.2 *Noncarbonaceous Electrodes*

The use of carbon as the positive electrode has many advantages including natural abundance, cost, porosity, conductivity, and most importantly attractive (low) molecular weight, which contributes to the high theoretical specific energy densities. However, in addition to the decomposition of the electrolyte, carbon-based electrodes have also been found to exhibit stability issues. The carbonaceous electrode itself is oxidized during discharge and recharge. An in situ ambient X-ray photoelectron spectroscopy (XPS) study using a reduced graphene oxide electrode and a solid electrolyte shows the possible nucleophilic addition of $O_2^-$ during discharge resulting in epoxy and carbonate functional groups on the carbonaceous electrode.[85] In addition, carbonaceous electrodes are not stable at >3.5 V regardless of whether the surface property is hydrophobic or hydrophilic, establishing that the carbonaceous electrode is main source for the production of $Li_2CO_3$ in DMSO or ether-based glyme solution.[68] The contribution of the carbonaceous electrode to side reactions is clearly demonstrated from in situ gas analysis with the detection of $^{13}CO_2$ evolution when isotope-labeled $^{13}C$ electrodes are employed.[7,68]

In response to the mounting evidence pointing to the instability of the carbonaceous electrode, there have been a variety of approaches that attempt to eliminate the carbon surface. Conductive metals or ceramic materials such as nanoporous Au and titanium carbide (TiC) have been reported.[7] However, McCloskey and coworkers brought into question the reported results of suppressed side reactions for these materials by performing quantitative in situ gas analysis, and showed similarly low $O_2$ evolved/consumed ratios in comparison to carbonaceous electrodes, irrespective of the electrolyte solution.[7] In addition, the possible irreversibility from the production of a passivating oxide layer along with the disadvantages of weight and cost should be considered for practical applications.[86]

Alternatively, coating the carbon electrode surface with a thin oxide layer including $Al_2O_3$ and ZnO via atomic layer deposition (ALD) has been investigated.[87,88] The partial or full coating of a nanometer thick oxide layer can curtail side reactions while avoiding the conductivity limitations of a thicker and bulk oxide layer. It is important to note that these efforts for noncarbon electrodes should be examined together with the development of electrolyte solutions to identify the optimum condition and beneficial synergetic effects to mitigate side reactions in Li–$O_2$ cells.

## 4 CONCLUSION AND FUTURE OUTLOOK, FROM $O_2$ TO AIR: REALIZING Li–AIR BATTERIES

The significant research that has been invested in developing next-generation Li–$O_2$ batteries has resulted not only in an extension of knowledge relating to the fundamentals of $O_2$ electrochemistry but also in exploring the great potential of Li–$O_2$ batteries in providing higher energy densities. The inherent challenges that must be surmounted to attain reversible $Li_2O_2$ formation and decomposition require thorough understanding of the discharge and recharge processes and the manipulation of electrodes, electrolytes, additives and cell conditions.

In addition, we mention the possible differences in Li–$O_2$ battery designs at the current laboratory scale versus future designs with the target of commercialization, which will have additional challenges. Fundamental studies on Li–$O_2$ batteries at the laboratory scale have largely been conducted using *dry and highly pure* $O_2$ gas in either a closed or open cell configuration. Significant care in the prevention of ambient air ingress including $N_2$, $CO_2$, $H_2O$, etc. has been necessary to simplify the $O_2$ electrochemistry, achieve greater reversibility and to avoid additional side reactions. This indicates that practical Li–$O_2$ batteries may need to be accompanied by the addition of pure $O_2$ storage, and indeed, reports on the feasibility of well-engineered battery devices based on Li–$O_2$ technologies have considered the design of cells wherein the oxygen is maintained in a storage vessel.[74]

However, the critical point for developing a battery system with high specific energy (i.e., stored energy content per unit mass) is the creation of a workable cell device that has minimal weight meaning that an open design with the *ingress of ambient air* is preferred. In this case, essential efforts need to be made to maintain suitable conditions inside the Li–$O_2$ cell, where membranes that filter gases from the air, inclusion of nonvolatile electrolyte solutions, and advanced separators or protection layers for the metallic Li electrode are integrated. It is also of note that the lower partial pressure of $O_2$ gas in ambient air can reduce the Li–$O_2$ cell performance in comparison to operation under a pure $O_2$ gas. Hydrophobic membranes may prevent the ingress of water, while there has been no development of a specialized membrane to prevent the ingress of $N_2$ and $CO_2$. The presence of $CO_2$ impurities can promote the formation of lithium carbonates. It has been found that operation under $O_2/CO_2$ gas mixtures results in predominantly $CO_2$ gas evolution during recharge.[89] Thus, the development of workable solutions to prevent contamination will need to be made.[90,91] Indeed, the rational design of the electrolyte composition can limit the ingress of water. IL, solvent-in-salt and molten salt electrolytes have negligible volatility, and thus would be unlikely to suffer the effects of evaporation upon operation in a cell with an open design. Moreover, the inherent water adsorption properties of the electrolyte can be tuned to provide minimal equilibrium water content in the electrolyte in the presence of ambient air.[92] In addition, an advanced separator and protection layer, which addresses the cross-over issue of all impurities, $N_2$ from ambient air, also $O_2$, and redox shuttling, can result in an enduring and robust metallic Li negative electrode. Although $N_2$ is one of the most stable gas molecules, it can be reduced upon encountering the metallic Li negative electrode to form lithium nitride ($Li_3N$). Several approaches have been considered with establishment of a protection layer consisting of solid inorganic materials, gel or dry polymers, which results in enhanced reversibility of Li stripping and deposition and suppressed dendrite formation.[93,94]

In closing, we mention that the knowledge gained from the Li–$O_2$ field can be utilized in other metal–$O_2$, and conversion-type systems (e.g., Li–S battery). Thus, whilst elegant engineering design may provide various solutions, mechanistic understanding of the fundamental electrochemistry that can be revealed by various in situ and ex situ analytical tools can be considered as prerequisites that can pave the way for the beyond-LIB era.

## 5 ABBREVIATIONS AND ACRONYMS

ALD = atomic layer deposition; AN = acceptor number; CNT = carbon nanotube; CV = cyclic voltammetry; DBBQ = 2,5-di-*tert*-butyl-1,4-benzoquinone; DEMS = differential electrochemical mass spectrometry; DMAc = *N*,*N*-dimethylacetamide; DMSO = dimethyl sulfoxide; DN = donor number; EC-AFM = electrochemical atomic force microscopy; EV = electric vehicle; HAADF-STEM = high-angle annular dark field scanning transmission electron microscopy; HOMO = highest occupied molecular orbital; HOPG = highly orientated pyrolytic graphite; IL = ionic liquid; OEMS = on-line electrochemical mass spectrometry; OER = oxygen evolution reaction; ORR = oxygen reduction reaction; PEMFC = proton exchange membrane fuel cell; PEO = poly(ethylene oxide); RM = redox mediator; RRDE = rotating ring-disk electrode; SERS = surface-enhanced Raman spectroscopy; SPE = solid-polymer electrolyte; $TBA^+$ = tetrabutylammonium; $TEA^+$ = tetraethylammonium; TEMPO = 2,2,6,6-tetramethylpiperidin-1-yl)oxyl; TiC = titanium carbide; TTF = tetrathiafulvalene; XPS = X-ray photoelectron spectroscopy; XRD = X-ray diffraction; 1-D = one-dimensional

## 6 REFERENCES


1. P. G. Bruce, S. A. Freunberger, L. J. Hardwick and J.-M. Tarascon, *Nat. Mater.*, 2012, **11**, 19.
2. D. Aurbach, B. D. McCloskey, L. F. Nazar and P. G. Bruce, *Nat. Energy*, 2016, **1**, 16128.
3. N. Imanishi, A. C. Luntz and P. Bruce, 'The Lithium Air Battery: Fundamentals', Springer, New York, 2014.
4. K. Abraham and Z. Jiang, *J. Electrochem. Soc.*, 1996, **143**, 1.
5. D. T. Sawyer, G. Chiericato, C. T. Angelis, E. J. Nanni and T. Tsuchiya, *Anal. Chem.*, 1982, **54**, 1720.
6. Z. Peng, S. A. Freunberger, L. J. Hardwick, Y. Chen, V. Giordani, F. Bardé, P. Novák, D. Graham, J. M. Tarascon and P. G. Bruce, *Angew. Chem. Int. Ed.*, 2011, **50**, 6351.
7. A. C. Luntz and B. D. McCloskey, *Chem. Rev.*, 2014, **114**, 11721.
8. L. G. Cota and P. De La Mora, *Acta Crystallogr. B*, 2005, **61**, 133.
9. M. K. Chan, E. L. Shirley, N. K. Karan, M. Balasubramanian, Y. Ren, J. P. Greeley and T. T. Fister, *J. Phys. Chem. Lett.*, 2011, **2**, 2483.
10. M. D. Radin and D. J. Siegel, *Energy Environ. Sci.*, 2013, **6**, 2370.
11. M. D. Radin, C. W. Monroe and D. J. Siegel, *Chem. Mater.*, 2015, **27**, 839.

12. F. Tian, M. D. Radin and D. J. Siegel, *Chem. Mater.*, 2014, **26**, 2952.
13. H. Lim, E. Yilmaz and H. R. Byon, *J. Phys. Chem. Lett.*, 2012, **3**, 3210.
14. R. A. Wong, C. Yang, A. Dutta, O. Minho, M. Hong, M. L. Thomas, K. Yamanaka, T. Ohta, K. Waki and H. R. Byon, *ACS Energy Lett.*, 2018, **3**, 592.
15. R. Wen, M. Hong and H. R. Byon, *J. Am. Chem. Soc.*, 2013, **135**, 10870.
16. P. Liu, J. Han, X. Guo, Y. Ito, C. Yang, S. Ning, T. Fujita, A. Hirata and M. Chen, *Sci. Rep.*, 2018, **8**, 3134.
17. C. Yang, R. A. Wong, M. Hong, K. Yamanaka, T. Ohta and H. R. Byon, *Nano Lett.*, 2016, **16**, 2969.
18. Z. Lyu, Y. Zhou, W. Dai, X. Cui, M. Lai, L. Wang, F. Huo, W. Huang, Z. Hu and W. Chen, *Chem. Soc. Rev.*, 2017, **46**, 6046.
19. N. B. Aetukuri, B. D. McCloskey, J. M. García, L. E. Krupp, V. Viswanathan and A. C. Luntz, *Nat. Chem.*, 2015, **7**, 50.
20. B. D. Adams, C. Radtke, R. Black, M. L. Trudeau, K. Zaghib and L. F. Nazar, *Energy Environ. Sci.*, 2013, **6**, 1772.
21. S. Lau and L. A. Archer, *Nano Lett.*, 2015, **15**, 5995.
22. V. Viswanathan, K. S. Thygesen, J. Hummelshøj, J. K. Nørskov, G. Girishkumar, B. McCloskey and A. Luntz, *J. Chem. Phys.*, 2011, **135**, 214704.
23. S. Ganapathy, B. D. Adams, G. Stenou, M. S. Anastasaki, K. Goubitz, X.-F. Miao, L. F. Nazar and M. Wagemaker, *J. Am. Chem. Soc.*, 2014, **136**, 16335.
24. B. D. McCloskey, D. S. Bethune, R. M. Shelby, G. Girishkumar and A. C. Luntz, *J. Phys. Chem. Lett.*, 2011, **2**, 1161.
25. H.-M. Kwon, M. L. Thomas, R. Tatara, Y. Oda, Y. Kobayashi, A. Nakanishi, K. Ueno, K. Dokko and M. Watanabe, *ACS Appl. Mater. Interfaces*, 2017, **9**, 6014.
26. M. Hong, C. Yang, R. A. Wong, A. Nakao, H. C. Choi and H. R. Byon, *J. Am. Chem. Soc.*, 2018, **140**(20), 6190–6193.
27. S. Kang, Y. Mo, S. P. Ong and G. Ceder, *Chem. Mater.*, 2013, **25**, 3328.
28. J. Varley, V. Viswanathan, J. Nørskov and A. Luntz, *Energy Environ. Sci.*, 2014, **7**, 720.
29. P. Albertus, G. Girishkumar, B. McCloskey, R. S. Sánchez-Carrera, B. Kozinsky, J. Christensen and A. C. Luntz, *J. Electrochem. Soc.*, 2011, **158**, A343.
30. B. D. McCloskey, C. M. Burke, J. E. Nichols and S. E. Renfrew, *Chem. Commun.*, 2015, **51**, 12701.
31. H.-M. Kwon, M. L. Thomas, R. Tatara, A. Nakanishi, K. Dokko and M. Watanabe, *Chem. Lett.*, 2017, **46**, 573.
32. D. Sharon, D. Hirsberg, M. Salama, M. Afri, A. A. Frimer, M. Noked, W. Kwak, Y. K. Sun and D. Aurbach, *ACS Appl. Mater. Interfaces*, 2016, **8**, 5300.
33. I. Gunasekara, S. Mukerjee, E. J. Plichta, M. A. Hendrickson and K. M. Abraham, *J. Electrochem. Soc.*, 2015, **162**, A1055.
34. K. U. Schwenke, M. Metzger, T. Restle, M. Piana and H. A. Gasteiger, *J. Electrochem. Soc.*, 2015, **162**, A573.
35. X. Gao, Y. Chen, L. Johnson and P. G. Bruce, *Nat. Mater.*, 2016, **15**, 882.
36. Y.-C. Lu, B. M. Gallant, D. G. Kwabi, J. R. Harding, R. R. Mitchell, M. S. Whittingham and Y. Shao-Horn, *Energy Environ. Sci.*, 2013, **6**, 750.
37. Y. Zhang, Q. Cui, X. Zhang, W. C. McKee, Y. Xu, S. Ling, H. Li, G. Zhong, Y. Yang and Z. Peng, *Angew. Chem. Int. Ed.*, 2016, **55**, 10717.
38. R. Tatara, D. G. Kwabi, T. P. Batcho, M. Tulodziecki, K. Watanabe, H.-M. Kwon, M. L. Thomas, K. Ueno, C. V. Thompson, K. Dokko, Y. Shao-Horn and M. Watanabe, *J. Phys. Chem. C*, 2017, **121**, 9162.
39. E. Yilmaz, C. Yogi, K. Yamanaka, T. Ohta and H. R. Byon, *Nano Lett.*, 2013, **13**, 4679.
40. R. A. Wong, A. Dutta, C. Yang, K. Yamanaka, T. Ohta, A. Nakao, K. Waki and H. R. Byon, *Chem. Mater.*, 2016, **28**, 8006.
41. A. Dutta, R. A. Wong, W. Park, K. Yamanaka, T. Ohta, Y. Jung and H. R. Byon, *Nat. Commun.*, 2018, **9**, 680.
42. Z. Chang, J. Xu and X. Zhang, *Adv. Energy Mater.*, 2017, **7**, article number 1700875.
43. B. D. McCloskey and D. Addison, *ACS Catal.*, 2016, **7**, 772.
44. R. Black, J. H. Lee, B. Adams, C. A. Mims and L. F. Nazar, *Angew. Chem. Int. Ed.*, 2013, **52**, 392.
45. K. U. Schwenke, S. Meini, X. Wu, H. A. Gasteiger and M. Piana, *Phys. Chem. Chem. Phys.*, 2013, **15**, 11830.
46. Y. C. Lu and Y. Shao-Horn, *J. Phys. Chem. Lett.*, 2013, **4**, 93.
47. J. Lu, Y. J. Lee, X. Luo, K. C. Lau, M. Asadi, H.-H. Wang, S. Brombosz, J. Wen, D. Zhai and Z. Chen, *Nature*, 2016, **529**, 377.
48. J. K. Papp, J. D. Forster, C. M. Burke, H. W. Kim, A. C. Luntz, R. M. Shelby, J. J. Urban and B. D. McCloskey, *J. Phys. Chem. Lett.*, 2017, **8**, 1169.
49. B. J. Bergner, A. Schurmann, K. Peppler, A. Garsuch and J. Janek, *J. Am. Chem. Soc.*, 2014, **136**, 15054.
50. W.-J. Kwak, D. Hirshberg, D. Sharon, M. Afri, A. A. Frimer, H.-G. Jung, D. Aurbach and Y.-K. Sun, *Energy Environ. Sci.*, 2016, **9**, 2334.
51. K. P. C. Yao, J. T. Frith, S. Y. Sayed, F. Bardé, J. R. Owen, Y. Shao-Horn and N. Garcia-Araez, *J. Phys. Chem. C*, 2016, **120**, 16290.
52. A. Nakanishi, M. L. Thomas, H.-M. Kwon, Y. Kobayashi, R. Tatara, K. Ueno, K. Dokko and M. Watanabe, *J. Phys. Chem. C*, 2018, **122**, 1522.
53. Y. Chen, S. A. Freunberger, Z. Peng, O. Fontaine and P. G. Bruce, *Nat. Chem.*, 2013, **5**, 489.

54. C. M. Burke, R. Black, I. R. Kochetkov, V. Giordani, D. Addison, L. F. Nazar and B. D. McCloskey, *ACS Energy Lett.*, 2016, **1**, 747.

55. Y. Chen, X. Gao, L. R. Johnson and P. G. Bruce, *Nat. Commun.*, 2018, **9**, 767.

56. S. H. Lee, J.-B. Park, H.-S. Lim and Y.-K. Sun, *Adv. Energy Mater.*, 2017, **7**, 1602417.

57. D. J. Lee, H. Lee, Y.-J. Kim, J.-K. Park and H.-T. Kim, *Adv. Mater.*, 2016, **28**, 857.

58. J. Wang, J. Liu, Y. Cai, F. Cheng, Z. Niu and J. Chen, *ChemElectroChem*, 2018, **5**, 1702.

59. J. Lu, L. Li, J.-B. Park, Y.-K. Sun, F. Wu and K. Amine, *Chem. Rev.*, 2014, **114**, 5611.

60. M. Hong, H. C. Choi and H. R. Byon, *Chem. Mater.*, 2015, **27**, 2234.

61. S. A. Freunberger, Y. Chen, Z. Peng, J. M. Griffin, L. J. Hardwick, F. Barde, P. Novak and P. G. Bruce, *J. Am. Chem. Soc.*, 2011, **133**, 8040.

62. Z. Peng, S. A. Freunberger, Y. Chen and P. G. Bruce, *Science*, 2012, **337**, 563.

63. V. S. Bryantsev, J. Uddin, V. Giordani, W. Walker, D. Addison and G. V. Chase, *J. Electrochem. Soc.*, 2012, **160**, A160.

64. F. Wang, H. Chen, Q. Wu, R. Mei, Y. Huang, X. Li and Z. Luo, *ACS Omega*, 2017, **2**, 236.

65. W. Walker, V. Giordani, J. Uddin, V. S. Bryantsev, G. V. Chase and D. Addison, *J. Am. Chem. Soc.*, 2013, **135**, 2076.

66. S. A. Freunberger, Y. Chen, N. E. Drewett, L. J. Hardwick, F. Bardé and P. G. Bruce, *Angew. Chem. Int. Ed.*, 2011, **50**, 8609.

67. N. Mahne, B. Schafzahl, C. Leypold, M. Leypold, S. Grumm, A. Leitgeb, G. A. Strohmeier, M. Wilkening, O. Fontaine, D. Kramer, C. Slugovc, S. M. Borisov and S. A. Freunberger, *Nat. Energy*, 2017, **2**, 17036.

68. M. M. Ottakam Thotiyl, S. A. Freunberger, Z. Peng and P. G. Bruce, *J. Am. Chem. Soc.*, 2012, **135**, 494.

69. J. Hassoun, F. Croce, M. Armand and B. Scrosati, *Angew. Chem. Int. Ed.*, 2011, **50**, 2999.

70. N. Bonnet-Mercier, R. A. Wong, M. L. Thomas, A. Dutta, K. Yamanaka, C. Yogi, T. Ohta and H. R. Byon, *Sci. Rep.*, 2014, **4**, 7127.

71. K. Ueno, R. Tatara, S. Tsuzuki, S. Saito, H. Doi, K. Yoshida, T. Mandai, M. Matsugami, Y. Umebayashi, K. Dokko and M. Watanabe, *Phys. Chem. Chem. Phys.*, 2015, **17**, 8248.

72. C. Austen Angell, Y. Ansari and Z. Zhao, *Faraday Discuss.*, 2012, **154**, 9.

73. K. Yoshida, M. Nakamura, Y. Kazue, N. Tachikawa, S. Tsuzuki, S. Seki, K. Dokko and M. Watanabe, *J. Am. Chem. Soc.*, 2011, **133**, 13121.

74. K. G. Gallagher, S. Goebel, T. Greszler, M. Mathias, W. Oelerich, D. Eroglu and V. Srinivasan, *Energy Environ. Sci.*, 2014, **7**, 1555.

75. M. Watanabe, M. L. Thomas, S. Zhang, K. Ueno, T. Yasuda and K. Dokko, *Chem. Rev.*, 2017, **117**, 7190.

76. S. Higashi, Y. Kato, K. Takechi, H. Nakamoto, F. Mizuno, H. Nishikoori, H. Iba and T. Asaoka, *J. Power Sources*, 2013, **240**, 14.

77. G. A. Elia, J. Hassoun, W. J. Kwak, Y. K. Sun, B. Scrosati, F. Mueller, D. Bresser, S. Passerini, P. Oberhumer, N. Tsiouvaras and J. Reiter, *Nano Lett.*, 2014, **14**, 6572.

78. U. Ulissi, G. A. Elia, S. Jeong, F. Mueller, J. Reiter, N. Tsiouvaras, Y.-K. Sun, B. Scrosati, S. Passerini and J. Hassoun, *ChemSusChem*, 2018, **11**, 229.

79. B. D. McCloskey, D. S. Bethune, R. M. Shelby, T. Mori, R. Scheffler, A. Speidel, M. Sherwood and A. C. Luntz, *J. Phys. Chem. Lett.*, 2012, **3**, 3043.

80. T. Zhang and Z.-Y. Wen, *J. Phys. Chem. C*, 2017, **121**, 5968.

81. S. Seki, Y. Kobayashi, H. Miyashiro, Y. Ohno, A. Usami, Y. Mita, N. Kihira, M. Watanabe and N. Terada, *J. Phys. Chem. B*, 2006, **110**, 10228.

82. S. Meini, S. Solchenbach, M. Piana and H. A. Gasteiger, *J. Electrochem. Soc.*, 2014, **161**, A1306.

83. V. Giordani, D. Tozier, H. Tan, C. M. Burke, B. M. Gallant, J. Uddin, J. R. Greer, B. D. McCloskey, G. V. Chase and D. Addison, *J. Am. Chem. Soc.*, 2016, **138**, 2656.

84. C. Xia, C. Y. Kwok and L. F. Nazar, *Science*, 2018, **361**, 777.

85. D. M. Itkis, D. A. Semenenko, E. Y. Kataev, A. I. Belova, V. S. Neudachina, A. P. Sirotina, M. Hävecker, D. Teschner, A. Knop-Gericke, P. Dudin, A. Barinov, E. A. Goodilin, Y. Shao-Horn and L. V. Yashina, *Nano Lett.*, 2013, **13**, 4697.

86. D. Kundu, R. Black, B. Adams, K. Harrison, K. Zavadil and L. F. Nazar, *J. Phys. Chem. Lett.*, 2015, **6**, 2252.

87. J. Lu, Y. Lei, K. C. Lau, X. Luo, P. Du, J. Wen, R. S. Assary, U. Das, D. J. Miller, J. W. Elam, H. M. Albishri, D. A. El-Hady, Y. K. Sun, L. A. Curtiss and K. Amine, *Nat. Commun.*, 2013, **4**, 2383.

88. Y. Bae, D.-H. Ko, S. Lee, H.-D. Lim, Y.-J. Kim, H.-S. Shim, H. Park, Y. Ko, S. K. Park, H. J. Kwon, H. Kim, H.-T. Kim, Y.-S. Min, D. Im and K. Kang, *Adv. Energy Mater.*, 2018, **8**, 1702661.

89. Y. Qiao, J. Yi, S. Wu, Y. Liu, S. Yang, P. He and H. Zhou, *Joule*, 2017, **1**, 359.

90. O. Crowther and M. Salomon, *Membranes*, 2012, **2**, 216.

91. U. R. Farooqui, A. L. Ahmad and N. A. Hamid, *Renew. Sustain. Energy Rev.*, 2017, **77**, 1114.

92. M. L. Thomas, Y. Oda, R. Tatara, H.-M. Kwon, K. Ueno, K. Dokko and M. Watanabe, *Adv. Energy Mater.*, 2017, **7**, 1601753.

93. X. B. Cheng, R. Zhang, C. Z. Zhao and Q. Zhang, *Chem. Rev.*, 2017, **117**, 10403.

94. S. S. Zhang, *ACS Appl. Mater. Interfaces*, 2018, **1**, 910.

# Structural Engineering of Cathode Materials for Lithium–Sulfur Batteries

**Ligui Li, Jingping Yu, Nan Wang, Jun Zhao, Bin Fan and Shuaibo Zeng**

*South China University of Technology, Guangzhou, China*

and

**Shaowei Chen**

*South China University of Technology, Guangzhou, China; University of California, Santa Cruz, CA, USA*



## 1 INTRODUCTION

Lithium–sulfur (Li–S) batteries have been recognized as a promising energy technology to power stationary and portable electronic devices and electrical vehicles, since the discovery in the 1960s.[1] The typical reactions entail $Li \rightarrow Li^+ + e^-$ at the anode and $S + 2e^- \rightarrow S^{2-}$ at the cathode, with an overall reaction of $2Li + S \rightarrow Li_2S$.[2] Li–S batteries in general exhibit multiple merits. (i) Li–S batteries can deliver a theoretical energy density of 2567 Wh kg$^{-1}$ that is nearly 10-fold higher than those of conventional Li-ion batteries (LIBs).[3,4] For instance, an energy density of about 180 Wh kg$^{-1}$ is usually observed for $LiCoO_2$/C battery, 210 Wh kg$^{-1}$ for $LiFePO_4$/C battery, 200 Wh kg$^{-1}$ for $LiMnO_4$/C battery, and about 260 Wh kg$^{-1}$ for $LiNi_{1/3}Mn_{1/3}Co_{1/3}$/C battery.[5,6] This is because of the high specific capacity of sulfur (1672 mAh g$^{-1}$), which can meet the ever-increasing demand of portable electronic devices and electrical vehicles.[7–9] (ii) Sulfur is one of the most abundant elements on the earth's crust, and more than 70 million tons of elemental sulfur are annually produced as a by-product from hydrodesulfurization in petroleum refining processes, coal desulfurization, etc.[10,11] (iii) Sulfur is a low-cost natural resource and readily accessible.[12,13] (iv) Sulfur is also environmentally compatible because of its low toxicity and low impacts to the environment.[9,14–16]

However, large-scale commercialization of the Li–S battery technology is greatly impeded by a series of challenging issues, which are mostly correlated with the cathode (Scheme 1). These include (i) poor electrical conductivity of sulfur and polysulfides ($\sim 5 \times 10^{-30}$ S cm$^{-1}$ at 25 °C),[17–19] resulting in low specific capacity, low Coulombic efficiency, low utilization of active sulfur, and low charge–discharge rate performance; (ii) rapid mechanical degradation of cathode due to large volume expansion (up to 80%) of sulfur after complete lithiation; and (iii) dissolution of lithium polysulfides ($Li_2S_x$, $4 \leq x \leq 8$) into organic electrolytes and the subsequent so-called "shuttle" effects, which results in uncontrollable sulfide deposition on the anode of lithium metal, lowering the Coulombic efficiency, and causing rapid capacity attenuation.[20–23]

To mitigate these issues, a number of strategies have been developed and reported in recent literatures, mostly based on physical and/or chemical confinement of sulfur in a conductive matrix (Scheme 1). Therefore, in this chapter, we will summarize recent progress in the design and engineering of effective cathodes for Li–S battery that entails mostly physical and chemical confinements.

*Inorganic Battery Materials*. Edited by Hailiang Wang and Boniface P.T. Fokwa
 ISBN 978-1-119-43199-2

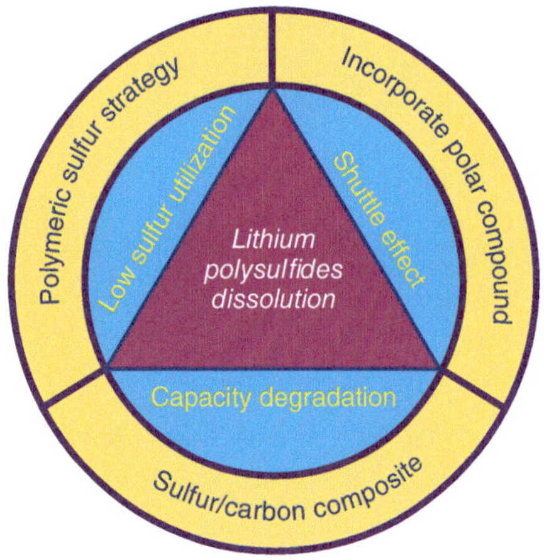


**Scheme 1** Summary of the challenges that limit the performance of Li–S battery and mitigation strategies

## 2 DESIGN AND ENGINEERING OF SULFUR CATHODES

### 2.1 Sulfur/Carbon Composites

In 1962, Herbert and Ulam were the first to demonstrate that sulfur might be used as a cathode material for lithium batteries.[1] Yet as the sulfur cathode exhibits only very low electrical conductivity,[24] sulfur is usually compounded with a conductive material. Among these, carbon materials are commonly used because of apparent electrical conductivity, high chemical stability, low density, and low costs. In early research, sulfur powders were simply mixed with toner in a physical manner, and the resulting cathodes usually showed only a poor electrochemical performance. Peled *et al.*[25] showed that by infilling sulfur into porous carbon, the specific capacity, and cycle stability of thus-prepared Li–S batteries could be markedly improved. An example for such a strategy was reported in 2009 by Ji *et al.*,[26] where sulfur was loaded into ordered mesoporous carbons, leading to significant enhancement of the cycling stability of the resulting Li–S cathode. A wider range of carbon materials of different morphologies have since been used as conductive supports in the preparation of S/C composites. These carbon materials also mitigate the issue of polysulfide dissolution into electrolyte by physical confinements and/or chemical adsorption. In addition, they minimize the negative impacts of poor electrical conductivity of sulfur and the large volume fluctuation during charge and discharge process because of the conductive carbon skeleton and porous structures.[27]

#### 2.1.1 *Carbon Nanoparticles*

Spherical or spheroidal carbon particles, due to their simple structure, small diameter, and large surface area, have been used rather extensively to effectively adsorb polysulfides through their surface functional groups. In fact, carbon particles have been widely utilized in the preparation of sulfur cathodes. In 2011, Archer and coworkers[28] utilized a simple silica hard-template method[29] to prepare mesoporous hollow carbon nanospheres. Experimentally, petroleum pitch was used as the carbon source, and after calcination at 1300 °C for 12 h under an argon flow, the silica templates were etched away by soaking in a HF solution, affording rather uniform hollow carbon nanospheres with a diameter of about 200 nm (Figure 1a). Molten sulfur was subsequently infiltrated into the hollow carbon nanospheres and used as the cathode for a Li–S battery, which retained a capacity of 850 mAh $g^{-1}$ after 100 charge–discharge cycles at 0.5 C. It should be recognized, however, that although a considerable device performance can be achieved by using the hollow carbon nanospheres, the hard-template procedure involves several complicated, time-/energy-consuming processes, such as synthesis of nanospheres templates, homogeneous coating of precursor materials on the surface of nanosphere templates, and post-synthesis removal of the templates, which hinder scale-up production.

In another study, Xu *et al.*[30] proposed a "soft template" method without the physical templates and post-synthesis removal process. Experimentally, they used aniline and pyrrole as monomers and Triton X-100 as emulsifier to prepare uniform hollow carbon nanospheres with an ultra-high specific surface area over 3000 $m^2\ g^{-1}$ by low-temperature polymerization. The inner diameter of the obtained hollow carbon sphere was only about 70 nm (Figure 1b). By controlling the carbonization conditions, they found that the micropores on the surface of the hollow carbon nanospheres could be tailored but the hollow structure was preserved. After the loading of sulfur, the resulting S@C composite electrode showed a specific discharge capacity of 1240 mAh $g^{-1}$ at 0.2 C. Note that although the encapsulation of sulfur in the mesopores hollow carbon nanospheres slowed down the dissolution of polysulfides, the weak interactions still allowed gradual diffusion of polysulfides out of these hollow carbon nanospheres during prolonged cycling. This suggests that physical confinement is not an effective strategy in achieving a stable performance of Li–S battery cathodes.

In 2013, Zhou *et al.*[31] prepared a sulfur-polyaniline composite and used it as the cathode material which exhibited excellent cycling stability. As shown in Figure 1(c), part of the hollow polyaniline nanospheres were loaded with sulfur, forming a unique yolk-shell structure. Because of the void space between the carbon shell and sulfur, volume expansion of sulfur during discharging was effectively alleviated. Using such a yolk-shell sulfur-polyaniline composite as the cathode material, the corresponding Li–S battery displayed a capacity of 765 mAh $g^{-1}$ at 0.2 C

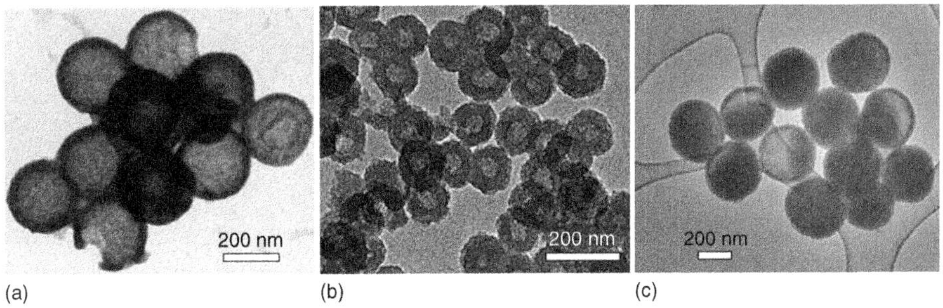


**Figure 1** TEM images of hollow carbon nanospheres prepared by the (a) hard template method; (b) soft template method. (c) TEM images for yolk-shell carbon nanospheres. [Panel (a) reproduced with permission from Ref. 28. © Wiley-VCH Verlag GmbH & Co. KGaA; panel (b) reproduced with permission from Ref. 30. © 2015, Nature Publishing Group; panel (c) reprinted with permission from Ref. 31. © 2013, American Chemical Society]

and retained a specific capacity of 628 mAh $g^{-1}$ after 200 charge–discharge cycles at 0.5 C.

### *2.1.2 Carbon Nanotubes*

Carbon nanotubes (CNTs) are another structural scaffolds used extensively for sulfur cathodes in Li–S battery, because the hybridization orbitals of the $\pi$ electrons are perpendicular to the plane of the carbon atoms, leading to excellent electrical conductivity along the tubular axis.[32,33] Carbon fibers have also been utilized widely because of their excellent mechanical properties, chemical stability, and electrical and thermal conductivity.[34,35] However, the main issue of these one-dimensional carbon structures is their low porosity and hence low specific surface area (typically less than 200 $m^2$ $g^{-1}$), which renders them ineffective in preventing the dissolution and diffusion of lithium polysulfides.[36,37]

To solve these problems, Wang and coworkers[38] developed a template wetting technique to prepare disordered carbon nanotubes (DCNTs). They used commercial anodic aluminum oxide (AAO) membranes as templates, into which sulfur was filled by physical vapor deposition (PVD). As shown in Figure 2(a), the vaporized sulfur diffused into the tiny voids of the graphitized carbon layers and defects in amorphous carbon. Thermal treatment of the sulfur powders at high temperatures led to the decomposition of $S_8$ molecules into smaller $S_6$ or $S_2$. These small molecules are able to pass through tiny pores in the shell and form more stable $S_8$ macromolecules. Note that these tiny channels can also effectively impede the penetration of liquid electrolyte and avoid the dissolution of polysulfides. The cathode material based on such sulfur-impregnated DCNTs exhibited excellent cycle stability and an unprecedented high Coulombic efficiency of 96% in 2011. Subsequently, Xin *et al.*[39] deposited a ~100 nm thick microporous carbon layer onto multi-walled carbon nanotubes (MWCNTs) to synthesize carbon nanotubes@microporous carbon (CNT@MPC) structures which were then heat-treated with the long-chain $S_8$ for a long time (155 °C, 20 h) to form short-chain $S_{2-4}$ allotropes, as shown in Figure 2(b). These metastable small $S_{2-4}$ molecules were confined within the MPC layers with a pore size of about 0.5 nm, which avoided the unfavorable transition from $S_4^{2-}$ to $S_8$ during charge and discharge and fundamentally solved the problem of dissolution and diffusion of polysulfides in Li–S batteries. A Li–S battery prepared with this material exhibited excellent cycle stability and rate performance, and the first discharge specific capacity at 0.1 C reached 1670 mAh $g^{-1}$, very close to the theoretical capacity of sulfur (1675 mAh $g^{-1}$). The capacity was 800 mAh $g^{-1}$ at 5 C, and after 200 cycle at 0.1 C, it still reached 1149 mAh $g^{-1}$. However, the sulfur loading remained too low. If too much sulfur was loaded into the hybrid carbon, the utilization rate would be reduced. In response to this situation, Xu and coworkers[36] reported a novel tube-in-tube carbon nanomaterial, as shown in Figure 2(c). They first treated the CNTs with acid and then coated the treated nanotubes with $SiO_2$ using an organosilicon compound (octadecyltrichlorosilane, $C_{18}$TMS) as the porogen agent and carbon precursors. After high-temperature calcination, $C_{18}$TMS was converted into carbon, and then $SiO_2$ was etched away with a NaOH solution to obtain a tube-in-tube carbon nanomaterial. Sulfur was then infiltrated into the gap between the tubes and into the tube by melt infiltration. This unique tube-in-tube structure not only improves the electrical conductivity of the CNTs but also increases the sulfur loading. As a cathode material for lithium–sulfur batteries, it exhibits an excellent electrochemical performance. At the current density of 500 mA $g^{-1}$, the discharge specific capacity still reached 918 mAh $g^{-1}$ after 500 cycles.

In addition to CNTs, carbon fibers have also been widely used in Li–S batteries. Cui and coworkers[40] used an AAO film as a template to obtain a hollow carbon nanofiber array by carbonization of polystyrene, as shown in Figure 2(d) and (e). The AAO film was not only a

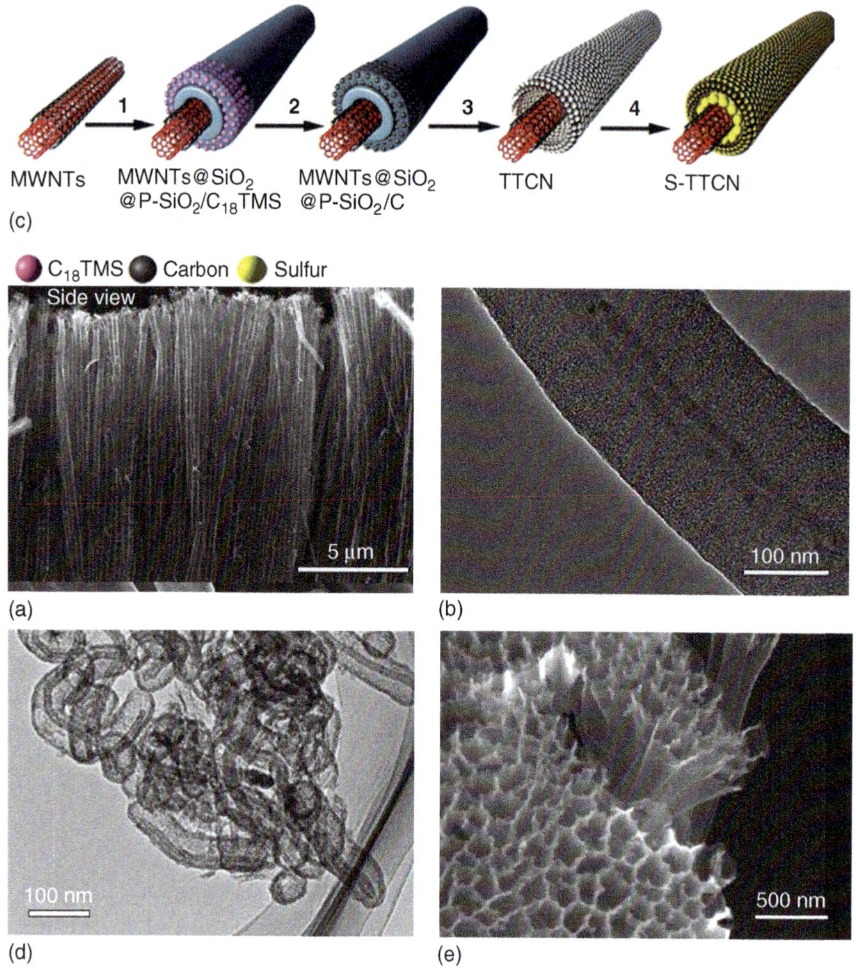


**Figure 2** SEM image of (a) disordered carbon nanotubes prepared by template-wetting method; (b) TEM image of microporous carbon layer-coated carbon nanotube; (c) schematic of the synthesis of tube-in-tube structured carbon nanomaterial; (d) TEM image of tube-in-tube structured carbon nanomaterials; (e) SEM image of hollow carbon nanofiber array. [Panel (a) was reprinted with permission from Ref. 38. © 2011, American Chemical Society; panel (b) was reprinted with permission from Ref. 39. © 2012, American Chemical Society; and panel (c) was reproduced with permission from Ref. 36. © 2014, Wiley-VCH Verlag GmbH & Co. KGaA; panels (d,e) were reprinted with permission from Ref. 40. © 2011, American Chemical Society]

template for the formation of carbon nanofibers but also prevented sulfur from depositing on the outer surface of the carbon fiber, thereby facilitating the penetration of sulfur into the interior of the carbon nanofiber. The hollow structure of carbon fibers effectively alleviated volume change of sulfur during charging and discharging. Owing to the thin carbon walls and large voids, lithium ions can penetrate easily. This hollow carbon fiber array can not only enable facile transport of electrons but also provide a large deposition area for $Li_2S_2$ and $Li_2S$. In fact, Li–S battery based on this material exhibited excellent electrochemical performance, with a high specific capacity of about 730 mAh $g^{-1}$ at 0.2 C rate after 150 cycles of charge/discharge. Further introduction of $LiNO_3$ additive into the electrolyte was proven to improve the Coulombic efficiency to over 99% at 0.2 C.

### *2.1.3 Graphene Derivatives*

Two-dimensional carbon nanomaterials have also been employed in the design and fabrication of sulfur cathode, such as graphene.[41,42] However, simple physical mixing of graphene and sulfur does not form an effective constraint on sulfur. During the charge and discharge processes, polysulfides can easily diffuse out of the sheet structure of graphene, compromising the battery performance. Cui and coworkers[43] developed a graphene–sulfur composite by coating sulfur particles with a surfactant and loading them onto carbon black-decorated graphene oxide (GO) sheets (Figure 3a–c). The graphene layer conducted electricity and confined the polysulfide intermediates, and the active agent layer accommodated volume change of sulfur during charging and discharging. This

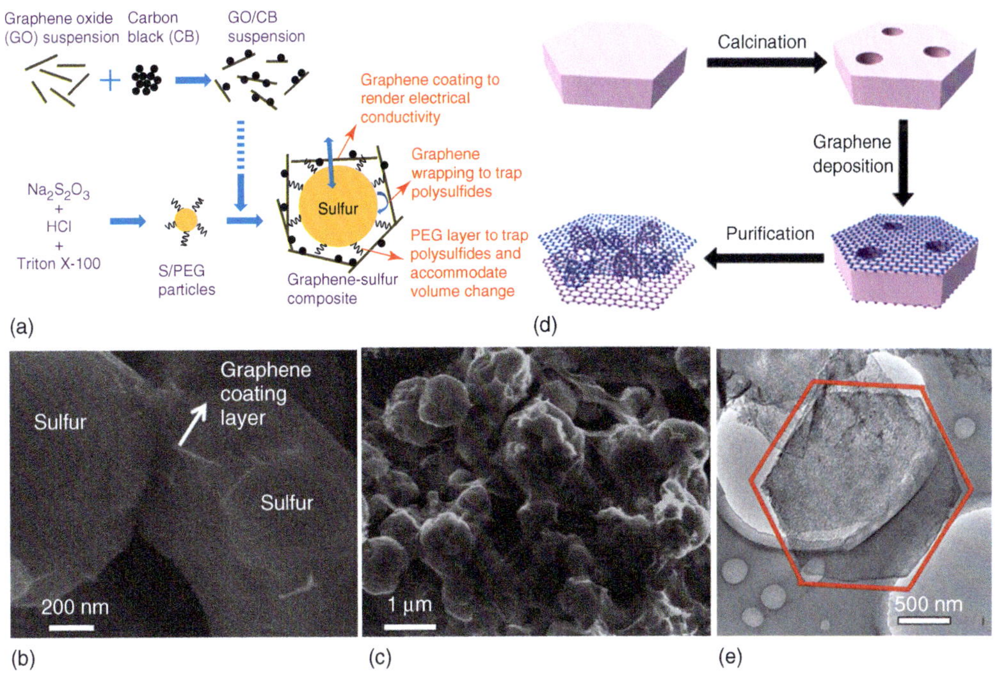


**Figure 3** (a) Schematic of the synthesis of a graphene–sulfur composite. (b,c) Representative SEM images of the graphene–sulfur composite. (d) Unstacked two-layer graphene synthesis scheme. (e) TEM image of unstacked two-layer graphene. [Panels (a–c) were reprinted with permission from Ref. 43. © 2011, American Chemical Society; panels (d,e) were reproduced with permission from Ref. 46. © 2014, Nature Publishing Group]

graphene–sulfur composite material exhibited an excellent electrochemical performance as a cathode material for Li–S batteries. The capacity was stable at 600 mAh $g^{-1}$ at 0.2 C and decayed only 15% after 100 cycles. In another study, Zhang *et al.*[44] synthesized graphene oxide sheets using the modified Hummers method, and then nitridized the graphene oxide (GO) sheets under an $NH_3$ atmosphere to obtain nitrogen-doped graphene (NG). Ultrafine sulfur particles were then deposited on the NG sheets by chemical reduction. A Li–S battery with this S@NG material as the positive electrode showed a specific discharge capacity of 1167 mAh $g^{-1}$ at 0.2 C, and the capacity decayed only 0.028% per cycle in 2000 cycles at 2 C. In a separate study, Yang and coworkers[45] recycled industrial waste gas $H_2S$ to prepare high-performance electrode materials for Li–S battery, whereby they reduced graphene oxide (rGO) with $H_2S$ to obtain a graphene/sulfur composite. The sulfur particles were uniformly dispersed on the graphene sheets, which were interconnected with a crimped structure.

It is well-known that graphene sheets are prone to stacking because of strong $\pi$–$\pi$ interaction. This will compromise the specific surface area and seriously affect the energy storage property. A number of strategies have been reported to inhibit the stacking of graphene. For instance, Zhao *et al.*[46] synthesized unstacked bilayer graphene by template-directed chemical vapor deposition. They used mesoporous SiC nanosheets as a template to deposit graphene and form graphene protrusions at the mesopores. These protrusions acted as a barrier to prevent stacking of graphene layers on both sides of the nanosheets, as shown in Figure 3(d)–(e). The graphene synthesized by this method showed high electrical conductivity; and loading of sulfur into the mesoporous structure effectively adsorbed polysulfides, inhibited the shuttle effect, and thus exhibited an excellent electrochemical performance. The specific discharge capacity reached 1034 and 734 mAh $g^{-1}$ at the high rates of 5 and 10 C, respectively. After 1000 cycles at 5 C, the specific discharge capacity still reached 530 mAh $g^{-1}$.

The Li–S batteries described above generally use an aluminum foil as a current collector, which accounts for 5–9% of the total weight of the battery, and thereby reduces the energy density of the battery.[47] Cheng and coworkers[48] proposed a unique sandwich structure (Figure 4), where they used graphene as the cathode current collector and membrane, forming a unique graphene–sulfur–graphene-separator structure between the cathode active material and the separator. The graphene layer was used as a protective layer, conductive layer, and also separator for the battery, which suppressed the shuttle effects of lithium polysulfides. The use of graphene as a positive current collector not only simplifies the electrode preparation process but also increases the energy density of the battery.

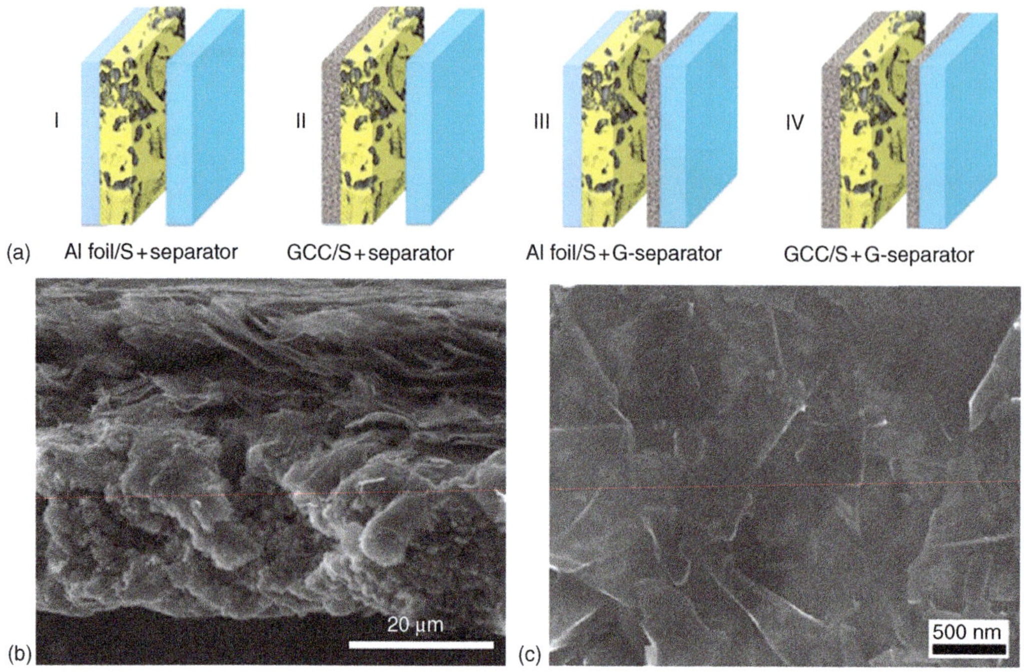


**Figure 4** (a) Diagram of an electrode structure formed by different materials. (b) SEM image of a graphene–sulfur–graphene sandwich structure. (c) SEM image of graphene nanosheets on a graphene separator. [Reproduced with permission from Ref. 48, © 2014, Wiley-VCH Verlag GmbH & Co. KGaA]

### 2.1.4 *Three-Dimensional Carbon Scaffolds*

Three-dimensional (3D) carbon scaffolds, consisting of a network of carbon nanoparticles and graphene, generally show a high specific surface area and a rich pore structure that can facilitate electron and ion transport.[49] Such a structure is also advantageous to increase sulfur loading, which is typically low with carbon nanoparticles or graphene alone. For instance, Zheng and coworkers[50] prepared hierarchical porous carbon rods (HPCR) decorated with vertically oriented porous graphene-like nanosheets. As depicted in Figure 5(a), MgO microrods were first hydrolyzed at room temperature to produce $Mg(OH)_2$ microrods decorated with vertically oriented $Mg(OH)_2$ nanosheets. Graphene-like carbon layers were then grown by chemical vapor deposition onto these $Mg(OH)_2$ microrods in a $CH_4$ atmosphere. After the removal of MgO by etching with a HCl aqueous solution, a carbon rod composed of vertically aligned porous carbon sheets was obtained (Figure 5b). The obtained 3D porous carbon showed a high specific surface area of $2226\,cm^2\,g^{-1}$ and pore volume of $4.9\,cm^3\,g^{-1}$. The skeleton consisted of interconnected vertical porous carbon sheets, a unique feature favorable for ion and electron transport. A Li–S battery with this material as the positive sulfur carrier exhibited an excellent electrochemical performance, and the reversible capacity reached $700\,mAh\,g^{-1}$ after 300 cycles at 1 C.

In another study, Hoffmann *et al.*[51] used silica nanospheres as templates and immersed them in a polycarbosilane precursor to obtain a $SiC/SiO_2$ composite. After removal of the silica template, a porous carbon material having monodisperse pores, pore size-adjustable and microporous carbon wall was obtained, as shown in Figure 5(c) and (d). This material showed an extremely high specific surface area of $2450\,m^2\,g^{-1}$ and pore volume of $5\,cm^3\,g^{-1}$. The large pores provided a high sulfur loading of 80 wt%, and the first specific discharge capacity of the battery with this material as the positive electrode reached $1165\,mAh\,g^{-1}$ at 0.1 C. In a separate study, Yuan *et al.*[52] prepared carbon nanopaper with a 3D interconnected conductive frame. CNTs of different lengths were alternately connected in the carbon paper. Short CNTs were employed as the short-range electrical conductive framework for sulfur accommodation, while long CNTs formed the basis of conductive networks and mechanical scaffolds. Such a material exhibited a sulfur loading of $17.3\,mg\,cm^{-2}$ and an area capacity of $15.1\,mAh\,cm^{-2}$.

In the above studies, the electrochemical performance of the sulfur/carbon composite electrodes was improved due to an increase of the specific surface area and pore volume of the electrode, but the influence of the structural characteristics on the electron-transfer kinetics was inadequately explained. Huang and coworkers[53] pointed out that smaller sulfur molecules have better cycle stability and electrolyte compatibility, but their discharge

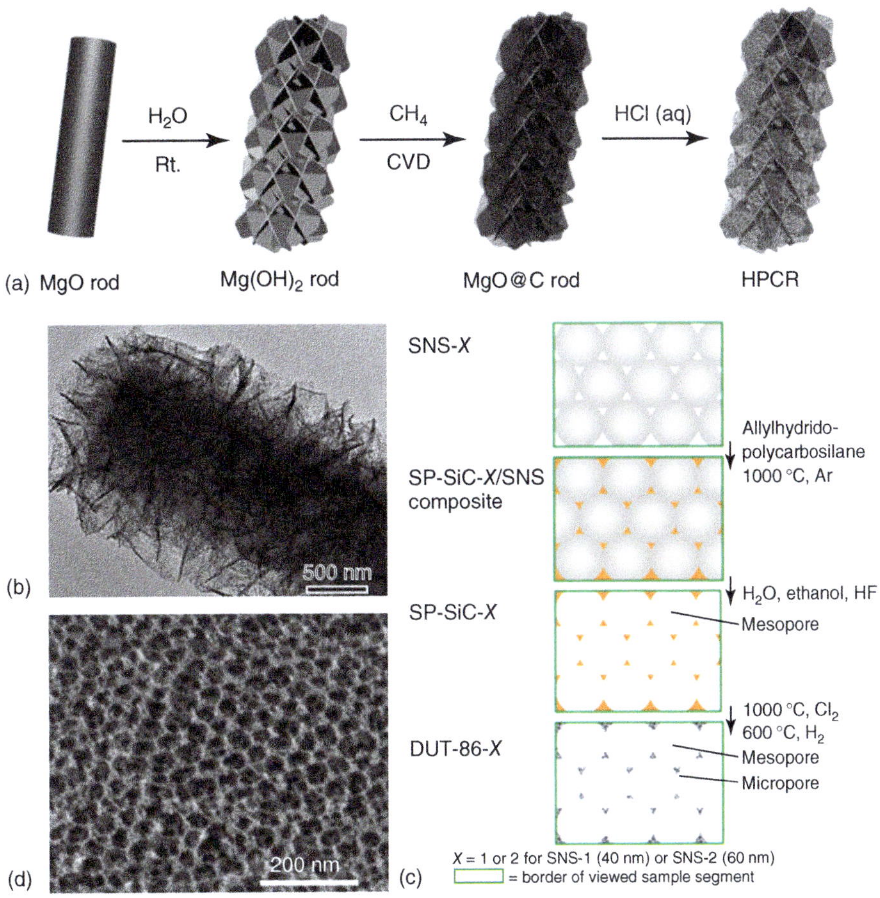


**Figure 5** (a) Schematic of the synthetic process for HPCR. (b) TEM image of HPCR. (c) Synthesis scheme from parent silica nanospheres (SNS-*X*) to nanoporous silicon carbides (SP-SiC-*X*) and further to carbide-derived carbon materials (DUT-86-*X*) with $X = 1$ and 2 for silica sphere size of 40 and 60 nm, respectively. White parts symbolize pores. (d) SEM image of microporous carbon wall porous carbon material. [Panels (a,b) were reproduced with permission from Ref. 50, © 2016, Wiley-VCH Verlag GmbH & Co. KGaA; panels (c,d) were reprinted with permission from Ref. 51. © 2014, American Chemical Society]

voltage platform is too low, thus reducing the energy density of the battery. Although larger sulfur molecules have a higher discharge platform, they can only exist in larger pores, and the electrolyte solution can easily penetrate into these large pores and cause dissolution and diffusion of polysulfides. If the advantages of the smaller sulfur molecules of $S_{2-4}$ and the larger $S_8$ molecule can be combined, it is possible to further improve the electrochemical performance of the battery. This will necessitate the development of new effective strategies in electrode design and engineering.

## 2.2 Polymeric Sulfur Strategy

The strategies described in the earlier section are primarily focused on physical methods to suppress polysulfide dissolution and the shuttle effects. The corresponding batteries often suffer a rapid decrease of capacity during charge–discharge cycles and hence have limited applications. Note that at ambient conditions, elemental sulfur exists in the form of an eight-membered ring ($S_8$). When elemental sulfur is heated up to 159 °C, equilibrium ring-opening polymerization (ROP) can start and form linear polysulfane with diradical chain ends. Chemical copolymerization strategies can, therefore, be employed to avoid the dissolution of long-chain polysulfides by copolymerization with different organic monomers, forming strong covalent bonds between the carbon frameworks and sulfur. Herein, we highlight three copolymers that have been used in Li–S batteries.

### 2.2.1 *Alkene-Derived Organosulfur Polymers*

Copolymerization between elemental sulfur and unsaturated alkenes has been widely studied to prepare sulfur-enriched copolymers. When alkenes are copolymerized with ring-opened sulfur forming C—S bonds, the resulting copolymers can be used as active cathodes for Li–S batteries. Cross-bonding and strong chemical

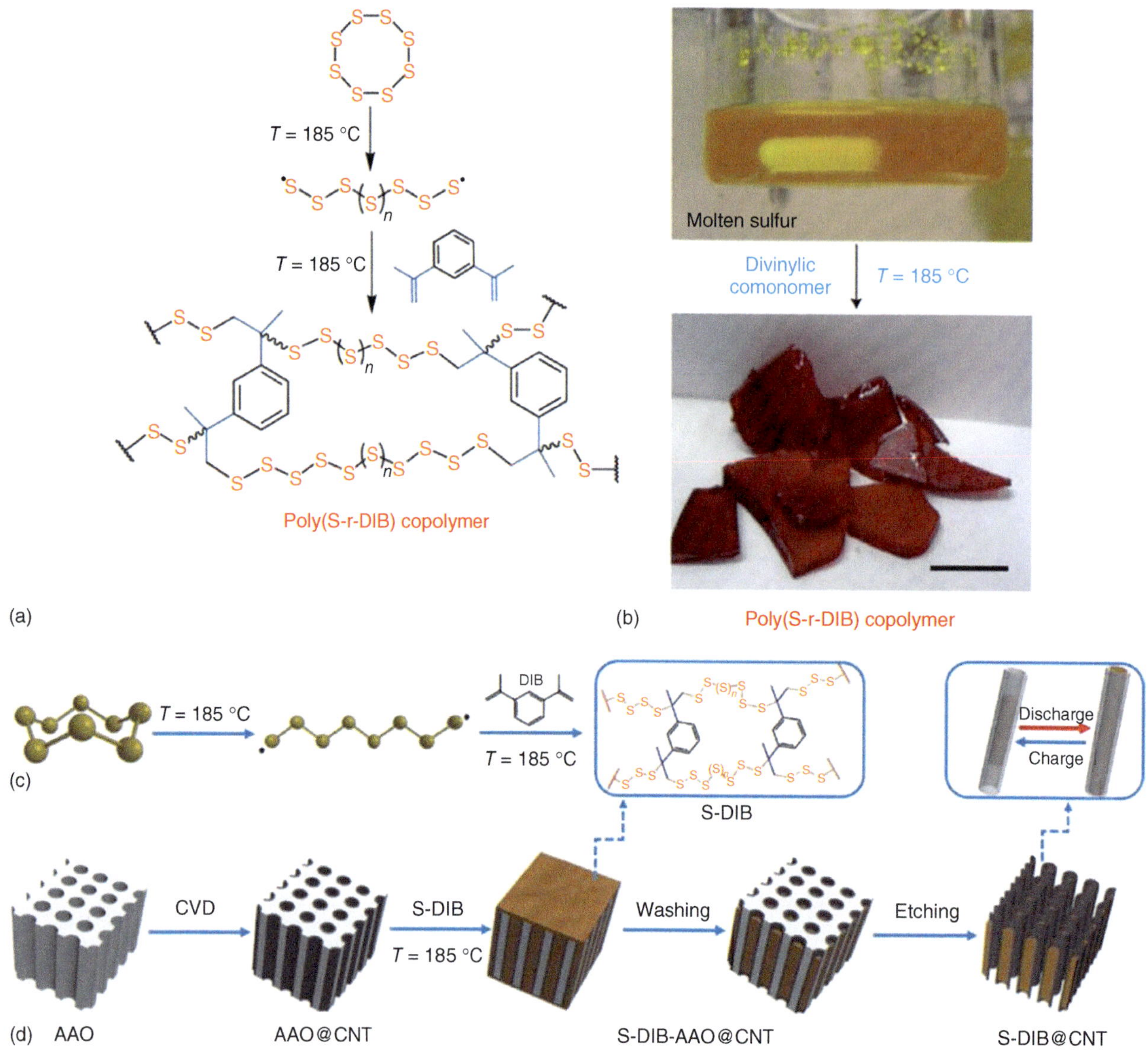


**Figure 6** (a) Synthetic scheme for the copolymerization of $S_8$ with DIB to form chemically stable sulfur copolymers. (b) Images of liquid sulfur and poly(S-r-DIB) glass. (c) Synthesis scheme of S–DIB copolymer. (d) Schematic of the fabrication process of the S-DIB@CNT hybrid. [Panels (a,b) were reproduced with permission from Ref. 10. © 2013, Nature Publishing Group; panels (c,d) were reproduced with permission from Ref. 54, © 2017, Wiley-VCH Verlag GmbH & Co. KGaA]

interactions of sulfur with carbon can effectively suppress the dissolution of polysulfides. For instance, Chung *et al.*[10] described an effective method to prepare sulfur-rich chemically stable polymeric materials by copolymerizing sulfur with 1,3-diisopropenylbenzene (DIB) (Figure 6a). The general strategy involved adding sulfur to a glass vial under magnetic stirring in an oil bath which was heated to 185 °C in order to open the sulfur rings (Figure 6b). Then the copolymer (S-DIB) was cooled to room temperature after adding DIB directly under stirring for 8–10 min. The resulting red-color S-DIB, was used as an electrochemically active sulfur reservoir material, which exhibited a high discharge capacity of 1100 mAh $g^{-1}$ and long-term cycling performance of 823 mAh $g^{-1}$ after 100 cycles at 0.1 C. Hu *et al.*[54] further improved the performance by infiltrating S–DIB into the AAO@CNT (Figure 6c), which was then sequentially washed by $CS_2$ and etched by NaOH (Figure 6d), affording a novel S-DIB@CNT hybrid cathode. The combination of physical and chemical confinement helped resolve the conductivity problem of S-r-DIB's, leading to enhanced cycling and rate performance with 880 mAh $g^{-1}$ at 1 C after 100 cycles and a capacity retention of over 98%.

In another study, our group[55] directly vulcanized a thiourea aldehyde resin (TAR) and formed TAR backbone structures that were highly crosslinked with

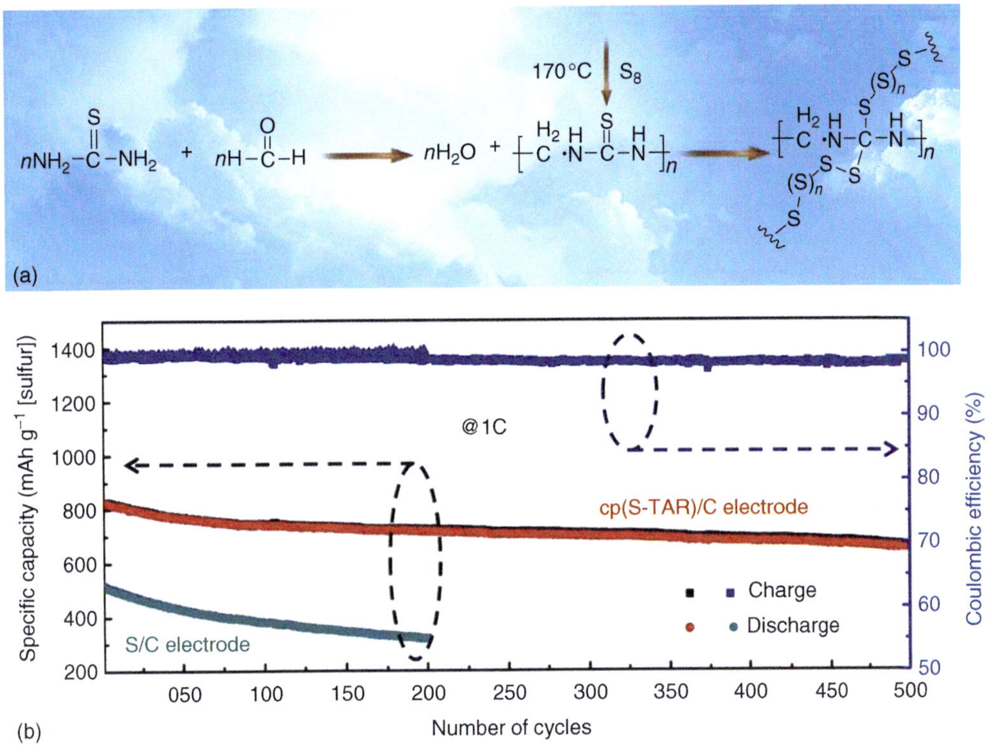


**Figure 7** (a) Synthetic procedure of cp(S–TAR). (b) Long-term cycling performance and Coulombic efficiency for cp(S–TAR)/C and S/C cathodes at 1 C. [Reproduced with permission from Ref. 55, © 2018, Elsevier]

sulfur (Figure 7a). The resulting sulfur-rich copolymer cp(S–TAR) comprised abundant mesopores, which provided short ion channels (good for $Li^+$ diffusion), leading to a high initial capacity of 1285 mAh $g^{-1}$. Moreover, the formation of C—S covalent bonds enhanced the cycling stability with an ultralow fading rate of 0.045% per cycle and a Coulombic efficiency of 99% during the 500 deep discharge–charge cycles (Figure 7b).

The polymer matrices used above are electrically insulating, not ideal for maximizing the utilization of active sulfur and hence the rate performance. To alleviate these problems, Zentel and coworkers[56] directly copolymerized sulfur with allyl-terminated P3HT (poly(3-hexylthiophene-2,5-diyl)) (Figure 8a and b). The covalent bonds endowed S-P3HT with an enhanced battery performance, as compared to a simple mixture of sulfur and P3HT which were not covalently linked. Notably, uniform incorporation of P3HT with elemental sulfur can constitute an interpenetrating framework that is good for electron and ion transport in Li–S battery, which can also partially account for the enhanced battery performance. In another study, Fu *et al.* templated the vulcanization reaction by using an AAO array to synthesize vulcanized polyisoprene (PIP) nanowires (Figure 8c), which indeed facilitated $Li^+$ transport.[57] In addition, it was found that the electrochemical lithiation behavior of covalently bonded sulfur chains in polyisoprene (SPIP) were apparently different from that of isolated elemental sulfur molecules or macromolecules. Depending on the length of sulfur chain, different lithium sulfides, such as $Li_2S$, $Li_2S_2$, and other lower-order polysulfides (both reversible and irreversible) could be directly formed because lithiation induced the cleavage of S—S bonds between the first and second S atoms in the sulfur chain. They also pointed out that the practical lithiation-delithilation reversibility might be contributed by the redistribution of the unanchored lithium sulfide species and the spatial confinement effects resulting from the crosslinked network, while the gradual capacity decay of SPIP cathodes was most likely due to the formation of irreversible C=S groups as well as thiol species during long-term charge–discharge cycling.

### 2.2.2 *Thiol-Derived Organosulfur Polymers*

Theoretically, thiol-containing molecules can also form covalent bonds with sulfur. For example, in practical applications, sulfur is widely used as an effective dehydrogenating reagent. Therefore, ring-opened sulfur can substitute the hydrogen in -SH to form S—S bonds, and the reaction mechanism for the synthesis of thiol-derived organosulfur polymers is somewhat similar to that for the alkene counterparts.[58] For instance, Park and coworkers[59] used trithiocyanuric acid (TTCA) crystals as soft templates

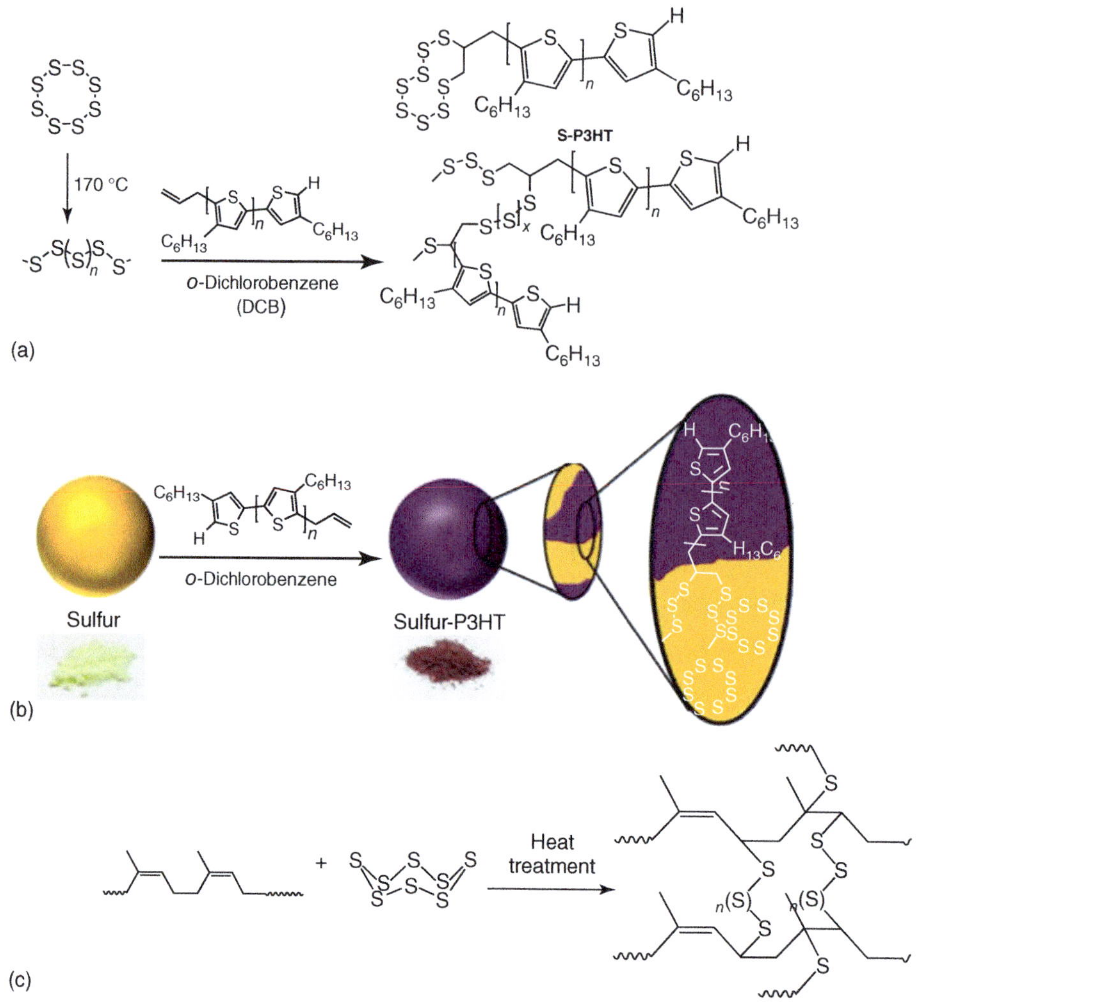


**Figure 8** (a) Synthetic approach for the copolymerization of allyl-terminated P3HT and sulfur. (b) Proposed microstructure of the copolymer-containing sample. (c) Schematic of the vulcanization of PIP. [Panels (a,b) were reprinted with permission from Ref. 56. © 2015, American Chemical Society; panel (c) was reprinted with permission from Ref. 57. © 2016, American Chemical Society]

to copolymerize with ring-opened sulfur through vulcanization. During heat treatment process, the crystallized TTCA was changed to porous frameworks, into which was impregnated with ring-opened sulfur followed by vulcanization (Figure 9a). Owing to the organized amine groups and covalent bonds in TTCA, the S-TTCA polymer showed an excellent discharge capacity of 945 $mAh\,g^{-1}$ after 100 cycles at 0.2 C and a rate performance of 1210 $mAh\,g^{-1}$ at 0.1 C, which was one of the highest specific capacities at that time. In order to enhance the performance of S-TTCA copolymer, Zou and coworkers[60] put TTCA into hollow carbon nanospheres (HCNs) with a high surface area of 2330 $m^2\,g^{-1}$ and large pore volume of 1.75 $cm^3\,g^{-1}$ and then vulcanized (Figure 9b). The HCNs–TTCA–S composite showed a high initial discharge capacity of 1430 $mAh\,g^{-1}$ at 0.1 C, but a restricted rate and cycle performance. Hence, in order to enhance the distribution and electroconductibility, Chen's group[61] activated polymerization of sulfur and TTCA onto the surface of highly conductive rGO nanosheets (Figure 9c). Owing to the highly crosslinked organosulfur and excellent electrical conductivity of rGO, the composite showed a remarkable rate performance of 1341 $mAh\,g^{-1}$ at 0.1 C and an enhanced battery cycling performance at a decay rate of only 0.0404% per cycle in 500 deep charge–discharge cycles, markedly better than those of similar copolymers. It is worth mentioning that the (S-TTCA)@rGO composite cathode had a higher loading of active sulfur at 2.0 $mg\,cm^{-2}$ than those of other copolymers (~1.0 $mg\,cm^{-2}$), which is a suitable material for commercialization.

Inspired by Park's work,[59] Je *et al.* in 2016 reported an orthogonal, one-pot synthetic approach to secure elemental sulfur through thiol groups.[62] The sulfur chains are covalently linked to benzoxazine polymer (BOP) in a quantitative yield by reacting 4,4-dihydroxydiphenyldisulfide and hexahydro-1,3,5-triphenyl-1,3,5-triazine (Figure 10a).

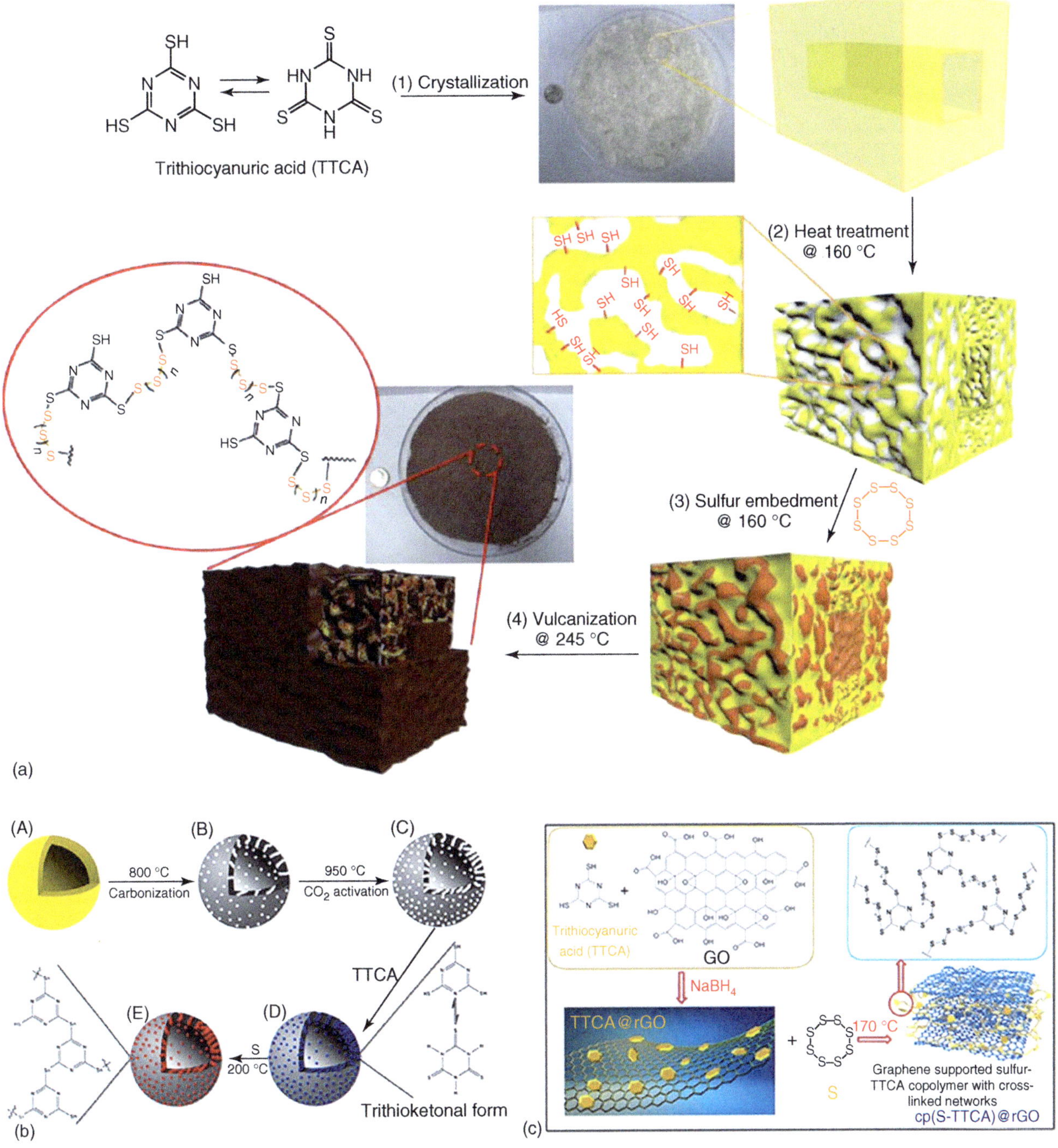


**Figure 9** (a) Synthetic route of sulfur-rich polymers. Schematic drawings describing the synthetic procedures of sulfur-rich polymers with controllable morphology. (b) Schematic illustration of the preparation of HCNs and HCNs-TTCA-S. (c) Schematic illustration of the synthesis of graphene-supported highly crosslinked sulfur copolymer nanoparticles, cp(S-TTCA)@rGO. [Panel (a) was reproduced with permission from Ref. 59, © 2015, Nature Publishing Group; panel (b) was reproduced with permission from Ref. 60, © 2017, Elsevier; and panel (c) was reproduced with permission from Ref. 61, © 2017, Elsevier]

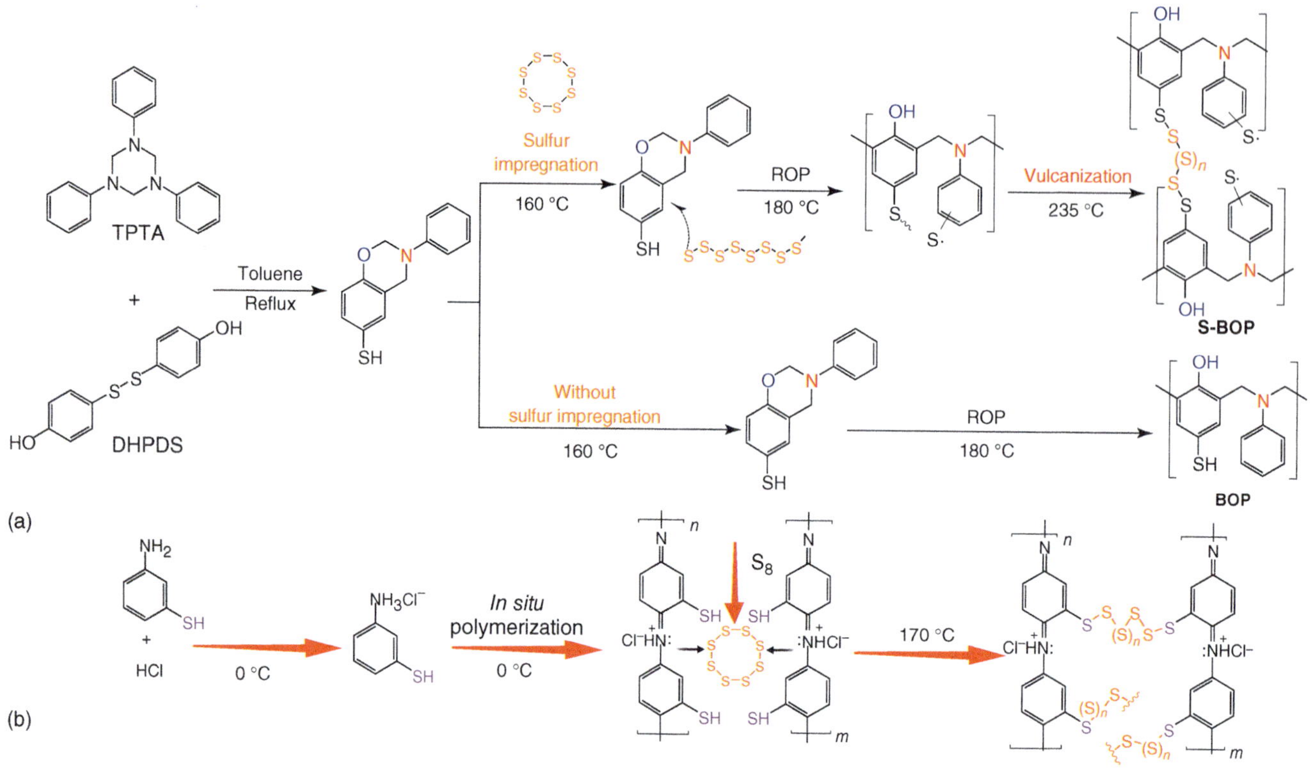


**Figure 10** Synthetic procedures of (a) S-BOP copolymer and (b) cp(S-PMAT) copolymer. [Panel (a) was Reprinted with permission from Ref. 62. © 2016, American Chemical Society; panel (b) was reproduced with permission from Ref. 63. © 2017, Wiley-VCH Verlag GmbH & Co. KGaA]

Owing to the homogeneous distribution of sulfur (in situ formation) and the C—S bonds, the S-BOP electrode delivered an appropriate cycling performance of 92.7% retention after 1000 cycles. Recently, Chen and coworkers[63] utilized m-aminothiophenol (MAT) to form conducting poly(m-aminothiophenol) (PMAT) with rich thiol groups. The resulting PMAT was then impregnated with molten elemental sulfur at 170 °C (Figure 10b) to form crosslinked sulfur side chains. Thanks to the highly crosslinked structure and high electrical conductivity of PMAT backbone, the cp(S–PMAT)/C cathode exhibited a high specific capacity (1240 mAh $g^{-1}$ at 0.1 C, 880 mAh $g^{-1}$ at 1 C and 600 mAh $g^{-1}$ at 5 C) and robust cyclability (66.9% retention, where a capacity of 495 mAh $g^{-1}$ was retained after 1000 cycles at a relatively high rate of 2 C).

### 2.2.3 *Nitrile-Derived Organosulfur Polymers*

Long-chain polysulfides can cause severe shuttling effects and affect the stability of Li–S batteries. One useful method to mitigate this issue is to decompose $S_8$ into small $S_{2-4}$ molecules. Wang and coworkers prepared sulfur–polyacrylonitrile (SPAN) composites (sulfurized polyacrylonitrile) and observed good charge and ion transport behaviors that were needed for Li–S battery. Yet the chemical structure and electrochemical performance of SPAN remained under active debate. Wang and co-workers[64] argued that elemental sulfur was embedded in the pyrolyzed PAN (Figure 11a), whereas Yu *et al.*[65] and Fanous *et al.*[66] believed sulfur was covalently bonded onto the PAN backbones in the form of short -$S_x$- chains (Figure 11b and c). In 2014, Zhang *et al.*[67] demonstrated that the structures reported by Yu and Fanous were inaccurate because of inconsistency of the resulting of *C*/*H* ratio and the correct one was the one shown in Figure 11d.

Recently, Wang's group[52] synthesized PAN nanoparticles, less than 100 nm in size, on the surface of graphene nanosheets (GNS) by a simple in situ method, followed by sulfurization (Figure 12a). Because the as-synthesized pPAN–S particles had an intimate contact with GNS and could be well-dispersed on GNS, the electrochemical measurements of the pPAN–S/GNS composite cathode showed an excellent discharge capacity of ca.1500 mAh $g^{-1}$ (corresponding to ca. 90% sulfur utilization) and a competitive capacity of ca. 800 mAh $g^{-1}$, even at up to 6 C. More recently, a highly ordered mesoporous sulfurized polyacrylonitrile (MSPAN) was obtained by directly heating a SBA-15 template embedded with sulfur followed by removal of SBA-15 templates with acidic etching (Figure 12b).[68] The resulting composite MSPAN

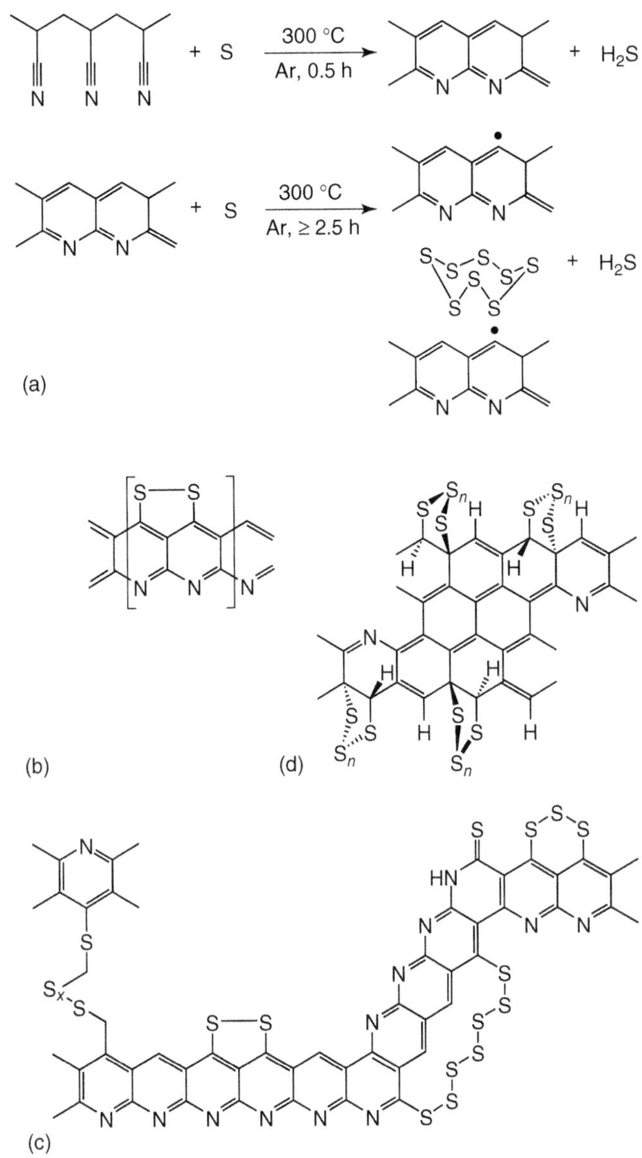


**Figure 11** (a) Schematic of the thermo-reaction between PAN and sulfur. Possible molecular structures of SPAN according to Yu's (b) and (c) Fanous's reports. (d) Revised structure of SPAN proposed by Zhang, matching the results of *C*/*H* ratio in SPAN. [Panel (a) was reproduced with permission from Ref. 64, © 2003, Wiley-VCH Verlag GmbH & Co. KGaA; panel (b) was reproduced with permission from Ref. 65, © 2004, Elsevier; panel (c) was reprinted with permission from Ref. 66, © 2011, American Chemical Society; panel (d) was reproduced with permission from Ref. 67, © 2014, MDPI]

provided a large surface area and abundant porous structures. The corresponding nanocomposite cathode exhibited a stable cycling property with 610 mA h $g^{-1}$ after 900 cycles at 2 C (Figure 12c). Even at 5 C, the MSPAN-based cathode could also display reasonable capacity retention.

In 2015, Coskun and coworkers[69] prepared S-CTF-1 (covalent triazine framework) through a catalyst- and solvent-free method without using any external templates where elemental sulfur was incorporated onto 1,4-dicyanobenzene at 400 °C. The resulting copolymer S-CTF-1 constituted a macrocycle backbone with multiple sulfur side chains (Figure 13a). When used as cathode in Li–S batteries, it exhibited an apparent electrochemical performance, with 85.8% capacity retention after 300 cycles along with a high initial Coulombic efficiency (ICE) of ca. 94.4% at 0.05 C. In order to further improve the electrochemical performances, Xu *et al.*[70] introduced electronegative fluorine onto the CTF (Figure 13b). The polar fluorine species was argued to serve as anchoring sites for chemical adsorption of polysulfides, and hence suppress the dissolution and accelerate the conversion of polysulfides. As a result, the fluorinated porous triazine-based frameworks (FCTF)-S showed a superior performance as compared to CTF–S without fluorine,

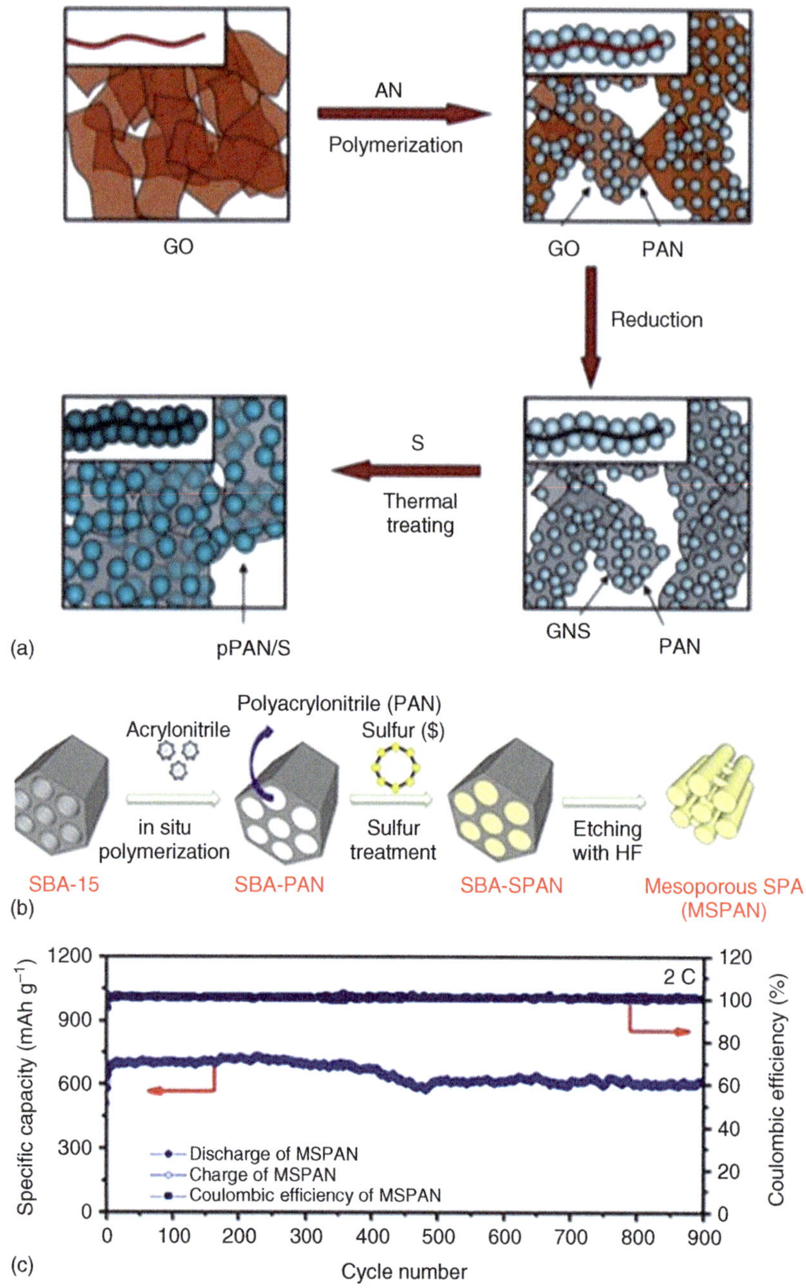


**Figure 12** (a) Illustration of in situ polymerization and synthesis of pPAN-S/GNS. (b) Schematic diagram of the sequential fabrication steps for MSPAN. (c) Cycling performance and Coulombic efficiency of MSPAN cell at 2 C rate. [Panel (a) was reproduced with permission from Ref. 52, © 2014, Wiley-VCH Verlag GmbH & Co. KGaA; panels (b,c) were reprinted with permission from Ref. 68. © 2017, American Chemical Society]

with an initial discharge capacity of 1296 mA h $g^{-1}$ at 0.1 C and a retention of 833 mA h $g^{-1}$ after 150 cycles at 0.5 C.

## 2.3 Incorporation of Polar Compounds

In addition to the above physical confinement and covalent bonding strategies, recent studies have also shown that chemical binding of polysulfides is an important, novel strategy to improve the performance of Li–S batteries. In theory, an ideal chemical binding material should be able to selectively control the shuttling of polysulfide anions via strong bonding interactions between them, while not disturbing $Li^+$ ion transfer. Developing a lightweight and chemically selective host materials is therefore of urgent importance. In the following section, we will summarize the

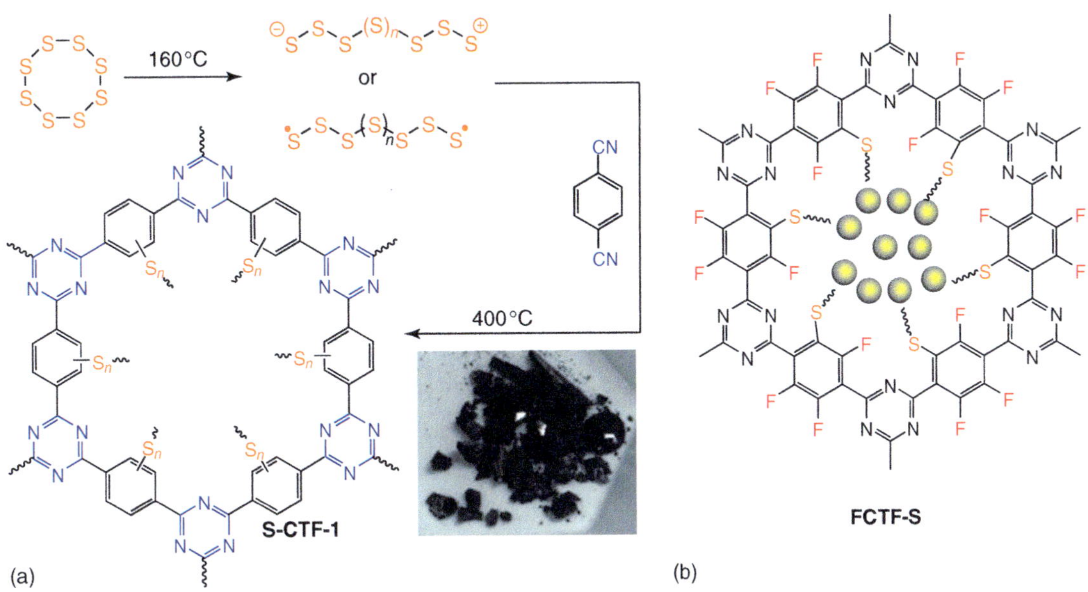


**Figure 13** (a) Synthesis procedure of S-CTF-1 and an optical image of S-CTF-1. (b) Chemical structure of FCTF-S. [Panel (a) was reproduced with permission from Ref. 69, © 2016, Wiley-VCH Verlag GmbH & Co. KGaA; panel (b) was reprinted with permission from Ref. 70. © 2017, American Chemical Society]

commonly used polar anchoring materials for tailoring the performance of Li–S batteries.

### 2.3.1 *Heteroatom-Doped Nanocarbon*

It was first recognized by Zhang *et al.*[71] that the weak Van der Waals' (vdW) interaction between the intrinsic nonpolar carbon-based matrix materials and polar $Li_2S_x$ species could only provide weak confinement, hence detachment of $Li_2S_x$ species from the carbon matrix surface usually occurred, and their further diffusion into electrolyte caused rapid degradation in capacity and rate performance. Intuitively, heteroatom doping should change the electronic state of carbon surface and hence modify the polarity of carbon matrix. A systematic density functional theory (DFT) calculation suggested that chemical modification with N or O dopant significantly enhanced the interaction between the carbon matrix and $Li_2S_x$ species via electrostatic attraction and thereby effectively prevented the appearance of polysulfides shuttle, enabling a high discharge capacity and a high Coulombic efficiency.[72] Moreover, both N and O heteroatoms with extra pairs of electrons were considered as electron-rich donors that naturally served as a Lewis base sites to effectively interact with strong Lewis acid of Li terminal atoms in $Li_2S_x$ species via dipole–dipole electrostatic interaction. In contrast, the monodoping of B, F, S, P, or Cl into the carbon matrix was unsatisfactory, in terms of the strength of electrostatic interaction with Li terminal atoms in $Li_2S_x$ species.

Subsequently, theoretical calculations were conducted to determine the binding energy ($E_b$) of the series of doping elements in different doping configurations. From the results showed in Figure 14, the principles for the rational design of doped carbon scaffolds to achieve strong coupling with $Li_2S_x$ species in Li–S batteries can be summarized as follows: (i) doping atom has a lone pair electron; (ii) doping atom exhibits a higher electronegativity than carbon and a small radius that matches Li; (iii) doping atom forms a delocalized $\pi$ bond with the conjugated system; and (iv) doping atom forms a stable bond to the carbon plane. With the fulfillment of these conditions, the carbon matrix could form a strong electrostatic dipole–dipole interaction with $Li_2S_x$ species, coinciding with the theoretical and experimental results that N or O doping is a better choice to relieve the shuttle effect.

In another work, Pang *et al.*[73] developed a strategy based on strong chemisorption of polysulfides utilizing an ordered nanoporous carbon dual-doped with nitrogen and sulfur (Figure 15). The doped carbon was synthesized from self-templating of biomaterial-derived cellulose nanocrystals and exhibited a tunable hierarchically porous structure as well as high surface area. The dual doping with N and S atoms of the porous carbon remarkably enhanced the chemisorption of lithium polysulfides. XPS studies revealed a synergistic N–Li and S–S interactions. The binding configuration was also confirmed by ab initio DFT calculations, where $Li^+$ and doped N atoms formed Li–N chemical bond while the polysulfide anions and doped S atoms formed S–S. In addition, the electrical conductivity of the porous carbon was greatly improved,

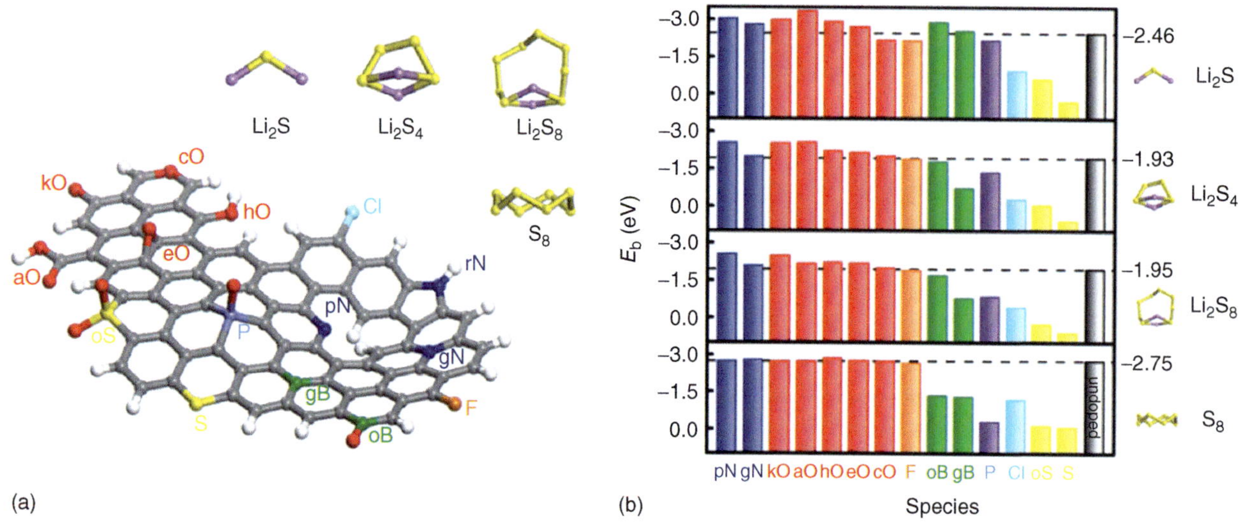


**Figure 14** (a) Schematic diagram of $X$-doped nanocarbon materials ($X$ = N, O, F, B, P, S, Cl) and PBE level optimized structure of $Li_2S$, $Li_2S_4$, $Li_2S_8$, and $S_8$ molecules, which are key intermediates in the charge/discharge process of Li–S battery. (b) Binding energy $E_b$ (eV) of $Li_2S$, $Li_2S_4$, $Li_2S_8$, and $S_8$ interacting with $X$-doped GNRs with zigzag edge, and binding energies of $Li_2S$, $Li_2S_4$, $Li_2S_8$, and $S_8$ interacting with pristine (undoped) GNR are shown as grey bars and dashed lines for reference. Carbon, hydrogen, nitrogen, oxygen, fluorine, boron, phosphorus, chlorine, lithium, and sulfur elements are represented by grey, white, blue, red, orange, green, lavender, cyan, purple, and yellow, respectively. [Reprinted with permission from Ref. 72, © 2016, Wiley-VCH Verlag GmbH & Co. KGaA]

as compared to undoped carbon. Using this dual-doped carbon as a sulfur host, the sulfur electrode was able to deliver a high capacity of 1370 mAh $g^{-1}$ at 0.05 C and discharge/charge for 1100 cycles at 2 C rate with a very low capacity fading rate of 0.052% per cycle. These results demonstrated that a synergistic functionalization of intrinsic carbons with nitrogen and sulfur heteroatoms dramatically modified the electron density distribution of the host carbons substrate, leading to much stronger polysulfide binding than that for undoped or nitrogen mono-doped carbons.

Besides heteroatom doping, direct incorporation of nitrogen-rich polar materials is also proven to be effective for polysulfides anchoring/binding. Pang and Nazar[74] reported that introduction of a light-weight nanoporous graphitic carbon nitride (high-surface-area g-$C_3N_4$) into sulfur cathode enabled an ultralow capacity fading rate of only 0.04% per cycle over 1500 long-term cycles at a practical rate of 0.5 C. On the basis of the results from spectroscopic studies and first-principles theoretical calculations, they pointed out that the ultrahigh concentration of accessible pyridinic nitrogen (up to 53.5 at%) of g-$C_3N_4$ acted as strong chemisorption sites for lithium polysulfides, which accounted for the greatly improved cycling performance, as compared to that of N-doped or undoped carbons.

### 2.3.2 *Transition Metal Sulfides*

With the advancement in the synthesis methods of 2D dichalcogenides in recent years, various transition metal sulfides have been explored for Li–S batteries.[75] As compared with carbon-based materials, transition metal sulfides demonstrate several advantages, such as (i) low lithiation voltages vs Li/$Li^+$, which is outside of the working voltage window of Li–S batteries; (ii) a strong sulfiphilic property toward the $Li_2S_x$ species; and (iii) there are a large number of metallic or half-metallic phases of metal chalcogenides, such as pyrite, spinel, and NiAs structures, which are highly electronically conductive. Many transition metal sulfides have been proposed as polysulfide-anchoring materials, considering their polarity and electrical conductivity. For example, $TiS_2$ was the earliest light-weight intercalating cathode material to be used in secondary lithium batteries,[76] and exhibited a high diffusion rate of lithium ions and high electronic conductivity. With its polar surface a relatively stable sulfur cathode comprised of $TiS_2$ host was prepared by Archer and coworkers,[77] showing a marked capacity of about 17 mAh $cm^{-2}$ after 100 cycles with a high sulfur loading of 40 mg $cm^{-2}$. They also performed DFT calculations and found that the binding energy between lithium polysulfides and $TiS_2$ was about 2.60 eV, which is nearly three times greater than that between lithium polysulfides and polyacrylonitrile and around 10 times higher than that between lithium polysulfides and carbon-based graphene.[78] This strong chemical interaction between lithium polysulfides and $TiS_2$ host significantly alleviated the dissolution of polysulfides into electrolytes and hence enabled an improved cycling performance.

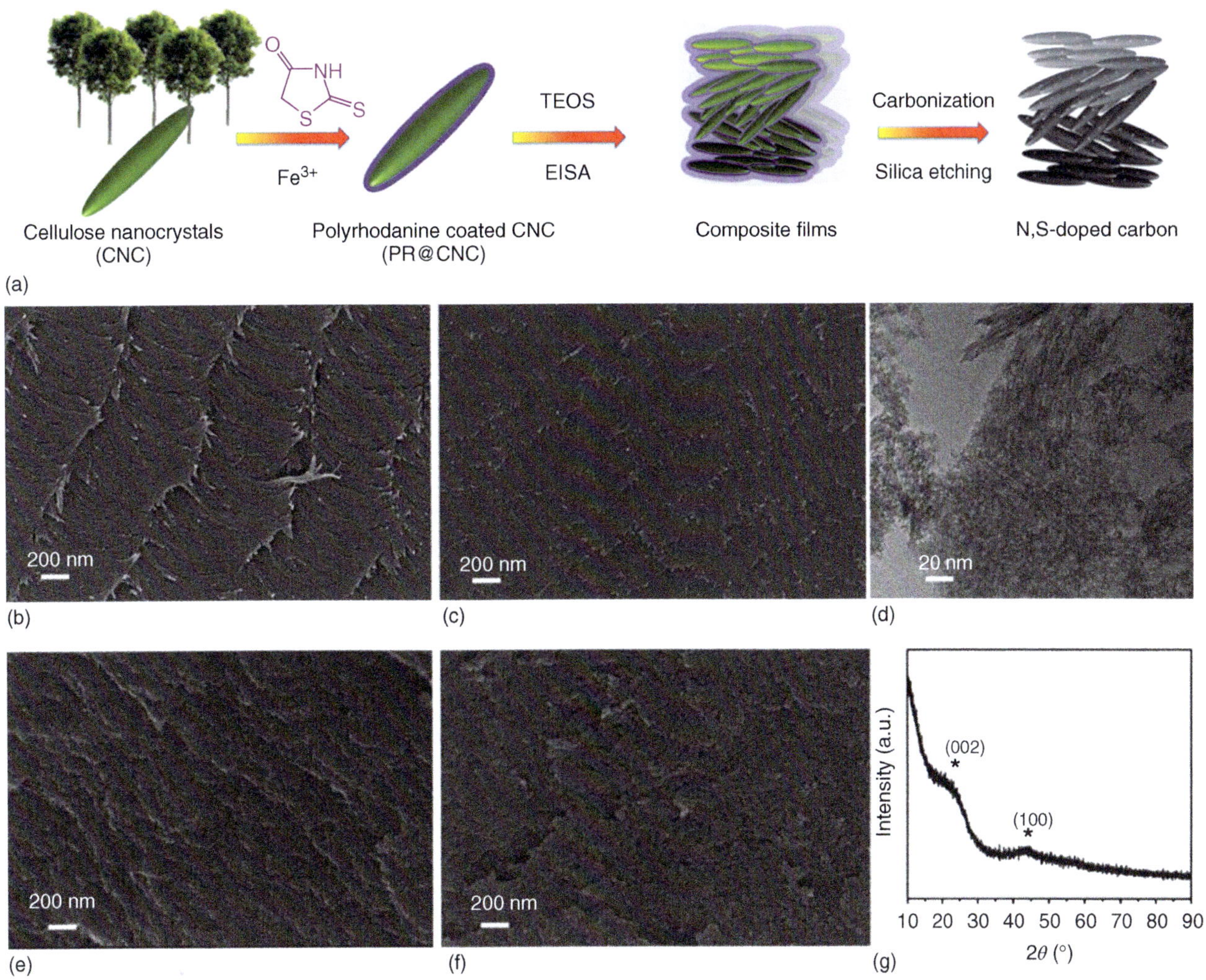


**Figure 15** (a) Schematic illustration of the synthesis of nitrogen/sulfur-doped cellulose (NSC); (b) SEM images of periodic self-assembled cellulose nanocrystals (CNCs) and (c) PR@CNC precursors for the synthesis of CDC and NSC, respectively; (d) TEM and (e) SEM images of NSC; (f) SEM image of the NSC/S-70 composite; (g) XRD pattern of the NSC. [Reprinted with permission from Ref. 73, © 2015, Wiley-VCH Verlag GmbH & Co. KGaA]

Actually, $TiS_2$ was theoretically proven to be an ideal chemical anchoring material for polysulfides by Cui's group.[71] They used a first-principles method with vdW interaction included to systematically investigate the adsorption of different $Li_2S_x$ species on transition metals-based 2D layered materials (e.g., oxides, sulfides, and chlorides), and study the detailed interaction and electronic structure, including binding strength, configuration distortion, and charge transfer. The resulting simulated binding energies ($E_b$) with vdW correction at different lithiation stages are shown in Figure 16, where one can find that for the adsorption of pristine $S_8$, all the investigated materials show a small binding energy in the range of 0.75–0.85 eV, which is very comparable to that for graphene, regardless of crystalline graphene (c-G) and amorphous graphene (a-G) (Figure 16a). However, when lithiation starts, the binding energy for the different materials varies (Figure 16b). For $V_2O_5$ and $MoO_3$, the binding energies of both oxides are higher than 2.0 eV over the whole lithiation process. For the chlorides ($TiCl_2$ and $ZrCl_2$), their $E_b$ is lower than 0.80 eV. For most of the sulfides, including $MoS_2$, $TiS_2$, $ZrS_2$, $NbS_2$, and $VS_2$, their binding energies are moderate, not so high as that for oxides, but much higher than that for chlorides. Specifically, except for $MoS_2$, all the sulfides shown in Figure 16 demonstrate an $E_b$ value higher than 1.00 eV. Figure 16 also reveals that the evolution trend of $E_b$ with lithiation process for transition metal chlorides is similar to that of graphene, but opposite to that for both transition metal oxides and sulfides during lithiation. Because the $E_b$ continuously increases for both transition metal oxides and sulfides as lithiation proceeds, the anchoring effect of these two kinds of materials should mainly derive from their chemical interaction with the Li atoms by forming

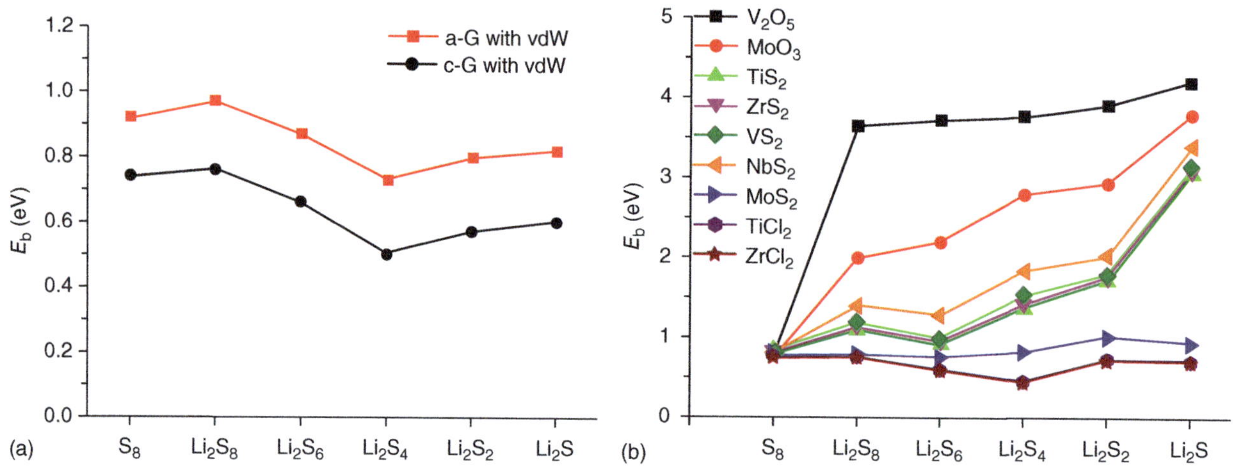


**Figure 16** Variation of binding energies of lithium sulfides at four different lithiation stages ($S_8$, $Li_2S_6$, $Li_2S_4$, and $Li_2S_2$) on various anchoring materials. (a) a-G and c-G; and (b) different layered transitions metal compounds. [Reprinted with permission from Ref. 71. © 2015, American Chemical Society]

Li—S and Li—O bonds on transition metal sulfides and oxides, respectively. In contrast, the weak anchoring effect between chlorides and $Li_2S_x$ species is mainly attributed to the vdW interaction.

From the viewpoint of anchoring effect, a higher $E_b$ enables a stronger binding strength for $Li_2S_x$ species. However, anchoring materials with too strong binding strength (such as $V_2O_5$ and $MoO_3$) cause the break of Li—S bonds in $Li_2S_x$ species, leading to the formation of isolated $Li^+$ and $S^{2-}$ ions, which can be readily dissolved into the electrolyte. Moreover, the decomposition of Li—S bonds in $Li_2S_x$ species will impair the effect of sulfur, which is the key functional material that imparts the high capacity to the cathode of Li–S batteries. Therefore, too strong anchoring materials such as $V_2O_5$ and $MoO_3$, with a binding energy higher than 2.0 eV, might not be good candidates, while those with a moderate binding energy can maintain a balance between binding strength and integrity of the $Li_2S_x$ species, overcoming the inherited disadvantages of both too strong and too weak anchoring materials. These findings can explain very well why anchoring materials with a medium binding energy, including $TiS_2$, $ZrS_2$, and $VS_2$, can show superior cycling performances to many oxide materials.[79]

In a subsequent experimental study,[80] Zhou *et al.* confirmed that transition metal sulfides based moderate-strength anchoring materials such as $TiS_2$, $VS_2$, and $CoS_2$, could effectively anchor $Li_2S_6$. As shown in Figure 17(a), after prolonged contact with Li2S6, nonpolar G/CNT had no apparent effect on adsorbing polysulfides as the color of the solution remained the same as that of the control sample, indicating only weak physical adsorption for G/CNT sample. In contrast, FeS and $SnS_2$ both demonstrated higher adsorption capability of $Li_2S_6$, as compared to G/CNT,[81] whereas $Ni_3S_2$ exhibited lower adsorption, as demonstrated by the lack of any significant color change (Figure 17b–d). Notably, the originally yellow-colored polysulfide solution became colorless after the addition of $TiS_2$ or $VS_2$, and became much lighter in color for $CoS_2$, suggesting a strong interaction between $Li_2S_6$ and these sulfide hosts (the binding energy was about 1.0 eV for $Li_2S_6$) (Figure 17e–g).

In addition to the above anchoring effect, Zhou *et al.*[80] discovered that some of the transition metal sulfides were able to serve as an activation catalyst to facilitate the oxidation of the discharge product, $Li_2S$, back to the charge product, $S_8$. Specifically, from the cyclic voltammetry (CV) curves of a sulfur cathode employing $SV_2$ as the anchoring materials (Figure 18a), one can see two cathodic peaks at around 2.30 V (denoted as $I_{C1}$) and 1.95 V (denoted as $I_{C2}$), due to the reduction of $S_8$ to long-chain lithium polysulfides ($S_8 \rightarrow Li_2S_x$) and the subsequent formation of short-chain $Li_2S_2/Li_2S$ ($Li_2S_x \rightarrow Li_2S_2/Li_2S$), respectively. Accordingly, the anodic peak at around 2.50 V (denoted as $I_A$) derives from the conversion of short-chain $Li_2S_2/Li_2S$ to long-chain polysulfides and finally $S_8$ ($Li_2S_2/Li_2S \rightarrow S_8$). The heights of these three current peaks ($I_{C1}$, $I_{C2}$, $I_A$) show a linear correlation with the square root of potential scanning rate of all the investigated electrodes employing various anchoring materials (Figure 18b–d), signifying a diffusion-limited process. Therefore, the lithium diffusion process can be approximately described with the classical Randles–Sevcik equation,[82,83] $I_p = (2.69 \times 10^5) n^{3/2} S (D_{Li^+})^{1/2} C_{Li^+} v^{1/2}$, where $I_p$ is the corresponding peak current of the CV curve, $n$ corresponds to the charge transfer number, $S$ is the geometric surface area of the active part of the working electrode, $D_{Li^+}$ is the diffusion coefficient of $Li^+$, $C_{Li^+}$ is the concentration of $Li^+$

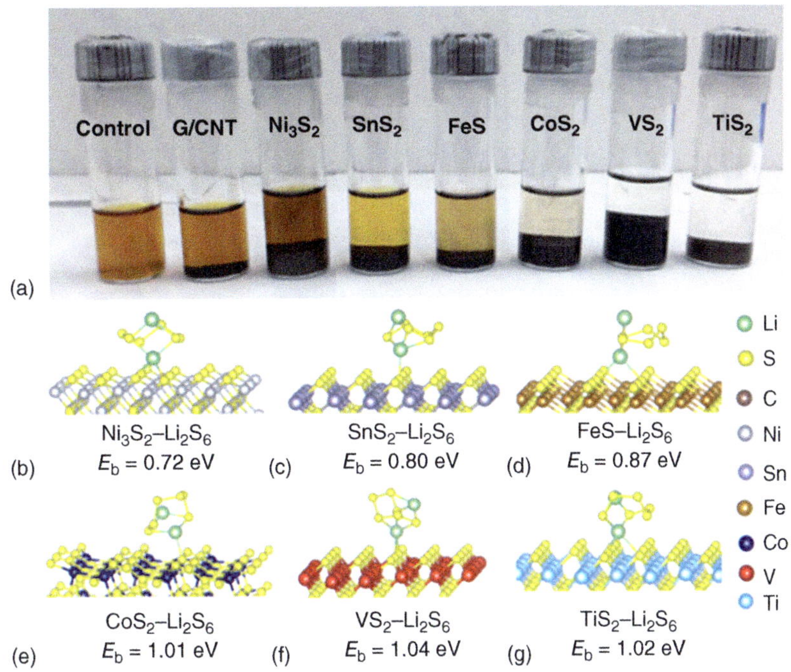


**Figure 17** Lithium polysulfide ($Li_2S_6$) adsorption by carbon and metal sulfides and corresponding simulation of $Li_2S_6$ adsorbed on the surface of metal sulfides. (a) Digital image of the $Li_2S_6$ (0.005 M) captured by carbon and metal sulfides in a DOL/DME solution. Atomic conformations and binding energy for $Li_2S_6$ species adsorption on (b) $Ni_3S_2$, (c) $SnS_2$, (d) FeS, (e) $CoS_2$, (f) $VS_2$, and (g) $TiS_2$. Here, green, yellow, gray, purple, brown, blue, red, and cyan balls represent lithium, sulfur, nickel, tin, iron, cobalt, vanadium, and titanium atoms, respectively. [Reprinted with permission from Ref. 80, © 2017, National Academy of Sciences]

in sulfur cathode, and $v$ is the electrode potential scanning rate. From the linear plots depicted in Figure 18(b)–(d), one can find that the carbon-based G/CNT anchoring material displays the smallest slopes and hence shows the lowest lithium ion diffusivity, which is attributed to the weak adsorption of $Li_2S_x$ species and inefficiency in catalyzing $Li_2S$ conversion, probably caused by the high viscosity of $Li_2S_x$ in electrolyte, or the deposition of a thick nonconductive layer on electrode surface. By contrast, the electrodes with $VS_2$, $CoS_2$, or $TiS_2$ exhibit much faster $Li^+$ diffusion than G/CNT and also better redox kinetics than those electrodes with $Ni_3S_2$, $SnS_2$, and FeS, indicating that the introduction of moderate-strength polar metal sulfide hosts can efficiently catalyze the redox conversion of sulfur. The high $Li^+$ diffusivity on $CoS_2$, $VS_2$, and $TiS_2$ is also supported by their simulation results, where graphene showed the highest $Li^+$ diffusion barrier of about 0.30 eV, while for $Ni_3S_2$, $SnS_2$, and FeS, the $Li^+$ diffusion barriers were lower but about 0.1 eV larger than those for $CoS_2$, $VS_2$, and $TiS_2$. Therefore, the requirements for a suitable anchoring material in the cathode should include: (i) a moderate binding energy with $Li_2S_x$ species; (ii) capability in rational control of $Li_2S$ deposition; (iii) fast $Li^+$ diffusion ability; and (iv) efficient catalytic conversion of the sulfur redox. These requirements can help decrease electrode polarization, increase active sulfur utilization, and improve rate performance and the charge–discharge cycling stability.

Moreover, pyrite $CoS_2$ was also proven to have high catalytic activity in polysulfide reduction by Jin and coworkers.[84] The pyrite type $CoS_2$ crystal has marked electronic conductivity of $6.7\times10^3\ S\ cm^{-1}$ at 300 K. Yuan *et al.* incorporated the sulfiphilic half-metallic pyrite-type $CoS_2$ into the cathode of Li–S battery to facilitate the conversion process of polysulfide redox[85] by mixing $CoS_2$ with graphene and sulfur at different ratios. CV measurements showed that the current density significantly increased by an order of magnitude as the content of $CoS_2$ was increased from 0 to 30 wt%, demonstrating that the interactions of $CoS_2$–polysulfides not only statically existed but also dynamically accelerated the electrochemical reactions of $Li_2S_x$. Electrochemical impedance spectroscopic (EIS) tests revealed that charge transfer at the $CoS_2$–polysulfide interface was much faster than that at the graphene–polysulfide interface, and the redox kinetics of polysulfides in liquid phase ($Li_2S_8 \leftrightarrow Li_2S_6 \leftrightarrow Li_2S_4$) were also improved after introduction of sulfiphilic $CoS_2$ hosts. Along this line, hierarchically porous $CoS_2$/carbon composite paper was developed and applied as an interlayer for capturing lithium polysulfides through physical absorption and also chemical binding in a working Li–S battery.[86] The corresponding composite $CoS_2$/C/S cathode was able to deliver a high initial sulfur utilization of 74% at 0.2 C and demonstrate a long cycling life.

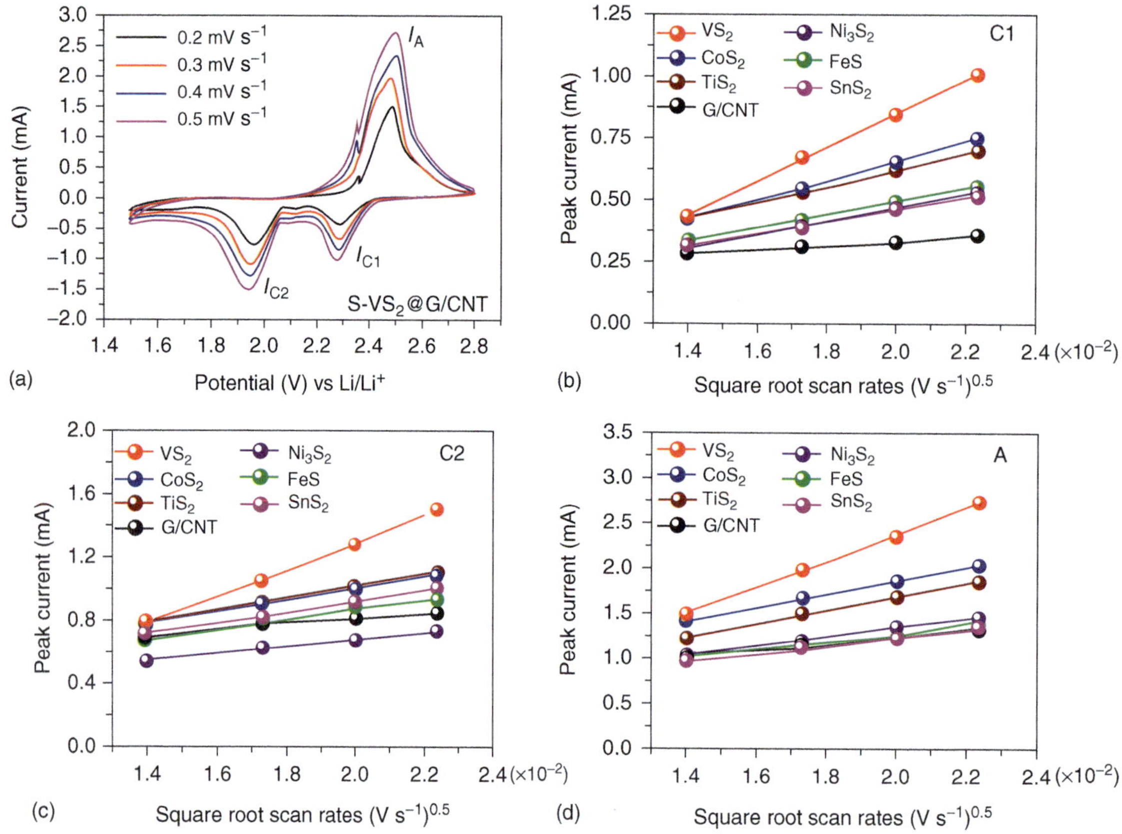


**Figure 18** (a) CV curves of the sulfur cathode employing various anchoring materials at different scan rates. CV peak current for the (b) first cathodic reduction process ($I_{C1}$ : $S_8 \rightarrow Li_2S_x$); (c) second cathodic reduction process ($I_{C2}$ : $Li_2S_x \rightarrow Li_2S_2/Li_2S$); and (d) anodic oxidation process ($I_A$ : $Li_2S_2/Li_2S \rightarrow S_8$) against the square root of scan rates. [Adapted with permission from Ref. 80, © 2017, National Academy of Sciences]

In addition to the transition metal sulfides discussed above, recently 3D interconnected graphene-like $Co_9S_8$ nanosheets were synthesized through a microwave-assisted solvothermal method by Pang *et al.*[87] $Co_9S_8$ is a member of the Pentlandite family, with a particularly high room-temperature electrical conductivity of 290 S cm$^{-1}$. In contrast to results in earlier studies where only Li–S (or Li–O) binding between the Li atoms in $Li_2S_x$ species and the layered metal sulfide (oxide) was observed, a strong synergistic binding effect from both Li and S ions in $Li_2S_x$ species was confirmed. The chemical binding of positively charged $Li^+$ to $S^{2-}$ of $Co_9S_8$ and the terminal sulfur to Co (if available) completely dominated the interaction between $Co_9S_8$ and $Li_2S_4$, where the binding energy of 6.93 eV (vdW interactions included) was the highest ever reported for $Li_2S_x$ on an anchoring material surface to date.

### *2.3.3 Transition Metal Nitrides*

Sun *et al.*[88] prepared a conductive porous vanadium nitride nanoribbon/graphene composite (VN/G) as the cathode material of a Li–S battery, where the shuttle effect of lithium polysulfides was drastically diminished. The fabrication process of a VN/G composite and cell assembly were schematically shown in Figure 19. The 3D free-standing structure composed of a graphene network facilitated the electrolyte absorption and the transport of electrons and ions. In addition, VN showed a strong anchoring effect for polysulfides and its high conductivity also accelerated the polysulfide conversion. The VN/G-based sulfur cathode delivered an excellent initial discharge capacity of 1471 mAh g$^{-1}$, and more importantly, it was able to maintain a stable cycling performance with a Coulombic efficiency above 99.5% for 100 charge–discharge cycles at 0.2 C, indicating that dissolution of polysulfides into the organic electrolyte was largely mitigated in the VN/G electrode.

In addition, Cui *et al.*[89] reported the use of a mesoporous TiN–S cathode to improve the performance of Li–S batteries. The TiN–S composite was synthesized by a melt-diffusion method that encapsulated the sublimed sulfur in mesoporous TiN that was produced by a solid–solid phase separation method with zinc titanate

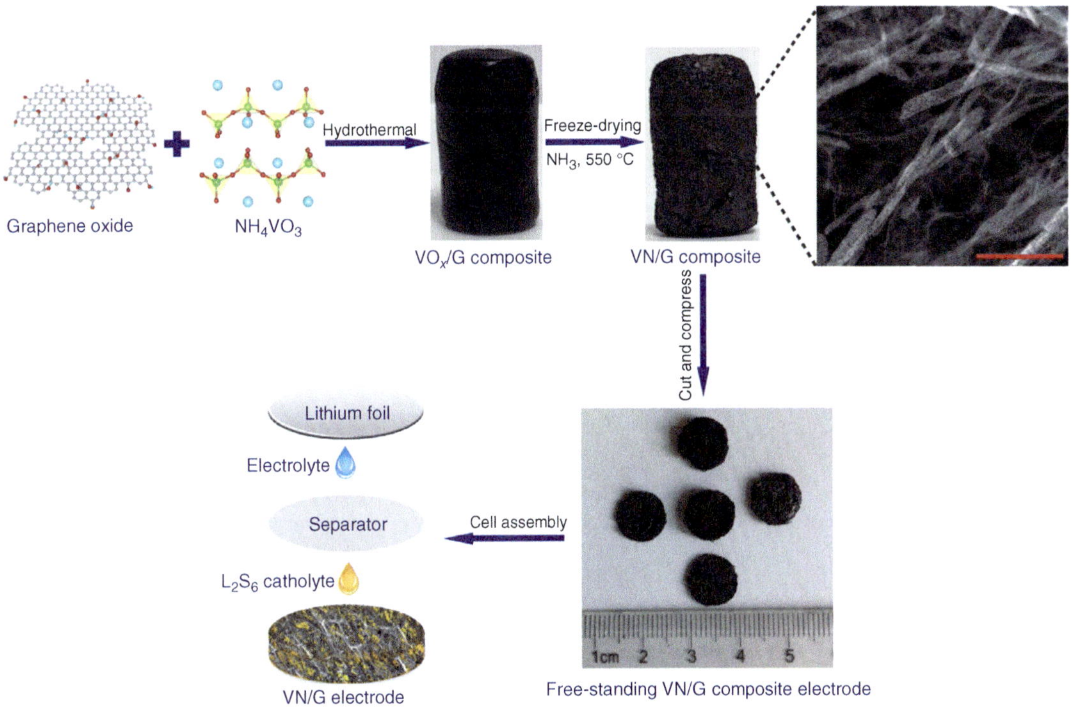


**Figure 19** Schematic of the fabrication of a porous VN/G composite and the cell assembly with corresponding optical images of the material obtained. Scale bar 500 nm. [Reprinted with permission from Ref. 88. © Nature Publishing Group, 2017]

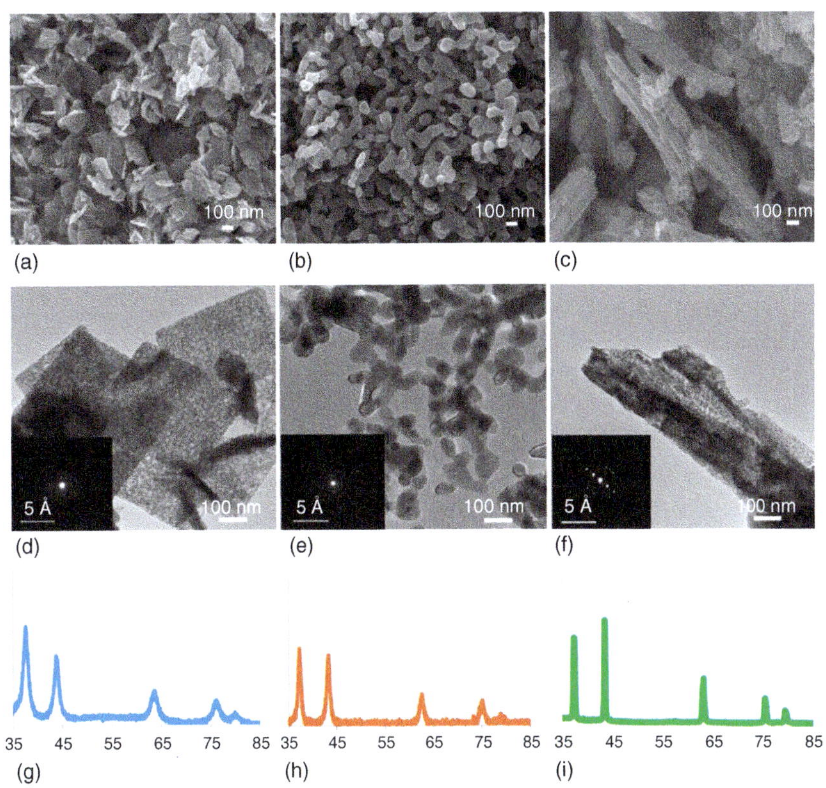


**Figure 20** SEM image of (a) WN, (b) VN, and (c) $Mo_2N$. TEM images and SAED patterns of (d) WN, (e) VN, and (f) $Mo_2N$. Powder XRD patterns of (g) WN, (h) VN, and (i) $Mo_2N$. [Reprinted with permission from Ref. 90. © Elsevier, 2017]

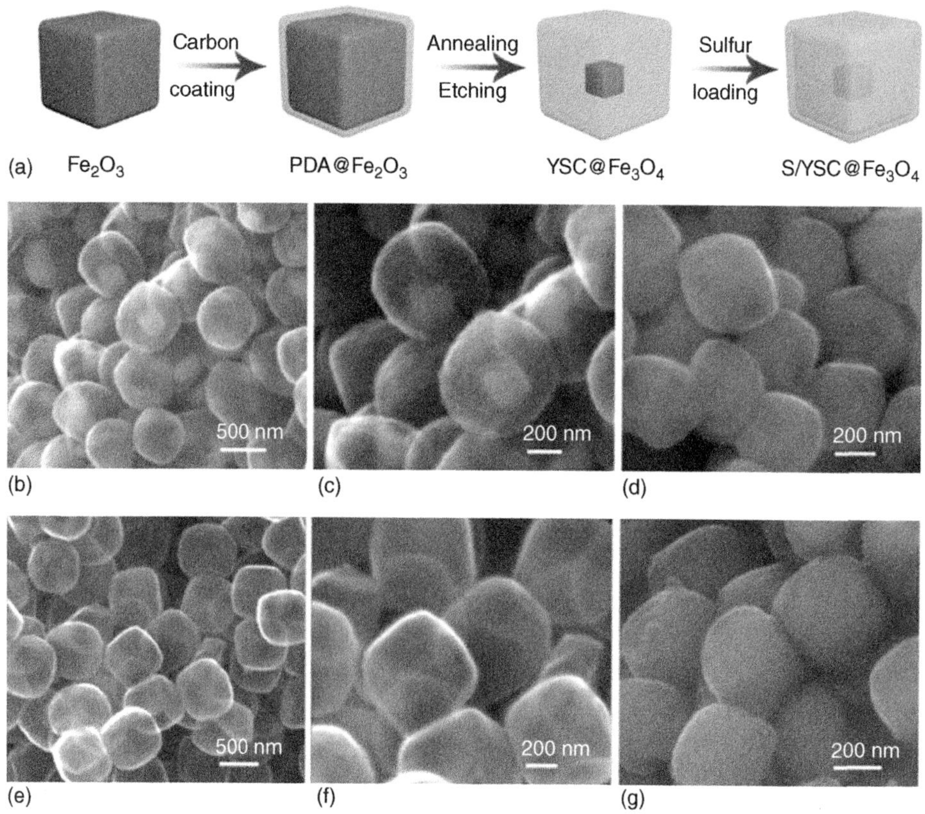


**Figure 21** (a) Schematic of the synthetic procedure of the S/YSC@$Fe_3O_4$ composite. SEM images of (b,c) YSC@$Fe_3O_4$, (d) S/YSC@$Fe_3O_4$, (e,f) C nanobox, and (g) S/C nanobox. [Reprinted with permission from Ref. 91. © Wiley-VCH Verlag GmbH & Co. KGaA, 2017]

as a starting material. Benefiting from the high electrical conductivity, robust porous frameworks and favorable adsorption properties of TiN, the TiN–S composite cathode exhibited a high specific capacity and excellent rate capability. More importantly, compared with mesoporous $TiO_2$–S and Vulcan C–S composite cathodes, the TiN–S composite cathode displayed the best cycling stability, with a capacity degradation rate of only 0.07% per cycle over 500 charge/discharge cycles. The excellent electrical conductivity and the chemical binding of the soluble intermediate species of mesoporous TiN allow for high rate of charge/discharge.

In another study, Mosavati *et al.*[90] prepared three different transition metal nitrides (WN, VN, and $Mo_2N$) as cathode materials for Li–S batteries (Figure 20). Among these materials, WN demonstrated the most promising cycling performance and a high capacity of 700 mAh $g^{-1}$ after 100 cycles. By increasing the WN cathode loading to 9.5 and 12.5 mg $cm^{-2}$, an improved capacity of 980 and 1283 mAh $g^{-1}$ after 100 cycles was observed. The superior performance of WN was attributed to the existence of S–WN bonding at the electrode surface, which indicated a strong interaction between lithium polysulfides and WN. Although these results indicated that WN was a highly promising candidate for high-performance Li–S batteries, delicate structure modification was required to further improve its cycling stability.

### 2.3.4 *Transition Metal Oxides*

Recently, He *et al.*[91] reported the preparation of well-defined yolk-shell-like carbon@$Fe_3O_4$ (YSC@$Fe_3O_4$) nanoboxes and their utilization as sulfur hosts for Li–S batteries (Figure 21). Thanks to both physical confinement of carbon shells and strong chemical interaction with $Fe_3O_4$ cores, this unique nanoarchitecture effectively immobilized the active material and inhibited diffusion of polysulfide intermediate species. Moreover, due to their high conductivity, the carbon shells and the polar $Fe_3O_4$ cores facilitated fast electron/ion transport and promoted continuous reactivation of the active material during the charge/discharge process, resulting in improved electrochemical utilization and reversibility. The resulting S/YSC@$Fe_3O_4$ cathode, at a sulfur loading of 2.2 mg $cm^{-2}$, delivered a high initial specific capacity of 1366 mAh $g^{-1}$. After 200 cycles, the capacity was

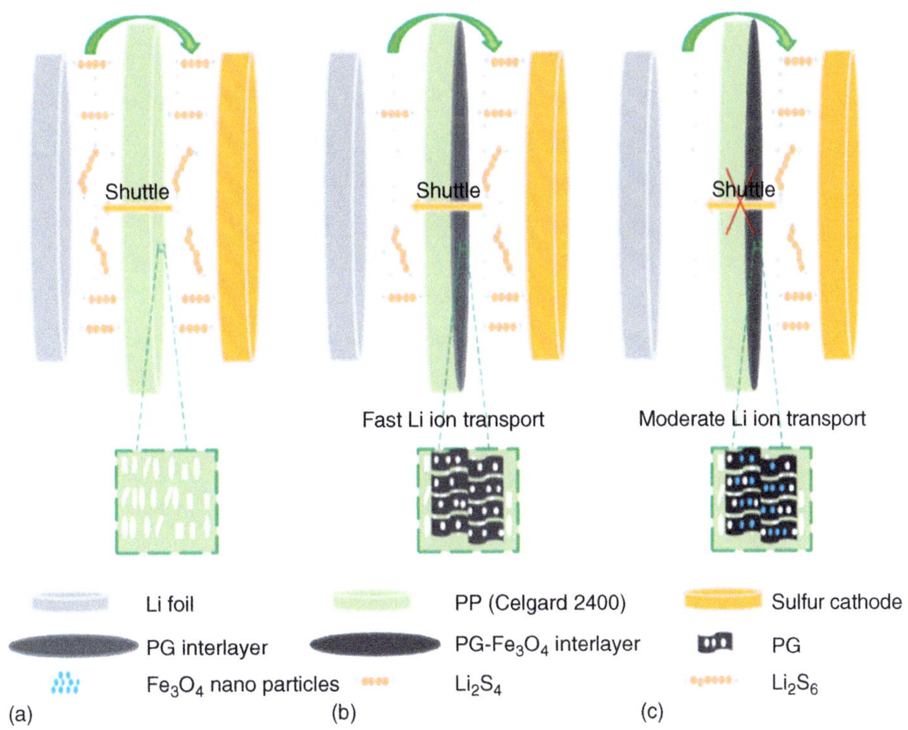


**Figure 22** Schematic configuration of Li–S batteries with (a) a PP (polypropylene, Celgard 2400) separator; (b) porous graphene (PG)@PP; and (c) PG-$Fe_3O_4$@ PP interlayer. [Reprinted with permission from Liu, Y.; Qin, X.; Zhang, S.; Liang, G.; Kang, F.; Chen, G.; Li, B. $Fe_3O_4$-decorated porous graphene interlayer for high-performance lithium-sulfur batteries. ACS applied materials & interfaces 2018. © 2018, American Chemical Society]

still stabilized at 1165 mAh $g^{-1}$, corresponding to a high capacity retention of 85.3% and a capacity fading rate of only 0.07% per cycle. For a comparison, the S/C nanobox-based cathode suffered from a severe capacity fade with a low capacity of 532 mAh $g^{-1}$ after 200 cycles. These results clearly demonstrate that polysulfide dissolution was effectively mitigated in the S/YSC@$Fe_3O_4$ cathode.

In another study, Liu *et al.*[92] constructed a multifunctional interlayer on a separator by depositing $Fe_3O_4$ nanoparticles on a porous graphene film to immobilize polysulfides via strong chemical interaction (Figure 22). The graphene layer acted as a physical barrier of polysulfides and guaranteed good Li-ion transport, whereas $Fe_3O_4$ nanoparticles were able to chemically immobilize polysulfides. After optimizing the interlayer, the Li–S batteries showed an excellent cycling performance (732 mAh $g^{-1}$ after 500 cycles and 356 mAh $g^{-1}$ after total 2000 cycles at 1 C with a 49.0% capacity retention and capacity decay rate of 0.02%) and a superior rate capability (589 mAh $g^{-1}$ at 2 C and 1423 mAh $g^{-1}$ at 0.1 C).

In addition, a previous report demonstrated that coating a common S–C cathode with a thin layer of mesoporous $TiO_2$ promoted the interaction between $TiO_2$ and S, which was believed to be an electrostatic attraction (S–Ti–O)[93] and hence improved the adsorption of lithium polysulfides on $TiO_2$ surface. Similarly, Xiao *et al.*[94] prepared an integrated selective interlayer simply by coating the surface of a C–S cathode with a commercial graphene/$TiO_2$ film, which accounted for only ca. 7.8 wt% of the whole cathode. In this composite cathode, the porous graphene afforded an additional electrically conductive network, and physically trapped S and the intermediate lithium polysulfides; while $TiO_2$ in the graphene/$TiO_2$ barrier film further chemically suppressed the dissolution of lithium polysulfides, and hence greatly alleviated the undesirable shuttle effect. The thus-prepared cathode coated with a graphene/$TiO_2$ film was able to deliver a reversible specific capacity of ca. 1040 mAh $g^{-1}$ over 300 cycles at 0.5 C. Ultralow capacity decay rates of 0.01 and 0.018% per cycle were observed at 2 and 3 C, respectively, over 1000 cycles.

Luo *et al.*[95] reported a novel strategy for decorating the separator with an interwoven $MoO_3$@CNT scaffold as an interlayer for trapping polysulfides in Li–S batteries (Figure 23). The conductive CNT networks facilitated electron transport, resulting in high rate capabilities. The polar $MoO_3$ nanorods tightly enfolded by CNTs (Figure 24) provided potent chemical interaction with polysulfides, thus mitigating the shuttle effect and giving rise to a longer cycle life. Moreover, such a dense

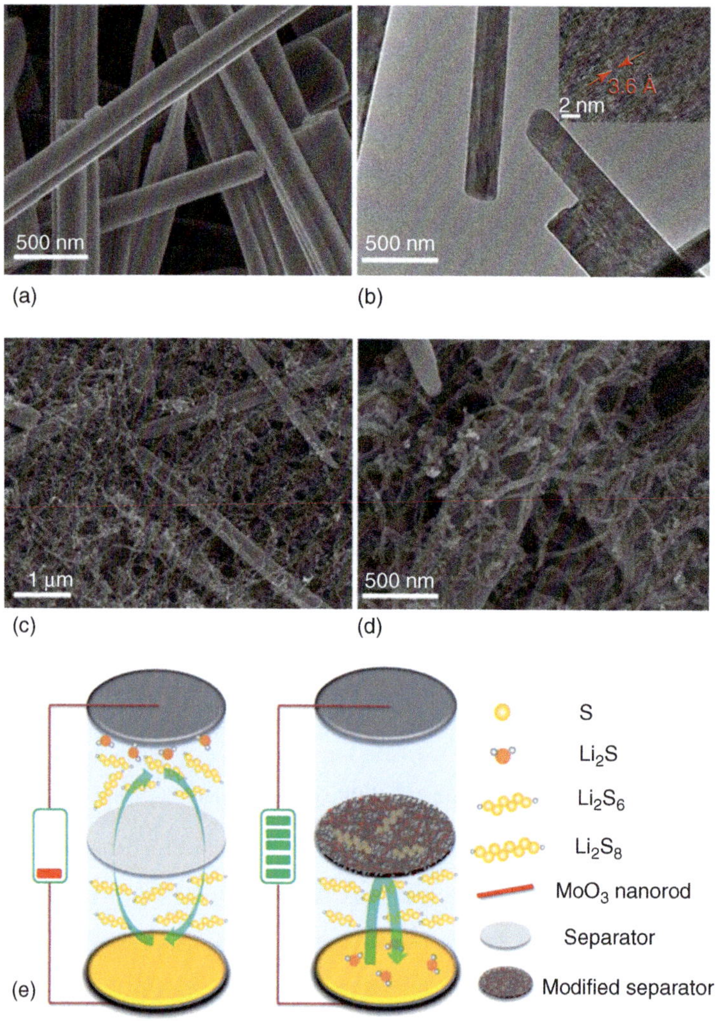


**Figure 23** (a) SEM and (b) TEM images of $MoO_3$ nanorods; (c–d) SEM images of the interwoven $MoO_3$@CNT framework. (e) Schematic of Li–S batteries with original and $MoO_3$@CNT modified separators. [Reprinted with permission from Ref. 95. © Royal Society of Chemistry, 2018]

coating of $MoO_3$@CNTs but with a porous structure also reserved abundant ion transfer channels. Therefore, the thin $MoO_3$@CNT hybrid interlayer could immobilize polysulfides without sacrificing fast lithium-ion transport. As a result, the Li–S batteries with the $MoO_3$@CNT interlayer showed remarkable electrochemical performances, with specific capacities of 1251 and 655 mAh $g^{-1}$ at 0.3 C and 3 C, respectively. After 200 cycles at 0.3 C, the batteries still retained a high capacity of 755 mAh $g^{-1}$. Furthermore, the facilely synthesized $MoO_3$@CNT hybrid layer was also able to buffer the large volume fluctuation of sulfur cathode during cycling process.

### 2.3.5 *Transition Metal Carbides*

Metal carbides as a cathode material for Li–S battery have been widely studied since the discovery of $Ti_3C_2$ in 2011. Compared to other metal compounds such as sulfide, oxide, and phosphide, carbides have many unique characteristics: (i) high electrical conductivity; (ii) excellent chemical and physical stability; and (iii) high electrocatalytic activity for the redox processes of lithium polysulfides. In 2015, Nazar and coworkers[96] first reported a new class of sulfur host materials the so-called delaminated MXene phases that capitalized on a combination of inherently high conductivity and highly active 2D surfaces to chemically bond intermediate polysulfides by metal–sulfur interactions, providing a very stable cycling performance and high capacity even at 70 wt% S. With the aid of XPS analysis, the authors verified that sulfur/sulfide species replaced the hydroxy groups on $Ti_2C$ surface, forming a strong Ti–S interaction with the core, as shown in Figure 24. Specifically, the optimized 70S/d-$Ti_2C$-based cathode delivered a discharge capacity of 1090 mAh $g^{-1}$ at

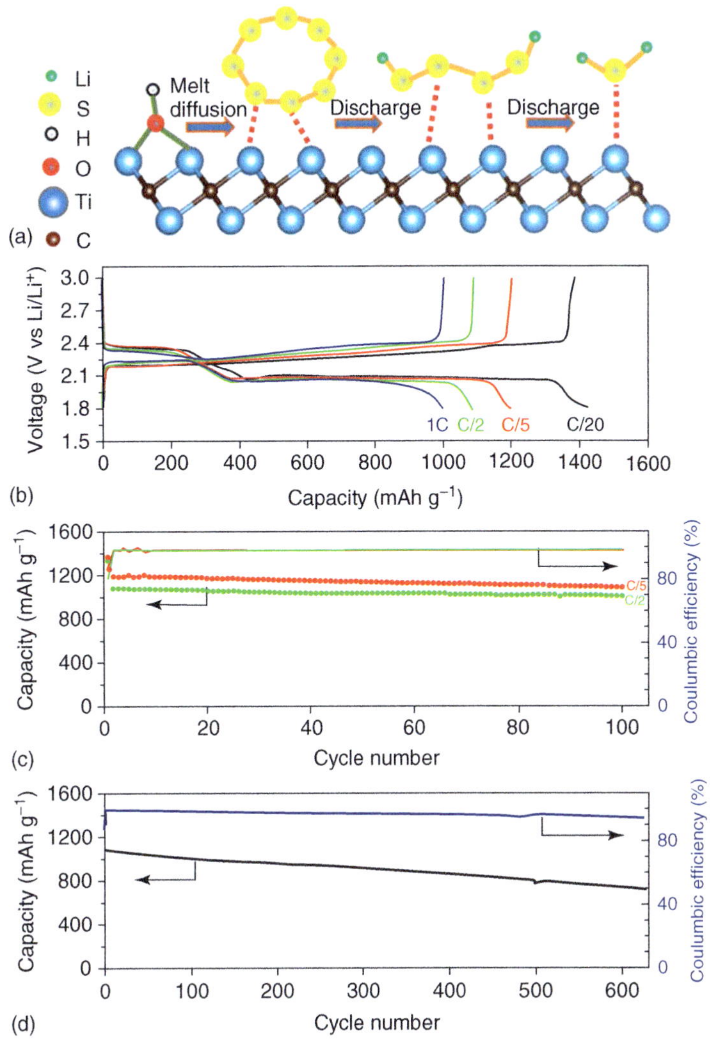


**Figure 24** (a) Replacement of the Ti—OH bond on the MXene surface with a S—Ti—C bond by heat treatment or by contact with polysulfides. (b) Voltage profiles of 70S/d-$Ti_2C$ at various rates ranging from 0.05 to 1 C. (c) Cycling performance of 70S/d-$Ti_2C$ at 0.2 and 0.5 C. (d) Long-term cycling at 0.5 C. Cells are conditioned for the first cycle at 0.05 C to facilitate electrode wetting. The increase in rate on the 2nd cycle results in a change in capacity. [Reprinted with permission from Ref. 96. Wiley-VCH Verlag GmbH & Co. KGaA, 2015]

a 0.5 C rate (837 mA $g^{-1}$) and while at 1 C (1675 mA $g^{-1}$) the discharge capacity was still as high as1000 mAh $g^{-1}$ and showed excellent cycling stability at various rates in Figure 24(b)–(d).

Among the metal carbides discovered to date, niobium carbide (NbC) delivers the highest electrical conductivity up to $2.9 \times 10^6$ S $m^{-1}$ and excellent mechanical and chemical stabilities, while its polar molecular structure offers strong chemical interactions with polysulfide anions. Hence NbC can serve as a potential interlayer material for Li–S batteries. Zhu and coworkers[97] reported a facile autoclave method to synthesize nanocrystalline NbC through a magnesiothermic reaction at 600 °C, as shown in Figure 25, and the sulfur cathode employing the NbC-coated membrane exhibited outstanding cyclability over 1500 cycles, excellent rate capability up to 5 C, and high areal capacity of 3.6 mAh $cm^{-2}$ with a relatively high sulfur loading of 4 mg $cm^{-2}$ in a coin cell, as well as a reliable operation in soft package.

In a recent study, Wang *et al.*[98] infiltrated unexfoliated $Ti_3C_2$ stacks with molten sulfur, which was then coated with a thin layer of porous polydopamine. The electrically conductive MXene sheets, that is, $Ti_3C_2$, contained a number of Lewis-acid Ti-sites which enabled the Lewis acid–base interaction with $Li_2S_x$. At the same time, the polar N- and O-atom in the polydopamine layer bound strongly with polar $Li^+$, further immobilizing $Li_2S_x$ within the cathode. Such a combined $Li_2S_x$ immobilization effect offered by the MXene matrix and polydopamine layer helped maximize the $Li_2S_x$ immobilization and allowed

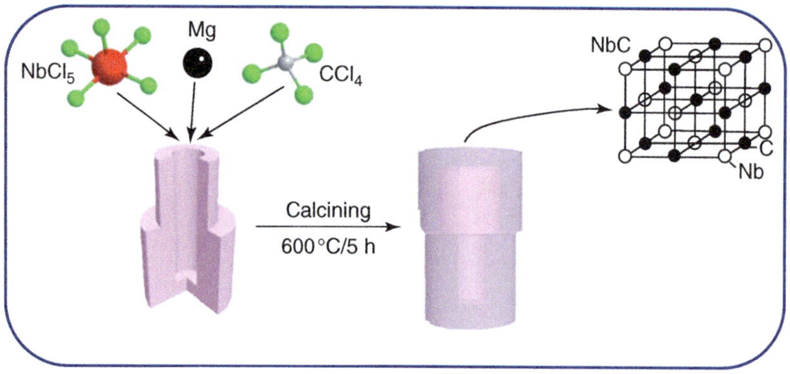


**Figure 25** Schematic illustration of the preparation of NbC. Many other carbides have been used as host materials for Li–S batteries, such as MoC and $Wo_2C$. For examples, Yu's group prepared MoC nanoparticles encapsulated in N-doped carbon nanofibers (NCF), theoretical calculations, adsorption experiments, and cyclic voltammetry analysis proved that MoC is active sites on the electrode with double functionality: to improve both the cycling stability and rate capability of Li–S batteries via the dual effect of providing adsorption and catalytic sites. [Reprinted with permission from Ref. 97. Wiley-VCH Verlag GmbH & Co. KGaA, 2018]

the new S-cathodes to exhibit high and stable energy/rate capacity even at a high S loading. Specifically, the capacity decay was only 0.035% per cycle, which was achieved at a sulfur loading of 5 mg cm$^{-2}$. Meanwhile, the electrode kinetics of cathode with a high sulfur loading was also promoted by the efficient deposition of $Li_2S$ driven by the dual $Li_2S_x$ immobilization and the conductive network stabilized by polydopamine.[20,98,99]

## 3 SUMMARY AND OUTLOOK

Dissolution of intermediate lithium polysulfides from cathode to organic electrolyte and their subsequent diffusion toward the anode are two main reasons for rapid capacity degradation, low Coulombic efficiency, and poor sulfur utilization. A number of methods, such as physical confinement, covalent bonding to form polymeric sulfur, and incorporation of polar anchoring materials, have been developed for cathode engineering to effectively mitigate or even suppress these two processes, so as to maximize the potential of Li–S battery and realize its high capacity and energy density, as compared to traditional lithium-ion batteries. An ideal material for shielding polysulfide dissolution and diffusion should have sufficiently strong bonding/binding interactions with the $Li_2S_x$ species, high electron and ion conductivity, good catalytic activity for the polysulfides redox process as well as light-weight, low cost, high electrochemical and mechanical stability. In addition, the traditional composite materials based on an active sulfur yolk and a conductive shell having strong binding strength toward polysulfides, as well as the vast family of metal–organic framework hosts[100] will demonstrate high potential in the preparation of efficient polysulfides immobilizers for long-life sulfur cathodes.[101–103] These will be the foci of future research.

## 4 ACKNOWLEDGMENTS

This work was supported by the National Natural Science Foundation of China (NSFC Grant No. 21528301 and 51402111), and the Fundamental Research Funds for the Central Universities (SCUT Grant No. 201522/05).

## 5 ABBREVIATIONS AND ACRONYMS

3D = Three-dimensional; AAO = anodic aluminum oxide; BOP = benzoxazine polymer; CNT@MPC = carbon nanotubes@microporous carbon; CNT = Carbon nanotube; CV = cyclic voltammetry; DCNT = disordered carbon nanotube; DFT = density functional theory; EIS = Electrochemical impedance spectroscopic; FCTF = fluorinated porous triazine-based framework; GNS = graphene nanosheets; GO = graphene oxide; HCN = hollow carbon nanosphere; HPCR = hierarchical porous carbon rod; ICE = initial Coulombic efficiency; Li–S = Lithium–sulfur; LIBs = Li-ion batteries; MAT = m-aminothiophenol; MSPAN = mesoporous sulfurized polyacrylonitrile; MWCNT = multi-walled carbon nanotube; NbC = niobium carbide; NG = nitrogen-doped graphene; PIP = polyisoprene; PMAT = poly(m-aminothiophenol); PVD = physical vapor deposition; rGO = reduced graphene oxide; ROP = ring-opening polymerization; PIP = polyisoprene; TAR = thiourea aldehyde resin; TTCA = trithiocyanuric acid; vdW = Van der Waals

## 6 REFERENCES

1. H. Danuta, U. Juliusz, Electric Dry Cells and Storage Batteries, US, Patent 3,043,896[P], 1962-7-10.

2. Y. Yang, G. Zheng and Y. Cui, *Chem. Soc. Rev.*, 2013, **42**, 3018.
3. J. Song, M. L. Gordin, T. Xu, S. Chen, Z. Yu, H. Sohn, J. Lu, Y. Ren, Y. Duan and D. Wang, *Angew. Chem.*, 2015, **127**, 4399.
4. M. A. Pope and I. A. Aksay, *Adv. Energy Mater.*, 2015, **5**, 1500124.
5. B. Ji, F. Zhang, M. Sheng, X. Tong and Y. Tang, *Adv. Mater.*, 2017, **29**, 1604219.
6. C.-X. Zu and H. Li, *Energy Environ. Sci.*, 2011, **4**, 2614.
7. H. S. Kang and Y. K. Sun, *Adv. Funct. Mater.*, 2016, **26**, 1225.
8. X. Gu, C.-j. Tong, C. Lai, J. Qiu, X. Huang, W. Yang, B. Wen, L.-m. Liu, Y. Hou and S. Zhang, *J. Mater. Chem. A*, 2015, **3**, 16670.
9. L. Borchardt, M. Oschatz and S. Kaskel, *Chem. Eur. J.*, 2016, **22**, 7324.
10. W. J. Chung, J. J. Griebel, E. T. Kim, H. Yoon, A. G. Simmonds, H. J. Ji, P. T. Dirlam, R. S. Glass, J. J. Wie and N. A. Nguyen, *Nat. Chem.*, 2013, **5**, 518.
11. V. Etacheri, R. Marom, R. Elazari, G. Salitra and D. Aurbach, *Energy Environ. Sci.*, 2011, **4**, 3243.
12. N. Osada, C. B. Bucur, H. Aso and J. Muldoon, *Energy Environ. Sci.*, 2016, **9**, 1668.
13. Y. Xiang, J. Li, J. Lei, D. Liu, Z. Xie, D. Qu, K. Li, T. Deng and H. Tang, *ChemSusChem*, 2016, **9**, 3023.
14. J. Wang, Y. S. He and J. Yang, *Adv. Mater.*, 2015, **27**, 569.
15. H. Wang, W. Zhang, H. Liu and Z. Guo, *Angew. Chem. Int. Ed.*, 2016, **55**, 3992.
16. H. Cheng and S. Wang, *J. Mater. Chem. A*, 2014, **2**, 13783.
17. X. Ji and L. F. Nazar, *J. Mater. Chem.*, 2010, **20**, 9821.
18. G. C. Li, G. R. Li, S. H. Ye and X. P. Gao, *Adv. Energy Mater.*, 2012, **2**, 1238.
19. F. Wu, J. Qian, R. Chen, T. Zhao, R. Xu, Y. Ye, W. Li, L. Li, J. Lu and K. Amine, *Nano Energy*, 2015, **12**, 742.
20. A. Rosenman, E. Markevich, G. Salitra, D. Aurbach, A. Garsuch and F. F. Chesneau, *Adv. Energy Mater.*, 2015, **5**, 1500212.
21. Z. Li, Y. Huang, L. Yuan, Z. Hao and Y. Huang, *Carbon*, 2015, **92**, 41.
22. G. Li, J. Sun, W. Hou, S. Jiang, Y. Huang and J. Geng, *Nat. Commun.*, 2016, **7**, 10601.
23. M. Yan, Y. Zhang, Y. Li, Y. Huo, Y. Yu, C. Wang, J. Jin, L. Chen, T. Hasan and B. Wang, *J. Mater. Chem. A*, 2016, **4**, 9403.
24. N. A. Lange ed., 'Lange's Handbook of Chemistry', McGraw-Hill Education, 2016.
25. E. Peled, A. Gorenshtein, M. Segal and Y. Sternberg, *J. Power Sources*, 1989, **26**, 269.
26. X. Ji, K. T. Lee and L. F. Nazar, *Nat. Mater.*, 2009, **8**, 500.
27. A. Manthiram, Y. Fu and Y. S. Su, *Acc. Chem. Res.*, 2013, **46**, 1125.
28. N. Jayaprakash, J. Shen, S. S. Moganty, A. Corona and L. A. Archer, *Angew. Chem. Int. Ed.*, 2011, **50**, 5904.
29. G. Buechel, K. K. Unger, A. Matsumoto and K. Tsutsumi, *Adv. Mater.*, 1999, **10**, 1036.
30. F. Xu, Z. Tang, S. Huang, L. Chen, Y. Liang, W. Mai, H. Zhong, R. Fu and D. Wu, *Nat. Commun.*, 2015, **6**, 7221.
31. W. Zhou, Y. Yu, H. Chen, F. J. Disalvo and H. D. Abruña, *J. Am. Chem. Soc.*, 2013, **135**, 16736.
32. R. H. Baughman, A. A. Zakhidov and W. A. D. Heer, *Science*, 2002, **297**, 787.
33. R. B. Heimann, S. E. Evsvukov and Y. Koga, *Carbon*, 1997, **35**, 1654.
34. P. Morgan, 'Carbon Fibers and Their Composites', Marcel Dekker, 2005.
35. K. Shirvanimoghaddam, S. U. Hamim, M. K. Akbari, S. M. Fakhrhoseini, H. Khayyam, A. H. Pakseresht, E. Ghassali, M. Zabet, K. S. Munir and S. Jia, *Compos. A Appl. Sci. Manuf.*, 2017, **92**, 70.
36. Y. Zhao, W. Wu, J. Li, Z. Xu and L. Guan, *Adv. Mater.*, 2014, **26**, 5113.
37. D. W. Wang, Q. Zeng, G. Zhou, L. Yin, F. Li, H. M. Cheng, I. R. Gentle and G. Q. M. Lu, *J. Mater. Chem. A*, 2013, **1**, 9382.
38. J. Guo, Y. Xu and C. Wang, *Nano Lett.*, 2011, **11**, 4288.
39. S. Xin, L. Gu, N. H. Zhao, Y. X. Yin, L. J. Zhou, Y. G. Guo and L. J. Wan, *J. Am. Chem. Soc.*, 2012, **134**, 18510.
40. G. Zheng, Y. Yang, J. J. Cha, S. S. Hong and Y. Cui, *Nano Lett.*, 2011, **11**, 4462.
41. K. S. Novoselov, A. K. Geim, S. V. Morozov, D. Jiang, Y. Zhang, S. V. Dubonos, I. V. Grigorieva and A. A. Firsov, *Science*, 2004, **306**, 666.
42. H. Wang, H. S. Casalongue, Y. Liang and H. Dai, *J. Am. Chem. Soc.*, 2010, **132**, 7472.
43. H. Wang, Y. Yang, Y. Liang, J. T. Robinson, Y. Li, A. Jackson, Y. Cui and H. Dai, *Nano Lett.*, 2011, **11**, 2644.
44. Y. Qiu, W. Li, W. Zhao, G. Li, Y. Hou, M. Liu, L. Zhou, F. Ye, H. Li, Z. Wei, S. Yang, W. Duan, Y. Ye, J. Guo and Y. Zhang, *Nano Lett.*, 2014, **14**, 4821.
45. C. Zhang, W. Lv, W. Zhang, X. Zheng, M. B. Wu, W. Wei, Y. Tao, Z. Li and Q. H. Yang, *Adv. Energy Mater.*, 2013, **4**, 1301565.
46. M. Q. Zhao, Q. Zhang, J. Q. Huang, G. L. Tian, J. Q. Nie, H. J. Peng and F. Wei, *Nat. Commun.*, 2014, **5**, 3410.
47. L. Hu, J. W. Choi, Y. Yang, S. Jeong, F. La Mantia, L.-F. Cui and Y. Cui, *Proc. Natl. Acad. Sci.*, 2009, **106**, 21490.
48. G. Zhou, S. Pei, L. Li, D. W. Wang, S. Wang, K. Huang, L. C. Yin, F. Li and H. M. Cheng, *Adv. Mater.*, 2014, **26**, 625.
49. D.-W. Wang, F. Li, M. Liu, G. Q. Lu and H.-M. Cheng, *Angew. Chem. Int. Ed.*, 2008, **47**, 373.

50. Z. M. Zheng, H. C. Guo, F. Pei, X. Zhang, X. Y. Chen, X. L. Fang, T. H. Wang and N. F. Zheng, *Adv. Funct. Mater.*, 2016, **26**, 8952.
51. C. Hoffmann, S. Thieme, J. Brückner, M. Oschatz, T. Biemelt, G. Mondin, H. Althues and S. Kaskel, *ACS Nano*, 2014, **8**, 12130.
52. Z. Yuan, H.-J. Peng, J.-Q. Huang, X.-Y. Liu, D.-W. Wang, X.-B. Cheng and Q. Zhang, *Adv. Funct. Mater.*, 2014, **24**, 6105.
53. Z. Li, L. X. Yuan, Z. Q. Yi, Y. M. Sun, Y. Liu, Y. Jiang, Y. Shen, Y. Xin, Z. L. Zhang and Y. H. Huang, *Adv. Energy Mater.*, 2014, **4**, 1301473.
54. G. Hu, Z. Sun, C. Shi, R. Fang, J. Chen, P. Hou, C. Liu, H. M. Cheng and F. Li, *Adv. Mater.*, 2017, **29**, 1603835.
55. S. Zeng, L. Li, J. Yu, N. Wang and S. Chen, *Electrochim. Acta*, 2018, **263**, 53.
56. B. Oschmann, J. Park, C. Kim, K. Char, Y.-E. Sung and R. Zentel, *Chem. Mater.*, 2015, **27**, 7011.
57. C. Fu, G. Li, J. Zhang, B. Cornejo, S. S. Piao, K. N. Bozhilov, R. C. Haddon and J. Guo, *ACS Energy Lett.*, 2016, **1**, 115.
58. J. Liu, M. Wang, N. Xu, T. Qian and C. Yan, *Energy Storage Mater.*, 2018, **15**, 53.
59. H. Kim, J. Lee, H. Ahn, O. Kim and M. J. Park, *Nat. Commun.*, 2015, **6**, 7278.
60. S.-Z. Zeng, Y. Yao, X. Zeng, Q. He, X. Zheng, S. Chen, W. Tu and J. Zou, *J. Power Sources*, 2017, **357**, 11.
61. S. Zeng, L. Li, L. Xie, D. Zhao, N. Zhou, N. Wang and S. Chen, *Carbon*, 2017, **122**, 106.
62. S. H. Je, T. H. Hwang, S. N. Talapaneni, O. Buyukcakir, H. J. Kim, J.-S. Yu, S.-G. Woo, M. C. Jang, B. K. Son and A. Coskun, *ACS Energy Lett.*, 2016, **1**, 566.
63. S. Zeng, L. Li, L. Xie, D. Zhao, N. Wang and S. Chen, *ChemSusChem*, 2017, **10**, 3378.
64. J. Wang, J. Yang, C. Wan, K. Du, J. Xie and N. Xu, *Adv. Funct. Mater.*, 2003, **13**, 487.
65. X.-g. Yu, J.-y. Xie, J. Yang, H.-j. Huang, K. Wang and Z.-s. Wen, *J. Electroanal. Chem.*, 2004, **573**, 121.
66. J. Fanous, M. Wegner, J. Grimminger, A. n. Andresen and M. R. Buchmeiser, *Chem. Mater.*, 2011, **23**, 5024.
67. S. S. Zhang, *Energies*, 2014, **7**, 4588.
68. Y. Liu, A. K. Haridas, K.-K. Cho, Y. Lee and J.-H. Ahn, *J. Phys. Chem. C*, 2017, **121**, 26172.
69. S. N. Talapaneni, T. H. Hwang, S. H. Je, O. Buyukcakir, J. W. Choi and A. Coskun, *Angew. Chem. Int. Ed.*, 2016, **55**, 3106.
70. F. Xu, S. Yang, G. Jiang, Q. Ye, B. Wei and H. Wang, *ACS Appl. Mater. Interfaces*, 2017, **9**, 37731.
71. Q. Zhang, Y. Wang, Z. W. Seh, Z. Fu, R. Zhang and Y. Cui, *Nano Lett.*, 2015, **15**, 3780.
72. T. Z. Hou, X. Chen, H. J. Peng, J. Q. Huang, B. Q. Li, Q. Zhang and B. Li, *Small*, 2016, **12**, 3283.
73. Q. Pang, J. Tang, H. Huang, X. Liang, C. Hart, K. C. Tam and L. F. Nazar, *Adv. Mater.*, 2015, **27**, 6021.
74. Q. Pang and L. F. Nazar, *ACS Nano*, 2016, **10**, 4111.
75. X. Liu, J. Q. Huang, Q. Zhang and L. Mai, *Adv. Mater.*, 2017, **29**, 1601759.
76. X. Y. Yu, L. Yu and X. W. Lou, *Adv. Energy Mater.*, 2016, **6**, 1501333.
77. L. Ma, S. Wei, H. L. Zhuang, K. E. Hendrickson, R. G. Hennig and L. A. Archer, *J. Mater. Chem. A*, 2015, **3**, 19857.
78. L. Ma, H. Zhuang, Y. Lu, S. S. Moganty, R. G. Hennig and L. A. Archer, *Adv. Energy Mater.*, 2014, **4**, 1400390.
79. Z. W. Seh, J. H. Yu, W. Li, P.-C. Hsu, H. Wang, Y. Sun, H. Yao, Q. Zhang and Y. Cui, *Nat. Commun.*, 2014, **5**, 5017.
80. G. Zhou, H. Tian, Y. Jin, X. Tao, B. Liu, R. Zhang, Z. W. Seh, D. Zhuo, Y. Liu and J. Sun, *Proc. Natl. Acad. Sci.*, 2017, **114**, 840.
81. M. Q. Zhao, X. F. Liu, Q. Zhang, G. L. Tian, J. Q. Huang, W. Zhu and F. Wei, *ACS Nano*, 2012, **6**, 10759.
82. R. E. Jones, F. S. Gittleson, J. A. Templeton and D. K. Ward, *J. Electrochem. Soc.*, 2017, **164**, A6422.
83. X. Tao, J. Wang, C. Liu, H. Wang, H. Yao, G. Zheng, Z. W. Seh, Q. Cai, W. Li and G. Zhou, *Nat. Commun.*, 2016, **7**, 11203.
84. M. S. Faber, R. Dziedzic, M. A. Lukowski, N. S. Kaiser, Q. Ding and S. Jin, *J. Am. Chem. Soc.*, 2014, **136**, 10053.
85. Z. Yuan, H. J. Peng, T. Z. Hou, J. Q. Huang, C. M. Chen, D. W. Wang, X. B. Cheng, F. Wei and Q. Zhang, *Nano Lett.*, 2016, **16**, 519.
86. S. Rehman, S. Guo and Y. Hou, *Adv. Mater.*, 2016, **28**, 3167.
87. Q. Pang, D. Kundu and L. F. Nazar, *Mater. Horiz.*, 2016, **3**, 130.
88. Z. Sun, J. Zhang, L. Yin, G. Hu, R. Fang, H.-M. Cheng and F. Li, *Nat. Commun.*, 2017, **8**, 14627.
89. Z. Cui, C. Zu, W. Zhou, A. Manthiram and J. B. Goodenough, *Adv. Mater.*, 2016, **28**, 6926.
90. N. Mosavati, S. O. Salley and K. S. Ng, *J. Power Sources*, 2017, **340**, 210.
91. J. He, L. Luo, Y. Chen and A. Manthiram, *Adv. Mater.*, 2017, **29**, 1702707.
92. Y. Liu, X. Qin, S. Zhang, G. Liang, F. Kang, G. Chen and B. Li, *ACS Appl. Mater. Interfaces*, 2018, **10**, 26264.
93. Z. Liang, G. Zheng, W. Li, Z. W. Seh, H. Yao, K. Yan, D. Kong and Y. Cui, *ACS Nano*, 2014, **8**, 5249.
94. Z. Xiao, Z. Yang, L. Wang, H. Nie, M. Zhong, Q. Lai, X. Xu, L. Zhang and S. Huang, *Adv. Mater.*, 2015, **27**, 2891.
95. L. Luo, X. Qin, J. Wu, G. Liang, Q. Li, M. Liu, F. Kang, G. Chen and B. Li, *J. Mater. Chem. A*, 2018, **6**, 8612.

96. X. Liang, A. Garsuch and L. F. Nazar, *Angew. Chem.*, 2015, **127**, 3979.

97. W. Cai, G. Li, K. Zhang, G. Xiao, C. Wang, K. Ye, Z. Chen, Y. Zhu and Y. Qian, *Adv. Funct. Mater.*, 2018, **28**, 1704865.

98. X. Wang, C. Yang, X. Xiong, G. Chen, M. Huang, J.-H. Wang, Y. Liu, M. Liu and K. Huang, *Energy Storage Mater.*, 2019, **16**, 344.

99. A. Manthiram, Y. Fu, S.-H. Chung, C. Zu and Y.-S. Su, *Chem. Rev.*, 2014, **114**, 11751.

100. H. Jiang, X. C. Liu, Y. Wu, Y. Shu, X. Gong, F. S. Ke and H. Deng, *Angew. Chem.*, 2018, **130**, 3980.

101. Z. W. Seh, W. Li, J. J. Cha, G. Zheng, Y. Yang, M. T. McDowell, P.-C. Hsu and Y. Cui, *Nat. Commun.*, 2013, **4**, 1331.

102. A. S. Arico, P. Bruce, B. Scrosati, J.-M. Tarascon and W. V. Schalkwijk, in 'Materials For Sustainable Energy: A Collection of Peer-Reviewed Research and Review Articles from Nature Publishing Group', ed. W. Van Schalkwijk, World Scientific, 2011, p.148.

103. Y. G. Guo, J. S. Hu and L. J. Wan, *Adv. Mater.*, 2008, **20**, 2878.

# PART 3
# Materials and Chemistry of Non-Lithium Batteries

# How to Maximize the Potential of Zn–Air Battery: Toward Acceptable Rechargeable Technology with or without Electricity

**Joohyuk Park, Jang-Soo Lee and Jaephil Cho**

*Ulsan National Institute of Science and Technology (UNIST), Ulsan, Republic of Korea*



## 1 INTRODUCTION

Demands for better alternative energy storage and conversion systems are increasing owing to fluctuating oil prices and the effects of global warming. To date, Li-ion batteries have played an important role in portable electronic equipment and devices. However, its maximum energy density is limited owing to a rocking chair principle, and thus it has intrinsic limitation in achieving a practical application of electric vehicles. Therefore, possible strategy to increase its energy density is free from energy storage within limited sized batteries and thus to use an energy conversion. In this regard, a metal–air battery with a mixed concept between traditional battery and fuel cells has been involved to address these problems. Therefore, it is possible to maximize its energy density by coupling lithium as anode, the lightest metal, and unlimited oxygen ($O_2$) from atmosphere as cathode, and its theoretical energy density is $\approx 5200\,Wh\,kg^{-1}$ (including $O_2$), higher than that of current Li-ion battery (about $200–250\,Wh\,kg^{-1}$), as shown in Figure 1.[1]

Therefore, the next-generation batteries such as lithium–oxygen or sulfur batteries, based on the lithium metal, the lightest metal, have been attracting attention.[2,3] However, their potential has been much more overestimated because of excess use of lithium metal.[4,5] In addition, although artificial solid electrolyte interface (SEI) layer can partially suppress lithium dendrite formation during cycling, the safety problems related to lithium are still critical issues.[6] Furthermore, it is worthy to note that advanced nanotechnologies such as highly loaded silicon nanoarchitectures enabled current Li-ion batteries to have more energy density, and their values are even comparable to those of the aforementioned Li–$O_2$ and Li–S batteries.[7] However, the highly loaded cathode electrodes should also be realized to practically increase the energy density of the state-of-the-art Li-ion battery. This is why it is extremely difficult to meet the ever-increasing high-energy demand of Li-ion battery. Therefore, it is required to develop the batteries with high energy density which are not based on intercalation or conversion chemistry and lithium metal coupled organic electrolyte. It follows that aqueous electrolyte-based metal–air, not oxygen, batteries have attracted much attention as a possible alternative because their energy density is extremely high compared to that of other rechargeable batteries, and it is intrinsically safe compared to other lithium-based batteries.

*Inorganic Battery Materials*. Edited by Hailiang Wang and Boniface P.T. Fokwa
 ISBN 978-1-119-43199-2

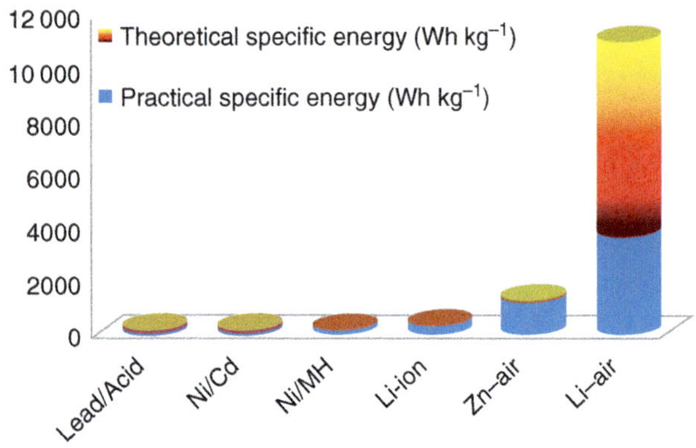


**Figure 1** Theoretical and practical energy densities of various types of rechargeable batteries.[1] [Reproduced with permission from Ref. J.-S. Lee, S. Tai Kim, R. Cao, N.-S. Choi, M. Liu, K. T. Lee, J. Cho, Adv. Energy Mater. 2011, 1, 34. © Wiley, 2011]

Among aqueous-based metal–air batteries, aluminum and zinc batteries are practically appropriate for the aqueous system. For example, the theoretical energy density of Al–air battery is around 6000 Wh kg$^{-1}$, and its application as a primary cell looks very promising and have already come close to commercialization for electric vehicle in spite of the lack of its rechargeability resulting from instability of aluminum metal under alkaline electrolyte (http://www.alcoa.com). Therefore, Zn–air batteries with a rechargeable possibility have been a powerful potential candidate as the next-generation energy storage and conversion devices. Besides their potential as secondary batteries, zinc metal batteries have several advantages such as low cost, abundance, low equilibrium potential, environmental benignity, flat discharge voltage, long shelf-life, and economical aspects (http://www.mee-t.com/default/00/01.php). However, some technical problems, which will be discussed in this chapter, have prevented Zn–air batteries from maximizing its potential beyond just hearing aids.

Recently, some papers provided a great insight into all about Zn–air battery with an overview of either individual components or the whole system.[8–10] Therefore, in this chapter, we briefly introduce Zn–air battery and discuss the efforts that should be devoted to maximize its tremendous potential as obvious next-generation battery and then a practical perspective view of a further Zn–air battery.

## 2 Zn–AIR BATTERY—OVERVIEW

Primary Zn–air batteries are composed of fully immersed zinc anode into alkaline electrolyte, a separator, and partially immersed air cathode, and its battery electrochemistry including side reactions is described in Figure 2. During reactions, the only consumed material is zinc because oxygen can be supplied continuously. Therefore, its voltage profile is flat based on the Gibbs phase rule.[11] During discharge, zinc electrochemistry, at anode side, in alkaline solution can be summarized as charge (soluble species) and then mass transfer reaction (dissolution/precipitation).[12]

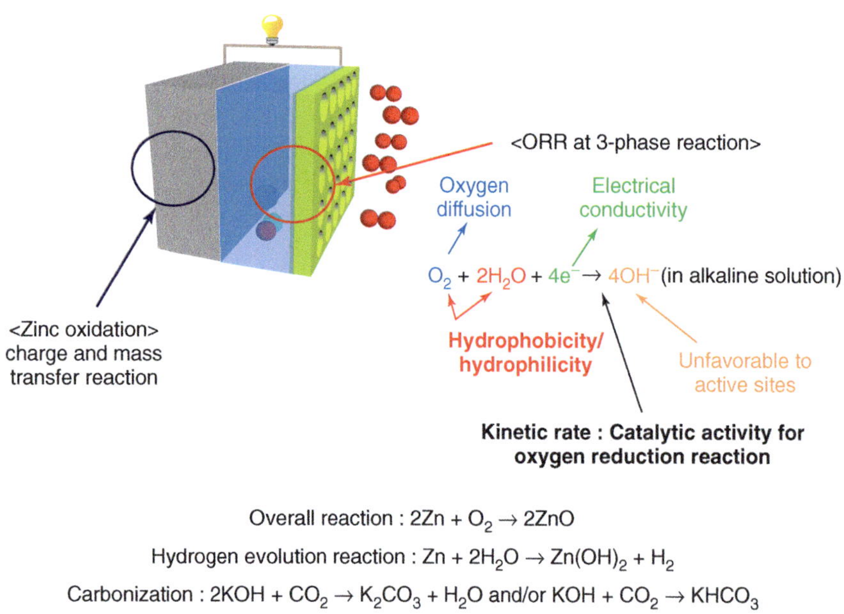


**Figure 2** Schematic principle of operation for alkaline electrolyte-based Zn–air battery

$$Zn\,(0.140\,cm^3\,g^{-1}) + 4OH^- \rightarrow Zn(OH)_4^{2-}(\text{soluble}) + 2e^-$$
(via charge transfer reaction)

$$Zn(OH)_4^{2-} \rightarrow ZnO\,(0.178\,cm^3\,g^{-1}) + 2OH^- + H_2O$$
(via mass transfer reaction)

During oxidation of zinc, electrons move to the cathode and zinc ion is dissolved as zincate ions in alkaline electrolyte. On reaching its saturation point, zincate ions precipitate to zinc oxide (ZnO) with ≈27% of its volume expansion from Zn ($0.140\,cm^3\,g^{-1}$) to ZnO ($0.178\,cm^3\,g^{-1}$).[13,14] Besides this oxidation reaction of zinc, hydrogen evolution reaction (HER) can also be possible with water consumed because standard reference potential of $Zn/Zn^{2+}$ in alkaline solution is lower than −1.0 V versus NHE. Eventually, this undesired parasitic reaction can lower the utilization of zinc and ionic conductivity, thus decreasing its capacity.

$$Zn + 2H_2O \rightarrow Zn(OH)_2 + H_2 \uparrow \quad (\text{HER})$$

On the other hand, at the cathode side, oxygen gas diffuses into porous gas diffusion electrode due to a difference in a pressure of oxygen between the outside and inside of the cell. This permeated oxygen gas is mainly reduced at the surface of the catalysts' contact with aqueous electrolyte due to more easily available gas-phase oxygen, not dissolved oxygen in solution.[15]

$$O_2\,(g) + 2H_2O\,(l) + 4e^-\,(s) \rightarrow 4OH^-\,(l) \quad (\text{ORR})$$

Therefore, it is worthy to note that most oxygen reduction reactions (ORRs) in aqueous electrolyte occur at the three phases of the reaction sites: gas, liquid, and solid.

On the basis of the aforementioned half-cell reaction, Zn–air full-cell reactions show 1.65 V of an equilibrium potential, and its practical potential is typically less than 1.4 V at desired discharge current densities.[11] Proper electrocatalysts should be required to increase its potential because major potential loss results from sluggish ORR. Besides both the electrodes, using alkaline electrolyte can cause carbonization problems, which can decrease ionic conductivity or block a pore within the air electrode.

$$2KOH + CO_2 \rightarrow K_2CO_3 + H_2O \text{ and/or } KOH + CO_2 \rightarrow KHCO_3 \quad (\text{carbonization})$$

Therefore, for building better Zn–air batteries, the aforementioned two side reactions, HER and carbonization, as well as two main reactions, oxidation of zinc and ORR, should be considered.

## 3 PERSPECTIVE VIEW OF A POSSIBILITY AS A PRIMARY AND MECHANICALLY RECHARGEABLE Zn–AIR BATTERY

The theoretical energy density of Zn–air batteries (ca. $1084\,Wh\,kg^{-1}$) is much higher than the state-of-art Li-ion batteries (ca. $200–250\,Wh\,kg^{-1}$) when assuming 100% utilization of zinc metal. However, generally, zinc utilization is around less than 60% so that its practical energy density is lower than its theoretical value, but it has still larger energy density than that of the current Li-ion batteries. For example, when assuming that Zn occupies 50–70% of the total battery pack, the specific energy of Zn–air battery with just 20–30% of depth of discharge (DOD) is comparable to commercial Li-ion batteries.[16] Therefore, if zinc utilization can be increased and zinc refiling station is possible, primary Zn–air batteries can still have a high potential based on the various aforementioned benefits described in Section 1. Furthermore, a dynamic response time of Zn–air batteries for current step-up or current step-down is in milliseconds, and thus it can be excellently applied to vehicles with rapid dynamic response demands.[17]

To maximize the potential of primary Zn-air batteries, increasing zinc utilization and decreasing polarization loss at cathode, can be the most practical approach. To address these two key problems, it is significant to know that Zn–air battery is an open system and to correlate each electrode with aqueous electrolyte so that understanding each electrode with solution can be a starting point to build better Zn–air batteries.

### 3.1 Fully Immersed Zinc Electrode in Alkaline Electrolyte

Zinc metal has been widely used as an anode material in alkaline electrolyte and fundamentally studied and reviewed.[18] Similar to other primary Zn-based batteries, zinc metal is fully immersed in electrolyte. The desired reaction is oxidation of zinc at anode, which produces the cationic compound, $Zn^{2+}$, which can permeate through the separator to air electrode, increasing polarization and decreasing capacity, which is one of the problems in Zn-based battery system. To prevent this crossover, a modified separator with sulfonium cation[19] and $Mn(OH)_2$[20] is prepared and studied. For example, sulfonium-treated separators effectively suppress cation permeation and increase the capacity sixfold.[19] Another problem related with zinc metal results from the fact that equilibrium potential for zinc is lower than that of hydrogen. Therefore, HER at zinc surface is thermodynamically possible although its kinetic rate is slow because of high overpotential of HER of zinc metal. This parasitic reaction can consume zinc metal as well as water, decreasing both the capacity of zinc and the ionic conductivity of the electrolyte. Therefore, a surface coating,[21,22] additive,[23,24] and alloy[25] have been applied to increase the stability of zinc metal. Furthermore, at the end of the discharge of zinc metal, zinc oxide is precipitated in the vicinity of separator and can be accelerated under high current discharge. Once this passivation layer is fully covered onto pristine zinc surface, zinc metal at the current collector cannot participate in the electrochemical oxidation process, causing the utilization of zinc metal and then decreasing the capacity. In addition, about 27% of volume expansion from zinc to zinc oxide in alkaline electrolyte occurs and can increase inner cell pressure, resulting in electrolyte flooding through porous air electrode. Recently, Parker *et al.*[16] developed a zinc sponge with a porous, monolithic, three-dimensional (3D) aperiodic architecture and they demonstrated that zinc utilization approaches 90% (728 mAh $g^{-1}{}_{Zn}$) as a primary Zn–air battery.

### 3.2 Partially Immersed Air Electrode in Alkaline Electrolyte

The ORR at cathode is a vital reaction, and especially it is worthy to note that it occurs at triple phase boundary. This means that the air electrode is partially immersed in the electrolyte to maximize the utilization of gas-phase oxygen. Obviously, using oxygen gas can significantly increase energy density, but the way to accelerate this sluggish reaction and then minimize the overall voltage loss require highly efficient electrocatalysts as well as engineered air electrode architectures. ORR in alkaline electrolyte has better kinetics than in acid because of lower overpotential for first electron transfer step ($O_2/O_2 \cdot^-$)[26] and weaker anions adsorption.[27] As a result, various nonprecious metal catalysts can catalyze this reaction and have been studied.[28–30] Almost all efforts, however, have focused to prepare catalysts and test it under minimized mass transfer controlled experimental condition using a thin film rotating disk electrode technique[31–33] and limited evaluation parameters such as a half wave potential, the number of transfer electrons, and peroxide yield. Of course, highly efficient electrocatalysts can critically decrease polarization loss, but it is worthy to note that this ideal experimental condition cannot fully describe an electrochemical behavior of actual air electrode operated under static state air pressure, not artificial oxygen pressure. Let us consider ORR equation in alkaline electrolyte, and it indicates an importance of three-phase reaction, which could be maximized using hydrophilic carbon in catalysts layer and hydrophobic carbon in gas-diffusion layer.[34] Therefore, how to facilitate oxygen gas diffusion into solid catalysts contacted with electrolyte is a significant issue and can be directly connected to better air electrode design. On the basis of the aforementioned discussion, understanding why air electrode is flooded is a key step to develop better air electrode. During discharge, generated heat affects electroosmosis of electrolyte and polytetrafluoroethylene (PTFE) hydrophobic binder, causing air electrode polarization due to electrolyte infiltration.[35] Especially, this decrease in hydrophobicity of PTFE can be accelerated in alkaline KOH electrolyte, causing decrease in three-phase reaction sites.[36] Besides this electrolyte infiltration in aqueous solution although it is slower than in organic electrolyte, 8a alkaline-based Zn–air battery is vulnerable to $CO_2$ concentration due to its open system. This carbonization leads to formation of carbonates within a pore of air electrodes as well as decrease in ionic conductivity of alkaline electrolyte, causing degradation of air electrode.[37]

### 3.3 How to Maximize a Primary and Mechanically Rechargeable Zn–Air Battery

When considering only consumed materials in zinc metal during discharge, how to regenerate zinc efficiently can be a most practical approach to maximize the potential as a primary and mechanically rechargeable Zn–air battery application when aforementioned problems can be addressed. Especially, this process requires additional devices because zinc is not regenerated using electric charging of Zn–air battery. For example, an electrowinning and solzinc process,[38] nonelectrical route, can be used for zinc regeneration. Especially, compared to other zinc regeneration process, in solzinc process, carbon consumption can be 1/5 under solzinc process and it can be environmental friendly. In addition, these processes work at lower temperature, 1200 °C than original thermal decomposition of zinc oxide, 1750 °C, thanks to addition of carbon.

Another promising possibility of primary Zn–air battery is to combine it with conventional Li-ion battery, as shown in Figure 3.[39] By controlling the charging

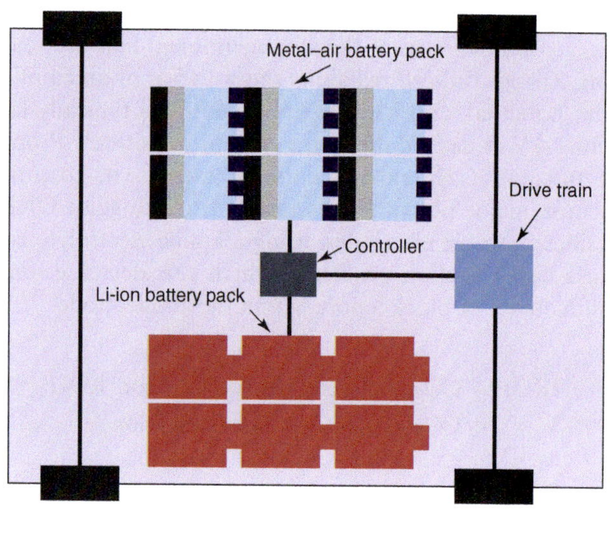


**Figure 3** Efficient dual-source battery pack system for an electric vehicle

method and the electrical engineering method, this dual source battery system could complement each other, and this proposal could maximize a primary and mechanically rechargeable Zn–air battery technology.

## 4 PERSPECTIVE VIEW OF A POSSIBILITY AS AN ELECTRICALLY RECHARGEABLE Zn–AIR BATTERY

For making electrically rechargeable Zn–air batteries, zinc metal should be plated uniformly onto current collector and oxygen should be evolved efficiently without air electrode damage. Nevertheless, ≈2 V of significant charging voltage is required because of sluggish oxygen evolution reaction (OER), resulting in a low round trip energy efficiency of ≈60%. For example, for zinc anode, it is extremely difficult for zincate to fully deposit on the same position at the electrode so that this irreversible process can yield electrode morphology change or dendrite formation during charging process. Therefore, zinc electrode is gradually degraded, decreasing the overall battery performance. Furthermore, for air electrode, oxygen evolution as well as ORR should be possible. This bifunctionality requires both highly efficient and durable electrocatalysts and air electrode architectures. To meet these strong criteria, recently, more advanced zinc electrode,[16] new design of air electrode,[40,41] new charging method,[42] and flow type battery concept (http://www.revolttechnology.com/; http://www.eosenergystorage.com) have been studied, which presented a more practical possibility for rechargeable Zn–air battery, as shown in Figure 4.

### 4.1 Fully Immersed Zinc Electrode in Alkaline Electrolyte

Zinc electrodes undergo shape change due to uneven redistribution of the zinc metal during cycling because of high local current density resulting from a loss of interparticle connectivity. To decrease this local current density to prepare reversible zinc electrode, porous and well-wired interconnected zinc sponge architecture is developed. This monolithic three-dimensional structure can increase the electrified active area so that its local current density effectively is 100 times lowered than that required to form dendrites (typically

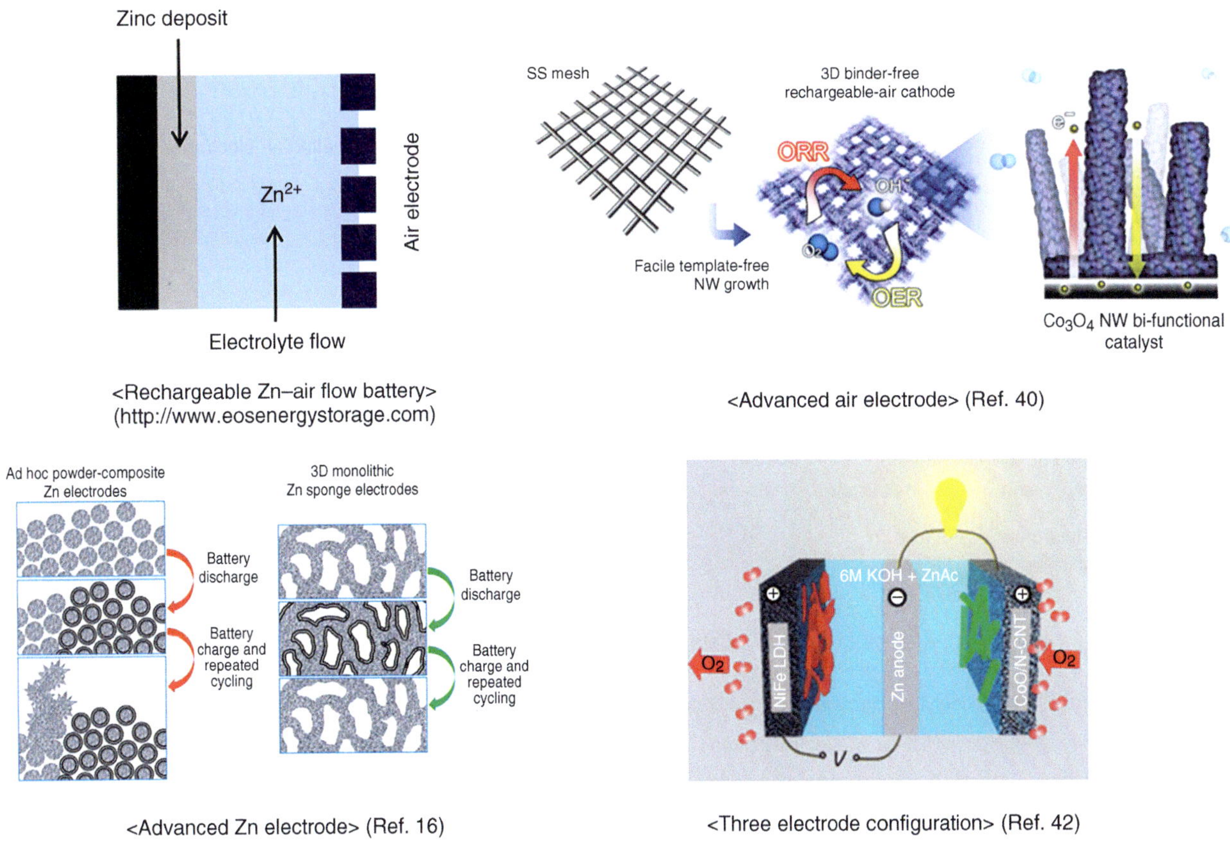


**Figure 4** State-of-the-art possible solution for electrically rechargeable Zn–air battery.[1] [Reproduced with permission from Ref. J.-S. Lee, S. Tai Kim, R. Cao, N.-S. Choi, M. Liu, K. T. Lee, J. Cho, Adv. Energy Mater. 2011, 1, 34. © Wiley, 2011]

>10 mA cm$^{-2}$).[16] Another possible strategy is to flow alkaline electrolyte to reduce zinc dendrite formation and carbonization from $CO_2$. By circulating fresh KOH or using natural electrolyte, it is possible to minimize $CO_2$ effect as well as zinc dendrite formation as demonstrated by industries (http://www.revolttechnology.com/; http://www.eosenergystorage.com). Compared to vanadium redox flow battery (VRFB), flow-type Zn–air battery has much higher power and energy density. Using flowing electrolyte with high alkali content can prevent zinc passivation and reduce dendrite formation, and thermal management issues are also minimized due to electrolyte circulating.

### 4.2 Partially Immersed Air Electrode in Alkaline Electrolyte

For preparing reversible air electrode, bifunctional electrocatalysts for ORR as well as OER are required. During charging, applied high oxidation potential could damage carbon materials composed of air electrode, and increase hydrophilicity of carbon-based electrode and then polarization.[43] Furthermore, oxygen gas is evolved during charging and can degrade air electrode architecture physically.[44] Achieving two functions, ORR and OER, however, is highly extremely difficult using just one air electrode. To address these problems, Dai group separated these two functions using three-electrode structures and they showed significantly improved rechargeability and durability although these systems can be more complicated compared to conventional two-electrode systems of Zn–air battery architectures.[42] To avoid carbon corrosion problems, Chen and Jaramilio[40,41] groups independently developed carbon and binder-free reversible air electrode by deposition of metal oxides onto metal mesh directly. They showed highly improved durability by avoiding carbon corrosion.

## 5 CONCLUSION

Despite the high potential of Zn–air battery, its practical application has been limited to hearing aids

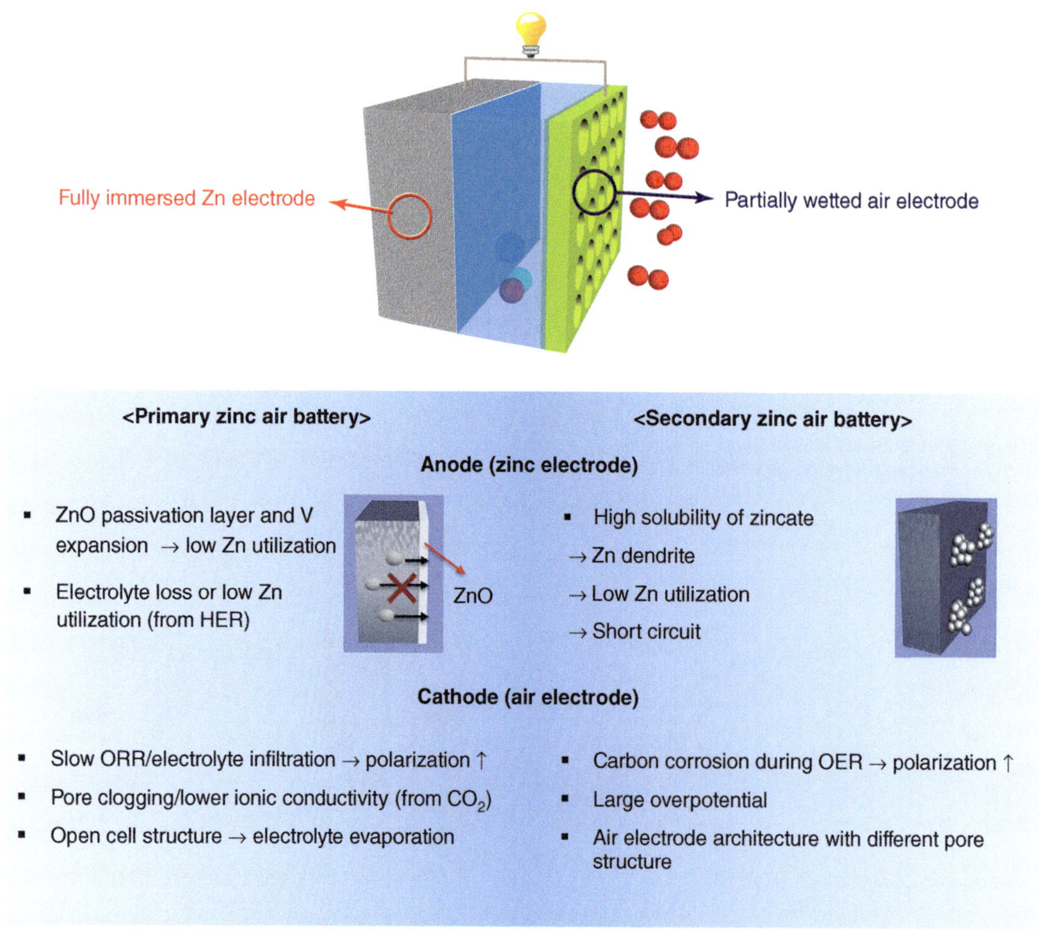


**Figure 5** Summary of current stage problems for Zn–air battery

because of fully unsolved aforementioned problems to date, as shown in Figure 5. Toward maximizing the potential of Zn–air technology, firstly, we need to consider two side reactions, HER and carbonization, as well as zinc oxidation and ORR, and know that these four reactions are loosely related to one another. It indicates that simultaneous approach should be required for building better batteries. In addition, it is possible to select a primary and mechanically rechargeable or electrically rechargeable battery option once addressing the aforementioned problems. The objective of both the systems is to regenerate zinc metal as anode. For making practically available primary and mechanically rechargeable Zn–air batteries, their potentials can be maximized by increasing zinc utilization and developing durable air electrode. Firstly, more efforts to develop air electrode, not just catalysts, should be devoted, indicating that with even highly efficient and durable catalysts, we could not fully utilize their performance with poor engineered air electrode architectures. Secondly, it can be possible to maximize their potentials by combining Li-ion battery with a well-established infra for zinc regeneration and fuel supply. For making electrically rechargeable Zn–air batteries, circulating electrolyte can bring more practical approach than static state charging although additional devices should be further required. With the aforementioned two strategies, Zn–air battery technology is not century-old battery technology anymore and can be really possible next-generation energy storage and conversion system in the near future.

## 6 ACKNOWLEDGMENT

This work was supported by the 2018 Research Funds (1.180019.01) of Ulsan National Institute of Science and Technology (UNIST).

## 7 RELATED ARTICLE

Li–$CO_2$ Batteries

## 8 ABBREVIATIONS AND ACRONYMS

DOD = depth of discharge; HER = hydrogen evolution reaction; OER = oxygen evolution reaction; ORRs = oxygen reduction reactions; PTFE = polytetrafluoroethylene; SEI = solid electrolyte interface; VRFB = vanadium redox flow battery

## 9 REFERENCES

1. J.-S. Lee, S. Tai Kim, R. Cao, N.-S. Choi, M. Liu, K. T. Lee and J. Cho, *Adv. Energy Mater.*, 2011, **1**, 34.
2. P. G. Bruce, S. A. Freunberger, L. J. Hardwick and J.-M. Tarascon, *Nat. Mater.*, 2012, **11**, 172.
3. S. Evers and L. F. Nazar, *Acc. Chem. Res.*, 2012, **46**, 1135.
4. J. Dahn, 'Scalable Energy Storage: Beyond Lithium Ion', 2009.
5. W. Xu, J. Wang, F. Ding, X. Chen, E. Nasybulin, Y. Zhang and J.-G. Zhang, *Energy Environ. Sci.*, 2014, **7**, 513.
6. G. Zheng, S. W. Lee, Z. Liang, H.-W. Lee, K. Yan, H. Yao, H. Wang, W. Li, S. Chu and Y. Cui, *Nat. Nanotechnol.*, 2014, **9**, 618.
7. N. Liu, Z. Lu, J. Zhao, M. T. McDowell, H.-W. Lee, W. Zhao and Y. Cui, *Nat. Nanotechnol.*, 2014, **9**, 187.
8. Y. Li and H. Dai, *Chem. Soc. Rev.*, 2014, **43**, 5257.
9. P. Pei, K. Wang and Z. Ma, *Appl. Energy*, 2014, **128**, 315.
10. V. Caramia and B. Bozzini, *Mater. Renew. Sustain. Energy*, 2014, **3**, 1.
11. R. A. Huggins, 'Materials for Lithium-Ion Batteries', Springer, Netherlands, 2000, Vol. **85**, p. 21.
12. M. Geng and D. O. Northwood, *Int. J. Hydrog. Energy*, 2003, **28**, 633.
13. T. Arlt, D. Schroder, U. Krewer and I. Manke, *Phys. Chem. Chem. Phys.*, 2014, **16**, 22273.
14. S. Siahrostami, V. Tripkovic, K. T. Lundgaard, K. E. Jensen, H. A. Hansen, J. S. Hummelshoj, J. S. G. Myrdal, T. Vegge, J. K. Norskov and J. Rossmeisl, *Phys. Chem. Chem. Phys.*, 2013, **15**, 6416.
15. K. E. Gubbins and R. D. Walker, *J. Electrochem. Soc.*, 1965, **112**, 469.
16. J. F. Parker, C. N. Chervin, E. S. Nelson, D. R. Rolison and J. W. Long, *Energy Environ. Sci.*, 2014, **7**, 1117.
17. P. Pei, Z. Ma, K. Wang, X. Wang, M. Song and H. Xu, *J. Power Sources*, 2014, **249**, 13.
18. F. R. McLarnon and E. J. Cairns, *J. Electrochem. Soc.*, 1991, **138**, 645.
19. E. L. Dewi, K. Oyaizu, H. Nishide and E. Tsuchida, *J. Power Sources*, 2003, **115**, 149.
20. Y. Kiros, *J. Power Sources*, 1996, **62**, 117.
21. Y.-D. Cho and G. T.-K. Fey, *J. Power Sources*, 2008, **184**, 610.
22. S.-M. Lee, Y.-J. Kim, S.-W. Eom, N.-S. Choi, K.-W. Kim and S.-B. Cho, *J. Power Sources*, 2013, **227**, 177.
23. C. W. Lee, K. Sathiyanarayanan, S. W. Eom, H. S. Kim and M. S. Yun, *J. Power Sources*, 2006, **160**, 161.
24. M. Yano, S. Fujitani, K. Nishio, Y. Akai and M. Kurimura, *J. Power Sources*, 1998, **74**, 129.
25. C. W. Lee, K. Sathiyanarayanan, S. W. Eom and M. S. Yun, *J. Power Sources*, 2006, **160**, 1436.

26. N. Ramaswamy and S. Mukerjee, *J. Phys. Chem. C*, 2011, **115**, 18015.
27. J. S. Spendelow and A. Wieckowski, *Phys. Chem. Chem. Phys.*, 2007, **9**, 2654.
28. R. Cao, J.-S. Lee, M. Liu and J. Cho, *Adv. Energy Mater.*, 2012, **2**, 816.
29. F. Cheng and J. Chen, *Chem. Soc. Rev.*, 2012, **41**, 2172.
30. V. Neburchilov, H. Wang, J. J. Martin and W. Qu, *J. Power Sources*, 2010, **195**, 1271.
31. J. Suntivich, H. A. Gasteiger, N. Yabuuchi and Y. Shao-Horn, *J. Electrochem. Soc.*, 2010, **157**, B1263.
32. Y. Garsany, I. L. Singer and K. E. Swider-Lyons, *J. Electroanal. Chem.*, 2011, **662**, 396.
33. S. S. Kocha, J. W. Zack, S. M. Alia, K. C. Neyerlin and B. S. Pivovar, *ECS Trans.*, 2013, **50**, 1475.
34. L. Jörissen, *J. Power Sources*, 2006, **155**, 23.
35. Y. Xu, X. Xu, G. Li, Z. Zhang, G. Hu and Y. Zheng, *Int. J. Electrochem. Sci.*, 2013, **8**, 11805.
36. M. Balaish, A. Kraytsberg and Y. Ein-Eli, *ChemElectroChem*, 2014, **1**, 90.
37. J. F. Drillet, F. Holzer, T. Kallis, S. Muller and V. M. Schmidt, *Phys. Chem. Chem. Phys.*, 2001, **3**, 368.
38. C. Wieckert, M. Epstein, G. Olalde, S. Santén and A. Steinfeld, *Encycl. Electrochem. Power Sources*, 2009, **5**, 469.
39. K. R. Kelty, V. H. Mehta and J. B. Straubel, 'Google Patents', 2012.
40. D. U. Lee, J.-Y. Choi, K. Feng, H. W. Park and Z. Chen, *Adv. Energy Mater.*, 2014, **4**, 1301389.
41. J. W. D. Ng, M. Tang and T. F. Jaramillo, *Energy Environ. Sci.*, 2014.
42. Y. Li, M. Gong, Y. Liang, J. Feng, J.-E. Kim, H. Wang, G. Hong, B. Zhang and H. Dai, *Nat. Commun.*, 2013, **4**, 1805.
43. X. Wang, W. Li, Z. Chen, M. Waje and Y. Yan, *J. Power Sources*, 2006, **158**, 154.
44. G. Toussaint, P. Stevens, L. Akrour, R. Rouget and F. Fourgeot, *ECS Trans.*, 2010, **28**, 25.

# Solid State and Materials Chemistry for Sodium-Ion Batteries

**Divya Sehrawat and Neeraj Sharma**

*University of New South Wales (UNSW Sydney), Sydney, Australia*

and

**Jennifer H. Stansby**

*University of New South Wales (UNSW Sydney), Sydney, Australia;*
*Australia Nuclear Science and Technology Organisation, Sydney, Australia*



## 1 INTRODUCTION

### 1.1 A Battery and the Rise of Lithium-Ion Batteries

An electrochemical cell, where a battery can be multiple or single electrochemical cells, stores energy and releases it to an external circuit. Chemical energy is converted into electrical energy, and the humble battery has revolutionized our world. Primary batteries turn chemical energy into electrical energy for typically one discharge and then are discarded. A typical example is the so-called alkaline batteries that are found in children's toys delivering about 1.4 V or so. Rechargeable or secondary batteries can have the chemical reactions inside the battery reversed numerous times and therefore can store, release, and recharge in cycles with typical examples being lead-acid and lithium-ion batteries (LIBs).

In all cases, electrodes are present in the battery, a cathode and an anode or a positive and negative electrode, respectively, as often referred to in the LIB literature. The electrodes act as the reaction centers, where reduction and oxidation reactions occur transforming the chemistry of the electrodes. In the first commercial LIBs introduced in the 1990s,[1] graphite was used as the anode and $LiCoO_2$ as the cathode. During charge, lithium-ions were extracted from the cathode and Co oxidized from $3^+$ to $4^+$ and the lithium-ions were inserted into the anode forming $Li_xC_6$[2–4] and reducing graphite. These reactions are called (de)intercalation reactions, a type of (de)insertion reactions. These intercalation or insertion reactions can proceed via different pathways, for example, solid solution or two-phase reactions that, in solid-state chemistry, are second-and first-order phase transitions, respectively. Other insertion mechanisms have been explored in the literature, including conversion and alloying reactions.

The potential difference between the cathode and anode reaction results in the cell potential or voltage, the higher the voltage and the higher the energy density. The larger the amount of Li that can be reversibly inserted or removed from both electrodes equates to a larger

*Inorganic Battery Materials*. Edited by Hailiang Wang and Boniface P.T. Fokwa
 ISBN 978-1-119-43199-2

energy storage density or capacity. Minimal change in the electrodes during ion insertion/extraction typically results in better "reversibility," which often leads to longer lifetimes. Compounded with these considerations is the need to choose the cheapest and most nontoxic components for the electrodes and the need for binders and additional conductive materials such as carbon black to make up the electrode paste. The result is that electrodes are a mix of 3 or more components in varying ratios, with oxidation state, crystal structures, and typically particle dimensions changing during function, in addition to surface-based reactions. All of these lead to massive performance considerations, and they need to function together in order to achieve safe and stable operation.

Although the electrodes control much of the electrochemistry, other components are also critical in the cell. Cu and Al current collectors are used in the case of LIBs. Ideally, Al current collectors would be preferred due to cost, but at low potentials Li alloys with Al and therefore Al cannot be used as the anode current collector. A separator is used to physically separate the electrodes, as contact results in shorting and the cell voltage would reduce to zero. An electrolyte solution is added to the battery, often a salt mixed in an organic solution, noting that using water-based solutions limits the cell voltage to around 1.5 V due to water participation in oxidation and reduction. The electrolyte in the case of LIBs has a number of additives added to it to allow the formation of solid–electrolyte interphase (SEI) layers. These SEI layers are very challenging to characterize and do lead to a large proportion of degradation of rechargeable batteries with cycling.

Taking these facets together, the LIB has been commercialized and optimized for a variety of mobile applications and is finding significant use in electric vehicles and as energy storage solutions for renewable energy generation technologies. The question remains whether LIBs are the best option for all applications, and this question is driven by their perceived safety issues, cost, toxicity (especially with Co-containing electrodes), and performance.

### 1.2 The Sodium-Ion Battery Equivalent of a Lithium-Ion Battery

Electrodes that insert/extract sodium from their crystal structures, much like commercial LIBs, were also explored in the 1980s, for example, Ref. 5. The lithium-ion-based system began to dominate, and the sodium equivalents were largely left at this stage until about the 2000s when interest in this chemistry was re-invigorated.[6] Using the same principle as a LIB described earlier, the intention was reversible insertion/extraction from electrodes at room temperature using $Na^+$ as the charge carrier rather than $Li^+$. Almost immediately two aspects were different, the ionic radii of $Na^+$ being about 1.34 times larger than $Li^+$, and therefore resulting in the need for modified or different hosts, and the response of the host typically being larger for the removal and insertion of the larger cation. Furthermore, the standard electrochemical potential of $Na^+/Na$ differs by 0.3 V compared to $Li^+/Li$. However, the abundance of sodium and the focus on cheap electrodes has resulted in significant research uptake and development of a few commercial prototypes. Another advantage has been the ability to use Al current collectors for both electrodes, as Na does not alloy with Al within the potential windows used.

A number of cathode options exist that are Co-free with cheap transition metals and show moderate energy densities, and the electrolyte behaves with similar electrochemistry to that of LIB systems but the anode is slightly troublesome. Graphite that works well with LIBs (for a number of applications) does not work in sodium-ion batteries, and this has resulted in significant attention on the anode.

### 1.3 High-Temperature and Other Sodium-Based Battery Configurations

It is worthwhile to note that sodium-based batteries have been commercialized and do exist. These are typically run at higher temperatures, above 200 °C and are often referred to as molten salt batteries and are typically rechargeable.[7,8] New electrolytes are being pursued that might bring the operation temperature down to 90 °C and the structure is often composed of Na/NaCl, $Ni/NiCl_2$ or $Fe/FeCl_2$ electrodes and a combination of $\beta$ alumina and $NaAlCl_4$ as electrolytes. It should be noted these batteries need appropriate casing and insulation when operating. The molten salt acts similar to the liquid electrolytes used in conventional LIBs. Other chemistries currently being investigated include the equivalent to lithium systems of $Na–O_2$ and Na–S batteries.[9] These face a number of challenges similar to lithium systems of polysulfide solubility in electrolyte and the ability to reversibly deposit and dissolve $Na_xO_2$ precipitates during cycling. The concept of Na–S has been examined since the 1970s. Protection of the sodium electrode is critical in these systems. It is interesting to note, that many next-generation lithium-based systems are now routinely being investigated with sodium and even potassium-based configurations.

## 2 ELECTRODES AND THEIR MECHANISMS

Electrode materials are critical in determining the cell voltage, capacity, energy density, and hence overall performance of a battery. Like LIBs, sodium-ion batteries typically function through the insertion of $Na^+$ into the cathode upon charge and subsequent extraction of Na from the cathode and insertion into the anode during discharge. Although the desired characteristic for sodium-ion battery electrodes are generally the same as those for LIBs (good structural stability, high capacity, nontoxic), the larger size of $Na^+$ compared to $Li^+$ mediates the choice of electrodes. Typically, wider diffusion pathways and larger interlayer distances are requirements for cathode and anode materials, respectively, or for both electrodes, which create challenges regarding long-term structural stability. Furthermore, a greater emphasis is placed on cost, environmental friendliness, and lifetime for sodium-ion batteries due to their prospective use in large-scale energy storage applications. In contrast, LIBs are primarily used in portable electronic devices for which a high voltage, capacity, and therefore energy density are of foremost importance. Some critical parameters that should be considered when designing and testing new electrodes are detailed in the following sections.

### 2.1 Voltage

The higher the voltage, the higher the attainable energy density. Since the cell voltage is the potential difference between the cathode and anode, it is desirable to have a high-voltage cathode material, that is, efficient oxidizing agent and low-voltage anode material, that is, efficient reducing agent.

### 2.2 Specific Capacity

High-capacity materials are ideal for both electrodes. Often materials that achieve high capacities show a compromise between coulombic efficiency, capacity retention, and lifetime. High specific capacity translates into a structure or composite electrode that accommodates the reversible insertion of large amounts of $Na^+$, the larger the amount of $Na^+$ insertion/extraction, the higher the capacity.

### 2.3 Fast Reaction Kinetics and High Electronic/Ionic Conductivity

These properties can be improved through nanosizing. However, nanosizing can lead to a large capacity loss after the initial cycles, which are associated with irreversible reactions between the electrolyte and electrode material.[10] Factors such as tap density and active surface areas need to be considered.

### 2.4 Cyclability

Good cyclability, or in other words structural stability (minimal structural change) and mechanistic reversibility at the electrodes, is important to produce a battery with a good lifetime. Capacity fade may be caused by irreversible structural changes or mechanical fractures within the bulk electrode material during cycling, surface reactions, electrolyte oxidation, insertion of electrolyte solvent molecules, or decomposition of cell components at high voltages.[11]

### 2.5 Other Factors

Ideally, materials used to synthesize electrode materials are low cost, abundant, and environmentally friendly. Moreover, the synthetic route should be simple and scalable. It is also important to consider the battery safety, which is influenced by the stability of the electrodes and the types of reactions that occur on the electrode surface. Finally, the electrode materials should be compatible with the electrolyte material to minimize dendrite formation, dissolution of the electrode material into the electrolyte, and oxidation of the electrolyte. For stable electrode–electrolyte interfaces, ideally the cathode electrochemical potential lies below the electrolyte HOMO while the anode potential lies above the LOMO of the electrolyte. For LIBs, the operating voltage is often limited to 4.5 V, while for sodium-based systems the conventional electrolyte will become unstable above 4.2 V because the equilibrium potential of $Na^+/Na$ is 0.3 V higher than that of the $Li^+/Li$ redox couple.[11]

### 2.6 Cathodes

The cathode is arguably the greatest bottleneck for the commercialization of sodium-ion batteries due to its influence on the electrochemical performance in terms of specific energy, cycle life, and specific power. Furthermore, it is also the component that offers the greatest cost reduction in comparison to the lithium counterparts. Research in this area is often driven by two approaches: (i) families of compounds that are analogous to LIB electrodes or (ii) new families. For example, the widely used cathode material $LiCoO_2$ outperforms $NaCoO_2$ in terms of operating voltage (3.9 V vs 2.5–3.0 V) and specific capacity (140 mAh $g^{-1}$ vs ~121 mAh $g^{-1}$).[12,13] Both materials possess layers of $CoO_6$ edge-sharing octahedra. However, the larger ionic radius of Na (1.02 Å) versus Li (0.76 Å) and lower electrochemical potential restrict the performance of $NaCoO_2$ systems. In contrast, the $Fe^{3+/4+}$ redox couple is not active in $LiFeO_2$ while electrochemical activity has been reported for $NaFeO_2$ in Na cells.

Promising classes of sodium-ion battery cathodes include layered oxides, polyanionic-based systems, framework-based materials including metal–organic frameworks (MOFs) and more recently organic compounds.

### 2.6.1 *Layered Oxides*

The larger size and screening effect of Na compared to Li give rise to the tendency of Na to favor the formation of layered materials.[14] $Na_xMO_2$ compounds, M = Ti, V, Cr, Mn, Fe, Co, Ni, and a combination of two or more of these elements, are cathode candidates due to their relatively high capacity, use of inexpensive transition metals, and facile synthesis.[14] Their structure can be described as $MO_2$ layers consisting of edge-sharing $MO_6$ octahedra with Na-ions residing on sites between the layers. The Na-ions share either an edge ($Na_e$) or a face ($Na_f$) with the adjacent $MO_6$ octahedra whereby $Na_e$ is often preferentially occupied.[15] Using notation developed by Delmas *et al.*,[5] they are classified based on the Na coordination geometry (O: octahedral, P: prismatic) and the number of transition-metal-oxide layers within the unit cell; a prime symbol (′) indicates distortion from the ideal phase. Typical phases include O3-, P3-, P2-, and P′2-. Na-ions diffuse in and out of the layers, leading to structural changes during operation. P2 to OP4, O2 or P3 to O3 are examples of phase transitions that may take place upon cycling while preserving M—O bonding. In contrast, P2 to O3 transitions rarely take place due to the high input of energy required to break M—O bonds. These phase transitions manifest themselves as potential plateaus in charge/discharge curves. The structural evolution (stability) and Na-ion diffusion kinetics upon cycling has a direct influence on the electrochemical performance of the material. Ideally, minimal phase transitions will occur during cycling (smooth charge/discharge curve) to enable long-term stability and an enhanced battery lifetime.

*P2-Type* In the 1980s, Na intercalation was identified in P2-$Na_xCoO_2$ (Figure 1), and it became one of the first materials to be incorporated into a full sodium-ion battery. However, interest in this system diminished as it was outperformed by the $LiCoO_2$/graphite system. For comparison, commercially used O3-$LiCoO_2$ produces a capacity of 140 mAh $g^{-1}$ over cycles,[16] while P2-$Na_xCoO_2$ produces a reversible specific capacity of 126 mAh $g^{-1}$ at a current rate of C/10.[13] P2-$Na_xCoO_2$ regained interest in 2010 and was found to form for $0.46 < x < 0.83$, while O3-$Na_xCoO_2$ forms when $0.83 < x < 1.00$.[17] In general, P2-type layered oxides are typically stable, and therefore synthesized, with a nonstoichiometric quantity of Na ($0.6 < x < 0.7$).[18] Many differences were found between P2-$Na_xCoO_2$ and lithium analogues. The larger size of $Na^+$ leads to different behavior of the material upon cycling; multiple plateaus in the charge/discharge curves (phase transitions) are observed for P2-$Na_xCoO_2$, while for O3-$Li_xCoO_2$ a single and stable plateau around 4.0 V with low polarization is observed (Figure 2).[18] The presence of multiple plateaus indicates the presence of both single-phase (solid solution) and two-phase mechanisms.

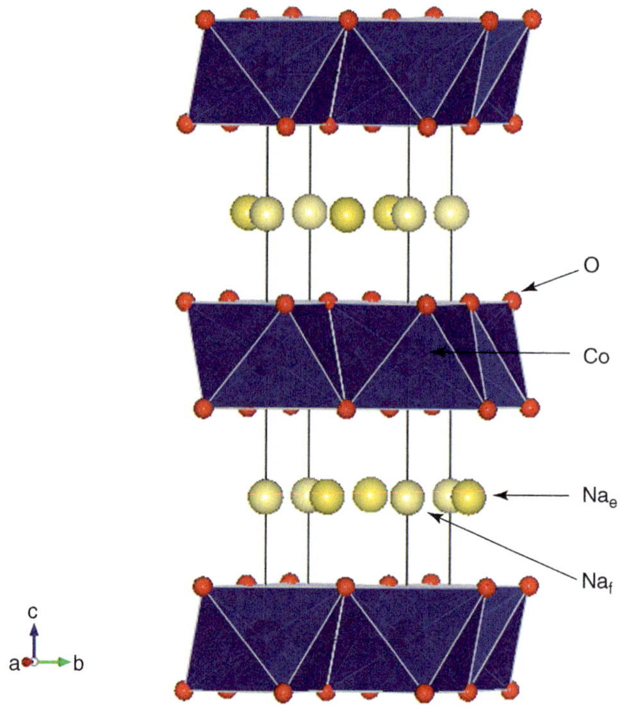


**Figure 1** The P2-$Na_xCoO_2$ structure

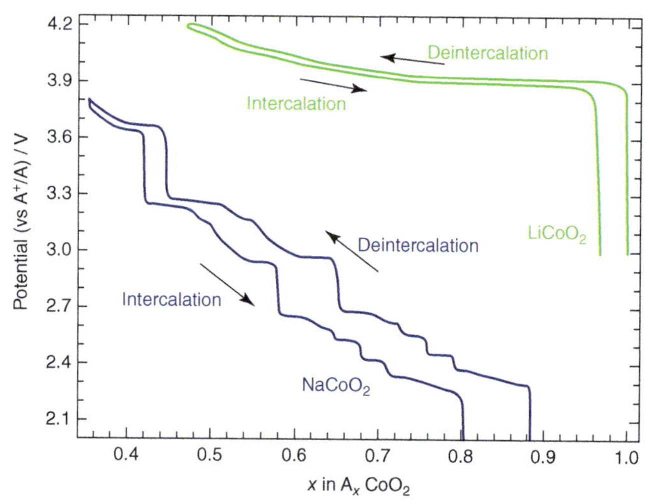


**Figure 2** Voltage profiles of $LiCoO_2$ and $NaCoO_2$. [Reproduced with permission from Ref. 19 © Wiley, 2017]

Consequently, the sodium counterpart achieves a much lower energy density despite its faster diffusion coefficients ($1.5 \times 10^{-10}$ cm$^2$ s$^{-1}$ vs $1 \times 10^{-11}$ cm$^2$ s$^{-1}$).[20]

A wide range of metals (Ti, Ni, Co, V, Fe, Mn, Mg, Cu) can be incorporated onto the transition metal site of sodium-based layered oxides. Altering the chemical composition of layered oxides influences the structure and hence properties and performance of the material. A major

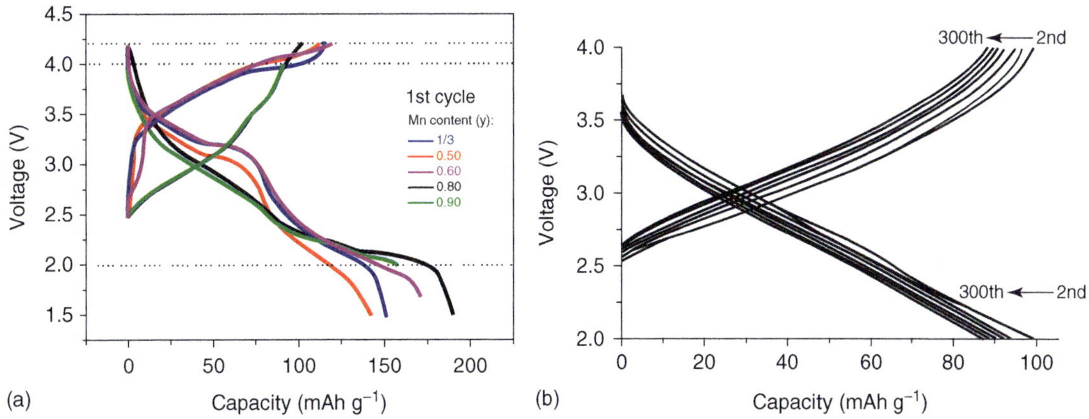


**Figure 3** Comparative electrochemical performance across the (a) P2-$Na_{2/3}Fe_{1-y}Mn_yO_2$ series for the 1st charge–discharge curves [Reproduced with permission from Ref 23. © Ameican Chemical Society, 2018] and (b) P2-$Na_{2/3}Mn_{0.8}Fe_{0.1}Ti_{0.1}O_2$ within the voltage range 4.0–2.0 V at 1C for 300 cycles. [Reproduced with permission from Ref 24. © Ameican Chemical Society, 2016.]

advancement was made when the earth-abundant P2-$Na_{2/3}[Fe_{1/2}Mn_{1/2}O_2]$ was reported to deliver 190 mAh g$^{-1}$ of reversible capacity.[21] Layered cathode materials typically deliver specific capacities between 120 and 200 mAh g$^{-1}$ with an average operating voltage between 2.6 and 3.2 V. Limiting the operating voltage to below 4.0 V helps to minimize electrolyte oxidation.[11]

Manganese-based P2-type layered oxides are environmentally friendly cathode materials. P2-$Na_xMnO_2$ is prone to rapid capacity fade due to Jahn–Teller distortion associated with high spin $Mn^{3+}$ $((t_{2g})^3\,(e_g{}^*)^1)$.[22] The structural stability of the material can be improved through substitution of Mn for different metals such as Fe. Further investigations into the P2-$Na_xFe_{1-y}Mn_yO_2$ series have highlighted the sensitivity of the structural evolution and hence electrochemical performance toward Mn content (Figure 3a). Greater capacities and long-term capacity retention is achieved with higher Mn content ($y \geq 0.8$).[23]

Substitution of Mn for Mg to produce the $Na_{0.67}Mn_{1-x}Mg_xO_2$ series also improves the electrochemical performance relative to the parent material, P2-$Na_xMnO_2$. The $Na_{0.67}Mn_{0.8}Mg_{0.2}O_2$ composition displays the best performance in terms of reversible capacity and capacity retention. The substitution of a small amount of electrochemically inactive elements into the oxide layer is a promising method to improve the stability and hence electrochemical performance of Mn-rich cathode materials. To this end, ternary layered oxides such as P2-$Na_{2/3}Mn_{0.8}Fe_{0.1}Ti_{0.1}O_2$ have been investigated (Figure 3b). In situ X-ray powder diffraction (XRD) data for P2-$Na_{2/3}Mn_{0.8}Fe_{0.1}Ti_{0.1}O_2$ during cycling shows that a solid solution reaction is mainly observed within 4.0–2.0 V, except for a small biphasic mechanism at or below 2.2 V.[24] Structural stability is reflected in the electrochemical stability, whereby P2-$Na_{2/3}Mn_{0.8}Fe_{0.1}Ti_{0.1}O_2$ displays enhanced capacity retention after 50 cycles of 95%[24] compared to 73% for P2-$Na_{2/3}Mn_{0.8}Fe_{0.2}O_2$.[25] For completeness, comparison with unsubstituted P2-$Na_{2/3}MnO_2$, the capacity retention after the 20 cycles is 87%[26] while it is 73% for the Mn poor phase, P2-$Na_{2/3}Mn_{0.5}Fe_{0.5}O_2$.[23] Overall, research in this area has highlighted P2-$Na_{2/3}Mn_{0.8}M_{0.2}O_2$ as the ideal composition to date with M shown to be a single or multielement mix. Current work involves tuning this composition further through incorporation of additional low-cost metals (Al, Cu, Zn) for M.

The cycling performance and energy density of layered Mn-based cathode can be improved through the addition of a stable interface.[27] For example, coating the surface with nanoparticles or applying a carbon coating prevents leaching of the transition metal ions during cycling.[17] The use of aluminum oxide as a surface coating to produce $Al_2O_3$/P2-$Na_{2/3}Ni_{1/3}Mn_{2/3}O_2$ increases the materials' cyclability by suppressing unfavorable side reactions and exfoliation of the transition-metal-oxide layer.[28] Coating of aluminum oxide via atomic layer deposition onto P2-$Na_{2/3}Mn_{0.54}Ni_{0.13}Co_{0.13}O_2$ reduces the mechanical stresses exerted on the material during cycling, which improves the structural stability and hence improves electrochemical performance.[29]

*O3-Type* O3-type $NaMO_2$ (M = Co, Ni, Mn, Fe) are also appealing cathode candidates due to their good electrochemical activity and simple synthesis.[30] Furthermore, in some analogous systems, the diffusion of Na-ions is faster than Li-ions.[31] O3-$NaFeO_2$ and O3-$NaCrO_2$ are examples of cathodes for which the lithium counterpart is electrochemically inactive.[32]

Similar to P2-type compounds, Mn-containing materials are some of the most promising O3-type cathodes known to date. For example, O3-$NaNi_{0.5}Mn_{0.5}O_2$ can deliver more than 185 mAh g$^{-1}$ when charged to

4.5 V, which is comparable to O3-$NaMnO_2$.[22] At 1C the material displays a capacity of $105\,mAh\,g^{-1}$ within the voltage range 2.2–3.8 V and retains 75% of the initial reversible capacity after 50 cycles.[33] Often the phase transitions observed for O3-type structures are more complex than for P2-type materials. During sodium extraction, O3-$NaNi_{0.5}Mn_{0.5}O_2$ transforms into the O′3, P3, P′3, and P″3 phases.[33] Again, many doping regimes have been explored, including quaternary O3-type compounds such as $NaTi_{0.5}Fe_{0.5}Co_{0.5}Ni_{0.5}O_2$, which delivers $163\,mAh\,g^{-1}$ at C/20 with 89% capacity retention over 20 cycles.[34]

Due to their greater Na stoichiometry, the O3 phase normally displays greater reversible capacity within the same voltage range compared to its P2 counterpart. However, O3-type materials often display a poor cycle life, low thermal stability, slow diffusion kinetics, and high moisture sensitivity.[35] The poor air stability associated with these materials can be improved by reducing the interlayer spacing and simultaneously increasing the valence state of the transition metals. Cu/Ti codoping is one method that can be used to achieve this as demonstrated through the synthesis and characterization of O3-$NaNi_{0.45}Cu_{0.05}Mn_{0.4}Ti_{0.1}O_2$.[30] Density functional theory (DFT) calculations suggest that structural modulation toward greater air stability requires cosubstitution with one atom of similar electronegativity ($Cu^{2+}$) and one with a substantially different Fermi level ($Ti^{4+}$).[36]

The first direct comparison between two phases with the same stoichiometry was carried out on P2- and O3-$Na_{2/3}Mn_{1/3}Fe_{2/3}O_2$. These materials produced initial charge capacities of 114.7 and $134.01\,mAh\,g^{-1}$, respectively, when cycled between 4.2 and 1.5 V at C/10. It is generally believed that P2-type layered cathode materials have better structural stability and reversibility compared to O3-type structures.[37] However, similar electrochemical performances were observed for both structures, indicating that stoichiometry is the more influential factor.

*P2/O3-Type* Mixed P2/O3 phases are relatively unexplored in the literature and present interesting research opportunities in the field of cathodes. Multiphase systems such as P2/O3-$Na_{2/3}Li_{0.18}Mn_{0.8}Fe_{0.2}O_2$ and P2/O3-$Na_{0.66}Li_{0.18}Mn_{0.71}Ni_{0.21}Co_{0.08}O_{2+\delta}$, have been reported to display enhanced electrochemical performance relative to similar single-phase systems.[38,39] The improved performance is thought to arise through exploitation of the high sodium content of the O3 phase and fast Na diffusion kinetics of the P2 phase.[40]

In summary, the main advantages of layered oxides include their high capacity, often ease of synthesis, environmentally friendly and inexpensive nature of the materials. One of the drawbacks of sodium-ion batteries that use layered oxides is their relatively low energy density and cycle life. If energy density and cycling stability of these cathode materials continue to improve, it is likely that this class of materials will be used in commercial prototypes.

### 2.6.2 *Polyanionic Compounds*

High-voltage 3D polyanionic compounds (phosphates, pyrophosphates, fluorophosphates, oxychlorides) have the general formula $Na_xMZ$, M = Ti, V, Fe, F and $Z = (PO_4)^{3-}$ or $(SiO_4)^{4-}$ and display significant thermal stability.[41] Their rigid structures contain open 3D channels through which $Na^+$ can easily diffuse in and out of the material.

*Phosphates* The olivine $LiFePO_4$ electrode is used in commercial LIBs.[42] However, the Na counterpart, $NaFePO_4$, crystallizes into the more thermodynamically stable maricite structure and is considered to be electrochemically inactive.[43] As shown in Figure 4, the olivine and maricite structures possess very similar phosphate frameworks, but the M1 and M2 site occupancy in

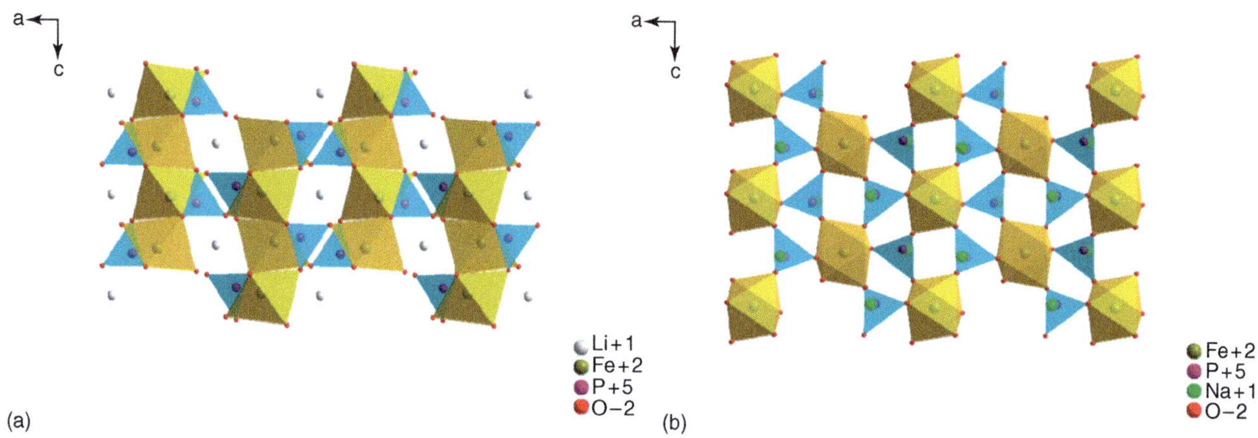


**Figure 4** (a) Olivine and (b) maricite $NaFePO_4$ structures shown along the 101 plane. [Reprinted with permission from K. T. Lee, T. N. Ramesh, F. Nan, G. Botton and L. F. Nazar, *Chem. Mater.*, 2011, **23**, 3593–3600. © 2011 American chemical society]

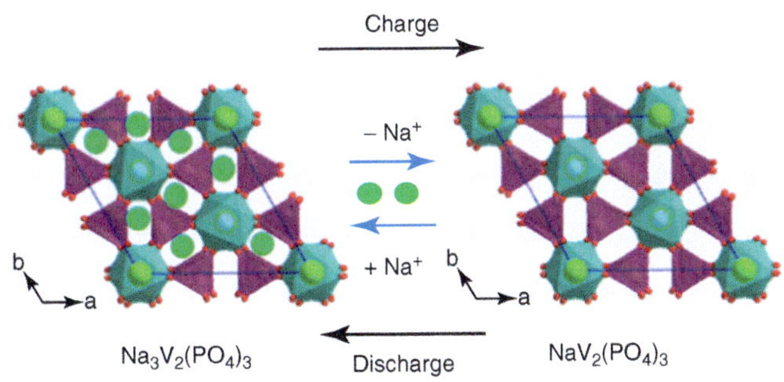


**Figure 5** $Na_3V_2(PO_4)_3$ structure at charged and discharged states. [Reproduced with permission from Ref. 53. © Elsevier, 2018]

maricite is the opposite to that in olivine. The resulting Fe and Na connectivity within the maricite structure gives rise to a closed network that does not allow for (de)intercalation of Na-ions.[45]

In 2015, the ability of nanosized $NaFePO_4$ to function as a cathode was reported, and in this case the Na-ion hopping activation barrier was found to be approximately four times lower. The electrode displays a capacity of 142 mAh g$^{-1}$ and 95% retention after 200 cycles.[46] Additional electrochemically active maricite $NaFePO_4$ type cathodes have been developed more recently. $NaFePO_4$/C/graphene comprises maricite $NaFePO_4$ nanoparticle entwined graphene sheets and retains 98% (142 mAh g$^{-1}$) capacity after 300 cycles.[47]

An electrochemically active $NaFePO_4$ phase can be also synthesized via ion exchange from the olivine $LiFePO_4$ phase. This $NaFePO_4$ has the highest capacity among phosphate polyanion cathode materials of 125 mAh g$^{-1}$, retaining 110 mAh g$^{-1}$ after 50 cycles.[48] Different potential profiles are observed upon charge and discharge, indicating two different mechanisms for the insertion and extraction of Na-ions.[17] On charge, a two-stage process is observed: complete conversion from $NaFePO_4$ to $Na_{0.7}FePO_4$ followed by formation of $FePO_4$. On discharge, a single-step process takes place.[49]

Metastable, electrochemically active mixed metal olivine-type phases $Na[Mn_{1-x}M_x]PO_4$ (M = Fe, Ca, Mg), $0 \leq x \leq 0.5$, have been synthesized by a low-temperature ($\leq$100 °C) method using topotactic reactions.[44] Unusual solid solution behavior was observed for these materials, and only a single-phase reaction took place on extraction of Na-ions. Overall, the lack of a straightforward synthetic route makes $NaFePO_4$ type materials a less attractive choice for use in sodium-ion batteries.

$Na_3V_2(PO_4)_3$ adopts a NASICON (Na superionic conductor) type framework comprising octahedral V sites corner-linked to $(PO_4)^{3-}$ tetrahedra and $Na^+$ occupying either 6- or 8-coordinate sites. This particular material can be used for either the cathode or anode due to the two potential plateaus at 3.4 and 1.6 V corresponding to $V^{3+}/V^{4+}$ and $V^{3+}/V^{4+}$ redox couples.[50] Furthermore, the possibility to incorporate NASICON framework compounds into solid-state batteries with materials such as $Na_3Zr_2Si_2PO_{12}$ as electrolytes make them an appealing research focus. During operation as a cathode material, ~2 Na-ions can be reversibly extracted/inserted yielding a capacity of 117 mAh g$^{-1}$. A volume change of 8.26% takes place during the biphasic mechanism between $Na_3V_2(PO_4)_3$ and $NaV_2(PO_4)_3$.[51] The first exploration of Na-ion migration pathways in $Na_3V_2(PO_4)_3$ was carried out in 2014, which produced an estimate of $10^{-11}$ cm$^2$ s$^{-1}$ for the diffusion constant.[52] The DFT calculations carried out indicate that 3D Na-ion migration can take place through channels between $PO_4$ tetrahedra along the $x$ direction, through voids between $PO_4$ tetrahedra and $VO_6$ octahedra along the $y$ direction, or via a curved route along the $z$ direction (Figure 5).[52]

*Pyrophosphates* $Na_2FeP_2O_7$ has a triclinic structure and belongs to the $P\bar{1}$ space group; see Figure 6. The 3D framework structure consists of corner-sharing $P_2O_7$ groups and $FeO_6/FeO_5$ polyhedra giving rise to open Na diffusion pathways along the [011] direction.[18] Na-ions reside on one of eight sites within the resulting channels. Due to the high activation energy associated with the deintercalation of multiple Na-ions, $Na_2FeP_2O_7$ displays low theoretical capacities (<100 mAh g$^{-1}$).[56] Mechanistic studies into $Na_2FeP_2O_7$ cycled between 2.0 and 4.5 V at C/20 revealed two distinct plateaus in the voltage profile corresponding to a single-phase region followed by two-phase reactions.[55] The material also displays high thermal stability, with no thermal decomposition and/or oxygen evolution up to 600 °C.[54] When doped with Mn to form $Na_2Fe_{0.5}Mn_{0.5}P_2O_7$, different Na coordination and site occupancy occurs, resulting in a single-phase reaction mechanism.[57] The material displays a reversible capacity of ~80 mAh g$^{-1}$ with an average redox potential of 3.2 V. For comparison, $Na_2MnP_2O_7$ displays a reversible capacity of ~90 mAh g$^{-1}$ with an average redox potential of 3.8 V.

*Mixed Phosphate–Pyrophosphates* Mixed phosphate–pyrophosphates, such as $Na_4Fe_3(PO_4)_2(P_2O_7)$,

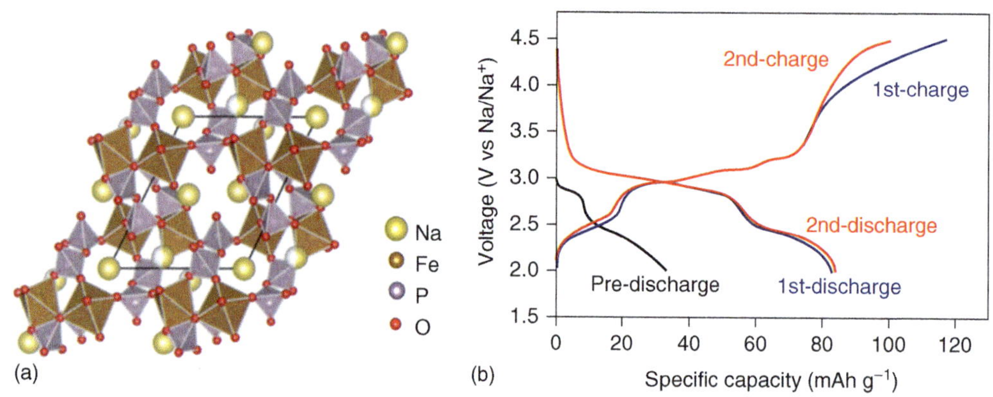


**Figure 6** (a) Structure of $Na_2FeP_2O_7$ compound and (b) voltage profile of $Na_2FeP_2O_7$ at C/20. [(a) Reprinted with permission from P. Barpanda, G. Liu, C. D. Ling, M. Tamaru, M. Avdeev, S.-C. Chung, Y. Yamada and A. Yamada, *Chem. Mater.*, 2013, **25**, 3480–3487. © 2013 American chemical society]

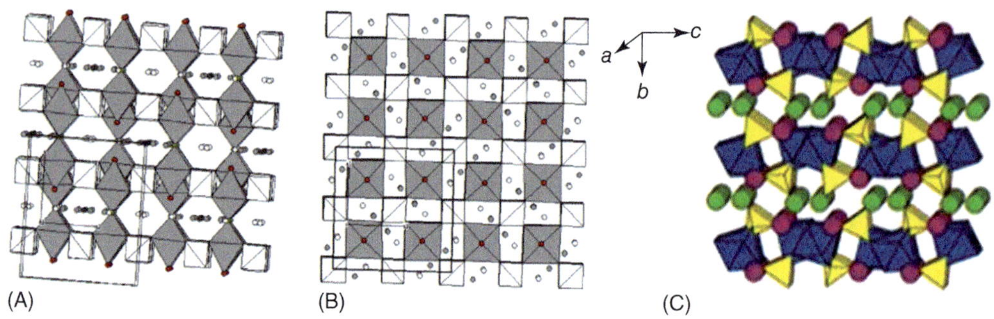


**Figure 7** Structure of $Na_3V_2(PO_4)_2F_3$ projected along the (A) *a* and (B) *c* axes and (C) $Na_2FePO_4F$ structure. In (C) iron octahedra are shown in blue and the phosphate tetrahedra in yellow; the two alkali sites are distinguished as Na(1) in green and Na(2) in pink. [(C) Reproduced with permission from Ref. 61 © Elsevier, 2007 and Ref. 60 © Nature Publishing Group, 2006]

are appealing cathodes due to their highly stable framework structure and low cost of their constituents.[58] $Na_4Fe_3(PO_4)_2(P_2O_7)$ displays a capacity of 100 mAh g$^{-1}$ at C/20 and small volume changes (<4%), which are attractive for good long-term cyclability.[59] Mössbauer spectroscopy revealed activity of the $Fe^{2+}/Fe^{3+}$ redox couple and deinsertion of up to 3 Na-ions.[59]

*Fluorophosphates* $Na_3V_2(PO_4)_2F_3$ contains three distinct Na sites from which Na can be reversibly (de)intercalated (Figure 7). Carbon-coated $Na_3V_2(PO_4)_2F_3$ nanoparticles deliver a capacity of 130 mAh g$^{-1}$ with 50 % capacity retention over 3000 cycles.[62] During cycling, up to 2 Na-ions per formula unit are extracted and minimal volume changes (<2%) are observed. In operando studies reveal three biphasic reactions upon sodium deintercalation involving intermediate phases as well as solid-solution-type behavior for $Na_xV_2(PO_4)_2F_3$ $(1.8 \leq x \leq 1.3)$.[63]

The structure of $Na_2FePO_4F$ exhibits two-dimensional ion conduction pathways (Figure 7C).[64] Unlike other phosphates, which display two-phase behavior, this material primarily shows solid-solution-type behavior. $Na_2FePO_4F$ displays a reversible capacity of 110 mAh g$^{-1}$ with an average operating voltage of 3.0 V. Interestingly, with the composition (c) $Li_xNa_{2-x}FePO_4F$, it can also be utilized as a Li-ion cathode.[61]

*Sulfates* Introduction of the highly electronegative sulfate ion, $SO_4^{2-}$, increases the redox potential in comparison to phosphate-based materials.[65] The highest known $Fe^{3+}/Fe^{2+}$ redox potential (~3.8 V) among sulfate-containing materials is displayed by $Na_2Fe_2(SO_4)_3$ (Figure 8A). The material crystallizes to the alluaudite structure with the $P2_1/c$ space group and displays a reversible capacity of ~100 mAh g$^{-1}$.[66] The structure consists of corner-sharing $Fe_2O_{10}$ dimers and $SO_4^{2-}$ units with Na-ions located on sites within the resulting large tunnels along the c direction. Ab initio and bond valance calculations showed that all Na-ions are accessible for (de)intercalation, which takes place via 1D conduction channels along the *c* axis (Figure 8B and C).[65]

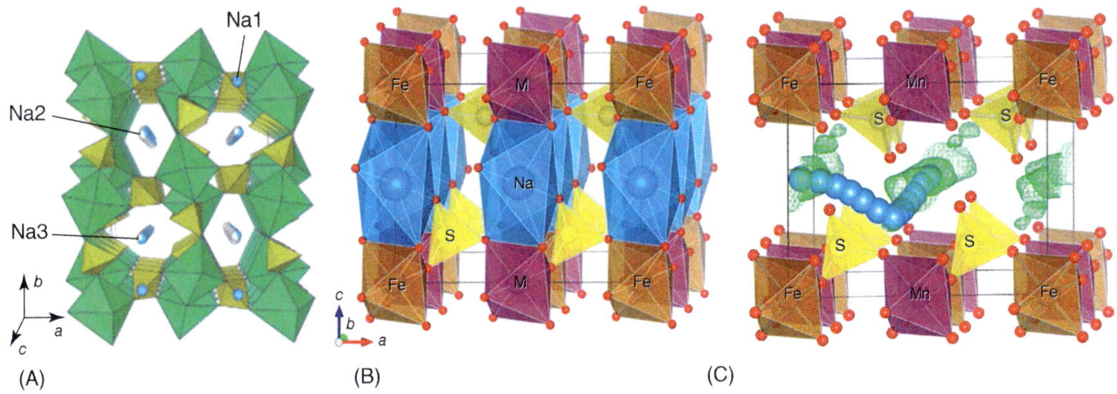


**Figure 8** (A) Structure of $Na_2Fe_2(SO_4)_3$ projected along the *c* axis and (B) layered structure of $Na_x(Fe_{1/2}M_{1/2})(SO_4)_2$ (M = Mn, Co, Ni) with (C) an isosurface plot from bond valence sum difference analysis suggesting possible Na migrations obtained from DFT calculations. [(C) Reproduced with permission from Ref. 66 © Nature Publishing Group, 2014]

A new family of sulfate-based cathodes, $Na_x(Fe_{1/2}M_{1/2})(SO_4)_2$ (M = Mn, Co, Ni), has recently been proposed. Higher electrode potentials (~4.37 V) are expected from $Na_x(Fe_{1/2}M_{1/2})(SO_4)_2$ (M = Mn, Co, Ni) based on DFT calculations.[65]

### 2.6.3 *Metal–Organic Frameworks*

Due to the larger size of Na-ions compared to Li-ions, a variety of materials such as MOFs, which possess sufficiently large channels to accommodate the diffusion of Na-ions, have been explored.

*Prussian Blue Analogues* Prussian blue derivatives ($Na_xM[M'(CN)_6]{\cdot}nH_2O$, M,M′ = K, Mn, Fe, Co, Ni, Zn, In) are typically easy to synthesize via precipitation method from cheap starting materials.[67] A detailed structural study of $Na_xFe[Fe(CN)_6]_{1-x}{\cdot}yH_2O$[68] showed an open host framework, which contains large interstitial sites on which, in theory, up to 2 Na-ions per formula unit can be found.[56] Prussian blue analogues are known to absorb water, so it is likely that water may occupy some of these sites. As-synthesized, the dry $Na_xFe[Fe(CN)_6]_{1-x}{\cdot}yH_2O$ powder exists as a single phase, which adopts the $Fm\bar{3}m$ space group. When placed in a battery, the material transforms into a sodium-free minor phase and sodium-containing phase. For application as a cathode, a balance between the ability of a material to store Na-ions and the ease of insertion/extraction is required. During battery construction, $Na_xFe[Fe(CN)_6]_{1-x}.yH_2O$ splits into a major sodium-containing phase and minor sodium-free phase allowing for Na storage, while its open framework structure leads to a low activation energy for $Na^+$ diffusion allowing for reversible (de)intercalation during cycling.[58] The presence of crystal water in the inactive minor phase blocks the pores, restricting diffusion of Na-ions. Therefore, anhydrous Prussian Blue analogues such as $Na_{2-x}Fe[Fe(CN)_6]$ $(0 < x < 1)$ show improved performance (discharge capacity of 160 mAh $g^{-1}$, retaining 80 mAh $g^{-1}$ after 20 cycles and average operating voltage of 3.15 V).[62] In contrast, layered oxides such as P2-$Na_{2/3}Mn_{0.8}Fe_{0.1}Ti_{0.1}O_2$ display a stable electrochemical performance even after moisture exposure.[24]

### 2.6.4 *Organic Compounds*

Organic compounds provide a promising low-cost, sustainable alternative to the widely studied transition-metal-containing inorganic materials. They are appealing due to their structural flexibility and hence ability to accommodate volume changes upon insertion and extraction of Na-ions. $Na_4C_8H_2O_6$ is an interesting candidate because it displays potential windows between 1.6–2.8 V (Figure 9) and 0.1–1.8 V, so it can be used as both an anode and cathode. As such, an all-organic sodium-ion battery with $Na_4C_8H_2O_6$ has been developed.[69] The battery functions through the $Na_2C_8H_2O_6/Na_4C_8H_2O_6$ redox couple at the cathode and $Na_4C_8H_2O_6/Na_6C_8H_2O_6$ at the anode. Although a reversible capacity of 187 mAh $g^{-1}$ can be achieved, the nature of the redox reactions means the material displays a low operating voltage of 1.8 V.[69]

A suitable material should be lightweight, display a high working voltage and reversible capacities, which should be compatible with an appropriate electrolyte. Layered oxides display high specific capacities but often suffer from poor long-term structural stability. Phosphates display high operating voltages, yet low capacities. Significant capacity degradation is observed for MOFs and organic compounds. Overall, improvements to cathodes are still required and significant research opportunities exist.

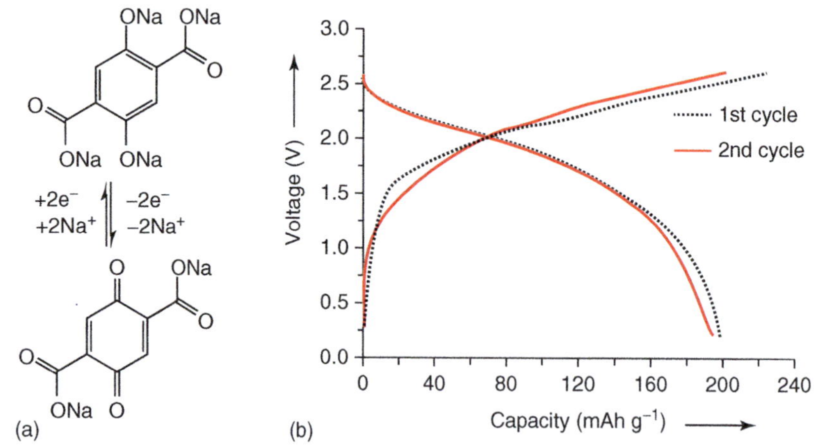


**Figure 9** (a) $Na_4C_8H_2O_6$ redox couple at 2.3 V and (b) voltage profile for the first two cycles. [Reproduced with permission from Ref. 69 © Wiley, 2014]

## 2.7 Anodes

During charge, Na-ions are inserted into the host lattice (anode) and are subsequently extracted during discharge, ideally with minimal structural change. Graphite and silicon are commonly used for LIBs because Li can reversibly insert/extract into/from these structures to form $LiC_6$ and $Li_{4.4}Si$, respectively. In contrast, Na cannot be reversibly intercalated into graphite on a suitable timescale, and the capacity achieved for Na with graphite is <20 mAh g$^{-1}$. This has led to the need to develop a suitable anode, and based upon their mechanism, proposed sodium-ion battery anodes can be divided into two main families, insertion and conversion/alloying materials.

### 2.7.1 *Insertion Materials*

*Hard Carbons* Hard and soft carbons are examples of amorphous nongraphitic carbons with short-range order.[56] Their structure consists of disordered cross-linked layers of planar hexagonal, sp$^3$ hybridized carbon networks. Soft carbons possess weak crosslinking between the layers, while strong crosslinking, which immobilizes the layers, is characteristic of hard carbons.[70] They are often prepared by the pyrolysis of organic polymers or hydrocarbons at high temperatures (<1500 °C).[71]

The larger interlayer spacing in hard carbon, compared to graphite, makes it an attractive anode, and they have been shown to produce an initial capacity of up to 300 mAh g$^{-1}$.[72] As suggested by wide-angle in situ X-ray scattering and solid-state $^{23}$Na NMR (nuclear magnetic resonance), Na-ions are inserted between parallel layers as well as into nanopores within the structure during operation.[73,74] These two mechanisms also manifest themselves in the voltage profile of hard carbon, shown in Figure 10, as the sloping region above ~0.1 V and long plateau region below 0.1 V respectively.[76] However, this mechanism is still disputed in the literature.[77] More recent work has suggested that filling of nanopores during the plateau region does not take place. It has been suggested that the sloping region corresponds to the interaction of

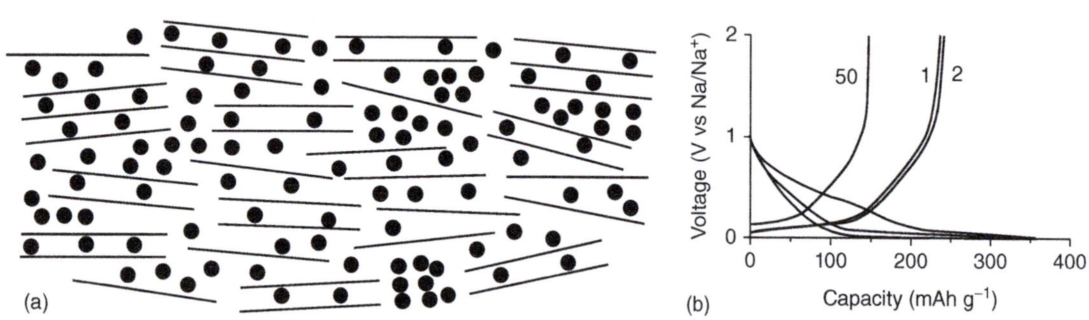


**Figure 10** (a) Schematic showing Na-filled hard carbon and (b) voltage profile for sugar hard carbon (SHC). [Reproduced with permission from Ref. 75 © ECS, 2000 and Ref. 76 © ECS, 2014]

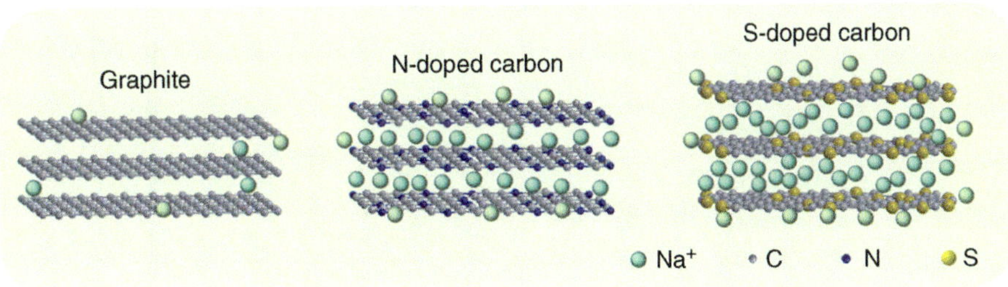


**Figure 11** Illustrations of graphite, N-doped carbon, and S-doped carbon. [Reproduced with permission from Ref. 83 © Wiley]

the Na-ions with the carbon edge defects while the plateau region is related to the insertion of Na-ions into the layered structure.[77]

The main limitations of hard carbon are its poor rate capability and cyclability as well as low charging potential leading to poor stability usually associated with electrodeposition of Na or SEI formation.[78] Therefore, strategies to modify hard carbon anodes as well as alternative carbon-based anodes have been investigated. The electrolyte plays an important role in the stability of the anode material. Inclusion of a protective SEI formed from an ester-based electrolyte improves the electrochemical performance whereby a capacity of 200 mAh $g^{-1}$ at a rate of 500 mA $g^{-1}$ can be retained over 1000 cycles.[79]

*Heteroatom-Doping* Nitrogen-doped, boron-doped, phosphorous-doped, and sulfur-doped graphene are examples of graphene-based materials that have been investigated.[80–83] Doping can increase the number of active sites and surface defects without inducing too much structural deformation, reducing the adsorption energy for Na-ions.[84] N-doping introduces in-plane hole defects, which increase the surface adsorption capacity. In contrast, P-doping creates protrusions, which facilitate both the absorption and insertion of Na-ions, allowing a high capacity of 374 mAh $g^{-1}$ after 120 cycles at 25 mA $g^{-1}$ to be achieved.[82] In theory, B-doped graphene can realize a maximum capacity of 762 mAh $g^{-1}$, which is ~2.54 times that of a hard-carbon anode. However, the development of a suitable synthetic route is still required. The larger covalent radius of sulfur (1.02 Å) compared to carbon (0.77 Å) suggests that S-doping can enlarge the interlayer distance, which may facilitate the insertion/extraction of Na-ions (Figure 11).

*Reduced Graphene Oxide and Expanded Graphite* Reduced graphene oxide has been shown to deliver a reversible capacity of 217.2 mAh $g^{-1}$ at 20 mA $g^{-1}$ [78] and can be prepared in large amounts by slight modification of the Hummers method, making it suitable for mass production.[85] Although reduced graphene oxide enables the insertion of Na-ions, it is limited by steric hindrance associated with large quantities of oxygen-containing groups. To address this, expanded graphite that contains fewer oxygen-containing groups while retaining a suitable interlayer spacing was investigated. It is prepared via a two-step oxidation–reduction process of graphite, which results in an enlarged interlayer spacing of 4.3 Å compared to the interlayer distance of graphite (~3.4 Å). Theoretical calculations suggest that a minimum interlayer spacing of 3.7 Å is required for the insertion of Na-ions.[86] At a current density of 100 mA $g^{-1}$, expanded graphite can deliver a capacity of 184 mAh $g^{-1}$ and retain a capacity of 136 mAh $g^{-1}$ even after 2000 cycles.[86] The enhanced capacity retention is reflected in the Na-ion insertion mechanism whereby insertion into the expanded graphite interlayer spacing, with reversible volume changes, takes place (Figure 12).

*Nanostructured Materials* Hollow carbon nanostructures are promising because they provide a large surface area and short diffusion distance to aid diffusion of Na-ions.[87] They are synthesized via hydrothermal decomposition of glucose onto a latex template followed by nonoxidative thermal treatment at 1000 °C to remove the template and release hollow carbon spheres (~12 mm thickness).[87] Compared to carbon spheres, hollow carbon nanospheres display enhanced electrochemical performance; see Figure 13. The thin shell thickness means there is a short diffusion distance for Na-ions. Furthermore, the large surface area of the hollow carbon nanospheres provides a large number of active sites for insertion, which, in combination with the large interlayer spacing (~4 Å), facilitates the diffusion and storage of Na-ions.[87] Similar results are observed for hollow carbon nanowires (HCNWs) for which a reversible capacity of 149 mAh $g^{-1}$ at 500 mA $g^{-1}$ can be achieved.[88]

The low cost and ease of large-scale synthesis of graphene nanosheets (GNSs) make it an appealing material for practical applications,[78] and it delivers reversible capacities of 220 mAh $g^{-1}$ at 30 mA $g^{-1}$ with ~80% capacity retention after 300 cycles.[78] During the insertion, two mechanisms have been shown to take place—surface adsorption/desorption at high potential and bulk insertion/deinsertion at low potential.[78] It has also been

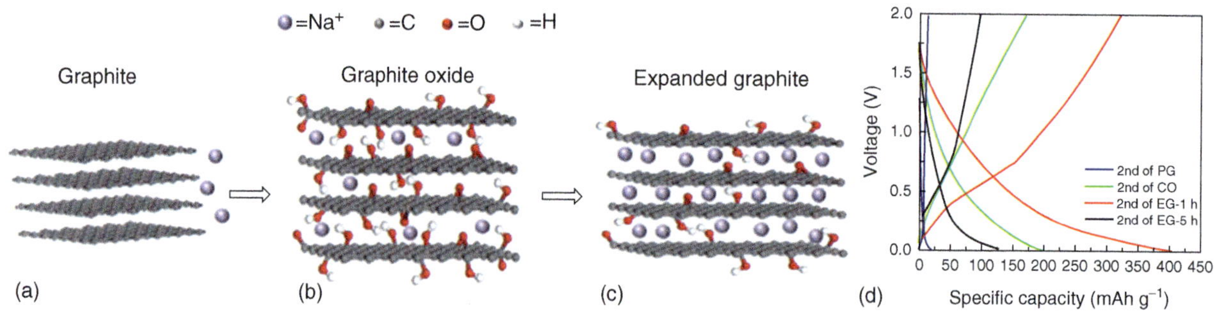


**Figure 12** Illustrations of sodium storage in (a) graphite, (b) graphite oxide, (c) expanded graphite, and (d) charge/discharge curves for the second cycle of pristine graphite (PG), reduced graphene oxide (GO), expanded graphene with thermal reduction times of 1 h (EG-1 h) and 5 h (EG-5 h). [Reproduced with permission from Ref. 86 © Nature Publishing Group, 2014]

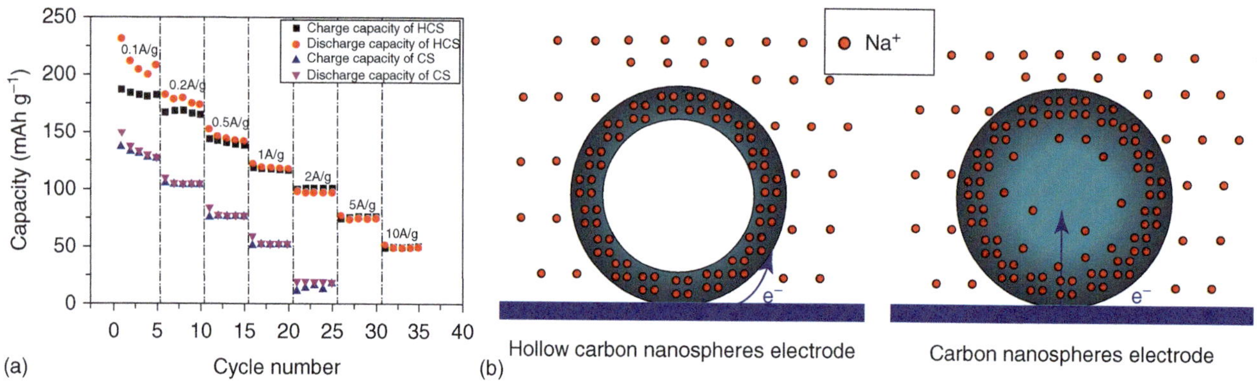


**Figure 13** (a) Rate capability of hollow carbon nanospheres (HCS) and carbon spheres (CS). (b) Schematic of the electrochemical mechanism. [Reproduced with permission from Ref. 87 © Wiley, 2012]

proposed that the capacity fade is associated with trapping of Na-ions at defects within the carbon layers, which are therefore not released during the next cycle.

Compared to GNSs, carbon nanotubes (CNTs) display a much lower reversible capacity of 82 mAh $g^{-1}$ at 30 mA $g^{-1}$. Although a similar mechanism for Na-ion storage is observed, insufficient active sites for insertion/extraction and an insufficient interlayer spacing limit the reversible capacity.[78] Activated carbons (ACs) and mesocarbon microbeads (MCMBs) have also been investigated but appear to be unsuitable. On charging, irreversible oxidation takes place for ACs while the low surface area of MCMBs prevents the storage of significant amounts of Na-ions.[78]

The microstructure, texture, and surface functionality and therefore performance of carbon nanofibers (CNFs) can be tuned through the carbonization temperature used.[89] CNFs carbonized above 2000 °C displayed the best performance, delivering a capacity of 200 mAh $g^{-1}$ and a single plateau at ~0.1 V in the voltage profile. Transmission electron microscopy (TEM) images reveal the formation of a more ordered structure comprising of graphitic layers following carbonization at temperatures above 2000 °C, which are thought to facilitate the uptake of Na-ions.[89] The performance of CNFs can be improved by embedding Sb nanoparticles homogenously into the structure via electrospinning followed by calcination.[90] Sb-C nanofibers can deliver a reversible capacity of 631 mAh $g^{-1}$ at 40 mA $g^{-1}$, with 90% capacity retention after 400 cycles.[90] Other nanocomposites such as SbSn-C can also deliver high reversible capacities (~500 mAh $g^{-1}$).[91] In addition to the Na-ion (de)insertion mechanism, Na-Sb and Na-Sn (de)alloying reactions take place during cycling, yielding higher capacities.

### 2.7.2 *Alloy-Based Anodes*

Alloying-based materials are comprised of elements that can alloy with sodium, often yielding higher capacities than carbonaceous materials.[92] Since Si, which is widely studied in the context of LIBs, does not alloy with Na, alternative materials such as those containing Sn,

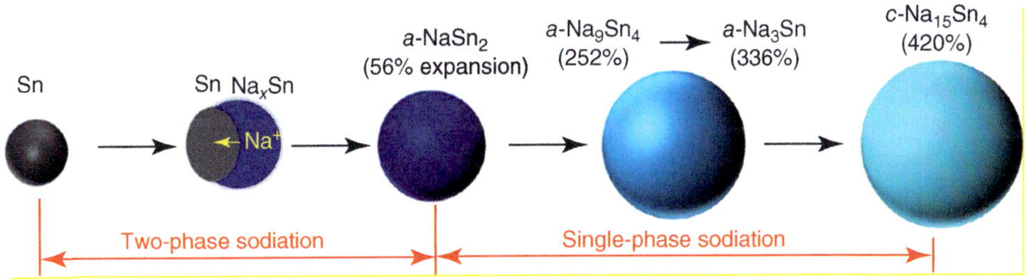


**Figure 14** Schematic for the structural evolution of Sn nanoparticles upon discharge. [Reproduced with permission from Ref. 88 © 2012 American Chemical Society]

Ge, P, or Sb are of greater interest for use in sodium-ion batteries. High theoretical capacities make them attractive anode materials, while the major drawback is the significant volume changes that take place upon cycling, which results in poor long-term stability.[93] In part, nanostructures can help to minimize the large volume changes that take place upon cycling.[94] Furthermore, embedding nanoparticles into a carbon network can alleviate expansion/contraction stress during insertion/extraction.[95]

Sn-based alloying anodes display particularly high theoretical capacities (up to 1137 mAh $g^{-1}$) and have received the most research attention.[94] The sodiation process for metallic Sn nanoparticles is thought to take place via a two-phase mechanism, during which an amorphous $Na_xSn$ alloy ($x \sim 0.50$) forms, followed by a single-phase mechanism to form the final crystalline phase, $Na_{15}Sn_4$. A huge volume change of 420% is associated with the process (Figure 14).[88]

Tin oxides and sulfides both display higher theoretical capacities than metallic Sn. They operate initially via a conversion reaction and subsequently via an alloying reaction. For example, the mechanism for $SnS_2$ is given by Equations (1) and (2) and interestingly upon desodiation, $SnS_2$ does not reform.[96]

$$SnS_2 + 2Na \rightarrow Sn + Na_2S \quad (1)$$

$$Sn + 3.75Na \rightarrow Na_{15}Sn_4 \quad (2)$$

### 2.7.3 *Selected Conversion Anodes*

Spinel $NiCo_2O_4$ cannot accommodate Na-ions on its empty $T^d$ sites and so upon discharge it undergoes a conversion reaction to form $Na_2O$ as shown in Equation (3). A reversible capacity of ~200 mAh $g^{-1}$ is observed, which is much lower than that of 884 mAh $g^{-1}$ observed versus Li and this is suspected to be due to the loss of long-range order and reduction of transition metals during the first discharge.[97]

$$NiCo_2O_4 + 8Na \rightarrow Ni + 2Co + 4Na_2O \quad (3)$$

Normally, iron oxide materials display large volume changes during Na (de)intercalation, resulting in large capacity losses during cycling. 3D porous $\gamma$-$Fe_2O_3$ on carbon nanocomposites can be prepared by aerosol spray pyrolysis and appears to be an ideal candidate due to its abundance, low cost, nontoxicity, and high theoretical capacity (~1007 mAh $g^{-1}$).[98] At high current densities of 2000 mA $g^{-1}$, a reversible capacity of 358 mAh $g^{-1}$ after 1400 cycles is observed.[98] Equation (4) shows the electrochemical conversion reaction, which is thought to take place upon discharge.

$$Fe_2O_3 + 6Na^+ + 6e^- \rightarrow 2Fe^0 + 3Na_2O \quad (4)$$

A one-pot synthesis has been developed for $Co_3O_4$ intertwined with CNTs. This material displays a reversible capacity of 487 mAh $g^{-1}$ and delivers a capacity of 184 mAh $g^{-1}$ at a high current rate of 3200 mA $g^{-1}$.[99] The electrochemical performance has been attributed to the presence of mixed electron and ion conducting networks within the structure as well as to the ability for full reduction of $Co^{2+}/Co^{3+}$ cations by Na.[100]

### 2.7.4 *Titanium-Containing Materials*

$TiO_2$-based materials are extremely stable, inexpensive, nontoxic, abundant and are known to (de)insert Li-ions.[101] As such, studies on these materials, particularly the anatase phase, have been extended to sodium-ion batteries.[102–104] Detailed studies using in situ and ex situ XRD, ex situ X-ray photoelectron spectroscopy (XPS), Raman spectroscopy, and GC-MS indicate the immediate reaction mechanism involves the formation of metallic Ti, sodium oxide, and amorphous sodium titanate (Figure 15). Insertion of ~0.41 Na per $TiO_2$, which corresponds to a capacity of ~140 mAh $g^{-1}$, can be delivered by the amorphous sodium titanate phase.[105]

$Na_2Ti_3O_7$ can deliver a capacity of 200 mAh $g^{-1}$ and has the lowest operating voltage of 0.3 V among oxide insertion materials.[106] The structure consists of

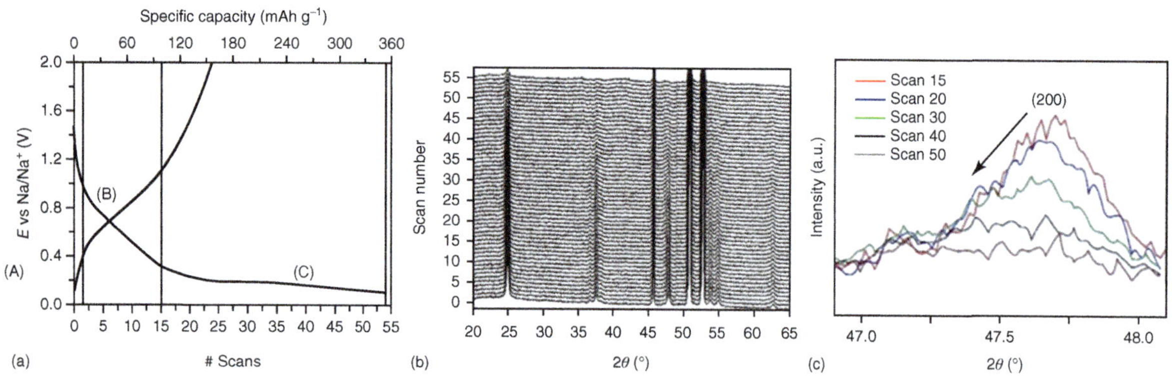


**Figure 15** The (a) potential profile, (b) in situ XRD upon discharge, and (c) evolution of the (200) reflection for the $TiO_2$ electrode. [Reproduced with permission from Ref. 105 ©, Wiley 2015]

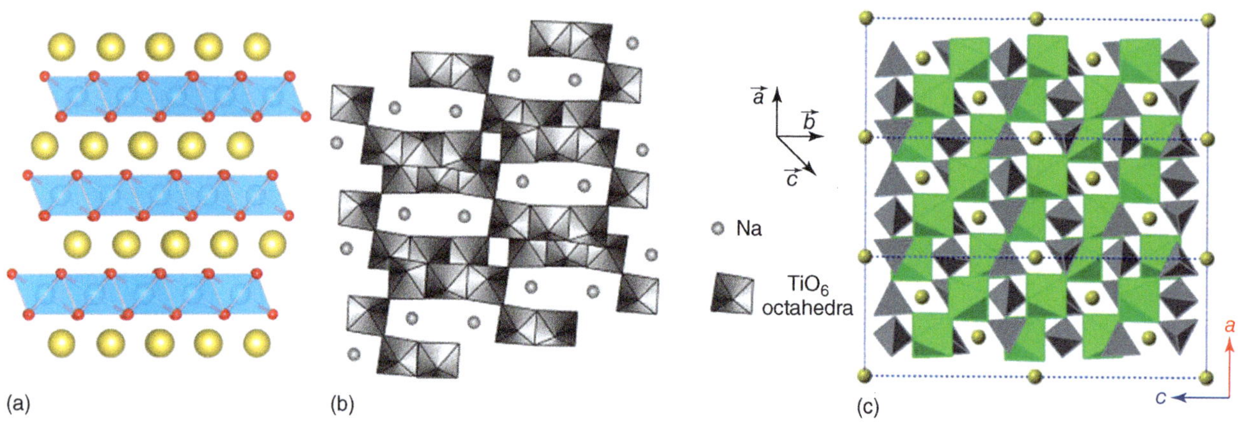


**Figure 16** DFT optimized structure for stoichiometric (a) $NaTiO_2$, (b) $Na_2Ti_6O_{13}$ structure, and (c) $NaTi_2(PO_4)_3$ NASICON structure. [Reproduced with permission from Ref. 104 © ACS, 2017, Ref. 111 © Elsevier, 2006 and Ref. 112 © ECS, 2014 respectively]

corrugated $(Ti_3O_7)^{2-}$ layers with Na-ions located in the interlayer space.[107] During cycling, reversible insertion of two additional Na-ions takes place to form $Na_4Ti_3O_7$, leading to the reduction of Ti(IV) to Ti(III).[108] The performance is limited by the mechanical strain that takes place upon cycling as well as irreversible insertion of Na-ions. In an attempt to further improve the electrochemical performance and electrode stability, nanoscale $Na_2Ti_3O_7$ has been investigated and improved cyclability was found for $Na_2Ti_3O_7$ nanoarrays.[109]

$Na_2Ti_6O_{13}$ has been shown to deliver good long-term performance, retaining 85% of its capacity after 5000 cycles.[110] The monoclinic structure (*C2/m* space group) consists of edge-sharing $TiO_6$ octahedra with Na-ions located on the interstitial sites between the $TiO_6$ units (Figure 16). Insertion and extraction of Na-ions is thought to take place along the tunnel space parallel to the *b* axis.[113] Based upon first-principle calculations on $Na_2Ti_6O_{13}$ anomalously high Na-ion and low Li-ion mobility was found.[114] This was attributed to strengthened Na—O bonding at the transition state, which compensates for repulsion between Na-ions within the structure.

Another promising Ti-containing material is $NaTi_2(PO_4)_3$, which adopts a NASICON type structure and displays a reversible capacity of 130 mAh g$^{-1}$.[115] The open 3D framework structure with large interstitial spaces for Na-ions results in small volume changes during cycling. However, the material is limited by its poor electronic conductivity and therefore slow transport of Na-ions.[116] Various forms of the material have been synthesized and tested, including graphene-supported $NaTi_2(PO_4)_3$, carbon-intercalated porous $NaTi_2(PO_4)_3$ spheres, and CNT-decorated $NaTi_2(PO_4)_3$/C nanocomposites.[112,117,118] Superior electrochemical properties are found for these carbon-coated materials because the resulting network provides a pathway for rapid insertion/extraction of Na-ions.[119]

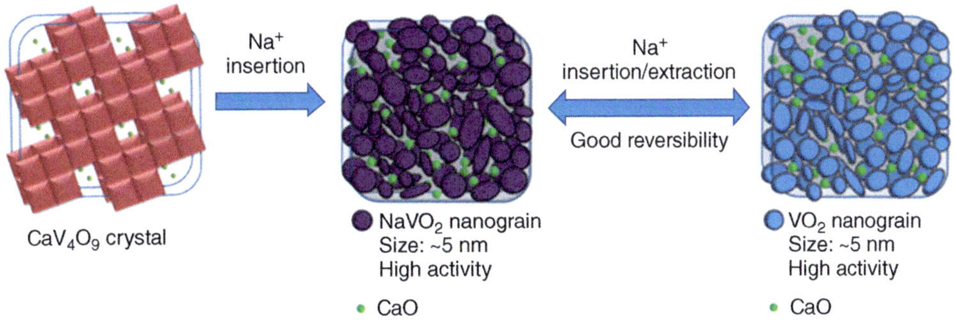


**Figure 17** Schematic of the sodium storage mechanism of $CaV_4O_9$ during cycling. [Reproduced with permission from Ref. 120 © Nature Publishing Group, 2017]

#### 2.7.5 *Other Anode Materials*

As mentioned in Section 2.6.4, $Na_4C_8H_2O_6$ has the potential to act as both a cathode and anode material. Reversible capacities of ~250 mAh $g^{-1}$ have been reported, which are associated with the (de)sodiation of a carbonyl group of $Na_4C_8H_2O_6/Na_6C_8H_2O_6$ at ~0.3 V.[69] Alkaline earth metal vanadates, namely, $CaV_4O_9$ nanowires (NWs), have been identified as promising anodes[120] and a novel mechanism has been reported. This involves the formation of spectator CaO nanograins (Figure 17), which leads to good reversibility and negligible volume changes upon cycling. Such work opens up new ideas and directions for research into anodes.

Overall, nanoscale anode materials show the greatest potential for application in sodium-ion batteries. The challenge remains to find a low cost anode material displaying good long-term cycling stability as well as high capacity and rate capability.

## 3 ELECTROLYTES

### 3.1 Liquid Electrolytes

Carbonate sodium-based salt solutions still dominate the electrolytes used for rechargeable room-temperature sodium-ion batteries. In terms of salts used, $NaPF_6$, the sodium-equivalent of the commercially used lithium salt, is widely adopted and $NaClO_4$ is also used, but there are concerns of this salt's stability especially with respect to shock resistance. Ionic liquids have also been considered and field is elegantly reviewed in Ref. 121. However, if sodium-metal-based batteries were to be developed, parallel to the implementation of lithium-metal batteries, then ionic liquids are likely to be investigated in more detail. This leads to an interesting aside, with the push for lithium-metal-based batteries, there has not been a similar push in sodium-metal systems, most likely purely from a reactivity and safety point of view. Lithium, although reactive in air/moisture, often forms nitrides and slowly darkens or stops functioning. Since sodium is significantly more reactive, it has to be protected and handled significantly more carefully than lithium. Furthermore, as lithium dendrites are known to cause failure in conventional LIBs (without lithium-metal anodes), significant attention has been devoted to assess formation and growth characteristics compared to their sodium-based counterparts, with an exception of a few studies.[122] If aqueous cells were to be developed, various conventional sodium salts are applicable, for example, $Na_2SO_4$ and NaCl.

### 3.2 Solid-State Electrolyte and Solid-State Batteries

All solid-state rechargeable sodium-ion batteries have attracted significant attention as an alternative for large-scale energy storage applications due to their potential benefits in terms of safety, low cost, and high energy density.[58,123] For such systems to be a reality, it is necessary to develop a suitable solid-state sodium electrolyte that allows fast ionic conduction at room temperature. The electrolyte should show high sodium transport numbers at and around room temperature, be able to conduct in a wide potential window, be chemically, air/moisture and thermally stable, and form beneficial interfaces with the anode and cathode active materials. The key advantage is no leakage and a completely safe and potentially stable battery. The challenges of the electrolyte include ionic conductivities at room temperature, air stability, and depending on the choice of electrolyte, accessible potential windows. In terms of a whole solid-state battery, challenges include volume change of the components during battery function, choice of battery construction materials, architecture of the battery, and the ionic/electronic connectivity between the components and format of the battery (e.g., thin film or bulk). Arguably, the most challenging are the interfaces in the device such as between electrolyte–electrode, electrode

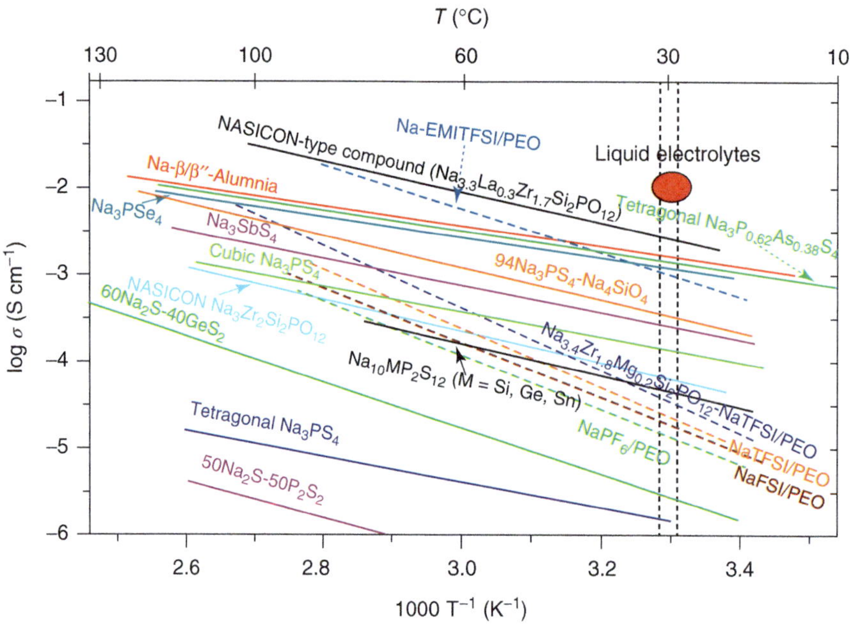


**Figure 18** A summary of the ionic conductivities reported in the literature for various sodium-ion conductors. [Adopted from Ref. 124 © Wiley 2018]

and current collectors, and the sheer number of interfaces. Focusing on the electrolytes, Figure 18 shows conductivity comparisons of sodium-based solid electrolytes relative to conventional carbonate-based liquid systems.[124] There are a number of materials classes actively investigated that show appreciable conductivity and these will be discussed in the following text.

The disadvantage of all solid-state sodium batteries is the reactivity of sodium metal, and this is where all solid-state lithium batteries can potentially overpower sodium equivalents. The use of metallic lithium provides a high energy density option, whereas it is highly unlikely metallic sodium would be used due to its reactivity and safety concerns. Alternatives including safer sodium-rich alloys might still make such batteries viable.

Various formats (Figure 19) can be used in all solid-state batteries, from research-scale bulk batteries that can be heated (for measurement and performance tests), to thin-film format cells that could be integrated in microelectronics and coin cells with pellet type arrangements of the components. There are significant differences in the interaction between components at the particle level, for example, in thin film compared bulk solid-state battery architectures. One of the major differences is the sheer number of interfaces and how connectivity has to be considered in such systems. In a Na–$O_2$ battery, a solid electrolyte could protect the metallic sodium electrode and work in combination with a liquid electrolyte.

### 3.3 Selected Solid-State Electrolytes

Various classes of materials exist ranging from ceramics, glassy ceramics, and glasses to plastic crystals.[121] Plastic crystals are a predominantly organic option for solid-state electrolytes, with other polymeric and gel-like electrolytes also being reported.[123,124] Significant efforts are also being directed toward computer-based approaches for materials discovery and ion conduction mechanism determination.[125]

#### *3.3.1 Na Superionic Conductor*

One of the most promising classes of materials is the Na SuperIonic CONductor or NASICON family of compounds with the general formula $Na_{1+x}Zr_2Si_xP_{3-x}O_{12}$ ($0 \leq x \leq 3$), which in addition to its high ionic conductivity, also has excellent air and moisture stability.[126–130] NASICON's structure consists of alternating $ZrO_6$ octahedra and Si/$PO_4$ tetrahedra, which form a three-dimensional network, essentially providing the framework with wide channels that allow the fast diffusion of the $Na^+$ ions.[131] At temperatures below 160 °C, NASICON undergoes a reversible phase transition from a rhombohedral ($R\bar{3}c$) to a monoclinic ($C2/c$) phase.[132–134] Electrochemical impedance measurements have shown that monoclinic NASICON has a higher ionic conductivity than its rhombohedral phase. Moreover, this can be significantly improved through chemical doping,[129,135–138] with a recent report of Mg-doped NASICON achieving

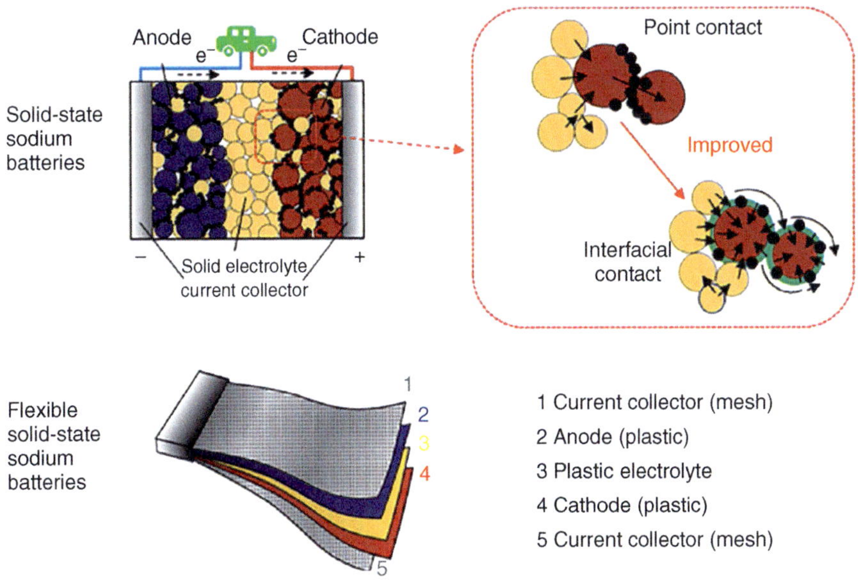


**Figure 19** Examples of two architectures of all solid-state batteries, with the insert showing some of the interactions that take place[124]

a total ionic conductivity ($2.7\,mS\,cm^{-1}$) comparable to some conventional liquid sodium electrolytes such as NaOTf.[138,139]

In these and related solid-state ionic conductors, recent research has taken more of a focus on the grain boundary influence, the role of surface conductivity versus bulk conductivity and whether the formation of secondary phases can assist in conduction pathways. The formation of secondary phases rather than direct chemical doping into the NASICON structure is being brought into light with high-resolution diffraction studies showing minute quantities of such phases on par with dopant concentration levels. In some cases, the combination of secondary phases and some doping can lead to improvement in conductivity of the whole sample or pellet. This provides an impetus to study the precise doping scheme. In combination with diffraction, solid-state NMR for local environment (especially around Na), and activation energies, computational calculations and impedance measurements lay the foundation for characterization methods of solid-state electrolytes generally. For more glassy substances, pair distribution function (PDF) analysis in combination with X-ray absorption spectroscopy (XAS) might be more appropriate.

### 3.3.2 *Sulfides*

Sulfide glasses and ceramics in some systems equivalent to lithium-based systems have been explored, for example, with $Na_2S$-$P_2S_5$ glasses and ceramics generated and determined using various heating regimes.[124,140–142] The $Na_3PS_4$ crystalline phase shows promise due to its stability with sodium metal, ionic conductivity of $10^{-4}\,S\,cm^{-1}$ at room temperature, and voltage window up to 5 V. Various isovalent substitution regimes have also been applied to generate related phases.[124]

A recent discovery in sulfides is $Na_{11}Sn_2PS_{12}$, which has reports of ionic conductivity ranging from $1.4\,mS\,cm^{-1}$ [143] to $4\,mS\,cm^{-1}$ [144] and a reported energy barrier for Na-ion mobility of 0.25 eV. Work has been carried out on both single crystals and powders with simulations using ab initio molecular dynamics simulations[143] and bond valence[144] maps indicating potential pathways. These studies show the elegant combination of precise structural details with calculation even though measured conductivities can vary.

An interesting demonstration of an all solid-state sodium-ion battery at room temperature used $Na_{2+2\delta}Fe_{2-\delta}(SO_4)_3|Na_{3+x}M_xP_{1-x}S_4|Na_2Ti_3O_7$ where M = Ge, Ti, or Sn.[145] These batteries were cycled at relatively high C-rates, within 30 min achieving capacities larger than $80\,mAh\,g^{-1}$ after 100 cycles. This demonstration shows the practical and feasible nature of such all solid-state devices.

### 3.3.3 *Other Chemistries*

Recent work on $Na_2M_2TeO_6$, where M = Ni, Co, Zn, Mg, is interesting as these are layered type compounds.[146] They show structural similarities to hexagonal layered P2-type cathodes discussed earlier and can be considered superstructures of these with ordering of M and Te in the octahedral layer. The ionic conductivity is found to be $4–11\,S\,m^{-1}$ at 300 °C with the highest electronic conductivity reported for the Co-related sample due to mixed valence nature, and it only accounts for 0.1% of the ionic conductivity. Another example of a layered solid-state electrolyte is $Na_xCr_xTi_{1-x}O_2$, which at selected compositions shows ionic conductivity of $12\,S\,m^{-1}$ at room temperature.[147] The versatility in composition allows studies of diffusion and whether vacancies are critical for such conduction. The question of whether such a material may act as a cathode or anode needs to be explored. These studies indicate that a wholly layered system, comprised of cathode, electrolyte, and anode, maybe feasible for an all solid-state battery.

The humble $\beta$ or $\beta''$ alumina has been the cornerstone for high-temperature sodium-based batteries, and a number of studies have investigated ordering of vacancies, growth of crystals on various substrates, doping ratios, and temperatures.[148,149] The $\beta''$ alumina composition is $NaAl_5O_8$ and differs from $\beta$ in the layering and stacking sequence of the spinel-like subunits and has higher ionic conduction but is thermodynamically less stable. To avoid mixtures of $\beta''$ and $\beta$, various compounds are used as stabilizers or Na-rich compositions are used to produce phase pure $\beta''$. During the 1960s and 1970s, the concept of the solid state or similar sodium–sulfur cell was already discussed as shown in Figure 20, with $NaAl_{11}O_{17}$ $\beta$ alumina as the electrolyte.[150] It is worthy to note that early measurements of conductivity included the use of liquid sodium and cation exchange at elevated temperatures.

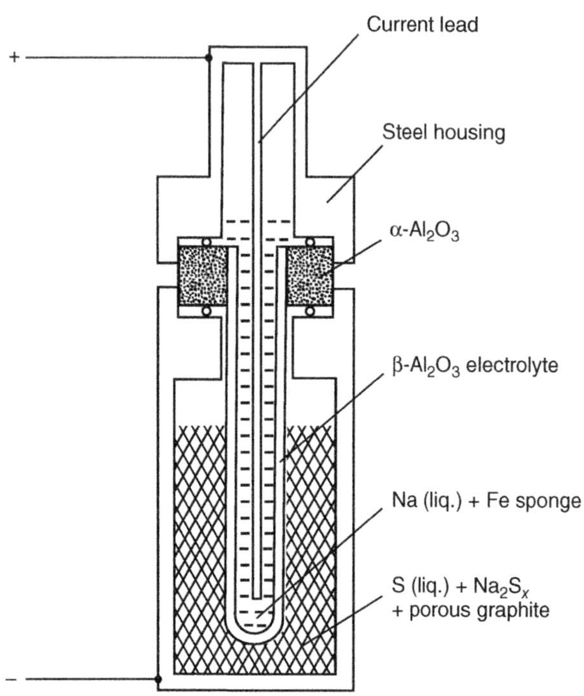


**Figure 20** A schematic of an early sodium–sulfur cell. [Reproduced with permission from Ref. 150 © Wiley, 1978]

## 4 MECHANISM—SURFACE CHEMISTRY DURING ELECTROCHEMICAL CYCLING

### 4.1 Surface Chemistry of Lithium-Ion Batteries

In addition to the major charge storage reactions, in most cases critical or significant irreversible reactions also take place on electrodes in LIBs. These reactions occur between the electrode and electrolyte, which ultimately lead to the formation of layers on the electrode surface (cathode or anode) in the solid state termed as the "solid–electrolyte interface (SEI)" by Peled in 1979.[151] Ideally, this layer is permeable to ions but not to electrons. The electron-insulating nature of the layer chokes the direct electron supply between the electrodes to electrolyte. This helps to limit irreversible reactions and can minimize the degradation of the electrolyte and electrode materials. These layers help in the improvement of battery kinetics in terms of cell performance, stability, and safety. To distinguish between the interface layer of the anode and cathode, the term "surface permeable interface (SPI)" or "cathodic electrolyte interface (CEI)" for the interface layer toward the cathode was introduced in 2004.[152]

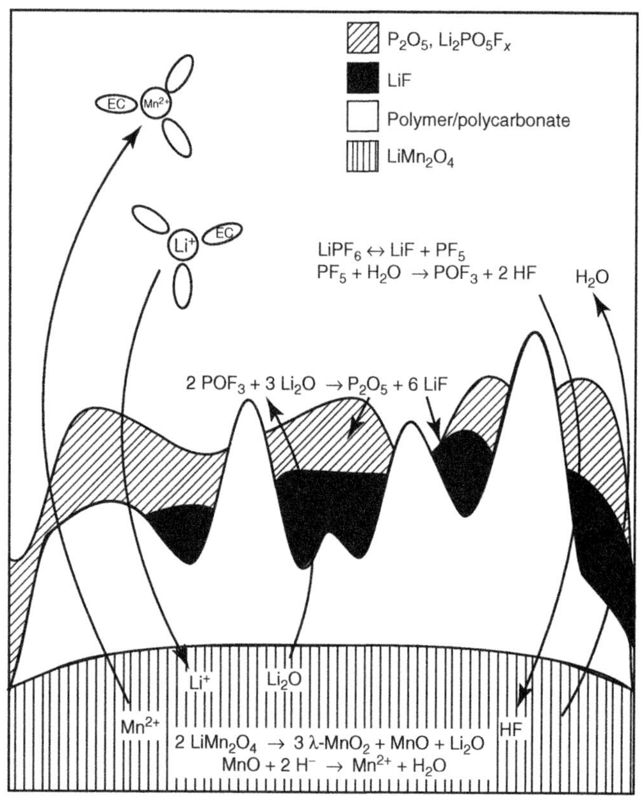


**Figure 21** A proposed model for the SPI layer formed on a $LiMn_2O_4$ electrode. [Reproduced with permission from Ref. 152 © Elsevier, 2004]

Researchers indicate that a good SEI should be an electronic resistor, uniform ionic conductor (with cation transference number close to unity) to minimize the overpotential, have a uniform morphology, and be able to manage the electrode volume change during electrochemical cycling. The latter suggests that the SEI layer should be both mechanically stable once formed but flexible when forming. The chemical composition of the SEI consists of inorganic and organic constituents, which are the breakdown products of the electrolyte salt, solvents, additives, occasionally surface layers of the electrode, and contaminants such as water and oxygen.[153] Generally, the inner level (near the SEI/electrode interface) of the SEI is comprised of inorganic components and the outer layer (near the SEI/electrolyte interface) by predominantly organic constituents.[154] A proposed model for the formation of SPI surface layer on the $LiMn_2O_4$ is shown in Figure 21. The failure of the SEI or failure in SEI formation results in battery degradation.[155]

At the anode, upon charge–discharge, volume expansion in the electrode materials results in fatigue, cracking, and structural changes in both the SEI and electrode materials. The metallic dendrite (needlelike) growth of lithium at a lower potential, also known as lithium plating, toward the anode can be generated on the SEI or pierce the SEI as well as the separator, which can result in cell failure. Gas generation, metal dissolution and deposition, electrochemical degradation of the SEI components, and cathode material lattice oxygen removal in dual redox electrodes also affect the stability of the SEI.[155] To solve some of these issues, researchers have designed electrode materials with a premeditated coating of inorganic,[156–158] organic,[159] polymeric[160] or composite[161] layers often mimicking natural SEI components over active electrode materials. These coatings are known as the "artificial solid–electrolyte interface". They show similar properties to the naturally (or electrochemically) formed SEI but are designed to be mechanically and chemically more robust than natural SEI. Hence, a carefully selected artificial SEI can change the course of interfacial reactions and minimize electrolyte degradation.[155] For example, an artificial SEI was created on $Li_4Ti_5O_{12}$ by coating 5–10 nm of p-type semiconductor $NiO_x$ (where Ni = $2^+$, $3^+$).[162] In situ XRD confirmed that the cell stability improved as no structural change in the lattice was observed. FT-IR showed a lower amount of $Li_2CO_3$ and almost a constant amount of ROCOOLi on the surface of $Li_4Ti_5O_{12}$ with the $NiO_x$ coating in comparison to the bare $Li_4Ti_5O_{12}$. At the cathode, MgO coating on $LiCoO_2$ was shown to improve the stability of cathode by minimizing the reduction of Co on the surface, especially in the absence of Li in crystal lattice at the charged state.[163] Researchers have also explored organic coating layers, for example, 5% methyl viologen (MV) in the ether electrolyte was shown to form a stable and uniform conductive interfacial coating on Li-metal after it is electrochemically reduced during battery cycling (Figure 22), that is, in situ generated during battery function.[159] Electrolyte additives have been widely explored and perfected for their ability to form SEI layers on electrodes.

Since the mid-1950s graphite was found to be reactive with lithium but it showed very poor electrochemical performance with commonly used propylene carbonate (PC)-based electrolytes.[164] In LIBs, the organic electrolyte has oxidation potentials around 4.7 V versus $Li^+/Li$ and reduction potentials close to 1.0 V versus

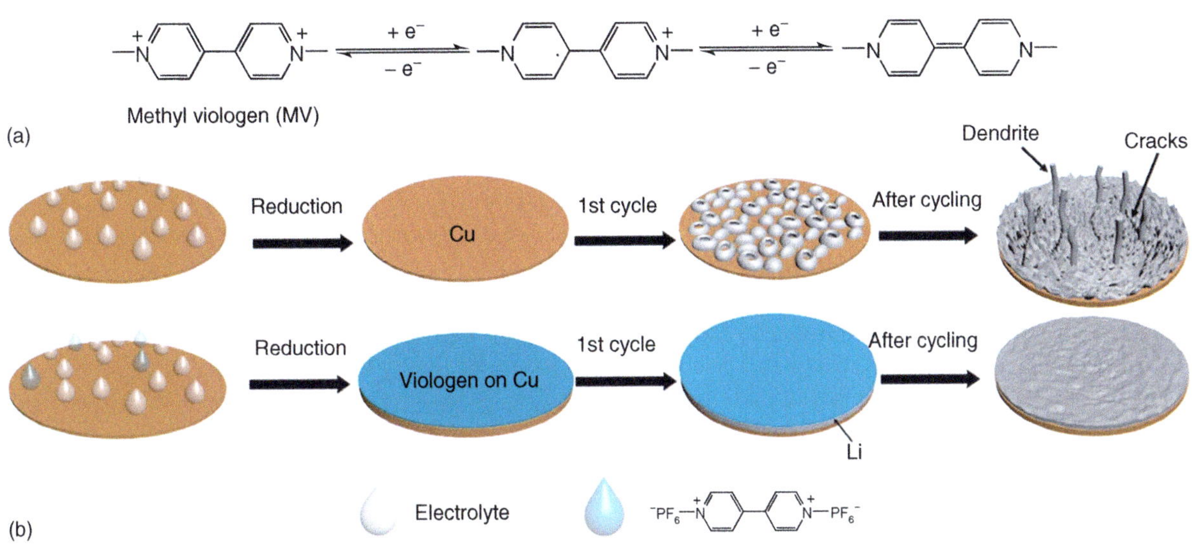


**Figure 22** (a) Two-step reduction process of methyl viologen (MV) and (b) cartoon of the proposed function of MV-containing electrolyte relative to MV-free electrolyte. [Reproduced with permission from Ref 159. © American Chemical Society, 2017]

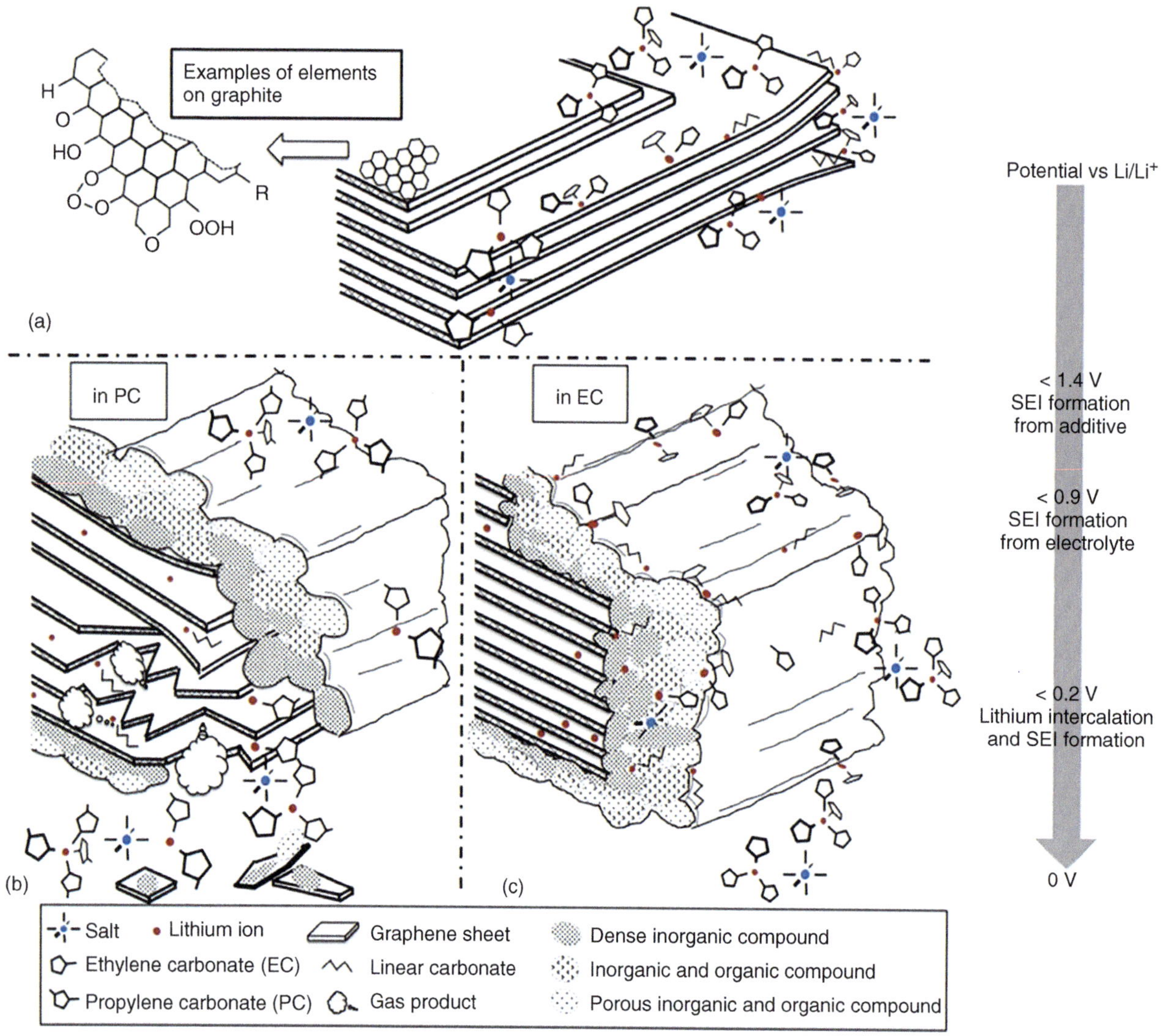


**Figure 23** Schematic of formation of the anode SEI showing (a) graphene layers surrounded by electrolyte salts and solvents above 1.4 V versus $Li/Li^+$, (b) propylene-carbonate (PC) intercalation with lithium ions into graphene layers resulting exfoliation below 0.9 V versus $Li/Li^+$ and (c) stable SEI formation in ethylene-carbonate (EC)-based electrolyte below 0.9 V versus $Li/Li^+$; plane side with thinner SEI and edge side with thicker SEI. [Reproduced with permission from Ref. 165 © Elsevier, 2016]

$Li^+/Li$. The intercalation potential of Li into graphite is below the reduction potential of the electrolyte (between 0 and 0.25 V vs $Li^+/Li$). Thus, the potential of graphite electrodes falls below the stability window of the electrolyte during charging, and it decomposes at the graphite surface forming the SEI.[165] In 1990, Dahn and coworkers studied the role of SEI in the reversibility of carbonaceous electrodes and the effect of ethylene carbonate (EC) with PC (50:50 mixture with 1 M $LiAsF_6$) as an electrolyte on Li/graphite and Li/petroleum coke cells.[166] These cells exhibit excellent reversibility and can be recycled without any significant capacity loss and was related to the ability of EC to assist in the formation of an SEI layer.[166] Soon thereafter the $LiPF_6$ salt was commercially adopted, which after reduction precipitates to form LiF or $Li_xPF_y$ and carbonates from the electrolyte solvent precipitate with lithium ions to form $Li_2CO_3$, lithium alkyl carbonate ($ROCO_2Li$), or other organic compounds. LiF, $Li_2O$, $Li_2CO_3$, and other insoluble products remain as components of the SEI, whereas some soluble products may diffuse back into the electrolyte upon cycling; see Figure 23. The combination of using an SEI forming component in the electrolyte and stabilizing the SEI, particularly at the graphite anode, led to the successful commercialization of a LIB system by Sony Corporation based on petroleum coke and $LiCoO_2$.[1]

It has been generally accepted that the formation of SEI on graphite is a two-step process, in the first step the graphite electrode is polarized and new chemical species form due to the reductive decomposition of organic electrolyte components typically near the surface oxygen sites[164] and in the second step the decomposition products precipitate on the graphite surface sites to form the SEI layer.[165] The SEI forms mostly during the first charge, but the formation continues with cycling until the SEI layer is fully developed[165] and ideally stable. However, it is not easy to control the formation and growth of SEI on graphite or alternative electrodes because the chemical composition, morphology, and stability depend on the type and morphology of graphite (or electrode), electrolyte composition, cell temperature, and electrochemical composition.[165] Furthermore, the type of insertion reaction can have a drastic effect on the SEI, for example, alloying reactions may feature large volume expansion/contraction and this may require an SEI to be continuous formed and broken.

## 4.2 Surface Chemistry of Sodium-Ion Batteries

Graphite is a well-known anode for LIBs as it reversibly forms $Li_xC_6$ on insertion. However, graphite does not reversibly insert/extract $Na^+$ in significant quantities and is found not to form Na–C alloys.[167] The majority of Na atoms are rather found to adsorb onto the carbon surface or nanoscopic pores throughout hard carbon particles.[72,168] Therefore, alternatives to graphite are widely considered, with a range of them focusing on carbon-based materials due to their cost advantages such as hard carbon,[169] pitch-cokes[170] and modified graphite.[92] However, very few studies have explored the SEI in these systems, and a better understanding of the SEI in carbon-based electrodes for NIBs is critically required as it can inform electrode design and electrolyte choice, which, in turn, leads to performance enhancement.

In terms of the surface chemistry, an example is the comparison of commercially available single-wall carbon nanotubes (SW-CNTs) and multiwall carbon nanotubes (MW-CNTs) with SW-CNTs producing first discharge capacity of 951 and 65 mAh $g^{-1}$ after 100 cycles, while MW-CNTs show an initial discharge capacity of 346 mAh $g^{-1}$ that drops below 30 mAh $g^{-1}$ by the 40th cycle. $^{23}$Na NMR was used to quantify the amount of sodium intercalated in carbon materials and suggest their distribution. The lower capacity delivery by MW-CNTs was attributed to sodium ions trapped in the MW-CNT electrodes (unable to be cycled) while for SW-CNTs a larger fraction was accessible, and this was associated with Na inserted into the tubes or distributed among the surface that correlated to a higher surface area. The surface-dominated reaction suggests the SEI is critical in these systems and access to the surface is essential.

One of the major considerations in carbon-based electrode performance for NIBs appears to be the link between surface area of the active material, surface chemistry (SEI formation), and performance. When high-surface-area carbons are used, clearly SEI formation is arguably even more pertinent than in graphite. Examples of surface-based studies and correlated phenomena in NIBs are appearing. $^{23}$Na solid-state nuclear magnetic resonance (SS-NMR) was used to study the interaction between inserted (and extracted) Na-ions and solvothermal synthesized graphene (STSG) and thermally exfoliated graphite oxide (TEGO) in an ex situ manner, after charge/discharge cycles.[171] Although typically considered a bulk technique, the local information accessible by solid-state NMR can be used to indirectly propose whether Na is in a mobile surface-type layer or in more bound sites within SEI layers or in the electrodes. In this work, at least three distinct Na chemical environments were identified, which were active during charge/discharge and tentatively considered to be mobile, but the majority of Na was found to be predominantly located on an immobile site, assigned to the SEI layer. A growth of the immobile sites and the loss of mobile sites are observed with cycling with TEGO, which correlates with a decrease in capacity. STSG shows evidence of changing Na local environments with cycling and this suggests a less stable SEI layer with respect to Na and potentially a reason for the lower capacity observed.[171] The correlation of local phenomenon with surface chemistry and electrochemistry is critical in design; comparative work like that shown above may be a means to elucidate the reasons behind performance.

An alternative approach with graphite investigates the intercalation of solvated Na-ions, which has shown the ability for the electrode to periodically "breathe" by about 70–100% during battery cycling and maintains excellent reversibility.[172,173] This is speculated to be because the graphite particles exfoliate to crystalline platelets rather than delaminate or clog-up at the surface. Online electrochemical mass spectrometry (OEMS) gas analysis was used to show that electrolyte decomposition is largely restricted to the first cycle only and hence SEI formation is associated with the first cycle.[172]

Additives have also been considered; an example is the work with sugar pyrolysis derived hard carbon electrodes[169] in $NaClO_4$ in EC:PC electrolytes with or without 2 vol% fluoroethylene carbonate (FEC) additive. Cells with the FEC polymerizing agent exhibit much poorer conductivity, which results in low specific capacity and coulombic efficiency[169] than the neat EC:PC mixture[174] and this was related to the SEI formation. Furthermore, FEC concentrations larger than 2% were found to have detrimental effects on the reversible capacity as it prevented fully accessing the low potential plateau.[169] This work contrasts studies on hard carbon (nongraphitizable carbon) where addition of 2 vol% FEC with $NaClO_4$ in PC showed remarkable improvements in capacity retention.[175] The improved cycle performance was thought to originate from the modification of the surface passivation layer including SEI as FEC here is suppressing the electrolyte decomposition.[175] The key difference in these appears to be the use of EC and FEC as an SEI layer formation component or simply using just FEC.

### 4.3 Surface Chemistry of Non-Carbon-Based Anodes in Sodium-Ion Batteries

$Na_2Ti_3O_7$ a potential negative electrode shows poor capacity retention, and the stability of SEI is considered as one of the reasons for its underperformance. Ex situ XPS confirms the formation of SEI layers composed of carbonates ($Na_2CO_3$), alkyl carbonates ($NaCO_3R$), fluoride (NaF), chloride (NaCl), and polyethylene oxides (PEO) upon discharge.[176] Though the SEI is partially dissolved upon every charge, they do not appear to be completely stable. $Na_2CO_3$, $NaCO_3R$, NaCl, and PEO are formed from the decomposition of electrolyte salt while NaF is formed during the electrode preparation (or initial cycles) having its origin in the polyvinylidene difluoride (PVDF) dehydrofluorination reactions.

### 4.4 Characterizing the Surface

The SEI layer is very sensitive toward $CO_2$, oxygen, and $H_2O$ present in atmosphere. Hence, while transferring the SEI specimens from the inert atmosphere of the glove box to an analytical instrument or even opening an electrochemical cell, careful handling, sample specific chambers or conditions are required to minimize SEI layer contamination or modification. The question of whether the SEI layer that is found in a cell is the same as explored in an ex situ technique is notable and requires caveats. A number of tools and techniques do not require cell opening and can be used to interrogate the SEI, for example, electrochemical impedance spectroscopy (EIS) and cyclic voltammetry (CV). To image the morphology and surface features of an extracted electrode, scanning electron microscopy (SEM), TEM and atomic force microscopy (AFM) can be used, while chemical identification, bonding environment can be probed with XPS and Fourier transform infrared spectroscopy (FTIR). However, some of these techniques require high vacuum environments that can pull off or modify the SEI layer, but some cells are available to avoid this. Other techniques used to probe the SEI either directly or indirectly include Raman spectroscopy, X-ray diffraction (in situ or ex situ), NMR, neutron diffraction (ND), and XAS. Overall, significant care needs to be taken when interpreting data associated with the SEI, especially with respect to sample handling.

## 5 STATE-OF-THE-ART CHARACTERIZATION TOOLS

### 5.1 In Situ Methods

Electrode material characterization has relied heavily on ex situ measurements; however, recently in situ and *in operando* studies have become more accessible and used. They allow researchers to explore the structure–function relationships in an electrode material or entire device during operation.

#### *5.1.1 X-Ray Diffraction*

In situ XRD is a nondestructive analytical tool to monitor the structural change and phase evolution of electrochemically active electrode materials (crystalline materials). Laboratory and synchrotron X-ray-based sources can be used where the latter provides the choice of wavelength and a higher intensity source. For an in situ setup when rapid measurement is not required during battery cycling, then a multiple-cell arrangement can be employed, examples of which include translation stages or the eight-position modified coin cell battery rotation stage (shown in Figure 24).[177] These setups allow electrochemistry to be controlled for each cell simultaneously while each position is cycled through the beam and a dataset collected. An example of an in situ synchrotron XRD study using an in situ multiple-cell configuration is the structural evolution of the P2 $Na_{2/3}Fe_{0.2}Mn_{0.8}O_2$ [25] with data collection every 20.7 min during charge–discharge for a period of 2.5 min in transmission geometry. Here the current rates applied were relatively slow to ensure that subtle changes in the structure are not missed or overlapped in collected data. This work showed at this Fe/Mn composition, the P2 structure is maintained throughout and neither disordered Z or OP4 phases were formed at charged state nor orthorhombic P′2 phase formed in the discharged state[178] under the electrochemical conditions used.[25] For rapid measurement, when using high current rates and to see fast phase transitions, *in operando* XRD experiments are typically performed where XRD datasets are continuously collected while cycling the cell. *In operando* synchrotron XRD was used to understand the origin of capacity fade in P2 $Na_{2/3}Fe_{1/2}Mn_{1/2}O_2$ when cycling between 1.5 and 4.2 V[179] with data collected continuously every ~3 min. Notably selected studies are able to model the sodium site occupancies in crystal structures and their evolution as a function of charge and discharge. Currently in the literature there are over 100 studies (Web of Science) using in situ or *in operando* XRD to investigate the structure evolution of possible electrode materials in sodium-ion batteries.

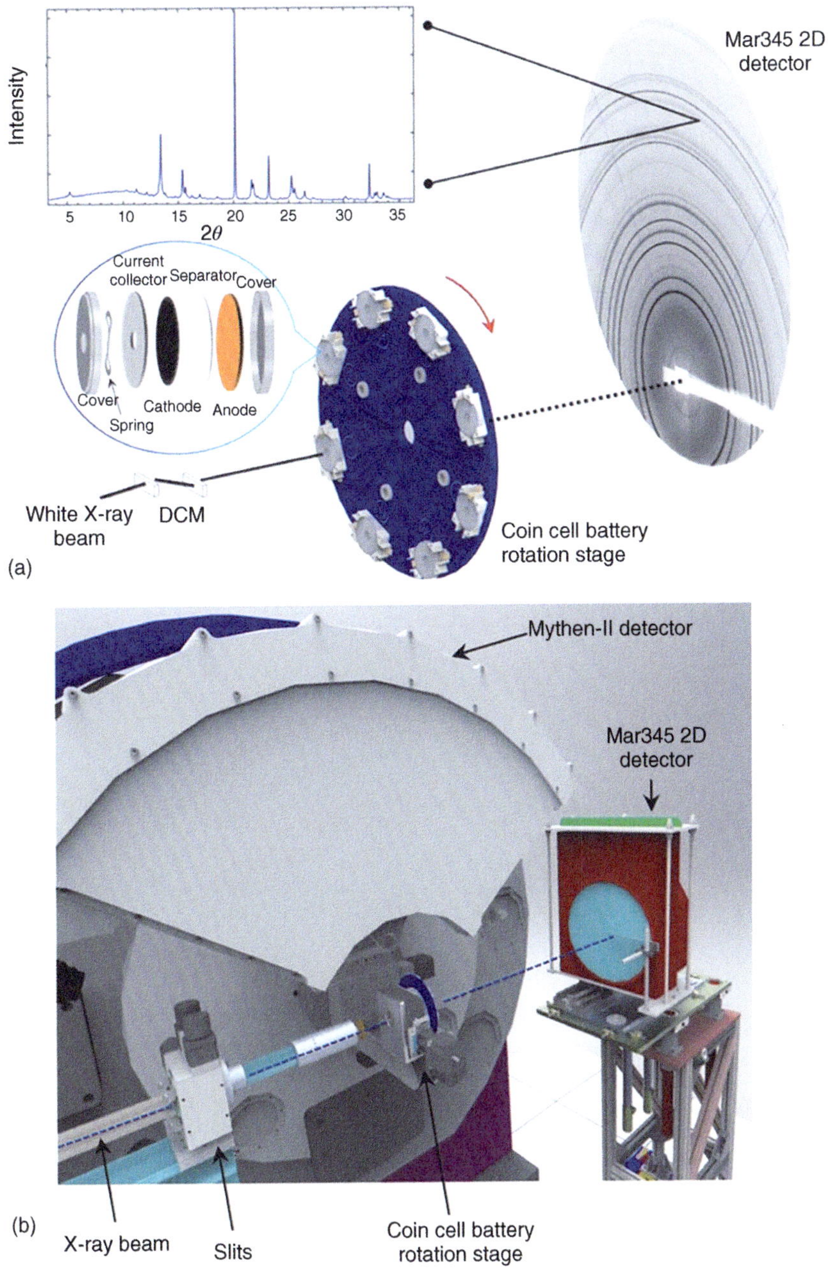


**Figure 24** (a) Schematic of a typical modified coin cell in situ SXPD experimental setup at the powder diffraction beamline, Australian synchrotron. (b) An eight-position modified coin cell battery rotation stage for high-throughput in situ experiments. [Reproduced with permission from Ref. 177 © Wiley, 2017]

### *5.1.2 Neutron Diffraction*

In situ ND is another tool that has recently been used heavily for the study of LIBs. The electrochemical cell design is more important for ND studies due to the interaction of the neutrons with atomic nuclei, particularly with hydrogen, which increases the background dramatically in as collected data (see Figure 25 as an example). Furthermore, deuterated electrolytes are often used for ND studies. The advantages of using ND include the large penetration depth, different scattering factors relative to XRD (e.g., between Fe and Mn), and contrast offered by isotopic substitutions, while disadvantages include relative amount of sample required, contribution of hydrogen, and limited facilities available worldwide. In LIBs, full commercial 18 650 batteries have been probed, allowing the structural evolution of both electrodes to be determined,[180–187] and similar experiments are likely to be conducted for NIBs.

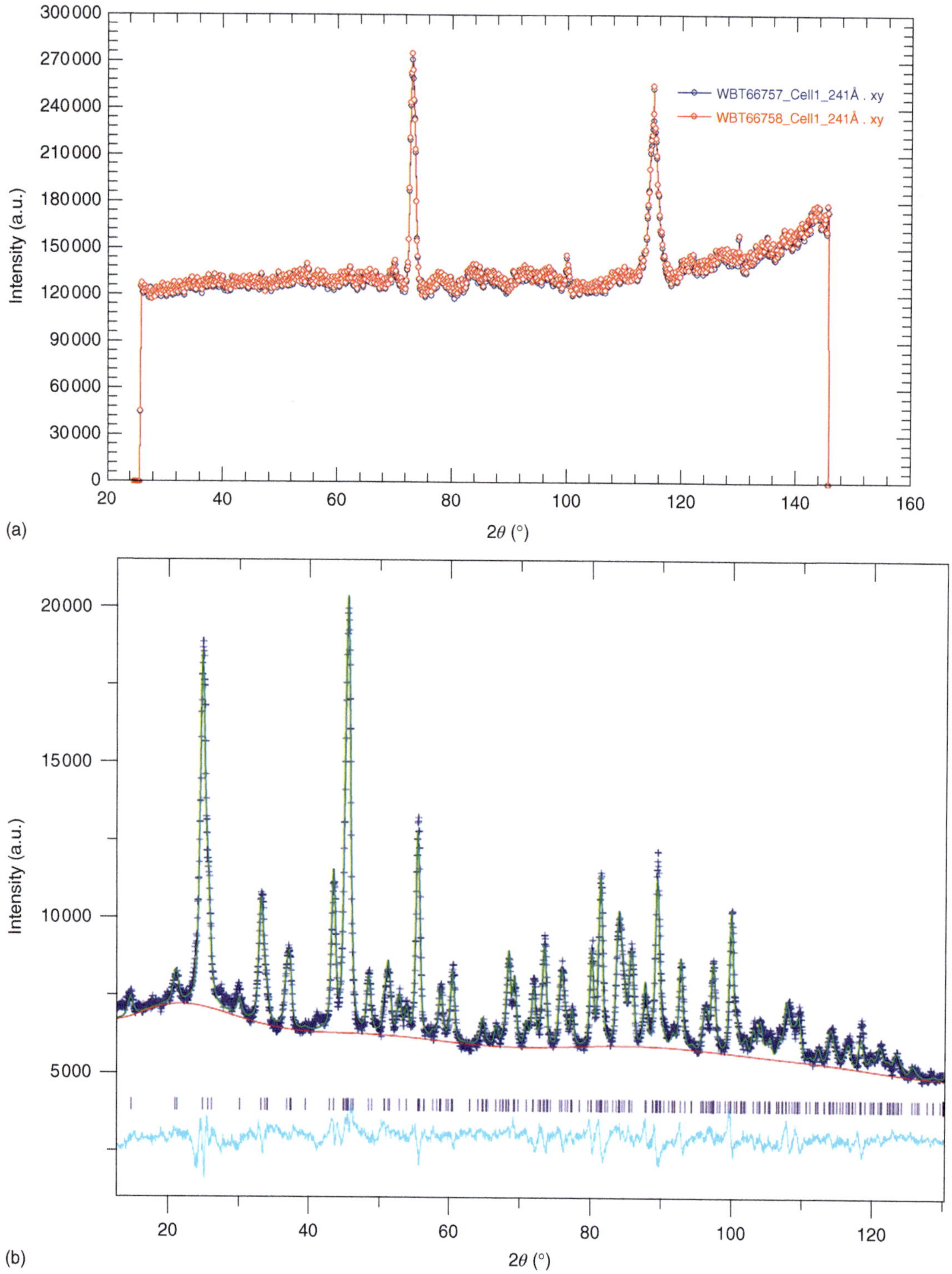


**Figure 25** An example of a ND pattern (a) of a 18 650 sodium-ion battery on the WOMBAT neutron diffractometer at the Australian Nuclear Science and Technology Organisation where several milliliters of electrolyte are used in comparison with the Rietveld analysis of the pure cathode powder (b) (unpublished work)

ND has been applied to study solid electrolytes ($Na_3OBr$ and the modified layered antiperovskite $Na_4OI_2$) as a function of temperature while impedance data were collected.[188] The intention was to elucidate the Na ionic conductivities and the potential transport pathways, and the authors found in both cubic $Na_3OBr$ and layered $Na_4OI_2$ the main sodium transport pathway is between the nearest $Na^+$ ions in the $NaO_6$ octahedra building block of the antiperovskite. The polarized nuclear densities of $I^-$ ions bridges the $Na^+$ ions jumping pathway between two isolated octahedra along the *c* axis of the layered $Na_4OI_2$ antiperovskite, but $Br^-$ ions are not involved in $Na_3OBr$.

### 5.1.3 *X-Ray Absorption Spectroscopy*

XAS is a powerful technique to measure the atomic structure and identify the oxidation states of metal ions (typically) in electrode materials at different stages of battery cycling. It is element specific and sensitive to very minimal concentrations and can be used for short-range ordered compounds, that is, amorphous and nanoparticles. For in situ studies, the absorption by other elements in a battery is minimized by tuning the X-ray energy close to absorption edge of the element being analyzed. XAS spectra are divided into two parts: X-ray absorption near-edge spectroscopy (XANES), which covers the energy range up to 50–100 eV near the absorption edge, while extended X-ray absorption fine structure (EXAFS) covers several hundred electron volts, starting about 30 eV above the absorption edge. In situ and ex situ XANES spectroscopy at the Fe K edge confirms the presence of electrochemically active $Fe^{2+}/Fe^{3+}$ and $Fe^{3+}/Fe^{4+}$ redox couples in $Na_3FePO_4CO_3$ nanoplates, suggesting the existence of a two-electron intercalation reaction during charge/discharge and contribution to the capacity.[189] This is a very powerful technique for nanocrystalline electrode materials.

### 5.1.4 *Raman Spectroscopy*

Raman spectroscopy can be used for obtaining structural information at the atomic scale for a range of inorganic and organic compounds having Raman-active vibrational modes. In situ Raman measurements have been employed to evaluate the structural evolution of $Na_2Ti_{0.94}Cr_{0.06}O_{0.97}$,[190] $Na_{2/3}[Fe_{1/2}Mn_{1/2}]O_2$,[191] crystalline few-layered graphene (FLG) materials,[192] and $Li_{0.1}Na_{0.7}Co_{0.5}Mn_{0.5}O_2$[193] upon (de)sodiation. In situ Raman spectroscopy was performed on sol–gel synthesized (Figure 26b) and commercial antimony (Figure 26c) to study their different reversibility and the Raman

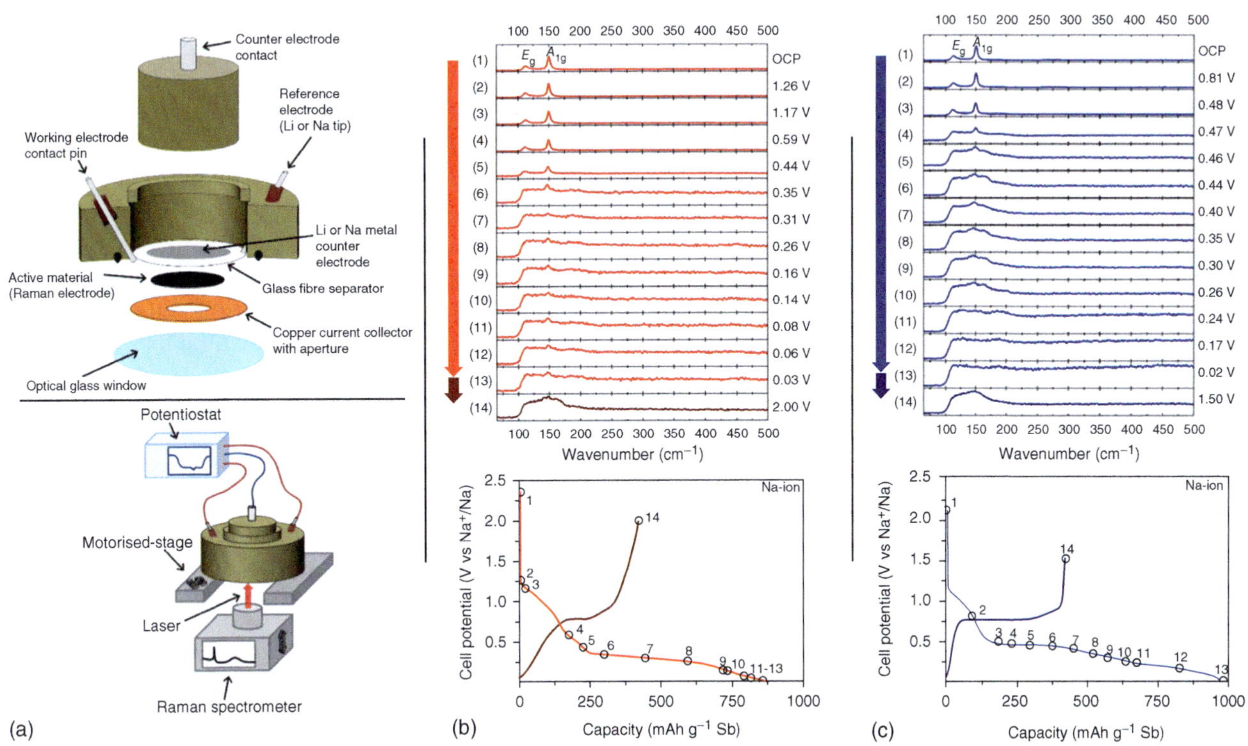


**Figure 26** (a) Schematic of Raman cell setup, in situ Raman spectroscopy data for (b) sol–gel synthesized and (c) commercial antimony versus $Na^+/Na$ over 1 charge/discharge cycle at 60 μA $g^{-1}$. [Reproduced with permission from Ref. 194 © Elsevier, 2017]

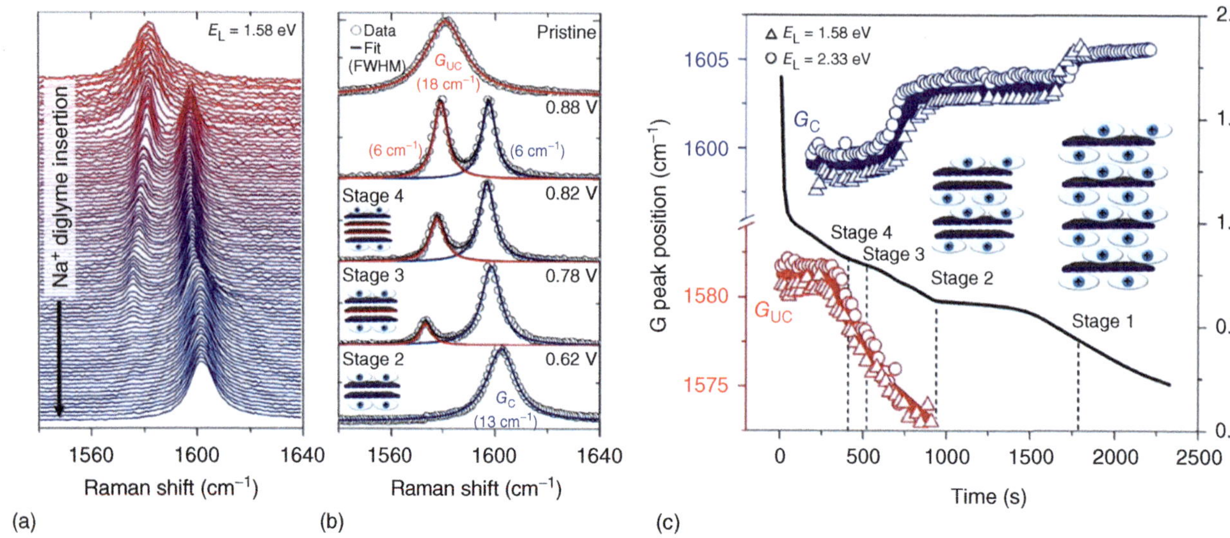


**Figure 27** (a) In situ Raman spectra (normalized) of FLG showing the highly ordered staging reaction as measured using a 1.58 eV laser with (b) selected spectra and Lorentzian fits of GC (blue line) and GUC (red line) components. (c) Tracking the positions of the Raman G peak components (GC shown in blue and GUC shown in red) measured in situ with the 1.58 eV laser (triangles) and the 2.33 eV laser (circles) during the electrochemical intercalation reaction with the corresponding galvanostatic discharge (~0.2 A g$^{-1}$) profile shown with respect to right *y*-axis (black line). [Reproduced with permission from Ref 192 © Ameican Chemical Society, 2016]

cell assembly is shown in Figure 26(a).[194] For sol–gel synthesized antimony material, no peak shift is observed during cell discharge, suggesting that phase is stable until 0.59 V versus $Na^+$/Na while below 0.59 V versus $Na^+$/Na, long-range features for antimony start to disappear. Specifically, the $E_g$ band disappears and there is the weak presence of $A_{1g}$ bands, which suggests some order remains at the end of discharge. This contrasts to the commercial antimony; the $A_{1g}$ band completely disappears, indicating the completion of sodiation and formation of $Na_3Sb$. The authors speculate that the carbon-coated sol–gel synthesized antimony restricts the diffusion of the larger sodium ions into the antimony particles, which would also explain the reduction in capacity compared to commercial antimony. Upon charge, both materials do not return to a crystalline state, but instead to an amorphous phase represented by a broad feature centered at ca. 140 cm$^{-1}$.[194] In situ Raman measurement was also used to identify the $Na^+$ staging sequence into highly crystalline FLG with which the Fermi energy of stage 1 and stage 2 intercalation compounds were determined to be 1.2 and 0.8 eV, respectively (Figure 27).[192] Raman spectroscopy will play a critical role in the development of anode materials in particular for sodium-ion batteries.

### 5.1.5 *Electron Microscopy*

SEM provides topographical images and compositional information of a sample by scanning the surface with typically a focused electron beam. SEM needs high vacuum working conditions and an electronically conductive sample, which limit its application for in situ SEM studies with a conventional liquid electrolyte. Due to the electron-insulating nature of the electrolyte, it is difficult to study the dendritic structures immersed in the electrolyte. For in situ SEM visualization, either solid electrolyte or low vapor pressure electrolytes, for example, ionic liquids, that can be sustained under ultrahigh vacuum conditions should be used. *In operando* SEM was used to study the morphological changes, such as crack generation and volume expansion, in a tin electrode during discharge of a LIB[195] followed up by studies of the Si anode and an ionic liquid electrolyte.[196] There is a potential for parallel studies for Na-based systems.

In situ TEM is an especially useful tool for obtaining an insight into a range of phase transformation behavior, including site-specific sodiation events and decomposition of electrolytes on electrode surfaces.[197] In situ TEM together with ex situ time-of-flight, secondary-ion mass spectrometry (TOF-SIMS), and focused ion beam–helium ion scanning microscope (FIB-HIM) analysis have been used to study the time-dependent structural and compositional changes in nanometer-scale (50 and 100 nm) Sb thin films (single crystal or twinned Ge NW grown on Si wedge, which was coated with TiN layer) during sodiation/desodiation (Figure 28).[197] In situ TEM shows time-dependent buckling (shown by white arrow in Figure 28f–h) and localized separation of the sodiated films from the TiN-Ge NW support. No direct

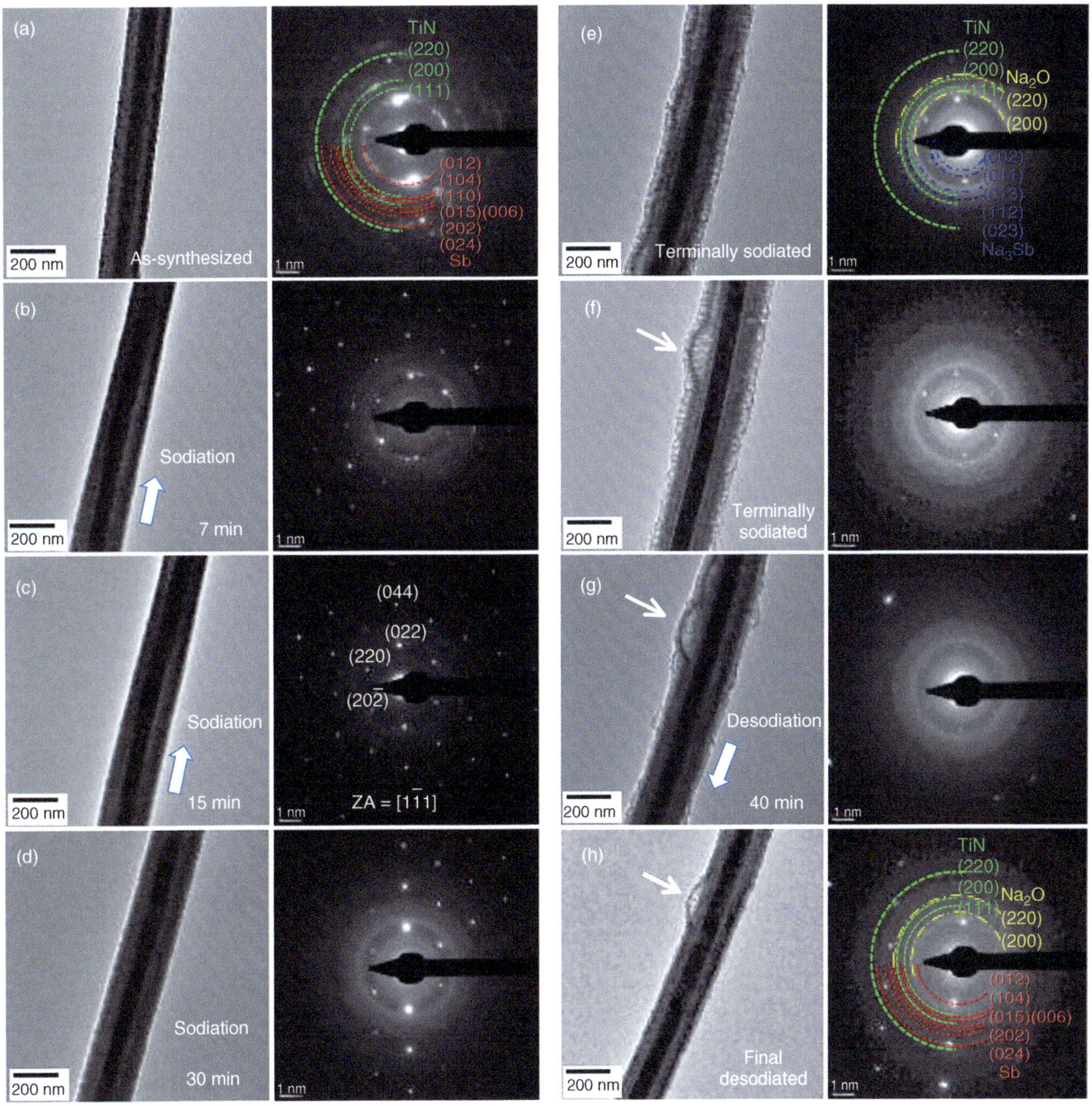


**Figure 28** In situ TEM analysis and indexed SAED of a single 50 nm Sb/NW for a full sodiation–desodiation cycle under a beam-blank experimental condition: (a) in the as-synthesized state, (b) after 7 min of sodiation, (c) after 15 min of sodiation, (d) after 30 min of sodiation, (e) in the terminally sodiated state after 7 h, (f) another region in the terminally sodiated state, (g) after 40 min of desodiation, and (h) in the final desodiated state. The supporting nanowire is oriented near the [1$\bar{1}$1] symmetric zone axis. [Reproduced with permission from Ref 197. © Ameican Chemical Society, 2015]

localized horizontal fracture was observed at the interface, but rather at a short distance away within the bulk of the Sb.[197] The real-time sodiation/desodiation behavior of α-$MoO_3$ was also probed using in situ TEM[198] where these nanobelts were found to undergo a unique multistep phase transformation. Upon the first sodiation, α-$MoO_3$ nanobelts initially form amorphous $Na_xMoO_3$, then intermediate crystalline $NaMoO_2$, and finally crystallized Mo nanograins embedded within the $Na_2O$ matrix. During the first desodiation process, Mo nanograins are reoxidized into the intermediate $NaMoO_2$ that is further transformed into amorphous $Na_2MoO_3$, resulting in an irreversible phase transformation. Finally, in situ TEM experiments can also be used to reveal the mechanical degradation of CNFs[199] during sodiation and potassiation because of the formation of longitudinal cracks near the crystalline carbon/disordered carbon interface. Such work provides significant microstructural insight into electrode function.

### 5.1.6 *UV–Visible Spectroscopy*

In situ UV–Visible spectroscopy was used to study the reason of capacity fade in a metal–organic cathode material, cuprous 7,7,8,8-tetracyanoquinodimethane (CuTCNQ).[200] This compound consists of both cationic ($Cu^{II} \leftrightarrow Cu^{I}$) and anionic ($TCNQ^0 \leftrightarrow TCNQ^- \leftrightarrow TCNQ^{2-}$) reversible redox reactions and delivers a high capacity, but dissolution of TCNQ causes rapid capacity fade. The TCNQ chromophore is sensitive to UV/Visible radiation hence, the dissolved TCNQ species can be identified by in situ UV/Visible absorption spectroscopy (Figure 29c) under various conditions. An in situ cell assembled in a

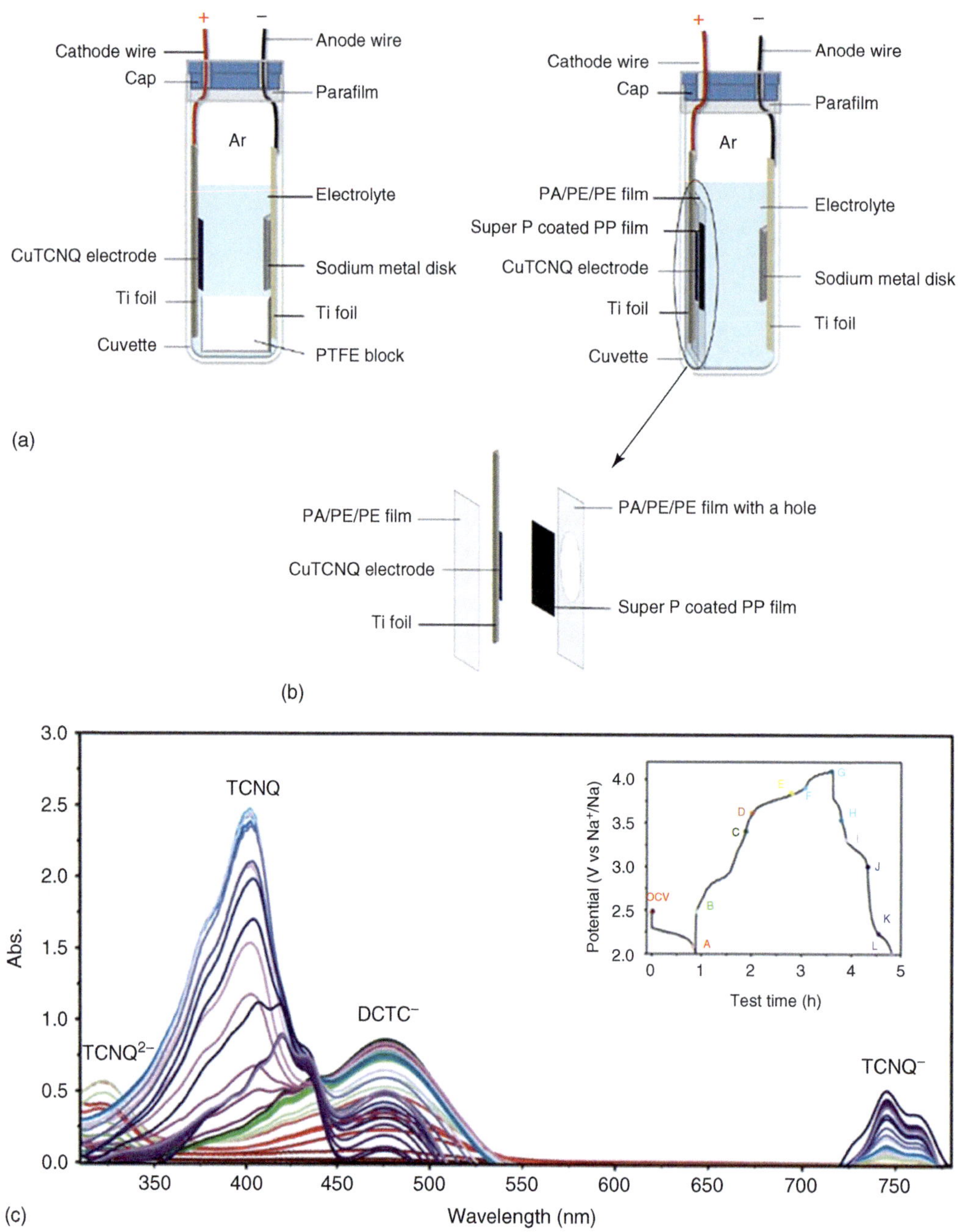


**Figure 29** Cell assembly for in situ UV–Visible absorption spectroscopy measurement in a cuvette (a) CuTCNQ electrode assembled without a carbon film, (b) CuTCNQ electrode encapsulated with a Super P-coated PP film between two pieces of PA/PE/PE films. The magnification of PA/PE/PE films encapsulated CuTCNQ electrode with a carbon film and (c) in-situ UV–Visible absorption spectra collected during charge–discharge. [Reproduced with permission from Ref. 200 © Wiley, 2017]

cuvette is shown in Figure 29(a) and (b) and the results at various states of charge/discharge.[200]

## 5.2 Atomic Resolution

Imaging techniques provide useful information about morphology and chemical composition (when coupled with elemental analysis). These data can be analyzed from selected regions or mapped for the entire electrode (or another component, e.g., separator). The investigation of SEI on electrodes is difficult due to their instability and potential to decompose outside a battery, when exposed to vacuum or under high-energy radiation or light. Most electron microscopy studies are carried out ex situ or post mortem; however, significant examples of in situ TEM studies exist (Section 5.1.5). TEM has been extensively used in batteries for crystal orientation analysis, preferred lithiation/sodiation direction, morphological changes both in ex situ and in situ conditions.[155] However, one of the most promising applications is atomic-scale visualization.

Detailed atomic-scale structural information on P2-$Na_{2/3}Ni_{1/3}Mn_{2/3}O_2$ and P2-$Na_{2/3}Ni_{1/3}Mn_{1/3}Ti_{1/3}O_2$ were confirmed by scanning transmission electron microscopy (STEM) with annular bright field (ABF) and high-angle annular dark field imaging (HAADF) techniques (Figure 30).[201] The position of transition metals (TM = Ni, Mn, Ti) columns are revealed in ABF-STEM images (Figure 30d and f). The distance between the adjacent layer $d_c$ is calculated in HAADF-STEM images, which is consistent with the refined crystallographic data. Further, energy dispersive spectroscopy (EDX) spectroscopy maps demonstrate the uniform elemental distribution of Na, Ni, Mn, Ti, and O in the P2-$Na_{2/3}Ni_{1/3}Mn_{1/3}Ti_{1/3}O_2$ particles.[201] Such atomic details provide amazing opportunities for direct quantification of structure and properties.

## 5.3 Surface Characterization

The SEI layer consists of both amorphous and crystalline layers (lower molecular weight inorganic compounds); see Section 5 and examples therein. XRD is one technique that can be used, but in comparison to electrode materials, the thickness of SEI with significant crystallinity is much thinner, and amorphous and organic compounds are major constituents of SEI, which restricts its conclusive analysis. Still, there are few noteworthy synchrotron-based in situ XRD studies of SEI formation. For example, the graphene/SiC surface in LIBs[202] with 1 M $LiPF_6$/(EC: DMC) electrolyte was used to study the formation and growth of LiF crystallites at the graphene surface. The crystallites were shown to increase in size with lithiation and were serendipitously textured such that the LiF (002) planes were approximately parallel to the graphene sheets and easily observed in the XRD data. In a full cell containing $LiNi_{1/3}Mn_{1/3}Co_{1/3}O_2$ (cathode) and graphite (anode) with $LiPF_6$ (EC: EMC) electrolyte, in situ XRD was used to quantify the amount of lithium used in SEI formation.[203] The irreversible Li loss during the charge–discharge process was calculated by quantifying the amount of lithium used in lattice expansion of the electrode. The lithium loss was correlated with the SEI formation on the graphite electrode surfaces. Parallel work can be undertaken here. Extending XRD-based analysis, PDF analysis uses Bragg with diffuse scattering. New phases of sodium antimony ($Na_{3-x}Sb$) were observed; depending on the step of sodiation/desodiation, amount of Na trapped in the SEI during battery cycling using a combination PDF analysis and $^{23}Na$ NMR spectroscopy.[204] This is an elegant example of using multiple techniques to address an open question.

The role of FEC additive, morphological change after cycling, and chemical composition of SEI films on the Sb anode and $NaNi_{0.35}Mn_{0.35}Fe_{0.3}O_2$ cathode have been recently examined.[205] After 10 cycles, the Sb-based anode in FEC-free 0.8 M $NaPF_6$/TMP (trimethyl phosphate) electrolyte showed no or indiscernible SEI film. In contrast, the surface of the Sb-based anode in FEC-containing electrolyte suggests the presence of SEI coating with FEC decomposition products confirmed by XPS. For the $NaNi_{0.35}Mn_{0.35}Fe_{0.3}O_2$ cathode in electrolyte with or without FEC, a distinct peak at 687.7 eV in F 1 s spectra is observed in XPS, corresponding to some organic fluorides from the PVDF binder. A new peak appears in FEC-containing electrolyte that can be attributed to NaF. This indicates that FEC decomposition also contributes to the SEI formation on the cathode. Thus, XPS results demonstrate that FEC in TMP electrolyte can form stable SEI layers on both cathode and anode, which effectively suppresses TMP decomposition, improves the initial efficiency and reversible capacity of the electrodes.[205] An elegant ex situ study coupled SEM and XPS to study the formation of SEI on Sn particles where the SEI polymeric layer was found to completely cover the particle at the end of sodiation and helps to limit the damage to particle structure, that is, fracturing of the particles due to large volume change.[206]

AFM or scanning probe microscopy (SPM) is a scanning-based technique for surfaces (with 3D capability) and can be tailored for a battery system, where the probe acts as an electrode in some cases. AFMs can be used to measure the forces between the probe and the sample as a function of their mutual separation (in the simplest setup) providing a high-resolution image of the "3-D" shape of a sample surface. In situ AFM and optical images were used to analyze the electrochemical deposition of Na metal at the anode attempting to characterize the three stages—nucleation, growth, and appearance of dendrites.[122] The addition of FEC to the electrolyte, which promotes the growth of a NaF-containing SEI with

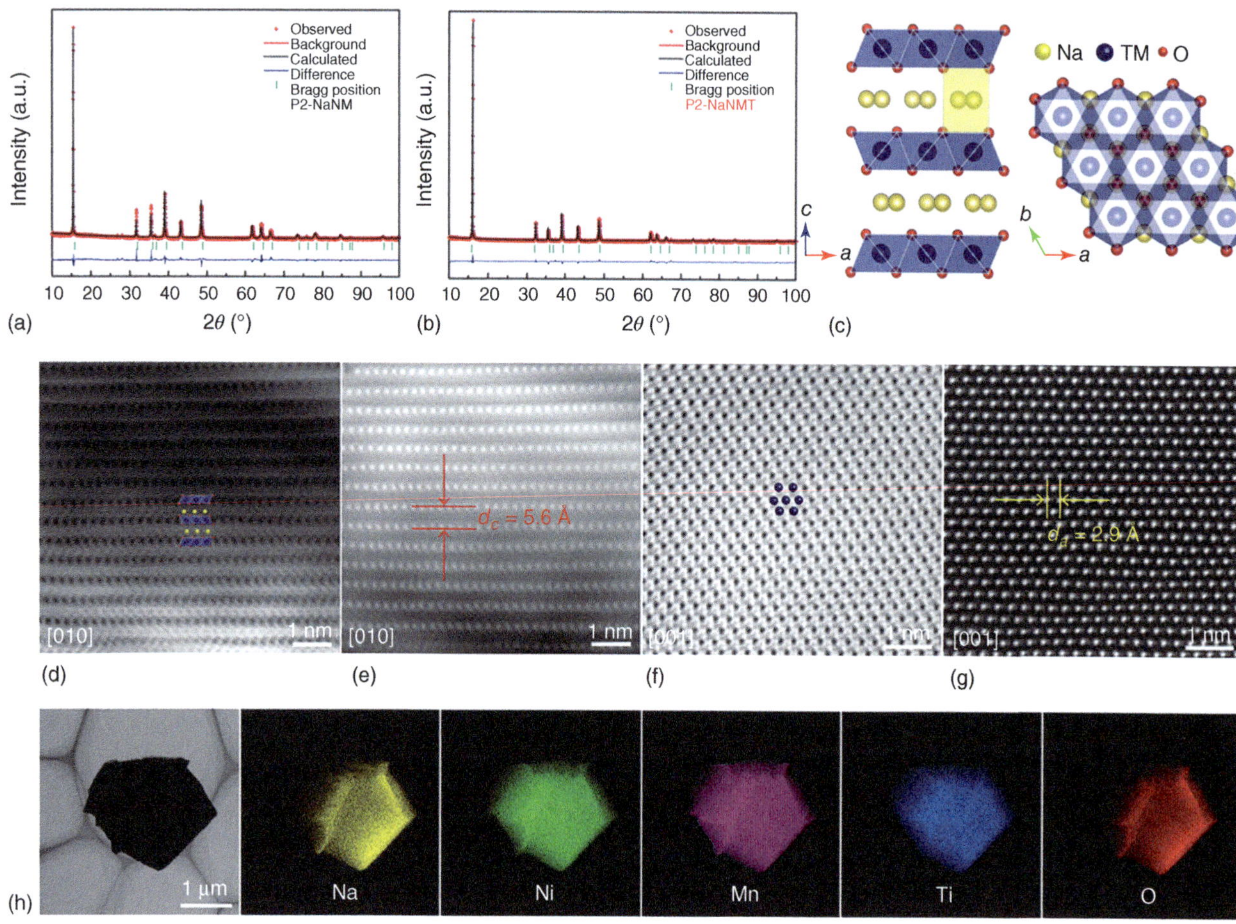


**Figure 30** XRD and Rietveld plots of (a) P2-$Na_{2/3}Ni_{1/3}Mn_{2/3}O_2$ and (b) P2-$Na_{2/3}Ni_{1/3}Mn_{1/3}Ti_{1/3}O_2$ with (c) crystal structure of the P2 phase samples along the *a* and *c* axes. (d) ABF and (e) HAADF-STEM image of P2-$Na_{2/3}Ni_{1/3}Mn_{1/3}Ti_{1/3}O_2$ along the [010] zone axis. (f) ABF and (g) HAADF-STEM image of P2-$Na_{2/3}Ni_{1/3}Mn_{1/3}Ti_{1/3}O_2$ along the [001] zone axis. (h) TEM image and EDX spectroscopy maps of P2-$Na_{2/3}Ni_{1/3}Mn_{1/3}Ti_{1/3}O_2$. [Reproduced with permission from Ref. 201 © Science Advances 2018]

homogenous morphology and high modulus, leads to stable sodium deposition and high coulombic efficiency (88% over 50 cycles) of the sodium-metal anode.[122] This directly shows the mechanism by which the deposition of sodium can be controlled and is highly applicable to Na-metal batteries and future Na–$O_2$ or Na–air batteries. *Operando* AFM probed the SEI formation on $MoS_2$ flakes in NIBs. Prior to Na insertion, two layers of SEI layer were found to form with thickness ranging between 8 and 54 nm (20.4 ± 10.9 nm on average, where quantitative force spectroscopy was used to evaluate the thickness).[207]

## 6 CONCLUSIONS

There is significant potential to realize commercial sodium-ion batteries that may compete with LIBs and other battery technologies for specialized applications. This is underpinned by materials discovery and a mechanistic understanding of the function of each component inside a battery and holistically of the entire battery itself. This chapter provides a snapshot on the materials used for various components, electrodes and electrolytes and how these components evolve and interact. The key to sodium-ion battery development lies in mechanistic understanding, and this was elaborated throughout the chapter. The future is exciting for battery technology, and sodium-ion batteries are likely to play an important part in the battery mix for the future.

## 7 ACKNOWLEDGMENTS

This work was financially supported by the Australian Research Council DECRA (DE160100237) and DP (DP170100269) programs. JS would like to thank the

Australian Nuclear Science and Technology Organisation (ANSTO) and UNSW for PhD supporting funds. We would also like to thank Emily Cheung and Damian Goonetilleke for support during this work.

## 8 ABBREVIATIONS AND ACRONYMS

ABF = annular bright field; AC = Activated carbon; AFM = atomic force microscopy; CEI = cathodic electrolyte interface; CNF = carbon nanofiber; CNT = carbon nanotube; CV = cyclic voltammetry; DFT = Density functional theory; EC = ethylene carbonate; EDX = energy dispersive spectroscopy; EIS = electrochemical impedance spectroscopy; EXAFS = extended X-ray absorption fine structure; FEC = fluoroethylene carbonate; FIB-HIM = focused ion beam–helium ion scanning microscope; FLG = few-layered graphene; FTIR = Fourier transform infrared spectroscopy; GNS = graphene nanosheet; HAADF = high-angle annular dark field; HCNW = hollow carbon nanowire; LIBs = lithium-ion batteries; MCMB = mesocarbon microbead; MOF = metal–organic framework; MV = methyl viologen; NASICON = Na superionic conductor; ND = neutron diffraction; NMR = nuclear magnetic resonance; NW = nanowire; OEMS = Online electrochemical mass spectrometry; O = octahedral; PC = propylene carbonate; PDF = pair distribution function; PEO = polyethylene oxide; PVDF = polyvinylidene difluoride; P = prismatic; SEI = solid–electrolyte interphase; SEM = scanning electron microscopy; SPI = surface permeable interface; SPM = scanning probe microscopy; SS-NMR = solid-state nuclear magnetic resonance; STEM = scanning transmission electron microscopy; STSG = solvothermal synthesized graphene; TEGO = thermally exfoliated graphite oxide; TEM = Transmission electron microscopy; TMP = trimethyl phosphate; TOF-SIMS = time-of-flight, secondary-ion mass spectrometry; XANES = X-ray absorption near-edge spectroscopy; XAS = X-ray absorption spectroscopy; XPS = X-ray photoelectron spectroscopy; XRD = X-ray powder diffraction

## 9 REFERENCES


1. T. Nagaura and K. Tozawa, *Progr. Batt. Solar Cells*, 1990, **9**, 209.
2. M. L. Foo, T. Klimczuk and R. J. Cava, *Mater. Res. Bull.*, 2005, **40**, 665.
3. S. Park, Y. Lee, W. Si and T. Vogt, *Solid State Commun.*, 2005, **134**, 607.
4. K. Takada, H. Sakurai, E. Takayama-Muromachi, F. Izumi, R. A. Dilanian and T. Sasaki, *J. Solid State Chem.*, 2004, **177**, 372.
5. C. Delmas, C. Fouassier and P. Hagenmuller, *Physica B+C*, 1980, **99**, 81.
6. M. L. Foo, R. E. Schaak, V. L. Miller, T. Klimczuk, N. S. Rogado, Y. Wang, G. C. Lau, C. Craley, H. W. Zandbergen, N. P. Ong and R. J. Cava, *Solid State Commun.*, 2003, **127**, 33.
7. R. Ishikawa, Y. Ono, Y. Miyazaki and T. Kajitani, *Jpn. J. Appl. Phys.*, 2002, **41**, L337.
8. M. S. Whittingham, *Chem. Rev.*, 2004, **104**, 4271.
9. J. Jorgensen, M. Avdeev, D. Hinks, J. Burley and S. Short, *Phys. Rev. B*, 2003, **68**, 1.
10. S. Wenzel, T. Hara, J. Janek and P. Adelhelm, *Energy Environ. Sci.*, 2011, **4**, 3342.
11. Y. You and A. Manthiram, *Adv. Energy Mater.*, 2018, **8**, 1701785.
12. R. Berthelot, D. Carlier and C. Delmas, *Nat. Mater.*, 2011, **10**, 74.
13. B. Venkata, R. Reddy, R. Ravikumar, C. Nithya and S. Gopukumar, *J. Mater. Chem. A*, 2015, **3**, 18059.
14. M. Sathiya, K. Hemalatha, K. Ramesha, J. M. Tarascon and A. S. Prakash, *Chem. Mater.*, 2012, **24**, 1846.
15. Q. Huang, M. L. Foo, R. A. Pascal, J. W. Lynn, B. H. Toby, T. He, H. W. Zandbergen and R. J. Cava, *Phys. Rev. B Condens. Matter*, 2004, **70**, 184110.
16. J. M. Tarascon and M. Armand, *Nature*, 2001, **414**, 359.
17. L. Chen, M. Fiore, J. E. Wang, R. Ruffo, D.-K. Kim and G. Longoni, *Adv. Sustainable Syst.*, 2018, 1700153. DOI: 10.1002/adsu.201700153.
18. Y. Fang, Z. Chen, L. Xiao, X. Ai, Y. Cao and H. Yang, *Small*, 2018, **14**, 1703116.
19. P. K. Nayak, L. Yang, W. Brehm and P. Adelhelm, *Angew. Chem. Int. Ed.*, 2017, **57**, 102.
20. T. Shibata, Y. Fukuzumi, W. Kobayashi and Y. Moritomo, *Sci. Rep.*, 2015, **5**, 9006.
21. N. Yabuuchi, M. Kajiyama, J. Iwatate, H. Nishikawa, S. Hitomi, R. Okuyama, R. Usui, Y. Yamada and S. Komaba, *Nat. Mater.*, 2012, **11**, 512.
22. X. Ma, H. Chen and G. Ceder, *J. Electrochem. Soc.*, 2011, **158**, A1307.
23. W. M. Dose, N. Sharma, J. C. Pramudita, M. Avdeev, E. Gonzalo and T. Rojo, *Chem. Mater.*, 2018. DOI: 10.1021/acs.chemmater.8b02456
24. M. H. Han, E. Gonzalo, N. Sharma, J. M. López Del Amo, M. Armand, M. Avdeev, J. J. Saiz Garitaonandia and T. Rojo, *Chem. Mater.*, 2016, **28**, 106.
25. W. M. Dose, N. Sharma, J. C. Pramudita, H. E. A. Brand, E. Gonzalo and T. Rojo, *Chem. Mater.*, 2017, **29**, 7416.
26. J. Billaud, G. Singh, A. R. Armstrong, E. Gonzalo, V. Roddatis, M. Armand, T. Rojo and P. G. Bruce, *Energy Environ. Sci.*, 2014, **7**, 1387.
27. S. Guo, Q. Li, P. Liu, M. Chen and H. Zhou, *Nat. Commun.*, 2017, **8**, 135.

28. F. Wang, Y. Lin, L. Suo, X. Fan, T. Gao, C. Yang, F. Han, Y. Qi, K. Xu and C. Wang, *Energy Environ. Sci.*, 2016, **9**, 3666.
29. K. Kaliyappan, J. Liu, A. Lushington, R. Li and X. Sun, *ChemSusChem*, 2015, **8**, 2537.
30. H. R. Yao, P. F. Wang, Y. Gong, J. Zhang, X. Yu, L. Gu, C. OuYang, Y. X. Yin, E. Hu, X. Q. Yang, E. Stavitski, Y. G. Guo and L. J. Wan, *J. Am. Chem. Soc.*, 2017, **139**, 8440.
31. E. de la Llave, V. Borgel, K. J. Park, J. Y. Hwang, Y. K. Sun, P. Hartmann, F. F. Chesneau and D. Aurbach, *ACS Appl. Mater. Interfaces*, 2016, **8**, 1867.
32. V. Duffort, E. Talaie, R. Black and L. F. Nazar, *Chem. Mater.*, 2015, **27**, 2515.
33. S. Komaba, N. Yabuuchi, T. Nakayama, A. Ogata, T. Ishikawa and I. Nakai, *Inorg. Chem.*, 2012, **51**, 6211.
34. P. Vassilaras, S. T. Dacek, H. Kim, T. T. Fister, S. Kim, G. Ceder and J. C. Kim, *J. Electrochem. Soc.*, 2017, **164**, A3484.
35. P.-F. Wang, Y. You, Y.-X. Yin and Y.-G. Guo, *J. Mater. Chem. A*, 2016, **4**, 17660.
36. P.-F. Wang, Y. You, Y.-X. Yin and Y.-G. Guo, *Adv. Energy Mater.*, 2018, **8**, 1701912.
37. H. Wang, R. Gao, Z. Li, L. Sun, Z. Hu and X. Liu, *Inorg. Chem.*, 2018, **57**, 5249.
38. D. H. Lee, J. Xu and Y. S. Meng, *Phys. Chem. Chem. Phys.*, 2013, **15**, 3304.
39. S. Guo, P. Liu, H. Yu, Y. Zhu, M. Chen, M. Ishida and H. Zhou, *Angew. Chem. Int. Ed.*, 2015, **54**, 5894.
40. M. Keller, D. Buchholz and S. Passerini, *Adv. Energy Mater.*, 2016, **6**, 1501555.
41. J.-Y. Hwang, S.-T. Myung and Y.-K. Sun, *Chem. Soc. Rev.*, 2017, **46**, 3529.
42. J. Wang and X. Sun, *Energy Environ. Sci.*, 2015, **8**, 1110.
43. J. N. Bridson, S. E. Quinlan and P. R. Tremaine, *Chem. Mater.*, 1998, **10**, 763.
44. K. T. Lee, T. N. Ramesh, F. Nan, G. Botton and L. F. Nazar, *Chem. Mater.*, 2011, **23**, 3593.
45. S. P. Ong, V. L. Chevrier, G. Hautier, A. Jain, C. Moore, S. Kim, X. Ma and G. Ceder, *Energy Environ. Sci.*, 2011, **4**, 3680.
46. J. Kim, D.-H. Seo, H. Kim, I. Park, J.-K. Yoo, S.-K. Jung, Y.-U. Park, W. A. Goddard Iii and K. Kang, *Energy Environ. Sci.*, 2015, **8**, 540.
47. M. M. Rahman, I. Sultana, S. Mateti, J. Liu, N. Sharma and Y. Chen, *J. Mater. Chem. A*, 2017, **5**, 16616.
48. S.-M. Oh, S.-T. Myung, J. Hassoun, B. Scrosati and Y.-K. Sun, *Electrochem. Commun.*, 2012, **22**, 149.
49. M. Casas-Cabanas, V. V. Roddatis, D. Saurel, P. Kubiak, J. Carretero-González, V. Palomares, P. Serras and T. Rojo, *J. Mater. Chem.*, 2012, **22**, 17421.
50. Y. Zhang, H. Zhao and Y. Du, *J. Mater. Chem. A*, 2016, **4**, 7155.
51. Z. Jian, W. Han, X. Lu, H. Yang, Y.-S. Hu, J. Zhou, Z. Zhou, J. Li, W. Chen, D. Chen and L. Chen, *Adv. Energy Mater.*, 2013, **3**, 156.
52. W. Song, X. Ji, Z. Wu, Y. Zhu, Y. Yang, J. Chen, M. Jing, F. Li and C. E. Banks, *J. Mater. Chem. A*, 2014, **2**, 5358.
53. D. Guo, J. Qin, Z. Yin, J. Bai, Y.-K. Sun and M. Cao, *Nano Energy*, 2018, **45**, 136.
54. P. Barpanda, G. Liu, C. D. Ling, M. Tamaru, M. Avdeev, S.-C. Chung, Y. Yamada and A. Yamada, *Chem. Mater.*, 2013, **25**, 3480.
55. H. Kim, R. A. Shakoor, C. Park, S. Y. Lim, J.-S. Kim, Y. N. Jo, W. Cho, K. Miyasaka, R. Kahraman, Y. Jung and J. W. Choi, *Adv. Funct. Mater.*, 2013, **23**, 1147.
56. L. Yu, L. P. Wang, H. Liao, J. Wang, Z. Feng, O. Lev, J. S. C. Loo, M. T. Sougrati and Z. J. Xu, *Small*, 2018, **14**, 1703338.
57. R. A. Shakoor, C. S. Park, A. A. Raja, J. Shin and R. Kahraman, *Phys. Chem. Chem. Phys.*, 2016, **18**, 3929.
58. V. Palomares, M. Casas-Cabanas, E. Castillo-Martínez, M. H. Han and T. Rojo, *Energy Environ. Sci.*, 2013, **6**, 2312.
59. H. Kim, I. Park, D. H. Seo, S. Lee, S. W. Kim, W. J. Kwon, Y. U. Park, C. S. Kim, S. Jeon and K. Kang, *J. Am. Chem. Soc.*, 2012, **134**, 10369.
60. R. Gover, A. Bryan, P. Burns and J. Barker, *Solid State Ionics*, 2006, **177**, 1495.
61. B. L. Ellis, W. R. Makahnouk, Y. Makimura, K. Toghill and L. F. Nazar, *Nat. Mater.*, 2007, **6**, 749.
62. Q. Liu, D. Wang, X. Yang, N. Chen, C. Wang, X. Bie, Y. Wei, G. Chen and F. Du, *J. Mater. Chem. A*, 2015, **3**, 21478.
63. M. Bianchini, F. Fauth, N. Brisset, F. Weill, E. Suard, C. Masquelier and L. Croguennec, *Chem. Mater.*, 2015, **27**, 3009.
64. R. Tripathi, S. M. Wood, M. S. Islam and L. F. Nazar, *Energy Environ. Sci.*, 2013, **6**, 2257.
65. G.-C. Ri, S.-H. Choe and C.-J. Yu, *J. Power Sources*, 2018, **378**, 375.
66. P. Barpanda, G. Oyama, S. Nishimura, S. C. Chung and A. Yamada, *Nat. Commun.*, 2014, **5**, 4358.
67. A. Kumar, S. M. Yusuf and L. Keller, *Phys. Rev. B Condens. Matter*, 2005, **71**, article ID: 054414.
68. J. C. Pramudita, S. Schmid, T. Godfrey, T. Whittle, M. Alam, T. Hanley, H. E. A. Brand and N. Sharma, *Phys. Chem. Chem. Phys.*, 2014, **16**, 24178.
69. S. Wang, L. Wang, Z. Zhu, Z. Hu, Q. Zhao and J. Chen, *Angew. Chem. Int. Ed.*, 2014, **53**, 5892.
70. J. R. Dahn, T. Zheng, Y. Liu and J. S. Xue, *Science*, 1995, **270**, 590.
71. E. Irisarri, A. Ponrouch and M. R. Palacin, *J. Electrochem. Soc.*, 2015, **162**, A2476.
72. S. Komaba, W. Murata, T. Ishikawa, N. Yabuuchi, T. Ozeki, T. Nakayama, A. Ogata, K. Gotoh and K. Fujiwara, *Adv. Funct. Mater.*, 2011, **21**, 3859.

73. D. A. Stevens and J. R. Dahn, *J. Electrochem. Soc.*, 2001, **148**, A803.
74. K. Gotoh, T. Ishikawa, S. Shimadzu, N. Yabuuchi, S. Komaba, K. Takeda, A. Goto, K. Deguchi, S. Ohki, K. Hashi, T. Shimizu and H. Ishida, *J. Power Sources*, 2013, **225**, 137.
75. D. A. Stevens and J. R. Dahn, *J. Electrochem. Soc.*, 2000, **147**, 1271.
76. R. Suresh Babu and M. Pyo, *J. Electrochem. Soc.*, 2014, **161**, A1045.
77. C. Matei Ghimbeu, J. Górka, V. Simone, L. Simonin, S. Martinet and C. Vix-Guterl, *Nano Energy*, 2018, **44**, 327.
78. X.-F. Luo, C.-H. Yang, Y.-Y. Peng, N.-W. Pu, M.-D. Ger, C.-T. Hsieh and J.-K. Chang, *J. Mater. Chem. A*, 2015, **3**, 10320.
79. P. Bai, Y. He, P. Xiong, X. Zhao, K. Xu and Y. Xu, *Energy Storage Mater.*, 2018, **13**, 274.
80. J. Xu, M. Wang, N. P. Wickramaratne, M. Jaroniec, S. Dou and L. Dai, *Adv. Mater.*, 2015, **27**, 2042.
81. C. Ling and F. Mizuno, *Phys. Chem. Chem. Phys.*, 2014, **16**, 10419.
82. Y. Yang, D.-M. Tang, C. Zhang, Y. Zhang, Q. Liang, S. Chen, Q. Weng, M. Zhou, Y. Xue, J. Liu, J. Wu, Q. H. Cui, C. Lian, G. Hou, F. Yuan, Y. Bando, D. Golberg and X. Wang, *Energy Environ. Sci.*, 2017, **10**, 979.
83. L. Qie, W. Chen, X. Xiong, C. Hu, F. Zou, P. Hu and Y. Huang, *Adv. Sci.*, 2015, **2**, 1500195.
84. Y. Liu, V. I. Artyukhov, M. Liu, A. R. Harutyunyan and B. I. Yakobson, *J. Phys. Chem. Lett.*, 2013, **4**, 1737.
85. W. S. Hummers and R. E. Offeman, *J. Am. Chem. Soc.*, 1958, **80**, 1339.
86. Y. Wen, K. He, Y. Zhu, F. Han, Y. Xu, I. Matsuda, Y. Ishii, J. Cumings and C. Wang, *Nat. Commun.*, 2014, **5**, 4033.
87. K. Tang, L. Fu, R. J. White, L. Yu, M.-M. Titirici, M. Antonietti and J. Maier, *Adv. Energy Mater.*, 2012, **2**, 873.
88. Y. Cao, L. Xiao, M. L. Sushko, W. Wang, B. Schwenzer, J. Xiao, Z. Nie, L. V. Saraf, Z. Yang and J. Liu, *Nano Lett.*, 2012, **12**, 3783.
89. B. Zhang, C. M. Ghimbeu, C. Laberty, C. Vix-Guterl and J.-M. Tarascon, *Adv. Energy Mater.*, 2016, **6**, 1501588.
90. L. Wu, X. Hu, J. Qian, F. Pei, F. Wu, R. Mao, X. Ai, H. Yang and Y. Cao, *Energy Environ. Sci.*, 2014, **7**, 323.
91. L. Xiao, Y. Cao, J. Xiao, W. Wang, L. Kovarik, Z. Nie and J. Liu, *Chem. Commun. (Camb.)*, 2012, **48**, 3321.
92. H. Kang, Y. Liu, K. Cao, Y. Zhao, L. Jiao, Y. Wang and H. Yuan, *J. Mater. Chem. A*, 2015, **3**, 17899.
93. J. Conder, C. Marino, P. Novák and C. Villevieille, *J. Mater. Chem. A*, 2018. DOI: 10.1039/c7ta10622j.
94. M. Lao, Y. Zhang, W. Luo, Q. Yan, W. Sun and S. X. Dou, *Adv. Mater.*, 2017, **29**, 1700622.
95. Y. Liu, N. Zhang, L. Jiao, Z. Tao and J. Chen, *Adv. Funct. Mater.*, 2015, **25**, 214.
96. Z. Li, J. Ding and D. Mitlin, *Acc. Chem. Res.*, 2015, **48**, 1657.
97. R. Alcántara, M. Jaraba, P. Lavela and J. L. Tirado, *Chem. Mater.*, 2002, **14**, 2847.
98. N. Zhang, X. Han, Y. Liu, X. Hu, Q. Zhao and J. Chen, *Adv. Energy Mater.*, 2015, **5**, 1401123.
99. Z. Jian, P. Liu, F. Li, M. Chen and H. Zhou, *J. Mater. Chem. A*, 2014, **2**, 13805.
100. F. Klein, B. Jache, A. Bhide and P. Adelhelm, *Phys. Chem. Chem. Phys.*, 2013, **15**, 15876.
101. Z. Yang, D. Choi, S. Kerisit, K. M. Rosso, D. Wang, J. Zhang, G. Graff and J. Liu, *J. Power Sources*, 2009, **192**, 588.
102. L. Wu, D. Buchholz, D. Bresser, L. Gomes Chagas and S. Passerini, *J. Power Sources*, 2014, **251**, 379.
103. J. H. Kim, W. Choi, H.-G. Jung, S. H. Oh, K. Y. Chung, W. I. Cho, I. H. Oh and I. W. Nah, *J. Alloys Compd.*, 2017, **690**, 390.
104. W. Li, M. Fukunishi, B. J. Morgan, O. J. Borkiewicz, K. W. Chapman, V. Pralong, A. Maignan, O. I. Lebedev, J. Ma, H. Groult, S. Komaba and D. Dambournet, *Chem. Mater.*, 2017, **29**, 1836.
105. L. Wu, D. Bresser, D. Buchholz, G. A. Giffin, C. R. Castro, A. Ochel and S. Passerini, *Adv. Energy Mater.*, 2015, **5**, 1401142.
106. P. Senguttuvan, G. Rousse, V. Seznec, J.-M. Tarascon and M. R. Palacín, *Chem. Mater.*, 2011, **23**, 4109.
107. M. A. Tsiamtsouri, P. K. Allan, A. J. Pell, J. M. Stratford, G. Kim, R. N. Kerber, P. C. M. M. Magusin, D. A. Jefferson and C. P. Grey, *Chem. Mater.*, 2018, **30**, 1505.
108. G. Rousse, M. E. Arroyo-de Dompablo, P. Senguttuvan, A. Ponrouch, J.-M. Tarascon and M. R. Palacín, *J. Mater. Chem.*, 2013, **25**, 4946.
109. J. Ni, S. Fu, C. Wu, Y. Zhao, J. Maier, Y. Yu and L. Li, *Adv. Energy Mater.*, 2016, **6**, 1502568.
110. A. Rudola, K. Saravanan, S. Devaraj, H. Gong and P. Balaya, *Chem. Commun. (Camb.)*, 2013, **49**, 7451.
111. R. Dominko, E. Baudrin, P. Umek, D. Arčon, M. Gaberšček and J. Jamnik, *Electrochem. Commun.*, 2006, **8**, 673.
112. X. Li, X. Zhu, J. Liang, Z. Hou, Y. Wang, N. Lin, Y. Zhu and Y. Qian, *J. Electrochem. Soc.*, 2014, **161**, A1181.
113. J. C. Perez-Flores, F. Garcia-Alvarado, M. Hoelzel, I. Sobrados, J. Sanz and A. Kuhn, *Dalton Trans.*, 2012, **41**, 14633.
114. C. Ling and R. Zhang, *Phys. Chem. Chem. Phys.*, 2017, **19**, 10036.
115. K. Kubota and S. Komaba, *J. Electrochem. Soc.*, 2015, **162**, A2538.

116. G. Pang, C. Yuan, P. Nie, B. Ding, J. Zhu and X. Zhang, *Nanoscale*, 2014, **6**, 6328.
117. H. Geng, J. Yang, H. Yu, C. Li and X. Dong, *Mater. Chem. Front.*, 2017, **1**, 1435.
118. L. Wang, B. Wang, G. Liu, T. Liu, T. Gao and D. Wang, *RSC Adv.*, 2016, **6**, 70277.
119. Y. Jiang, J. Shi, M. Wang, L. Zeng, L. Gu and Y. Yu, *ACS Appl. Mater. Interfaces*, 2016, **8**, 689.
120. X. Xu, C. Niu, M. Duan, X. Wang, L. Huang, J. Wang, L. Pu, W. Ren, C. Shi, J. Meng, B. Song and L. Mai, *Nat. Commun.*, 2017, **8**, 460.
121. A. Basile, M. Hilder, F. Makhlooghiazad, C. Pozo-Gonzalo, D. R. MacFarlance, P. C. Howlett and M. Forsyth, *Adv. Energy Mater.*, 2018, **8**, 1703491.
122. M. Han, C. Zhu, T. Ma, Z. Pan, Z. Tao and J. Chen, *Chem. Commun. (Camb.)*, 2018, **54**, 2381.
123. V. Palomares, P. Serras, I. Villaluenga, K. B. Hueso, J. Carretero-Gonzalez and T. Rojo, *Energy Environ. Sci.*, 2012, **5**, 5884.
124. C. Zhao, L. Liu, X. Qi, Y. Lu, F. Wu, J. Zhao, Y. Yu, Y. S. Hu and L. Chen, *Adv. Energy Mater.*, 2018, **8**, 1703012.
125. G. Avall, J. Mindemark, D. Brandell and P. Johansson, *Adv. Energy Mater.*, 2018, **8**, 17030036.
126. J. B. Goodenough, H. Y. P. Hong and J. A. Kafalas, *Mater. Res. Bull.*, 1976, **11**, 203.
127. H. Y. P. Hong, *Mater. Res. Bull.*, 1976, **11**, 173.
128. F. Lalère, J. B. Leriche, M. Courty, S. Boulineau, V. Viallet, C. Masquelier and V. Seznec, *J. Power Sources*, 2014, **247**, 975.
129. S. Song, H. M. Duong, A. M. Korsunsky, N. Hu and L. Lu, *Sci. Rep.*, 2016, **6**, 32330.
130. R. O. Fuentes, F. M. Figueiredo, F. M. B. Marques and J. I. Franco, *Solid State Ionics*, 2001, **140**, 173.
131. J. S. Thorne, S. Chowdhury, R. A. Dunlap and M. N. Obrovac, *J. Electrochem. Soc.*, 2014, **161**, A1801.
132. K. Yubuta, Y.-K. Hasegawa, Y. Miyazaki and T. Kajitani, *Jpn. J. Appl. Phys.*, 2007, **46**, 712.
133. K. Knížek, Z. Jirák, J. Hejtmánek, P. Brázda, J. Buršík, M. Soroka and P. Beran, *J. Solid State Chem.*, 2015, **229**, 160.
134. E. Gonzalo, M. H. Han, J. M. López del Amo, B. Acebedo, M. Casas-Cabanas and T. Rojo, *J. Mater. Chem. A*, 2014, **2**, 18523.
135. A. G. Jolley, G. Cohn, G. T. Hitz and E. D. Wachsman, *Ionics*, 2015, **21**, 3031.
136. Z. Khakpour, *Electrochim. Acta*, 2016, **196**, 337.
137. Q. Ma, M. Guin, S. Naqash, C.-L. Tsai, F. Tietz and O. Guillon, *Chem. Mater.*, 2016, **28**, 4821.
138. M. Samiee, B. Radhakrishnan, Z. Rice, Z. Deng, Y. S. Meng, S. P. Ong and J. Luo, *J. Power Sources*, 2017, **347**, 229.
139. A. Bhide, J. Hofmann, A. Katharina Durr, J. Janek and P. Adelhelm, *Phys. Chem. Chem. Phys.*, 2014, **16**, 1987.
140. M. Tatsumisago, M. Nagao and A. Hayashi, *J. Asian Ceram. Soc.*, 2013, **1**, 17.
141. K. Noi, A. Hayashi and M. Tatsumisago, *J. Power Sources*, 2014, **269**, 260.
142. A. Hayashi, K. Noi, A. Sakuda and M. Tatsumisago, *Nat. Commun.*, 2012, **3**, 856.
143. Z. Zhang, E. Ramos, F. Lalère, A. Assoud, K. Kaup, P. Hartman and L. F. Nazar, *Energy Environ. Sci.*, 2018, **11**, 87.
144. M. Duchardt, U. Ruschewitz, S. Adams, S. Dehnen and B. Roling, *Angew. Chem.*, 2018, **57**, 1351.
145. R. P. Rao, H. Chen, L. L. Wong and S. Adams, *J. Mater. Chem. A*, 2018, **5**, 3377.
146. M. A. Evstigneeva, V. B. Nalbandyan, A. A. Petrenko, B. S. Medvedev and A. A. Kataev, *Chem. Mater.*, 2011, **23**, 1174.
147. M. Y. Avdeev, V. B. Nalbandyan and B. S. Medvedev, *Inorg. Mater. (Engl. Transl.)*, 1997, **33**, 500.
148. H. Engstrom, J. B. Bates, W. E. Brundage and J. C. Wang, *Solid State Ionics*, 1981, **2**, 265.
149. K. K. Kim, J. N. Mundy and W. K. Chen, *J. Phys. Chem. Solids*, 1979, **40**, 743.
150. H. Rickert, *Angew. Chem. Int. Ed. Engl.*, 1978, **17**, 37.
151. E. Peled, *J. Electrochem. Soc.*, 1979, **126**, 2047.
152. K. Edström, T. Gustafsson and J. O. Thomas, *Electrochim. Acta*, 2004, **50**, 397.
153. D. Aurbach, *J. Power Sources*, 2000, **89**, 206.
154. S. Shi, P. Lu, Z. Liu, Y. Qi, L. G. Hector, H. Li and S. J. Harris, *J. Am. Chem. Soc.*, 2012, **134**, 15476.
155. A. M. Tripathi, W.-N. Su and B. J. Hwang, *Chem. Soc. Rev.*, 2018, **47**, 736.
156. S.-K. Hu, G.-H. Cheng, M.-Y. Cheng, B.-J. Hwang and R. Santhanam, *J. Power Sources*, 2009, **188**, 564.
157. B. J. Hwang, C. Y. Chen, M. Y. Cheng, R. Santhanam and K. Ragavendran, *J. Power Sources*, 2010, **195**, 4255.
158. T. Takamura, K. Sumiya, J. Suzuki, C. Yamada and K. Sekine, *J. Power Sources*, 1999, **81–82**, 368.
159. H. Wu, Y. Cao, L. Geng and C. Wang, *Chem. Mater.*, 2017, **29**, 3572.
160. B. D. Assresahegn, B. D. Ossonon and D. Bélanger, *J. Power Sources*, 2018, **391**, 41.
161. C.-F. Lin, M. Noked, A. C. Kozen, C. Liu, O. Zhao, K. Gregorczyk, L. Hu, S. B. Lee and G. W. Rubloff, *ACS Nano*, 2016, **10**, 2693.
162. M. R. Jo, G.-H. Lee and Y.-M. Kang, *ACS Appl. Mater. Interfaces*, 2015, **7**, 27934.
163. Y. Orikasa, D. Takamatsu, K. Yamamoto, Y. Koyama, S. Mori, T. Masese, T. Mori, T. Minato, H. Tanida, T. Uruga, Z. Ogumi and Y. Uchimoto, *Adv. Energy Mater.*, 2014, **1**, 1400195.

164. J. Yan, B.-J. Xia, Y.-C. Su, X.-Z. Zhou, J. Zhang and X.-G. Zhang, *Electrochim. Acta*, 2008, **53**, 7069.

165. S. J. An, J. Li, C. Daniel, D. Mohanty, S. Nagpure and D. L. Wood, *Carbon*, 2016, **105**, 52.

166. R. Fong, U. von Sacken and J. R. Dahn, *J. Electrochem. Soc.*, 1990, **137**, 2009.

167. M. K. Datta, R. Epur, P. Saha, K. Kadakia, S. K. Park and P. N. Kumta, *J. Power Sources*, 2013, **225**, 316.

168. M. M. Doeff, Y. Ma, S. J. Visco and L. C. De Jonghe, *J. Electrochem. Soc.*, 1993, **140**, L169.

169. A. Ponrouch, A. R. Goñi and M. R. Palacín, *Electrochem. Commun.*, 2013, **27**, 85.

170. P. Thomas, J. Ghanbaja and D. Billaud, *Electrochim. Acta*, 1999, **45**, 423.

171. J. C. Pramudita, A. Rawal, M. Choucair, D. Pontiroli, G. Magnani, M. Gaboardi, M. Riccò and N. Sharma, *ACS Appl. Mater. Interfaces*, 2017, **9**, 431.

172. M. Goktas, C. Bolli, E. J. Berg, P. Novák, K. Pollok, F. Langenhorst, M. V. Roeder, O. Lenchuk, D. Mollenhauer and P. Adelhelm, *Adv. Energy Mater.*, 2018, **8**, 1702724.

173. L. Seidl, N. Bucher, E. Chu, S. Hartung, S. Martens, O. Schneider and U. Stimming, *Energy Environ. Sci.*, 2017, **10**, 1631.

174. A. Ponrouch, E. Marchante, M. Courty, J.-M. Tarascon and M. R. Palacin, *Energy Environ. Sci.*, 2012, **5**, 8572.

175. S. Komaba, T. Ishikawa, N. Yabuuchi, W. Murata, A. Ito and Y. Ohsawa, *ACS Appl. Mater. Interfaces*, 2011, **3**, 4165.

176. M. A. Muñoz-Márquez, M. Zarrabeitia, E. Castillo-Martínez, A. Eguía-Barrio, T. Rojo and M. Casas-Cabanas, *ACS Appl. Mater. Interfaces*, 2015, **7**, 7801.

177. Q. Gu, J. A. Kimpton, H. E. A. Brand, Z. Wang and S. Chou, *Adv. Energy Mater.*, 2017, **7**, 1602831.

178. B. Mortemard de Boisse, D. Carlier, M. Guignard, L. Bourgeois and C. Delmas, *Inorg. Chem.*, 2014, **53**, 11197.

179. W. K. Pang, S. Kalluri, V. K. Peterson, N. Sharma, J. Kimpton, B. Johannessen, H. K. Liu, S. X. Dou and Z. Guo, *Chem. Mater.*, 2015, **27**, 3150.

180. N. Sharma and V. K. Peterson, *J. Power Sources*, 2013, **244**, 695.

181. W. R. Brant, S. Schmid, G. Du, H. E. A. Brand, W. K. Pang, V. K. Peterson, Z. Guo and N. Sharma, *J. Vis. Exp.*, 2014, 52284. DOI: 10.3791/52284.

182. N. Sharma, G. Du, A. J. Studer, Z. Guo and V. K. Peterson, *Solid State Ionics*, 2011, **199–200**, 37.

183. M. Roberts, J. J. Biendicho, S. Hull, P. Beran, T. Gustafsson, G. Svensson and K. Edström, *J. Power Sources*, 2013, **226**, 249.

184. S. Taminato, M. Yonemura, S. Shiotani, T. Kamiyama, S. Torii, M. Nagao, Y. Ishikawa, K. Mori, T. Fukunaga, Y. Onodera, T. Naka, M. Morishima, Y. Ukyo, D. S. Adipranoto, H. Arai, Y. Uchimoto, Z. Ogumi, K. Suzuki, M. Hirayama and R. Kanno, *Sci. Rep.*, 2016, **6**, 28843.

185. X.-L. Wang, K. An, L. Cai, Z. Feng, S. E. Nagler, C. Daniel, K. J. Rhodes, A. D. Stoica, H. D. Skorpenske, C. Liang, W. Zhang, J. Kim, Y. Qi and S. J. Harris, *Sci. Rep.*, 2012, **2**, 747.

186. J. Li, R. Petibon, S. Glazier, N. Sharma, W. K. Pang, V. K. Peterson and J. R. Dahn, *Electrochim. Acta*, 2015, **180**, 234.

187. N. Sharma, M. V. Reddy, G. Du, S. Adams, B. V. R. Chowdari, Z. Guo and V. K. Peterson, *J. Phys. Chem. C*, 2011, **115**, 21473.

188. J. Zhu, Y. Wang, S. Li, J. W. Howard, J. Neuefeind, Y. Ren, H. Wang, C. Liang, W. Yang, R. Zou, C. Jin and Y. Zhao, *Inorg. Chem.*, 2016, **55**, 5993.

189. W. Huang, J. Zhou, B. Li, J. Ma, S. Tao, D. Xia, W. Chu and Z. Wu, *Sci. Rep.*, 2014, **4**, 4188.

190. S. Song, M. Kotobuki, Y. Chen, S. Manzhos, C. Xu, N. Hu and L. Lu, *Sci. Rep.*, 2017, **7**, 373.

191. G. Singh, J. M. Lopez del Amo, M. Galceran, S. Perez-Villar and T. Rojo, *J. Mater. Chem. A*, 2015, **3**, 6954.

192. A. P. Cohn, K. Share, R. Carter, L. Oakes and C. L. Pint, *Nano Lett.*, 2016, **16**, 543.

193. K. Mukai, T. Inoue, Y. Kato and S. Shirai, *ACS Omega*, 2017, **2**, 864.

194. N. E. Drewett, I. M. Aldous, J. Zou and L. J. Hardwick, *Electrochim. Acta*, 2017, **247**, 296.

195. P. R. Raimann, N. S. Hochgatterer, C. Korepp, K. C. Möller, M. Winter, H. Schröttner, F. Hofer and J. O. Besenhard, *Ionics*, 2006, **12**, 253.

196. C.-Y. Chen, T. Sano, T. Tsuda, K. Ui, Y. Oshima, M. Yamagata, M. Ishikawa, M. Haruta, T. Doi, M. Inaba and S. Kuwabata, *Sci. Rep.*, 2016, **6**, 36153.

197. Z. Li, X. Tan, P. Li, P. Kalisvaart, M. T. Janish, W. M. Mook, E. J. Luber, K. L. Jungjohann, C. B. Carter and D. Mitlin, *Nano Lett.*, 2015, **15**, 6339.

198. W. Xia, F. Xu, C. Zhu, H. L. Xin, Q. Xu, P. Sun and L. Sun, *Nano Energy*, 2016, **27**, 447.

199. Y. Liu, F. Fan, J. Wang, Y. Liu, H. Chen, K. L. Jungjohann, Y. Xu, Y. Zhu, D. Bigio, T. Zhu and C. Wang, *Nano Lett.*, 2014, **14**, 3445.

200. F. Chun, H. Ying, Y. Lixia, L. Yaojun, C. Weilun, H. Yangyang, C. Kongyao, H. Jiantao, L. Qingju and H. Yunhui, *Angew. Chem. Int. Ed.*, 2017, **56**, 6793.

201. P.-F. Wang, H.-R. Yao, X.-Y. Liu, Y.-X. Yin, J.-N. Zhang, Y. Wen, X. Yu, L. Gu and Y.-G. Guo, *Sci. Adv.*, 2018, **4**, eaar6018.

202. S. Chattopadhyay, A. L. Lipson, H. J. Karmel, J. D. Emery, T. T. Fister, P. A. Fenter, M. C. Hersam and M. J. Bedzyk, *Chem. Mater.*, 2012, **24**, 3038.

203. I. Buchberger, S. Seidlmayer, A. Pokharel, M. Piana, J. Hattendorff, P. Kudejova, R. Gilles and H. A. Gasteiger, *J. Electrochem. Soc.*, 2015, **162**, A2737.

204. P. K. Allan, J. M. Griffin, A. Darwiche, O. J. Borkiewicz, K. M. Wiaderek, K. W. Chapman, A. J. Morris, P. J. Chupas, L. Monconduit and C. P. Grey, *J. Am. Chem. Soc.*, 2016, **138**, 2352.

205. Z. Ziqi, J. Xiaoyu, L. Ran, Y. Dingding, A. Xinping, Y. Hanxi and C. Yuliang, *Adv. Sci.*, 2016, **3**, 1600066.

206. L. O. Vogt, M. El Kazzi, E. Jämstorp Berg, S. Pérez Villar, P. Novák and C. Villevieille, *Chem. Mater.*, 2015, **27**, 1210.

207. S. D. Lacey, J. Wan, A. W. Cresce, S. M. Russell, J. Dai, W. Bao, K. Xu and L. Hu, *Nano Lett.*, 2015, **15**, 1018.

# Multivalent Metallic Anodes for Rechargeable Batteries

**Jennifer L. Schaefer and Laura C. Merrill**

*University of Notre Dame, Notre Dame, IN, USA*



## 1 INTRODUCTION

Multivalent metallic anode materials are sought after to serve as an alternative to lithium-ion anodes or lithium metal. The materials discussed in this chapter are less reactive compared to the alkali metal anodes (lithium, sodium, and potassium), making them less facile in electrolyte decomposition and, in some respects, safer. The state of art, challenges, and prospects of magnesium, aluminum, calcium, zinc, and iron anodes will be reviewed in this chapter. Because many of these technologies encompass decades of research, only the most important historical aspects and latest modern work will be discussed.

The draw of multivalent anodes lies in the capacities and abundances of the materials. The theoretical capacity of the anode material, $Q$ (Ah $g^{-1}$), can be calculated as shown in Equation (1).

$$Q = \frac{nF}{\mathrm{MW}} \times \frac{1}{3600} \quad (1)$$

where $n$ is the number of electrons, $F$ is Faraday's constant, and MW is the molecular weight. Therefore, a multivalent metal will then double or triple the capacity for a compound with the same molecular weight. Another parameter to consider is the reduction potential of each metal. Because power, $P$ (W), and energy, $E$ (J), are proportional to the voltage as reflected in Equations (2) and (3), it is desired to have a more negative potential from the anode. This will increase the working potential as in Equation (4), and therefore increase the energy and power of the battery.

$$P = I V_{\mathrm{cell}} \quad (2)$$

$$E = Q\, V_{\mathrm{cell}} \quad (3)$$

$$V_{\mathrm{cell}} = V_{\mathrm{cathode}} - V_{\mathrm{anode}} \quad (4)$$

Figure 1 depicts these theoretical properties for the metals mentioned compared to lithium.

Another important parameter to consider is the natural abundance of the element and the locations of the reserves. As mentioned, lithium has much lower natural abundance compared to the metals discussed in this chapter. Despite the large concentrations of lithium in seawater, the extraction process is both energy intensive and costly.[1] Materials such as iron, magnesium, and aluminum are greatly abundant and the processes to extract these materials from ore (iron, magnesium, and aluminum) and seawater (magnesium) are already in practice.[2,3] Furthermore these materials are readily recycled. For the case of iron, so much of it is recycled that it is considered a separate commodity from iron ore. Figure 2(a) summarizes the average cost in USD $kg^{-1}$ of each material from July 2017 to July 2018; however, the authors note that the prices are that of commodity materials and that electrochemical grade metal is typically more expensive due to high purification costs.[4] The identified reserves of the materials are summarized in the Figure 2(b). Although calcium

*Inorganic Battery Materials*. Edited by Hailiang Wang and Boniface P.T. Fokwa
 ISBN 978-1-119-43199-2

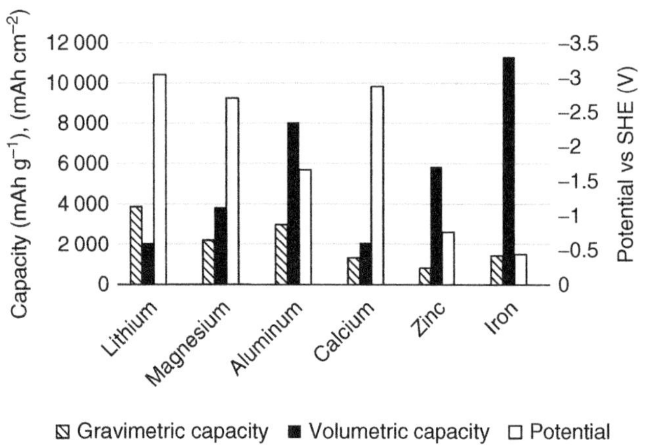


**Figure 1** Comparisons of the capacities and the reduction potentials of the multivalent metals compared to lithium

is not included on this table, it is considered to be "virtually unlimited" due to the large amount of calcium in living creatures.[3] Lithium and calcium are much higher in price as they do not occur naturally in their metallic states and are not air stable.

Although these metals are appealing for use as anodes based purely on their charge capacities and abundances, energy storage devices based on multivalent ions are less developed than lithium-ion batteries. This is, in part, due to the commercial success of the lithium-ion battery technology in 1991 and subsequent optimization from this original design. Rechargeable batteries based on multivalent metal anodes exhibit chemistry distinctly different than in lithium-ion. Herein we present the multivalent technologies in two categories: aqueous (Zn, Fe) and nonaqueous batteries (Mg, Al, and Ca). Zinc and iron are typically used as anodes in metal–air batteries with alkaline aqueous solutions, and the focus of these materials in this chapter will be on these chemistries. Both aqueous and nonaqueous materials have different challenges relating to the reversible electrodeposition of metal, and the specific problems relating to each material will be discussed in more detail in each section.

## 2 ZINC ANODES

Zinc has been widely used as an anode material for hundreds of years; however, it recently has gained attention for application in grid-level storage.[5–8] Several companies already offer commercialized or pilot grid-level zinc–air batteries including, EOS Energy Storage, Fluidic Energy, and ZincNyx Energy Systems.[5,7] Many energy storage platforms using zinc as an electrode have been developed, including zinc–bromine flow batteries and nickel–zinc batteries. Representative energy density values for select zinc battery chemistries are shown in Table 1 for comparison. The energy density of flow batteries is limited by the solubility of the active species in solution, thus they are inherently lower in energy density compared to zinc–air or nickel–zinc batteries.

The focus of this section, for the purpose of conciseness, is on zinc–air batteries, specifically relating to how metallic zinc and the electrolytes have been altered to extend battery life and improve cell performance. Bifunctional air electrodes (BAEs) that are capable of catalyzing both the oxygen reduction reaction (ORR) and the oxygen evolution reaction (OER) are a large area of research that will not be covered in this section, but more information on the current direction of research can be found in the cited reference.[7]

Zinc–air batteries typically use an aqueous alkaline electrolyte composed of concentrated KOH. Equation (5) shows the reaction at the zinc electrode, and Equation (6) shows the reaction at the air electrode.[5,9]

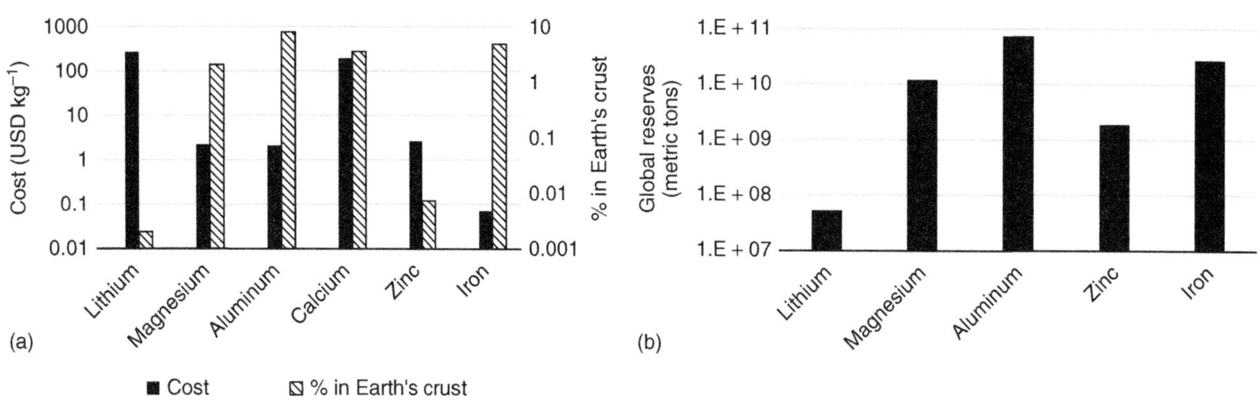


**Figure 2** (a) Cost in USD $kg^{-1}$ of commodity metals and amount in earth's crust (b) global reserves of commodity metals

**Table 1** Comparison of zinc battery chemistries

| Cell chemistry | Nominal voltage (V) | Theoretical engery density[(a)] ($Wh\,kg^{-1}$) | Practical energy density ($Wh\,kg^{-1}$) |
|---|---|---|---|
| Zinc–cerium (hybrid-flow) | 2.2 | 1800 | 20 |
| Zinc–bromine (hybrid-flow) | 1.85 | 1520 | 50 |
| Zinc–air | 1.66 | 1090 | 400 |
| Nickel–zinc | 1.73 | 930 | 100 |

[(a)]Based on mass of anode material and electrochemical reaction at anode.

$$Zn + 2OH^- \leftrightarrow ZnO + H_2O + 2e^-, \quad E = -1.26\,V \text{ vs SHE} \quad (5)$$

$$O_2 + 2H_2O + 4e^- \leftrightarrow 4OH^-, \quad E = -0.4\,V \text{ vs SHE} \quad (6)$$

This results in a theoretical cell voltage of 1.66 V, from the overall reaction in Equation (7).

$$2Zn + O_2 \leftrightarrow 2ZnO, \quad E = 1.66\,V \text{ vs SHE} \quad (7)$$

The oxidation of zinc to form zinc oxide at the anode is composed of two mechanisms. The first, described by Equation (8), is the oxidation of zinc to zincate in the electrolyte.

$$Zn + 4OH^- \leftrightarrow Zn(OH)_4^{2-} + 2e^- \quad (8)$$

Once the electrolyte becomes saturated, zinc oxide will precipitate on the surface, shown by Equation (9).[9] This form of zinc oxide is typically referred to as "Type 1" zinc oxide.[5,9]

$$Zn(OH)_4^{2-} \leftrightarrow ZnO + H_2O + 2OH^- \quad (9)$$

The zinc oxide layer is passivating to the $OH^-$ ions and can ultimately cause battery failure.[10,11] This happens when the initial zinc oxide layer grows, due to a change in pH at the interface, to form "Type 2" zinc oxide. Some studies speculate the Type 2 zinc oxide grows from a nucleation and growth mechanism whereas others suggest it grows through an electrochemical adsorption mechanism.[5,7,12] This can be controlled by changing the cutoff voltages in cell cycling.

Besides electrode passivation, the zinc anode also suffers from corrosion due to the hydrogen evolution reaction (HER) during charging and dendrite growth/shape changes of the anode.[5,9,13] Hydrogen evolution is caused because the standard reduction of zinc, −1.26 V versus SHE, is more negative than the potential that causes water splitting, −0.87 V versus SHE. Thus in order to charge aqueous zinc batteries, it is inevitable that hydrogen evolution will occur. Zinc–air batteries also can form dendrites, or fingerlike growths of metal, which can pierce a separator and short circuit a battery; in zinc–air batteries dendrites are caused by concentration driven electrodeposition of zinc.[7] The applied potential or current can bring the battery into this regime and dendrites can form from nucleation sites on the surface.[5,7,14] Finally, shape change is caused from the dissolution and redeposition of zinc onto the electrode; due to partial passivation, the zinc will deposit differently from the initial state of the anode.

Present day research of the zinc anode focuses on approaches to mitigate these problems by using 3-D electrode structure, using electrode or electrolyte additives, and coating the electrode. In some cases, studies have employed multiple methods simultaneously in order to improve cell performance. The nature of the alkaline electrolyte is what leads to many of these problems and low reversibility with the zinc anode; nevertheless, the alkaline electrolyte is necessary for the performance of the BAE as it allows for a four-electron reduction pathway which ultimately increases cell capacity.[5,7,9,15]

### 2.1 Zinc Electrode

#### 2.1.1 *Electrode Additives*

One of the primary ways electrodes have been modified to improve performance has been by inclusion of additives. The different additives include both inorganic and organic materials, and use of heavy metals is one of the more common approaches. Heavy metal additives, such as Bi or In, are often added to the electrode in the form of an oxide or hydroxide.[5,7,16–18] The additive can alter the surface to create a more compact deposit, thus lowering the rate of hydrogen evolution.[5,7,17] The decreased rate of hydrogen evolution is believed to be due to the additive increasing the overpotential required for evolution. Another approach to incorporate heavy metal additives is to coat the electrode as opposed to just physically mixing it.[5,7,19] Guo and coworkers saw significant improvement in terms of discharge capacity when the electrode was film coated with a Bi compound to give a modified electrode.[19] Despite the success of heavy metal additives, it is not desired to have widespread use of these additives due to health and environmental reasons.

Another common approach is to use additional inorganic salts in the electrolyte such as $Ca(OH)_2$ or

$Ba(OH)_2$.[5,20] The calcium case, in particular, is able to react, as in Equation (10), with the zincate ion ($Zn(OH)_4{}^{2-}$) to form a calcium–zinc crystal.[20]

$$2Zn(OH)_4^{2-} + Ca(OH)_2 + 2H_2O \leftrightarrow Ca(OH)_2 \cdot 2Zn(OH)_2 \cdot 2H_2O + 4OH^- \quad (10)$$

This complex is insoluble and forms within the crystal of the electrode. Like the heavy metal additives, coating the calcium salt onto the electrode surface versus physical mixing the additive in results in improved cycling behavior. This is known as a discharge trapping electrode additive.[5,7] The electrode is then able to deposit zinc while minimizing dendrite growth due to concentration-dependent deposition.[5,7,8,21] Generally, the reduction of these inorganic materials to form metallic compounds helps to improve the electronic conductivity throughout the electrode, which can lead to better current distribution. The better current distribution, in turn, can mitigate the risks of shape change and dendrite formation. Secondary zinc–air batteries typically have higher quantities of ZnO compared to their primary analogs.[5,7,22] Incorporation of ZnO with metallic zinc helps to control the volume changes of the zinc anode throughout cycling.[5,23]

Finally, different organic additives such as polymers or surfactants have been used to improve performance.[5,15,24] These additives are not mixed in as with inorganic additives but are coated onto the electrode. The organic materials can adsorb onto the surface of the zinc and help to prevent corrosion of the anode. Furthermore, the hydrophobic tails of surfactants and polymers can prevent the water in the electrolyte from reaching the surface of the zinc and generating hydrogen. An in-depth review of corrosion prevention due to polymer coatings has been reported by Chen and coworkers.[24]

### *2.1.2 Electrode Morphology*

Three-dimensional and high surface area zinc electrodes are used to mitigate dendrite formation while increasing zinc utilization. This method has traditionally been done by loading zinc onto high surface area metal foams.[7,23,25] Though this approach allows for more contact between the electrolyte and the electrode and a more even current distribution, these electrodes can be more susceptible to passivation, corrosion, and hydrogen evolution. These electrodes also tend to be more fragile and cannot reach deep charge and discharge.[25] Interestingly, the use of a copper foam support appeared to decrease the rate of hydrogen evolution compared to nickel foam due to the large overpotential for hydrogen evolution on copper. The foam supported electrodes have a lowered theoretical capacity and energy density.

Recent work from the United States Naval Research Laboratory showed that zinc sponges can be made and used in zinc–air batteries.[26,27] These sponges are monolithic in nature and reduced under alkaline conditions to remove the native ZnO layer.[27] The zinc sponge exhibits greater discharge capacities due to the good electrical conductivity within the structure (compared to zinc particles dispersed within a binder). Furthermore, the zinc sponges show little to no structural change after cycling. The small length within the pores of the anode, along with the well-distributed current, helped to prevent the conditions that cause zinc dendrite formation. The original reports of the 3D electrode were without additives, but in a follow-up study, Parker and coworkers integrated different electrode and electrolyte additives to improve the performance to make the anode more suitable to NiZn type applications. The report found a combination of KOH, KF, and $K_2CO_3$ to be the most promising with regards to cycling behavior and the electrode integrity after cycling; these additives were used to decrease zincate solubility as demonstrated in previous studies.[28,29] With each of the electrolyte formulations, the anode maintained at least a 90% efficiency at 20% depth of discharge.

## 2.2 Electrolyte

While concentrated aqueous KOH solutions are the most common electrolytes used for zinc batteries, these solutions have been altered through use of additives or completely changed in order to improve cell performance. One common approach of modifying the electrolyte is to add $Zn(OH)_4{}^{2-}$ ions.[23] This is used to precipitate solid ZnO on the electrode to minimize shape change. In the case of NiZn batteries, inorganic compounds are added including KF, $K_2CO_3$, $K_3BO_3$, and $K_3PO_4$.[22,28,30] These compounds are used such that the concentration of KOH can be lowered to reduce the extent of shape change while maintaining good ionic conductivity of the electrolyte. The optimized performance of a NiZn battery used a combination of 3.2–4.5 M KOH, 2 M KF, 2 M $K_2CO_3$, saturated ZnO, and 0.5 M LiF.[30] The incorporation of carbonate anions, however, is not practical for metal–air batteries as it can lead to precipitation of $K_2CO_3$ on the air electrode.[5,7,15] Aside from preventing shape change, other additives have been used to prevent dendrite growth and/or hydrogen evolution by the adsorption of the additive onto the electrode surface. These include weak acids, fluorinated surfactants, and ammonium hydroxides.[15]

While KOH is one of the most commonly used electrolyte components, different salts have been used with zinc batteries to create neutral or acidic electrolytes. This alters the zinc electrochemical reactions as shown in Table 2.[15]

The pH of the electrolyte changes the air cathode reaction even more. The four electron reaction in the

**Table 2** Electrochemical reaction at the anode and the corresponding electrolyte pH

| Electrolyte pH | Reaction at anode |
|---|---|
| Basic | $Zn + 2OH^- \leftrightarrow ZnO + H_2O + 2e^-$ |
| Neutral | $Zn \leftrightarrow Zn^{2+} + 2e^-$ |
| Acidic | $Zn \leftrightarrow Zn^{2+} + 2e^-$ |

basic electrolyte moves to a two-electron reaction in the acidic electrolyte. The acidic and neutral electrolytes have slightly higher theoretical potentials. The authors recommend readers review Blazquez and coworkers' comprehensive review of electrolytes for zinc–air batteries; this review covers acidic, neutral, basic, and nonaqueous electrolyte systems.[15] Many of the commercial systems have strayed away from basic electrolyte. For instance, EOS Energy Storage uses a neutral chloride-based electrolyte based on zinc chloride and ammonium chloride, and fluidic energy appears to offer a unit with a nonaqueous electrolyte.[15]

### 2.3 Future Directions

Ongoing research in zinc–air batteries focuses on the development of the BAE, prevention of the phenomena that limit the zinc's performance (passivation, dendrites, etc.), and new electrolyte materials. There is an increasing interest in moving away from the traditional alkaline electrolyte. This is likely related to issues with the alkaline electrolyte including the solubility of metallic zinc in the alkaline solution. Another method of improving zinc–air batteries is related to the cell configuration; planar, flow/hybrid-flow, and flexible batteries are different configurations that are currently established within the literature. Zinc anodes are the most developed of the anode materials discussed in this chapter, and there are multiple pilot-scale operations of zinc–air chemistries.

## 3 IRON ANODES

Iron has been used in batteries since late 1800s; however, the rechargeable iron–air battery was not developed until the mid-1900s. NASA pioneered the work on the Fe–air battery in the late 1960s, and during the 1970s Matsushita, Siemens, Swedish National Development Corp., and Westinghouse Electric Corp. developed their own versions of the iron–air cell.[31] Iron–nickel and iron–air batteries operate similarly in terms of the electrochemistry at the iron anode surface, but they employ different cathode materials. In recent years (2010 to present), the iron–air battery has significantly improved due to advances in nanotechnology and government funding programs.[31,32]

Iron is an attractive anode material for rechargeable batteries due to its low cost and multiple oxidation states; furthermore, iron is easily oxidized and can be electrodeposited from aqueous solutions.[31] The open circuit potential of the iron–air battery is 1.28 V whereas its predecessor, nickel–iron (FeNi), has an open circuit potential (OCP) of 1.41 V. The low cost and ease of obtaining iron makes it an attractive option for grid-level storage. Equations (11) and (12) show the discharge reactions at the anode. First, the metallic iron is oxidized to $Fe^{II}$ as follows, and once formed, $Fe^{II}$ can be further oxidized to form $Fe^{III}$.

$$Fe + 2OH^- \rightleftharpoons Fe(OH)_2 + 2e^-, \ E = -0.88\,V \text{ vs SHE} \tag{11}$$

$$3\,Fe(OH)_2 + 2OH^- \rightleftharpoons Fe_3O_4 + 4H_2O + 2e^-, \quad E = -0.76\,V \text{ vs SHE} \tag{12}$$

The process is reversed during charging, but $H_2$ evolution can occur. The reduction of water into hydroxide ions and hydrogen gas occurs at −0.87 V versus SHE. This potential is between the values of the initial reduction of Fe to $Fe^{II}$ and the second reduction of $Fe^{II}$ to $Fe^{III}$. Hydrogen evolution leads to a lowered charging efficiency, shown in Equation (13).

$$\text{Charging Efficiency} = \frac{Q_{charging} - Q_{hydrogen}}{Q_{charging}} \tag{13}$$

Aside from hydrogen evolution, layers of $Fe(OH)_2$ can passivate the surface of the metallic iron.[32–34] The combination of these phenomena causes a decrease in discharge capacity during cycling. Recent research efforts have focused on changing the electrode materials or adding materials to the electrode/electrolyte to mitigate these problems.[32,35–37] Because the iron–air chemistry is similar to that of zinc–air, many approaches of the approaches used to improve cycling behavior are analogous to the methods used in zinc–air batteries. To date, the highest performing battery has been able to cycle 2000 times with an efficiency of 50%. For grid-level storage, the Department of Energy (DOE) of the United States has set long-term goals of 5000 cycles at 80% efficiency for 100 USD $kWh^{-1}$.[31]

### 3.1 Iron Electrode

#### *3.1.1 Electrode Kinetics and Morphology*

Yushin and coworkers studied the morphology and phase changes in carbonyl iron electrodes during cycling.[38] The authors identified four stages: I, development; II, retention; III, fading; and IV, failure. The initial stage involves the formation of higher surface area particles, where the cell performance is increasing until the performance plateaus (which is then stage II, retention). The conversion of Fe to $\gamma$-$Fe_2O_3$, identified from X-ray diffraction (XRD), scanning electron microscopy (SEM), and Raman studies of the electrodes, was attributed to the fading and failure of the cells. Most studies have attributed the failure of cells to be due to the passivation by electrically insulating $Fe(OH)_2$.[31,39] Similar analyses have been completed using in situ atomic force microscopy (AFM) on ingot iron and ex situ SEM on carbonyl iron.[39] Initially, the electrode's surface area increases, due to the reduction of $Fe(OH)_2$ to elemental iron.[38,39] Eventually, the discharge capacity begins to plateau, at which point the $Fe(OH)_2$ is not completely reduced, and the surface area can no longer increase. Eichel and coworkers studied the effect of increasing the capacity throughout charging. By increasing the charge capacity once the cell reached Stage II, the passivated surface was completely reduced; the discharge capacity immediately increased from 50 to 80 mAh $g^{-1}$ and then plateaued around 100 mAh $g^{-1}$, with a charge efficiency of 80%.[39]

Many researchers have shown dramatic improvement by adding sulfide species into alkaline iron batteries. Early kinetic studies of alkaline iron batteries compared the iron and hydrogen evolution redox reactions for sulfide-modified and sulfide-free electrodes.[33,34] It is suggested that the $S^{2-}$ anion stably adsorbs onto the electrode surface, and remains throughout charge and discharge cycling. Thus the sites normally available for $OH^-$ are taken by $S^{2-}$, which prevents the formation of the passivation film.[33] The $Fe(OH)_2$ films have low ionic conductivity, making it difficult to completely reduce the discharge products, but the HER can still occur since the film is electronically conductive. The sulfide ion changes the film characteristics and changes the resistances throughout the pores of the electrode.[33,34]

### 3.1.2 *Anode Materials*

Anode materials can either consist of an oxidized form of iron (such as $Fe_3O_4$ or $Fe_2O_3$) or metallic iron (typically carbonyl iron). The oxidized forms of iron exhibit different first cycle performance as the iron oxides are reduced to metallic iron. The oxides are supported on conductive materials such as carbon to increase surface area and conductivity. To improve discharge capacity retention and charge efficiency, different additives are incorporated into the electrode or the electrolyte

Narayanan and coworkers first used carbonyl iron, a highly pure form of α-Fe, to create the highest performance iron–air cell to date.[32] It is hypothesized that the impurities in the electrode can increase the ease of adsorption, thus the higher purity iron increased the overpotential required for hydrogen evolution. Without additives, the charging efficiency was 90%. To improve the battery performance further, Narayanan and coworkers added 5% $Bi_2S_3$ to the electrode. Upon charging the $Bi_2S_3$ is reduced to bismuth, described by Equation (14).

$$Bi_2S_3 + 6e^- \rightleftharpoons 2Bi + 3S^{2-} \tag{14}$$

The bismuth further increased the overpotential for hydrogen evolution, making "electro-sorption" energetically unfavorable, and the sulfide anion reacted with the $Fe(OH)_2$ to form FeS and $Fe_3S_4$. This disrupts the passivating $Fe(OH)_2$ layer and creates an electronically conductive layer. From Tafel analysis, it was identified that the kinetics for hydrogen evolution were an order of magnitude lower for the electrodes with the $Bi_2S_3$ additive. The electrode had a 96% charging efficiency.[32] To further investigate the effect of additives, Narayan and coworkers studied the impact of bismuth oxide on cell performance with and without sulfur additives to the electrolyte, FeS and $Na_2S$.[40] The electrochemical performance was lower than the electrodes with the $Bi_2S_3$ additive, but improved compared to the carbonyl iron alone.

Magnetite (magnetic $Fe_3O_4$) particles have been studied both on their own and in conjunction with metallic iron particles.[37,41,42] The magnetite alone showed no voltage plateaus following the first discharge cycle; this is attributed to complete passivation of the iron particles upon the first cycle.[37] Magnetic $Fe_3O_4$ and $\alpha$-Fe have been used in tandem for iron battery anodes.[41,42] The ratios of the two species was determined from XPS, and the highest performance electrode had negligible magnetite. An ultrafast FeNi battery that was developed used $FeO_x$/graphene as the anode material, and the active material was identified to be a combination of metallic Fe and $Fe_3O_4$.[42]

Ferric oxide ($Fe_2O_3$) is commonly used as an alternative to carbonyl iron in iron–air batteries. During the first charge, $Fe_2O_3$ is reduced to $Fe(OH)_2$ and then to metallic iron.[43] Often, the iron is mixed with or loaded onto some form of carbon to increase conductivity. These electrode designs varied the carbon morphology, particle size/shape, loading to optimize performance, and additives to improve performance. Hang and coworkers studied the effect of different carbon morphologies with and without sulfide additives for $Fe_2O_3$-based anodes.[36,44,45] Generally, $Fe_2O_3$ loaded carbon exhibited higher capacities due to the more even distribution of iron (as opposed to physically mixing the ferric oxide with the carbon), and increasing the carbon to iron ratio improved performance.[36,44] Performance was further improved through the use of sulfide additives.[44] Most studies of $Fe_2O_3$ anodes use synthesized materials as opposed to buying commercial microparticles and nanoparticles because synthesized particles typically have a narrower particle size distribution and improved electrochemical performance.[43]

### 3.2 Electrolyte

Iron–air batteries typically use a concentrated aqueous KOH solution (6 or 8 M).[31,32] Different additives to the electrolyte are used to mitigate hydrogen evolution. Commonly, the additive is $K_2S$, $Na_2S$, or another sulfur-containing compound, and can result in a 10-fold decrease in hydrogen evolution.[36,37,40,44,46] From the electrode performance previously summarized, it is clear that the addition of the sulfur-containing compounds to the electrolyte can drastically improve the behavior of the iron battery.

Thiols as additives to the electrolyte can create a monolayer on the surface of the iron electrode.[47,48] The monolayer caused decreased cathodic current densities; due to the decreased surface area of the electrode, and hindered electron transfer process for hydrogen evolution. The alkane chains were used to exclude water from the iron surface. The longer chain *n*-alkanethiols suppressed the HER almost completely, and the other chemistries also showed suppression, and improved performance compared to the electrode without any additives.

### 3.3 Future Directions

Ongoing research of iron–air batteries focuses on the development of a suitable air cathode catalyst and on the reversibility problems of the iron anode (relating to hydrogen evolution and passivation). At this point, it is unclear if there is an advantage to using high purity carbonyl iron over in-house synthesized iron oxides and this is a point that should be addressed. Furthermore, although the $Bi_2S_3$ additive showed the highest performance, the use of heavy metals in batteries is still not advised for the reasons mentioned previously in the zinc section. Thus, future research may focus on alternatives to this material that could give comparable performance.

## 4 MAGNESIUM ANODES

Magnesium batteries are a more recent type of multivalent battery technology which first gained attention in the early 1990s with the work of Thomas Gregory from Dow Chemical Company.[49] This led to research on Grignard type solutions where several researchers attempted to achieve reversible deposition of magnesium metal.[49–52] In the early 2000s, the first prototype was developed by Aurbach et al.[50] This prototype complexed di-*n*-butyl magnesium ($Bu_2Mg$) and ethylaluminum dichloride ($EtAlCl_2$) to form the dichloro complex (DCC). A full cell was developed using this electrolyte and a Chevrel Phase (CP) $Mo_6S_8$ cathode that was able to cycle reversibly. This electrolyte was later improved upon this design by varying the Grignard reagent and the Lewis Acid used.[51,52] Because this technology is still fairly new, no commercial magnesium batteries have been released.

The magnesium surface is readily passivated in common electrolyte components such as carbonates, acetonitrile, and simple salts; this layer, analogous to a solid electrolyte interface (SEI), is impermeable to the magnesium ion due to the high charge density of the magnesium cation.[51–54] Many simple salts (e.g., $(MgClO_4)_2$ or $Mg(TFSI)_2$) are thought to be unstable against the magnesium surface, and most high-performance electrolytes are magnesium complexes. These complexes result from the exchange of ligands between two salts to form an electrochemically active cation species.

Because Grignard reagents in ether were known to reversibly deposit magnesium, electrolytes were developed that used a Grignard reagent, such as di-*n*-butyl magnesium or phenyl magnesium chloride complexed with a strong Lewis acid.[50–52] The addition of the Lewis acid ($AlCl_3$) to the electrolyte allowed for a wider electrochemical window along with reversible deposition. Since the advent of the first magnesium battery, research has focused on developing less corrosive and flammable electrolytes, a suitable cathode material, and on the deposition mechanism of magnesium.[52–54] This section will focus on the electrolyte composition and deposition mechanism of magnesium electrolytes.

The vast majority of magnesium electrolytes have reported a magnesium–chloride complex as the active species in solution. In tetrahydrofuran (THF), a six-coordination site magnesium cation, $[Mg_2Cl_3 \cdot 6THF]^+$ and $[MgCl \cdot 5THF]^+$, is what has typically been reported as the active species.[50,54–56] The former structure has been identified via single crystal XRD and is depicted in Figure 3.[50,56]

Although the single crystal is not necessarily what forms in solution, several complementary studies using alternative methods agree with this speciation.[57–59] Analogous cations have been reported with the glymes as the solvent instead of THF.[58] Muldoon and coworkers showed that the single crystal was electrochemically active after re-dissolving it in THF.[56] Some studies

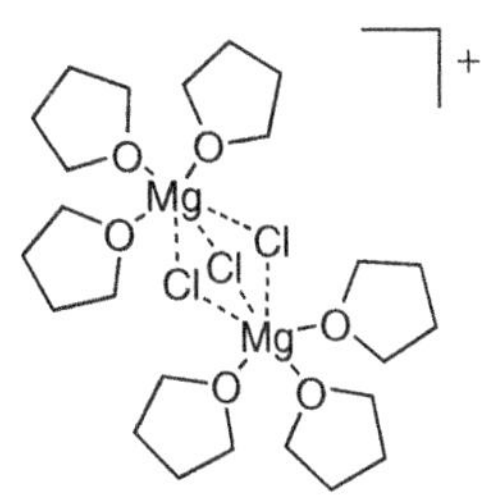


**Figure 3** Structural schematic of the tetrahydrofuran solvated $Mg_2Cl_3^+$ active cation

report a four coordination site magnesium cation such as $[MgCl \cdot 3THF]^+$, specifically this has been identified from computational studies.[60] Boron-based electrolytes and other halide free electrolytes report a monovalent magnesium-anion-solvent species as the active cation, the case for simple salt systems.[51,54,61]

### 4.1 Magnesium Electrodes

Like the other nonaqueous multivalent batteries, metallic magnesium is used as the electrode. The native oxide layer is typically mechanically removed from the magnesium foil, either by a corundum suspension or mechanically scraping.[55,62] Recently, studies have reported the use of alloys and pretreatments, but the vast majority of the literature uses metallic magnesium as an anode.[63,64] The use of $I_2$ was recently reported by Wang and coworkers to create an SEI on the surface of the magnesium.[63,65] It was reported that this layer decreased the overpotential/underpotential of chloride-free $Mg(TFSI)_2$-based electrolytes. Thus far, no rechargeable magnesium battery has been demonstrated where the SEI serves to protect the anode and maintain long-term cycling stability like in the lithium-ion systems. Bismuth–magnesium alloys have also gained attention for use with chloride-free $Mg(TFSI)_2$ systems; however, this will ultimately lower the theoretical capacity of the electrode; thus it is desired to use metallic magnesium as the anode.[64,66]

#### *4.1.1 Electrodeposition Mechanism*

Early studies by Aurbach focused on the deposition mechanism of magnesium from Grignard reagents of the form RMgX.[55,67] This mechanism was determined via in situ Fourier transform infrared spectroscopy (FTIR) studies. The deposition mechanism is described by Equations (15) and (16), and the dissolution mechanism is shown in Equations (17) and (18).

Deposition Mechanism

$$2RMg^+ + 2e^- \leftrightarrow 2RMg\cdot_{ad} \quad (15)$$

$$2RMg\cdot_{ad} \leftrightarrow Mg + MgR_{2sol} \quad (16)$$

Dissolution Mechanism

$$Mg + MgR_2 \leftrightarrow 2e^- + 2RMg^+ \quad (17)$$

$$Mg + 2RMgX_2^- \leftrightarrow 2e^- + 2RMgX + MgX_2 \quad (18)$$

The Grignard reagent undergoes both a Schlenk and an ionization equilibrium to form an electrochemically active species. Then magnesium deposition occurs via the reduction of the active magnesium species in solution which results in an adsorbed Mg–R (such as Mg–Cl) species.[55,67] This adsorbed species is what ultimately deposits magnesium. More details on the magnesium deposition mechanism can be found in Refs 55, 67.

Electrochemical quartz crystal microbalance (EQCM) measurements helped to further study the mechanism of magnesium deposition. It was hypothesized that $MgCl^+$ and neutral $R_xAlCl_{3-x}$ act as $Cl^-$ acceptors from the active species in solution.[55,67] The mass from EQCM measurements showed that the electrodeposition of Mg metal is more complicated than just the $Mg^{2+}$ reaction. It is hypothesized that magnesium deposition is preceded by the adsorption of a Mg–Cl-THF cation and/or the neutral aluminum species. The initial adsorption process is fast and is followed by a slow electron transfer to form a neutral MgCl species. Then a second fast electron transfer process occurs to form metallic Mg which then can diffuse to form Mg crystals. A more recent study of the $Mg(TFSI)_2$–$MgCl_2$ system suggests that the electrochemical deposition of magnesium is preceded by a chemical equilibrium step, reflecting the previously predicted processes.[68] The chemical equilibrium step is suggested due to the increase in current with decreasing scan rate in cyclic voltammetry measurements. The rate of deposition will determine the morphology of the deposit. This will bring the system into either a diffusion or kinetically limited regime as described by Matsui for a Grignard system.[69]

### 4.2 Electrolytes

Aside from reversible magnesium deposition, it is also desired to have electrolytes that are more practical for eventual commercial use. This includes qualities such as noncorrosiveness, wider electrochemical stability windows, and increased thermal stability. Recent work on electrolytes has focused on magnesium–aluminum chloride complexes (MACC) and MMAC, alkoxide/phenoxide, nonnucleophilic materials, $Mg(TFSI)_2$, and boron-based electrolytes. This section focuses only on the liquid mixture electrolytes for the purposes of conciseness; however, more information on polymer or ionic liquid-based electrolytes can be found at other references.[51–54]

The all-inorganic salt-based MACC electrolytes are formed by complexing magnesium chloride and aluminum chloride in ether (THF, glymes).[57,58,70,71] These electrolytes need to be conditioned in order to achieve high electrochemical performance. Prior to conditioning, cyclic voltammograms of the electrolyte show large amounts of reductive current and little to no overpotentials.[71] In-depth studies from Gewirth and coworkers showed that the Mg/Al ratio increased after conditioning the electrolyte.[71] Prior to reaching the optimized Mg/Al ratio, both magnesium and aluminum irreversibly deposit, reflected in the reductive currents seen in early cycling. Liu and coworkers modified this system by adding metallic magnesium powder to the electrolyte, termed MMAC.[72] They hypothesized that magnesium acted as a scavenger to for the species that caused irreversible deposition. These electrolytes did not require conditioning the way that MACC electrolytes did and were able to achieve efficiencies up to 100%.

The mid-2010s saw a surge in bulkier magnesium salts for application in magnesium batteries. Magnesium hexamethyldisilazide ($Mg(HMDS)_2$, MgHMDSCl)—$AlCl_3$ complexes were used to make a nonnucleophilic electrolyte that was compatible with sulfur.[56,62,73] This particular electrolyte has had varying degrees of success. Liao et al.[62] showed that the $Mg(HMDS)_2$ could be complexed with $MgCl_2$, but this formulation had lowered oxidative stability and showed little to no deposition in glymes due to low solubility of $MgCl_2$. Grignard derivatives, such as alkoxide and phenoxide salts (ROMgCl or $(RO)_2Mg$), have been widely studied to determine the effect of substituent groups on the electrochemical properties of the electrolyte.[74–76] These types of electrolytes could be complexed with either $AlCl_3$ or $MgCl_2$ to form an active species in solution; however, the $MgCl_2$ complexation resulted in a higher oxidative stability.[76,77]

Another class of electrolytes is based on different magnesium–boron salts. Some forms of these electrolytes can reversibly deposit without the addition of chloride ions. The $Mg(BH_4)_2$ system is an example of a chloride-free system, but it has a low oxidative stability (1.7 V vs $Mg^{2+}/Mg^0$).[78] Owing to the lack of chlorides in solution, the active species will be different from the predicted $Mg_2Cl_3^+$ dimer; proposed structures include $Mg(BH_4)_2 \cdot$ Diglyme, $Mg_2(BH_4)_4 \cdot 3DME$, and $Mg(BH_4)_2 \cdot 3THF$ in each respective solvent.[61] The authors predict that the performance of the electrolyte in diglyme is greater due to the active species being more favorable to form due to steric restraints. The chlorine-free carborane-based electrolyte, $Mg(CB_{11}H_{12})$, showed reversible deposition with increased oxidative stability compared to $Mg(BH_4)_2$.[79] In 2017, Zhao-Karger et al.[80] reported a fluorinated alkoxyborate-based electrolyte. The electrolyte is formed by reacting $Mg(BH_4)_2$ and a fluorinated alcohol in the electrolyte solvent to form $Mg(B(OR)_4)_2$. The electrolyte reported in depth was $[B(hfip)_4]^-$, or tetrakis(hexafluoroisopropyloxy) borate. A single crystal was precipitated out of solution and was identified as $Mg[B(hfip)_4]_2 \cdot 3DME$. The authors hypothesized that $[Mg \cdot 3DME]^{2+}$ was the active species in solution.

Recently, there has been an increased interest in $Mg(TFSI)_2$ electrolytes.[65,81–85] The TFSI anion, though highly dissociable, has low reversibility in magnesium systems.[81,84] Some attribute the lowered reversibility to the instability of the $MgTFSI^+$ complex against the magnesium metal whereas others attribute this to presence of residual water that comes with the salt.[54,81,85] Complexing the $Mg(TFSI)_2$ salt with $MgCl_2$ helps to improve performance, likely due to a surface effect of the chlorine anion.[85,86] The electrolyte typically requires conditioning in order to reach high performance. The majority of the Mg–S battery literature uses this electrolyte.

The use of sulfones as a solvent in magnesium systems is also a recent approach.[87,88] Lee and coworkers showed that $MgCl_2$ could deposit magnesium in sulfone/THF mixtures.[87] To our knowledge, this is the only work that has demonstrated the reversible electrodeposition of Mg from $MgCl_2$ without a supporting salt. The electrolyte needs to be conditioned in order to achieve high 90 + % efficiencies and exhibit low deposition underpotentials.

### 4.3 Future Directions

Magnesium battery research has moved on from synthesizing complex magnesium salts for the electrolyte to focusing on cathode materials. The intercalation of magnesium into traditional cathode materials such as transition metal oxides proves to be difficult due to the high charge density of the magnesium cation and has led to interest in alternative cathode materials, such as sulfur.[51,52,56,86] There is also interest in noncorrosive electrolytes, such that much research is focused on improving the $Mg(TFSI)_2$ chemistry through use of bimetallic anodes and surface coatings.[63,64,66] Finally, there is also an interest in replacing THF as the electrolyte solvent with higher order glymes and sulfones. Although the field of magnesium batteries has advanced beyond the original prototype developed by Aurbach, they are still far from commercialization.

## 5 ALUMINUM ANODES

The earliest nonaqueous aluminum batteries date back to the late 1960s and early 1970s when Tyco Labs in partnership with NASA developed an Al–$Cl_2$ battery using an $AlCl_3$–KCl–NaCl eutectic.[89,90] Despite the widespread abundance and high volumetric capacity of aluminum, fundamental problems including poor anodic reversibility and lack of suitable cathode materials, along with the development of the lithium-ion battery, slowed the research progress of aluminum batteries.[89,90]

Early research of electrolytes established that chloroaluminate electrolytes can be considered basic (<50 mol% $AlCl_3$), neutral (50 mol% $AlCl_3$), or acidic (>50 mol% $AlCl_3$) depending on the amount of aluminum chloride in the electrolyte. Only acidic electrolytes are able to facilitate electrodeposition and dissolution. In acidic solutions, the primary anions are $AlCl_4^-$ and $Al_2Cl_7^-$. This results in the following reversible electrodeposition reaction, Equation (19).

$$4Al_2Cl_7^- + 3e^- \rightleftharpoons Al + 7AlCl_4^- \quad (19)$$

In later research, electrolytes were developed based on chloride salt eutectics and incorporating ionic liquids, and different cathode materials were developed to counter the impracticalities of chlorine gas.[89,90] Contemporary research focuses on developing suitable cathode materials and improving electrolyte design.

## 5.1 Electrolytes

### *5.1.1 Ionic Liquid Electrolytes*

Electrolytes based on haloaluminate ionic liquids were developed to lower the operating temperature of the varying chloride salt eutectics. The earliest haloaluminate ionic liquid electrolytes were based on mixtures of $AlCl_3$ and 1-ethyl-3-methylimidazolium chloride (EMIMCl) or 1-butylpyridinium chloride (BPC). These mixtures were liquids at room temperature and were able to achieve reversibilities of 99% at room temperature with an oxidative stability of 2.5 V.[89–91] Alternatives to EMIMCl and BPC were studied due to the exothermic heat of mixing that occurs between $AlCl_3$ and these ionic liquids, which can lead to thermal decomposition of the electrolyte. Nevertheless, acidic solutions of $AlCl_3$ and 1-butyl-3-methylimidazolium chloride (BMIMCl) or EMIMCl are still the standard electrolyte in the aluminum literature.[92–96]

To lessen the corrosiveness of the typical ionic liquid electrolytes, halide-free electrolytes have been studied. Johansson and Mandai developed fully solvated aluminum chloride complexes, $[Al(L)_6]X_3$ (where L is a ligand and X is the anion) that exhibited semi-reversible behavior. The imidazole ligand and a TFSI anion showed deposition and dissolution at low reversibility without the addition of a solvent.[97] Wu and coworkers developed an electrolyte based on BMIM(TfO) and $Al(TfO)_3$.[98] For this study, the aluminum anode had to be treated with acidic $AlCl_3$/BMIMCl to remove the oxide layer. Despite the stable capacities achieved in the full Al-$V_2O_5$ cell, the report did not confirm reversible aluminum electrodeposition.

### *5.1.2 Non-IL-Based Electrolytes*

Despite the high performance of the haloaluminate ionic liquid electrolytes, the electrolytes themselves are corrosive, viscous, and hygroscopic. Thus, other approaches to achieve reversible electrodeposition have been studied, specifically involving aluminum salts in different organic solvents. Only certain combinations of aluminum salts and organic solvents can show reversible behavior, for instance $Al(TfO)_3$ in propylene carbonate (PC)/THF mixtures is not electrochemically active and $Al(TFSI)_3$ in acetonitrile shows little to no reversibility despite observed aluminum electrodeposition.[93,99]

Despite previous efforts to deposit aluminum from a salt/solvent mixture, reversible deposition could be achieved with high concentrations of $AlCl_3$ in diglyme (1–5 molar ratio).[100] Despite the success with diglyme, aluminum electrodeposition could not be achieved using higher order glymes.[101] The active species is identified to be the six-coordination site aluminum cation $[AlCl_2(diglyme)_2]^+$. This cation is elucidated from the formation of $AlCl_4^-$ and not $Al_2Cl_7^-$ in Raman spectra. Menke and coworkers completed a study of $Al(TfO)_3$ in glymes, however, reversible deposition of aluminum was not achieved.[96]

The use of sulfones in aluminum batteries dates back to the 1990s when formulations of $AlCl_3$ in di-alkyl sulfones were developed.[89,90] Early research proposed the following solution equilibria, Equation (20).

$$4AlCl_3 + 3DXSO_2 \rightarrow Al(DXSO_2)_3^{3+} + 3AlCl_4^- \qquad (20)$$

Similar electrolytes were developed by Nakayama et al.,[102] but with lower melting point sulfones. The electrolyte was compared to acidic $AlCl_3$/EMICl using $^{27}Al$ NMR, XAS, and cyclic voltammetry. Through X-ray absorption near edge structure (XANES) and extended X-ray absorption fine structure (EXAFS), it was identified that $Al_2Cl_7^-$ is the active species in solution that is responsible for aluminum deposition, as opposed to $[Al(Sulfone)_3]^{3+}$.[102] This was further confirmed when $AlCl_3$ was replaced by $Al(BF_4)_3$ in solution, and no electrochemical activity was present.

Mandai and Johansson[103] developed a deep eutectic melt of $Al(TfO)_3$, NMA, and urea to create a room temperature electrolyte with enhanced stability in ambient conditions that facilitates aluminum deposition. This approach was used to create an electrolyte that was more stable in air. The combination of NMA and urea allow $Al(TfO)_3$ to dissociate. The electrolyte was able to deposit and strip, albeit with greater overpotentials and lower reversibilities compared to acidic $AlCl_3$/EMIMCl. Dai and coworkers showed that this formulation could be used to make a high-efficiency electrolyte, with the addition of $AlCl_3$. $Al_2Cl_7^-$ was again identified as the active species in solution.[104]

## 5.2 Future Directions

Current research on aluminum batteries focuses on the development of a suitable cathode material. The trivalency of aluminum creates a high charge density that can make intercalation into different host matrices difficult, and the strong interactions between aluminum and organic solvents can lead to high energetic barriers for desolvation. For example, vanadium oxide cathode materials show evidence of deformation of the crystal structure upon aluminum intercalation/deintercalation,[92,98,105] and other studies have reported the intercalation of $AlCl_4^-$ into graphite as opposed to $Al^{3+}$.[106,107] Of the intercalation materials, the one with the most success was CP $Mo_6S_8$; this material was able to reversibly intercalate aluminum with stable cycling over 50 cycles.[94] Sulfur has become an increasingly attractive option as a conversion cathode material due to its high theoretical energy density, but polysulfide dissolution can cause capacity fading over time.[92,108,109]

Aside from cathode materials, aluminum battery research focuses on the development of a more suitable electrolyte. The electrolytes that are able to facilitate reversible deposition often contain high concentrations of $AlCl_3$ and thus are very corrosive, and efforts to remove $AlCl_3$ from the cell have achieved varying success in terms of reversibility.

## 6 CALCIUM ANODES

Metallic calcium is another alkaline earth metal that could serve as a possible anode material. Until recently, calcium metal was only been demonstrated in primary cells, due to low reversibility at the anode. Similar to the majority of nonaqueous multivalent systems, the anode is passivated in the presence of common electrolyte components.[53,110] Despite metallic calcium's low reversibility, it is predicted that $Ca^{2+}$ has faster solid state diffusion compared to $Mg^{2+}$ due to its lower charge density.[54] This suggests that $Ca^{2+}$ may be compatible with different intercalation cathode materials. Early research by Aurbach suggested that $TBABF_4$ is stable against calcium, therefore, $Ca(BF_4)_2$ could be stable with *N*-methyl-2-pyrrolidone (NMP) or THF.[111] This study confirmed the formation of a SEI at the calcium anode and eludes to the possibility of reversible calcium electrodeposition with a stable surface layer. This chapter does not demonstrate reversible calcium plating.

Owing to the difficulties in achieving calcium electrodeposition, various primary cells and calcium ion cells have been developed. Studies have demonstrated $Ca-SOCl_2$ and Ca–S cells, but electroplating, or charging, was hindered in many studies due to the formation of a passivating film on the electrode surface.[53,110,112,113] The calcium–sulfur battery, in a proof-of-concept work, showed a high discharge capacity but no reversibility.[113] Calcium-ion batteries use electrolytes that are calcium conducting but use intercalation anodes or alloys as the anode. Sn–Ca and Mg–Ca alloys have shown reversible anodic behavior.[112,114,115] Prussian blue and $V_2O_5$ cathodes are the most studied cathode materials for calcium and calcium-ion batteries; however, a full cell with a metallic calcium anode has not been developed.[110,116–118]

Quasi-reversible calcium electrodeposition was demonstrated from electrolytes composed of $Ca(BF_4)_2$ in PC/ethylene carbonate (EC) at moderate temperatures (50–100 °C).[119] The formation of a $Ca^{2+}$ conducting SEI was reported. Organic carbonate solvents were chosen due to their high dielectric constants and ability to form even surface layers.[111] Electrolytes made of $Ca(ClO_4)_2$ and $Ca(TFSI)_2$ showed irreversible reductive currents from cyclic voltammetry. Studies confirmed metallic calcium in the deposit from the $Ca(ClO_4)_2$ and $Ca(BF_4)_2$ electrolytes. The deposit from the $Ca(BF_4)$ electrolyte had $CaF_2$ present. The deposit from the $Ca(ClO_4)_2$ electrolyte had $Ca(OH)_2$ present, this is consistent with previous studies.[111,119] A computational study of a similar system, $Ca(ClO_4)$ in EC, shows that the solvent preferentially decomposes to form CO and $C_2H_4O^{2-}$ both alone and in the presence of the salt.[120]

A breakthrough study by Bruce and coworkers in 2018 showed that highly reversible calcium electrodeposition can be achieved at room temperature.[121] Using $Ca(BH_4)_2$ in THF, calcium was able to be plated and stripped at 95% efficiency with an oxidative stability around 3 V versus $Ca/Ca^{2+}$. The underpotential for calcium deposition was around −200 mV, attributed to the strong reducing nature of the borohydride anion. Studies of the calcium deposits showed no decomposition products aside from the formation of $CaH_2$, from THF. The $CaH_2$ layer prevented electrolyte decomposition at open circuit potential but the thickness of the $CaH_2$ grew with each cycle. Therefore, the authors stress the importance of forming an SEI that can conduct $Ca^{2+}$ but prevent the reduction of solvent species at the electrode surface in order to achieve 99 + % coulombic efficiencies.

### 6.1 Future Directions

In summary, calcium batteries are still in their infancy and a suitable electrolyte still needs to be developed. Such an electrolyte would support reversible calcium electrodeposition, without continual solvent decomposition. Given the low reduction potential of calcium, and the large deposition underpotential required to induce calcium electrodeposition, it is necessary that a stable, $Ca^{2+}$ conducting SEI is formed on the calcium anode to support a high-efficiency rechargeable battery.[111,121]

## 7 CONCLUSIONS

Each multivalent anode material discussed is at a different stage of development and faces different challenges. Not discussed in this chapter in detail are the challenges with cathode material development; having a functional cathode material is necessary for a full cell. Currently, one type of multivalent battery chemistry is unlikely to be suitable for every battery application (i.e., grid-level storage, automotive, and electronic devices) and completely replace lithium-ion chemistries. However, multivalent metallic anodes provide a path by which more abundant materials can be used while maintaining electrochemical performance. The future directions of these technologies, aside from cathode development, will continue to focus on reversibility at the anode.

## 8 RELATED ARTICLE

Solid State and Materials Chemistry for Sodium-Ion Batteries

## 9 ABBREVIATIONS AND ACRONYMS

AFM = atomic force microscopy; BAE = Bifunctional air electrodes; BMIMCl = 1-butyl-3-methylimidazolium chloride; BPC = 1-butylpyridinium chloride; CP = Chevrel Phase; DCC = dichloro complex; DOE = Department of Energy; EC = ethylene carbonate; EMIMCl = 1-ethyl-3-methylimidazolium chloride; EQCM = Electrochemical quartz crystal microbalance; EXAFS = extended X-ray absorption fine structure; FTIR = Fourier transform infrared spectroscopy; HER = hydrogen evolution reaction; MACC = magnesium–aluminum chloride complexes; NMP = *N*-methyl-2-pyrrolidone; OCP = open circuit potential; OER = oxygen evolution reaction; ORR = oxygen reduction reaction; PC = propylene carbonate; SEI = solid electrolyte interface; SEM = scanning electron microscopy; THF = tetrahydrofuran; XANES = X-ray absorption near edge structure; XRD = X-ray diffraction

## 10 REFERENCES


1. U. Bardi, *Sustainability*, 2010, **2**, 980.
2. U.S. Geological Survey, 'Mineral Commodity Summaries 2018', U.S. Geological Survey, 2018, p. 1.
3. Minor Metals Trade Association. Calcium – MMTA; https://mmta.co.uk/metals/ca/ (accessed Dec 2018).
4. InvestmentMine. Commodity and Metal Prices, Metal Price Charts – InvetmentMine; http://www.infomine.com/investment/metal-prices/ (accessed Dec 2018).
5. A. R. Mainar, L. C. Colmenares, J. A. Blázquez and I. Urdampilleta, *Int. J. Energy Res.*, 2018, **42**, 903.
6. V. Caramia and B. Bozzini, *Mater. Renew. Sustain. Energy*, 2014, **3**, 1.
7. J. Fu, Z. P. Cano, M. G. Park, A. Yu, M. Fowler and Z. Chen, *Adv. Mater.*, 2017, **29**, 1.
8. P. Gu, M. Zheng, Q. Zhao, X. Xiao, H. Xue and H. Pang, *J. Mater. Chem. A*, 2017, **5**, 7651.
9. A. R. Mainar, O. Leonet, M. Bengoechea, I. Boyano, I. De Meatza, A. Kvasha, A. Guerfi and J. Alberto Blázquez, *Int. J. Energy Res.*, 2016, **40**, 1032.
10. L. M. Baugh and A. Higginson, *Electrochim. Acta*, 1985, **30**, 1163.
11. L. M. Baugh and A. R. Baikie, *Electrochim. Acta*, 1985, **30**, 1173.
12. M.-B. Liu, *J. Electrochem. Soc.*, 1981, **128**, 1663.
13. L. Binder and K. Kordesch, *J. Electroanal. Chem.*, 1984, **180**, 495.
14. J. W. Diggle, A. R. Despic and J. O. Bockris, *J. Electrochem. Soc.*, 1969, **116**, 1503.
15. A. R. Mainar, E. Iruin, L. C. Colmenares, A. Kvasha, I. de Meatza, M. Bengoechea, O. Leonet, I. Boyano, Z. Zhang and J. A. Blazquez, *J. Energy Storage*, 2018, **15**, 304.
16. Y. Li and H. Dai, *Chem. Soc. Rev.*, 2014, **43**, 5257.
17. J. McBreen and E. Gannon, *J. Power Sources*, 1985, **15**, 169.
18. R. Wang, Z. Yang, B. Yang, T. Wang and Z. Chu, *J. Power Sources*, 2014, **251**, 344.
19. Y. F. Yuan, L. Q. Yu, H. M. Wu, J. L. Yang, Y. B. Chen, S. Y. Guo and J. P. Tu, *Electrochim. Acta*, 2011, **56**, 4378.
20. R. Shivkumar, G. Paruthimal Kalaignan and T. Vasudevan, *J. Power Sources*, 1995, **55**, 53.
21. C. Zhang, J. M. Wang, L. Zhang, J. Q. Zhang and C. N. Cao, *J. Appl. Electrochem.*, 2001, **31**, 1049.
22. F. R. Mclarnon and E. J. Cairns, *J. Electrochem. Soc.*, 1991, **138**, 645.
23. S. Müller, F. Holzer and O. Haas, *J. Appl. Electrochem.*, 1998, **28**, 895.
24. T. K. A. Hoang, T. N. L. Doan, K. E. K. Sun and P. Chen, *RSC Adv.*, 2015, **5**, 41677.
25. Z. Yan, E. Wang, L. Jiang and G. Sun, *RSC Adv.*, 2015, **5**, 83781.
26. J. F. Parker, C. N. Chervin, I. R. Pala, M. Machler, M. F. Burz, J. W. Long and D. R. Rolison, *Science*, 2017, **356**, 415.
27. J. F. Parker, C. N. Chervin, E. S. Nelson, D. R. Rolison and J. W. Long, *Energy Environ. Sci.*, 2014, **7**, 1117.
28. T. C. Adler, F. R. Mclarnon and E. J. Cairns, *J. Electrochem. Soc.*, 1993, **140**, 289.
29. J. F. Parker, E. S. Nelson, M. D. Wattendorf, C. N. Chervin, J. W. Long and D. R. Rolison, *ACS Appl. Mater. Interfaces*, 2014, **6**, 19471.
30. T. C. Adler, F. R. McLarnon and E. J. Cairns, *Ind. Eng. Chem. Res.*, 1998, **37**, 3237.
31. R. D. McKerracher, C. Ponce de Leon, R. G. A. Wills, A. A. Shah and F. C. Walsh, *Chempluschem*, 2014, **80**, 323.

32. A. K. Manohar, S. Malkhandi, B. Yang, C. Yang, G. K. Surya Prakash and S. R. Narayanan, *J. Electrochem. Soc.*, 2012, **159**, A1209.

33. K. Vijayamohanan, A. K. Shukia and S. Sathyanarayana, *J. Electroanal. Chem.*, 1990, **295**, 59.

34. K. Vijayamohanan, A. K. Shukla and S. Sathyanarayana, *J. Electroanal. Chem.*, 1990, **289**, 55.

35. B. T. Hang, T. Watanabe, M. Egashira, I. Watanabe, S. Okada and J. I. Yamaki, *J. Power Sources*, 2006, **155**, 461.

36. B. T. Hang, T. Watanabe, M. Eashira, S. Okada, J. I. Yamaki, S. Hata, S. H. Yoon and I. Mochida, *J. Power Sources*, 2005, **150**, 261.

37. A. Ito, L. Zhao, S. Okada and J. I. Yamaki, *J. Power Sources*, 2011, **196**, 8154.

38. D.-C. Lee, D. Lei and G. Yushin, *ACS Energy Lett.*, 2018, **3**, 794.

39. H. Weinrich, J. Come, H. Tempel, H. Kungl, R. A. Eichel and N. Balke, *Nano Energy*, 2017, **41**, 706.

40. A. K. Manohar, C. Yang, S. Malkhandi, G. K. S. Prakash and S. R. Narayanan, *J. Electrochem. Soc.*, 2013, **160**, A2078.

41. A. S. Rajan, S. Sampath and A. K. Shukla, *Energy Environ. Sci.*, 2014, **7**, 1110.

42. H. Wang, Y. Liang, M. Gong, Y. Li, W. Chang, T. Mefford, J. Zhou, J. Wang, T. Regier, F. Wei and H. Dai, *Nat. Commun.*, 2012, **917**, 1.

43. B. T. Hang and D. H. Thang, *J. Alloys Compd.*, 2016, **655**, 44.

44. B. T. Hang, S. H. Yoon, S. Okada and J. I. Yamaki, *J. Power Sources*, 2007, **168**, 522.

45. B. T. Hang, D. H. Thang, N. T. Nga, P. T. L. Minh and E. Kobayashi, *J. Electrochem. Soc.*, 2013, **160**, A1442.

46. J. O. Gil Posada and P. J. Hall, *J. Power Sources*, 2014, **268**, 810.

47. S. Malkhandi, B. Yang, A. K. Manohar, G. K. S. Prakash and S. R. Narayanan, *J. Am. Chem. Soc.*, 2013, **135**, 347.

48. B. Yang, S. Malkhandi, A. K. Manohar, G. K. Surya Prakash and S. R. Narayanan, *Energy Environ. Sci.*, 2014, **7**, 2753.

49. T. D. Gregory, R. J. Hoffman and R. C. Winterton, *J. Electrochem. Soc.*, 1990, **137**, 775.

50. D. Aurbach, Z. Lu, A. Schechter, Y. Gofer, H. Gizbar, R. Turgeman, Y. Cohen, M. Moshkovich and E. Levi, *Nature*, 2000, **407**, 724.

51. C. B. Bucur, T. Gregory, A. G. Oliver and J. Muldoon, *J. Phys. Chem. Lett.*, 2015, **6**, 3578.

52. H. D. Yoo, I. Shterenberg, Y. Gofer, G. Gershinsky, N. Pour and D. Aurbach, *Energy Environ. Sci.*, 2013, **6**, 2265.

53. J. Muldoon, C. B. Bucur and T. Gregory, *Chem. Rev.*, 2014, **114**, 11683.

54. N. N. Rajput, T. J. Seguin, B. M. Wood, X. Qu and K. A. Persson, *Top. Curr. Chem.*, 2018, **376**, 1.

55. D. Aurbach, I. Weissman, Y. Gofer and E. Levi, *Chem. Rec.*, 2003, **3**, 61.

56. H. S. Kim, T. S. Arthur, G. D. Allred, J. Zajicek, J. G. Newman, A. E. Rodnyansky, A. G. Oliver, W. C. Boggess and J. Muldoon, *Nat. Commun.*, 2011, **2**, 1.

57. K. A. See, K. W. Chapman, L. Zhu, K. M. Wiaderek, O. J. Borkiewicz, C. J. Barile, P. J. Chupas and A. A. Gewirth, *J. Am. Chem. Soc.*, 2016, **138**, 328.

58. C. J. Barile, R. G. Nuzzo and A. A. Gewirth, *J. Phys. Chem. C*, 2015, **119**, 13524.

59. A. Benmayza, M. Ramanathan, T. S. Arthur, M. Matsui, F. Mizuno, J. Guo, P. A. Glans and J. Prakash, *J. Phys. Chem. C*, 2013, **117**, 26881.

60. P. Canepa, S. Jayaraman, L. Cheng, N. N. Rajput, W. D. Richards, G. S. Gautam, L. A. Curtiss, K. A. Persson and G. Ceder, *Energy Environ. Sci.*, 2015, **8**, 3718.

61. Y. Shao, T. Liu, G. Li, M. Gu, Z. Nie, M. Engelhard, J. Xiao, D. Lv, C. Wang, J. G. Zhang and J. Liu, *Sci. Rep.*, 2013, **3**, 1.

62. C. Liao, N. Sa, B. Key, A. K. Burrell, L. Cheng, L. A. Curtiss, J. T. Vaughey, J.-J. Woo, L. Hu, B. Pan and Z. Zhang, *J. Mater. Chem. A*, 2015, **3**, 6082.

63. X. Li, T. Gao, F. Han, Z. Ma, X. Fan, S. Hou, N. Eidson, W. Li and C. Wang, *Adv. Energy Mater.*, 2018, **8**, 1.

64. Y.-H. Tan, W.-T. Yao, T. Zhang, T. Ma, L.-L. Lu, F. Zhou, H.-B. Yao and S.-H. Yu, *ACS Nano*, 2018, **12**, 5856.

65. H. Tian, T. Gao, X. Li, X. Wang, C. Luo, X. Fan, C. Yang, L. Suo, Z. Ma, W. Han and C. Wang, *Nat. Commun.*, 2017, **8**, 1.

66. J. Niu, H. Gao, W. Ma, F. Luo, K. Yin, Z. Peng and Z. Zhang, *Energy Storage Mater.*, 2018, **14**, 351.

67. D. Aurbach, A. Schechter, M. Moshkovich and Y. Cohen, *J. Electrochem. Soc.*, 2001, **148**, A1004.

68. K. Ta, K. A. See and A. A. Gewirth, *J. Phys. Chem. C*, 2018, **122**, 13790.

69. M. Matsui, *J. Power Sources*, 2011, **196**, 7048.

70. I. Shterenberg, M. Salama, Y. Gofer, E. Levi and D. Aurbach, *MRS Bull.*, 2014, **39**, 453.

71. C. J. Barile, E. C. Barile, K. R. Zavadil, R. G. Nuzzo and A. A. Gewirth, *J. Phys. Chem. C*, 2014, **118**, 27623.

72. J. Luo, S. He and T. L. Liu, *ACS Energy Lett.*, 2017, **2**, 1197.

73. Z. Zhao-Karger, X. Zhao, O. Fuhr and M. Fichtner, *RSC Adv.*, 2013, **3**, 16330.

74. E. G. Nelson, J. W. Kampf and B. M. Bartlett, *Chem. Commun.*, 2014, **50**, 5193.

75. A. J. Crowe, K. K. Stringham and B. M. Bartlett, *ACS Appl. Mater. Interfaces*, 2016, **8**, 23060.

76. N. Sa, B. Pan, A. Saha-Shah, A. A. Hubaud, J. T. Vaughey, L. A. Baker, C. Liao and A. K. Burrell, *ACS Appl. Mater. Interfaces*, 2016, **8**, 16002.

77. B. Pan, J. Zhang, J. Huang, J. T. Vaughey, L. Zhang, S.-D. Han, A. K. Burrell, Z. Zhang and C. Liao, *Chem. Commun.*, 2015, **51**, 6214.

78. R. Mohtadi, M. Matsui, T. S. Arthur and S. J. Hwang, *Angew. Chem. Int. Ed.*, 2012, **51**, 9780.

79. O. Tutusaus, R. Mohtadi, T. S. Arthur, F. Mizuno, E. G. Nelson and Y. V. Sevryugina, *Angew. Chem. Int. Ed.*, 2015, **54**, 7900.

80. Z. Zhao-Karger, M. E. Gil Bardaji, O. Fuhr and M. Fichtner, *J. Mater. Chem. A*, 2017, **5**, 10815.

81. I. Shterenberg, M. Salama, H. D. Yoo, Y. Gofer, J.-B. Park, Y.-K. Sun and D. Aurbach, *J. Electrochem. Soc.*, 2015, **162**, A7118.

82. S. Hebié, H. P. K. Ngo, J. C. Leprêtre, C. Iojoiu, L. Cointeaux, R. Berthelot and F. Alloin, *ACS Appl. Mater. Interfaces*, 2017, **9**, 28377.

83. N. N. Rajput, X. Qu, N. Sa, A. K. Burrell and K. A. Persson, *J. Am. Chem. Soc.*, 2015, **137**, 3411.

84. S. Y. Ha, Y. W. Lee, S. W. Woo, B. Koo, J. S. Kim, J. Cho, K. T. Lee and N. S. Choi, *ACS Appl. Mater. Interfaces*, 2014, **6**, 4063.

85. H. D. Yoo, S.-D. Han, I. L. Bolotin, G. M. Nolis, R. D. Bayliss, A. K. Burrell, J. T. Vaughey and J. Cabana, *Langmuir*, 2017, **33**, 9398.

86. T. Gao, S. Hou, F. Wang, Z. Ma, X. Li, K. Xu and C. Wang, *Angew. Chem. Int. Ed.*, 2017, **56**, 13526.

87. S. J. Kang, S. C. Lim, H. Kim, J. W. Heo, S. Hwang, M. Jang, D. Yang, S. T. Hong and H. Lee, *Chem. Mater.*, 2017, **29**, 3174.

88. B. Yujing, J. Luo, S. He, K. Nielson and T. Liu, *ECS Trans.*, 2017, **80**, 343.

89. T. Schoetz, C. P. de Leon, M. Ueda and A. Bund, *J. Electrochem. Soc.*, 2017, **164**, A3499.

90. G. A. Elia, K. Marquardt, K. Hoeppner, S. Fantini, R. Lin, E. Knipping, W. Peters, J. F. Drillet, S. Passerini and R. Hahn, *Adv. Mater.*, 2016, **28**, 7564.

91. J. J. Auborn and Y. L. Barberio, *J. Electrochem. Soc.*, 1985, **132**, 598.

92. G. Cohn, L. Ma and L. A. Archer, *J. Power Sources*, 2015, **283**, 416.

93. N. Jayaprakash, S. K. Das and L. A. Archer, *Chem. Commun.*, 2011, **47**, 12610.

94. L. Geng, G. Lv, X. Xing and J. Guo, *Chem. Mater.*, 2015, **27**, 4926.

95. H. Wang, Y. Bai, S. Chen, X. Luo, C. Wu, F. Wu, J. Lu and K. Amine, *ACS Appl. Mater. Interfaces*, 2015, **7**, 80.

96. L. D. Reed, A. Arteaga and E. J. Menke, *J. Phys. Chem. B*, 2015, **119**, 12677.

97. T. Mandai and P. Johansson, *J. Phys. Chem. C*, 2016, **120**, 21285.

98. H. Wang, S. Gu, Y. Bai, S. Chen, F. Wu and C. Wu, *ACS Appl. Mater. Interfaces*, 2016, **8**, 27444.

99. M. Chiku, H. Takeda, S. Matsumura, E. Higuchi and H. Inoue, *ACS Appl. Mater. Interfaces*, 2015, **7**, 24385.

100. A. Kitada, K. Nakamura, K. Fukami and K. Murase, *Electrochemistry*, 2014, **82**, 946.

101. A. Kitada, K. Nakamura, K. Fukami and K. Murase, *Electrochim. Acta*, 2016, **211**, 561.

102. Y. Nakayama, Y. Senda, H. Kawasaki, N. Koshitani, S. Hosoi, Y. Kudo, H. Morioka and M. Nagamine, *Phys. Chem. Chem. Phys.*, 2015, **17**, 5758.

103. T. Mandai and P. Johansson, *J. Mater. Chem. A*, 2015, **3**, 12230.

104. M. Angell, C.-J. Pan, Y. Rong, C. Yuan, M.-C. Lin, B.-J. Hwang and H. Dai, *Proc. Natl. Acad. Sci.*, 2017, **114**, 834.

105. S. Gu, H. Wang, C. Wu, Y. Bai, H. Li and F. Wu, *Energy Storage Mater.*, 2017, **6**, 9.

106. M. S. Wu, B. Xu, L. Q. Chen and C. Y. Ouyang, *Electrochim. Acta*, 2016, **195**, 158.

107. S. C. Jung, Y. J. Kang, D. J. Yoo, J. W. Choi and Y. K. Han, *J. Phys. Chem. C*, 2016, **120**, 13384.

108. H. Yang, L. Yin, J. Liang, Z. Sun, Y. Wang, H. Li, K. He, L. Ma, Z. Peng, S. Qiu, C. Sun, H. M. Cheng and F. Li, *Angew. Chem. Int. Ed.*, 2018, **130**, 1916.

109. T. Gao, X. Li, X. Wang, J. Hu, F. Han, X. Fan, L. Suo, A. J. Pearse, S. B. Lee, G. W. Rubloff, K. J. Gaskell, M. Noked and C. Wang, *Angew. Chem. Int. Ed.*, 2016, **55**, 9898.

110. A. Ponrouch and M. Palacin Rosa, *Curr. Opin. Electrochem.*, 2018, **9**, 1.

111. D. Aurbach, *J. Electrochem. Soc.*, 1991, **138**, 3536.

112. R. J. Staniewicz, *J. Electrochem. Soc.*, 1980, **127**, 782.

113. K. A. See, J. A. Gerbec, Y. S. Jun, F. Wudl, G. D. Stucky and R. Seshadri, *Adv. Energy Mater.*, 2013, **3**, 1056.

114. M. Wang, C. Jiang, S. Zhang, X. Song, Y. Tang and H. M. Cheng, *Nat. Chem.*, 2018, **10**, 667.

115. T. Ouchi, H. Kim, B. L. Spatocco and D. R. Sadoway, *Nat. Commun.*, 2016, **7**, 1.

116. T. Tojo, Y. Sugiura, R. Inada and Y. Sakurai, *Electrochim. Acta*, 2016, **207**, 22.

117. M. Hayashi, H. Arai, H. Ohtsuka and Y. Sakurai, *J. Power Sources*, 2003, **119–121**, 617.

118. A. F. Sammells and B. Schumacher, *J. Electrochem. Soc.*, 1986, **133**, 235.

119. A. Ponrouch, C. Frontera, F. Bardé and M. R. Palacín, *Nat. Mater.*, 2016, **15**, 169.

120. J. Young and M. Smeu, *J. Phys. Chem. Lett.*, 2018, **9**, 3295.

121. D. Wang, X. Gao, Y. Chen, L. Jin, C. Kuss and P. G. Bruce, *Nat. Mater.*, 2018, **17**, 16.

# Redox-Active Inorganic Materials for Redox Flow Batteries

**Bo Hu, Jian Luo, Camden DeBruler, Maowei Hu, Wenda Wu and T. Leo Liu**

*Utah State University, Logan, UT, USA*



## 1 INTRODUCTION

Extensive consuming of fossil fuels within less than 200 years has resulted in the truth that concentration of $CO_2$ in the atmosphere increased dramatically from 280 ppm at the beginning of industrial revolution in the nineteenth century to about 408 ppm in 2018,[1] which has caused serious substantial environmental impacts. Utilization of renewable energy sources such as solar and wind energy represents a sustainable and environmentally benign strategy to alleviate the world's severe dependency on traditional fossil fuels.[2] It is estimated that 68% of today's electrical energy is supplied from fossil fuels while only 3% from renewable energy technologies.[3] However, under the pressure of global warming and climate change, more attention and efforts have been focused on renewable energy utilization. For example, global wind power generation capacity is expected to reach 474 GW in the year 2020, about five times increase as compared to 2007.[3] In order to manage the intermittent and fluctuating nature of solar and wind energy, cost-effective technologies for energy conversion and storage are in urgent need.[2–4] Among numerous energy storage technologies, redox flow batteries (RFBs, Figure 1) have been recognized as a promising technology to overcome the intermittency of renewable energy and supply reliable electricity to electricity grids with a scale up to MW/MWh.[3–5]

$$\text{Anode:}\quad A^{n} + xe^{-} \rightleftharpoons A^{n-x} \qquad (1)$$

$$\text{Cathode:}\quad C^{n} \rightleftharpoons C^{n+y} + ye^{-} \qquad (2)$$

RFBs operate in the general mechanism of a rechargeable battery by using redox-active chemicals dissolved in liquid supporting electrolyte solutions stored in two reservoirs (Figure 1). The redox-active liquid electrolytes are termed as anolyte in the anode side and catholyte in the cathode side, respectively. A separator (ion exchange membrane or porous membrane) sandwiched between two electrodes allows the crossover of supporting electrolyte while prohibiting the crossover of active materials. During charge/discharge processes, redox

*Inorganic Battery Materials*. Edited by Hailiang Wang and Boniface P.T. Fokwa
 ISBN 978-1-119-43199-2

reactions (Equations 1 and 2) happen on the surface of the electrode while supporting ions (such as $H^+$, $Na^+$, $K^+$) as charge carriers migrate through the ion exchange membrane to balance the charge. Thereby, a flow battery is able to store or release electrical energy. The unique cell architecture of RFBs empowers a number of attractive technological merits for large-scale energy storage in comparison to traditional static rechargeable batteries. First, RFBs can modulate their energy (the volume of the electrolyte reservoirs) and power (electrode surface area) independently, which makes the power supplying more facile to manage.[3–5] Second, RFBs can operate at high current and high power densities due to fast electrochemical kinetics and high conductivities of aqueous supporting electrolytes. Third but not last, RFBs represent a safe energy storage technology by using nonflammable aqueous electrolyte materials. In addition to centralized grid-scale energy storage, RFBs with their technological merits are suitable for relatively small-scale decentralized energy storage applications including residential and commercial power backups, remote micro-grids, charge stations for electric vehicles, and fuel productions (Figure 1). It is worth noting that it is promising to develop high energy density RFBs as direct power sources for electric vehicles.

Volumetric energy density as a key parameter for a RFB which stands for the amount of energy that can be stored in the battery per liter electrolyte (Wh $L^{-1}$). The volumetric energy density of a RFB is determined by its cell voltage (V) and charge capacity (Ah $L^{-1}$) as shown in Equation (3). Charge capacity is further determined by the concentrations of active materials, and the number of electrons involved in the redox reactions given by Equation (3), where $n$ is the number of electrons, $F$ is the faradic constant, $c$ is the concentration of the more soluble electrolyte material between the anolyte and catholyte, $\mu$ is the volume factor for the full cell and defined as $(1 + V_{larger}/V_{smaller})$ ($V_{smaller}$ is the volume for the more soluble electrolyte and $V_{larger}$ is the volume for the less soluble electrolyte).

$$\text{Energy density} = CV = \frac{nFc}{\mu}V \tag{3}$$

The efficiencies of RFBs are manifested by coulombic efficiency (CE), voltage efficiency (VE), and energy efficiency (EE) as defined subsequently. CE is the ratio between discharge capacity and charge capacity. CE is an important indicator that can reflect active material crossover and irreversible side reactions if it is apparently lower than 100%.[6] VE is the ratio between discharge voltage and charge voltage, which reflects the overpotential of the battery during charge/discharge process. The multiplication of CE and VE gives EE.

Active materials applied in RFBs are very diverse such as redox-active metal salt,[7,8] nonmetal inorganic materials,[9] organometallic compounds,[10–12] organic

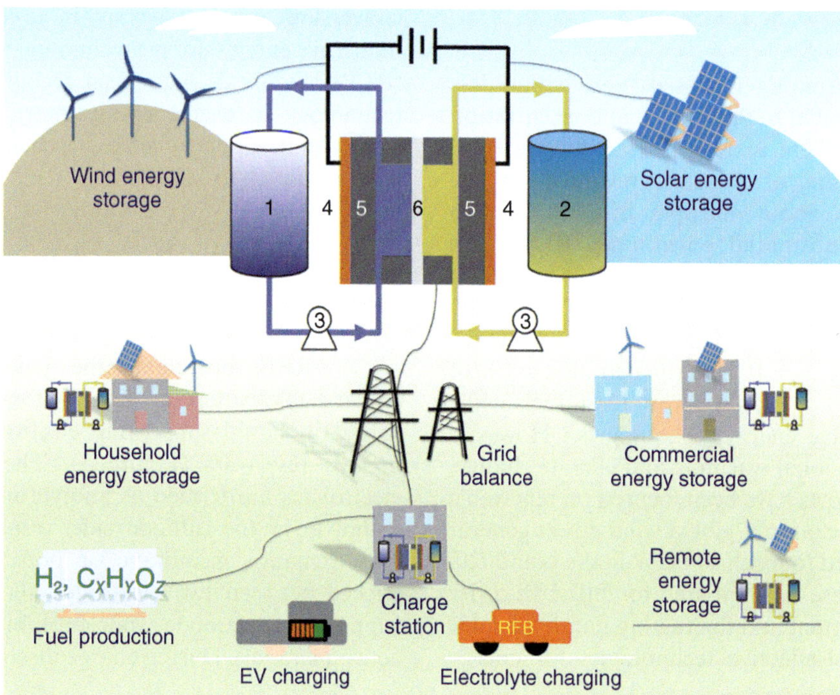


**Figure 1** Graphic representation of diverse applications and cell components of redox flow batteries (RFBs). 1. Anolyte reservoir; 2. catholyte reservoir; 3. pumps; 4. current collectors; 5. carbon electrodes; 6. ion exchange membrane

molecules,[13–16] and even polymers.[17] Among the various active materials, a few inorganic systems (Fe–Cr, $Zn-Br_2$, all-vanadium) have been demonstrated for large-scale application. Here, we present a review to discuss inorganic materials applied in RFBs with a primary focus of their most recent technological advances in aqueous inorganic RFBs. With an overview of the battery performance of different redox-active inorganic species, a brief perspective toward improved battery performance is also presented.

## 2 IRON–CHROMIUM REDOX FLOW BATTERY

Iron–chromium redox flow battery (ICRFB) stands for the first true redox flow battery (RFB) developed in the 1970s at National Aeronautics and Space Administration (NASA) as a possible energy storage means for deep-space missions.[18] $Fe^{2+}/Fe^{3+}$ and $Cr^{2+}/Cr^{3+}$ couples in hydrochloric acid are employed in an ICRFB providing a cell voltage of 1.18 V (Equations 1–3).

$$\text{Anode:}\quad Cr^{3+} + e^- \underset{\text{discharging}}{\overset{\text{charging}}{\rightleftharpoons}} Cr^{2+}\qquad E^0 = -0.41\ \text{V vs SHE} \tag{4}$$

$$\text{Cathode:}\quad Fe^{2+} \underset{\text{discharging}}{\overset{\text{charging}}{\rightleftharpoons}} Fe^{3+} + e^-\qquad E^0 = +0.77\ \text{V vs SHE} \tag{5}$$

$$\text{Overall:}\quad Cr^{3+} + Fe^{2+} \underset{\text{discharging}}{\overset{\text{charging}}{\rightleftharpoons}} Cr^{2+} + Fe^{3+}\qquad E_{\text{cell}} = 1.18\ \text{V} \tag{6}$$

Although ICRFBs have been extensively studied in the last decades, two issues remain to be overcome before it fully meets the performance and cost requirement matrices for broad market penetration. First, unlike the fast $Fe^{2+}/Fe^{3+}$ redox reaction on bare carbon felt, $Cr^{2+}/Cr^{3+}$ redox reaction shows poor kinetics.[19] Although high-temperature operation condition could facilitate the $Cr^{2+}/Cr^{3+}$ redox reaction, it would also result in significant parasitic energy loss, complexity of the system, and cost increase.[19] Hydrogen evolution reaction (HER) is severe for acidic ICRFBs, which not only reduces the CE, but also causes the state of charge (SOC) of positive and negative electrolytes to be imbalanced over prolonged cycles, eventually causing capacity decay.[20] Therefore, in order to operate ICRFBs efficiently at room temperature, catalysts are required to facilitate the $Cr^{2+}/Cr^{3+}$ redox reaction and suppress the HER simultaneously. To counter this issue, catalysts, such as Bi and Au–Ti alloy, are deposited on the electrode surface to enhance the electrochemical kinetics of the $Cr^{2+}/Cr^{3+}$ redox reaction while alleviate HER.[21,22] Second, active materials crossover is another problem for ICRFB, which can cause irreversible capacity decay because of unsatisfactory ion selectivity of membranes. This issue could be solved by using a mixed electrolyte of Fe and Cr salts for both half-cells. The capacity decay during long-term operation can be recovered by simply remixing the positive and negative electrolyte.[20] This strategy is also used for other RFB systems, such as all-vanadium RFB.[23]

In addition to the fundamental study, extensive efforts have also been made to install the ICRFB system for the purpose of large-scale energy storage. EnerVault demonstrated the world's largest ICRFB system up to 250 kW/1 MWh in California, USA.[24]

## 3 VANADIUM REDOX FLOW BATTERY

### 3.1 All-Vanadium Redox Flow Battery

Since the invention of the all-vanadium redox flow battery (VRFB) by Skyllas-Kazacos *et al.* in 1980s, the all-vanadium RFBs have been receiving massive studies and continuous commercialization.[25] Several VRFB systems have been installed, including the world's largest redox flow battery system rated at 60 MWh (15 MW for 4 h) in the Minami-Hayakita substation of Hokkaido Electric Power Co., Inc. (HEPCO) by Sumitomo Electric Industries, Ltd.[26] An even larger system is under construction by Rongke Co. in Dalian (China), with a scale of 200 MW or 800 MWh.[27] The all-VRFB represents the most mature RFB system hitherto for large-scale energy storage in terms of well-developed battery chemistry and system components.

$$\text{Anode:}\quad V^{3+} + e^- \underset{\text{discharging}}{\overset{\text{charging}}{\rightleftharpoons}} V^{2+}\qquad E^0 = -0.26\ \text{V vs SHE} \tag{7}$$

$$\text{Cathode:}\quad VO^{2+} + H_2O \underset{\text{discharging}}{\overset{\text{charging}}{\rightleftharpoons}} VO_2^+ + 2H^+ + e^-\qquad E^0 = +1.00\ \text{V vs SHE} \tag{8}$$

$$\text{Overall:}\ V^{3+} + VO^{2+} + H_2O \underset{\text{discharging}}{\overset{\text{charging}}{\rightleftharpoons}} V^{2+} + VO_2^+ + 2H^+\qquad E_{\text{cell}} = 1.26\ \text{V} \tag{9}$$

The VRFB use the $V^{2+}/V^{3+}$ redox couple (−0.26 V vs standard hydrogen electrode (SHE)) for the anode side and the $V^{4+}/V^{5+}$ redox couple (+1.0 V vs SHE) for the cathode side, giving a theoretical cell voltage of 1.26 V

(Equations 7–9). Crossing through a Nafion membrane, protons work as the shuttle to balance the charge of the electrolytes. In order to operate the VRFB with $V^{4+}/V^{5+}$ and $V^{2+}/V^{3+}$ redox reactions, a pre-charging step is required in most cases.[28] Particularly, the same $VOSO_4$ (in $H_2SO_4$) electrolyte is distributed to catholyte and anolyte reservoirs with a volume ratio of 2 to 1. Then the cell is pre-charged with a SOC of 100%. At this point, $VO^{2+}$ ($V^{4+}$) is fully oxidized to $VO_2^+$ ($V^{5+}$) in the cathode side and $V^{4+}$ is fully reduced to $V^{2+}$ in the anode side. Subsequently, half of the fully charged catholyte ($VO_2^+$) is removed for charge balancing and a fully charged VRFB with $V^{5+}$ and $V^{2+}$ species on each side is obtained.[28] $V_2O_5$ is also being used as the raw materials to replace the pricy $VOSO_4$. To solve the solubility issue of $V_2O_5$ in sulfuric acid, two methods were initially developed to produce VRFB electrolytes from $V_2O_5$ powder.[25] The first method involves chemical dissolution of the vanadium pentoxide powder by reaction with sulfur dioxide to produce a solution of $V^{4+}$ in sulfuric acid. The second method involves the electrolysis of suspended $V_2O_5$ powder in the negative half-cell of an electrolysis cell to produce a mixture of $V^{3+}$ and $V^{4+}$ species (also defined as $V^{3.5+}$).[25]

From the intrinsic properties of vanadium redox chemistry, VRFBs have displayed many advantages for large-scale energy storage. (i) Long lifetime: due to particularity of single element configuration, the cross-contamination could be solved by electrolyte remixing.[5] By doing this, the capacity loss could be restored effectively. This advantage helps VRFB to display the long calendar and cycle lives with more than 200 000 cycles already demonstrated in a NEDO funded demonstration system in Japan, and over 10 years of operation already completed for another VRFB system in Europe.[29] (ii) High EE: electrolytes used in VRFBs are highly conductive because of using strong acidic supporting electrolytes. Typically, VRFBs can achieve more than 80% EE with a current of $100\,mA\,cm^{-2}$. (iii) High power performance: because of high electrolyte conductivity and a high cell voltage, VRFBs can deliver high power density up to several hundred $MW\,cm^{-2}$, representing one of the most powerful RFBs known to date. Despite these salient features, challenges are also existing, which hinder VRFBs' practical applications. Vanadyl (IV) sulfate ($VOSO_4$) is the most common active material for VRFBs, which is highly soluble in water (up to 4 M). However, adding sulfuric acid, as the supporting electrolyte to provide higher conductivity, would dramatically decrease its solubility (1 M in 7 M $H_2SO_4$) due to the common ion effect. With increasing $H_2SO_4$ concentration, the total sulfate concentration increases, which shifts the equilibrium to lower dissociation of vanadyl sulfate.[30] On the other hand, the V(V) ions in the positive half-cell electrolyte can undergo thermal precipitation to form $V_2O_5$ at elevated temperatures of 40 °C (Equations 10 and 11).[19,31] Therefore, to balance the stability and the solubility of the vanadium active species, the concentration of vanadium and total $SO_4^{2-}$ is usually controlled at less than 2 and 5 M, respectively.[4] This operation concentration determines an energy density of less than $25\,Wh\,L^{-1}$.[5]

$$[VO_2(H_2O)_3]^+ \xrightarrow{\text{heat}} H_3VO_4 + H_3O^+ \tag{10}$$

$$2H_3VO_4 \longrightarrow V_2O_5 + 3H_2O \tag{11}$$

In order to improve the solubility and stability of the active materials, strategies have been conducted such as using mixed supporting electrolyte and adding additives. In 2011, scientists at Pacific Northwest National Laboratory reported sulfate–chloride mixed acid electrolytes for all-vanadium RFB application.[7] Wang and coworkers found that the mixed electrolyte supported the vanadium ion dissolution with a high concentration up to 2.5 M in the presence of 2.5 M $SO_4^{2-}$ and 6 M $Cl^-$, which indicated an energy density increase of 70% compared to traditional pure $H_2SO_4$ supported VRFBs with a wider operation window from −5 to 50 °C. In addition, using density functional theory, and $^{51}V$&$^{35}Cl$ NMR, $V^{5+}$ was proved to be existing as a neutral complex as $VO_2Cl(H_2O)_2$ is more stable than $VO_2(H_2O)_3^+$ cation owing to the prohibition of the formation reaction from $VO_2(H_2O)_3^+$ to $H_3VO_4$ (Figure 2). This dimer complex has a higher thermal stability due to its resistance to the deprotonation reaction which is the initial step of $V_2O_5$ formation (Equation 10), thus it can stabilize the $V^{5+}$ electrolyte. Moreover, the vanadium redox couples in mixed electrolytes demonstrated improved reaction kinetics while remaining as highly reversible as in the pure sulfate supporting electrolyte. Following the mixed $HCl/H_2SO_4$ electrolyte study, Kim *et al.*[32] suggested a vanadium chloride solution for all-vanadium RFBs. For the chloride-based electrolyte, thermally stable ions such as the vanadium dinuclear species of $[V_2O_3{\cdot}4H_2O]^{4+}$ or dinuclear-chloro complex of $[V_2O_3ClH_2O]^{3+}$ also display high solubility (2.3 M) within a large temperature range from 0 to 50 °C. Besides HCl, other forms of acids, such as phosphoric acid, EDTA, and $CH_3SO_3H$, have been also examined as VRFBs' supporting electrolyte.[33,34] However, they are not as effective as HCl. Organic and inorganic additives have also been explored to increase the solubility and stability of vanadium active species. Skyllas-Kazacos *et al.* investigated additives including sodium hexametaphosphate (SHMP), $K_2SO_4$, $Li_2SO_4$, and urea.[33] Supersaturated 4 M vanadyl sulfate solutions were prepared and the stability was evaluated. In 2016, they reported a high energy density vanadium redox battery employing a 3 M vanadium electrolyte using $K_3PO_4$, $H_3PO_4$, $(NH_4)_2SO_4$ as additives.[33] The VRFB was subjected to 90 charge/discharge cycles and no precipitation or capacity loss was observed in the presence of 1 wt% $H_3PO_4$ and 2 wt% ammonium sulfate.

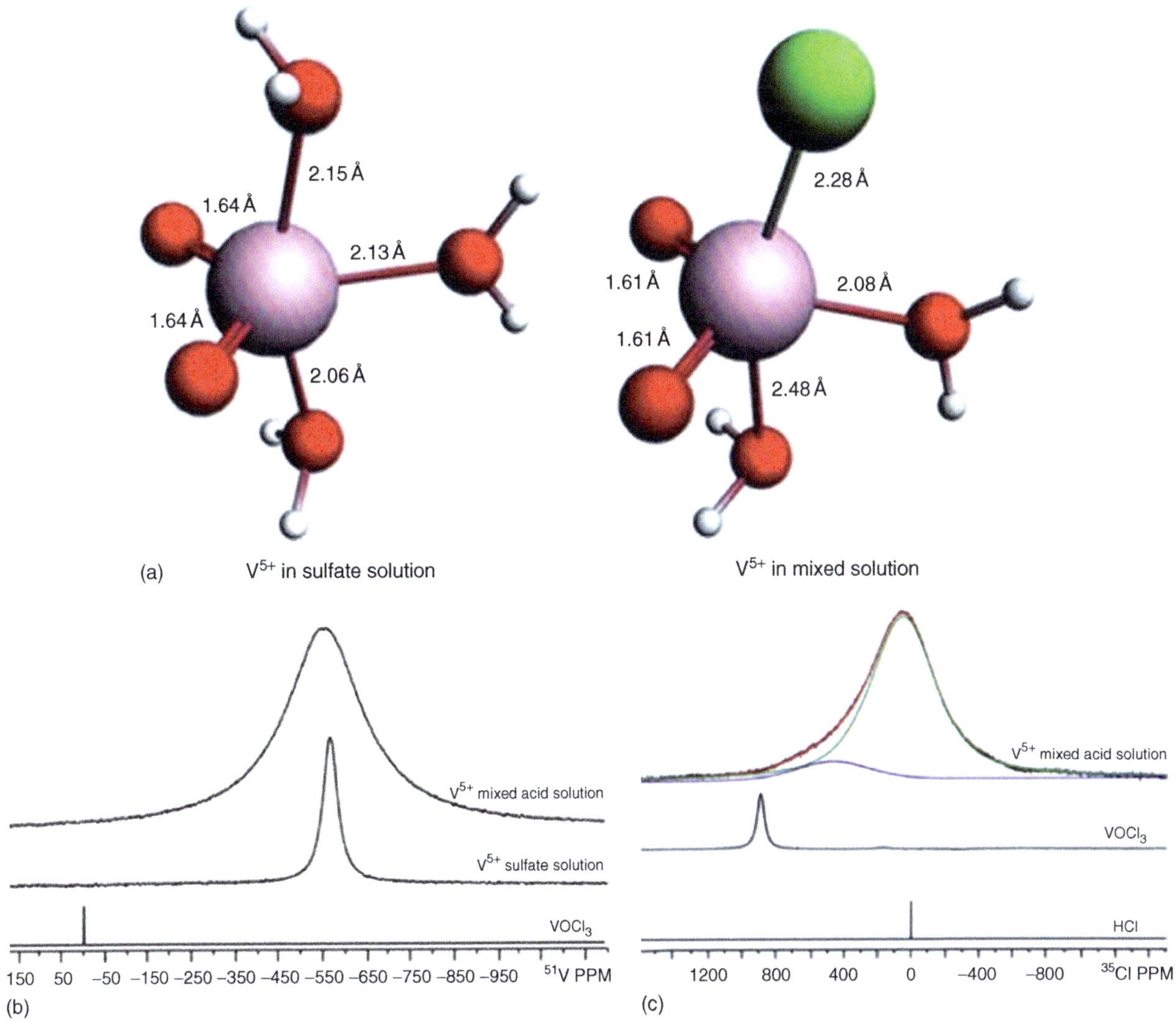


**Figure 2** (a) The structures of $V^{5+}$ monomer complexes in the sulfate and the sulfate–chloride mixed solutions. (b) $^{51}V$ NMR spectra of a mixed sulfate–chloride solution. (c) $^{35}Cl$ NMR spectra of a mixed sulfate–chloride solution. [Adapted with permission from Ref. 7 © Wiley-VCH, 2011]

Actually, additives cannot only serve as solubility boosters and stabilizers for vanadium redox-active materials but also catalysts for the redox reactions, thus improving the VRFBs' EE. Some inorganic ions such as $Sb^{3+}$ and $Bi^{3+}$ have been added into the anolyte to improve kinetics for $V^{2+}/V^{3+}$ and $V^{4+}/V^{5+}$ redox reactions.[35,36] The improved electrochemical performance proved to be ascribed to the prominent catalytic effect of metal particles, which are simultaneously electrodeposited onto the surface of graphite felts (GFs) during operation of a flow cell and facilitate charge transfer process. Wang *et al.* employed electrolytes containing $Bi^{3+}$ to improve the performance of a VRFB.[35] With the adding 0.01 M $Bi^{3+}$, the EE was significantly improved by about 11%. A surface image of GF modified with Bi nanoparticles (NPs) at different magnifications after cycling was obtained from field emission scanning electron microscope (FESEM), scanning transmission electron microscopy (STEM), and high-resolution transmission electron microscopy (HRTEM) as shown in Figure 3. The formed Bi NPs were only found at the negative electrode and facilitate the redox reaction of the $V^{2+}/V^{3+}$ redox couple which was confirmed by the minimizing of the peak–peak separation of the $V^{2+}/V^{3+}$ redox couple in cyclic voltammetry (CV) experiments (Figure 3d). In addition, the charge/discharge overpotential was also dramatically reduced, and the specific capacity was doubled.

Shi *et al.* reported that the EE was increased from 57.5% to 67.1% at a current density of $120\,mA\,cm^{-2}$ by introducing 5 mM $Sb^{3+}$ into the negative electrolyte.[36] $In^{3+}$, $Cr^{3+}$, and $Mn^{2+}$ ions were applied as additives to improve redox kinetics of $V^{4+}/V^{5+}$ couple. In these cases,

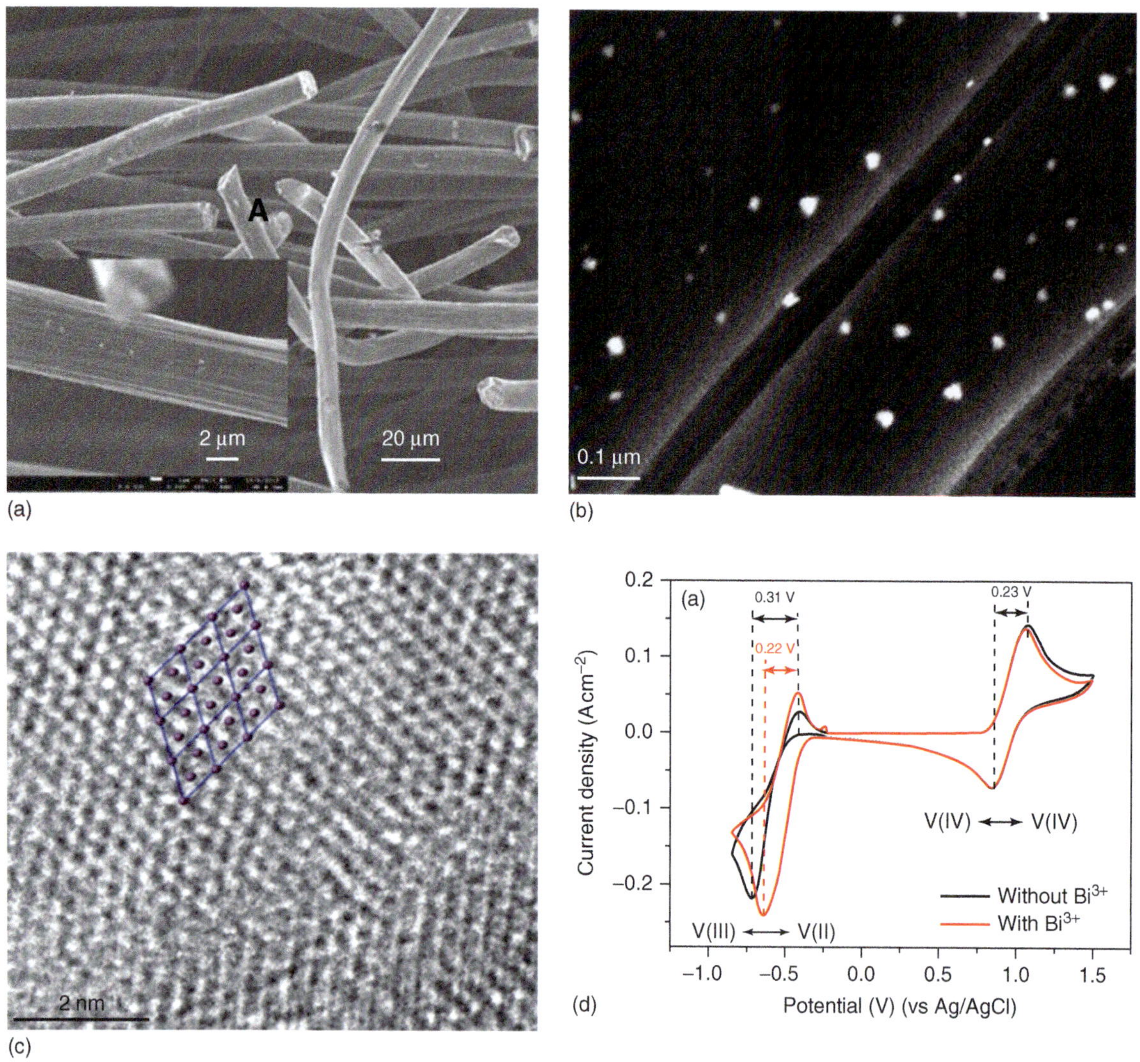


**Figure 3** (a) FESEM images of GFs modified with Bi nanoparticles at different magnifications after cycling; (b) STEM image of Bi nanoparticles in the anolytes resulting after cycling; (c) HRTEM images of Bi nanoparticles; (d) cyclic voltammograms with glassy carbon as working electrode in solutions vanadium electrolyte with or without $BiCl_3$. [Reprinted with permission from Li, B.; Gu, M.; Nie, Z.; Shao, Y.; Luo, Q.; Wei, X.; Li, X.; Xiao, J.; Wang, C.; Sprenkle, V.; Wang, W., Bismuth Nanoparticle Decorating Graphite Felt as a High-Performance Electrode for an All-Vanadium Redox Flow Battery. Nano Lett. 2013, 13 (3), 1330–1335. © American Chemical Society 2013]

metal cations, other than metal NPs, promote the charge transfer, redox reaction reversibility, and diffusion coefficient of the $V^{4+}$ and $V^{5+}$ species. As we can see, by adding $Bi^{3+}$ and $Sb^{3+}$ additives, electrodes were modified with metal catalysts.

In fact, extensive studies have also been conducted to modify the electrode with electrochemical catalysts, such as carbon materials, transition metal oxides, and redox mediators to meliorate performance for VRFBs.[37–40] In addition to poor redox reaction kinetics, the sluggish interfacial charge transfer from electrode surface to vanadium species is also responsible for EE loss of a VRFB.[39] Wang *et al.* introduced prussian blue (PB) and a prussian blue analog (PBA) with identical redox potentials to $VO^{2+}/VO_2^+$ and $V^{2+}/V^{3+}$ onto cathode and anode electrode as electron mediators to facilitate the charge transfer process.[40] During the charging process, the oxidized PB reacts with $VO^{2+}$ to generate $VO_2^+$ on the cathode electrode, and $V^{2+}$ is reduced to $V^{3+}$ by reacting with the reduced PBA on the electrode. Tafel plots shows that the PB-modified glass carbon electrode has an exchange current density ($3.8 \times 10^{-4}$ A cm$^{-2}$) two orders of magnitude as high as the pristine electrode ($3.0 \times 10^{-6}$ A cm$^{-2}$), implying a considerably enhanced kinetics of the modified electrode. This strategy boosts power density of a VRFB from 178 mW cm$^{-2}$ with a bare carbon electrode to 373 mW cm$^{-2}$ with a modified electrode.

### 3.2 Hybrid Vanadium Redox Flow batteries

As we mentioned previously, $V^{5+}$ species in the catholyte for the VRFBs tends to form precipitation. Moreover, due to the high oxidation ability and corrosiveness of $V^{5+}$ in the VRFB, the availability of compatible separators is very limited.[19] The high-cost Nafion membranes are still the best choice for an all-vanadium RFB. In order to overcome the problem of high-cost separators and avoid the issue of cathodic precipitation, researchers have made efforts to replacing $V^{4+}/V^{5+}$ contained catholyte with some less oxidative materials while $V^{2+}/V^{3+}$ solution is still used as the anolyte, thus resulting hybrid vanadium RFBs. Skyllas-Kazacos[41] employs a polyhalide solution in the positive half-cell electrolyte and a $V^{2+}/V^{3+}$ redox couple as the negative half-cell electrolyte. During charging, the bromide ions in the positive half-cell are considered to undergo oxidation to form the polyhalide ion $Br_2Cl^-$. The formal potential of this couple is about 0.8 V versus SCE. When combined with a formal potential of around −0.5 V versus SCE for the $V^{2+}/V^{3+}$couple in the chloride supporting electrolyte, an overall cell potential of approximate 1.3 V was obtained. It was also revealed that in the electrolyte with higher $Cl^-$ concentration, $V^{2+}/V^{3+}$couple redox kinetics is better.

Fe–V hybrid RFBs were first designed by the scientist at PNNL to overcome both the anodic issues of traditional Fe–Cr RFBs as we discussed previously and cathodic problem of the all-vanadium RFB. The combination of $Fe^{2+}/Fe^{3+}$ and $V^{2+}/V^{3+}$ redox couples gives a cell voltage of 1.02 V which is lower than the VRFB (1.26 V).[19] However, rapid fading in the cell capacity (70% capacity decay within 50 cycles) was observed caused by the crossover of the Fe/V ions through the membrane during the charge/discharge processes, which also caused a lower columbic efficiency (93%) throughout the cycling. Fortunately, this problem was effectively solved by using a mixed electrolyte of $Fe^{2+}$ and $V^{3+}$ ions for both anode and cathode. When 1.25 M Fe/V in 2.3 M HCl mixed solution was used in each half-cell, 95% capacity was retained within 50 cycles. In the following study, Wang and coworkers explored an optimized condition using 1.5 M $FeCl_2$/1.5 M $VOSO_4$ in 3.0 M HCl. Combined with low-cost microporous separator, the Fe–V RFB showed stable cycling performance and high EE (62% at −5 °C).[42] In addition, HCl–$H_2SO_4$ mixed acid was also examined as a supporting electrolyte for Fe–V RFBs.[43] With the combination of 1.5 M Fe and 1.5 M V in 1.5 M sulfate and 6.8 M total chloride solution, the redox flow cell achieved an EE greater than 80% at 50 mA cm$^{-2}$ and no capacity fading over 100 cycles employing Nafion 212 as the membrane. Impressively, even with a piece of low-cost polyethylene microporous separator, the Fe–V RFB also delivered satisfactory cell efficiencies and stable energy density over 50 cycles.

## 4 ZINC-BASED INORGANIC REDOX FLOW BATTERIES

VRFBs have been well developed and successfully commercialized. However, the total system capital cost of VRFBs is above \$400 per kWh where the cost of electrolyte accounts for 55%, which is way beyond the target \$150 per kWh set by U.S. Department of Energy (DOE).[44] In order to reduce the cost of the flow battery system, it is necessary to utilize low-cost active materials. Zinc has a much larger production up to 12 million tons per year, comparing to 76 thousand tons/year for vanadium. Meanwhile, the price of zinc metal is only one-tenth of $V_2O_5$,[45] which makes zinc very attractive as an anode material in RFBs. Another advantage of zinc electrolytes is the high solubility of zinc salts. For a typical VRFB, the operation concentration of vanadium is normally less than 2 M. In a zinc-based RFB, the concentration of zinc salts can be higher than 5 M which brings in a superior energy density compared to VRFBs. Different from VRFBs which have liquid phase redox reactions for both cathode and anode, in a Zn-based RFB, zinc deposition/stripping happens on the surface of the anode electrode, defining zinc-based RFB as a hybrid RFB system.[4] By minimizing the volume of anolyte, the volumetric energy density of the zinc RFB can be boosted up to a very high level. However, the Zn electrode has limitations of low current and power densities due to the formation of Zn dendrites. In addition, in contrast to all-liquid RFBs (such as VRFB), power, and energy of such hybrid systems are not fully decoupled.[4]

For a zinc-based inorganic RFB, a zinc salt ($ZnCl_2$, $ZnBr_2$, and $ZnI_2$) is dissolved in water and used as anolyte. The catholyte materials for zinc RFBs can be halogen salt ($Br^-$ and $I^-$),[46,47] low valent metal salt ($Ce^{3+}$, $Fe^{2+}$),[48,49] and organic molecules and polymer materials ((2,2,6,6-tetramethylpiperidin-1-yl)oxyl (TEMPO)).[50] It is worth noting that separator selection is crucial for Zn-based RFB. Anion exchange membranes (AEMs) feature quaternary ammonium species on the pore surface which lead to a side reaction/precipitation of $Zn^{2+}$.[51] Therefore, AEMs could lose their ion conductivity in the presence of $Zn^{2+}$. In this part, we introduce the Zn-based RFB systems which have been extensively investigated.

### 4.1 Zn–Halogen Redox Flow Batteries

As the first zinc–halogen battery example, a Zn–$Cl_2$ RFB was demonstrated in 1921, while the system received only a few development until 1980s due to the high-cost challenge of chlorine storage.[4] Comparing to Zn–$Cl_2$ RFBs, Zn–$Br_2$, and Zn–$I_2$ systems have been realized to be much more practical due to easier storage, lower volatility, and better electrochemical redox kinetics of $Br_2$ and $I_2$ than $Cl_2$.

### 4.1.1 *Zn–Br₂ Redox Flow Battery*

Along with vanadium RFBs, Zn–$Br_2$ system is another commercialized RFB platform with a kW/kWh$^{-1}$ scale.[52] The battery has a cell voltage more than 1.8 V and energy density as high as 60 Wh L$^{-1}$. Bipolar $ZnBr_2$ electrolyte is used for both anode and cathode in a Zn–$Br_2$ RFB. The electrochemical reactions are described in Equations (12)–(14).

$$\text{Anode:}\quad Zn^{2+} + 2e^- \underset{\text{discharging}}{\overset{\text{charging}}{\rightleftharpoons}} Zn \qquad E^0 = -0.76\ \text{V vs SHE} \tag{12}$$

$$\text{Cathode:}\quad 2Br^- \underset{\text{discharging}}{\overset{\text{charging}}{\rightleftharpoons}} Br_2 + 2e^- \qquad E^0 = +1.06\ \text{V vs SHE} \tag{13}$$

$$\text{Overall:}\quad Zn^{2+} + 2Br^- \underset{\text{discharging}}{\overset{\text{charging}}{\rightleftharpoons}} Zn + Br_2 \qquad E_{cell} = 1.82\ \text{V} \tag{14}$$

During battery charging process, zinc deposits on the cathode electrode while $Br^-$ anions are oxidized into $Br_2$ in the anode side. Although $ZnBr_2$ has a solubility of more than 4 M in water, liquid $Br_2$ is slightly soluble in water (0.35 mg mL$^{-1}$). As a serious safety concern, $Br_2$ vapor pressure will keep increasing during batteries' charging process, which is very dangerous as bromine is well known to be extremely toxic and corrosive to human bodies.[4] In addition, the low solubility of bromine would also limit the practical energy density of the full RFB.[53] In order to suppress the $Br_2$ vapor releasing, researchers have been exploring complexing reagents, such as quaternary ammonium bromide, to reduce the amount of free liquid or gas type bromine.[54,55] Zn–$Br_2$ RFBs are operated at acidic conditions ca. pH 4. At an acidic pH, zinc anode corrosion and HER become more dominant and lead to lower CE.[52] Besides, zinc dendrite formation is also severe, especially at high charge/discharge current densities. Thus, the operation current density (<20 mA cm$^{-2}$) is much lower than VRFB.[52]

### 4.1.2 *Zn–I₂ Redox Flow Battery*

The emerging of Zn–$I_2$ RFB is regarded as an alternative system to address the highly toxic and corrosive issue and performance drawbacks of Zn–$Br_2$ RFBs. Comparing to Zn–$Br_2$ RFBs, Zn–$I_2$ batteries have several impressive advantages. (i) In aqueous solution, $ZnI_2$ has a better solubility (7 M) than $ZnBr_2$ (4 M) which results in a higher theoretical energy density of Zn–$I_2$ RFBs than Zn–$Br_2$ RFB.[47] (ii) $I_2$ is solid at room temperature with a much lower vapor pressure than liquid bromine. Besides, $I_2$ can react with $I^-$ fast and generate highly soluble polyiodide $I_3^-$, which ensure a good capacity utilization of Zn–$I_2$ RFBs although 1 equiv of $I^-$ need to consumed as the complexing reagent. (iii) The $I_3^-/I^-$ redox couple has a better redox kinetic than the $Br_3^-/Br^-$ redox couple.

As an analogue of the Zn–$Br_2$ RFB, a Zn–iodine RFB employing $ZnI_2$ as a bipolar electrolyte was first reported by Wang *et al.* as shown in Figure 4.[47] The resulted battery displays an open circuit voltage (OCV) of about 1.3 V while it is dependent on the concentration of the active materials (decreases from 1.43 V to 1.22 V with the increasing concentration of $ZnI_2$ from 0.5 to 5.0 M). By minimizing the volume of the anolyte, the battery has a superb theoretical energy density about 322 Wh L$^{-1}$ which surpassed the $LiFePO_4$-based lithium ion batteries (ca. 223 Wh L$^{-1}$ based on cell level). A Nafion membrane was applied as the separator which proved to be effective to minimize $I^-$ anions crossover as a high CE about 99% was obtained. It is worth noting that the electrolyte showed a slightly acidic pH 3–4 at 0% state-of-charge which worked for suppressing both dramatic hydrogen evolution and $Zn(OH)_2$ precipitate formation. The flow cell was charge–discharge cycled within a range of current densities between 5 and 20 mA cm$^{-2}$. The energy efficiencies decreased from ca. 91% to ca. 76% as the concentration of zinc iodide increased from 0.5 to 3.5 M, which was attributed to the decrease of electrolyte conductivity. Using 5 M $ZnI_2$ as the electrolyte, the battery demonstrated an energy density as high as 166.7 Wh L$^{-1}$ under a real flow condition for 40 stable cycles. One of the most challenging issue for Zn-based RFBs is the dendrite formation which could penetrate the separator and cause serious self-discharge. In order to alleviate the dendrite formation during the Zn–$I_2$ RFB's charging process, ethanol was applied as an additive. The addition of an alcohol could induce ligand formation between oxygen on the hydroxyl group of an alcohol and zinc ions, which ameliorates the zinc dendrite issue. Meanwhile, the ethanol–water mixed solvent expanded the stable electrolyte temperature window from −20 to 50 °C. In addition, to further improve the dendrite, a small current density of 10 mA cm$^{-2}$ was also applied to cycle the battery which is much lower than most of the aqueous RFB systems. From this study, we can recognize that it is impossible to utilize all the $I^-$ active materials in Zn–$I_2$ RFB systems with pure bipolar $ZnI_2$ electrolyte due to the sacrificed 1 equiv $I^-$ for the formation of polyiodide species.

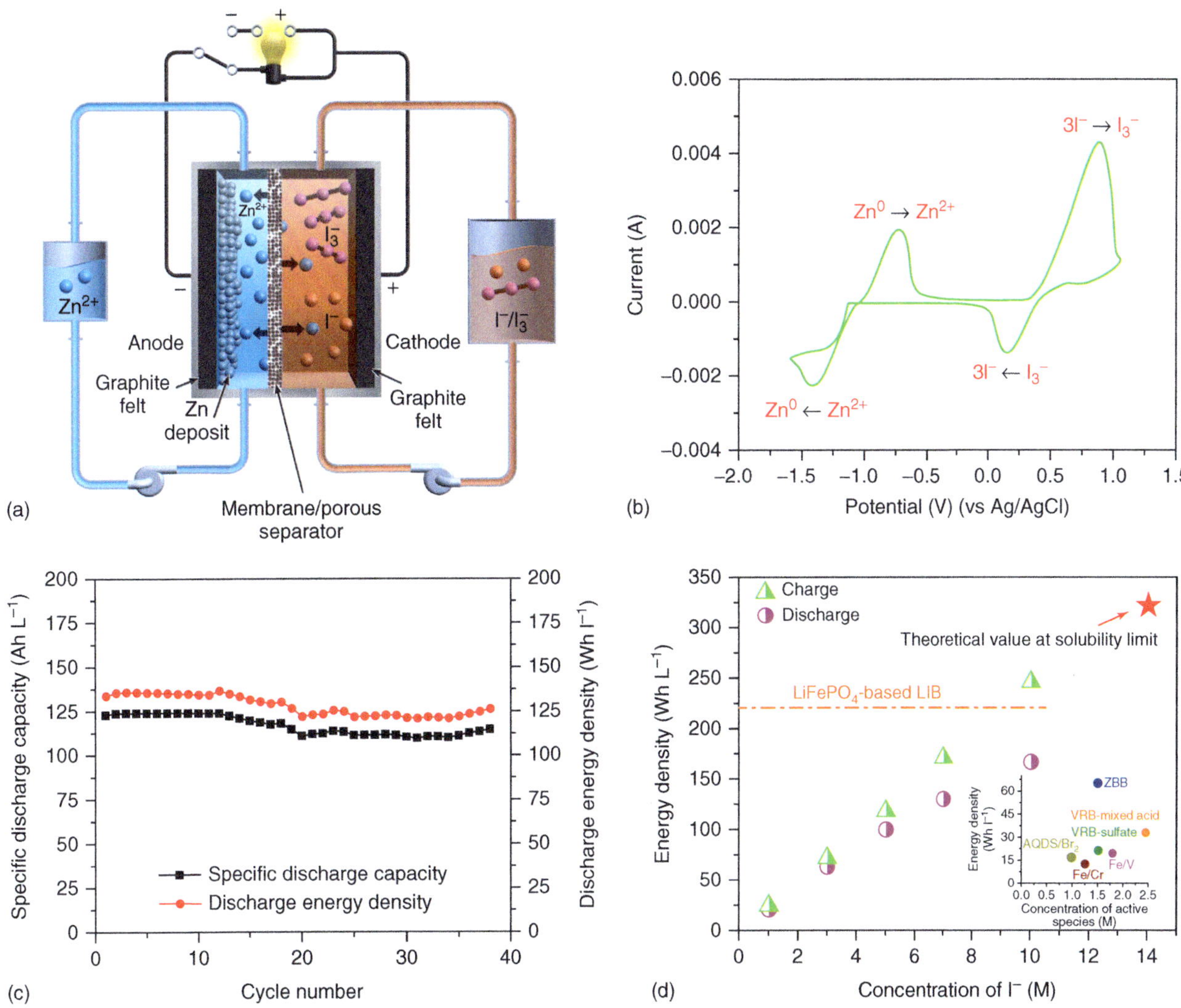


**Figure 4** (a) Schematic representation of the proposed Zn-iodine RFB system; (b) CV of 0.085 M $ZnI_2$ on a glassy carbon electrode; (c) cycling performances for discharge energy densities and capacities of the cell with 3.5 M $ZnI_2$ and Nafion 115 as membrane under the current density of $10\,mA\,cm^{-2}$; (d) the charge and discharge energy densities as a function of the concentration of $I^-$. The inset lists concentration versus energy density of several current aqueous RFB chemistries for comparison. [Source: 47, https://www.nature.com/articles/ncomms7303?page=2. Licensed under CC BY 4.0]

As we stated earlier, aiming to improve the energy density of RFB, strategies are required either increasing the concentration of redox-active material or enhance the cell voltage. Following Wang and coworkers' study on $Zn-I_2$ chemistry, Lu *et al.* applied bromide ions ($Br^-$) as a complexing agent to stabilize the iodine, forming iodine–bromide ions ($I_2Br^-$) as shown in the electrochemical reactions in Equations (15)–(17), which frees up a portion of iodide ions and increases the capacity.[56]

$$\text{Anode:}\quad Zn^{2+} + 2e^- \underset{\text{discharging}}{\overset{\text{charging}}{\rightleftharpoons}} Zn$$
$$E^0 = -0.76\ \text{V vs SHE} \qquad (15)$$

$$\text{Cathode:}\quad 2I^- + Br^- \underset{\text{discharging}}{\overset{\text{charging}}{\rightleftharpoons}} I_2Br^- + 2e^-$$
$$E^0 = +0.594\ \text{V vs SHE} \qquad (16)$$

$$\text{Overall:}\quad Zn^{2+} + 2Br^- \underset{\text{discharging}}{\overset{\text{charging}}{\rightleftharpoons}} Zn + Br_2$$
$$E_{cell} = 1.354\ \text{V} \qquad (17)$$

Applying this strategy, they demonstrated a novel zinc/iodine–bromide battery to achieve an energy density of 101 Wh $L^{-1}$ for the full cell (or 202 Wh $L^{-1}$ for the cathode half-cell), which is among the highest energy densities achieved for aqueous flow batteries to date. As shown in Figure 5, both centrosymmetric $I_3^-$ and asymmetric $I_2Br^-$

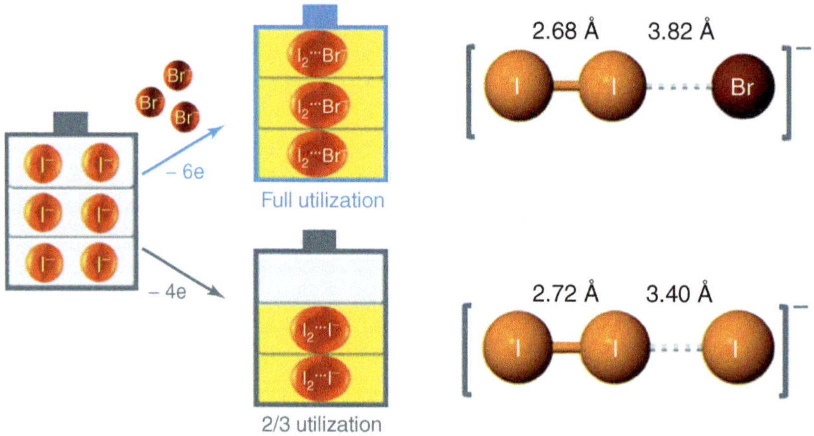


**Figure 5** Concept illustration of bromide as the complexing agent to stabilize iodine. Bonding length of the polyhalide ion of zinc polyhalide is obtained from first-principles density functional theory calculations. [Reproduced with permission from Weng, G.-M.; Li, Z.; Cong, G.; Zhou, Y.; Lu, Y.-C., Unlocking the capacity of iodide for high-energy-density zinc/polyiodide and lithium/polyiodide redox flow batteries. Energy Environ. Sci. 2017. © Royal Society of Chemistry 2017]

tend to have a linear (or nearly linear) trihalide structure, which is thermodynamically stable. Despite the high solubility of the iodide redox species, the overall capacity of this battery is still limited by the zinc negative electrode due to the dendrite formation.[4] Since the same electrolyte is used for both half-cells, the electrolyte crossover is mitigated.

Intuitively, the second strategy to enhance Zn–$I_2$ RFBs energy density is boosting the cell voltage. Zn-base redox couples exhibit dramatical redox potential difference between acidic and alkaline conditions. In acidic electrolyte, $Zn^{2+}$/Zn redox pair shows a redox potential of −0.76 V versus SHE, whereas in alkaline solution $Zn(OH)_4{}^{2-}$/Zn redox pair displays a redox potential at −1.26 V versus SHE. Meanwhile, the redox potential of $I_3{}^-/I^-$ couple is not pH dependent. Keeping the $I_3{}^-/I^-$ couple as the catholyte, Chen and coworkers simply replaced the $Zn^{2+}$ anolyte of Zn–$I_2$ RFBs with the alkaline $Zn(OH)_4{}^{2-}$ electrolyte and increased the cell voltage by 0.47 V as shown in the redox reaction (Equation 18) and Figure 6.[57] Started with the charged state, the designed batteries applied a Zn plate and KOH solution in anode side, and prepared neutral potassium polyiodide solution as catholyte by simply dissolving KI and $I_2$ in water. Nafion 117 was employed as the separator to mitigate the crossover of $Zn(OH)_4{}^{2-}$, $I^-$, $I_3{}^-$ and other species. During discharging process, zinc metal oxidation reaction occurs at the anode in an alkaline KOH anolyte, yielding soluble potassium zincate ($K_2Zn(OH)_4$) (Equation 19).

$$\text{Anode:}\quad Zn(OH)_4^{2-} + 2e^- \underset{\text{discharging}}{\overset{\text{charging}}{\rightleftharpoons}} Zn + 4OH^-$$
$$E^0 = -1.260\ \text{V vs SHE} \tag{18}$$

$$\text{Cathode:}\quad 2I^- + Br^- \underset{\text{discharging}}{\overset{\text{charging}}{\rightleftharpoons}} I_2Br^- + 2e^-$$
$$E^0 = +0.536\ \text{V vs SHE} \tag{19}$$

$$\text{Overall:}\quad Zn(OH)_4^{2-} + 3I^- + 4OH^- \underset{\text{discharging}}{\overset{\text{charging}}{\rightleftharpoons}} Zn + I_3^-$$
$$E_{cell} = 1.796\ \text{V} \tag{20}$$

$K^+$ cation instead of $Zn^{2+}$ shuttles between two electrodes to balance the charge. A small amount of zinc and iodide species were also detected at cathode and anode sides, respectively, which indicates slow active materials crossover. The high-energy-density Zn–$I_2$ RFB was demonstrated with a catholyte concentration of 6 M, which resulted in an energy density of 330.5 Wh $L^{-1}$. However, the battery was only cycled with 44% depth of the state of discharge for 10 cycles. The RFBs with 2 M and 4 M $I_3{}^-$ anolyte faced fast capacity and energy density decay within 10 cycles which might be due to long-term compatibility between the alkaline anolyte and the neutral $I_3{}^-/I^-$ catholyte. We hypothesize that during the cycling, there would be a pH change in both anolyte and catholyte. The pH decrease of the anolyte could result in the formation of $Zn(OH)_2$ precipitate while the pH increase of the catholyte could lead to the chemical decomposition of $I^-$ to iodate ($IO_3{}^-$ ).[58] Besides, dendrite formation was another issue for the demonstrated RFB which caused a fluctuation of EE from 65% to 80% at the current density of 20 mA $cm^{-2}$.

Overcoming the dendrite effect in Zn–$I_2$ RFBs is crucial to meliorate batteries' CE and prolong batteries' lifetime. By means of adding additives such as alcohol, and applying low current density (<20 mA $cm^{-2}$), the

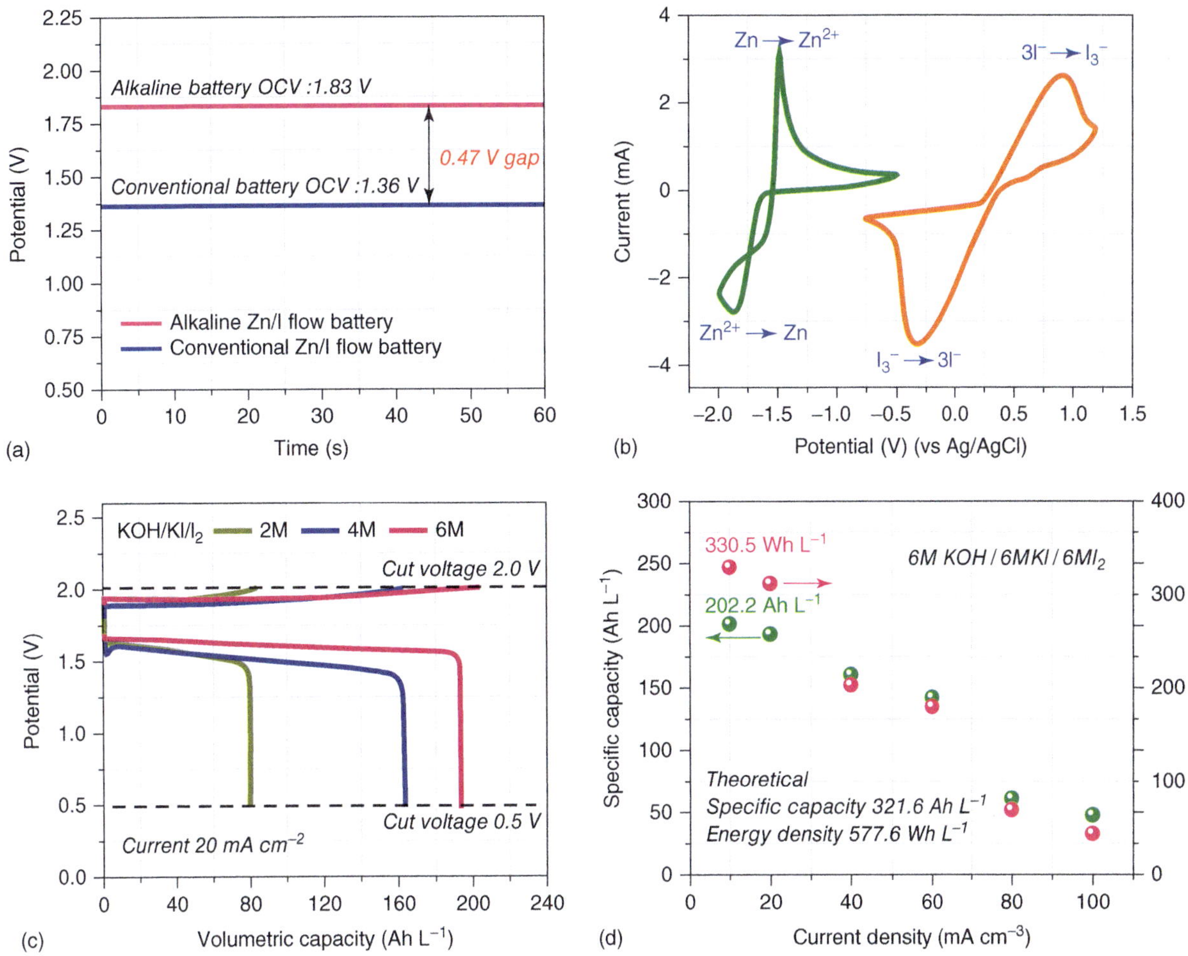


**Figure 6** (a) The open-circuit-voltage between the conventional and designed alkaline Zn–$I_2$ RFB; (b) cyclic voltammograms of 0.1 M $ZnAc_2$ with 3 M KOH (green curve) and 0.1 M KI (yellow curve) on a glassy carbon electrode; (c) representative charge/discharge curves at different electrolyte concentrations; (d) influence of the discharging current density on the achievable specific capacity and energy density. [Reproduced with permission from McCulloch, W. D.; Yu, M.; Wu, Y., pH-Tuning a Solar Redox Flow Battery for Integrated Energy Conversion and Storage. ACS Energy Lett. 2016, 1 (3), 578–582. © Royal Society of Chemistry 2017]

dendrite formation can be suppressed yet not be avoided especially after long-term cycling. Very recently, Li *et al.* reported a concept of a zinc–iodine RFB with dendrite self-healing capability which ensures the battery's long cycle life (1000 cycles) and high power density (Figure 7).[59] A piece of porous polyolefin membrane was employed as the separator. To eliminate the cross-contamination and improve the cycling life, a mixed electrolyte of KI and $ZnBr_2$ was used as both catholyte and anolyte. The pores in the membrane can be filled with a solution containing $I_3^-$ that can react with zinc dendrite. Therefore, by consuming zinc dendrite, the battery can self-recover from micro-short-circuiting resulting from overcharging, although the CE was sacrificed to a certain extent. Owing to the excellent conductivity of both electrolyte and separator, the battery was operated at a high current density (80 mA $cm^{-2}$) and still could maintain an EE of 82%. More impressively, a stack with 700 W output was demonstrated for 300 charge/discharge cycles, which displayed a promising future for practical application. However, long cycling stability is a big concern for this system, which might be attributed to hydrogen evolution in the anode side.

### 4.2 Zn–Fe Redox Flow Batteries

Because of high abundance and low cost of both zinc and iron metals, Zn–Fe RFBs are one of the most promising systems which could meet the capital cost target of \$150 per kWh for grid-scale energy storage by 2030 projected by the Department of Energy, United States. The first demonstrated Zn–Fe RFB was based on the Zn–ferrocyanide chemistry in alkaline solution.[60] However, further development of the alkaline Zn–ferrocyanide

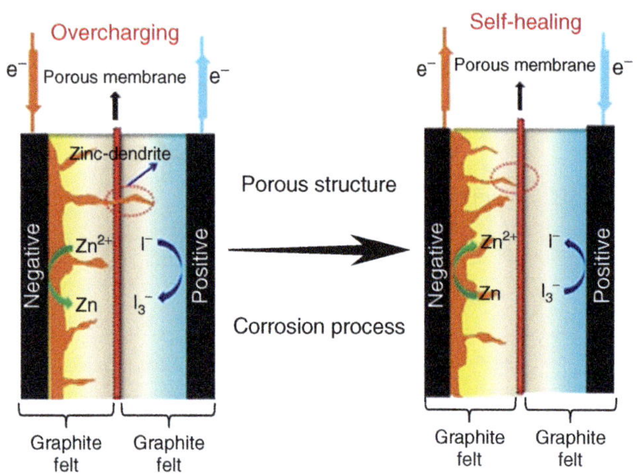


**Figure 7** Self-healing concept for a $Zn-I_2$ RFB with a porous membrane. [Reproduced with permission from Xie, C.; Zhang, H.; Xu, W.; Wang, W.; Li, X., A Long Cycle Life, Self-Healing Zinc-Iodine Flow Battery with High Power Density. Angew. Chem. 2018, 57 (35), 11171–11176. © Wiley-VCH 2018]

RFB was hindered due to the challenges of high cost of Nafion membrane, zinc dendrite formation, and ZnO precipitate generation from the supersaturated $Zn(OH)_4{}^{2-}$ electrolyte. Early in 2018, Li *et al.* developed a nonfluorinated ion conductive membrane and 3D carbon felt electrodes for the Zn–ferrocyanide RFB.[61] The low-cost ion selective membrane showed an ultra-high selectivity for hydroxide anion and a good mechanic resistance against zinc dendrite. The 3D carbon felt electrodes worked as a guidance for Zn deposition/stripping which suppressed the formation of dendrite. The battery displayed a good cycling stability and an impressive energy density of 82.8% at 160 mA cm$^{-2}$. Although Zn–ferrocyanide system is very energy efficient, this demonstrated system only utilized capacity of 65% and less than 50% at 100 and 160 mA cm$^{-2}$, respectively. The low active materials utilization could also contribute to the stable cycling stability by compensating the chemical decomposition of ferrocyanide compounds at alkaline conditions, bringing concerns for their applications in alkaline RFBs.[62] Half-cell studies of ferrocyanide compounds have been conducted by Liu *et al.*[61] to examine their chemical and electrochemical stability at different pH conditions. It is revealed that the redox couple functions best at neutral or near neutral conditions (Figure 8). Dramatic capacity decay of the redox couple was observed in a strong alkaline KOH electrolyte (pH = 14), which is due to the chemical decomposition associated with the $CN^-$ ligand dissociation. It is believed that the ligand dissociation can eventually lead to the formation of $Fe(OH)_X$ (X = 2 or 3).

Comparing to ferrocyanide compounds, iron salts (such as $FeCl_2$, $FeSO_4$) have much higher solubility in water. $Fe^{3+}/Fe^{2+}$ redox couples display a good redox reversibility and a redox potential of +0.77 V versus normal hydrogen electrode (NHE) which is 400 mV more positive than ferrocyanide. These merits make iron salts promising candidates as cathode materials in Zn–Fe RFBs. Since iron cations are not compatible with basic electrolytes, acid conditions are needed to avoid the formation of iron hydroxide precipitate and improve $Fe^{3+}/Fe^{2+}$ redox reaction kinetics.[63] In recent years, a few acidic Zn–Fe batteries have been developed.[63,64] However, all of them showed poor cycling stability due to active materials crossover and HER. Aiming to overcome these obstacles, Gong and coworkers put forward a hybrid Zn (alkaline)–Fe (acidic) three channel system with a low capital cost of $100 per Wh as shown in Figure 9.[65] Originated from the concept of the multiple ion-exchange membrane design, the alkaline Zn anolyte and NaCl solution in the middle chamber was separated by a piece of Nafion 211 membrane while an A901 AEM was applied

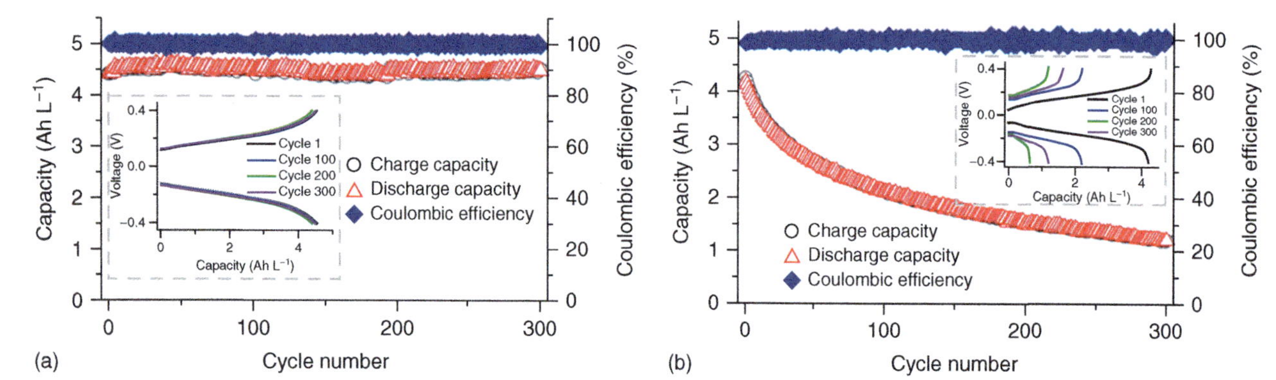


**Figure 8** $K_4Fe(CN)_6/K_3Fe(CN)_6$ half-cell cycling performance at different pH conditions. (a) pH 7.2; (b) pH 14. [Reproduced with permission from Gong, K.; Ma, X. Y.; Conforti, K. M.; Kuttler, K. J.; Grunewald, J. B.; Yeager, K. L.; Bazant, M. Z.; Gu, S.; Yan, Y. S., A zinc-iron redox-flow battery under $100 per kW h of system capital cost. Energy Environ. Sci. 2015, 8 (10), 2941–2945. © Royal Society of Chemistry 2015]

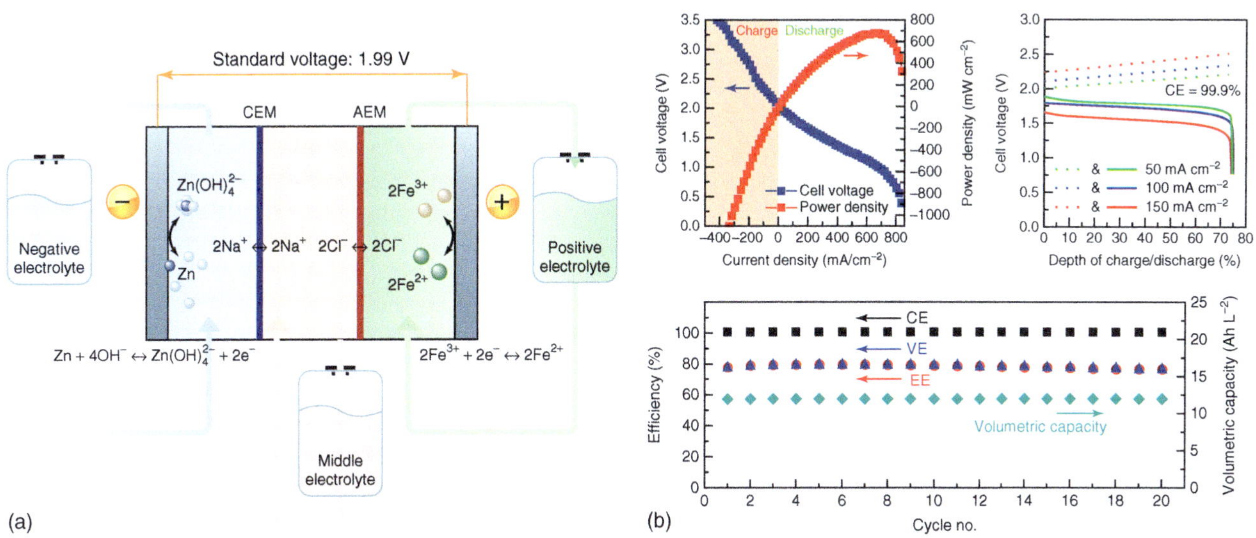


**Figure 9** (a) Schematic of the Zn–Fe RFB; (b) power density, rate dependent and cycling performance of the demonstrated Zn–Fe RFB. [Reproduced with permission from Ref. 68. © 2011 Elsevier]

between the cathode side and the middle chamber. This assembling of a two-separator-three-chamber system enables different pH values in the cathode side ($FeCl_2$ in HCl) and the anode side ($K_2Zn(OH)_4$ in KOH), which effectively solved the problems of both active materials crossover and the HER side reaction. The demonstrated system showed a low resistance of 2.3 Ω cm$^2$. When the flow rate was increased to 100 mL min$^{-1}$, battery's overall EE can be boost up to 74% at 80 mA cm$^{-2}$. With a high voltage of 1.99 V, the battery was capable of delivering an impressive power density of 676 mW cm$^{-2}$. However, only 20 charging/discharging cycles were reported. Long-term cycling reliability of this system still needs to be verified. It is believed that the hybrid alkaline/acidic Zn–Fe RFB suffers to similar issues as the hybrid alkaline/acidic Zn/$I_2$ RFB (Figure 6).

Owing to the issue of iron cation hydrolysis at pH neutral condition, a few studies have investigated pH neutral Zn–Fe RFBs. The first neutral Zn–Fe system was reported by Li *et al.*[66] Glycine was applied as ligand reagent to complex free iron cation thus not only alleviate iron hydrolysis but also improve the redox kinetic of $Fe^{3+}/Fe^{2+}$ couples. A piece of polybenzimidazole (PBI) porous membrane was employed as a separator which reduced active materials crossover. The resulted RFB displayed a CE of 97.75%, and an EE of 86.66% which was stable for 100 cycles. However, fast capacity (energy density) decay was observed during the cycling, which might come from iron complexes crossover and chemical decomposition. For this particular system, future work should focus on ligand design for iron cation stabilization and more ion-selective membranes for better active materials retention.

### 4.3 Zn–Ce Redox Flow Batteries

Zinc–cerium (Zn–Ce) RFBs were first developed by Plurion Inc. (UK) in 2004 with a 2 KW system installed.[67] The Zn–Ce system displays the highest cell voltage of 2.43 V among all aqueous RFBs due to the very positive redox potential (1.28–1.72 V vs SHE) of $Ce^{4+}/Ce^{3+}$ pair. In this battery, $Ce(CH_3SO_3)_3$ and $Zn(CH_3SO_3)_2$ were employed as cathode and anode active materials, respectively.[49] A piece of Nafion membrane was applied as the separator. Different from vanadium base sulfate compounds which display good solubility in sulfuric acid, the $Ce^{3+}$ active species are much less soluble (0.3 M) in the presence of sulfate. Methanesulfonic acid was found to be the best supporting electrolyte which enables good solubility of Ce species. High acid strength facilitates the solubility of $Ce^{4+}$ active species. However, the solubility of $Ce^{3+}$ decreases at higher acid concentrations which results in the highest operation concentration for $Ce^{3+}$ active material of less than 1 M (Figure 10).[68]

In the last 15 years, a number studies have been conducted to investigate the chemical stability, solubility, and battery charging/discharging overpotential of Zn–Ce RFBs.[52,69] Lots of efforts have also been made to protect zinc anode and improve the CE.[70] Based on the intrinsic chemistry and configuration, and the research progress of the Zn–Ce RFBs, several main challenges for Zn–Ce RFBs are summarized as follows: (i) relatively low energy density (25–35 Wh L$^{-1}$) due to the low solubility of active materials;[68] (ii) the poor redox kinetics of $Ce^{4+}/Ce^{3+}$ pair;[68] (iii) low CE due to Zn corrosion in the strong acid electrolyte and active materials crossover;[70] and (iv) water splitting side reaction because of the high operation potential window.[49]

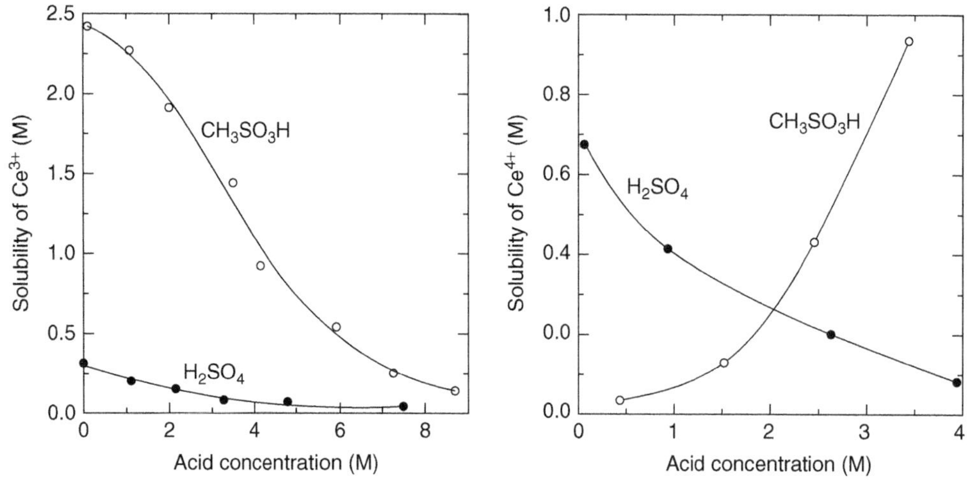


**Figure 10** Solubility of $Ce^{3+}$ and $Ce^{4+}$ species in $H_2SO_4$ and $CH_3SO_3H$ supporting electrolytes with different concentrations. [Reprinted with permission from Gong, K.; Xu, F.; Grunewald, J. B.; Ma, X.; Zhao, Y.; Gu, S.; Yan, Y., All-Soluble All-Iron Aqueous Redox-Flow Battery. ACS Energy Lett. 2016, 1 (1), 89–93. © 2016 American Chemical Society]

### 4.4 Zn–Organic Redox Flow Batteries

In addition to Zn–inorganic RFBs, zinc–organic hybrid aqueous systems have also been drawing many attentions recently. Comparing to inorganic redox-active materials, organic molecules have displayed several advantages for RFB application, such as great elemental abundance and potential low costs, excellent redox kinetics, and tunable properties (solubility, charge state, and redox potential). Several zinc–organic hybrid systems, based on zinc/TEMPO and zinc/polymer, have been developed and demonstrated to be promising for highly energy dense and low-cost energy storage.[50,51]

## 5 ALL-IRON REDOX FLOW BATTERY

The redox chemistry of all-iron RFBs is based on the $Fe^{3+}/Fe^{2+}$ redox couple at the cathode and the $Fe^{2+}/Fe^0$ couple at the anode (Equations 21–23),[71] which are semi-flow flow batteries similar as Zn-based RFBs. During the discharging process, $Fe^{3+}$ is reduced to $Fe^{2+}$ in the cathode side while metallic Fe is dissolved into the electrolyte and oxidized into $Fe^{2+}$. The processes are reversed during battery charging.

$$\text{Anode:}\quad Fe^{2+} + 2e^- \underset{\text{discharging}}{\overset{\text{charging}}{\rightleftharpoons}} Fe^0$$
$$E^0 = -0.44 \text{ V vs SHE} \tag{21}$$

$$\text{Cathode:}\quad 2Fe^{2+} \underset{\text{discharging}}{\overset{\text{charging}}{\rightleftharpoons}} 2Fe^{3+} + 2e^-$$
$$E^0 = +0.77 \text{ V vs SHE} \tag{22}$$

$$\text{Overall:}\quad 3Fe^{2+} \underset{\text{discharging}}{\overset{\text{charging}}{\rightleftharpoons}} 2Fe^{3+} + Fe^0$$
$$E_{\text{cell}} = 1.21 \text{ V} \tag{23}$$

Using the inexpensive and highly soluble $FeCl_2$ as the bipolar active materials for both cathode and anode, an aqueous all-iron redox flow rechargeable battery can deliver cell voltage of 1.21 V and theoretical energy density of 76 Wh $L^{-1}$ which stands for a promising, low-cost, durable, and eco-friendly energy storage system for large-scale applications. The very first all-iron RFB was reported by Hruska and Savinell in 1981.[71] Although it has been studied for almost 30 years, the all-iron RFB is still limited to lab research level due to several challenges such as low CE, low EE, and poor cycling stability. The sacrificed CE is mainly due to the hydrogen evolution and $Fe^{3+}$ crossover.[4] There is an intrinsic conflict between suppressing HER side reaction and hindering the iron cation hydrolysis in all-iron RFBs. The pH values of catholyte and anolyte are kept around 1 and 4, respectively. At this condition, HER competition in anode side could be thermodynamically favorable. Moreover, during charging process of the battery, the electro-deposition of iron presents substantial overpotential thereby facilitating the HER further. All-iron RFBs use Nafion membranes to minimize the resistance of the whole battery. In this circumstance, maintaining the pH difference is difficult. Once the acidity decreases in the anode side, precipitation of iron hydroxide would occur.[72] Meanwhile, $Fe^{3+}$ could cross through the membrane and react with $Fe^0$ which causes serious self-discharge. Efforts has been made to preserve the pH difference and minimize the Fe species crossover by employing AEMs.[72] However, the high resistance of the membrane resulted in a low EE (50%) and a low power density (50 mW cm$^{-2}$). The low EE of the all-iron RFBs is mainly owing to poor redox reaction kinetics of iron deposition/stripping process and low ion conductivity of the electrolyte.

To suppress iron cation hydrolysis and improve $Fe^{3+}/Fe^{2+}$ redox couple kinetics, a series of organic ligands were evaluated such as citrate, dimethyl sulfoxide (DMSO), glycerol, glycine, malic acid, malonic acid, and xylitol.[73] Among the studied ligands, glycine turned out to be the best for improving the reversibility of $Fe^{3+}/Fe^{2+}$ couple redox reaction. It was reported that a stable Fe cathode electrolyte (1:1 glycine to iron, pH = 2) displayed a solubility of 0.5 M and a redox potential of 0.690 versus SHE, suitable for the use as a positive redox couple in the all-iron flow battery. This ligand stabilized Fe electrolyte was also applied in Zn–Fe RFBs, which we introduced above. Glycine as a ligand shows a weak electronic effect to the iron metal center, as the $Fe^{2+}_{glycine}/Fe^{3+}_{glycine}$ couple shows a very close redox potential with free $Fe^{2+}/Fe^{3+}$. However, other ligands can play an important role of tuning Fe-ligand complexes' redox potential as shown in Figure 11.

Noteworthy, when triethanolamine (TEA) acts as the ligand, the resulted $[Fe(TEA)OH]^{-}/[Fe(TEA)OH]^{2-}$ couple shows a redox potential as low as −0.86 V versus SHE in alkaline solution (Figure 10), which is very attractive for high-voltage RFBs. The redox reaction is highly reversible which indicates excellent kinetics. The properties of the Fe–TEA complexes were first studied by Yang *et al.*[74] With an increase of the concentration of TEA, the solubility of Fe(III)–TEA can be increased to 0.6 M, and the solubility of Fe(II)–TEA is up to 0.4 M. Pairing the Fe(II)–TEA electrolyte (in 1 M NaOH) solution with neutral NaBr electrolyte, the resulted Fe–TEA/$Br_2$ RFB delivered a high OCV up to 2 V. Owing to the high viscosity of the anolyte, relatively low current density (20 mA cm$^{-2}$) was applied to charge and discharge the battery. An EE of 70% was obtained. However, long-term cycling stability is poor due to the active material ($Br_2$) crossover and pH change of the electrolyte. In addition, there was no post-cell analysis to verify stability of the Fe–TEA complexes.

Following Yang and coworkers' study,[74] Gong *et al.*[75] demonstrated an all soluble all iron alkaline RFB

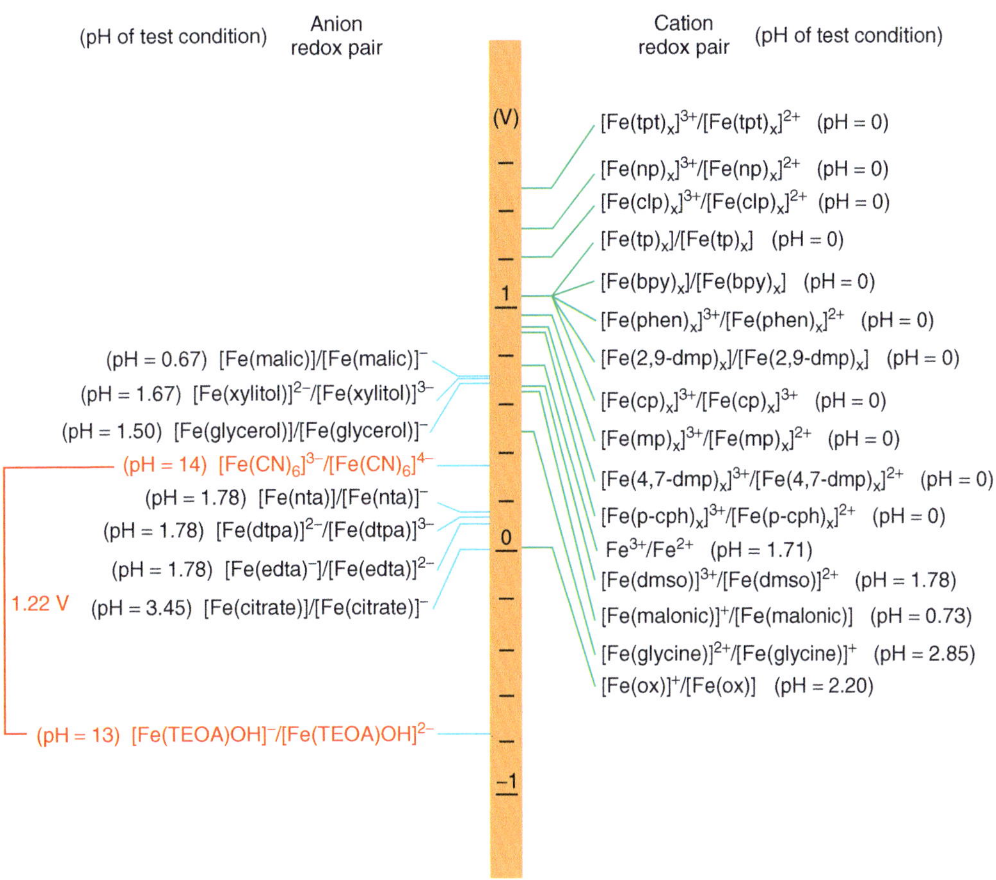


**Figure 11** Redox pairs of iron complexes that have been tested for RFB applications. [Reprinted with permission from Gong, K.; Xu, F.; Grunewald, J. B.; Ma, X.; Zhao, Y.; Gu, S.; Yan, Y., All-Soluble All-Iron Aqueous Redox-Flow Battery. ACS Energy Lett. 2016, 1 (1), 89–93. © 2016 American Chemical Society]

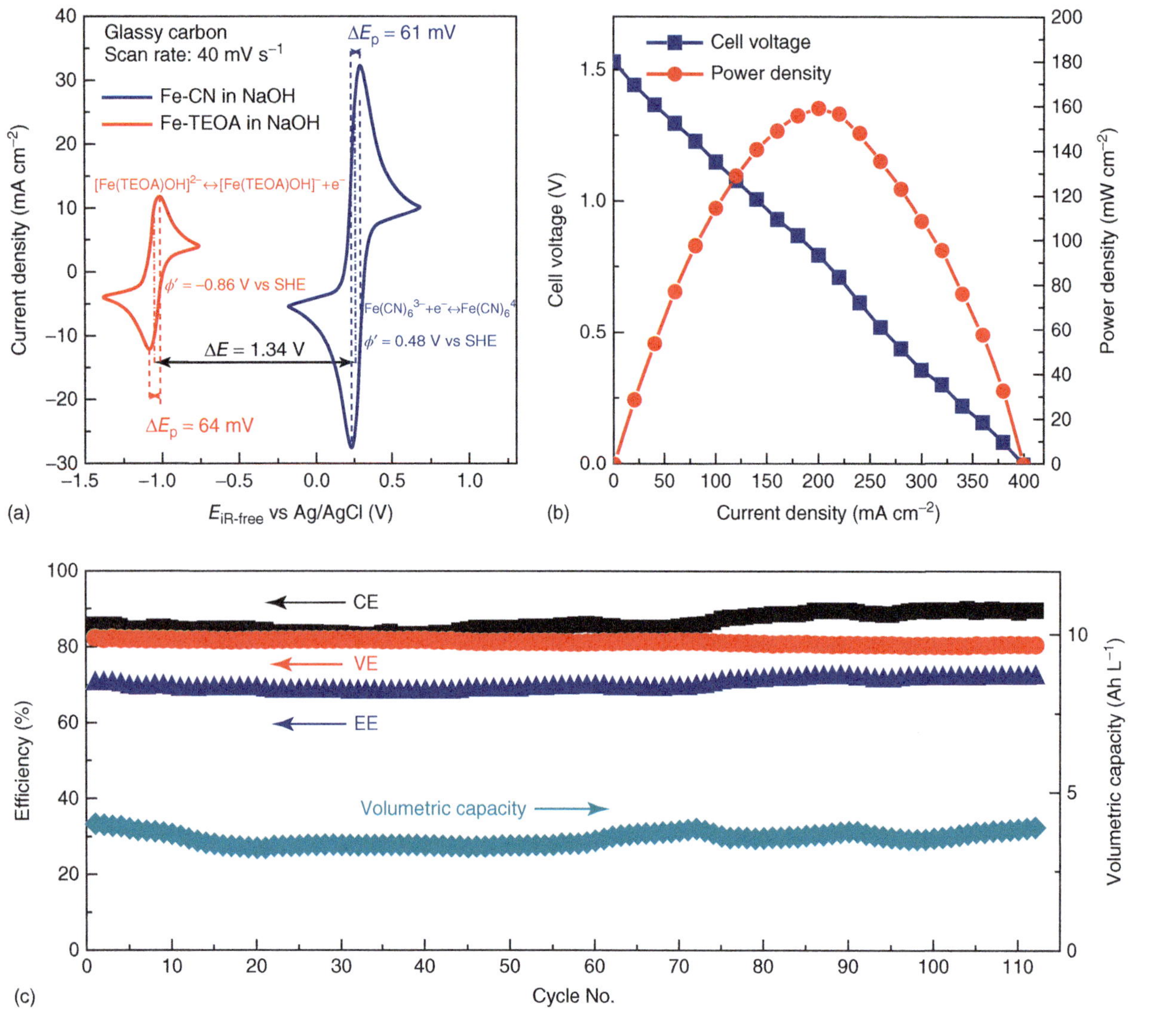


**Figure 12** (a) Cyclic voltammograms of $K_2Fe(CN)_6$ and Fe–TEA in NaOH solution; (b) power density of the all-iron RFB; (c) cycling performance of the all-iron RFB. [Reproduced with permission from Ref. 85. © Wiley VCH 2016]

based on an Fe(II)–TEA anolyte and an $K_4Fe(CN)_6$ catholyte. The reported all-iron RFB showed a cell voltage of 1.34 V and could deliver a power density of 160 mW cm$^{-2}$. With the solubility limit of the Fe–TEA complex, the RFB long-term cycling was conducted at the concentration of 0.2 M at 40 mA cm$^{-2}$, which indicated a low energy density of less than 3 Wh L$^{-1}$. A low CE of 80–90% was obtained, which was due to ligand dissociation and free TEA crossover (Figure 12).

Along with aqueous all-iron RFBs, concepts for nonaqueous all-iron RFBs were also reported in ionic liquids.[76,77] In the nonaqueous condition, HER side reaction and $Fe(OH)_X$ formation could be overcome. Ionic liquids were prepared by mixing $FeCl_X$, choline chloride, and ethylene glycol and used as electrolyte. The ionic liquids have acceptable intrinsic conductivities (without the need for added supporting electrolyte), high concentrations of iron (up to 6.3 M), large voltage windows, and enhanced safety due to the inert nature of the precursor materials as well as their low volatility. However, no battery cycling data was reported.

## 6 POLYOXOMETALATE AND HETEROPOLYACID REDOX FLOW BATTERY

A polyoxometalate (POM) is an oxoanion containing more than one metal atom formed by the linkage of $d^0$ metal-centered polyhedral with oxygen atoms located at the vertices. The versatility and accessibility of POMs have led to many applications such as catalysis,

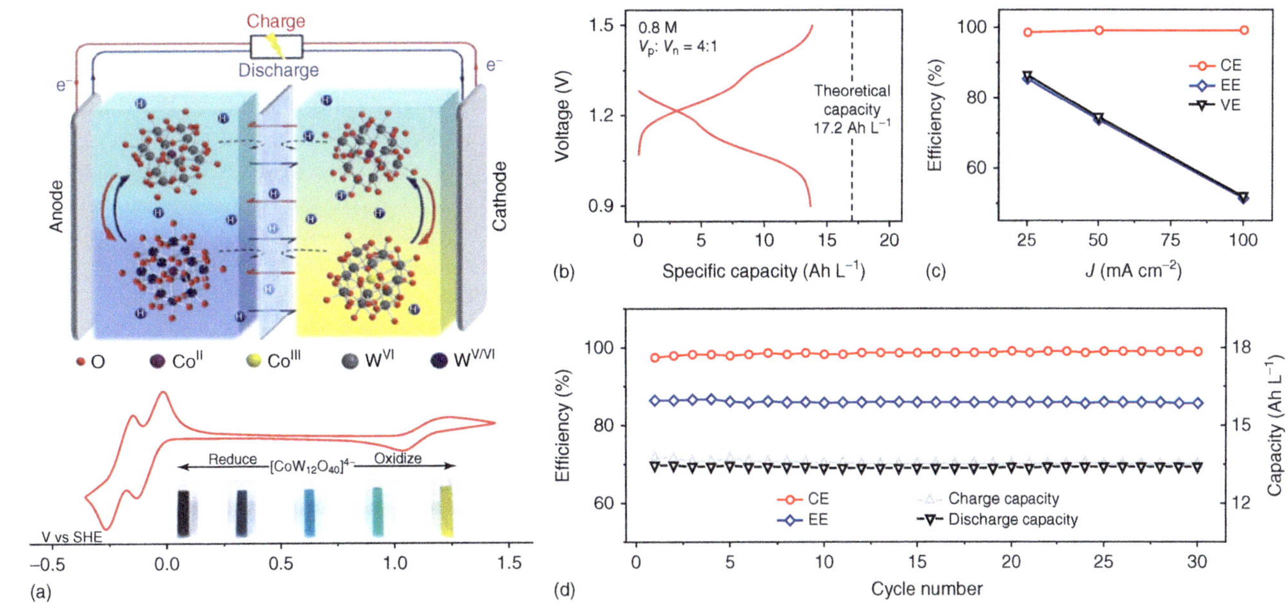


**Figure 13** (a) Cyclic voltammograms of $H_6CoW_{12}O_{40}$; (b) representative charge/discharge curve for the battery; (c) efficiency performance of the demonstrated RFB at various current density; (d) cycling performance of the RFB. [Reproduced with permission from Ref. 105. © Cell Press 2017]

luminescence, electronic conductors, and potential medicinal applications.[78] In addition, a POM has the capability to undergo reversible multi-electron redox processes due to multiple redox-active centers.[79] This desired property is very attractive for energy storage application because of the chance to largely improve energy density (Equation 3). In the recent years, POMs have been widely applied in lithium ion batteries, supercapacitors, and other electrochemical devices.[80]

Besides the merits of POMs mentioned above, POMs are easy to synthesize, stable, and soluble in water or organic solvents.[81] These advantages make POMs promising candidates for RFB applications, especially for using them as bipolar active materials (the same material for both cathode and anode) in a RFB, which overcomes the CE loss and capacity decay resulted from active materials crossover. Pratt *et al.* reported a series of bipolar electrolytes that contain POMs such as $[SiV_3W_9O_{40}]^{10-}$, $[PV_3W_9O_{40}]^{9-}$, and $[(SiFe_3W_9(OH)_3O_{34})_2(OH)_3]^{14-}$.[82–84] The RFBs based on these POMs displayed low cell voltages from 0.4 to 0.6 V due to the small redox potential difference between two redox events. Very low energy density (less than 1 Wh L$^{-1}$) was demonstrated because of the low concentrations of active materials. Owing to the poor redox reaction kinetics, all the three POM-based RFBs showed poor power densities and energy efficiencies (lower than 50%) even though strong acids were used as supporting electrolytes. In 2016, Yan and coworkers developed an acid form of POM material, $H_6CoW_{12}O_{40}$ (Tungsten–Cobalt Heteropolyacid), as a bipolar active material for an acidic aqueous RFB.[85] This heteropolyacid can undergo five electron redox processes including two two-electron redox reactions (for a total of four electrons) used the anode side, and a one-electron redox reaction used as the cathode side, giving a cell voltage of 1.2–1.4 V, as shown in Figure 13 and electrochemical reactions.

Anode:

$$Co^{II}W_{12}O_{40}^{6-} + 2e^- + 2H^+ \underset{\text{discharging}}{\overset{\text{charging}}{\rightleftharpoons}} H_2Co^{II}W_{12}O_{40}^{6-}$$
$$E^0 = -0.074 \text{ V vs SHE} \tag{24}$$

Anode:

$$H_2Co^{II}W_{12}O_{40}^{6-} + 2e^- + 2H^+ \underset{\text{discharging}}{\overset{\text{charging}}{\rightleftharpoons}} H_2Co^{II}W_{12}O_{40}^{6-}$$
$$E^0 = -0.191 \text{ V vs SHE} \tag{25}$$

Cathode:

$$Co^{II}W_{12}O_{40}^{6-} \underset{\text{discharging}}{\overset{\text{charging}}{\rightleftharpoons}} Co^{II}W_{12}O_{40}^{5-} + e^-$$
$$E^0 = +1.103 \text{ V vs SHE} \tag{26}$$

Along with a solubility of 1.0 M in water at room temperature, $H_6CoW_{12}O_{40}$ is very suitable for RFB application. As the $[CoW_{12}O_{40}]^{6-}$ anion is highly negatively charged and has a very large size, Nafion separator can effectively avoid the active materials crossover, which ensure a high CE. When one $H_6CoW_{12}O_{40}$ molecule is dissolved in water, it dissociates to give the redox-active anion and 6 protons which can be used as the charge shuttle. Therefore, no supporting electrolyte was required for this system. The all-$H_6[CoW_{12}O_{40}]$ RFB exhibited a high energy density of 86% with a GF electrode at a current density of 25 mA cm$^{-2}$ with a volumetric specific energy of 15.4 Wh L$^{-1}$. An extended cycling study was tested for only 30 cycles at 25 mA cm$^{-2}$ displaying a good capacity retention. The follow-up work for the $H_6CoW_{12}O_{40}$ RFB by the same group studied the effect of Nafion membrane thickness on performance of the battery.[86] By replacing Nafion 212 to Nafion 211 membrane, both EE and power density were improved dramatically, while CE was sacrificed. The obtained highest power density of this system was 0.56 W cm$^{-2}$, which represents one of the most powerful aqueous RFBs. It is important to note that, in order to perform best capacity utilization, the volume ratio between the catholyte and the anolyte was tuned to be 4:1 according to the amount of electrons involving in the anode and cathode reactions.

Stimming *et al.* reported two POMs that could be used as active materials for an asymmetric aqueous RFB.[87] Based on the trend that the redox potential decreases with the increasing electronegativity ($\chi$) of POM species, $[SiW_{12}O_{40}]^{4-}$ ($SiW_{12}$, $\chi_W = 2.36$) and $[PV_{14}O_{42}]^{9-}$ ($PV_{14}$, $\chi_V = 1.63$) were selected as anode materials and cathode materials, respectively. An in situ NMR test during an electrochemical experiment revealed excellent chemical and electrochemical stability of $PV_{14}$. Both $SiW_{12}$ and $PV_{14}$ can undergo multi-electron process. PV14 can be reduced by as many as seven electrons, whereas during battery operation the molecule was reduced by only four electrons. $SiW_{12}$ can be reduced by four electrons. With concentrations of 0.1 M for $PV_{14}$ and 0.2 M for $SiW_{12}$, the RFB had a theoretical capacity of 10.7 Ah L$^{-1}$ and was able to reach this value during actual battery operation. Capacity decay was clearly observed in 155 cycles and verified to be due to the ingress of atmospheric oxygen instead of the instability of the redox-active POMs. However, the 4 out of 7 electron utilization limitation of the $PV_{14}$ POM still needs to be further explored and elucidated.

Very recently, Cronin and coworkers reported a polyanion, $[P_2W_{18}O_{62}]^{6-}$, which displays notable proton-coupled electron redox activity and can undergo a series of total 18-electron redox processes.[88] Interestingly, with the increase of concentration of this POM materials, the amount of charge in terms of protons and electrons that can be reversibly stored also increased (4 electrons at the concentration of 2 mM, 16 electrons at 50 mM, and ca.17.2 electrons at 100 mM). Another salient feature is the high solubility $[P_2W_{18}O_{62}]^{6-}$ POM in water. The $Li^+$ counter cation-based POM displays a solubility of 1.9 M, which indicates a theoretical capacity of more than 900 Ah L$^{-1}$. When paired with $Br_2$, the resulted RFB gives an extremely high theoretical energy density up to 1000 Wh L$^{-1}$, three times of Zn–$I_2$ RFB. The demonstrated 1.25 V POM–$Br_2$ RFB with a POM concentration of 0.5 M in acidic electrolytes delivered an energy density of 225 Wh L$^{-1}$, a power density of 0.5 W cm$^{-2}$ and stable capacity retention for 20 cycles. However, the chemical stability and long-term cycling stability of POM molecules need to be further studied. In addition, the deeply reduced materials can also react with protons to generate hydrogen even at pH neutral condition in the absence of Pt/C catalyst. Therefore, the inevitable HER side reaction could jeopardize the long-term stability of $[P_2W_{18}O_{62}]^{6-}$ based acidic flow battery.

## 7 POLYSULFIDE–POLYHALIDE REDOX FLOW BATTERIES

Polysulfide has been recognized as a promising candidates as anolyte materials of a RFB due to its high abundance, low cost, and good solubility in aqueous media.[67] Polysulfide–polyhalide RFBs received the most attention and commercialization has been reported up to a 1 MW system. Polysulfide–bromine RFB system is developed by Regenesys Ltd.[89] For a polysulfide–bromine redox flow battery (SBRFB), during the charging cycle, the bromide is oxidized to bromine and complexed as tribromide ions in the cathode side. In the anode side, the sulfur present as soluble polysulfide anion is reduced as shown in Equations (27)–(29).[67] Normally, sodium or potassium cations are applied as charge carriers, thus cation exchange membrane is required for the SBRFB systems.

$$\text{Anode:}\quad S_4^{2-} + 2e^- \underset{\text{discharging}}{\overset{\text{charging}}{\rightleftharpoons}} 2S_2^{2-} \qquad E^0 = -0.256\ \text{V vs SHE} \tag{27}$$

$$\text{Cathode:}\quad 3Br^- \underset{\text{discharging}}{\overset{\text{charging}}{\rightleftharpoons}} Br_3^- + 2e^- \qquad E^0 = +1.09\ \text{V vs SHE} \tag{28}$$

$$\text{Overall:}\quad 3Br^- + S_4^{2-} \underset{\text{discharging}}{\overset{\text{charging}}{\rightleftharpoons}} 2S_2^{2-} + Br_3^- \qquad E_{\text{cell}} = 1.346\ \text{V} \tag{29}$$

The SBRFB suffers from the same cathode issues as other bromine base systems, such as Zn–$Br_2$ RFB discussed above. Besides, the poor kinetics of $S_2^{2-}/S_4^{2-}$ redox reaction also restrains the EE of the flow battery. Therefore, catalysts are essential for the polysulfide anode to improve battery performance. In recent years, metal sulfide-based catalysts have been studied to facilitate the polysulfide redox reaction.[90] For example, Yi and coworkers first employed nickel foam pretreated with $Na_2S_4$ solution as an electrocatalytic anode current collector in a SBRFB. EE was obtained as high as 77.2% at a current density of 40 mA $cm^{-1}$, which attributes to the generated $NiS_X$ catalyst on the surface of Ni foam.[90]

As an analog of the SBRFB system, polysulfide–iodine redox flow batteries (SIRFBs) also stand for a cost-effective system that are promising to meet the \$150 per kWh price target set by DOE. SIRFBs have been drawing increasing attention than SBRFB for fundamental studies due to much better solubility and redox kinetics of $I^-/I^{3-}$ species than $Br^-/Br_3^-$. Without any catalyst, CV shows that $I^-/I_3^-$ redox pair displays much smaller peak–peak separation (240 mV) than $Br^-/Br_3^-$ (730 mV) at the scan rate of 100 mV $s^{-1}$ at a carbon electrode.[9] Lu *et al.* reported a high-energy and low-cost SIRFB using nickel foam as anode electrode to facilitate $S_2^{2-}/S_4^{2-}$ redox reaction.[9] 3.3 M $K_2S_2$ in 1 M KOH and 6 M KI aqueous solution were employed as an anolyte and an catholyte, respectively. This combination gives a cell voltage of 1.05 V, which is lower than SBRFB because of the relatively lower redox potential of $I^-/I^{3-}$ redox couples than $Br^-/Br_3^-$. With a high concentration of active materials, the achieved energy density reaches 43.1 Wh $L^{-1}$. However, the high viscosity and low conductivity of the concentrated electrolyte resulted in low EE, only 55% at the current density of 20 mA $cm^{-2}$. In 2017, Brushett group also demonstrated a SIRFB using $Na^+$ as the charge shuttle.[91] The battery gives consistent cycling performance for more than 200 cycles. However, the EE is still not satisfactory. In addition, it is worth noting that polysulfides are not chemically stable under alkaline conditions,[92] meaning they should be used at pH neutral conditions for more stable cycling performance.

Besides the above-discussed RFBs, other inorganic-based aqueous RFBs have also been developed, such as hydrogen–halogen RFBs, soluble lead RFBs, and copper RFBs.[4] However, they are less attractive because of different technological limitations and will not be discussed here.

## 8 OTHER NONAQUEOUS INORGANIC REDOX FLOW BATTERIES

### 8.1 Lithium-Based Semi-Flow Redox Flow Batteries

Several lithium-based semi-flow RFBs have been reported using lithium metal as anode paired with polysulfide, iodide, and organic materials cathode materials.[93–96] Cui *et al.* first demonstrated the concept of a membrane-free Li/polysulfide hybrid battery with a theoretical energy density of 105 Wh $L^{-1}$.[93] However, the battery test under real flow condition was not conducted. An aqueous/nonaqueous hybrid $Li/I_2$ RFB was reported Byon and coworkers.[94] The LiI aqueous cathode served as a highly soluble and stable $I^-/I_3^-$ redox couple in the RFB, providing high energy and power densities (up to 130 mW $cm^{-2}$). 1 M $LiPF_6$ in ethylene carbonate (EC)/dimethyl carbonate (DMC) was used as anolyte. Solid electrolyte (LAPT) was applied as the separator to allow the Li ion conducting and avoid the crossover of any other materials and solvent. The addition of the flow device and aqueous electrolyte reservoir resulted in a further increase in capacity up to 90% of the theoretical capacity. The demonstrated $Li/I_2$ RFB displayed a total 100 mAh of battery performance without significant capacity loss and potential degradation in 20 cycles.

Although high energy density can be achieved with these systems, they sacrifice the capability of decoupling power and energy of a RFB. In order to keep both advantages, a solid-suspension approach was proposed using Li-intercalation materials as cathode materials which are suspended in solvent phase homogeneously.[97] As most of the solid active materials are insulated, conductive additives are required to improve the ionic conductivity of the slurry electrolyte. Duduta *et al.*[97] first applied this concept using $LiCoO_2$ slurry (20 vol%, 10.2 M and 1.5% ketjenblack) and $Li_4Ti_5O_{12}$ (10 vol%, 2.3 M and 2% ketjenblack) slurry as cathode and anode, respectively. The following study on silicon anode material, polysulfide, $LiTi(PO_4)_3$, $LiFePO_4$, and other material using similar methods were also conducted. The solid-suspension electrolyte-based RFBs can achieve a high theoretical energy density. For example, Duduta *et al.*[97] estimated a high energy density up to 397 Wh $L^{-1}$ for the $LiCoO_2$–$Li_4Ti_5O_{12}$ if the volume loading is as high as 40%. However, the low conductivity and high viscosity of the electrolyte result in low capacity utilization of these RFBs. Meanwhile, the low volume ratio (normally 20 vol%) of active material also limits further improvement of batteries' energy density.[98]

This issue could be overcome with a redox targeting approach which was first proposed by Wang *et al.*[99] The main idea of the redox targeting is to apply soluble redox-active molecules to shuttle electrons between insoluble active materials and electrodes, which allows the insoluble active materials to be oxidized/reduced without using a significant amount of conductive additives. Different from a typical RFB, in redox targeting RFBs, electron mediator solution flows through the cell while solid active materials such as $LiFePO_4$ are maintained statically in the anode and cathode reservoirs.[98] Using this strategy, several systems reported by Wang *et al.* have been demonstrated with very high energy density and active materials utilization.[100–104] In these RFBs, the redox potentials of redox mediators are required to match those of the solid active materials. This prerequisite requires the careful selection of redox targeting molecules. In the first reported redox targeting RFBs, two redox mediators with a single redox event[101,104] or one redox mediator with two redox events[102,103] were used to bracket the redox potential of

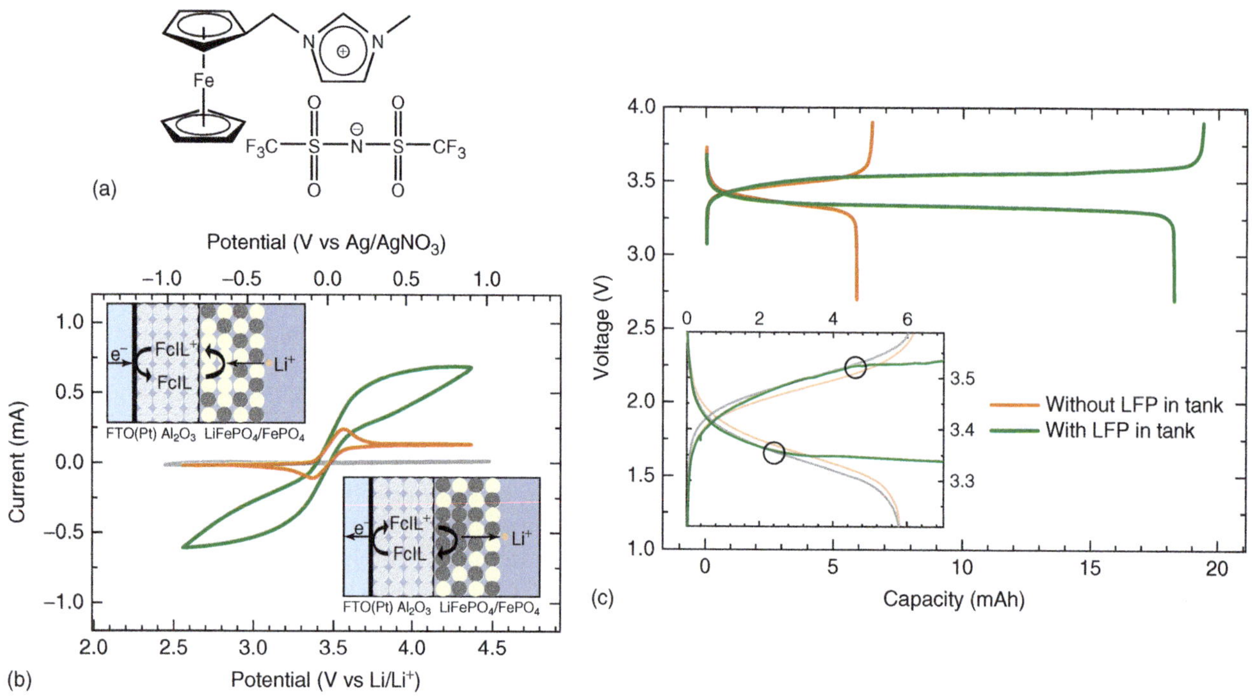


**Figure 14** (a) Molecular structure of FcIL. (b) Cyclic voltammograms of 50 mM FcIL on a double-layer electrode in the absence (orange) and presence (green) of $FePO_4/LiFePO_4$ (1:1). For comparison, the gray curve shows the CV of a $FePO_4/LiFePO_4$-coated double-layer electrode in the absence of FcIL in the electrolyte. The electrolyte was 1MLiTFSI/PC. The scan rate was $2\,mV\,s^{-1}$. The insets illustrate the reaction of FcIL on the double-layer electrode. (c) Voltage profiles of flow cells with 0.20 M FcIL in the catholyte and 0.37 M equivalent $LiFePO_4$ granules in the tank. The inset is the enlarged voltage profiles of the flow cells after IR correction. The electrolyte was 1 M LiTFSI/PC. The current density was $0.025\,mA\,cm^{-2}$. [Reproduced with permission from Ref. 108. © Cell Press 2018]

solid active materials, in which the lower potential process reduces the solid material for lithiation and the higher potential process oxidizes the solid material for delithiation. These two redox systems, however, inevitably induces voltage loss to the system as a result of the relatively large potential difference between the two redox events. More recently, the team reported a single redox mediator-based $LiFePO_4$/Li redox targeting RFB (Figure 14).[105] In this redox targeting RFB, an imidazolium functionalized ferrocene (FcIL, Figure 14a) was ingeniously selected as a single redox mediator. Because of its perfect redox matching with $LiFeO_4$ (Figure 14b), the ferrocene redox mediator through its single redox event at 3.42 V versus Li could promote Nernstian redox-driven reversible charge/discharge of the solid $LiFePO_4$ cathode material at $0.025\,mA\,cm^{-2}$ with a demonstrated energy density of $330\,Wh\,L^{-1}$. The FcIL single redox mediator-based redox targeting RFB clearly outperforms previously reported redox targeting RFBs in terms of VE and capacity utilization of solid cathode materials.

In addition, redox targeting RFBs need to operate at a low current density ($<1\,mA\,cm^{-2}$) to ensure effective redox mediating process to achieve the high charge capacities of solid active materials. In this regard, it is believed that redox targeting RFBs have potential to meet long-duration energy storage needs (hours to days) where power performance is not critically important. Moreover, the redox targeting RFBs also suffers from low CE due to the lack of effective separators in organic solvent which is needed to selectively mitigate the crossover of redox mediators. To overcome the separator issue while achieving higher current and power performance, it appears reasonable to develop redox targeting RFBs using well-developed ion exchange membranes in aqueous environment.

### 8.2 Deep-Eutectic-Solution-Based Inorganic RFBs

Deep eutectic solution (DES) is a mixture of electroactive metal salts, solvation additives, and organic solvent with unique physical properties, such as low melting points and high conductivities, and has been recognized as a distinct class of low-cost catholytes or anolytes that potentially have high concentration of electroactive species for RFB applications. The first concept of proof demonstration of DES redox-active electrolytes was reported by Zhou *et al.* in an Fe–DES/Li RFB.[106] The deep brown Fe-based DESs were simply prepared by mixing $FeCl_3{\cdot}6H_2O$ and urea and achieved a concentration of 5.4 M for $FeCl_3$ (a charge capacity of $144.72\,Ah\,L^{-1}$) with a freezing point at about −7.1 °C. Such Fe–DES is attractive

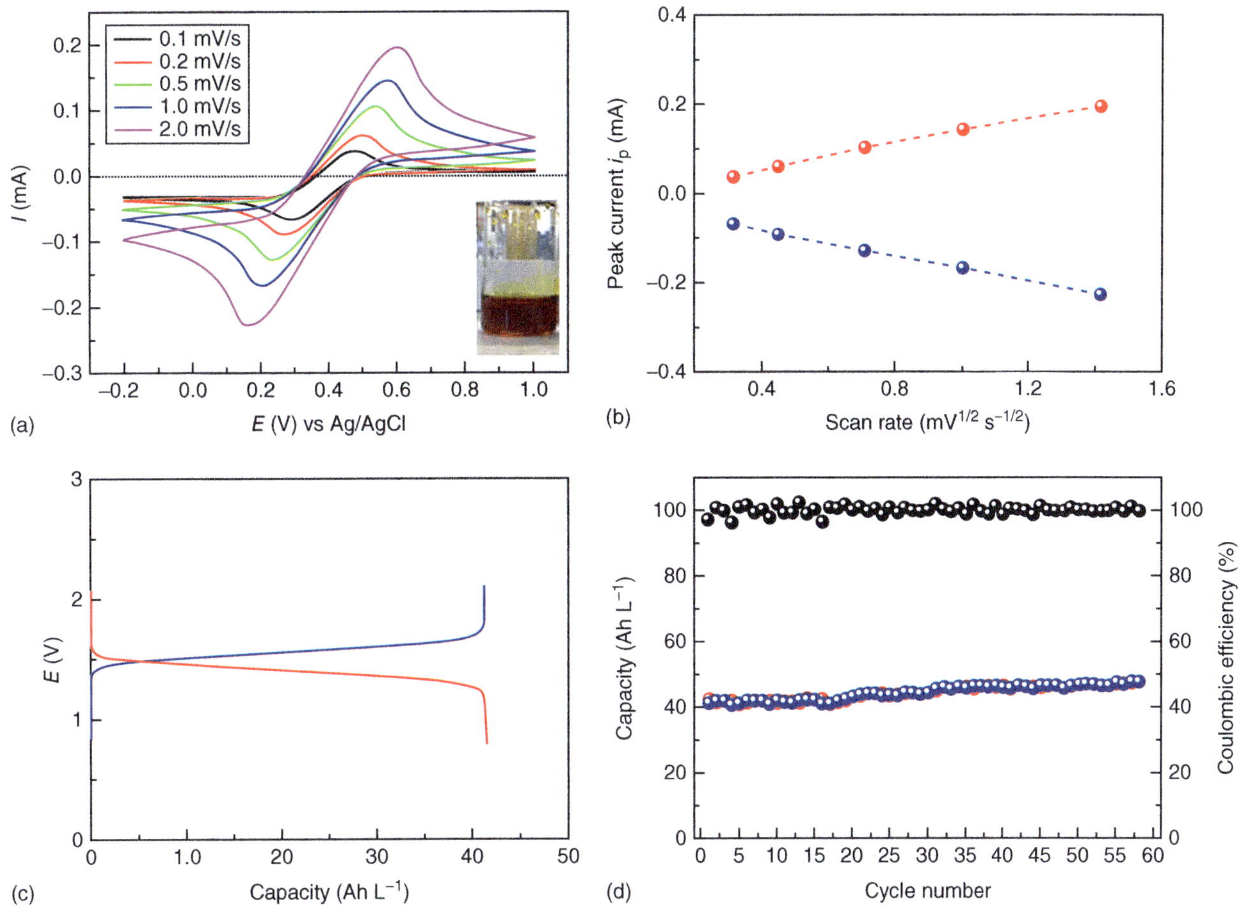


**Figure 15** (a) CV curves of Fe(126) DES at different sweeping rates. (b) Linear relationships between the oxidation and reduction peak current with the square root of the sweeping rate. (c) Charge/discharge profile of Fe(126)–DES/EC–DMC (1.7 M, 0.05 mL) paired with Al–DES/DCE (3.2 M, 0.1 mL) at current density of 0.1 mA $cm^{-2}$. (d) The charge and discharge capacity with corresponding coulombic efficiency of an Fe(126)–Al battery at room temperature over cycling. [Reproduced with permission from Ref. 107. © Cell Press 2017]

cathode-active materials because of their stability, high earth abundance, and high redox potential ($Fe^{3+/2+}$ at 0.77 V vs NHE or 3.4 V vs Li). At 40 °C, 97% capacity of the Fe–DES was achieved in a Li semi-flow battery, corresponding to an exceptionally high energy density of ca. 480 Wh $L^{-1}$. Subsequently, Yu *et al.* developed a low-cost Al–DES anolyte (2.2 V vs Li) to demonstrate a semi-flow battery with $I_3^-/I^-$ and $Br_3^-/Br^-$ with a high energy density of up to 200 Wh $L^{-1}$.[107] More recently, Yu group combined a new Fe(126)–DES catholyte (where 126 stands for the 1:2:6 ratio of $FeCl_3{\cdot}6H_2O$/urea/ethylene glycol) and the Al–DES anolyte to demonstrate a 1.4 V full DES RFB with a demonstrated energy density of 58.1 Wh $L^{-1}$ (Figure 14).[108] In spite of a low energy density compared to the reported DES/Li RFBs, the Fe–DES/Al–DES RFB delivered an impressive cycling stability for ca. 1500 h at 0.1 mA $cm^{-2}$ (Figure 14d). Low current performance (1 mA $cm^{-2}$) and low power densities (<5 mW $cm^{-2}$) of the reported DES RFBs are largely due to the use of a solid electrolyte separator with a low ion conductivity which is necessary to avoid the crossover of active materials. Thus, similar as redox targeting RFBs, DES RFBs with their high energy densities are more suitable for long duration energy storage applications (hours to days). In addition, it is important to develop a cost model for DES RFBs regarding their capital costs by considering both low-cost DES electrolytes and expensive solid electrolyte separator (Figure 15).

## 8.3 Polyoxometalate-Based Nonaqueous Inorganic RFBs

Not limited to aqueous systems, POMs-based nonaqueous RFBs have also been developed. To meet the requirement of nonaqueous RFB application, Matson *et al.* developed a series of POMs based on Ti and V metal elements and applied them as bipolar active materials.[109,110] Alkyl oxide groups, –OR (R represents for methyl or ethyl group), were introduced to improve the cluster's solubility in organic solvents. The designed $[V_6O_7(OR)_{12}]$ nonaqueous RFB showed a cell voltage of 1.6 V. The incorporation of one ($[TiV_5(OCH_3)_{13}]$) or two ($[Ti_2V_4(OCH_3)_{14}]$) titanium ions within the Lindqvist core ($[V_6O_7(OR)_{12}]$) significantly increases the cell voltage of the system (from 1.60 V, to 2.30 and 2.74 V, respectively) while the solubility and redox stability observed for cluster complexes is retained.

## 9 PERSPECTIVE

Over the past several decades, aqueous inorganic RFBs based on various chemistries have been proposed and developed in both academic institutions and industries. However, these systems are still facing several representative challenges regarding chemical stability of active materials, side reactions, and electrolyte crossover, which impede the cycling stability and efficiency of inorganic RFBs. Herein, we would like to point out considerations and possible solutions to address these issues.

1. Chemical stability. Metal salt-based RFBs are facing a problem of metal hydroxide or metal oxide formation under pH neutral or alkaline conditions, which substantially affects battery stability and efficiency (due to the increase of battery resistance). For nonmetal redox-active species, such as polysulfide and iodide, pH neutral condition is crucial as they are unstable under acidic or alkaline conditions as discussed above. The chemical nature of redox-active inorganic compounds leaves very limited space for structural modulation to improve their chemical stability as well as other physicochemical properties. When combining different inorganic redox couples to form a flow battery, one needs to carefully consider the best pH condition under which both anolyte and catholyte can be chemically stable in their charged and discharged states. A viable option is to use a suitable organic anolyte or catholyte to replace an unsuitable inorganic one to construct a hybrid inorganic/organic RFB, which can largely increase the design space to meet the chemical stability requirement.[15,111,112]
2. Side reactions. Simple metal salt-based RFBs usually need acidic electrolytes to prevent metal hydrolysis to form metal hydroxide precipitates or metal oxides. However, under acidic conditions, HER is thermodynamically favorable and competitive to occur if an anode active materials' redox potential is more negative than the HER potential or a flow battery is charged with an overpotential beyond the HER potential. HER may represent a major capacity decay mechanism for acidic RFBs. For example, the HER problem has been well documented for highly acidic all vanadium RFBs.[113,114] In addition to HER, water and supporting electrolyte anion oxidation (such as $Cl^-$) could also happen in RFBs with a catholyte with a very positive redox potential (such as $Ce^{3+}/Ce^{4+}$). When choosing an anode or cathode material and electrolyte conditions, these sides reactions need to be carefully considered. A possible option is developing low cost and efficient electrocatalysts to facilitate the redox chemistry of active materials to avoid overpotential to trigger the side reactions. Alternatively, it is also possible to develop engineering solutions to convert side products such as hydrogen gas back to protons and electrons to remedy the charge imbalance between an anolyte and an catholyte, which may lead to the increase of the system capital cost.
3. Active material crossover. Most of the redox inorganic compounds including metal salts or nonmetal salts typically have small solvation sphere sizes and can still crossover an ion exchange membrane leading to capacity decay. Again, due to their simple chemical structures, structural modification is not feasible to improve their membrane compatibility. Especially, the crossover issue is even more challenging for nonaqueous RFBs including both inorganic and organic RFBs. It means that we have to rely on the development of advanced membranes or separators to address this problem. As discussed above, it is possible to apply solid-state electrolyte separators for inorganic RFBs to achieve good capacity retention.

In addition, we would like to advocate to use a half-cell flow battery approach to evaluate the chemical stability and cycling performance of individual redox-active inorganic compounds. In the half-cell flow battery approach which adopts the same flow battery design as shown in Figure 1, both the charged state and discharged state of an individual redox-active compound will be applied in each of their respective electrode sides. The half-cell approach allows a systematic evaluation of the electrochemical performance of a single electrolyte material-based electrolyte under actual full flow battery conditions while decoupling the complications from the performance of another electrolyte material, as occurs in a full flow battery. By monitoring CE and cycling stability of the half-cell, this method can effectively diagnose side reactions and chemical decomposition through other spectroscopic tools such as NMR, UV–vis, and FTIR. This approach has been recently applied to elucidate the pH-dependent cycling stability of ferrocyanide by us.[62] Aziz and co-workers also applied this approach to determine the cycling stability of a number of individual redox-active organic molecules.[115]

## 10 CONCLUSION

Owing to the salient features of RFBs, they have a promising future to be the one of the key elements in large-scale grid-integrated energy storage systems because they hold promise to offer long cycle life and high energy efficiencies as well as low operational costs. Over the past few decades, RFBs based on various chemistries have been proposed and developed for both fundamental studies and commercialization attempts. In this chapter, we summarized the most important inorganic materials that have been applied in RFBs and noticed that only a few systems have been applied for grid-integrated energy storage with a scale from 1 MW to 15 MW. All-vanadium, Fe/Cr, and $Zn/Br_2$ systems share the same problem of hazardous and corrosive electrolytes, and they are not environmentally friendly electrochemical energy storage systems. To achieve a more advanced RFB system for sustainable, environmentally benign, low-cost and reliable energy storage, challenges need to be addressed for increasing energy density, improving cycling stability, boosting EE, reducing the capital cost, and using less hazardous electrolyte materials.

## 11 ACKNOWLEDGMENTS

We thank Utah State University for providing faculty startup funds to the PI (T. Leo Liu) and the Utah Science Technology and Research initiative (USTAR) UTAG award for supporting our flow battery project. Bo Hu is grateful to the China CSC Study Abroad program and Utah Energy Triangle Student Award for supporting his graduate program. Camden DeBruler is grateful for his USU Presidential Doctoral Research Fellowship (PDRF) supported by USU. Maowei Hu is grateful to the China CSC Study Abroad program for supporting her graduate program.

## 12 ABBREVIATIONS AND ACRONYMS

AEM = anion exchange membrane; CE = coulombic efficiency; CV = cyclic voltammetry; DES = deep eutectic solution; DOE = U.S. Department of Energy; DMC = dimethyl carbonate; DMSO = dimethyl sulfoxide; EC = ethylene carbonate; EE = energy efficiency; FESEM = field emission scanning electron microscope; GF = graphite felt; HER = hydrogen evolution reaction; HRTEM = high-resolution transmission electron microscopy; ICRFB= iron–chromium redox flow battery; NHE = normal hydrogen electrode; NP = nanoparticle; OCV = open circuit voltage; PBI = polybenzimidazole; POM = polyoxometalate; RFB = redox flow battery; SBRFB = polysulfide–bromine redox flow battery; SHE = standard hydrogen electrode; SHMP = sodium hexametaphosphate; SIRFB = polysulfide–iodine redox flow battery; STEM = scanning transmission electron microscopy; SOC = state of charge; TEA = triethanolamine; TEMPO = (2,2,6,6-tetramethylpiperidin-1-yl)oxyl; VE = voltage efficiency; VRFB = all-vanadium redox flow battery.

## 13 REFERENCES

1. https://www.esrl.noaa.gov/gmd/ccgg/trends/index.html, 2018.
2. B. Dunn, H. Kamath and J.-M. Tarascon, *Science*, 2011, **334**(6058), 928.
3. Z. Yang, J. Zhang, M. C. W. Kintner-Meyer, X. Lu, D. Choi, J. P. Lemmon and J. Liu, *Chem. Rev.*, 2011, **111**(5), 3577.
4. G. L. Soloveichik, *Chem. Rev.*, 2015, **115**(20), 11533.
5. W. Wang, Q. Luo, B. Li, X. Wei, L. Li and Z. Yang, *Adv. Funct. Mater.*, 2013, **23**(8), 970.
6. J. Winsberg, T. Hagemann, T. Janoschka, M. D. Hager and U. S. Schubert, *Angew. Chem. Int. Ed.*, 2017, **56**(3), 686.
7. L. Li, S. Kim, W. Wang, M. Vijayakumar, Z. Nie, B. Chen, J. Zhang, G. Xia, J. Hu, G. Graff, J. Liu and Z. Yang, *Adv. Energy Mater.*, 2011, **1**(3), 394.
8. B. Li, Z. M. Nie, M. Vijayakumar, G. S. Li, J. Liu, V. Sprenkle and W. Wang, *Nat. Commun.*, 2015, **6**, 6303.
9. Z. J. Li, G. M. Weng, Q. L. Zou, G. T. Cong and Y. C. Lu, *Nano Energy*, 2016, **30**, 283.
10. B. Hu, C. DeBruler, Z. Rhodes and T. L. Liu, *J. Am. Chem. Soc.*, 2017, **139**(3), 1207.
11. C. DeBruler, B. Hu, J. Moss, X. Liu, J. Luo, Y. Sun and T. L. Liu, *Chemistry*, 2017, **3**(6), 961.
12. E. S. Beh, D. De Porcellinis, R. L. Gracia, K. T. Xia, R. G. Gordon and M. J. Aziz, *ACS Energy Lett.*, 2017, **2**(3), 639.
13. T. B. Liu, X. L. Wei, Z. M. Nie, V. Sprenkle and W. Wang, *Adv. Energy Mater.*, 2016, **6**(3), 1501449.
14. T. Janoschka, N. Martin, M. D. Hager and U. S. Schubert, *Angew. Chem. Int. Ed.*, 2016, **55**(46), 14425.
15. K. X. Lin, Q. Chen, M. R. Gerhardt, L. C. Tong, S. B. Kim, L. Eisenach, A. W. Valle, D. Hardee, R. G. Gordon, M. J. Aziz and M. P. Marshak, *Science*, 2015, **349**(6255), 1529.
16. K. Lin, R. Gómez-Bombarelli, E. S. Beh, L. Tong, Q. Chen, A. Valle, A. Aspuru-Guzik, M. J. Aziz and R. G. Gordon, *Nat. Energy*, 2016, **1**, 16102.
17. T. Janoschka, N. Martin, U. Martin, C. Friebe, S. Morgenstern, H. Hiller, M. D. Hager and U. S. Schubert, *Nature*, 2015, **527**(7576), 78.

18. L.H. Thaller, 'Electrically Rechargeable Redox Flow Cells'. NASA TM X-71540, Lewis Research Centre, 1974.
19. W. Wang, S. Kim, B. Chen, Z. Nie, J. Zhang, G.-G. Xia, L. Li and Z. Yang, *Energy Environ. Sci.*, 2011, **4**(10), 4068.
20. Y. K. Zeng, T. S. Zhao, L. An, X. L. Zhou and L. Wei, *J. Power Sources*, 2015, **300**, 438.
21. C. D. Wu, D. A. Scherson, E. J. Calvo, E. B. Yeager and M. A. Reid, *J. Electrochem. Soc.*, 1986, **133**(10), 2109.
22. D. S. Cheng and E. Hollax, *J. Electrochem. Soc.*, 1985, **132**(2), 269.
23. M. Skyllas-Kazacos and M. Kazacos, *J. Power Sources*, 2011, **196**(20), 8822.
24. http://enervault.com/, 2018.
25. M. Skyllas-Kazacos, L. Cao, M. Kazacos, N. Kausar and A. Mousa, *ChemSusChem*, 2016, **9**(13), 1521.
26. DOE Global Energy Storage Database (2016) https://www.energystorageexchange.org/projects/2043.
27. https://electrek.co/2017/12/21/worlds-largest-battery-200mw-800mwh-vanadium-flow-battery-rongke-power/, 2017.
28. C. Choi, S. Kim, R. Kim, Y. Choi, S. Kim, H.-y. Jung, J. H. Yang and H.-T. Kim, *Renew. Sustain. Energy Rev.*, 2017, **69**, 263.
29. M. Ulaganathan, V. Aravindan, Q. Yan, S. Madhavi, M. Skyllas-Kazacos and T. M. Lim, *Adv. Mater. Interfaces*, 2016, **3**(1), 1500309.
30. G. Oriji, Y. Katayama and T. Miura, *Electrochim. Acta*, 2004, **49**(19), 3091.
31. F. Rahman and M. Skyllas-Kazacos, *J. Power Sources*, 2009, **189**(2), 1212.
32. S. Kim, M. Vijayakumar, W. Wang, J. Zhang, B. Chen, Z. Nie, F. Chen, J. Hu, L. Li and Z. Yang, *PCCP*, 2011, **13**(40), 18186.
33. S. Roe, C. Menictas and M. Skyllas-Kazacos, *J. Electrochem. Soc.*, 2016, **163**(1), A5023.
34. J. Zhang, L. Li, Z. Nie, B. Chen, M. Vijayakumar, S. Kim, W. Wang, B. Schwenzer, J. Liu and Z. Yang, *J. Appl. Electrochem.*, 2011, **41**(10), 1215.
35. B. Li, M. Gu, Z. Nie, Y. Shao, Q. Luo, X. Wei, X. Li, J. Xiao, C. Wang, V. Sprenkle and W. Wang, *Nano Lett.*, 2013, **13**(3), 1330.
36. J. Shen, S. Liu, Z. He and L. Shi, *Electrochim. Acta*, 2015, **151**, 297.
37. B. Li, M. Gu, Z. Nie, X. Wei, C. Wang, V. Sprenkle and W. Wang, *Nano Lett.*, 2014, **14**(1), 158.
38. W. Li, J. Liu and C. Yan, *Carbon*, 2011, **49**(11), 3463.
39. M. Zago and A. Casalegno, *Electrochim. Acta*, 2017, **248**, 505.
40. F. Zhang, S. Huang, X. Wang, C. Jia, Y. Du and Q. Wang, *Nano Energy*, 2018, **52**, 292.
41. M. Skyllas-Kazacos, *J. Power Sources*, 2003, **124**(1), 299.
42. B. Li, L. Li, W. Wang, Z. Nie, B. Chen, X. Wei, Q. Luo, Z. Yang and V. Sprenkle, *J. Power Sources*, 2013, **229**, 1.
43. W. Wang, Z. Nie, B. Chen, F. Chen, Q. Luo, X. Wei, G.-G. Xia, M. Skyllas-Kazacos, L. Li and Z. Yang, *Adv. Energy Mater.*, 2012, **2**(4), 487.
44. U.S. Department of Energy Grid Energy Storage, 2013; http://energy.gov/oe/downloads/grid-energy-storage-december-2013.
45. 2018, 2, 121, https://www.vanadiumprice.com/.
46. D. J. Eustace, *J. Electrochem. Soc.*, 1980, **127**(3), 528.
47. B. Li, Z. M. Nie, M. Vijayakumar, G. S. Li, J. Liu, V. Sprenkle and W. Wang, *Nat. Commun.*, 2015, **6**, p.ncomms7303.
48. F. R. McLarnon and E. J. Cairns, *J. Electrochem. Soc.*, 1991, **138**(2), 645.
49. F. C. Walsh, C. Ponce de Léon, L. Berlouis, G. Nikiforidis, L. F. Arenas-Martínez, D. Hodgson and D. Hall, *ChemPlusChem*, 2015, **80**(2), 288.
50. J. Winsberg, T. Janoschka, S. Morgenstern, T. Hagemann, S. Muench, G. Hauffman, J. F. Gohy, M. D. Hager and U. S. Schubert, *Adv. Mater.*, 2016, **28**(11), 2238.
51. J. Winsberg, C. Stolze, A. Schwenke, S. Muench, M. D. Hager and U. S. Schubert, *ACS Energy Lett.*, 2017, **2**(2), 411.
52. A. Khor, P. Leung, M. R. Mohamed, C. Flox, Q. Xu, L. An, R. G. A. Wills, J. R. Morante and A. A. Shah, *Mater. Today Energy*, 2018, **8**, 80.
53. M. Skyllas-Kazacos, M. H. Chakrabarti, S. A. Hajimolana, F. S. Mjalli and M. Saleem, *J. Electrochem. Soc.*, 2011, **158**(8), R55.
54. Q. Lai, H. Zhang, X. Li, L. Zhang and Y. Cheng, *J. Power Sources*, 2013, **235**, 1.
55. J.-D. Jeon, H. S. Yang, J. Shim, H. S. Kim and J. H. Yang, *Electrochim. Acta*, 2014, **127**, 397.
56. G.-M. Weng, Z. Li, G. Cong, Y. Zhou and Y.-C. Lu, *Energy Environ. Sci.*, 2017, **10**, 735.
57. J. Zhang, G. Jiang, P. Xu, A. Ghorbani Kashkooli, M. Mousavi, A. Yu and Z. Chen, *Energy Environ. Sci.*, 2018, **11**(8), 2010.
58. W. D. McCulloch, M. Yu and Y. Wu, *ACS Energy Lett.*, 2016, **1**(3), 578.
59. C. Xie, H. Zhang, W. Xu, W. Wang and X. Li, *Angew. Chem. Int. Ed.*, 2018, **57**(35), 11171.
60. G.B. Adams, Patent US4180623A, 1979.
61. Z. Yuan, Y. Duan, T. Liu, H. Zhang and X. Li, *iScience*, 2018, **3**, 40.
62. J. Luo, A. Sam, B. Hu, C. DeBruler, X. Wei, W. Wang and T. L. Liu, *Nano Energy*, 2017, **42**, 215.
63. Z. Xie, Q. Su, A. Shi, B. Yang, B. Liu, J. Chen, X. Zhou, D. Cai and L. Yang, *J. Energy Chem.*, 2016, **25**(3), 495.

64. S. Selverston, R. F. Savinell and J. S. Wainright, *J. Electrochem. Soc.*, 2017, **164**(6), A1069.

65. K. Gong, X. Y. Ma, K. M. Conforti, K. J. Kuttler, J. B. Grunewald, K. L. Yeager, M. Z. Bazant, S. Gu and Y. S. Yan, *Energy Environ. Sci.*, 2015, **8**(10), 2941.

66. C. Xie, Y. Duan, W. Xu, H. Zhang and X. Li, *Angew. Chem. Int. Ed.*, 2017, **56**(47), 14953.

67. C. Ponce de León, A. Frías-Ferrer, J. González-García, D. A. Szánto and F. C. Walsh, *J. Power Sources*, 2006, **160**(1), 716.

68. P. K. Leung, C. Ponce de León, C. T. J. Low and F. C. Walsh, *Electrochim. Acta*, 2011, **56**(5), 2145.

69. P. K. Leung, C. Ponce-de-León, C. T. J. Low, A. A. Shah and F. C. Walsh, *J. Power Sources*, 2011, **196**(11), 5174.

70. P. K. Leung, C. Ponce-de-León, F. J. Recio, P. Herrasti and F. C. Walsh, *J. Appl. Electrochem*, 2014, **44**(9), 1025.

71. L. W. Hruska and R. F. Savinell, *J. Electrochem. Soc.*, 1981, **128**(1), 18.

72. A. K. Manohar, K. M. Kim, E. Plichta, M. Hendrickson, S. Rawlings and S. R. Narayanan, *J. Electrochem. Soc.*, 2016, **163**(1), A5118.

73. K. L. Hawthorne, J. S. Wainright and R. F. Savinell, *J. Electrochem. Soc.*, 2014, **161**(10), A1662.

74. Y. H. Wen, H. M. Zhang, P. Qian, H. T. Zhou, P. Zhao, B. L. Yi and Y. S. Yang, *Electrochim. Acta*, 2006, **51**(18), 3769.

75. K. Gong, F. Xu, J. B. Grunewald, X. Ma, Y. Zhao, S. Gu and Y. Yan, *ACS Energy Lett.*, 2016, **1**(1), 89.

76. D. Lloyd, T. Vainikka, M. Ronkainen and K. Kontturi, *Electrochim. Acta*, 2013, **109**, 843.

77. M. A. Miller, J. S. Wainright and R. F. Savinell, *J. Electrochem. Soc.*, 2016, **163**(3), A578.

78. D. E. Katsoulis, *Chem. Rev.*, 1998, **98**(1), 359.

79. M. T. Pope and A. Müller, *Angew. Chem. Int. Ed. Engl.*, 1991, **30**(1), 34.

80. Y. Ji, L. Huang, J. Hu, C. Streb and Y.-F. Song, *Energy Environ. Sci.*, 2015, **8**(3), 776.

81. H. N. Miras, J. Yan, D.-L. Long and L. Cronin, *Chem. Soc. Rev.*, 2012, **41**(22), 7403.

82. H. D. Pratt Iii and T. M. Anderson, *Dalton Trans.*, 2013, **42**(44), 15650.

83. H. D. Pratt, N. S. Hudak, X. Fang and T. M. Anderson, *J. Power Sources*, 2013, **236**, 259.

84. H. D. Pratt, W. R. Pratt, X. Fang, N. S. Hudak and T. M. Anderson, *Electrochim. Acta*, 2014, **138**, 210.

85. Y. Liu, S. Lu, H. Wang, C. Yang, X. Su and Y. Xiang, *Adv. Energy Mater.*, 2016, **7**(8), 1601224.

86. Y. Liu, H. Wang, Y. Xiang and S. Lu, *J. Power Sources*, 2018, **392**, 260.

87. J. Friedl, M. V. Holland-Cunz, F. Cording, F. L. Pfanschilling, C. Wills, W. McFarlane, B. Schricker, R. Fleck, H. Wolfschmidt and U. Stimming, *Energy Environ. Sci*, 2018, **11**, 3010.

88. J.-J. Chen, M. D. Symes and L. Cronin, *Nat. Chem.*, 2018, **10**, 1042.

89. B. R. Chalamala, T. Soundappan, G. R. Fisher, M. R. Anstey, V. V. Viswanathan and M. L. Perry, *Proc. IEEE*, 2014, **102**(6), 976.

90. P. Zhao, H. Zhang, H. Zhou and B. Yi, *Electrochim. Acta*, 2005, **51**(6), 1091.

91. L. Su, A. F. Badel, C. Cao, J. J. Hinricher and F. R. Brushett, *Ind. Eng. Chem. Res.*, 2017, **56**(35), 9783.

92. I. Filpponen, A. Guerra, A. Hai, L. A. Lucia and D. S. Argyropoulos, *Ind. Eng. Chem. Res.*, 2006, **45**(22), 7388.

93. Y. Yang, G. Zheng and Y. Cui, *Energy Environ. Sci.*, 2013, **6**(5), 1552.

94. Y. Zhao and H. R. Byon, *Adv. Energy Mater.*, 2013, **3**(12), 1630.

95. F. R. Brushett, J. T. Vaughey and A. N. Jansen, *Adv. Energy Mater.*, 2012, **2**(11), 1390.

96. H. Chen, Y. Zhou and Y.-C. Lu, *ACS Energy Lett.*, 2018, **3**(8), 1991.

97. M. Duduta, B. Ho, V. C. Wood, P. Limthongkul, V. E. Brunini, W. C. Carter and Y.-M. Chiang, *Adv. Energy Mater.*, 2011, **1**(4), 511.

98. F. Pan and Q. Wang, *Molecules*, 2015, **20**(11), 19711.

99. Q. Wang, S. M. Zakeeruddin, D. Wang, I. Exnar and M. Grätzel, *Angew. Chem. Int. Ed.*, 2006, **45**(48), 8197.

100. Q. Huang, H. Li, M. Grätzel and Q. Wang, *PCCP*, 2013, **15**(6), 1793.

101. F. Pan, J. Yang, Q. Huang, X. Wang, H. Huang and Q. Wang, *Adv. Energy Mater.*, 2014, **4**(15), 1400567.

102. Q. Huang, J. Yang, C. B. Ng, C. Jia and Q. Wang, *Energy Environ. Sci.*, 2016, **9**(3), 917.

103. Y. G. Zhu, Y. Du, C. Jia, M. Zhou, L. Fan, X. Wang and Q. Wang, *JACS*, 2017, **139**(18), 6286.

104. F. Pan, Q. Huang, H. Huang and Q. Wang, *Chem. Mater.*, 2016, **28**(7), 2052.

105. M. Zhou, Q. Huang, T. N. Pham Truong, J. Ghilanc, Y. G. Zhu, C. Jia, R. Yan, L. Fan, H. Randriamahazaka and Q. Wang, *Chemistry*, 2017, **3**, 1036.

106. Y. Wang and H. Zhou, *Energy Environ. Sci.*, 2016, **9**(7), 2267.

107. C. Zhang, Y. Ding, L. Zhang, X. Wang, Y. Zhao, X. Zhang and G. Yu, *Angew. Chem. Int. Ed.*, 2017, **56**(26), 7454.

108. X. Zhou, F. Zhao, Y. Guo, Y. Zhang and G. Yu, *Energy Environ. Sci.*, 2018, **11**, 1985.

109. L. E. VanGelder, A. M. Kosswattaarachchi, P. L. Forrestel, T. R. Cook and E. M. Matson, *Chem. Sci.*, 2018, **9**(6), 1692.

110. L. E. VanGelder, *J. Mater. Chem. A*, 2018, **6**(28), 13874.

111. C. DeBruler, B. Hu, J. Moss, J. Luo and T. L. Liu, *ACS Energy Lett.*, 2018, **3**, 663.

112. A. Hollas, X. Wei, V. Murugesan, Z. Nie, B. Li, D. Reed, J. Liu, V. Sprenkle and W. Wang, *Nat. Energy*, 2018, **3**(6), 508.

113. A. A. Shah, H. Al-Fetlawi and F. C. Walsh, *Electrochim. Acta*, 2010, **55**(3), 1125.

114. C.-N. Sun, F. M. Delnick, L. Baggetto, G. M. Veith and T. A. Zawodzinski, *J. Power Sources*, 2014, **248**, 560.

115. M.-A. Goulet and M. J. Aziz, *J. Electrochem. Soc.*, 2018, **165**(7), A1466.

# Electrode and Electrolyte Interaction in Aqueous Electrochemical Energy Storage

**Xiaowei Teng**

*University of New Hampshire, Durham, NH, USA*



## 1 INTRODUCTION

Developing large-scale stationary electrochemical energy storage (EES) systems efficiently and economically storing the energy generated from renewable sources (e.g., solar or wind) are essential priorities to withstand the gaps between the peak production and peak consumption.[1,2] Rechargeable nonaqueous lithium-ion batteries (LIBs) have advanced rapidly in the past twenty years with much improved energy density, widely used from portable electronics to electrical vehicles.[3] However, high cost and safety concerns make the current LIBs less appealing for large-scale stationary EES systems, where cost, safety, and cycle life become important in addition to mere energy density.[4,5] The high safety risk of nonaqueous LIBs is attributed to the usage of flammable electrodes and organic electrolytes, as well as the hazard of thermal runaway reaction is attributed to the high reactivity of electrode materials and electrolytes.[6,7] Their high cost is due to the high price of transition metals, Li metal, organic electrolytes and the requirement of a dry room during the cell assembly processes. Moreover, limited supply of Li in terrestrial reserves could also hamper the large-scale implementation of LIB technology.[8] Rechargeable EES devices using aqueous electrolytes and Na ion as the charge carrier display several advantages over nonaqueous LIBs.[9–13] For example, using nonflammable and environmentally benign water-based electrolytes not only mitigates the safety issue of nonaqueous LIBs but also significantly lowers the production cost by avoiding rigorous manufacturing conditions. Meanwhile, the costs of Na salts are considerably lower than that of Li, mainly because the abundance of Na in the earth's crust is nearly 1000 times greater than Li and its sources are more geographically available. Moreover, compared with Li ions, hydrated Na ions possess the smaller Stokes radius, which is defined as the total radius of the ion and bound water molecules, and thus greater transport properties in aqueous electrolytes: the ionic conductivities of the aqueous electrolytes are usually two orders of magnitude higher than those of organic electrolytes.[14] The utilization of aqueous Na-ion storage could fulfill essential requirements of large-scale stationary EES systems, where low cost and environmental sustainability become imperative. However, the current aqueous Na-ion EES systems needs further improvement in the energy density in order to rival the performance of the nonaqueous LIBs. The energy density ($E$) of aqueous Na-ion EES systems is limited to its narrow voltage window ($V$) and low storage capacity ($C$) as Equation (1) indicates:

$$E = V \times C \tag{1}$$

*Inorganic Battery Materials*. Edited by Hailiang Wang and Boniface P.T. Fokwa
 ISBN 978-1-119-43199-2

## 2 ELECTRODE AND ELECTROLYTE INTERACTION IN EXTENDING THE POTENTIAL WINDOW OF AQUEOUS ENERGY STORAGE

### 2.1 Thermodynamics of Limited Voltage Window in Aqueous Energy Storage

In an EES device, the open-circuit voltage ($V_{OC}$), namely, the cell voltage when the current of a cell is kept infinitely small, is not only limited by the electrochemical potentials between the cathode ($\mu_c$) and anode ($\mu_a$) but also by the electrochemical window of the electrolyte ($E_g$). It is notable that $E_g$ is determined by the energy gap from the highest occupied molecular orbital (HOMO) to the lowest unoccupied molecular orbital (LUMO). In an electrochemical cell, the $\mu_a$ needs to be energetically lower than the LUMO, while the $\mu_c$ is greater than the HOMO; otherwise, the electrolyte will be on the anode or oxidized (to form oxygen gas) at the cathode. For a nonaqueous electrolyte, its LUMO and HOMO are usually above the $\mu_a$ and below the $\mu_c$, staying stable under a wide potential window (usually >3.0 V) during the charging and discharging processes. However, for an aqueous electrolyte, LUMO and HOMO of water are usually below the $\mu_a$ and above the $\mu_c$ of electrodes. Therefore, the working voltage window for an aqueous EES device is often determined by the thermodynamically stable voltage window of water, instead of the electrochemical potentials of the electrodes ($\mu_c$ or $\mu_a$).

Pourbaix diagram as shown in Figure 1 describes the dependence of an electrochemical cell voltage on pH of the electrolyte. Especially, the electrode potential of the hydrogen evolution reaction (HER) at the anode and oxygen evolution reaction (OER) at the cathode can be described by the Nernst equation as shown in Equations (2)–(5). The thermodynamically stable voltage window for an aqueous EES device is determined by the thermodynamically stable voltage window of water, which is always equal to 1.23 V at any given pH as shown in Equation (6).

$$\text{Anode: } 2H^+_{(liquid)} + 2e^- = H_{2(gas)} \tag{2}$$

$$E_{anode} = E^0_{anode} - \frac{RT}{nF} \ln \frac{\frac{P_{H_2}}{P^0}}{(C_{H^+})^2} = E^0 - \frac{2.3RT}{nF} \log \frac{P_{H_2}/P^0}{(C_{H^+})^2} = -0.059 \times pH_{anode} \tag{3}$$

$$\text{Cathode } \frac{1}{2}O_{2(gas)} + 2H^+_{(liquid)} + 2e^- = H_2O_{(liquid)} \tag{4}$$

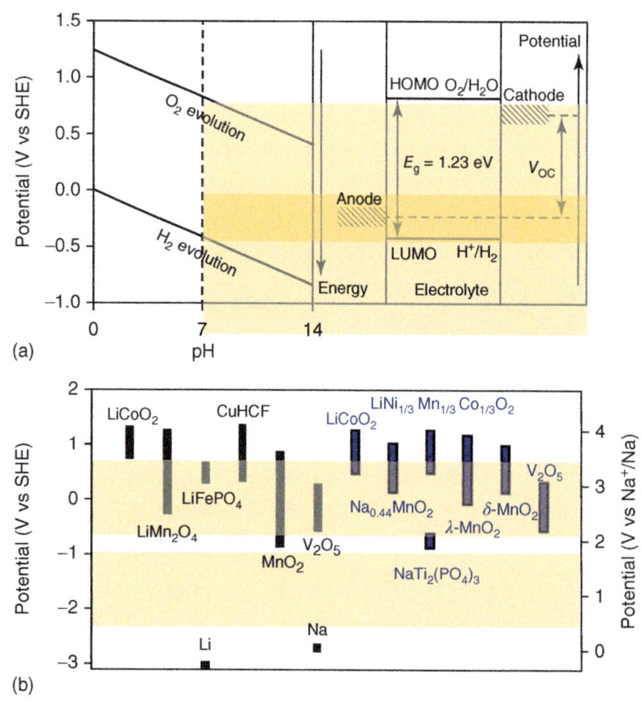


**Figure 1** (a) Electrochemical potential of water at different pH (Pourbaix diagram). The open-circuit voltage ($V_{OC}$) of an aqueous EES device is determined by electrochemical potentials between the cathode ($\mu_c$) and anode ($\mu_a$), and the potential window of the electrolyte ($E_g$); (b) the experimentally reported redox potential for electrode materials in nonaqueous (black) and aqueous (blue) electrolytes.[15] The thermodynamically stable potential window of water at pH = 7 is shadowed in gray, and SHE stands for standard hydrogen electrode

$$E_{\text{cathode}} = E^0_{\text{cathode}} - \frac{RT}{nF} \ln \frac{c_{\text{H}_2\text{O}}}{(P_{\text{O}_2})^{0.5}(C_{\text{H}^+})^2} = 1.23 - 0.059 \times \text{pH}_{\text{cathode}} \quad (5)$$

$$\text{Cell Voltage (V)} = E_{\text{cathode}} - E_{\text{anode}} = (1.23\ \text{V} - 0.059 \times \text{pH}_{\text{cathode}}) + 0.059 \times \text{pH}_{\text{anode}} \xrightarrow{\text{if pH}_{\text{anode}} = \text{pH}_{\text{cathode}} = \text{pH}} = (1.23 - 0.059 \times \text{pH}) + 0.059 \times \text{pH} \equiv 1.23\ \text{V} \quad (6)$$

## 2.2 Extending Voltage Window of Aqueous Energy Storage via pH Control

Extensive efforts have been made to break through the 1.23 V potential window limit in aqueous EES systems. One approach is to control the pH of the electrolyte to extend the voltage window. [9,16,17] As shown in Equation (6), when the pH of the electrolyte remains identical at the anode and cathode (which is a typical case due to the diffusion of $H^+$ and $OH^-$ across the porous separator), the cell voltage is always equal to 1.23 V. Simply using a basic electrolyte (e.g., pH = 15 as shown in Equation 7) instead of a neutral solution in the anode compartment, the electrode potential of the HER will be decreased from −0.413 (vs SHE) to −0.885 V. Analogously, when an acidic electrolyte (e.g., pH = −1) instead of a neutral solution is used in the cathode compartment, the electrode potential of the HER will be increased from 0.817 to 1.289 V. Therefore, the overall voltage for such an electrochemical cell expands from 1.23 to 2.174 V. When a more acidic or basic electrolyte is used, the cell voltage could be even further expanded.

$$\text{Cell Voltage (V)} = E_{\text{cathode}} - E_{\text{anode}} = (1.23 - 0.059 \times \text{pH}_{\text{cathode}}) - (-0.059 \times \text{pH}_{\text{anode}}) = (1.23 - 0.059 \times (-1)) - (-0.059 \times 15) = 2.174\ \text{V} \quad (7)$$

It is notable that such a voltage window expansion is completely driven by thermodynamics (described by the Nernst equation). Very often, even further expanded voltage has been experimentally observed due to the overpotential of the redox reactions (that is, the electrokinetic effect, which will be discussed later). However, such a cell design brings a great challenge for the separator that needs to withstand the drastic pH differences between the anode and cathode compartments. For example, Chen *et al.* reported an aqueous battery device with a potential window of 2.8 V. In their cell design, Zn metal, $KMnO_4$ salt, and 2 M KOH/2 M LiOH were used as anode, cathode, and electrolyte, respectively.[17] A commercially available ceramic lithium superionic conductor ($Li_{1+x+y}Al_xTi_{2-x}Si_yP_{3-y}O_{12}$, LATSP) thin film was used as the Li ion exchange membrane to separate the acidic and basic electrolytes. However, the cell appears the degradation after the prolonged cycles, primarily due to the drastic pH difference between two sides (Figure 2).

## 2.3 Extending Voltage Window of Aqueous Energy Storage via the Formation of Interphase between Electrode and Electrolyte

The potential window of aqueous energy storage can also be expanded by the introduction of water-based electrolytes into a conventional nonaqueous battery

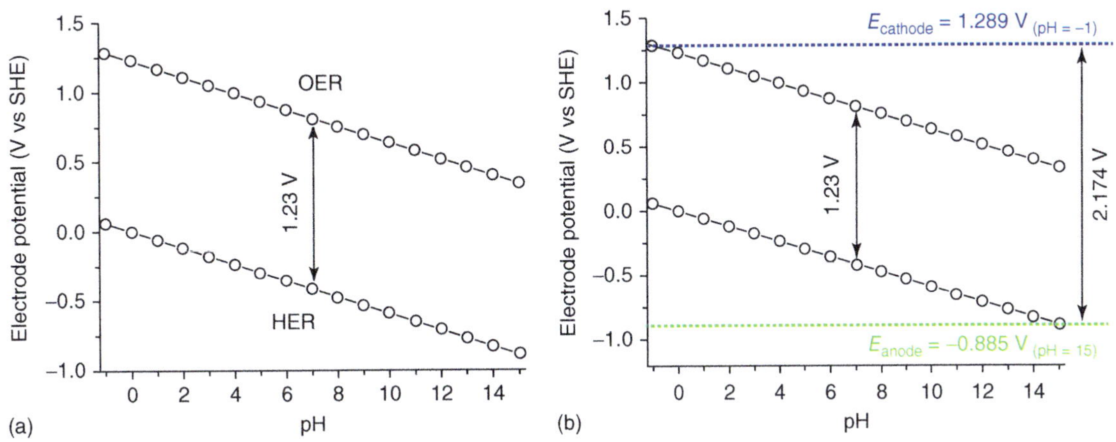


**Figure 2** (a) Potential–pH relationship (Pourbaix diagram), showing a thermodynamic stable potential window of 1.23 V at any pH value. (b) Pourbaix diagram showing that by controlling the pH values in the anodic (pH = −1) and cathodic (pH = 15) compartments, a thermodynamically stable potential window can reach 2.174 V

comprising water-sensitive metal or metal alloy anodes and Li–metal oxide cathodes. The electrode and electrolyte interaction at the interface is the key to understand battery electrochemistry, as it is where the electrons and Li ions combine and then get stored at the electrode, through intercalation, alloying, or simply as Li metal. This interface is further complicated by a passivation layer on the electrode where decomposition of organic electrolytes contributes to the formation of the solid-electrolyte interphase (SEI), a protective film, on the electrode surface (especially anode).[18–21] The understanding of this passivation layer on negative electrodes started from its observation on lithium metal immersed in a nonaqueous electrolyte by Dey in 1970.[22] Peled introduced the concept of the SEI in 1979 as an electronically insulating and ionically conducting passive layer and named it as the SEI because the passive layer formed between the electrode and electrolyte.[23] It is notable that the SEI formation blocks the interaction between the electron (from the electrode) and the electrolyte to avoid the further decomposition of the electrolyte. Meanwhile, the resulting SEI is still being ionically conductive for the transport of charge carriers between the electrolyte and electrode. In short, the SEI should be ionically conductive and electronically insulating. Finally, the SEI should be chemically and mechanically stable, that is, it should not react with electrolyte nor dissolve into the electrolyte and it should not crack during the fabrication process. The progressive understanding of the SEI for LIB applications has been recently summarized by Wang *et al.*, as shown in Figure 3.

The formation and properties of the SEI in a *nonaqueous* EES system have been studied for nearly five decades, which have been extensively reviewed previously[25–28] and therefore will not be the focus of

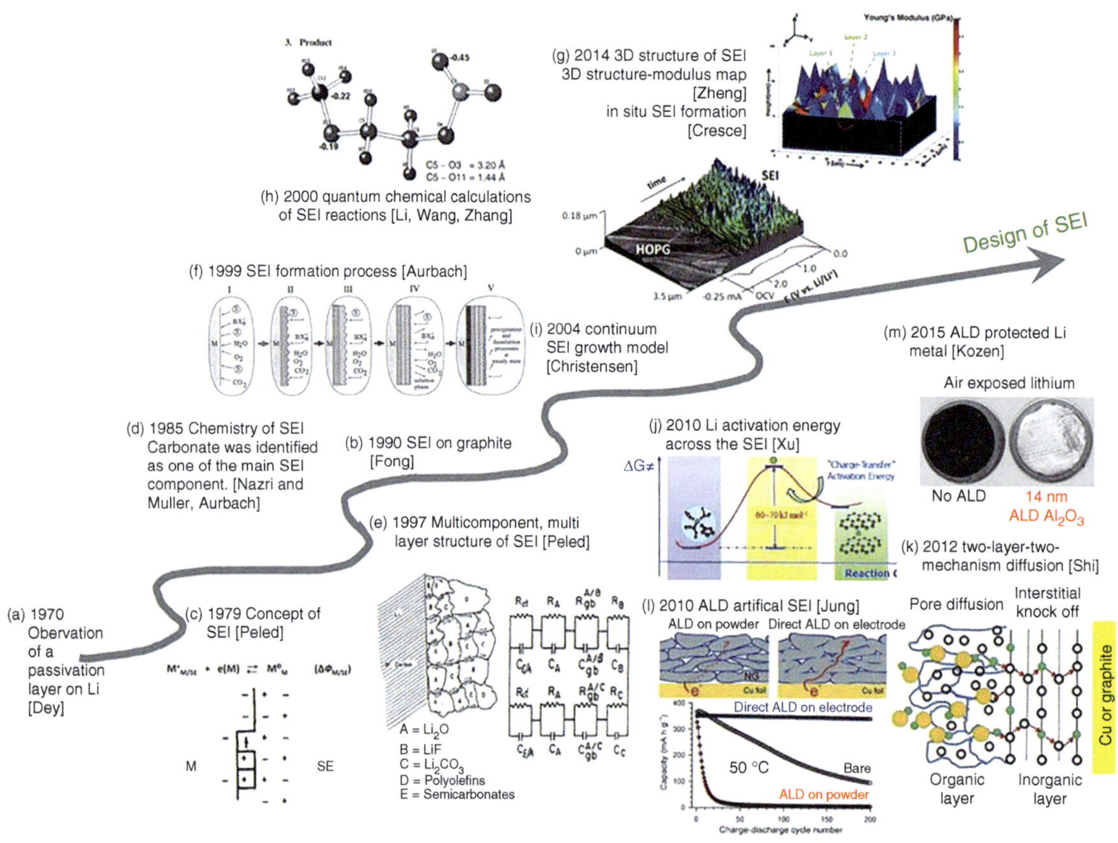


**Figure 3** Brief history of SEI from experimental and modeling efforts in the past four decades. (a) First observation of the passivation layer on lithium metal in the 1970s. (b) The report of effective SEI layer on graphite in 1990. (c) The introduction of the concept of SEI by Peled in 1979. (d) Experimental works on the chemistry of SEI and the formation mechanisms. (e) Peled pictured SEI as a mosaic structure and translated it into an equivalent circuit model in 1997. (f) Aurbach illustrated the formation process of SEI. (g) The report of the time-resolved evolution of the multicomponent and multilayer SEI formation by Cresce in 2014. (h) The work using quantum chemical calculations to simulate reaction pathways of SEI formation. (i) Continuum models developed to simulate SEI growth (assuming SEI is mainly $Li_2CO_3$) in 2004. (j) The measurement of the Li-ion transport energy barrier in 2010. (k) Calculations of Li-ion diffusion in $Li_2CO_3$. (g) The measurement of Young's modulus of SEI on a silicon anode. (h) Quantum chemical calculation of SEI formation. (l) The fundamental understanding of SEI leads to artificial SEI design. (m) Applying atomic layer deposition coating to protect a Li metal electrode. [Source: Wang *et al.*[24], https://www.nature.com/articles/s41524-018-0064-0. Licensed under CC BY 4.0]

this chapter. It is worth mentioning that an effective SEI for *aqueous* EES systems has been recently pioneered by Wang's and Xu's research groups. They reported the formation of a gel-like interphases on the surface of $Mo_6S_8$ anode in an aqueous "water-in-salt" electrolyte, which was obtained by dissolving lithium bis(trifluoromethane sulfonyl)imide (LiTFSI) at extremely high concentrations (molality >20 M).[29,30] The resulting interphase, similar to those observed in nonaqueous batteries, protected the $Mo_6S_8$ anode from direct contact with water. This novel electrolyte system introduced a desirable SEI that enables an aqueous Li ion full cell above 3.0 V without the HER.[31–33] Figure 4 compares cell voltage, capacity, and cycle life of the resulting water-in-salt battery with an $LiMn_2O_4$ cathode and an $Mo_6S_8$ anode, along with other previously investigated aqueous systems. The water-in-salt battery showed plausible cycle life, with more than 1000 stable charging–discharging cycles, and excellent energy density (~100 Wh kg$^{-1}$). It is notable that although electrochemical couples of $LiMn_2O_4$/acetylene black and $LiFePO_4/LiTi_2(PO_4)_3$ demonstrated a similar cycle life, these devices showed a relatively low potential window (<1.25 V) and therefore unsatisfactory energy density (<50 Wh kg$^{-1}$).

In comparison with SEI formed by the reduction of organic electrolytes typically occurring in nonaqueous LIBs, the formation of an interphase through the interaction between metal oxides and water has been investigated much less extensively. It is generally accepted that in conventional aqueous electrolytes containing water and strongly dissociated inorganic salts, a protective interphase resulting from the decomposition of water is very unlikely

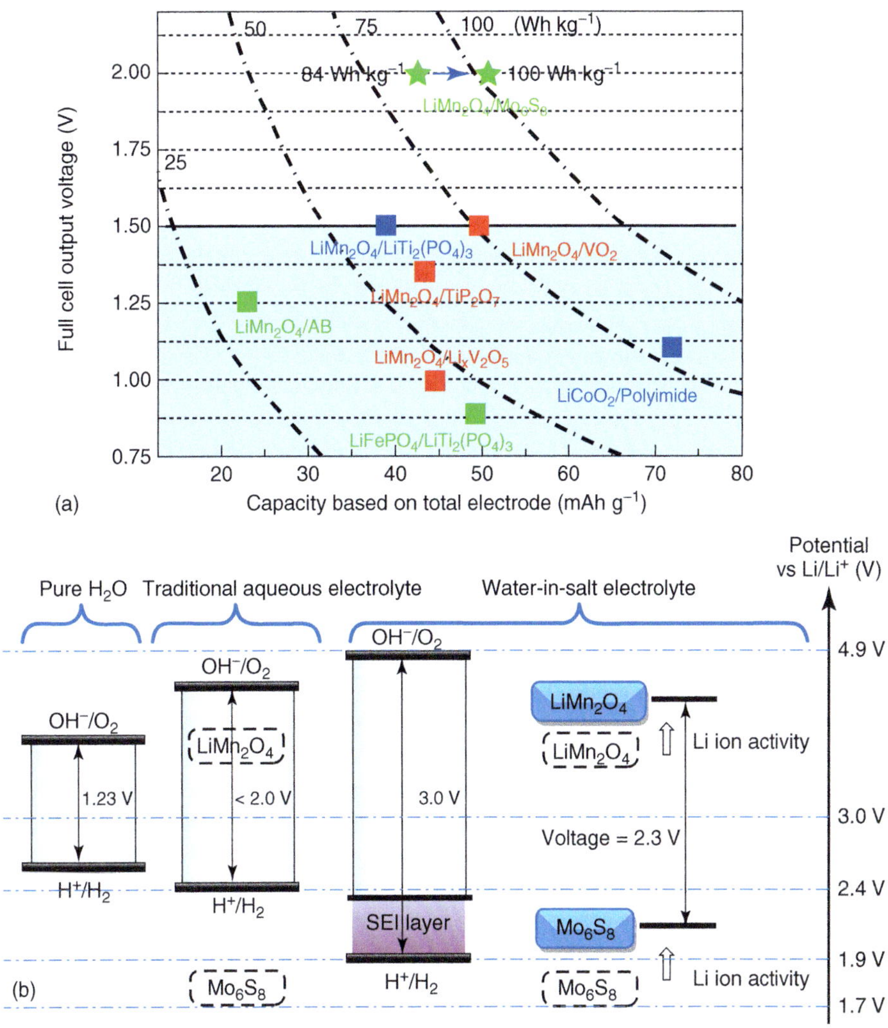


**Figure 4** The performance of aqueous Li-ion chemistries from the water-in-salt battery using a $LiMn_2O_4$ cathode and a $Mo_6S_8$ anode with a 3.0 V voltage window. (a) Comparison of performance between water-in-salt battery and other aqueous electrochemical systems. Color code for cycling stability: red, <100 cycles; blue, 100–200 cycles; green, >1000 cycles. (b) Schematic of the expanded electrochemical stability window for the water-in-salt battery

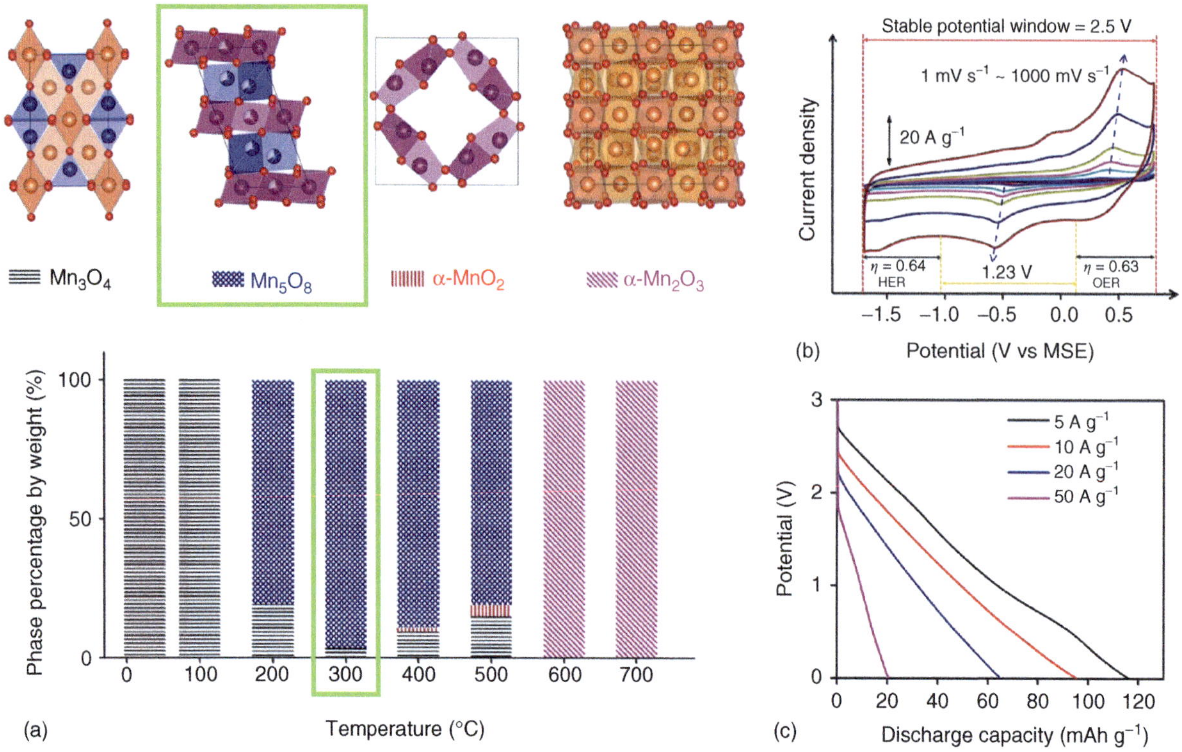


**Figure 5** (a) Illustration of formation of $Mn_5O_8$ through solid-state oxidation of $Mn_3O_4$ nanoparticles. Phase percentages and oxygen ratios (by weight) of manganese oxides analyzed by Rietveld XRD refinement with oxygen ratios of pure phase of $Mn_3O_4$, $Mn_5O_8$, $\alpha$-$MnO_2$ or $\alpha$-$Mn_2O_3$ indicated as dash lines and relative crystal structures provided above (blue, $Mn^{2+}$; yellow, $Mn^{3+}$; purple, $Mn^{4+}$; red, O; vacancy, white). (b) Cyclic voltammetry (CV) of $Mn_5O_8$ materials in 0.1 M $Na_2SO_4$ aqueous solution between −1.7 and 0.8 V (vs mercury sulfate electrode, MSE) in a half-cell, showing a stable potential window of 2.5 V at scan rates ranging from 1 to 1000 mV $s^{-1}$. (c) Discharge electrode capacities at current densities ranging from 5 to 50 A/g after 5000 galvanostatic cycles, showing a stable potential window of 3.0 V in two-electrode full cells

to form. This is because the decomposition products from water such as $H_2$ and $OH^-$ (the decomposition products from water reduction) or $O_2$ and $H^+$ (the decomposition products from water oxidation) cannot deposit on the surface of electrode materials in a dense solid state. However, the recent results reported by Shan *et al.* showed that interactions between the monoclinic manganese oxide ($Mn_5O_8$, $Mn^{2+}{}_2Mn^{4+}{}_3O_8$), the only layered and bivalent manganese oxide, and water during the electrochemical cycling could result in a new form of hydroxylated interphase on the surface of $Mn_5O_8$.[34] In such a bivalent manganese oxide, one-third of the cationic sites in the main octahedral sheets $[Mn_3{}^{4+}O_8]^{4-}$ were not fully occupied (Figure 5a); above and below the empty $Mn^{4+}$ sites were $Mn^{2+}$ cationic sites. The density function theory (DFT) calculation showed that $Mn^{2+}$ terminated surface had a lower surface energy (0.87 J $m^{-2}$) than $Mn^{4+}$ terminated surface (1.80 J $m^{-2}$) and suggested that $Mn_5O_8$ material was enclosed by the $Mn^{2+}$-rich surface, favoring the formation of a hydroxylated interphase upon the interaction between the $Mn^{2+}$-rich surface and water. The resulting hydroxylated interphase on the surface of $Mn_5O_8$ materials inhibited the hydrogen and OER and thus enabled a kinetically stable potential window of 2.5 V for aqueous Na-ion storage in half-cells (Figure 5).[34,35] An aqueous symmetric full cell using such electrodes demonstrated a stable potential window of 3.0 V and good energy and power performance. The results offered a new paradigm for developing electrode materials with a hydroxylated interphase for aqueous EES systems with a high voltage window.

It is worth mentioning that because such a hydroxylated interphase was very thin, conventional surface tools such as X-ray photoelectron spectroscopy (XPS) are not able to "identify" the structure of the interphase, which could only be identified by the oxygen K-edge soft X-ray absorption spectroscopy (sXAS) measurements. As shown in Figure 6(a), cycled $Mn_5O_8$ displayed peaks at the 535 and 537 eV, different from the pristine $Mn_5O_8$ material. Such features resembled the "pre-edge" and "main peak" fingerprints from water and thus indicated the formation of a hydroxylated interphase on the surface of $Mn_5O_8$. Additionally, the features between 528 and 534 eV represented the spectroscopic excitations to the hybridized state of O-2p and Mn-3d, which were split by the crystal field of

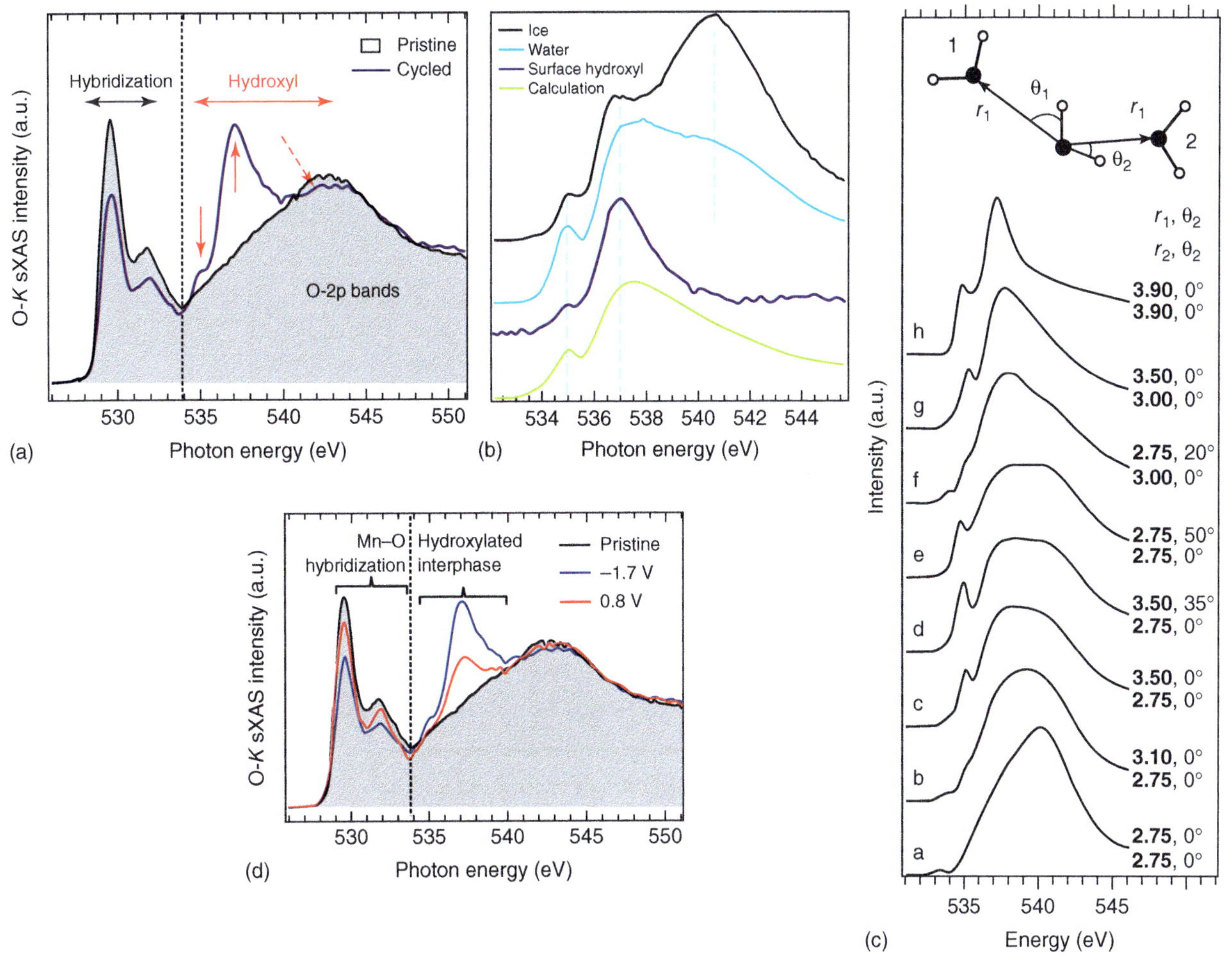


**Figure 6** (a) Oxygen K-edge sXAS of $Mn_5O_8$ (pristine) and $Mn_5O_8$ after two CV cycles. Features below 534 eV are from the hybridization of Mn-3d and O-2p states. The 535 and 537 eV peaks are fingerprints of water. (b) A comparison of the oxygen K-edge features of the hydroxylated interphase with water and ice, and (c) the reported calculation adapted from Ref. 36. (d) Oxygen K-edge sXAS spectra of pristine and cycled $Mn_5O_8$ at −1.7 and 0.8 V (vs MSE), showing strong hydroxyl features at −1.7 V

the local Mn–O coordination geometry.[37] A clear loss of intensity on this hybridization feature indicates the Mn–O coordination on the surface (consistent with the formation of a hydroxylated interphase). Figure 6(b) shows the oxygen K-edge spectra after subtracting the $Mn_5O_8$ (pristine) signal. It is clear that the hydroxylated interphase layer is different from ice, by missing the high-energy hump from the saturated H-bonds in ice. It reproduces the "prepeak" and "main peak" of liquid water, with the "main peak" at 537.5 eV dominating the spectrum.[36,38–40] The enhancement of this main peak was predicted by theoretical calculations, which stems from a combination of an extended O–O distance and perfectly aligned H-bonds along the O–O direction (Figure 6c).[36] Therefore, the sXAS not only evidently showed the formation of a hydroxylated interphase on the $Mn_5O_8$ surface after cycling but also a striking ordering of such an interphase. It is important to point out that the $Mn_5O_8$ material held at −1.7 V (vs MSE) showed more distinct hydroxyl features than the one at 0.8 V as shown in Figure 6(d). This result strongly indicated that the hydroxylated interphase can be stabilized at both high and low voltages, and more importantly, it suggested that the formation of the hydroxylated interphase might be strongly favored during the cathodic (reduction) process. At a lower voltage of −1.75 V, corresponding to over 0.6 V high overpotential toward the HER, the water is thermodynamically unstable and the HER happens as Equation (8) shows during the initial cyclic voltammetry (CV) cycling. Water decomposition (reduction) at low voltages could favor the formation of a hydroxylated interphase via the interaction between the resulting hydroxyl group (OH) and $Mn^{2+}$ on the $Mn_5O_8$ surface. It is also worth mentioning that the sXAS results also suggested that a strong bonding between $Mn^{2+}$ and OH could stand the consequent oxidative condition when the voltage was ramped to 0.8 V.

$$2H_2O\,(liq) + 2e^- \rightarrow H_2(g) + 2OH^-(liq) \qquad (8)$$

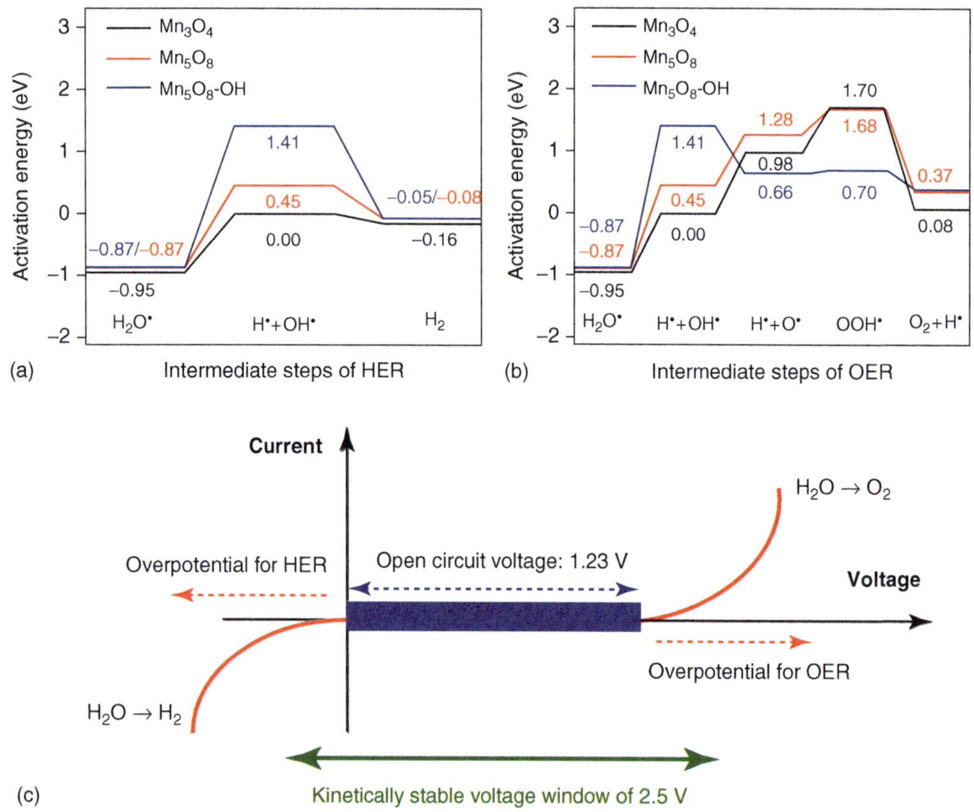


**Figure 7** The activation energies of intermediate reaction steps of (a) HER and (b) OER on $Mn_5O_8$ with a hydroxylated interphase, $Mn_5O_8$ without interphase, and $Mn_3O_4$ (the asterisk symbol denotes the adsorbed species on the surface). (c) Schematics of how to achieve a kinetically stable voltage window of 2.5 V by increasing the overpotential of the electrodes for HER and OER

Results from the DFT calculation further explained why hydroxylated interphase could result in a high overpotential toward water decomposition and thus afford a widened potential window beyond the thermodynamically limited window of 1.23 V. Figure 7 shows the activation energy of each intermediate reaction step during the HER and OER on surfaces of hydroxylated $Mn_5O_8$, pure $Mn_5O_8$, and $Mn_3O_4$.[34] The activation energy of the rate-determining step was substantially high for the OER, indicating that the OER was intrinsically difficult. Particularly, activation energy of water dissociation ($H_2O* \rightarrow H* + OH*$), the second catalytic step of gas evolution, followed the order of hydroxylated $Mn_5O_8$ (1.41 eV) > $Mn_5O_8$ (0.45 eV) > $Mn_3O_4$ (~0 eV). Hydroxylated $Mn_5O_8$ showed about 1 eV higher in activation energy than that of $Mn_5O_8$, implying that the reaction rate at 300 K was at least 17 orders of magnitude slower according to the Arrhenius equation. Therefore, the DFT calculation directly confirmed the inhibitive role of the hydroxylated interphase on the HER and OER. Therefore, the high overpotential toward the HER and OER resulted from the hydroxylated interphase offers a kinetically stable potential window of 2.5 V for Na-ion insertion and extraction without significant generation of hydrogen and oxygen.

## 3 ELECTRODE AND WATER INTERACTION IN IMPROVING THE KINETICS OF AQUEOUS ENERGY STORAGE

### 3.1 Understanding Kinetics of Charge Storage Process via Analysis of Experimental Data and Numerical Modeling

The charge storage process occurring in electrode materials is dependent on a relatively fast surface-controlled capacitive process and a relatively slow diffusion-controlled redox process. The diffusion-limited redox process is typically observed in battery-like materials and usually favors high-energy performance. The capacitive process is attributed to electrical double-layer capacitance (EDLC) and pseudocapacitance in capacitor-like materials via the separation or adsorption/desorption of ions at the near surface of the electrode and thus favors high-power performance of EES devices. The charge storage mechanism of the electrode materials could be understood and determined through analyzing the CV curves obtained experimentally at various scan rates. Once the dominated charge storage mechanism is known,

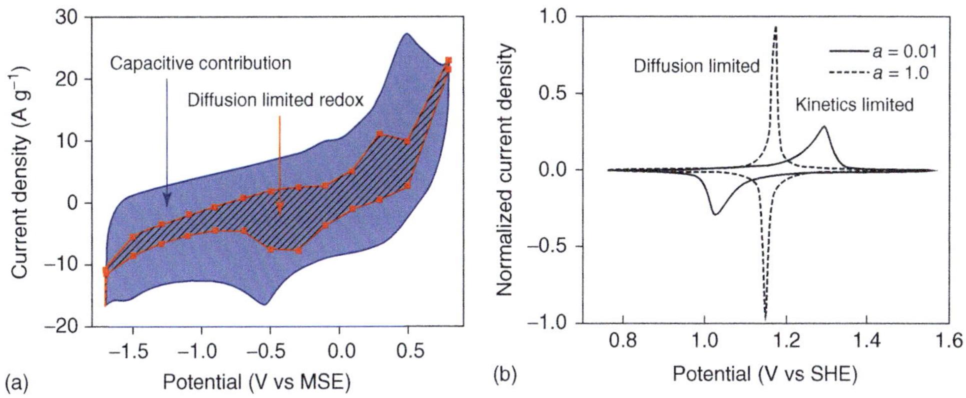


**Figure 8** (a) Electrokinetic analysis to reveal the different charge storage mechanisms by analyzing the CV data. (b) Simulated CV curves with different $a$ values using numerical modeling

operational parameters could be adjusted accordingly in order to achieve the desired energy or power performance, namely, one can promote capacitive process and/or demote diffusion-limited redox process if high power and/or low energy performance is needed and vice versa.

For a strictly diffusion-limited redox reaction, the current ($i_d$) is proportional to the square root of the scan rate ($v$), while for a strictly capacitive charge storage process, the capacitive current ($i_c$) from EDLC and pseudocapacitance has a linear dependence on the scan rate according to Equations 9 and 10), respectively.

$$i_d = k_d v^{0.5} \tag{9}$$

$$i_c = k_c v \tag{10}$$

Accordingly, the overall current ($i_{total}$) at a given potential can be expressed as the sum of two separate charge storage mechanisms:[41,42]

$$i_{total} = i_c + i_d = k_c v + k_d v^{0.5} \tag{11}$$

$$i_{total} = a v^b \tag{12}$$

where $a$ is an adjustable parameter and $b$ is a variable heavily dependent on the relative contribution from $i_c$ or $i_d$. It is apparent that the value of $b$ is equal to either 0.5 or 1 when the overall currents are strictly dominated by diffusion-limited redox current ($i_d$) or capacitive current ($i_c$), respectively.

$$\frac{i_{total}}{v^{0.5}} = k_c v^{0.5} + k_d \tag{13}$$

After rearranging Equations (11)–(13), the values of $k_c$ and $k_d$ can be determined by the slopes and intercepts via plotting $i/v^{0.5}$ as a function of $v^{0.5}$ at a given potential. Therefore, the current attributed to the diffusion-limited redox process ($i_d = k_d v^{0.5}$) or capacitive process ($i_c = k_c v$) at each scan rate can be determined, and the charge storage mechanisms (e.g., capacitive or diffusion-controlled redox) during the CV scans can be revealed. Figure 8(a) shows a typical CV of an electrode material at a scan rate of 500 mV s, where the distributions of diffusion-limited redox and capacitive processes are distinguished. Similar methods will be used to determine how experimental parameters (e.g., scan rate) affect the charge storage mechanism of the electrode materials (e.g., the capacitive or diffusion-limited redox charge storage mechanism) and understand how to achieve desired electrokinetics of the charge storage by manipulating these operational parameters. For example, by increasing the scan rate, the overall charge storage mechanism will tend to be dominated by the surface-controlled capacitive process and vice versa.

In addition to analyzing the experimental data, electrokinetics of the charge storage process can be modeled and estimated in a relatively accurate manner by unsteady state diffusion within the electrode materials using a numerical analysis, from which the interplay between the redox reaction and ion transport that determines electrokinetics of charge storage becomes obvious. Assuming charge carriers are homogenously mixed within an individual spherical electrode nanoparticle, the flux of the charge carrier ($N$), mass conservation of the charge carriers within the particle at unsteady state, and current density ($i$) on the surface of the particle can be expressed by Equations (14)–(16), respectively.

$$N = -D\nabla c \tag{14}$$

$$\frac{\partial c}{\partial t} = -\nabla \cdot N \tag{15}$$

$$\frac{i}{F} = -D\left(\left.\frac{\partial c}{\partial r}\right|_{r=r_0}\right) \tag{16}$$

Meanwhile, the following variables can be converted into dimensionless ones (time: $\tau = \frac{tD}{r_0^2}$; distance from center: $x = \frac{r}{r_0}$; concentration: $y = \frac{c}{c_0}$; current density: $j = \frac{ir_0}{FDc_0}$), where $t$ is time, $D$ is the diffusion coefficient, $r_0$

is the radius of the particle, $c_0$ is the initial concentration of charge carriers, $F$ is Faraday constant. Current density at the particle surface ($i$) is also governed by the Butler–Volmer equation:

$$\frac{i}{F} = k(c_1)^{1-\beta}(c_\theta)^{1-\beta}(c_s)^{\beta} \times \left\{ \exp\left(\frac{(1-\beta)F\eta}{RT}\right) - \exp\left(-\frac{\beta F\eta}{RT}\right) \right\} \quad (17)$$

where $k$ is the reaction rate constant, $C_1$ is the concentration of the charge carrier in the electrolyte (assumed to be a constant), $C_\theta$ is the surface concentration of vacant sites ready for cation intercalation, $c_s$ is the concentration of cations on the electrode surface, and $c_t$ is the concentration of total sites for seating cations ($C_\theta = C_t - C_s$). $R$ is the gas constant, $T$ is the temperature, and $\beta$ is the symmetry factor, representing the fraction of the electrical potential used to promote the cathodic reaction (assumed to be 0.5 in this study). $\eta$ is the overpotential (difference between the applied voltage and open-circuit voltage).

Combining Equations (14) and (15) gives the partial differential equation (Equation 18) that describes the conservation of charge carriers in spherical coordinates at unsteady state. Equation 18 can be solved with a set of initial condition (Equation 19) and boundary conditions at the center of the particle (Equation 20) and at the surface of the particle (Equation 22). As a result, the concentration gradient of charge carriers at different positions within the particles ($\partial y/\partial x$) can be obtained as a function of external voltage. Accordingly, the charging and discharging processes (e.g., CV and chronopotentiometry (CP) plots) can be simulated.[43–45]

$$\frac{\partial y}{\partial \tau} = \frac{\partial^2 y}{\partial x^2} + \frac{2}{x}\frac{\partial y}{\partial x} \quad (18)$$

$$y(x,0) = 1 \quad (19)$$

$$\left.\frac{\partial y}{\partial x}\right|_{x=0} = 0 \quad (20)$$

$$a = \frac{k r_0 c_1^{1-\beta}}{D} = \frac{k c_1^{2-\beta}}{D c_1/r_0} \quad (21)$$

$$\left.\frac{\partial y}{\partial x}\right|_{x=1} + a(y|_{x=1})^{\beta}(1-y|_{x=1})^{1-\beta} \times \left\{ \exp\left(\frac{(1-\beta)F\eta}{RT}\right) - \exp\left(-\frac{\beta F\eta}{RT}\right) \right\} = 0 \quad (22)$$

Figure 8(b) shows the results on simulating CV curves at a scan rate of 0.1 $mV\,s^{-1}$ using various values of $a$ using a partial differential equation-2D (PDE2D) solver.[46] It is notable that parameter $a$ is a dimensionless factor (Equation 21) and can be considered as the ratio of the interfacial reaction rate ($kc_1^{2-\beta}$) to the diffusion rate of the charge carriers on the surface ($Dc_1/r_0$), which strongly affects the shapes of the CV. A large value of $a$ reflects a fast interfacial reaction and/or a slow diffusion of charge carriers, and thus the charge transfer process is limited by solid-state diffusion of charge carriers. Accordingly, the solid particle functions more as a battery electrode material with a relatively high storage capacity and low charge transfer rate. If a stronger capacitive charge transfer is required to improve the power performance of an EES device (e.g., shortening the charging time), a smaller value of $a$ is needed. Therefore, (i) a smaller-sized electrode (smaller $r_0$) or (ii) an electrolyte with cations having a higher diffusivity (larger $D$) in the host electrode material is required or (iii) an electrode material with a large open framework for charge carrier insertion/extraction is required.[47]

It is also important to mention that the above mathematical model has apparent limitations such as only diffusion is considered while neglecting the migration of charge carriers in the electrode, only solid particles are considered while neglecting the possible porous nature of the electrode, and large numbers of parameters cannot be obtained experimentally. On the other hand, many structural parameters and operation conditions that are not achieved in practical cell operation can be considered in the numerical modeling. Hence, the mathematical model discussed here is very plausible, especially when experimental results and theoretical analyses are carefully compared.

### 3.2 Strategies to Improve Storage Kinetics of Electrode Materials: Design of Disordered Materials

Electrokinetics is usually slow in host materials (particularly cathodes), especially when such a material has a close-packed array of oxide ions which may not be able to accommodate the charge carriers for reversible insertion and extraction. Several design principles have been used to tackle the intercalation of charge carriers, especially the ones with a relatively large radius such as Na and K ions. The first one is to replace oxygen anions ($O^{2-}$) of host electrodes with other anions that have weaker bonding with $Na^+$. For example, electrode materials based on hexacyano ions $(C{\equiv}N)_6^{6-}$ have been intensively studied due to their weakened bonding between cyanide $(C{\equiv}N)^-$ and Na and/or K cations.[48–50] Cui and coworkers demonstrated that Prussian blue (potassium copper hexacyanoferrate) and its analogues could function as stable electrode materials for aqueous K- and Na-ion storage at a potential window of 1.5 V.[2,51–53] Similarly, sodium manganese hexacyanoferrate reported by Goodenough's group has demonstrated a high energy performance and reasonable cycle life in a nonaqueous electrolyte with a potential window up to 3.5 V.[48,49]

The second approach to improve storage kinetics and storage capacity of EES systems electrodes is to introduce a large interstitial host framework, especially layered structures.[54–57] These materials with planar or zigzag layers show different polymorphs (P2, P3, or O3 symmetry) with respective to the sites of the intercalated alkali ions by simply altering the stacking of transition metal–oxygen octahedra ($[MO_6]$). A number of recent studies on intercalation compounds have focused on earth-abundant transition metals, such as layered $Na_xMnO_2$ compounds, as promising cathode materials using Na as a charge carrier for its great potential to replace LIB technologies.

Moreover, designing the disordered electrode materials to improve the storage capacity and kinetics has drawn great research interest recently. The disordered electrodes demonstrated a great potential for enhancing storage capacity and cycle life of EES systems, especially in the nonaqueous battery system. Even for the close-packed metal–oxide electrode systems, electrodes with disordered cation or anion sites, accompanied accordingly with the changes of cation/anion vacancies and partial reduction of transition metals, showed a much enhanced storage capacity in $Na_{1.25+x}V_3O_8$, $VFe_2O_x$, $V_2O_5$, and $LaMnO_3$ electrode materials.[58–62] Lee *et al.* reported that the cation-disordered $Li_{1.211}Mo_{0.467}Cr_{0.3}O_2$ electrode materials showed a facile Li-ion transport due to a modified local environment for the transition metal ions and therefore a lower percolation threshold for ionic transport. As shown in Figure 9, in such disordered materials, Li ions could easily diffuse without directly contacting the transition metal framework on the atomic scale (so-called 0-TM channel for Li-ion diffusion pathway).[63] Lithium ions can diffuse much faster in such 0-TM pathways because of lower diffusion barrier rather than when contacting the framework (1-TM channel for Li-ion diffusion). It is worth noting that the relatively low average discharge voltage is a major disadvantage of this particular compound. However, this compound shed a light on a new design strategy for high-capacity electrodes by using Li-rich and disordered materials that can provide the percolation framework for Li-ion intercalation.

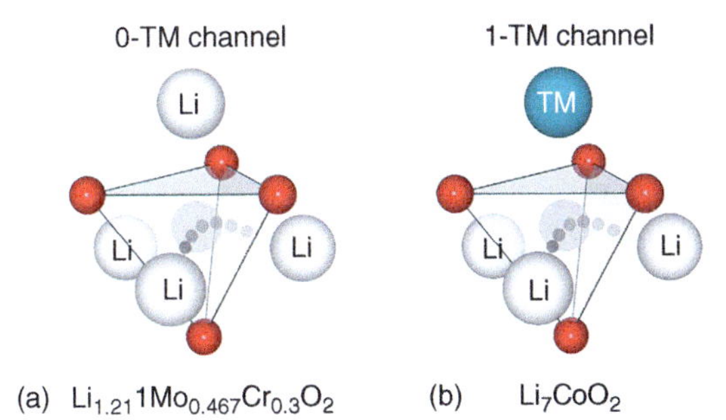


**Figure 9** Two possible environments for Li-ion transport in rock-salt-like Li-transition metal oxide, including (a) the activated state sharing faces with no octahedral transition metals (0-TM channel for Li-ion transport) and (b) one transition metal (1-TM channel)

Although rather extensively studied for non-aqueous EES systems, disordered electrode materials for aqueous EES systems have also been reported recently, showing that structural water could stabilize the disordered structure to improve storage kinetics and capacity. Charles *et al.* investigated the influence of structural water on the electrochemical performance of the disordered $V_2O_5$ nanostructures.[64] The CV curves of the disordered $V_2O_5$ contained multiple redox peaks at various potentials across the entirety of the one-volt potential window, as shown in Figure 10, during both anodic scans (0.24, 0.51, and 0.82 V vs Ag/AgCl) and cathodic scans (0.12, 0.43, and 0.79 V), indicating strong potential-dependent charge-storage mechanisms. Disordered $V_2O_5$ showed a much enhanced storage capacity for aqueous K-ion storage (183 mAh $g^{-1}$ at a scan rate of 5 mV $s^{-1}$, equivalent to 0.89 electrons transfer per vanadium atom) compared with its ordered counterpart (37 mAh $g^{-1}$) as shown in Figure 10(c) and (d).

Moreover, in situ X-ray diffraction collected during the CV measurements further confirmed that the disordered structure greatly contributed to the improved redox activity and capacity for aqueous K-ion storage. Figure 10(e) demonstrates that disordered $V_2O_5$ layered materials engaged with structural water showed more continuous expansion/contraction of (001) diffraction planes and a wider redox-active potential window during electrochemical cycling in a Contour plot, compared with highly crystalline materials (Figure 10f). Interestingly, a significant difference between the two materials is that the (001) peak shifting of the disordered $V_2O_5$ occurred within a much broader potential range than the highly crystalline $V_2O_5$. The results strongly suggest that K-ion transport between the V–O layers in the disordered structure has more lower energy barriers compared with those in the ordered structure, which is expected to expand the accessible potential window for K-ion insertion and extraction, and therefore to enhance the storage capacity. Another appealing feature is that the disordered $V_2O_5$ displayed continuous and reversible peak shifts of the (001) reflection during K-ion extraction/insertion without obvious structure transformation, compared to the highly crystalline $V_2O_5$, which undergoes a more distinct staged-like peak shifting.

Promotional effects of structural water on the stability of the disordered $V_2O_5$ were confirmed by neutron total scattering and pair distribution function (PDF) analysis. Neutron PDF was used due to the high sensitivity of neutrons to water (being particularly sensitive to both oxygen and hydrogen atoms) compared with X-ray and electron probes. Additionally, the lack of long-range order of the $V_2O_5$ nanomaterials and intercalated species (e.g., water and K ions) required PDF analysis so that the effect of hydration on the local structure can be studied. The neutron scattering experiments and PDF analysis showed

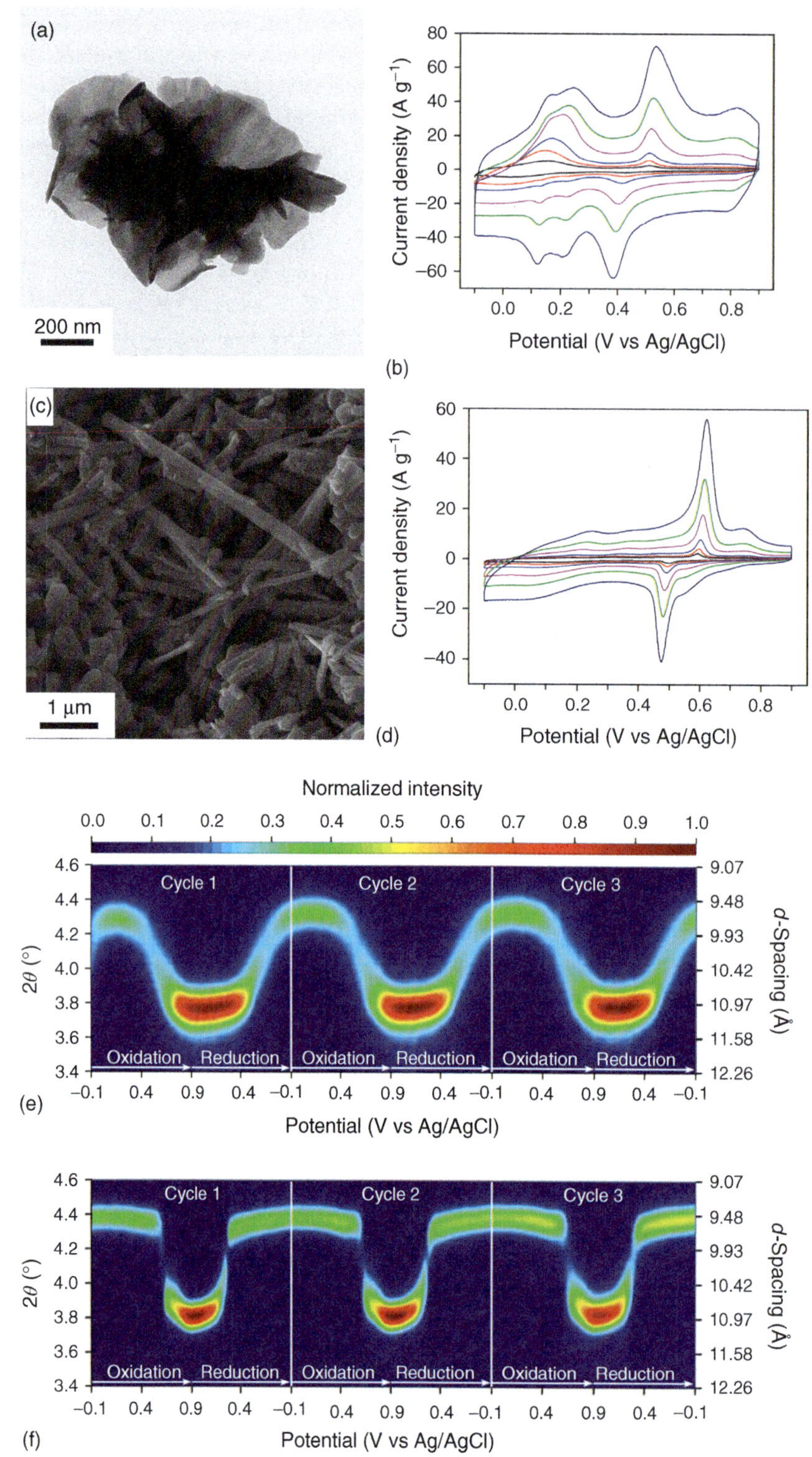


**Figure 10** The TEM image and CV curves of (a,b) disordered $V_2O_5$ nanomaterials and (c,d) ordered $V_2O_5$ nanomaterials. (e,f) In situ XRD spectra of the layered (001) plane during electrochemical cycling. The contour plot of in situ XRD spectra during the three CV cycles at a scan rate of 1 mV s$^{-1}$ for the (e) disordered $V_2O_5$ and (f) ordered $V_2O_5$ materials in a 3 M KCl electrolyte

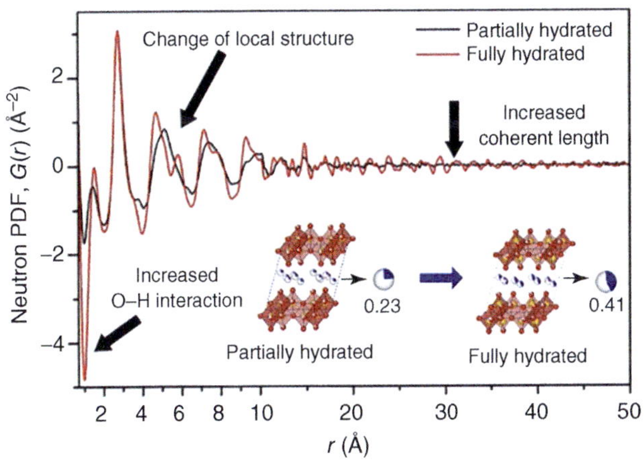


**Figure 11** Neutron PDFs of partially and fully hydrated and disordered $V_2O_5$. The fully hydrated material showed a much larger coherence length. The inset shows the rearrangement of the local structure upon hydration. The refined structures and occupancy of the water intercalated between the layers as quantified by fitting PDF analysis for the partially hydrated $V_2O_5$ have an $O_W$ (the oxygen atoms from the water molecules, blue) occupancy of 0.23 between V–O layers, while hydrated $V_2O_5$ has an $O_W$ occupancy of 0.41. In the refined crystal structure of $V_2O_5$ materials, the vanadium atoms are yellow, the potassium atoms are purple, the oxygen atoms from the V–O bilayers are red, and the vacancy sites are gray

that upon hydration the increase in structural water of the disordered $V_2O_5$ material was also accompanied by reordering the local structure and an increase in coherence. The refined structures of the partially hydrated and fully hydrated disordered $V_2O_5$ are shown in Figure 11. They possess a similar monoclinic structure having potassium and water intercalated within V–O bilayers consisting of $[VO_6]$ octahedral units. Albeit, a much larger coherence length is observed in the fully hydrated and disordered $V_2O_5$ (~50 Å) than that of the partially hydrated and disordered $V_2O_5$ (~28 Å). Many changes in the local structure are observed upon hydration of the disordered $V_2O_5$, including a large increase in water content evident by an increase in intensity of the O—H bond ($r = 1$ Å), as well as new O–O features are observed in the fully hydrated and disordered $V_2O_5$ at a radial distance of 4–6 Å in the PDF. Moreover, the disordered and hydrated $V_2O_5$ material showed a much larger occupancy of the water intercalated between the layers (0.41), compared with the partially hydrated $V_2O_5$ (0.41). All these results indicated a high structural stability of disordered $V_2O_5$ induced by the interaction between V–O layers and water, which also corroborated the excellent cycling stability and electrokinetics in electrochemical measurements. These results are particularly important for designing a new type of disordered electrode materials stabilized by structural water for aqueous energy storage with a large storage capacity.

## 4 FUTURE RESEARCH DIRECTIONS

The interactions between electrodes and electrolytes have been extensively described in EES systems, especially in nonaqueous LIBs.[18–21] In this chapter, the interaction between electrodes and electrolytes for aqueous EES systems was reviewed and discussed, and the comprehensive understanding of the SEI and its role in aqueous EES systems is still lacking. Here is the brief summary of the future research directions:

1. The exact mechanism contributing to the SEI formation in aqueous electrolytes is not fully understood. It is also unclear whether other transition metal oxides could form a similar hydroxylated interphase by interacting with water as the manganese oxides do and whether other types of interphases could be formed through interaction with species in the electrolyte besides water during the electrochemical process. Moreover, the structure of the SEI in aqueous electrolytes is still to be elucidated, so that establishing structure–property relationships is largely a trial-and-error process. Development of highly water-sensitive surface techniques to determine and analyze the structure of interphases resulted from the electrode/water interaction, such as soft X-ray absorption and neutron scattering, is highly demanded.
2. The nature of the electrode/electrolyte interactions, as well as the formation of SEI, is complex and dynamic during the charge-transfer process. Off-equilibrium operations involving the fast charge-transfer process and/or rapid change of the voltage frequently occur in a practical electrochemical process. It is expected that in aqueous electrolytes, the formation of SEI involves water reduction, electron transfer, and/or ion diffusion. Therefore, using in situ and operando tools to *observe* the formation of the interphases and study the influence of the interphases on ionic transport in electrodes is therefore extremely important to understand the SEI formation and its evolution. Moreover, through these in situ and operando tools, we can understand how the interphases influence the functionalities (e.g., kinetically stable voltage window, storage capacity, current density, and durability) of the electrode materials for aqueous EES systems at near-equilibrium and more importantly at off-equilibrium electrochemical conditions.
3. Although the influences of electrode/aqueous electrolyte interaction on EES systems are reviewed from experimental perspectives, modeling SEI in aqueous electrolytes is important and nontrivial, especially the formation and evolution of the SEI could occur at different lengths and time scales. In this context, developing a new modeling method, ranging from

electronic- to atomic-scale models and phenomenological models, is important and should be coordinated with experimental efforts in order to make qualitative and quantitative understanding of the SEI.

## 5 ACKNOWLEDGMENTS

This work was supported by the US Department of Energy (DOE), Office of Science, Basic Energy Sciences under Award No. DE-SC00 18922. The reported research used resources of the Advanced Photon Source (APS), a US DOE Office of Science User Facility operated for the DOE Office of Science by Argonne National Laboratory under Contract No. DE-AC02-06CH11357. The neutron scattering experiments were carried out at the Spallation Neutron Source, which is sponsored by the Scientific User Facilities Division, Office of Basic Energy Sciences, US Department of Energy, under Contract No. DE-AC05-00OR22725 with Oak Ridge National Laboratory.

## 6 ABBREVIATIONS AND ACRONYMS

CP = chronopotentiometry; DFT = density function theory; EDLC = electrical double-layer capacitance; EES = electrochemical energy storage; HER = hydrogen evolution reaction; HOMO = highest occupied molecular orbital; LIBs = lithium-ion batteries; LUMO = lowest unoccupied molecular orbital; OER = oxygen evolution reaction; PDF = pair distribution function; SEI = solid-electrolyte interphase; sXAS = soft X-ray absorption spectroscopy; XPS = X-ray photoelectron spectroscopy

## 7 REFERENCES

1. B. Dunn, H. Kamath and J. M. Tarascon, *Science*, 2011, **334**, 928.
2. M. Pasta, C. D. Wessells, R. A. Huggins and Y. Cui, *Nat. Commun.*, 2012, **3**, 7.
3. M. S. Whittingham, *Chem. Rev.*, 2004, **104**, 4271.
4. J. B. Goodenough and Y. Kim, *Chem. Mater.*, 2010, **22**, 587.
5. B. Scrosati and J. Garche, *J. Power Sources*, 2010, **195**, 2419.
6. P. G. Balakrishnan, R. Ramesh and T. P. Kumar, *J. Power Sources*, 2006, **155**, 401.
7. Q. S. Wang, P. Ping, X. J. Zhao, G. Q. Chu, J. H. Sun and C. H. Chen, *J. Power Sources*, 2012, **208**, 210.
8. J. M. Tarascon, *Nat. Chem.*, 2010, **2**, 510.
9. W. Li, J. R. Dahn and D. S. Wainwright, *Science*, 1994, **264**, 1115.
10. V. Palomares, P. Serras, I. Villaluenga, K. B. Hueso, J. Carretero-Gonzalez and T. Rojo, *Energy Environ. Sci.*, 2012, **5**, 5884.
11. N. Yabuuchi, K. Kubota, M. Dahbi and S. Komaba, *Chem. Rev.*, 2014, **114**, 11636.
12. L. M. Suo, O. Borodin, Y. S. Wang, X. H. Rong, W. Sun, X. L. Fan, S. Y. Xu, M. A. Schroeder, A. V. Cresce, F. Wang, C. Y. Yang, Y. S. Hu, K. Xu and C. S. Wang, *Adv. Energy Mater.*, 2017, **7**, 10.
13. Z. Li, D. Young, K. Xiang and W. Carter, *Energy Mater*, 2013, **3**, 290.
14. E. Glueckauf, *Trans. Faraday Soc.*, 1955, **51**, 1235.
15. H. Kim, J. Hong, K. Y. Park, H. Kim, S. W. Kim and K. Kang, *Chem. Rev.*, 2014, **114**, 11788.
16. J. Yu, C. Mu, X. Y. Qin, C. Shen, B. Y. Yan, H. G. Xue and H. Pang, *Adv. Mater. Interfaces*, 2017, **4**, 10.
17. L. Chen, Z. Y. Guo, Y. Y. Xia and Y. G. Wang, *Chem. Commun.*, 2013, **49**, 2204.
18. D. Aurbach, M. L. Daroux, P. W. Faguy and E. Yeager, *J. Electrochem. Soc.*, 1987, **134**, 1611.
19. D. Aurbach, *J. Power Sources*, 2000, **89**, 206.
20. P. Verma, P. Maire and P. Novak, *Electrochim. Acta*, 2010, **55**, 6332.
21. K. Xu, *Chem. Rev.*, 2014, **114**, 11503.
22. A. N. Dey and B. P. Sullivan, *J. Electrochem. Soc.*, 1970, **117**, 222.
23. E. Peled, *J. Electrochem. Soc.*, 1979, **126**, 2047.
24. A. P. Wang, S. Kadam, H. Li, S. Q. Shi and Y. Qi, *NPJ Comput. Mater.*, 2018, **4**, 222.
25. D. Aurbach, B. Markovsky, M. D. Levi, E. Levi, A. Schechter, M. Moshkovich and Y. Cohen, *J. Power Sources*, 1999, **81**, 95.
26. M. B. Pinson and M. Z. Bazant, *J. Electrochem. Soc.*, 2013, **160**, A243.
27. M. Gauthier, T. J. Carney, A. Grimaud, L. Giordano, N. Pour, H. H. Chang, D. P. Fenning, S. F. Lux, O. Paschos, C. Bauer, F. Magia, S. Lupart, P. Lamp and Y. Shao-Horn, *J. Phys. Chem. Lett.*, 2015, **6**, 4653.
28. X. B. Cheng, R. Zhang, C. Z. Zhao, F. Wei, J. G. Zhang and Q. Zhang, *Adv. Sci.*, 2016, **3**, 4653.
29. K. Xu and C. S. Wang, *Nat. Energy*, 2016, **1**, 2.
30. Y. Yamada, K. Usui, K. Sodeyama, S. Ko, Y. Tateyama and A. Yamada, *Nat. Energy*, 2016, **1**, 9.
31. L. Suo, O. Borodin, T. Gao, M. Olguin, J. Ho, X. Fan, C. Luo, C. Wang and K. Xu, *Science*, 2015, **350**, 938.
32. L. M. Suo, O. Borodin, W. Sun, X. L. Fan, C. Y. Yang, F. Wang, T. Gao, Z. H. Ma, M. Schroeder, A. von Cresce, S. M. Russell, M. Armand, A. Angell, K. Xu and C. S. Wang, *Angew. Chem. Int. Ed.*, 2016, **55**, 7136.

33. F. Wang, Y. X. Lin, L. M. Suo, X. L. Fan, T. Gao, C. Y. Yang, F. D. Han, Y. Qi, K. Xu and C. S. Wang, *Energy Environ. Sci.*, 2016, **9**, 3666.

34. X. Q. Shan, D. S. Charles, Y. K. Lei, R. M. Qiao, G. F. Wang, W. L. Yang, M. Feygenson, D. Su and X. W. Teng, *Nat. Commun.*, 2016, **7**, 8.

35. X. Q. Shan, F. H. Guo, W. Q. Xu and X. W. Teng, *Front. Energy*, 2017, **11**, 383.

36. P. Wernet, D. Nordlund, U. Bergmann, M. Cavalleri, M. Odelius, H. Ogasawara, L. A. Naslund, T. K. Hirsch, L. Ojamae, P. Glatzel, L. G. Pettersson and A. Nilsson, *Science*, 2004, **304**, 995.

37. F. M. F. de Groot, M. Grioni, J. C. Fuggle, J. Ghijsen, G. A. Sawatzky and H. Petersen, *Phys. Rev. B*, 1989, **40**, 5715.

38. J. D. Smith, C. D. Cappa, K. R. Wilson, B. M. Messer, R. C. Cohen and R. J. Saykally, *Science*, 2004, **306**, 851.

39. A. Nilsson, D. Nordlund, I. Waluyo, N. Huang, H. Ogasawara, S. Kaya, U. Bergmann, L. Å. Näslund, H. Öström, P. Wernet, K. J. Andersson, T. Schiros and L. G. M. Pettersson, *J. Electron Spectros. Relat. Phenomena*, 2010, **177**, 99.

40. J.-J. Velasco-Velez, C. H. Wu, T. A. Pascal, L. F. Wan, J. Guo, D. Prendergast and M. Salmeron, *Science*, 2014, **346**, 831.

41. T. C. Liu, W. G. Pell, B. E. Conway and S. L. Roberson, *J. Electrochem. Soc.*, 1998, **145**, 1882.

42. T. Brezesinski, J. Wang, J. Polleux, B. Dunn and S. H. Tolbert, *J. Am. Chem. Soc.*, 2009, **131**, 1802.

43. M. Doyle, J. Newman, A. S. Gozdz, C. N. Schmutz and J. M. Tarascon, *J. Electrochem. Soc.*, 1996, **143**, 1890.

44. D. Zhang, B. N. Popov and R. E. White, *J. Electrochem. Soc.*, 2000, **147**, 831.

45. R. Chandrasekaran, A. Magasinski, G. Yushin and T. F. Fuller, *J. Electrochem. Soc.*, 2010, **157**, A1139.

46. F. H. Guo, N. Gupta and X. Teng, in 'Supercapacitors—Theoretical and Practical Solutions', ed. L. Liudvinavičius, InTech, 2017.

47. B. E. Conway, *J. Electrochem. Soc.*, 1991, **138**, 1539.

48. Y. H. Lu, L. Wang, J. G. Cheng and J. B. Goodenough, *Chem. Commun.*, 2012, **48**, 6544.

49. L. Wang, Y. H. Lu, J. Liu, M. W. Xu, J. G. Cheng, D. W. Zhang and J. B. Goodenough, *Angew. Chem. Int. Ed.*, 2013, **52**, 1964.

50. A. Eftekhari, *J. Power Sources*, 2004, **126**, 221.

51. C. D. Wessells, R. A. Huggins and Y. Cui, *Nat. Commun.*, 2011, **2**, 5.

52. H. W. Lee, R. Y. Wang, M. Pasta, S. W. Lee, N. Liu and Y. Cui, *Nat. Commun.*, 2014, **5**, 6.

53. M. Pasta, C. D. Wessells, N. Liu, J. Nelson, M. T. McDowell, R. A. Huggins, M. F. Toney and Y. Cui, *Nat. Commun.*, 2014, **5**, 9.

54. C. Delmas, C. Fouassier and P. Hagenmuller, *Physica. B C*, 1980, **99**, 81.

55. A. Mendiboure, C. Delmas and P. Hagenmuller, *J. Solid State Chem.*, 1985, **57**, 323.

56. J. Billaud, R. J. Clement, A. R. Armstrong, J. Canales-Vazquez, P. Rozier, C. P. Grey and P. G. Bruce, *J. Am. Chem. Soc.*, 2014, **136**, 17243.

57. L. Chagas, *Chem. A*, 2014, **2**, 20263.

58. J. T. Mefford, W. G. Hardin, S. Dai, K. P. Johnston and K. J. Stevenson, *Nat. Mater.*, 2014, **13**, 726.

59. C. N. Chervin *et al.*, *J. Mater. Chem. A*, 2015, **3**, 12059.

60. E. Uchaker and G. Z. Cao, *Chem. Asian J.*, 2015, **10**, 1608.

61. E. Uchaker, H. Y. Jin, P. Yi and G. Z. Cao, *Chem. Mater.*, 2015, **27**, 7082.

62. O. B. Chae, J. Kim, I. Park, H. Jeong, J. H. Ku, J. H. Ryu, K. Kang and S. M. Oh, *Chem. Mater.*, 2014, **26**, 5874.

63. J. Lee, A. Urban, X. Li, D. Su, G. Hautier and G. Ceder, *Science*, 2014, **343**, 519.

64. D. S. Charles, M. Feygenson, K. Page, J. Neuefeind, W. Q. Xu and X. W. Teng, *Nat. Commun.*, 2017, **8**, 8.

# Na-Ion Batteries: Positive Electrode Materials

**Elizabeth H. Driscoll, Laura L. Driscoll and Peter R. Slater**

*University of Birmingham, Birmingham, UK*



## 1 INTRODUCTION

The major technological advances witnessed in the portable electronics industry and in the electrification of transport over recent decades can be attributed to the discovery and the successive commercialization of Li-ion batteries. Although Li-ion batteries are likely to maintain their dominant market share in such applications in the near future, future supply risks for lithium as a resource pose a threat, particularly with the forecasted rapid expansion in Li-ion battery use needed for worldwide targets for zero emission cars. In particular, easily accessible reserves have been estimated to run out within the century, with additional potential supply issues present due to over 90% of resources being found in just four countries. A further application that will add more strain on resources is the use of batteries for the storage of renewable energy in large-scale grid storage applications. Thus, efforts to find an alternative to Li have seen an increased interest in Na-ion batteries. This interest has been driven by the high natural abundance of Na globally and the intercalation chemistry of a similar nature to Li, meaning that these cells operate in a similar way to an Li-ion cell.

The architecture of an Na-ion cell (Figure 1) consists of two electrodes separated by an electrolyte, where Na ions move from the negative electrode to the positive electrode on discharge with electrons flowing through the circuit delivering power to the circuit (the reverse process occurs on charging the battery). Thus, a key feature of both electrodes is the ability to reversibly incorporate Na, as well as good electronic conductivity, while the electrolyte should possess high ionic conductivity with negligible electronic conductivity to prevent internal short circuiting.

While they are currently enjoying a renaissance in terms of research interest, it is important to note that Na-ion batteries have been around for over 50 years. High-temperature batteries such as Na–S and the Zebra (Na transition metal chloride) batteries have already been commercialized for electricity grid storage applications. However, the fact that both electrodes are molten in these batteries has raised safety concerns, leading to renewed interest in ambient-temperature Na-ion cells. Research on such ambient-temperature Na ion batteries initially ran parallel to the original research on Li-ion batteries in the 1970s–1980s, with a number of important initial studies on Na intercalation into layered Mn and Co oxide and titanium sulfide systems. However, due to additional issues encountered in comparison to the equivalent Li ion cells, such as increased moisture sensitivity, slightly lower cell voltage, increased size of $Na^+$ leading to increased volume changes, the inability of $Na^+$ to intercalate into graphite, research on these systems diminished as Li ion batteries became commercialised.

In this chapter, we focus on advances made in the development of positive electrode (cathode) materials for Na-ion batteries. As for Li-ion batteries, layered transition metal oxide systems have traditionally dominated research in this area, although following the development

*Inorganic Battery Materials*. Edited by Hailiang Wang and Boniface P.T. Fokwa
 ISBN 978-1-119-43199-2

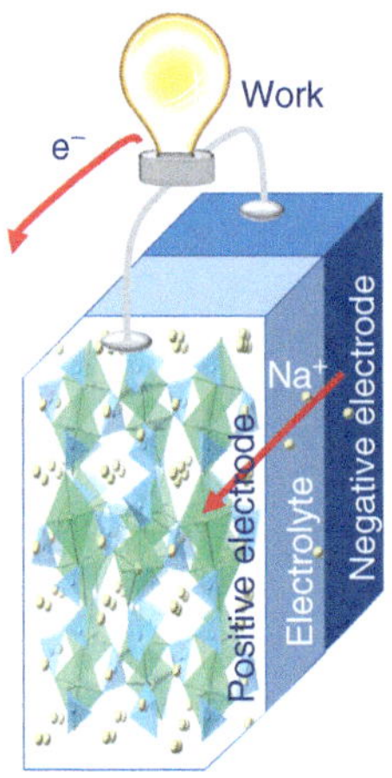


**Figure 1** Diagram depicting the movement of charges upon discharging of a $Na^+$-ion rechargeable battery

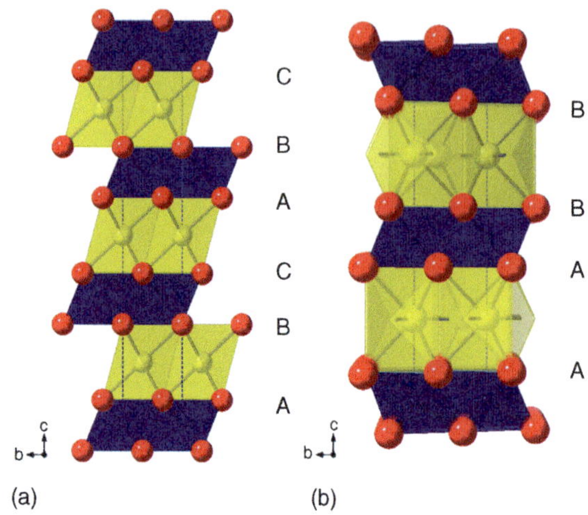


**Figure 2** Alternate layer stacking in $Na_xMO_2$. O3-type (a), P2-type (b). The letters show how the oxygen layers stack to form the layered material

of $LiFePO_4$ for Li ion batteries, there has been considerable interest in Na transition metal phosphate/sulfate/silicate systems among other materials, and we provide particular focus on these newer systems. Research is also continuing to look at Na–S batteries, with a focus on room temperature operation. Research in this area will not be part of this chapter, but a detailed review of progress can be found in a recent review.[1]

## 2 LAYERED TRANSITION METAL OXIDES ($Na_xMO_2$; M = TRANSITION METAL)

Extending work performed from the development of Li-ion batteries, layered transition metal oxides ($Na_xMO_2$; M = transition metal) were obvious candidates for exploration as potential positive electrode materials for Na-ion battery applications. Before delving into the numerous layered oxide compositions explored for this application and the various strategies used to improve their performance, the nomenclature (outlined by Delmas *et al.* in 1980) used to define how the layers stack will be briefly explained.[2] There are two main structure types that have been investigated extensively in the literature O3 and P2 (Figure 2). The letter corresponds to the coordination of the sodium in the layered structure (O—octahedral, P—prismatic) and the number refers to the number of $MO_2$ layers required to form the structure. Alternatively, the number can denote the stacking of the oxygen sublattice; for example, O3-type materials have an ABCABC stacking arrangement, whereas P2-type materials have AABB arrangement. If the letter is followed by (′) then this denotes that the structure contains a distortion within the $MO_2$ layers. The structure of the starting oxide (O3 vs P2) is extremely important as transformation from P2 to O3 or P3 would only be possible if an M—O bond was broken due to a $\pi/3$ rotation of ($MO_6$) octahedra, whereas the transformation from O3 to P3 only requires a relatively small displacement of the $MO_2$ layers. Therefore, it is common to observe O3 → P3 transitions during cycling, while P2-type structures are mostly retained through a wide composition range.

O3 materials usually possess high Na content, $x$ ($Na_xMO_2$) being close to 1, whereas the Na content for P-type materials can vary extensively where $x < 1$. O3-type materials typically show very good reversible sodium extraction/intercalation, although the materials are subject to voltage/capacity limitations to prevent irreversible structural changes, hence limiting the potential power density of these systems. P2-type materials are able to cycle within a wide voltage range although these materials undergo a P2–O2/OP4 phase transition at high voltages. Such transitions, caused by displacements of $MO_2$ layers and $Na^+$ ion/vacancy ordering, create large volume changes as the material is continually cycled, which leads to the eventual capacity fade after relatively few cycles. In addition, P2-type materials are usually sodium deficient and must be paired with a sodium metal anode in order to obtain relatively high capacities. Some of the most widely studied layered transition metal oxide materials will now be discussed.

O3-type α-$NaFeO_2$ forms a layered rock salt structure analogous to the commercial $LiCoO_2$/$LiNi_xMn_yCo_zO_2$ (NMC) layered materials used in Li-ion batteries. The structure is formed by alternating layers of $NaO_6$ and $FeO_6$ edge-sharing octahedra. $Na_{1-y}FeO_2$ displays a level charge curve up to $y = 0.4$; however, as more sodium is removed, the curve develops a slope indicative of a phase change within the material.[3] Similar to the lithium counterparts, sodium removal was limited in $Na_{1-y}FeO_2$

**Table 1** A summary of some of the most promising layered positive electrode materials reported in the literature

| Composition | Structure type | Voltage window (V) | Initial capacity (mAh $g^{-1}$) | Max. no. of cycles reported in paper | References |
|---|---|---|---|---|---|
| $NaCrO_2$ | O3 | 2.0–3.6 | 120 | 50 | 4 |
| $NaFeO_2$ | O3 | 2.5–4.5 | 100 | 3 | 5 |
| $Na[Fe_{0.5}Mn_{0.5}]O_2$ | O3 | 1.5–4.3 | 110 | 30 | 6 |
| $Na[Ni_{0.5}Mn_{0.5}]O_2$ | O3 | 2.0–4.0 | 140 | 100 | 7 |
| $Na[Ni_{0.5}Co_{0.5}]O_2$ | O3 | 2.0–4.5 | 115 | 20 | 8 |
| $Na[Ni_{0.5}Fe_{0.5}]O_2$ | O3 | 2.0–3.9 | 129 | 20 | 8 |
| $Na[Ni_{0.7}Fe_{0.3}]O_2$ | O3 | 2.5–3.8 | 135 | 30 | 9 |
| $Na[Ni_{0.3}Fe_{0.4}Mn_{0.3}]O_2$ | O3 | 2.0–3.8 | 130 | 50 | 10 |
| $Na[Ni_{0.33}Fe_{0.33}Mn_{0.33}]O_2$ | O3 | 2.0–4.0 | 100 | 150 | 11 |
| $Na[Ni_{0.4}Fe_{0.2}Mn_{0.2}Ti_{0.2}]O_2$ | O3 | 2.0–4.2 | 145 | 200 | 12 |
| $Na_{0.6}MnO_2$ | P2 | 2.0–3.5 | 140 | 8 | 13 |
| $Na_{0.66}MnO_2$ | P′2 | 1.5–4.4 | 216 | 25 | 14 |
| $Na_{0.74}CoO_2$ | P2 | 2.0–3.8 | 107 | 40 | 15 |
| $Na_{0.66}[Fe_{0.5}Mn_{0.5}]O_2$ | P2 | 1.5–4.0 | 120 | 50 | 16 |
| $Na_{1.0}[Co_{0.7}Mn_{0.3}]O_2$ | P2 | 2.0–4.1 | 100 | 225 | 17 |
| $Na_{0.78}[Ni_{0.23}Mn_{0.69}]O_2$ | P2 | 2.0–4.5 | 140 | 40 | 18 |
| $Na_{0.40}[Mn_{0.54}Co_{0.46}]O_2$ | P2 | 1.5–4.2 | 195 | 67 | 19 |
| $Na_{0.67}[Ni_{0.15}Co_{0.20}Mn_{0.65}]O_2$ | P2 | 2.0–4.4 | 140 | 50 | 20 |
| $Na_{0.45}[Ni_{0.22}Co_{0.11}Mn_{0.66}]O_2$ | P2 | 1.5–4.6 | 200 | 100 | 21 |
| $Na_{0.67}[Ni_{0.33}Ti_{0.17}Mn_{0.5}]O_2$ | P2 | 2.5–4.5 | 127 | 20 | 22 |
| $Na_{0.66}[Ni_{0.25}Cu_{0.083}Mn_{0.66}]O_2$ | P2 | 2.5–4.1 | 90 | 30 | 23 |
| $Na_{0.6}[Cr_{0.6}Ti_{0.4}]O_2$ | P2 | 2.5–3.85 | 70 | 200 | 24 |
| *Examples of Mg doped materials* | | | | | |
| $Na_{0.66}Mn_{0.95}Mg_{0.05}O_2$ | P2 | 1.8–3.8 | 140 | 50 | 25 |
| $Na_{0.67}Ni_{0.25}Mg_{0.1}Mn_{0.65}O_2$ | P2 | 1.5–4.2 | 140 | 30 | 26 |
| *Examples of Al doped materials* | | | | | |
| $Na_{0.66}Ni_{0.22}Al_{0.11}Mn_{0.66}O_2$ | P2 | 2.0–4.5 | 140 | 30 | 27 |
| $Na_{0.6}Ni_{0.22}Al_{0.11}Mn_{0.66}O_2$ | P2 | 1.5–4.6 | 160 | 50 | 28 |
| *Examples of Li-doping on the transition metal site* | | | | | |
| $Na_{0.8}[Li_{0.12}Ni_{0.22}Mn_{0.66}]O_2$ | P2 | 2.0–4.4 | 115 | 50 | 29 |
| $Na[Li_{0.07}Ni_{0.26}Mn_{0.4}Co_{0.26}]O_2$ | O3 | 1.5–4.5 | 147 | 50 | 30 |
| *Examples of electrochemical ion exchanged materials* | | | | | |
| $Na_{0.78}[Li_{0.18}Ni_{0.25}Mn_{0.58}]O_2$ | O3 | 1.5–4.2 | 190 | 50 | 31 |
| $Na_{0.95}Li_{0.15}(Mn_{0.55}Co_{0.10}Ni_{0.15})O_2$ | O3 | 1.0–4.6 | 170 | 40 | 32 |

to $y = 0.4–0.5$ in order to prevent the unwanted phase transition occurring (restricting the capacity of $NaFeO_2$ to $<100\,mAh\,g^{-1}$). Given the problem with this phase transition, most subsequent work has been targeted at stabilizing the structure and preventing this undesirable phase change.

Partial or complete substitution of Fe for other transition metals (such as Mn, Co, Cr, Ni) has been explored extensively in the literature (a summary table (Table 1) provides some of the most promising materials currently reported). The related Co materials are able to reversibly undergo a number of phase transitions when cycled, which is reflected in the multiple plateaus observed in the charge/discharge curves. Ceder *et al.* reported, through a mixed synthetic/electrochemical study of $Na_xCoO_2$ (for $0.5 \leq x \leq 1$), that O3-type $Na_xCoO_2$ undergoes a number of phase transitions and transforms from O3 → O′3 → P′3 → P3 → P′3 during cycling.[33] Delmas *et al.* reported that P2-type $Na_xCoO_2$ can undergo as many as nine phase transitions, with very small differences in sodium composition ($0.5 \leq x \leq 0.9$) due to $Na^+$ ion/vacancy ordering.[34]

$Na_xMnO_2$ has also been extensively researched in the literature as Mn is attractive because it is relatively inexpensive and much less toxic than Co. Two polymorphs of $NaMnO_2$ are known in the literature. α-$NaMnO_2$ can be synthesized at low temperatures and has a monoclinic O3-type structure with a minor distortion attributed to the Jahn–Teller effect ($Mn^{3+}$, $d^4$ high spin). High-temperature synthesis yields β-$NaMnO_2$, which adopts an alternative layered structure formed through zigzag layers of edge-sharing $MO_6$ octahedra separated by octahedral $Na^+$.[35] α-$NaMnO_2$ shows multiple plateaus in the electrochemical data, similar to those of $Na_xCoO_2$, and can deliver an initial capacity of ≈185 mAh $g^{-1}$, which drops very quickly to 132 mAh $g^{-1}$ after 20 cycles.[36] In comparison, β-$NaMnO_2$ has been reported to exhibit a capacity of 190 mAh $g^{-1}$ (C/20) but also shows a considerable drop to ≈140 mAh $g^{-1}$

after 100 cycles.[37] As with the Co systems, the layered Mn oxides can also form sodium-deficient $Na_xMnO_2$ phases, which fall into the P2-type category of materials. Komaba *et al.* synthesized two polymorphs of $Na_xMnO_2$ (where $x \approx 0.6$)—a distorted P′2 phase formed by quenching from high temperatures (1050 °C) and an undistorted P2 phase, synthesized by reheating the quenched phase up to 700 °C before slow cooling to 500 °C and then quenching. The distorted phase delivered a greater capacity of ≈215 mAh $g^{-1}$ compared to the undistorted phase (198 mAh $g^{-1}$) after the first charge/discharge cycle.[14] Furthermore, the undistorted P2 phase appears to display poorer cyclability compared to the distorted P′2 phase, with a considerable drop in reversible capacity after the first 20 cycles. Structural transformations in the low voltage region (at ≈2.2 V) appear to be the origin of this difference.[14]

Lower sodium contents in $Na_xMnO_2$ systems can also lead to nonlayered phases, such as $Na_{0.44}MnO_2$. This phase possesses a more open tunnel structure and displays interesting electrochemical behavior, with as many as five plateaus observed in the electrochemical data, indicative of multiple phase transitions.[38] At a rate of C/10, $Na_{0.44}MnO_2$ can deliver ≈80 mAh $g^{-1}$ although the capacity drops by 50% by the 40th cycle.

Overall, the work on $Na_xMnO_2$ systems has shown that, although they deliver some of the highest capacities reported in the literature, they suffer from large structural strain effects on cycling as a result of Jahn–Teller distortions associated with $Mn^{3+}$. In addition, Mn materials can also undergo disproportionation ($2Mn^{3+} \rightarrow Mn^{4+} + Mn^{2+}$) reactions, leading to the dissolution of $Mn^{2+}$ into the electrolyte and thus resulting in decomposition of the electrode material, which also contributes to capacity fade.

Other transition metal systems such as $Na_xNiO_2$ have also therefore been explored due to relatively high voltages for the $Ni^{2+} \rightarrow Ni^{3+} \rightarrow Ni^{4+}$ redox couples and the lower cost compared to Co. As with Co and Mn oxide materials, these Ni oxide materials also show a number of plateaus in the electrochemical data indicative of multiple phase transitions. When cycled in the voltage window of 1.25–3.75 V, monoclinic $NaNiO_2$ was shown to deliver an initial capacity of 123 mAh $g^{-1}$, which drops by 6% after 20 cycles. Further experiments expanding the voltage range to 4.5 V led to a higher initial capacity of 147 mAh $g^{-1}$, albeit at the cost of a larger reversible capacity decrease; in this case, the capacity decreased to 98 mAh $g^{-1}$ after the 20th cycle, representing a 33% decrease.[39] Other studies on $Na_xNiO_2$ have reported that this system undergoes a number of phase transitions ($O'3 \rightarrow P'3 \rightarrow P''3 \rightarrow O''3 \rightarrow O'''3$), which may be unfavorable for the performance.[40] Wang *et al.* suggested the origins for the two main causes for capacity fade in $Na_xNiO_2$; the first is the production of a new phase at high voltages, which does not reversibly undergo intercalation ($Na_{0.17}NiO_2$), and the second is attributed to changes in the interlayer spacing at low voltages.[41]

As a result of trying to balance the advantages and disadvantages of the above layered transition metal oxide systems, it is common to observe ternary or quaternary systems reported in the literature (Table 1). Common strategies include the substitution of $Mn^{3+}/Fe^{4+}$ by other elements to "dilute" the adverse Jahn–Teller distortions. Co and Ni are obvious choices for substitution but can be costly and/or toxic. Cr has also been investigated, but can undergo an irreversible layered to rock salt phase transition when too much sodium is removed from the structure. Vanadium systems have also been reported in the literature; however, the voltages reported for these systems are lower than for other systems,[42] and more detailed long-term cycling studies are needed. Other strategies to improve these layered materials include (i) substitution of transition metals for electrochemically inactive elements, such as Al and Mg, in order to maintain structural integrity while sacrificing capacity,[25–28] (ii) the doping of Li into the transition metal layer to maintain stability at low Na concentrations,[43] and (iii) the use of surface coatings (such as $Al_2O_3$) to reduce metal dissolution into the electrolyte during cycling.[44] Notable examples in the literature that have utilized mixed metal compositions include $Na[Ni_{0.4}Fe_{0.2}Mn_{0.2}Ti_{0.2}]O_2$ (121 mAh $g^{-1}$ after 200 cycles (2–3.8 V)),[12] proton exchanged $Na_{0.45}[Ni_{0.22}Co_{0.11}Mn_{0.66}]O_2$ (≈100 mAh $g^{-1}$ after 275 cycles (2.1–4.3 V)),[45] and $Na_{0.67}Co_{0.67}Mn_{0.23}Ti_{0.1}O_2$ (84 mAh $g^{-1}$ after 1000 cycles (2–4.2 V)).[46]

## 3 FLUORIDE SYSTEMS ($NaMF_3$)

Sodium fluoroperovskites, $NaMF_3$ (M = transition metal), have also been explored as potential positive electrodes following the reports of good performance of $FeF_3$/C composites in Li-ion cells. Interest in these systems has also been driven by the greater electronegativity of fluorine compared to oxygen, which should lead to higher cell voltages during cycling and hence higher power density. One of the first reports to evaluate the electrochemical performance of the sodium series was by Gocheva *et al.* in 2009.[47] They employed a mechanochemical method to synthesize the $NaMF_3$ series (Figure 3), involving high impact milling (750 rpm) of NaF and the transition metal fluorides, $MF_2$, in a dry Ar atmosphere. $NaFeF_3$ exhibited an initial capacity of ≈120 mAh $g^{-1}$ when cycled in the voltage range of 1.5–4.0 V and retained a reversible capacity of ≈100 mAh $g^{-1}$ after 20 cycles. The electrochemical properties were shown to improve with increasing milling time, which is attributed to a decrease in the particle size of the material. In the same study, the Mn and Ni analogues were also synthesized and tested but found

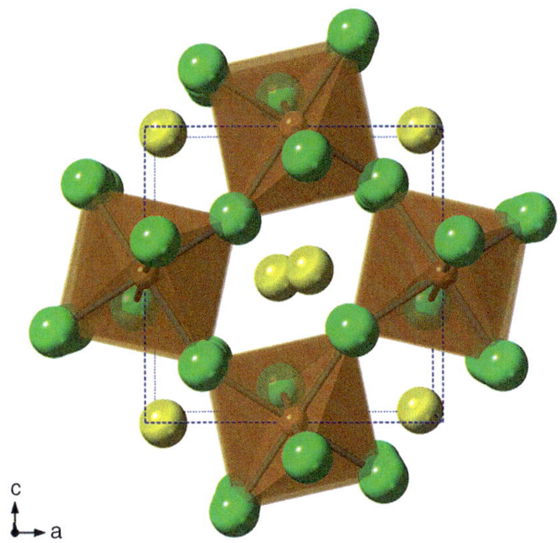


**Figure 3** Crystal structure of $NaMF_3$ (viewed down the *b*-axis)

to show poor electrochemical properties (≈40 mAh $g^{-1}$), possibly due to the cut off voltage of 4 V, which was utilized to prevent electrolyte decomposition.[47] As an alternative to the original mechanochemical method, the same group synthesized nanosized $NaFeF_3$ through a solution route utilizing organic acids and fluoroacetates, which led to improved performance at higher cycling rates.[48] The synthesis of $NaFeF_3$ nanoplates has also been reported using sodium iron hexafluoroacetylacetone ($NaFe(hfac)_3$) as a precursor. The $NaFeF_3$ nanoplates showed an initial capacity of 153 mAh $g^{-1}$, with a 50% capacity retention over 200 cycles.[49] Despite the promising performances of these fluorides, all of the above synthetic routes utilized to synthesize $NaFeF_3$ require the fluorides to be stored/handled/prepared using an inert atmosphere to prevent reaction with atmospheric moisture. Consequently, scaling up the synthesis of these systems for commercial applications would be difficult and costly compared to the layered transition metal oxide materials.

## 4 POLYANION SYSTEMS

Following on from the development of $LiFePO_4$ and other polyanion materials as positive electrode materials for Li ion batteries, research into similar polyanion systems for Na ion batteries has intensified. The interest in these systems stems from the ability to exploit changes in the bonding (ionic to covalent) to alter the potential of the $M^{n+1}/M^{n+}$ redox couple with respect to the alkali metal. For example, where the transition metal and counter anion are covalently bonded, the strong repulsion between the bonding and the antibonding orbitals lowers the $M^{n+1}/M^{n+}$ potential. The opposite is true if they are ionically bonded. Therefore, the use of polyanions (e.g., $XO_4^{n-}$; X = S, P, Si) offers the advantage of significantly influencing the energy of redox couples, which was first proposed by Goodenough *et al.*, thus allowing alternative redox couples, and hence alternative materials, to be exploited.[50] As a result, instead of the $Fe^{3+}/Fe^{4+}$ redox couple in $NaFeO_2$, for systems such as $NaFePO_4$, the $Fe^{2+}/Fe^{3+}$ redox couple is involved.

The use of a polyanion group within a cathode material originates from not only the interest in developing new materials but also from efforts to increase the stability and hence the operating safety of devices. In particular, concerns regarding the layered metal oxide system $LiCoO_2$ led the research to investigate olivine-type $LiFePO_4$ as an alternative safer electrode material. This latter system has proven to be one of the few non-layered transition metal oxide positive electrode materials that have been commercialized. The benefits of $LiFePO_4$ derive from the inexpensive and nontoxic nature of Fe, as well as the increased stability under cell operating conditions [with $FePO_4$ as a stable oxidation product] and a voltage of 3.5 V against Li metal.[50] Problems with such polyanion systems tend to be the more complex synthesis (commonly requiring carbothermal reduction) and the generally poor electronic (and ionic) conductivity, which requires nanosizing and carbon coating of materials to overcome.

### 4.1 Phosphate-Based Cathodes

#### *4.1.1 $NaFePO_4$ (Maricite Versus Olivine)*

Following on from the success of $LiFePO_4$, the Na analogue, $NaFePO_4$, has attracted significant attention in the battery community. $NaFePO_4$ is known to exhibit two crystal structures, one being the thermodynamically stable maricite structure and the other, like $LiFePO_4$, the olivine structure. The maricite structure contains edge-sharing $FeO_6$ units corner sharing to $PO_4$, resulting in no cationic channels. In contrast, the olivine structure contains a one-dimensional channel that allows for sodium ion diffusion (Figure 4). The generally poor electrochemical performance witnessed for the maricite variant has been attributed to these structural differences, which inhibit $Na^+$ ion diffusion in the maricite structure. Nevertheless, recent work has revealed that Na can be deintercalated from nanosized maricite $NaFePO_4$ and the resulting amorphous $FePO_4$ shows promising performance (capacity of 142 mAh$g^{-1}$ after the first cycle and retaining 95% of capacity at the slow rate of C/20).[51]

In order to synthesize olivine $NaFePO_4$, electrochemical insertion of Na into $FePO_4$ was initially used.[52] Further studies have shown the ability to form $NaMn_{1-x}Fe_xPO_4$ olivine solid solutions by a low-temperature topotactic molten salt ion exchange reaction

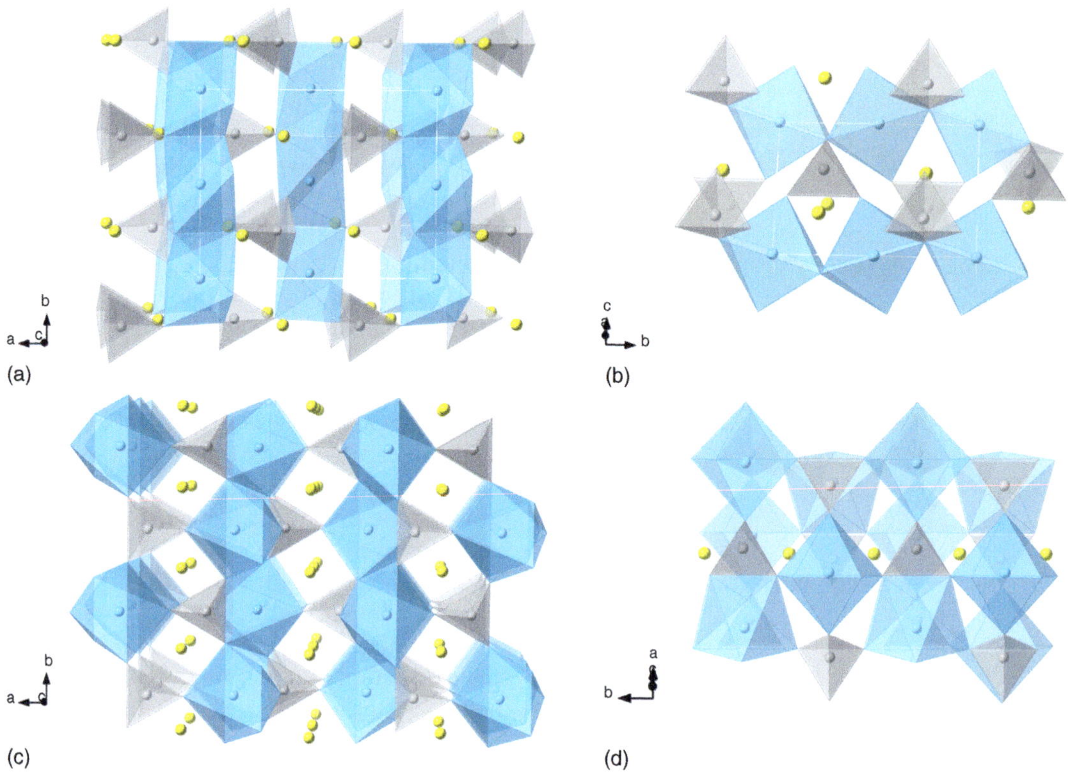


**Figure 4** $NaFePO_4$ shown in the maricite (a,b) and olivine (c,d) phases. (b) and (d) show the differences in framework: with edge-sharing $FeO_6$ units corner-sharing to $PO_4$ units and corner sharing $FeO_6$ units to edge- and corner-sharing $PO_4$, for the maricite and olivine, respectively

of $NH_4Mn_{1-x}Fe_xPO_4$.[53] However, despite the large interest in these systems, the electrochemical performances have not been able to compete with those of the layered transition metal oxides.

### *4.1.2 Nasicon-Type Materials*

Further interest into framework structures has resulted in developments deriving from the Naiscon framework. The name Nasicon—an abbreviation of Na SuperIonic CONductor—originates from initial compositions developed for $Na^+$ ion conducting solid electrolytes for high-temperature Na–S batteries. The general formulae of Nasicon systems is given by $A_xM_2(XO_4)_3$ ($0 \le x \le 4$), where A is an alkali-metal ions, M is a multivalent transition metal ion, while the polyanion substituent has the possibility to be a phosphate ($PO_4{}^{3-}$), silicate ($SiO_4{}^{4-}$), sulfate ($SO_4{}^{2-}$), or a combination of these.

The initial body of work for Nasicon related to the partial substitution of silicate units for phosphate in $NaZr_2(PO_4)_3$, which resulted in an increase in sodium ion conductivity through increased $Na^+$ content. The resulting high ionic conductivity is structurally related to interconnectivity of the $Na^+$ ion sites.[54] The 3D framework itself is built from connected $ZrO_6$ octahedra and $PO_4$ tetrahedra with $Na^+$ ions occupying cavities within this framework.[54]

In terms of applications of such systems as positive electrode materials, there has been significant interest in sodium vanadium phosphate systems. Nasicon-type $Na_3V_2(PO_4)_3$ (Figure 5), a material first reported by Delmas *et al.* in 1978,[55] has been shown to exhibit capacities of the order of $90\,mAh\,g^{-1}$ in the voltage range 2.7–3.8 V vs $Na^+/Na$, with good cycling performance. In this area, work has focused on improving the cycling performance through carbon coating and nanosizing the $Na_3V_2(PO_4)_3$.

A further aspect of interest is that $Na_3V_2(PO_4)_3$ has also been investigated as a negative electrode, with work showing that 1 Na could be inserted into the material at low voltage. This led to further work investigating the use of this general system in a symmetrical sodium ion battery. Initially, work by Noguchi *et al.* reported the use of $Na_3V_2(PO_4)_3$ as both positive and negative electrodes in an all Nasicon solid-state cell, utilizing $Na_3Zr_2Si_2PO_{12}$ as the electrolyte.[56] Further studies with a liquid electrolyte (1 M $NaClO_4$ in 1:1 ethyl carbonate/dimethyl carbonate with 5% fluoroethylene carbonate additive) and electrochemical pre-intercalation of Na in the negative

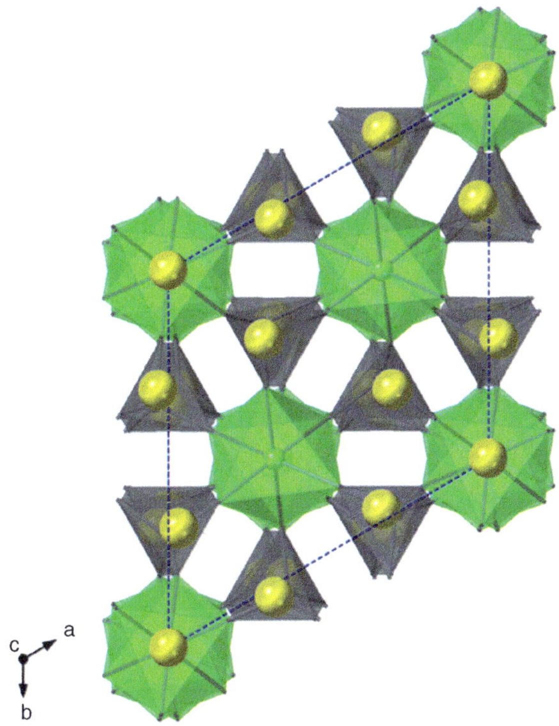


**Figure 5** The Nasicon-type structure exhibited by $Na_3V_2(PO_4)_3$

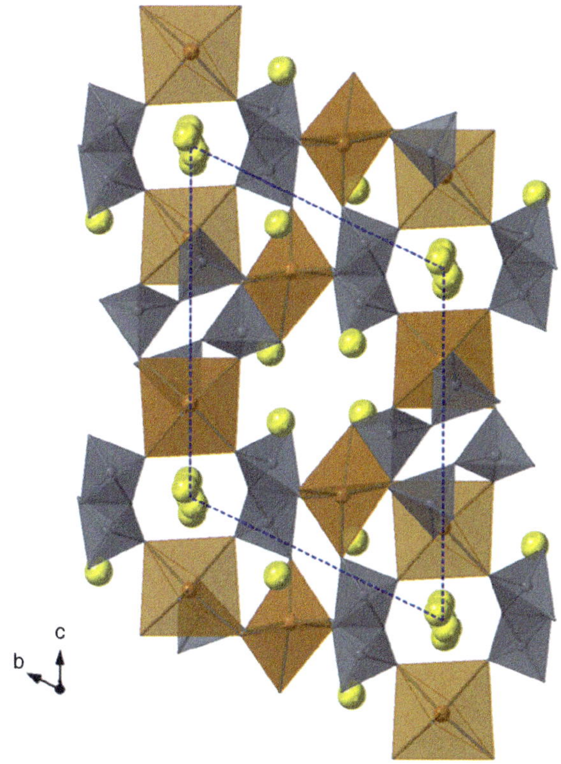


**Figure 6** Crystal structure of triclinic $Na_2FeP_2O_7$

electrode to give $Na_4V_2(PO_4)_3$, led to a cell with an overall voltage of 3 V and a higher cell level energy density (138 Wh $kg^{-1}$).[57]

Research has also investigated the sulfate Nasicon-type phase $Fe_2(SO_4)_3$ due to its low cost. A voltage of 3.2 V versus $Na^+/Na$ was reported with a capacity of 1 Na per formula unit, half the theoretical capacity.[58] Despite the reduced capacity, the material was shown to cycle relatively well with a capacity fade of ≈20% for 400 cycles. Further Galvanostatic Intermittent Titration Technique (GITT) studies showed that through using very slow rates, higher Na levels (up to 1.86 per formula unit) could be achieved, but the capacity decreased rapidly for higher currents.[59] Modeling studies suggested that this was due to the high activation energy for Na ion diffusion in comparison to related alluaudite sulfate systems (discussed later).

### *4.1.3 Pyrophosphate ($Na_2MP_2O_7$)*

Pyrophosphates have also been investigated as possible positive electrode materials. In particular, $Na_2FeP_2O_7$ (Figure 6) was shown to deliver a reversible capacity of 82 mAh $g^{-1}$ at a voltage of 3 V.[60] The material has a framework structure consisting of $FeO_6$ octahedra corner- and edge-sharing with the pyrophosphate groups in a staggered fashion, which results in tunnels occupied by $Na^+$ ions.[60]

Mn and Co analogues of this material have also been prepared. $Na_2MnP_2O_7$, like the Fe analogue, was shown to be electrochemically active with a capacity of 80 mAh $g^{-1}$ at a rate of C/20 and a voltage of ≈3.6 V.[61] Electrochemical activity has also been reported for the orthorhombic polymorph of the Co analogue, with a reversible capacity of ≈80 mAh $g^{-1}$ (C/20 rate).[62]

### *4.1.4 Carbonophosphates*

Mixed-polyanion systems have also begun to receive attention, in particular sodium transition metal carbonophosphates. Initial work was inspired by computational studies predicting compositions with the formula $A_xM(XO_3)(X'O_4)$, where A = Na and Li; M = Co, Mn, Fe, or Ni; X = B and C; X′ = P and Si.[63] Of these compositions, $Na_3FeCO_3PO_4$ and $Na_3MnCO_3PO_4$ (known as bonshtedtite and sidorenkite, respectively) were synthetically prepared, in addition to the lithium analogues formed through ion exchange of the sodium equivalent.[64] Both compositions are isostructural and have monoclinic unit cells. The transition metal octahedra, phosphate, and carbonate groups are connected such that they form 2D chains leading to a 3D framework with the alkali metals occupying channels within this framework.[64] Preliminary electrochemical results for $Na_3MnCO_3PO_4$ showed a high capacity (~125 mAh $g^{-1}$) and good cycling at an average voltage of 3.7 V.[65]

### 4.1.5 *Phosphate Fluorides*

There has also been a lot of interest in phosphate fluoride systems. $NaVPO_4F$ has been shown to operate in a sodium-ion cell with the use of hard carbon as the negative electrode, with the average discharge voltage reported to be 3.7 V, albeit with a significant capacity fade.[66] As noted earlier, a key issue with these polyanion materials is poor electronic conductivity, requiring the addition of carbon, and it has been shown that through optimizing the $NaVPO_4F$ to carbon composition, performance can be improved.[67]

The structure of $NaVPO_4F$ consists of a three-dimensional framework comprising of $VO_4F_2$ octahedra and phosphate tetrahedra. This material can be produced either via a hydrothermal route followed by a subsequent heat treatment or via a traditional ceramic route. Both routes use the intermediate $VPO_4$, which is produced through a carbothermal reduction process. Forming the intermediate via this route also offers the potential advantage of residual carbon aiding in the improvement of the electrochemical performance.

The $Na_3V_2(PO_4)_2F_3$ system has attracted particular interest following the reports by Gover *et al.* utilizing this material as a positive electrode in a Li-ion battery (specific capacity of 120 mAh g–[1] at an average voltage of 4.1 V versus Li).[68] As for $NaVPO_4F$, this compound can be prepared through heat treatment in Ar of NaF with $VPO_4$. The structure consists of a framework of $(V_2O_8F_3)$ bi-octahedra and $(PO_4)$ tetrahedra, with the Na located in channels within this framework. Following the demonstration of this material in a Li-ion cell, further research has investigated using this material in a Na-ion cell. This work has shown two voltage plateaus at ≈3.7 and ≈4.2 V with a capacity of 120 mAh $g^{-1}$ for a C/20 rate, with about 80% of this capacity achieved at a much higher 4C rate.[69] An interesting aspect noted by Broux *et al.* is the fact that there were a number of discrepancies in the literature concerning the results for this system. Through detailed studies, these researchers showed that this could be correlated with partial oxidation on synthesis giving some O/F exchange, that is, $Na_3V_2(PO_4)_2F_{3-x}O_x$, which highlights the need to ensure detailed characterization of these electrode materials.[70]

$Na_2FePO_4F$ and $Na_2Fe_{0.5}Mn_{0.5}PO_4F$ have also been investigated as positive electrode materials. $Na_2FePO_4F$ was shown to have a capacity of 100–110 mAh $g^{-1}$ at a cycling rate of C/20.[71] Through optimization of the preparation routes, $Na_2Fe_{0.5}Mn_{0.5}PO_4F$ was shown to have a similar capacity with a slightly higher average operating voltage.

## 4.2 Sulfate-Based Materials

### 4.2.1 *$Na_2M(SO_4)_2{\cdot}yH_2O$ (M = Transition Metal) and Their Dehydrated Products*

Out of all the polyanion systems, work in the Li-ion battery area has shown that sulfates provide the highest electrode voltages, and consequently there has been a significant interest in sodium transition metal sulfate systems. In this area, researchers have made use of low-temperature synthesis routes and structural water to devise new candidate electrode materials with different structures. For example, hydrated sodium transition metal sulfates have been explored including the tetrahydrate, $Na_2M(SO_4)_2{\cdot}4H_2O$, and the dihydrate, $Na_2M(SO_4)_2{\cdot}2H_2O$ systems. In this work, bloedite-type $Na_2Fe(SO_4)_2{\cdot}4H_2O$ was proposed as a potential positive electrode material.[72] The material is prepared through a facile route by simply dissolving $Na_2SO_4$ and $FeSO_4{\cdot}7H_2O$ in water and then evaporating it slowly to produce $Na_2Fe(SO_4)_2{\cdot}4H_2O$. Alternatively, it could be also prepared by precipitation through the addition of ethanol (so-called antisolvent synthesis) as well as solid-state milling of the precursor sulfates for 20 min.[72] The structure (Figure 7) consists of $FeO_2(H_2O)_4$ octahedra, with two sulfate tetrahedra bound to the metal octahedra. The octahedra are relatively isolated, and there is a strong hydrogen bonding network. A capacity of ≈60 mAh $g^{-1}$ was obtained at a voltage of ≈3.3 V versus $Na^+/Na$, albeit with a poor cycling performance attributed to electrolyte decomposition and amorphization of the cathode material.[72]

The second hydrate proposed as a cathode material is the dihydrate (krohnkite-type structure), $Na_2Fe(SO_4)_2{\cdot}2H_2O$.[73] This dihydrate is prepared through a similar route to that of the tetrahydrate but at a slightly higher temperature to facilitate water loss. The structures of krohnkite and bloedite differ significantly. The metal octahedra were isolated in the bloedite structure, while the krohnkite structure possesses chains of $FeO_4(H_2O)_2$ and sulfate tetrahedra along the *c*-axis. Parallel chains are connected through the hydrogen bonding network. Barpanda *et al.* showed that the material could achieve a reversible capacity of ≈70 mAh $g^{-1}$ at a voltage of ≈3.25 V, with improved cycling stability.[73]

However, as was seen in the bloedite-type material, removal of sodium can also lead to the co-removal of structural water from the system, the presence of which can lead to a number of safety issues when operating in a Na-ion battery. It has been shown that dehydration of the hydrated systems can yield a number of potential materials for sodium-ion batteries. Reynaud *et al.* reported the structure and performance of the dehydrated $Na_2Fe(SO_4)_2$ (212) system.[72] The material adopts a complicated monoclinic (space group: *C2/c*) structure in which transition metal octahedra corner and edge share with sulfate tetrahedra.

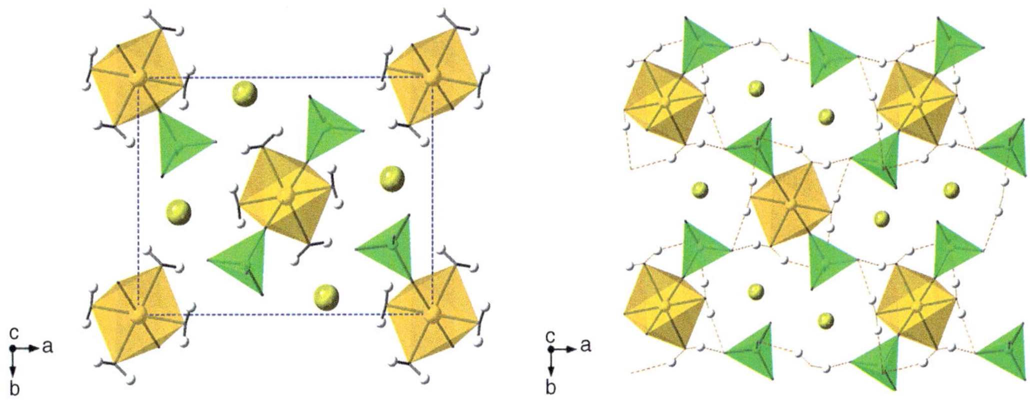


**Figure 7** Crystal structure of $Na_2Fe(SO_4)_2{\cdot}4H_2O$ (bloedite structure) shown down *c*-axis. The dashed orange lines show the hydrogen bond network that holds the framework together

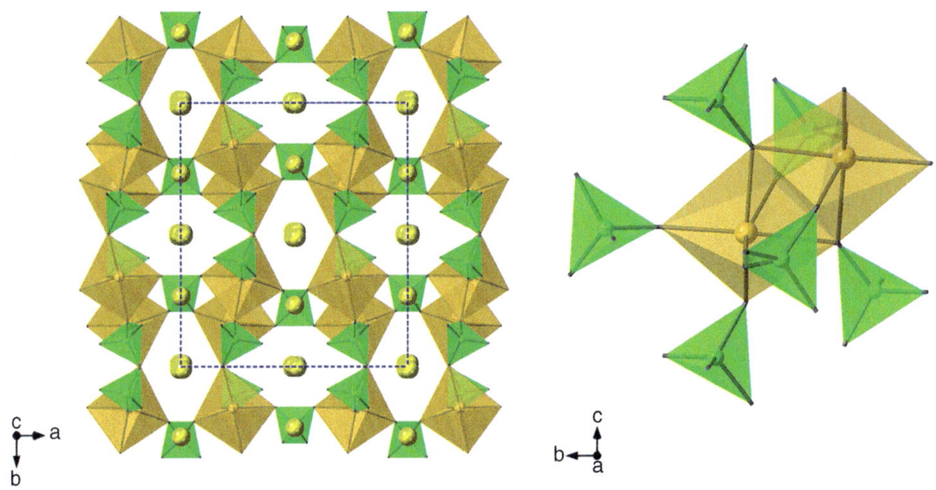


**Figure 8** Crystal structure of alluaudite-type $Na_2Fe_2(SO_4)_3$. A Fe cluster (Fe1) has been highlighted in order to show the edge-sharing octahedral biunits that form the basis of the structure

Sodium is distributed in channels throughout the framework. The electrochemical performance of this phase was evaluated, and it was shown to be limited to below 100 mAh $g^{-1}$, and in addition there was some evidence for electrolyte decomposition during cycling.

### 4.2.2 *Alluaudite-Type* $Na_{2+2x}Fe_{2-x}(SO_4)_3$

One of the most promising sulfate systems in recent years has been the alluaudite-type phase, $Na_{2+2x}Fe_{2-x}(SO_4)_3$. Although phosphate alluaudite systems have been known in the literature for some time, the sulfate equivalent was only discovered recently in 2014 by Barpanda *et al.*[74] The material was obtained by milling the anhydrous precursor sulfates (dried under vacuum at 200 °C for 12 h) for 4 h before annealing the mixture at 350 °C for 12 h under argon. The structure of alluaudite (Figure 8) consists of an open three-dimensional framework formed through the transition metal octahedra linked through edge sharing to form a bi-unit, which form chains along the *b*-axis, connected by corner-sharing sulfate tetrahedra. These chains then stack and alternate along the *a*-axis, separated by other crystallographically distinct sulfate tetrahedra. The sodium occupies channels in the structure with three sites available for occupation.

Barpanda *et al.* showed that $Na_2Fe_2(SO_4)_3$ was capable of delivering a capacity of 102 mAh $g^{-1}$ at ≈3.8 V vs $Na^+/Na$, which corresponds to ≈85% of the theoretical capacity.[74] The authors proposed that the high voltage observed for this alluaudite phase is due to the metal octahedra edge sharing, which creates shorter Fe–Fe distances leading to strong $Fe^{3+}$–$Fe^{3+}$ repulsion, hence raising the Gibbs free energy of the deintercalated framework.[74] These promising results have led to a large interest in materials with this structure type. Further research showed that these alluaudite materials display off-stoichiometric behavior, such that the formula should be written $Na_{2+2x}Fe_{2-x}(SO_4)_3$.[75] A number of groups have since reported the successful synthesis of the manganese and cobalt equivalents, $Na_{2+2x}Mn_{2-x}(SO_4)_3$ and $Na_{2.32}Co_{1.84}(SO_4)_3$, although no electrochemical data has been published with regard to these phases due to the restricted voltage window for present common electrolytes (maximum ≈4.5 V).[76,77] Other work related to these systems includes the stabilization of alluaudite type, $Na_3M_{1.5}(SO_4)_{1.5-x}(SeO_4)_{1.5x}$ (where M = Co, Mn and Ni), through selenate doping, which showed a clear relationship between the size of the transition metals selected and the amount of selenate required to achieve phase-pure alluaudite-type materials. In addition, this work was the first to stabilize Ni in an alluaudite framework as well as the highest reported sodium content for an alluaudite phase constructed of sulfate/selenate tetrahedra, which may have applications within Na-ion batteries.[78]

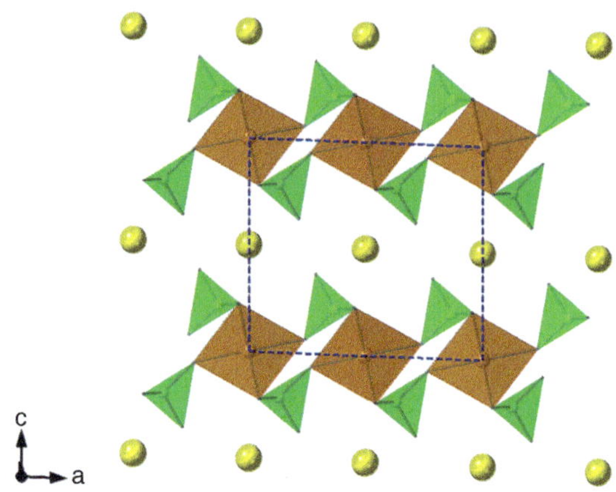


**Figure 9** Crystal structure of $NaFe(SO_4)_2$

### 4.2.3 *Eldfellite $NaFe(SO_4)_2$*

$NaFe(SO_4)_2$ was identified as a promising electrode material by Singh *et al.* in 2015.[79] The structure of $NaFe(SO_4)_2$ (Figure 9) differs from previously discussed framework materials as it is a pseudo-layered structure with sodium occupying the space in between the layers. Unlike the above systems, it contains $Fe^{3+}$ (rather than $Fe^{2+}$) and so requires the use of a Na-containing negative electrode (such as Na metal), which is a disadvantage of this system. Nevertheless, a capacity of 80 mAh $g^{-1}$ at ≈3 V vs $Na^+$/Na was obtained, with relatively good capacity retention.[79]

### 4.2.4 *Sulfate Fluorides, Hydroxy- and Oxysulfates*

As for the phosphate systems, sulfate fluorides have attracted interest, with $NaFeSO_4F$ (Figure 10) identified as a possible positive electrode material after reports on the Li-analogue $LiFeSO_4F$. $NaFeSO_4F$ can be synthesized hydrothermally using tetraethylene glycol as the solvent at 220 °C for 48 h.[80] Other routes have included using alternative solvents during synthesis, as well as ionothermal and topochemical solid-state reactions.[81] However, despite the promising performance of $LiFeSO_4F$, the electrochemical performance of $NaFeSO_4F$ is rather poor.[82] A number of factors are believed to contribute to the poor activity of $NaFeSO_4F$, in particular, low sodium ion mobility and large volume expansion.[82]

Hydroxy sulfates have also attracted interest, with $NaFe_3(SO_4)_2(OH)_6$ (Figure 10) identified by Pralong *et al.* as a potential candidate electrode material due to the possibility of incorporating additional sodium to form $Na_3Fe_3(SO_4)_2(OH)_6$.[83] The host phase can be synthesized through a precipitation route at 90 °C by dissolving $Na_2SO_4$ and $Fe_2(SO_4)_3{\cdot}5H_2O$ (ratio 1:4) in a weakly acidic solution (0.01 M $H_2SO_4$).[83] The structure consists of $FeO_4(OH)_2$ octahedra forming puckered layers in the *a*, *b* direction, with these layers connected through corner-sharing sulfate tetrahedra. This network

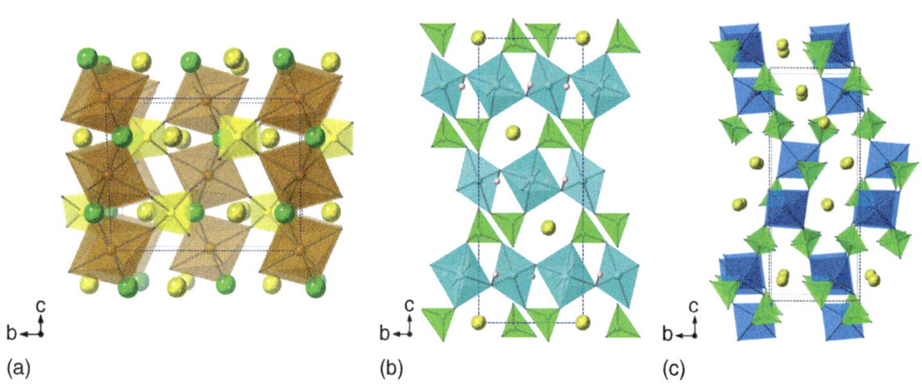


**Figure 10** Crystal structures of $NaFeSO_4F$ (a), Jarosite-type $NaFe_3(SO_4)_2(OH)_6$ (b), and $Na_2VO(SO_4)_2$ (c)

creates channels for sodium occupation down the *b*-axis. Electrochemical studies showed a capacity of 120 mAh $g^{-1}$ (lower than the expected 166 mAh $g^{-1}$ due to incomplete sodium removal from the material) after 20 cycles at an average potential of 2.72 V.[83] Although the material possesses a high capacity for a sodium-ion material, the requirement for a sodium metal anode, the need for slow charge rates (C/20), and low cell voltages may limit its future applications.

The oxy sulfate $Na_2VO(SO_4)_2$ (Figure 10) has also attracted significant interest. Although the material was initially identified in the 1990s as an impurity in sulfuric acid production, it was not until 2016 that $Na_2VO(SO_4)_2$ was proposed as a potential electrode material for Na-ion batteries.[84] Two routes have been proposed in the literature to synthesize $Na_2VO(SO_4)_2$: The first reports that the material could be synthesized by milling the precursor sulfates before heating under Ar for 12 h in the region of 400–415 °C. The second utilizes the reduction reaction of $V_2O_5$ with oxalic acid in solution to form $VOC_2O_4$ as an intermediate. Once reduced, $Na_2SO_4$ and $(NH_4)_2SO_4$ were then added, the water evaporated, and the precipitate heated in air at 350 °C for 4 h.[85] A reversible capacity of 60 mAh $g^{-1}$ was reported for this system, which corresponds to the removal of 0.7 $Na^+$.[84]

To summarize, sulfate materials are still in relative infancy compared to systems built from other polyanion units. Although the synthesis of these materials can all be carried out at relatively low temperatures, which gives potential benefits for future commercialization, electrolyte compatibility and stability appear to be crucial barriers limiting the use of the majority of these materials.

### 4.3 Silicate-Based Materials

In addition to the phosphate- and sulfate-based systems, silicates of the form $Na_2MSiO_4$ (M = Fe, Co and Mn) have attracted significant attention in recent years owing to high theoretical capacities (assuming the removal of 2 Na ions per formula unit).

Initial reports on the synthesis of the Fe analogue, $Na_2FeSiO_4$, made use of a solvothermal route, giving a capacity of 126 mAh $g^{-1}$ at a rate of C/40.[86] Further improvements have been shown through the synthesis via a sol–gel route, thus forming a carbon-coated $Na_2FeSiO_4$. In this case, a very high capacity of 181 mAh $g^{-1}$ was reported, with stable cycling retaining 88% of capacity up to 100 cycles.[87] For these synthesis routes, $Na_2FeSiO_4$ was shown to have triclinic symmetry, although similar to the related $Li_2FeSiO_4$, the Na analogues show a range of polymorphs.[88] Indeed, the cubic polymorph of $Na_2FeSiO_4$ has been shown to be successfully synthesized by both solid-state and sol–gel routes. Cycling the cubic phase achieved a reversible capacity of 106 mAh $g^{-1}$.[88] Density functional theory (DFT) calculations have been conducted for this system to further ascertain the processes occurring on the removal of sodium from this material—$Na_xFeSiO_4$ ($0 \geq x \geq 2$).[89] Corresponding to the removal of the two Na ions, two voltage plateaus have been determined with the first at ≈2.0 V ($2 \geq x \geq 1$) and the second at ≈4.5 V ($1 \geq x \geq 0$).[89] The oxidation of $Fe^{2+}$ to $Fe^{3+}$ was shown to occur upon the removal of the first Na ion to produce the stable intermediate $NaFeSiO_4$. With the removal of the second Na ion, oxidation of both $Fe^{3+}$ and $O^{2-}$ is observed.[89]

$Na_2CoSiO_4$ was initially prepared by a hydrothermal route,[90] with further progress achieved through a solvothermal process to directly prepare in situ carbon-coated $Na_2CoSiO_4$, the latter showing a capacity of 125 mAh $g^{-1}$ at a C/20 rate—a value twice that of the pristine $Na_2CoSiO_4$.[91]

$Na_2MnSiO_4$, produced via a sol–gel route, has also been investigated as a positive electrode, delivering a reversible capacity of 125 mAh $g^{-1}$ at a rate of C/10—when using an ionic liquid electrolyte at 90 °C.[92] The electrode performance was significantly improved with the aid of carbon coating and the elevated operating temperature.[92] Further improvements have been reported through alterations in cell fabrication, specifically within the electrolyte, which has led to excellent long-term cycling stability reported for this material over 500 cycles at 1C, with 90% of capacity retained.[93] Moreover, at a 5C rate a reversible discharge capacity of 100 mAh $g^{-1}$ was achieved. Other strategies to improve the electrochemical performance of the carbon-coated $Na_2MnSiO_4$ material have been reported including the formation of a nanocomposite $Na_2MnSiO_4$/carbon with the addition of graphene.[94] This $Na_2MnSiO_4$/carbon/graphene composite showed capacities of 182 mAh $g^{-1}$ (corresponding to the removal of 1.31 $Na^+$ insertion per $Na_2MnSiO_4$) and 60.7 mAh $g^{-1}$, at 0.1C and 5C rates, respectively.[94]

## 5 PRUSSIAN BLUE/WHITE MATERIALS

In the pursuit of other open-framework materials, Prussian blue complexes have gained considerable attention. The name Prussian blue relates to the compound $KFe_2(CN)_6$ but has become synonymous with analogous materials with the same structure. The cubic framework of Prussian blue is formed from alternating $Fe^{2+}$ and $Fe^{3+}$ octahedra bridged by linear cyano $(C{\equiv}N)^{1-}$ groups, where the high-spin $Fe^{2+}$ ions bond to the N atoms and the low-spin $Fe^{3+}$ to the C atoms.[95] Prussian blue–type electrode materials can be obtained through a low-temperature solution route at temperatures ≤100 °C.[96] In principle, these sodium iron hexacyanoferrates can utilize aqueous electrolytes: however, the performance of the material would be limited due to voltage restrictions. When organic electrolytes are utilized instead, the Prussian blue (with hard carbon as the negative electrode) has been shown to display a capacity of $140\,mAh\,g^{-1}$ when cycled in the region of 1.5–3.8 V while retaining 75% capacity after 200 cycles at a charge rate of 0.7C.[97] The related Prussian white phase, $Na_{1.92}Fe[Fe(CN)_6]$, has been reported to show an initial capacity of $\approx 160\,mAh\,g^{-1}$. Moreover, although the reversible capacity drops to $\approx 120\,mAh\,g^{-1}$ after the second discharge cycle, the material shows very little further change after 50 cycles (2.0–3.8 V).[98]

## 6 CONCLUSIONS

The growing need for large-scale stationary electrical power storage is driving research interest into Na-ion batteries, due to the greater abundance and lower cost of Na versus Li. A wide range of materials are attracting attention as positive electrodes, with layered transition metal oxide systems being the preferred materials as for Li-ion cells. The growing research on polyanion materials has, however, thrown up a number of new possibilities, with Nasicon, alluaudite, $Na_2MSiO_4$, Prussian blue systems attracting significant attention. Future trends include the investigation of all solid-state batteries. The use of a solid-state electrolyte is beneficial in terms of cell safety and would allow the potential use of Na metal, as well as leading to the potential to utilize higher voltage positive electrodes. In addition, as for Li ion batteries, there is strong interest in operating with an oxygen electrode in order to deliver higher electrode capacities. In such $Na-O_2$ batteries, sodium superoxide is formed as the discharge product.[99] However, as for $Li-O_2$ cells, a major issue is the need to separate oxygen from air. It is possible, however, that for a stationary $Na-O_2$ cell, this issue may possibly be overcome by utilizing oxygen in a closed-loop system. Overall, it is clear that the future prospects for Na-ion battery research looks bright, and there will continue to be a strive to discover new electrode systems through manipulation of the structural chemistry and synthetic approaches.

## 7 RELATED ARTICLE

Solid State and Materials Chemistry for Sodium-Ion Batteries

## 8 ABBREVIATIONS AND ACRONYMS

DFT = Density functional theory; GITT = Galvanostatic Intermittent Titration Technique; NMC = $LiCoO_2/LiNi_xMn_yCo_zO_2$

## 9 REFERENCES


1. W. Yun-Xiao, Z. Binwei, L. Weihong, X. Yanfei, C. Shu-Lei, L. Hua-Kun and D. Shi-Xue, *Adv. Energy Mater.*, 2017, **7**, 1602829.
2. C. Delmas, C. Fouassier and P. Hagenmuller, *Physica B+C*, 1980, **99**, 81.
3. Y. Takeda, K. Nakahara, M. Nishijima, N. Imanishi, O. Yamamoto, M. Takano and R. Kanno, *Mater. Res. Bull.*, 1994, **29**, 659.
4. S. Komaba, C. Takei, T. Nakayama, A. Ogata and N. Yabuuchi, *Electrochem. Commun.*, 2010, **12**, 355.
5. J. Zhao, L. Zhao, N. Dimov, S. Okada and T. Nishida, *J. Electrochem. Soc.*, 2013, **160**, A3077.
6. N. Yabuuchi, M. Kajiyama, J. Iwatate, H. Nishikawa, S. Hitomi, R. Okuyama, R. Usui, Y. Yamada and S. Komaba, *Nat. Mater.*, 2012, **11**, 512.
7. P.-F. Wang, Y. You, Y.-X. Yin and Y.-G. Guo, *J. Mater. Chem. A*, 2016, **4**, 17660.
8. P. Vassilaras, D. H. Kwon, S. T. Dacek, T. Shi, D. H. Seo, G. Ceder and J. C. Kim, *J. Mater. Chem. A*, 2017, **5**, 4596.
9. X. Wang, G. Liu, T. Iwao, M. Okubo and A. Yamada, *J. Phys. Chem. C*, 2014, **118**, 2970.
10. N. Yabuuchi, M. Yano, H. Yoshida, S. Kuze and S. Komaba, *J. Electrochem. Soc.*, 2013, **160**, A3131.
11. D. Kim, E. Lee, M. Slater, W. Lu, S. Rood and C. S. Johnson, *Electrochem. Commun.*, 2012, **18**, 66.
12. X. Sun, Y. Jin, C. Y. Zhang, J. W. Wen, Y. Shao, Y. Zang and C. H. Chen, *J. Mater. Chem. A*, 2014, **2**, 17268.
13. A. Caballero, L. Hernán, J. Morales, L. Sánchez, J. Santos Peña and M. A. G. Aranda, *J. Mater. Chem.*, 2002, **12**, 1142.
14. S. Kumakura, Y. Tahara, K. Kubota, K. Chihara and S. Komaba, *Angew. Chem. Int. Ed.*, 2016, **55**, 12760.
15. J. J. Ding, Y. N. Zhou, Q. Sun, X. Q. Yu, X. Q. Yang and Z. W. Fu, *Electrochim. Acta*, 2013, **87**, 388.
16. J. Xu, S. L. Chou, J. L. Wang, H. K. Liu and S. X. Dou, *ChemElectroChem*, 2014, **1**, 371.
17. Y. Shen, S. Birgisson and B. B. Iversen, *J. Mater. Chem. A*, 2016, **4**, 12281.
18. C. Ma, J. Alvarado, J. Xu, R. J. Clément, M. Kodur, W. Tong, C. P. Grey and Y. S. Meng, *J. Am. Chem. Soc.*, 2017, **139**, 4835.
19. X. Xu, S. Ji, R. Gao and J. Liu, *RSC Adv.*, 2015, **5**, 51454.
20. D. Yuan, W. He, F. Pei, F. Wu, Y. Wu, J. Qian, Y. Cao, X. Ai and H. Yang, *J. Mater. Chem. A*, 2013, **1**, 3895.
21. L. G. Chagas, D. Buchholz, L. Wu, B. Vortmann and S. Passerini, *J. Power Sources*, 2014, **247**, 377.
22. H. Yoshida, N. Yabuuchi, K. Kubota, I. Ikeuchi, A. Garsuch, M. Schulz-Dobrick and S. Komaba, *Chem. Commun.*, 2014, **50**, 3677.
23. L. Zheng, J. Li and M. N. Obrovac, *Chem. Mater.*, 2017, **29**, 1623.

24. Y. Wang, R. Xiao, Y. S. Hu, M. Avdeev and L. Chen, *Nat. Commun.*, 2015, **6**, 1.
25. R. J. Clément, J. Billaud, A. Robert Armstrong, G. Singh, T. Rojo, P. G. Bruce and C. P. Grey, *Energy Environ. Sci.*, 2016, **9**, 3240.
26. K. Hemalatha, M. Jayakumar, P. Bera and A. S. Prakash, *J. Mater. Chem. A*, 2015, **3**, 20908.
27. W. Zhao, H. Kirie, A. Tanaka, M. Unno, S. Yamamoto and H. Noguchi, *Mater. Lett.*, 2014, **135**, 131.
28. I. Hasa, S. Passerini and J. Hassoun, *J. Mater. Chem. A*, 2017, **5**, 4467.
29. J. Xu, D. H. Lee, R. J. Clément, X. Yu, M. Leskes, A. J. Pell, G. Pintacuda, X. Q. Yang, C. P. Grey and Y. S. Meng, *Chem. Mater.*, 2014, **26**, 1260.
30. J. Xu, H. Liu and Y. S. Meng, *Electrochem. Commun.*, 2015, **60**, 13.
31. H. Liu, J. Xu, C. Ma and Y. S. Meng, *Chem. Commun.*, 2015, **51**, 4693.
32. R. Kataoka, T. Mukai, A. Yoshizawa and T. Sakai, *J. Electrochem. Soc.*, 2013, **160**, A933.
33. Y. Lei, X. Li, L. Liu and G. Ceder, *Chem. Mater.*, 2014, **26**, 5288.
34. R. Berthelot, D. Carlier and C. Delmas, *Nat. Mater.*, 2011, **10**, 74.
35. A. Mendiboure, C. Delmas and P. Hagenmuller, *J. Solid State Chem.*, 1985, **57**, 323.
36. X. Ma, H. Chen and G. Ceder, *J. Electrochem. Soc.*, 2011, **158**, A1307.
37. J. Billaud, R. J. Clément, A. R. Armstrong, J. Canales-Vázquez, P. Rozier, C. P. Grey and P. G. Bruce, *J. Am. Chem. Soc.*, 2014, **136**, 17243.
38. F. Sauvage, L. Laffont, J. M. Tarascon and E. Baudrin, *Inorg. Chem.*, 2007, **46**, 3289.
39. P. Vassilaras, X. Ma, X. Li and G. Ceder, *J. Electrochem. Soc.*, 2012, **160**, A207.
40. M. H. Han, E. Gonzalo, M. Casas-Cabanas and T. Rojo, *J. Power Sources*, 2014, **258**, 266.
41. L. Wang, J. Wang, X. Zhang, Y. Ren, P. Zuo, G. Yin and J. Wang, *Nano Energy*, 2017, **34**, 215.
42. M. Guignard, C. Didier, J. Darriet, P. Bordet, E. Elkaïm and C. Delmas, *Nat. Mater.*, 2013, **12**, 74.
43. S. Zheng, G. Zhong, M. J. McDonald, Z. Gong, R. Liu, W. Wen, C. Yang and Y. Yang, *J. Mater. Chem. A*, 2016, **4**, 9054.
44. Y. Liu, X. Fang, A. Zhang, C. Shen, Q. Liu, H. A. Enaya and C. Zhou, *Nano Energy*, 2016, **27**, 27.
45. D. Buchholz, A. Moretti, R. Kloepsch, S. Nowak, V. Siozios, M. Winter and S. Passerini, *Chem. Mater.*, 2013, **25**, 142.
46. D. Zhou, W. Huang, X. Kang, F. Zhao, L. Zhao, Z. Deng, X. Yan, Y. Yu and M. Xiang, *Int. J. Electrochem. Sci.*, 2018, **13**, 2010.
47. I. D. Gocheva, M. Nishijima, T. Doi, S. Okada, J.-i. Yamaki and T. Nishida, *J. Power Sources*, 2009, **187**, 247.
48. Y. Yamada, T. Doi, I. Tanaka, S. Okada and J. I. Yamaki, *J. Power Sources*, 2011, **196**, 4837.
49. K. V. Kravchyk, T. Zünd, M. Wörle, M. V. Kovalenko and M. I. Bodnarchuk, *Chem. Mater.*, 2018, **30**, 1825.
50. A. K. Padhi, K. S. Nanjundaswamy, J. B. Goodenough and R. L. Batteries, *J. Electrochem. Soc.*, 1997, **144**, 1188.
51. J. Kim, D. H. Seo, H. Kim, I. Park, J. K. Yoo, S. K. Jung, Y. U. Park, W. A. Goddard and K. Kang, *Energy Environ. Sci.*, 2015, **8**, 540.
52. P. Moreau, D. Guyomard, J. Gaubicher and F. Boucher, *Chem. Mater.*, 2010, **22**, 4126.
53. K. T. Lee, T. N. Ramesh, F. Nan, G. Botton and L. F. Nazar, *Chem. Mater.*, 2011, **23**, 3593.
54. J. B. Goodenough, H. Y.-P. Hong and J. A. Kafalas, *Mater. Res. Bull.*, 1976, **11**, 203.
55. C. Delmas, R. Olazcuaga, F. Cherkaoui, R. Brochu and G. Leflem, *Comp. R. Hebd. Des Seances L Acad. DES Sci. Ser. C*, 1978, **287**, 169.
56. Y. Noguchi, E. Kobayashi, L. S. Plashnitsa, S. Okada and J. Yamaki, *Electrochim. Acta*, 2013, **101**, 59.
57. X. Yao, Z. Zhu, Q. Li, X. Wang, X. Xu, J. Meng, W. Ren, X. Zhang, Y. Huang and L. Mai, *ACS Appl. Mater. Interfaces*, 2018, **10**, 10022.
58. C. W. Mason, I. Gocheva, H. E. Hoster and D. Y. W. Yu, in 'Battery Chemistries Beyond Lithium Ion', eds. C. S. Johnson, G. M. Veith and S. Mukerjee, Electrochemical Soc Inc., Pennington, NJ, 2014, Vol. 58, p.41.
59. S.-C. Chung, J. Ming, L. Lander, J. Lu and A. Yamada, *J. Mater. Chem. A*, 2018, **6**, 3919.
60. P. Barpanda, T. Ye, S. I. Nishimura, S. C. Chung, Y. Yamada, M. Okubo, H. Zhou and A. Yamada, *Electrochem. Commun.*, 2012, **24**, 116.
61. P. Barpanda, T. Ye, M. Avdeev, S. C. Chung and A. Yamada, *J. Mater. Chem. A*, 2013, **1**, 4194.
62. P. Barpanda, J. Lu, T. Ye, M. Kajiyama, S.-C. Chung, N. Yabuuchi, S. Komaba and A. Yamada, *RSC Adv.*, 2013, **3**, 3857.
63. G. Hautier, A. Jain, H. Chen, C. Moore, S. P. Ong and G. Ceder, *J. Mater. Chem.*, 2011, **21**, 17147.
64. H. Chen, G. Hautier, A. Jain, C. Moore, B. Kang, R. Doe, L. Wu, Y. Zhu, Y. Tang and G. Ceder, *Chem. Mater.*, 2012, **24**, 2009.
65. H. Chen, Q. Hao, O. Zivkovic, G. Hautier, L. S. Du, Y. Tang, Y. Y. Hu, X. Ma, C. P. Grey and G. Ceder, *Chem. Mater.*, 2013, **25**, 2777.
66. J. Barker, M. Y. Saidi and J. L. Swoyer, *Electrochem. Solid-State Lett.*, 2003, **6**, A1.
67. Y. Lu, S. Zhang, Y. Li, L. Xue, G. Xu and X. Zhang, *J. Power Sources*, 2014, **247**, 770.

68. R. K. B. Gover, P. Burns, A. Bryan, M. Y. Saidi, J. L. Swoyer and J. Barker, *Solid State Ion.*, 2006, **177**, 2635.

69. R. A. Shakoor, D.-H. Seo, H. Kim, Y.-U. Park, J. Kim, S.-W. Kim, H. Gwon, S. Lee and K. Kang, *J. Mater. Chem.*, 2012, **22**, 20535.

70. T. Broux, T. Bamine, F. Fauth, L. Simonelli, W. Olszewski, C. Marini, M. Menetrier, D. Carlier, C. Masquelier and L. Croguennec, *Chem. Mater.*, 2016, **28**, 7683.

71. Y. Kawabe, N. Yabuuchi, M. Kajiyama, N. Fukuhara, T. Inamasu, R. Okuyama, I. Nakai and S. Komaba, *Electrochemistry*, 2012, **80**, 80.

72. M. Reynaud, G. Rousse, A. M. Abakumov, M. T. Sougrati, G. Van Tendeloo, J.-N. Chotard and J.-M. Tarascon, *J. Mater. Chem. A*, 2014, **2**, 2671.

73. P. Barpanda, G. Oyama, C. D. Ling and A. Yamada, *Chem. Mater.*, 2014, **26**, 1297.

74. P. Barpanda, G. Oyama, S. I. Nishimura, S. C. Chung and A. Yamada, *Nat. Commun.*, 2014, **5**, 1.

75. G. Oyama, S.-i. Nishimura, Y. Suzuki, M. Okubo and A. Yamada, *ChemElectroChem*, 2015, **2**, 1019.

76. D. Dwibedi, R. B. Araujo, S. Chakraborty, P. P. Shanbogh, N. G. Sundaram, R. Ahuja and P. Barpanda, *J. Mater. Chem. A*, 2015, **3**, 18564.

77. D. Dwibedi, R. Gond, A. Dayamani, R. B. Araujo, S. Chakraborty, R. Ahuja and P. Barpanda, *Dalton Trans.*, 2017, **46**, 55.

78. L. L. Driscoll, E. Kendrick, K. S. Knight, A. J. Wright and P. R. Slater, *J. Solid State Chem.*, 2018, **258**, 64.

79. P. Singh, K. Shiva, H. Celio and J. B. Goodenough, *Energy Environ. Sci.*, 2015, **8**, 3000.

80. R. Tripathi, T. N. Ramesh, B. L. Ellis and L. F. Nazar, *Angew. Chem. Int. Ed.*, 2010, **49**, 8738.

81. J. M. Tarascon, N. Recham, M. Armand, J. N. Chotard, P. Barpanda, W. Walker and L. Dupont, *Chem. Mater.*, 2010, **22**, 724.

82. P. Barpanda, J. N. Chotard, N. Recham, C. Delacourt, M. Ati, L. Dupont, M. Armand and J. M. Tarascon, *Inorg. Chem.*, 2010, **49**, 7401.

83. M. Gnanavel, O. I. Lebedev, P. Bazin, B. Raveau and V. Pralong, *Solid State Ion.*, 2015, **278**, 38.

84. M. Sun, G. Rousse, M. Saubanère, M.-L. Doublet, D. Dalla Corte and J.-M. Tarascon, *Chem. Mater.*, 2016, **28**, 6637.

85. L. L. Driscoll, A. J. Wright and P. R. Slater, *Dalton Trans.*, 2018, **47**, 13535.

86. Y. Kee, N. Dimov, A. Staykov and S. Okada, *Mater. Chem. Phys.*, 2016, **171**, 45.

87. W. Guan, B. Pan, P. Zhou, J. Mi, D. Zhang, J. Xu and Y. Jiang, *ACS Appl. Mater. Interfaces*, 2017, **9**, 22369.

88. S. Li, J. Guo, Z. Ye, X. Zhao, S. Wu, J.-X. Mi, C.-Z. Wang, Z. Gong, M. J. McDonald, Z. Zhu, K.-M. Ho and Y. Yang, *ACS Appl. Mater. Interfaces*, 2016, **8**, 17233.

89. S. Yu, J. Q. Hu, M. B. Hussain, S. Q. Wu, Y. Yang and Z. Z. Zhu, *J. Solid State Electrochem.*, 2018, **4**, 2237.

90. S. Gao, J. Zhao, Y. Zhao, Y. Wu, X. Zhang, L. Wang, X. Liu, Y. Rui and J. Xu, *Mater. Lett.*, 2015, **158**, 300.

91. V. S. Rangasamy, S. Thayumanasundaram and J. P. Locquet, *Electrochim. Acta*, 2018, **276**, 102.

92. C. Y. Chen, K. Matsumoto, T. Nohira and R. Hagiwara, *Electrochem. Commun.*, 2014, **45**, 63.

93. M. Law, V. Ramar and P. Balaya, *J. Power Sources*, 2017, **359**, 277.

94. H. Zhu, J. Wang, X. Liu and X. Zhu, *RSC Adv.*, 2017, **7**, 14145.

95. Y. Lu, L. Wang, J. Cheng and J. B. Goodenough, *Chem. Commun.*, 2012, **48**, 6544.

96. Y. You, X. Yu, Y. Yin, K. W. Nam and Y. G. Guo, *Nano Res.*, 2014, **8**, 117.

97. J. Song, L. Wang, Y. Lu, J. Liu, B. Guo, P. Xiao, J. J. Lee, X. Q. Yang, G. Henkelman and J. B. Goodenough, *J. Am. Chem. Soc.*, 2015, **137**, 2658.

98. L. Wang, J. Song, R. Qiao, L. A. Wray, M. A. Hossain, Y. De Chuang, W. Yang, Y. Lu, D. Evans, J. J. Lee, S. Vail, X. Zhao, M. Nishijima, S. Kakimoto and J. B. Goodenough, *J. Am. Chem. Soc.*, 2015, **137**, 2548.

99. P. Hartmann, C. L. Bender, M. Vracar, A. K. Duerr, A. Garsuch, J. Janek and P. Adelhelm, *Nat. Mater.*, 2013, **12**, 228.

# PART 4
# Electrolyte Chemistry for Rechargeable Batteries

# Solid-State Electrolyte

**Wei Luo**

*Tongji University, Shanghai, China*

**Chuang Yu**

*Delft University of Technology, Delft, The Netherlands*

and

**Liangbing Hu**

*University of Maryland, College Park, MD, USA*



## 1 INTRODUCTION

Since their commercialization in 1991, lithium-ion batteries (LIBs) have enabled the rapid development of portable electronics, such as smart phones and laptops, which have greatly changed our lifestyle. To promote the application of LIBs in electric vehicles (EVs), great efforts are being devoted to increasing the energy density of LIBs to provide EVs with enough range on a single charge. Recent studies have mainly focused on exploiting new electrode materials with higher specific energy density or new battery chemistries, such as those based on Li–S and $Li-O_2$, and as a result significant progress has been made in increasing the energy density over the past few decades.

However, safety concerns still represent a major technological challenge. Frequent fire-related accidents of LIBs have been reported, which were attributed to thermal runaway—a series of physiochemical reactions between the cathode, anode, electrolyte, and separator, which is exacerbated by the use of flammable organic liquid electrolyte. Moreover, internal short circuits occur in LIBs owing to the poor thermal and mechanical properties of polyolefin-type separators.

Recently, more research efforts have begun to shift toward the development of solid-state lithium-ion batteries (SSLIBs), in which inorganic solid-state electrolytes (SSEs) are employed to replace organic liquid electrolyte. In SSLIBs, SSEs not only act as the electrolyte but also function as the separator. The most critical advantage of SSEs is their nonflammable nature. Furthermore, the high mechanical strength of SSEs also largely reduces the risk of internal short circuit. Lithium metal is regarded as the "holy grail" anode in terms of specific capacity and redox potential. However, lithium metal anodes suffer from dendritic lithium growth and poor stability in liquid electrolyte systems. It is believed that SSEs can tackle nearly all the issues associated with the lithium metal anode, enabling its application in SSLIBs to provide high energy density. Moreover, SSEs may also be the best solution for addressing the dissolution and shuttle effect of polysulfides for Li–S batteries by physically blocking the polysulfides in the cathode side.

*Inorganic Battery Materials*. Edited by Hailiang Wang and Boniface P.T. Fokwa
 ISBN 978-1-119-43199-2

To date, a range of materials have been investigated as SSEs, including perovskites, antiperovskites, sodium super ionic conductor (NASICON)- and garnet-type lithium-ion conductors, lithium sulfide glasses, lithium sulfide glass-ceramics, crystalline lithium sulfides, thin-film-type lithium phosphorous oxynitride (LiPON), $Li_3N$, hydrides, and halides. An ideal SSE typically has a high ionic conductivity, excellent chemical stability, a wide electrochemical window, high mechanical strength, and is environmentally benign and of low cost. So far, many studies have focused on increasing ionic conductivity of SSEs and have achieved significant progress. For example, Kato *et al.*[1] reported a new SSE with the formula $Li_{9.54}Si_{1.74}P_{1.44}S_{11.7}Cl_{0.3}$, which exhibited an ionic conductivity of $25\,mS\,cm^{-1}$ at room temperature. This chapter aims to provide an overview of the reported SSEs and to survey their key parameters.

## 2 LITHIUM OXIDE SYSTEMS

Lithium oxide SSEs, such as lithium oxysalts ($Li_2SO_4$ and $Li_3PO_4$), as well as perovskite-type (Li, La)$TiO_3$, NASICON-type ($LiA_{2-x}B_x(PO_4)_3$), and garnet-type SSEs, have been widely studied and tend to either demonstrate high ionic conductivity or high electrochemical stability.

### 2.1 Lithium Oxysalts

The conductivity of $Li_4SiO_4$ is on the order of $10^{-3}\,S\,cm^{-1}$ at 300 °C. Over the past few decades, great efforts have been devoted to developing $Li_4SiO_4$ with high conductivity at room temperature.[2] Some studies have suggested that combining $Li_4SiO_4$ with $Li_3PO_4$ to form a solid solution can further enhance the ionic conductivity toward $10^{-2}\,S\,cm^{-1}$ at 300 °C with an activation energy of 0.55 eV.[3] Recently, Deng *et al.*[4] carefully investigated the diffusion mechanism of the $(1-z)Li_4SiO_4 - zLi_3PO_4$ system using AC impedance, NMR spectroscopy, and atomistic modeling, suggesting that the conductivity can be improved up to $\sim 10^{-3}\,S\,cm^{-1}$ at 200 °C when $z = 0.25$, 0.50, and 0.75.

### 2.2 Perovskite-Type SSEs

The general formula of perovskite-type SSEs is $ABO_3$, in which alkaline or rare earth metal ions are located at the $A$ sites and transition metal ions at the $B$ sites, with a 6-fold and 12-fold coordinated oxygen anion associated with the $B$ and $A$ cations, respectively.[5] To increase the ionic conductivity of perovskite-type SSEs, substituting $A$ with other cations has been shown to be effective since the substitution can affect the ion diffusion channels. To date, the most promising perovskite-type SSEs are $Li_{3x}La_{2/3-x}TiO_3$ (LLTO) and its analogs, which exhibit high bulk conductivity on the order of $10^{-3}\,S\,cm^{-1}$ at room temperature.[6]

Unfortunately, there are still several issues that limit the application of these materials. The first one is the loss of Li during the high-temperature (HT) sintering process that is used to obtain perovskite-type SSEs with high ionic conductivity. The second challenge is the large grain boundaries resistance.[7] Moreover, Ti-containing perovskite-type SSEs are electrochemically unstable against Li metal owing to the reduction of $Ti^{4+}$ in contact with Li metal.[5]

### 2.3 Anti-Perovskite-Type SSEs

Zhao and Daemen[8] recently reported a novel family of Li-rich anti-perovskite-type SSEs with the formula $A^-B^{2-}X_3^+$, in which $A^-$ is a monovalent anion at the dodecahedral center, $B^{2-}$ is a divalent anion at the octahedral center, and $X^+$ is a strongly electropositive monovalent cation located at the octahedral vertices. When the $A^-$ site is a halogen atom, such as F, Cl, Br, I, or a mixture of the halogens, oxygen in the $B^{2-}$ site and lithium is chosen as the monovalent anion element to produce materials such as $ClOLi_3$. This structure is similar to the high-temperature ionic superconductor $NaMgF_3$. According to the "cation-first" convention of ionic compounds, Zhao *et al.* recently reported the new lithium-rich antiperovskite $Li_3OX$ (X = Cl, Br, I). The synthesis of $Li_3OCl$ and $Li_3OCl_{0.5}Br_{0.5}$ antiperovskite materials can be accomplished in a simple high temperature route, using molten LiOH, LiCl, and LiBr as precursors, which can be readily scaled up. The conductivity of $Li_3OCl$ reaches $0.85\times10^{-3}\,S\,cm^{-1}$ at room temperature and increases up to $4.82\times10^{-3}\,S\,cm^{-1}$ at 250 °C. Recently, some studies have suggested that the as-prepared "$Li_3OX$" may in fact be $Li_2OHX$ antiperovskite. The Goodenough[9] group found that a fluorine-doped antiperovskite SSE $Li_2(OH)_{0.9}F_{0.1}Cl$ is an appealing SSE owing to its great stability against Li metal anode.

### 2.4 NASICON-Type SSEs

NASICON refers to variations of the compound $Na_{1+x}Zr_2Si_xP_{3-x}O_{12}$ $(0 < x < 3)$, which exhibits a high Na-ion conductivity.[10] Typically, NASICON-type Li-ion SSEs have a general formula $LiM_2(BO_4)_3$, in which tetravalent ions ($Ge^{4+}$, $Ti^{4+}$, and $Zr^{4+}$) or one trivalent ion ($B^{3+}$, $Al^{3+}$, and $Ga^{3+}$) occupies the $M$ site, and pentavalent ions ($P^{5+}$, $As^{5+}$, and $Sb^{5+}$) occupy the $B$ site. Its framework consists of $O$ shared $MO_6$ octahedra and $BO_4$ tetrahedra that form a three-dimensional (3D) structure.[5] Owing to this 3D open structure, Li ions can jump between octahedra and tetrahedra vacancies, and therefore the ionic conductivity of NASICON-type SSEs strongly depends on the size of this 3D framework. Thus, it is critical to match the channel size with the Li-ion radius. Partial occupancies of Li ions on the octahedral and tetrahedra positions are also an effective method to enhance the conductivity. Vacancies at the intersection of the ion conduction pathways in the 3D structure promote Li-ion diffusion. This provides us two important routes to improve the ionic conductivity of NASICON-type $Li_{1+x}A_{2-x}B_xIV(PO_4)_3$ structures. One is to tailor the framework size by choosing various $A$ ions with different sizes. The other is aliovalent substitution of $A$, such as using $M^{3+}$ ions, which can increase the concentration and mobility of Li ions.[5]

Cyclic voltammogram measurements confirmed that $Li_{1.5}Al_{0.5}Ge_{1.5}(PO_4)_3 - xLi_2O$ has a wide electrochemical window of up to 6 V versus $Li^+/Li$.[11] However, first-principle calculations suggested that the stability window of this NASICON-type electrolyte and its analogs are actually much smaller, that is, 2.9–4.2 V for $LiGe_2(PO_4)_3$ and $Li_{1.5}Al_{0.5}Ge_{1.5}(PO_4)_3$, and 2.17–4.2 V for $Li_{1.3}Ti_{1.7}Al_{0.3}(PO_4)_3$.[12,13] Moreover, similar to the perovskite-type LLTO, Ti-containing NASICON-type SSEs are also unstable with Li metal. It is worth noting that $Li_{1+x}Al_xGe_{2-x}(PO_4)_3$ has been used as the SSE for all-solid-state Li-air batteries, demonstrating excellent initial discharge and charge capacity properties of 1700 and 900 mAh $g^{-1}$ at room temperature.[14]

### 2.5 Garnet-Type Lithium Oxides

Garnet-type SSEs were first reported by Murugan *et al.*,[15] in which the nominal chemical composition is $Li_5La_3M_2O_{12}$ (M = Ta, Nb). In these structures, Li occupies the tetrahedral 24d position and the octahedral 48g position. The resulting materials displayed a conductivity of $10^{-6}$ S $cm^{-1}$ at room temperature and a wide potential window.[16] Recently, $Li_7La_3Zr_2O_{12}$ (LLZO) was discovered by replacing M with $Zr^{4+}$ in $Li_5La_3M_2O_{12}$ (M = Ta, Nb).[17] Surprisingly, LLZO showed a higher Li concentration and a high conductivity of $3 \times 10^{-4}$ S $cm^{-1}$ at room temperature with an activation energy of 0.3 eV.[16]

Interestingly, LLZO has two possible phases. One is the tetragonal phase with a space group of $I4_1/acd$ (no. 142) and the other is the cubic phase with a space group of $Ia3d$ (no. 230).[18] For the cubic LLZO, the ionic conductivity can reach $\sim 10^{-4}$ S $cm^{-1}$ at room temperature, while the tetragonal LLZO shows a much lower conductivity of $\sim 10^{-6}$ S $cm^{-1}$.[18] In tetragonal LLZO structure, Li ion occupancy over the tetrahedral and octahedral sites is ordered. In contrast, a complicated Li-vacancy disordering in the tetrahedral and octahedral sites is found for the cubic LLZO, as shown in Figure 1.[19] The phase transition between these two LLZO phases can be achieved as well. The tetragonal phase is stable at room temperature and can transform to the cubic phase between 100 and 150 °C.[21,22] However, this transformed cubic phase delivers a much lower ionic conductivity of $\sim 10^{-6}$ S $cm^{-1}$, which is about two orders magnitude smaller than that of the HT synthesized cubic phase.[23]

In the past several years, cubic LLZO has been intensively investigated owing to its acceptable conductivity and high stability against Li metal. To further improve the ionic conductivity of cubic LLZO, replacing $Zr^{4+}$ with pentavalent or hexavalent cations, such as $Ta^{5+}$, $Nb^{5+}$, and $Te^{6+}$, has been explored, which can increase the vacancy concentration and decrease the local Li ordering.[24–26] Several approaches have also been proposed to decrease the grain boundary resistance of cubic LLZO, such as using sintering additives and advanced synthesis technologies.[27,28]

Some reports have shown that LLZO has an extremely high decomposition voltage (>5.5 V), which has been confirmed by both experimental results and first-principle calculations.[17,29] As a result, many cubic LLZO-based all-solid-state batteries have been demonstrated. For example, Ohta *et al.*[30] combined an $LiCoO_2$ (LCO) cathode, Li metal anode, and Nb-doped LLZO electrolyte to assemble a solid-state cell that exhibited an excellent discharge capacity of 129 mAh $g^{-1}$ and stable cycling performance up to 100 cycles. The electrochemical performance of this LCO/LLZO/Li all-solid-state battery is even comparable to liquid cells using the same electrodes but liquid electrolyte.[31]

## 3 LITHIUM SULFIDE SYSTEMS

In this chapter, we classify various lithium sulfide electrolytes into three categories, including lithium sulfide glasses, lithium sulfide glass-ceramics, and crystalline lithium sulfides.

### 3.1 Lithium Sulfide Glasses

Lithium sulfide glassy electrolytes have notable advantages, such as high room temperature conductivities ($10^{-5}$–$10^{-3}$ S $cm^{-1}$), isotropic ionic conduction, relatively small grain boundary resistances, and a simple film manufacturing process, all of which make these electrolytes good candidates for all-solid-state batteries.[32] The high conductivity of the glassy sulfide electrolyte is due to the so-called open structure and the high polarizability of sulfur ions, which weakens the interaction between the anions and Li ions. The most popular sulfide glass materials are $Li_2S–SiS_2$, $Li_2S–P_2S_5$, and their analogs. Since

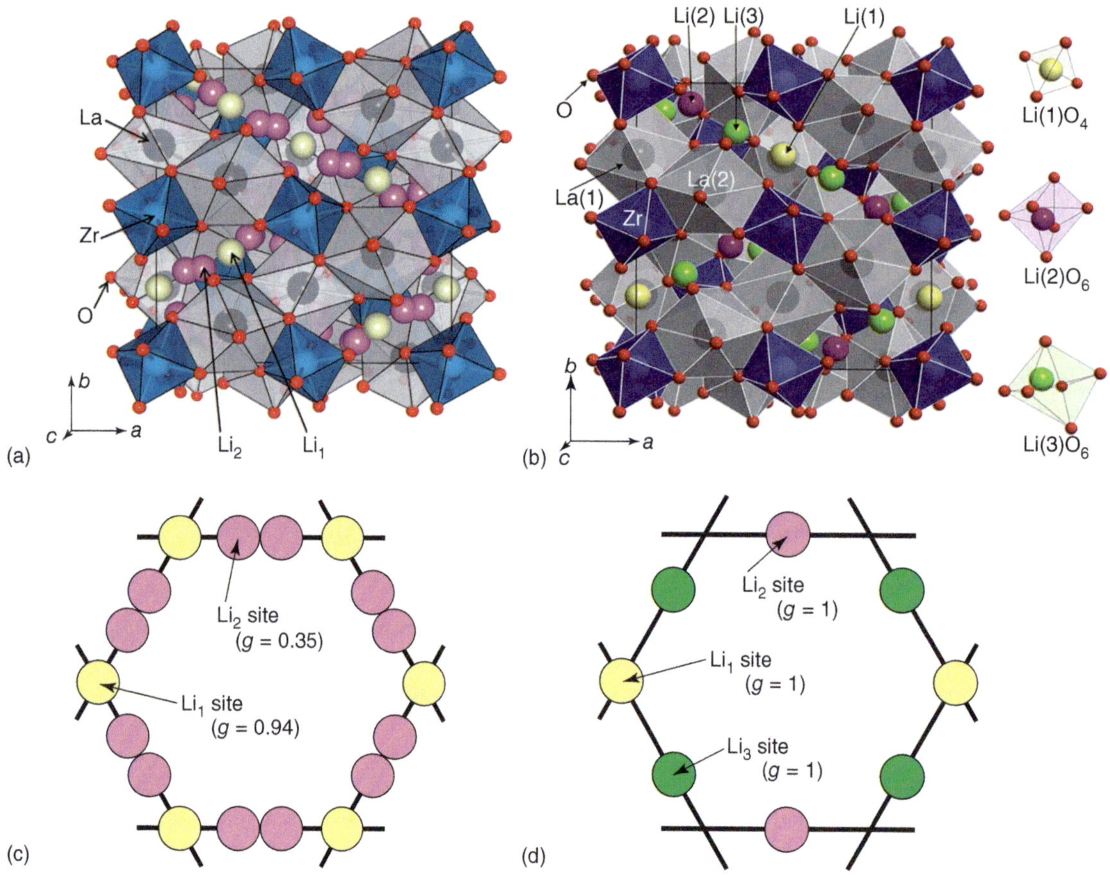


**Figure 1** Crystal structure of (a) cubic[19] and (b) tetragonal[20] $Li_7La_3Zr_2O_{12}$, and the corresponding loop structures constructed by Li atomic arrangement in (c) cubic and (d) tetragonal $Li_7La_3Zr_2O_{12}$. The occupancy value $g$ for each site is noted in parentheses. Note: The structural information of the low-temperature phase cubic $Li_7La_3Zr_2O_{12}$ is currently not clear. [(a,c,d) Reproduced with permission from Ref. Awaka, J.; Takashima, A.; Kataoka, K.; Kijima, N.; Idemoto, Y.; Akimoto, J., Crystal structure of fast lithium-ion-conducting cubic $Li_7La_3Zr_2O_{12}$. Chemistry letters 2010, 40 (1), 60–62 © The Chemical Society of Japan, 2010; (b) Reproduced with permission from Ref. Awaka, J.; Kijima, N.; Hayakawa, H.; Akimoto, J., Synthesis and structure analysis of tetragonal $Li_7La_3Zr_2O_{12}$ with the garnet-related type structure. J. Solid State Chem. 2009, 182 (8), 2046–2052. © Elsevier, 2009]

the crystallization process of glassy electrolytes is easy, glassy electrolytes are typically prepared by melting, followed by rapid quenching or a simple mechanical milling route. More importantly, lithium sulfide glasses are soft, which allows the solid-state batteries to be assembled by simply cold pressing lithium sulfide glasses with the anode and cathode materials (a very attractive prospect from a manufacturing perspective).

Unfortunately, like other sulfide materials, lithium sulfide glasses cannot be stored in air owing to the reaction between sulfides and moisture, yielding the smelly $H_2S$, which would lead to a high manufacturing cost. Furthermore, lithium sulfide glasses also show instability when they are coupled with oxide cathodes, such as $LiFePO_4$ and $LiCoO_2$. To improve this stability, partial replacement of sulfur in lithium sulfide glasses with oxygen atoms can suppress the generation of $H_2S$.[33] Mixing two different anions in the glassy system, in what is referred to as the "mixed anion effect", can further improve the conductivity of lithium sulfide glasses, as well.[34] Mixing lithium sulfide glasses with other lithium salts, such as lithium halides (LiX, X = Cl, Br, I),[35,36] lithium borohydride ($LiBH_4$),[37] and lithium *ortho*-oxosalts ($Li_3PO_4$)[38] can also enhance the ionic conductivity owing to the increased Li concentration and the decreased energy barrier for Li diffusion.

## 3.2 Lithium Sulfide Glass-Ceramics

Lithium sulfide glass-ceramics can be synthesized by annealing glassy lithium sulfide materials beyond their crystallization temperature, which are metastable phases but exhibit an extremely high room temperature conductivity ($>10^{-3}$ S cm$^{-1}$).[39] Similar to the sulfide glasses, the high ionic conductivity of lithium sulfide glass-ceramics is attributed to the large ionic radius and the high polarizability of the sulfur ions in the system.[32] Moreover, the crystalline domains are surrounded by amorphous phases in the glass-ceramics system, which can decrease the grain boundaries resistance and thus promote Li ion conduction.[32] $Li_7P_3S_{11}$, one of the most studied lithium sulfide glass-ceramic electrolytes in the $Li_2S-P_2S_5$ system, shows a conductivity of $3.2\times10^{-3}$ S cm$^{-1}$ at room temperature with an activation energy of 0.187 eV.[39] Previous research showed that $Li_7P_3S_{11}$ crystallizes in the triclinic unit cell indexed by the $P\bar{1}$ space group, which provides three types of Li positions located at the sites around the $P_2S_7$ and $PS_4$ groups and surrounded by 3, 4, or 5 sulfur atoms.[40] The atomic positions (especially $Li^+$) were previously determined by combined X-ray/neutron Rietveld analysis, showing a network of neighboring $Li^+$ cations (composed of Li–Li chains), demonstrating the Li-ion diffusion pathway.[41] The increment of the local conduction pathway of $Li^+$ ions from an $LiS_4$ unit to the neighboring $LiS_4$ units by the transformation from $(Li_2S)_{70}(P_2S_5)_{30}$ glass to $Li_7P_3S_{11}$ metastable crystal strongly contributes to the excellent conductivity.[42]

In addition to the diffraction and calculation methods,[40,43,44] solid-state NMR was also employed to investigate the diffusion mechanism of SSEs. For example, Hayamizu *et al.*[45] used pulsed-gradient spin-echo $^7Li$ NMR spectroscopy to identify the diffusion of Li ions in crystallized $(Li_2S)_7(P_2S_5)_3$, which showed that the observed $D_{Li}$ was consistent with the value evaluated from the conductivity using the Nernst–Einstein relation. Later on, Wohlmuth *et al.*[46] carried out spin-lattice relaxation (SLR) NMR experiments to quantitatively probe the short-range as well as long-range diffusion parameters of $Li_7P_3S_{11}$, the results of which suggested a multistep decay behavior and indicated the presence of at least three different relaxation rates in $Li_7P_3S_{11}$ glass-ceramics electrolytes. Moreover, $^{6/7}Li$ and $^{31}P$ solid-state NMR was applied in Murakami's study to confirm the original dynamics of high ionic conductivity of $Li_7P_3S_{11}$. Their results found that the significant motional fluctuation of the $P_2S_7$ ditetrahedra units plays a critical role for the observed high conductivity above 310 K.[47]

The conductivity of $Li_7P_3S_{11}$ can be further improved by optimizing the heat-treatment process. In Seino's experiment, a highly conductive $Li_7P_3S_{11}$ with a conductivity of $1.7\times10^{-2}$ S cm$^{-1}$ at room temperature was prepared by tailoring the heat-treatment process, which appears to have reduced the grain boundary resistance of $Li_7P_3S_{11}$.[48] However, the main disadvantages of $Li_7P_3S_{11}$ glass-ceramic is the instability in air and moisture, narrow electrochemical stability window,[13] as well as the incompatibility with Li metal negative electrodes.[49]

### 3.3 Crystalline Lithium Sulfide

Compared to the amorphous phases, crystalline lithium sulfide SSEs demonstrate higher conductivity and smaller activation energy.[50] Crystalline lithium sulfide conductors can be classified into thio-LISICON (thio-lithium superionic conductor)-type sulfides and argyrodite structured sulfides.

#### *3.3.1 Thio-LISICON-Type Sulfide Crystalline*

Kanno *et al.*[51] discovered that pure $Li_{3.25}Ge_{0.25}P_{0.75}S_4$, with a structure similar to LISICON (thio-LISICON), features a high ionic conductivity of $2.2\times10^{-3}$ S cm$^{-1}$ at 25 °C with negligible electronic conductivity in combination with a high electrochemical stability. Since its discovery, $Li_{3.25}Ge_{0.25}P_{0.75}S_4$ and its analogs have been intensively investigated owing to their high conductivity. Among these structures, the thio-LISICON compound based on Li silicon sulfide with a formula $Li_{3.4}Si_{0.4}P_{0.6}S_4$ is very attractive, displaying better phase stability but a slightly lower conductivity ($6.4\times10^{-4}$ S cm$^{-1}$ at 25 °C).[52] Currently, the most intensively researched sulfide crystalline electrolyte is $Li_{10}GeP_2S_{12}$ with a 3D framework that delivers an extremely high ionic conductivity of $1.2\times10^{-2}$ S cm$^{-1}$ at 25 °C, comparable to that of current liquid organic electrolytes.[53] Ab initio calculations for $Li_{10}GeP_2S_{12}$ showed that the conductivities in the *c*-direction and in the *ab*-plane are $4\times10^{-2}$ and $9\times10^{-4}$ S cm$^{-1}$, respectively, resulting in an overall conductivity of $9\times10^{-3}$ S cm$^{-1}$. The corresponding activation energies for the diffusion in the *c*-direction and in the *ab*-plane are 0.17 and 0.28 eV, respectively.[54] The agreement between the ionic conductivity obtained by calculations and experiments suggests a 3D conduction mechanism for $Li_{10}GeP_2S_{12}$ rather than a one-dimensional (1D) conduction scheme. The 1D conducting pathway along the *c*-direction is composed of sharing a common edge for the $LiS_4$ tetrahedra at the 16*h* and 8*f* sites. The *ab*-plane framework is constructed by connecting $(Ge/P)S_4$ tetrahedra to $LiS_6$ octahredra.[5,54] Experimental results suggested that $Li_{10}GeP_2S_{12}$ has a wide electrochemical stability window of up to 4 V versus $Li/Li^+$.[53] However, recent studies have questioned this reported electrochemical window for $Li_{10}GeP_2S_{12}$ owing to the reversible reduction oxidization at 0–1.7 and 2–2.5 V versus Li metal, suggesting an actual electrochemical window of 0.0–5.0 V.[55,56] The chemical stability and electrode compatibility of $Li_{10}GeP_2S_{12}$ are quite complicated owing to the potential oxidation of Ge and the instability toward atmospheric moisture. For its application in all-solid-state batteries, $Li_{10}GeP_2S_{12}$ appears to be suitable in combination with $LiCoO_2$ as a cathode and Li–In alloy as an anode.[57] In addition, all-solid-state batteries using $LiFePO_4$,[58] $LiMn_{1/3}Ni_{1/3}Co_{1/3}O_2$,[59] or $LiNi_{0.5}Mn_{1.5}O_4$[58] as the positive electrode in combination with $Li_{10}GeP_2S_{12}$ as electrolyte and Li metal as the negative electrode show promising capacities.

Although $Li_{10}GeP_2S_{12}$ appears to be an excellent solid electrolyte, the price of Ge is expected to limit its applications. Replacing Ge with other elements, such as Si, Sn, and Al, can promote the development of practical solid electrolytes in this structural family. As mentioned earlier, matching the diffusion channel size with the Li-ion radius is vital to achieve high ionic conductivities for Li-ion conductors.[18] Cation substitution is also a possible strategy to tailor the conductivity of $Li_{10}GeP_2S_{12}$. Replacing $Ge^{4+}$ with $Sn^{4+}$, which has a larger radius, results in large bulk conductivities amounting to $7\times10^{-3}$ and $4\times10^{-3}$ S cm$^{-1}$ for the total conductivity at 27 °C.[60] Replacing $Ge^{4+}$ by $Si^{4+}$, which has a smaller radius, can form $Li_{10}SiP_2S_{12}$. Calculations[56] have suggested that $Li_{10}SiP_2S_{12}$ has a conductivity of $2.3\times10^{-2}$ S cm$^{-1}$ at 300 K with an activation energy of 0.20 eV. Experimental results show that a maximum conductivity of up to $8.6\times10^{-3}$ S cm$^{-1}$ is achieved for a composition of $Li_{10}Ge_{0.95}Si_{0.05}P_2S_{12}$.[61] However, the high-pressure treatment and higher demands on sintering make the synthesis process more costly for Si substitution in $Li_{10}GeP_2S_{12}$, which limits its application. Calculations suggest that heterovalent substitution of $Ge^{4+}$ by $Al^{3+}$ will form $Li_{11}AlP_2S_{12}$, which is predicted to enhance the ionic conductivity and decrease the energy barrier for diffusion.[56] However, in reality, $Li_{11}AlP_2S_{12}$ shows a conductivity of $8.02\times10^{-4}$ S cm$^{-1}$ at 25 °C, which is much smaller than the result predicted by calculations ($3.3\times10^{-2}$ S cm$^{-1}$ at the same temperature).[62] The heterovalent cation substitution of $Ge^{4+}$ with $Al^{3+}$ increases the $Li^+$ concentration, but it has a small effect on the conductivity in this structure. Anion substitution is another strategy to modify the conductivity of $Li_{10}GeP_2S_{12}$. Oxygen-substituted $Li_{10}GeP_2O_{12}$ compounds that feature much lower conductivities ($3\times10^{-5}$ S cm$^{-1}$ at 300 K) are predicted to be unstable, while selenium-substituted $Li_{10}SnP_2Se_{12}$ compounds show a remarkable improvement of conductivity ($2.4\times10^{-2}$ S cm$^{-1}$ at 300 K), though at the cost of reduced electrochemical stability.[56]

Recently, two new members of this crystalline lithium sulfide family showing high ionic conductivity ($2.5\times10^{-2}$ S cm$^{-1}$ for $Li_{9.54}Si_{1.74}P_{1.44}S_{11.7}Cl_{0.3}$ at 25 °C) and excellent electrochemical stability (~0 V vs Li metal for $Li_{9.6}P_3S_{12}$) were reported by Kato *et al.*[1] The 3D conduction pathway of the $Li_{9.54}Si_{1.74}P_{1.44}S_{11.7}Cl_{0.3}$ electrolyte was revealed by analyzing the anisotropic thermal displacement of lithium and nuclear density distribution, revealing 1D diffusion along the *c*-axis and 2D diffusion in the *ab*-plane.

### 3.4 Argyrodite Electrolytes

Another important family of crystalline lithium sulfide solid electrolytes is the argyrodites described by the chemical formula $Li_6PS_5X$ (X = Cl, Br, I), which show promising conductivities in the range $10^{-3}$–$10^{-2}$ S cm$^{-1}$ at room temperature.[63] The conductivity of Li-argyrodites is slightly smaller than that of the $Li_{10}GeP_2S_{12}$ family, while their cost is much lower, making this family of electrolytes attractive for the application in all-solid-state batteries. The $Li_6PS_5X$ (X = Cl, Br, I) structures exhibit the high-symmetry argyrodite framework indexed by the $F\bar{4}3m$ space group. In the $Li_6PS_5X$ (X = Cl, Br, I) structure, phosphorus atoms occupy the tetrahedral interstices, forming a framework of isolated $PS_4$ tetrahedra, while lithium ions are randomly distributed over the remaining tetrahedral interstices (48*h* and 24*g* sites), as shown in Figure 2.[65] Among $Li_6PS_5X$ (X = Cl, Br, I), the highest conductivities have been achieved for $Li_6PS_5Br$ and $Li_6PS_5Cl$, reaching up to $10^{-3}$ S cm$^{-1}$ at room temperature.[66,67] $Li_6PS_5I$ shows a much lower conductivity owing to the different distribution of I and Br/Cl ions over the 4*a* and 4*c* sites.[64,68] The disorder of the $S^{2-}/Br^-$ and $S^{2-}/Cl^-$ distributions has been suggested to promote the Li-ion diffusivity, while the exchange between $S^{2-}$ and $I^-$ is impeded owing to its larger ionic radius, resulting in an anion-ordered $Li_6PS_5I$ with a low Li-ion conductivity.[68] The structural studies and bond-valence model results showed that the packing determined by the $PS_4^{\ 3-}$ anions has great influence on Li-ion transport in the argyrodites, $Li_6PS_5X$ (X = Cl, Br, I).[65]

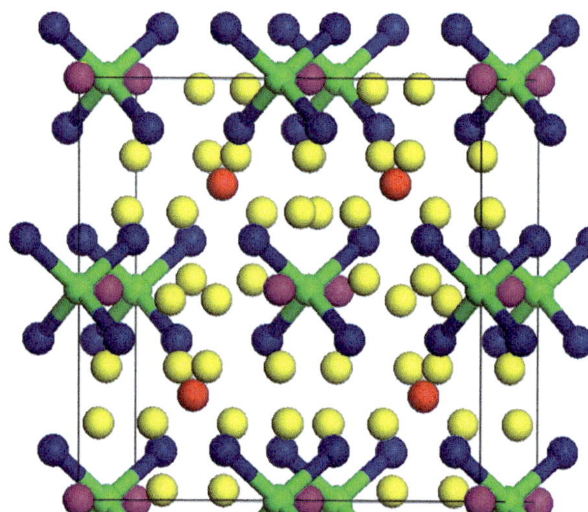

**Figure 2** Crystal structure of $Li_6PS_5X$ (X = Cl, Br, I). The yellow represents Li sites (48 h), green represents phosphorus, blue represents bonded sulfur, pink represents 4*a* sites, and red represents 4*c* sites.[64] [Reprinted with permission from de Klerk, N. J.; Rosłoń, I.; Wagemaker, M., Diffusion mechanism of Li argyrodite solid electrolytes for Li-ion batteries and prediction of optimized halogen doping: the effect of Li vacancies, halogens, and halogen disorder. Chemistry of Materials 2016, 28 (21), 7955–7963. © American Chemical Society 2016]

NMR results showed that there are a variety of jump processes in the Li-argyrodite,[69] which is associated with macroscopic Li-ion conductivity. Moreover, density functional theory (DFT) based on molecular dynamics (MD) simulations was applied to reveal the diffusion mechanism for Li-argyrodite $Li_6PS_5X$ (X = Cl, Br, I).[64] The calculation results showed that there are three types of Li jumps in Li-argyrodite: the transition between the paired 48*h* sites; the transition within the cages between the different 48*h* pairs; and the transitions interconnecting the four cages in each unit cell. These three transitions built up the 3D diffusion pathway through the whole crystal, and the slowest transition limits the macroscopic diffusion. Improvement of the ionic conductivity of Li-argyrodite $Li_6PS_5X$ (X = Cl, Br, I) can be achieved by optimization of the halogen distribution over the 4*a* and 4*c* sites and increasing the halogen content of the material.[64]

The common synthesis route of the argyrodite $Li_6PS_5X$ (X = Cl, Br, I) involves mechanical milling followed by annealing. In situ neutron diffraction and impedance measurement results showed that heat treatment is vital for achieving a high ionic conductivity.[66] To obtain an $Li_6PS_5X$ (X = Cl, Br, I) electrolyte with a high room temperature conductivity of $1 \times 10^{-3}\,S\,cm^{-1}$ and low activation energy of 0.16 eV, the samples need to be annealed over 250 °C.[66] Besides the excellent ionic conductivity, $Li_6PS_5X$ (X = Cl, Br) electrolytes initially were suggested to have a wide electrochemical window up to 7 V versus $Li/Li^+$.[67] However, recent computational studies indicated that the stability window of $Li_6PS_5Cl$ is about 2.0–2.5 V.[12] Other issues that may limit its application in all-solid-state batteries are the reactivity with air and moisture and the incompatibility with Li metal. Recently, various cathode materials ($LiCoO_2$,[67,70] S,[71,72] Cu–$Li_2S$,[73] CuS,[74] and $MoS_2$[75]) and anodes ($Li_4Ti_5O_{12}$,[70,76] Li–In,[71] In,[67] and Li–Al[76]) were combined together with the argyrodite $Li_6PS_5X$ (X = Cl, Br) solid electrolytes to investigate the electrochemical properties.

## 4 THIN-FILM SSEs

Thin-film lithium-ion-based solid-state microbatteries have been widely investigated as power sources in the microelectronics industry. Owing to the extremely small thickness, the conductivity requirements for SSEs in this application are not as high compared to the bulk SSLIBs. At the beginning, the thin-film SSE is sputter deposited, using lithium oxysalt ($Li_4SiO_4$–$Li_3PO_4$) pressed powder and $Li_2O$ pellets as targets. The resulting thin-film SSE has a conductivity of $5.0 \times 10^{-6}\,S\,cm^{-1}$ at room temperature. In 1993, thin-film LiPON was synthesized by magnetron sputtering an $Li_3PO_4$ target under an $N_2$ atmosphere.[77] This new thin-film SSE only exhibited a conductivity of $3.0 \times 10^{-6}\,S\,cm^{-1}$ at room temperature, but it had a high stability with both Li metal anodes and high voltage cathodes. In the thin-film SSE field, many groups are focusing on developing various deposition technologies to prepare dense thin-film SSEs with a relatively high growth rate. Currently, magnetron sputtering, chemical vapor deposition (CVD), and atomic layer deposition (ALD) are the three main methods to prepare thin-film SSEs for thin-film solid-state batteries.

## 5 HYDRIDES AND OTHER SSEs

In 2007, Orimo and coworkers[78] reported that $LiBH_4$ exhibited a relatively fast Li ion conductivity at its HT phase near 390 K, as well as a high electrochemical reaction rate at the $LiBH_4$–lithium metal interface. However, stabilizing $LiBH_4$ at its HT phase at room temperature presents a major challenge, which limits its practical applications. To stabilize the HT phase, Maekawa *et al.*[79] incorporated lithium halides (LiX, X = Cl, Br, I) into $LiBH_4$. The activation energy is particularly low when LiI is used, and the resulting $LiBH_4$–LiI solid solution demonstrates a high conductivity at room temperature. Furthermore, hydride-based SSEs with larger sized cage-like anions, such as $[B_{12}H_{12}]^{2-}$ and $[B_{10}H_{10}]^{2-}$, also provide high conductivities.

Lithium nitride ($Li_3N$), with a space group *P6/mmm*, is another solid electrolyte attracting interest owing to its high lithium content and open structure, described as a sequence of Li and $Li_2N$ layers perpendicular to the *c*-axis. The nitrogen atoms are located at the center of the elementary cell, which are surrounded by eight lithium atoms in the shape of a hexagonal bipyrdmid.[80,81] The material has a high lithium ion conductivity ($\sim 10^{-3}\,S\,cm^{-1}$) with negligible partial electronic conductivity at ambient temperature, stability with lithium metal up to its melting temperature of 813 °C, and a high decomposition voltage due to a strong kinetic hindrance of the decomposition reaction.[81] The lithium ion conductivity of $Li_3N$ can be improved by optimizing the synthesis conditions and providing suitable doping with the same family elements, such as P and Si. Moreover, the $Li_3N$ electrolyte can be used as a buffer layer for solid-state batteries for lithium ion conductors that feature chemical and electrochemical incompatibility with lithium metal.

A family of lithium oxide glasses, composed of a mixture of $Li_2O$ with $Al_2O_3$, $Ga_2O_3$, and $Bi_2O_3$, has been found to exhibit reasonably high ionic conductivities with low electronic conductivities for $Li_2O$ concentrations exceeding 50 mol%.[82] The ionic conductivity properties of this kind of lithium oxide glass can be enhanced by the introduction of a new oxide (MO) and/or a salt (LiX) due to the partial substation.[83]

## 6 SUMMARY/CONCLUSION

In summary, SSLIBs are promising energy devices with high safety properties. However, one of the fundamental challenges for a solid-state battery is exploring SSEs with fast ion mobility, high chemical and electrochemical stability, and good electrode material compatibility. The high stability of lithium oxide electrolyte makes it a promising candidate; however, the large grain boundary resistance and the poor processability limit its applications. Sulfide-based solid electrolytes are a good choice owing to their high room temperature lithium ion conductivity, even superior to that of current organic liquid electrolyte (25 mS $cm^{-1}$ at room temperature), and low density, providing the possibility to fabricate SSLIBs with high rate capability (even up to 100 °C). However, the chemical and electrochemical stability and the active material compatibility are still large obstacles that limit its commercial applications. Therefore, we expect that designing new inorganic solid electrolytes with ultrafast ionic conductivity and high chemical/electrochemical stability will continue to be a popular research topic in the near future.

## 7 ABBREVIATIONS AND ACRONYMS

ALD = atomic layer deposition; CVD = chemical vapor deposition; DFT = density functional theory; EV = electric vehicles; HT = high-temperature; LIB = lithium-ion batteries; $LiBH_4$ = lithium borohydride; LiPON = lithium phosphorous oxynitride; LLTO = $Li_{3x}La_{2/3-x}TiO_3$; MD = molecular dynamics; NASICON = sodium super ionic conductor; SLR = spin-lattice relaxation; SSE = solid-state electrolyte; SSLIB = solid-state lithium-ion batteries; thio-LISICON = thio-lithium superionic conductor; 1D = one-dimensional; 3D = three-dimensional

## 8 REFERENCES


1. Y. Kato, S. Hori, T. Saito, K. Suzuki, M. Hirayama, A. Mitsui, M. Yonemura, H. Iba and R. Kanno, *Nat. Energy*, 2016, **1**(4), 16030.
2. S. Pizzini, *J. Appl. Electrochem.*, 1971, **1**(3), 153.
3. Y. W. Hu, I. Raistrick and R. A. Huggins, *J. Electrochem. Soc.*, 1977, **124**(8), 1240.
4. Y. Deng, C. Eames, J.-N. Chotard, F. Lalère, V. Seznec, S. Emge, O. Pecher, C. P. Grey, C. Masquelier and M. S. Islam, *J. Am. Chem. Soc.*, 2015, **137**(28), 9136.
5. J. C. Bachman, S. Muy, A. Grimaud, H.-H. Chang, N. Pour, S. F. Lux, O. Paschos, F. Maglia, S. Lupart and P. Lamp, *Chem. Rev.*, 2015, **116**(1), 140.
6. E. Fortal'nova, O. Gavrilenkov, A. Belous and E. Politova, *Russ. J. Gen. Chem.*, 2009, **79**(9), 1987.
7. K. C. Alexander, P. Ganesh, M. Chi, P. Kent and B. G. Sumpter, *MRS Commun.*, 2016, **6**(4), 455.
8. Y. Zhao and L. L. Daemen, *J. Am. Chem. Soc.*, 2012, **134**(36), 15042.
9. Y. Li, W. Zhou, S. Xin, S. Li, J. Zhu, X. Lü, Z. Cui, Q. Jia, J. Zhou, Y. Zhao and J. B. Goodenough, *Angew. Chem. Int. Ed.*, 2016, **55**(34), 9965.
10. M. Bayard and G. Barna, *J. Electroanal. Chem. Interfacial Electrochem.*, 1978, **91**(2), 201.
11. X. Xu, Z. Wen, X. Wu, X. Yang and Z. Gu, *J. Am. Ceram. Soc.*, 2007, **90**(9), 2802.
12. W. D. Richards, L. J. Miara, Y. Wang, J. C. Kim and G. Ceder, *Chem. Mater.*, 2015, **28**(1), 266.
13. Y. Zhu, X. He and Y. Mo, *ACS Appl. Mater. Interfaces*, 2015, **7**(42), 23685.
14. H. Kitaura and H. Zhou, *Energy Environ. Sci.*, 2012, **5**(10), 9077.
15. R. Murugan, V. Thangadurai and W. Weppner, *Angew. Chem. Int. Ed.*, 2007, **46**(41), 7778.
16. H. Buschmann, J. Dölle, S. Berendts, A. Kuhn, P. Bottke, M. Wilkening, P. Heitjans, A. Senyshyn, H. Ehrenberg and A. Lotnyk, *Phys. Chem. Chem. Phys.*, 2011, **13**(43), 19378.
17. J. Awaka, A. Takashima, K. Kataoka, N. Kijima, Y. Idemoto and J. Akimoto, *Chem. Lett.*, 2010, **40**(1), 60.
18. J. Awaka, N. Kijima, H. Hayakawa and J. Akimoto, *J. Solid State Chem.*, 2009, **182**(8), 2046.
19. V. Thangadurai, H. Kaack and W. J. Weppner, *J. Am. Ceram. Soc.*, 2003, **86**(3), 437.
20. E. J. Cussen, *J. Mater. Chem.*, 2010, **20**(25), 5167.
21. M. Matsui, K. Sakamoto, K. Takahashi, A. Hirano, Y. Takeda, O. Yamamoto and N. Imanishi, *Solid State Ion.*, 2014, **262**, 155.
22. C. A. Geiger, E. Alekseev, B. Lazic, M. Fisch, T. Armbruster, R. Langner, M. Fechtelkord, N. Kim, T. Pettke and W. Weppner, *Inorg. Chem.*, 2010, **50**(3), 1089.
23. S. Toda, K. Ishiguro, Y. Shimonishi, A. Hirano, Y. Takeda, O. Yamamoto and N. Imanishi, *Solid State Ion.*, 2013, **233**, 102.
24. S. Adams and R. P. Rao, *J. Mater. Chem.*, 2012, **22**(4), 1426.
25. C. Deviannapoorani, L. Dhivya, S. Ramakumar and R. Murugan, *J. Power Sources*, 2013, **240**, 18.
26. S. Ohta, T. Kobayashi and T. Asaoka, *J. Power Sources*, 2011, **196**(6), 3342.
27. E. Rangasamy, J. Wolfenstine and J. Sakamoto, *Solid State Ion.*, 2012, **206**, 28.
28. J. Wolfenstine, J. Sakamoto and J. Allen, *J. Mater. Sci.*, 2012, **47**(10), 4428.
29. T. Thompson, S. Yu, L. Williams, R. D. Schmidt, R. Garcia-Mendez, J. Wolfenstine, J. L. Allen, E. Kioupakis, D. J. Siegel and J. Sakamoto, *ACS Energy Lett.*, 2017, **2**(2), 462.

30. S. Ohta, T. Kobayashi, J. Seki and T. Asaoka, *J. Power Sources*, 2012, **202**, 332.
31. S. Zhu, Y. Qin and J. Zhang, *Int. J. Electrochem. Sci.*, 2016, **11**, 6403.
32. C. Cao, Z.-B. Li, X.-L. Wang, X.-B. Zhao and W.-Q. Han, *Front. Energy Res.*, 2014, **2**, 25.
33. T. Ohtomo, A. Hayashi, M. Tatsumisago and K. Kawamoto, *J. Non-Cryst. Solids*, 2013, **364**, 57.
34. T. Minami, A. Hayashi and M. Tatsumisago, *Solid State Ion.*, 2006, **177**(26–32), 2715.
35. S. Ujiie, T. Inagaki, A. Hayashi and M. Tatsumisago, *Solid State Ion.*, 2014, **263**, 57.
36. S. Ujiie, A. Hayashi and M. Tatsumisago, *J. Solid State Electrochem.*, 2013, **17**(3), 675.
37. A. Yamauchi, A. Sakuda, A. Hayashi and M. Tatsumisago, *J. Power Sources*, 2013, **244**, 707.
38. K. Takada, N. Aotani and S. Kondo, *J. Power Sources*, 1993, **43**(1–3), 135.
39. F. Mizuno, A. Hayashi, K. Tadanaga and M. Tatsumisago, *Adv. Mater.*, 2005, **17**(7), 918.
40. H. Yamane, M. Shibata, Y. Shimane, T. Junke, Y. Seino, S. Adams, K. Minami, A. Hayashi and M. Tatsumisago, *Solid State Ion.*, 2007, **178**(15–18), 1163.
41. Y. Onodera, K. Mori, T. Otomo, A. C. Hannon, S. Kohara, K. Itoh, M. Sugiyama and T. Fukunaga, *J. Phys. Soc. Jpn.*, 2010, **79**(Suppl. A), 87.
42. Y. Onodera, K. Mori, T. Otomo, M. Sugiyama and T. Fukunaga, *J. Phys. Soc. Jpn.*, 2012, **81**(4), 044802.
43. Y. Onodera, K. Mori, T. Otomo, H. Arai, Y. Uchimoto, Z. Ogumi and T. Fukunaga, 'Structural Origin of Ionic Conductivity for $Li_7P_3S_{11}$ Metastable Crystal by Neutron and X-ray Diffraction', IOP Publishing, 2014, p.012021.
44. K. Mori, T. Ichida, K. Iwase, T. Otomo, S. Kohara, H. Arai, Y. Uchimoto, Z. Ogumi, Y. Onodera and T. Fukunaga, *Chem. Phys. Lett.*, 2013, **584**, 113.
45. K. Hayamizu and Y. Aihara, *Solid State Ion.*, 2013, **238**, 7.
46. D. Wohlmuth, V. Epp and M. Wilkening, *ChemPhysChem*, 2015, **16**(12), 2582.
47. M. Murakami, K. Shimoda, S. Shiotani, A. Mitsui, K. Ohara, Y. Onodera, H. Arai, Y. Uchimoto and Z. Ogumi, *J. Phys. Chem. C*, 2015, **119**(43), 24248.
48. Y. Seino, T. Ota, K. Takada, A. Hayashi and M. Tatsumisago, *Energy Environ. Sci.*, 2014, **7**(2), 627.
49. M. Tatsumisago, M. Nagao and A. Hayashi, *J. Asian Ceram. Soc.*, 2013, **1**(1), 17.
50. J. W. Fergus, *J. Power Sources*, 2010, **195**(15), 4554.
51. R. Kanno and M. Murayama, *J. Electrochem. Soc.*, 2001, **148**(7), A742.
52. M. Murayama, R. Kanno, M. Irie, S. Ito, T. Hata, N. Sonoyama and Y. Kawamoto, *J. Solid State Chem.*, 2002, **168**(1), 140.
53. N. Kamaya, K. Homma, Y. Yamakawa, M. Hirayama, R. Kanno, M. Yonemura, T. Kamiyama, Y. Kato, S. Hama and K. Kawamoto, *Nat. Mater.*, 2011, **10**(9), 682.
54. Y. Mo, S. P. Ong and G. Ceder, *Chem. Mater.*, 2011, **24**(1), 15.
55. F. Han, Y. Zhu, X. He, Y. Mo and C. Wang, *Adv. Energy Mater.*, 2016, **6**(8), 1501590.
56. S. P. Ong, Y. Mo, W. D. Richards, L. Miara, H. S. Lee and G. Ceder, *Energy Environ. Sci.*, 2013, **6**(1), 148.
57. Y. Kato, K. Kawamoto, R. Kanno and M. Hirayama, *Electrochemistry*, 2012, **80**(10), 749.
58. J. Hassoun, R. Verrelli, P. Reale, S. Panero, G. Mariotto, S. Greenbaum and B. Scrosati, *J. Power Sources*, 2013, **229**, 117.
59. J. M. Whiteley, J. H. Woo, E. Hu, K.-W. Nam and S.-H. Lee, *J. Electrochem. Soc.*, 2014, **161**(12), A1812.
60. P. Bron, S. Johansson, K. Zick, J. Schmedt auf der Günne, S. Dehnen and B. Roling, *J. Am. Chem. Soc.*, 2013, **135**(42), 15694.
61. Y. Kato, R. Saito, M. Sakano, A. Mitsui, M. Hirayama and R. Kanno, *J. Power Sources*, 2014, **271**, 60.
62. P. Zhou, J. Wang, F. Cheng, F. Li and J. Chen, *Chem. Commun.*, 2016, **52**(36), 6091.
63. H. J. Deiseroth, S. T. Kong, H. Eckert, J. Vannahme, C. Reiner, T. Zaiss and M. Schlosser, *Angew. Chem. Int. Ed.*, 2008, **47**(4), 755.
64. N. J. de Klerk, I. Rosłoń and M. Wagemaker, *Chem. Mater.*, 2016, **28**(21), 7955.
65. P. R. Rayavarapu, N. Sharma, V. K. Peterson and S. Adams, *J. Solid State Electrochem.*, 2012, **16**(5), 1807.
66. R. P. Rao and S. Adams, *Phys. Status Solidi (a)*, 2011, **208**(8), 1804.
67. R. P. Rao, N. Sharma, V. Peterson and S. Adams, *Solid State Ion.*, 2013, **230**, 72.
68. S. Boulineau, M. Courty, J.-M. Tarascon and V. Viallet, *Solid State Ion.*, 2012, **221**, 1.
69. V. Epp, O. Gün, H.-J. Deiseroth and M. Wilkening, *J. Phys. Chem. Lett.*, 2013, **4**(13), 2118.
70. S. Boulineau, J.-M. Tarascon, J.-B. Leriche and V. Viallet, *Solid State Ion.*, 2013, **242**, 45.
71. C. Yu, L. van Eijck, S. Ganapathy and M. Wagemaker, *Electrochim. Acta*, 2016, **215**, 93.
72. M. Chen and S. Adams, *J. Solid State Electrochem.*, 2015, **19**(3), 697.
73. M. Chen, R. P. Rao and S. Adams, *Solid State Ion.*, 2014, **262**, 183.
74. M. Chen, R. P. Rao and S. Adams, *Solid State Ion.*, 2014, **268**, 300.
75. M. Chen, X. Yin, M. Reddy and S. Adams, *J. Mater. Chem. A*, 2015, **3**(20), 10698.
76. F. Stadler and C. Fietzek, *ECS Trans.*, 2010, **25**(36), 177.

77. J. B. Bates, N. J. Dudney, G. R. Gruzalski, R. A. Zuhr, A. Choudhury, C. F. Luck and J. D. Robertson, *J. Power Sources*, 1993, **43**(1), 103.
78. M. Matsuo, Y. Nakamori, S.-i. Orimo, H. Maekawa and H. Takamura, *Appl. Phys. Lett.*, 2007, **91**(22), 224103.
79. H. Maekawa, M. Matsuo, H. Takamura, M. Ando, Y. Noda, T. Karahashi and S.-i. Orimo, *J. Am. Chem. Soc.*, 2009, **131**(3), 894.
80. U. V. Alpen, A. Rabenau and G. H. Talat, *Appl. Phys. Lett.*, 1977, **30**(12), 621.
81. U. Alpen, *J. Solid State Chem.*, 1979, **29**(3), 379.
82. A. Glass and K. Nassau, *J. Appl. Phys.*, 1980, **51**(7), 3756.
83. A. Magistris, G. Chiodelli and M. Villa, *J. Power Sources*, 1985, **14**(1–3), 87.

# Chemistry of Soft Matter Battery Electrolytes

**Jelena Popovic**

*Max Planck Institute for Solid State Research, Stuttgart, Germany*



## 1 INTRODUCTION

Electrolytes are key components of lithium and sodium ion batteries, enabling ion transport in broad electrochemical windows, and finally determining cell's power density as well as long-term performance and stability. Soft matter electrolytes have been used since the very early stages of battery development due to the beneficial properties such as high room temperature ionic conductivities, good wettability of the nanostructured electrodes, flexibility, and stable solid electrolyte interphase (SEI). Unlike their inorganic competitors, soft matter electrolytes rarely suffer from large volume changes upon electrochemical cycling or great charge-transfer resistances at electrode/electrolyte interfaces. The major issues, on the other hand, lie in the noncompatibility with high power electrodes and large concentration gradients due to low lithium (sodium) transference numbers ($t_{Li(Na)} < 0.5$), dendrite formation in contact with metal electrodes as a consequence of moderately electronically conductive or cracked SEI, thermal instability, and potential toxicity.

In the course of many decades, the ubiquity of specific chemistries of the electrolyte components allowed for the preparation of a very broad range of soft matter electrolytes with different consistencies (liquid, gel and solid), ionic conduction mechanisms and surface properties. This chapter deals specifically with conventional salt-in-solvent electrolytes (including ionic liquid (IL)-based electrolytes), polymer-based electrolytes, and inorganic/organic (hybrid) composite electrolytes. A number of valuable book chapters and reviews are already available on the general physical and electrochemical properties of the abovementioned materials classes and their components for lithium and sodium ion batteries.[1–9] This chapter, on the other hand will try to provide some general considerations on ionic conductivities and lithium (sodium) transference numbers as the most relevant criteria for evaluating soft matter battery electrolytes, as well as very recent material developments (last 5–10 years of research) in this field. Whenever indispensable, relevant characterization techniques are discussed, in particular, in the light of electrolyte/electrode interface stability investigation. At the end, some pertinent issues in soft matter electrolytes for multivalent, Li–S and Li–$O_2$ batteries are addressed.

## 2 IONIC CONDUCTIVITIES AND LITHIUM (SODIUM) TRANSFERENCE NUMBERS

In the quest for finding a suitable electrolyte for battery technologies, one of the most relevant parameters

*Inorganic Battery Materials*. Edited by Hailiang Wang and Boniface P.T. Fokwa
 ISBN 978-1-119-43199-2

to consider is the ionic conductivity. In soft matter, the ionic conductivity is proportional to the concentration of all ionic species $c$, their charge $z$, and mobility $u$:

$$\sigma \approx \Sigma zcu \tag{1}$$

The problem of improvement of the ionic conductivity in soft matter is nontrivial, as it is often so that the increase in the number of available ionic species leads to the increase in shear viscosity $\eta$, and thus decrease of $u$. In addition, concentration of ionic species frequently cannot be increased by the simple increase of the total salt concentration $c_{salt}$, since then neutral nonconductive species may form.

Ionic conductivity is commonly determined by ac electrochemical impedance spectroscopy (EIS) in a symmetrical cell with blocking electrodes such as stainless steel following the Ohm's law for the steady state conditions:

$$\sigma = \frac{l}{R_b A} \tag{2}$$

where $l$ is the thickness of a thin electrolyte layer, $R_b$ is the bulk electrolyte resistance, and $A$ is the interfacial surface area.

In many cases, soft matter electrolytes display empirically determined exponential Vogel–Fulcher–Tamman dependency of ionic conductivity vs temperature:

$$\sigma(T) = \frac{A}{\sqrt{T}} \exp\left[-\frac{B}{T - T_0}\right] \tag{3}$$

where $A$ and $B$ are constants and $T_0$ is the reference temperature (typically related to the glass transition temperature). Deviations of the VFT behavior due to the nonlinearity of the molecular volume vs $T$ relationship have also been repeatedly reported.

For composite soft matter electrolytes with two ionically conductive phases and an ionically conductive interface, total conductivity may be defined by a parallel switching model as:

$$\sigma_m = X\varphi_1\sigma_1 + Y\varphi_2\sigma_2 + Z\varphi_{\text{interface}}\sigma_{\text{interface}} \tag{4}$$

$X$, $Y$, $Z$ being the percolation parameters with values $\leq 1$ (if $X = 1$, phase 1 is fully percolating), $\varphi_1$, $\varphi_2$, $\varphi_{\text{interface}}$ being the volume fractions of the three phases ($\varphi_1 + \varphi_2 + \varphi_3 = 1$), and $\sigma_1$, $\sigma_2$, $\sigma_{\text{interface}}$ their conductivities. The percolation parameters and volume fractions are fixed for a specific composite morphology. The dominant phase will be directing the dominant conductivity–temperature behavior—if for example an inorganic phase is dominant, conductivity response will be Arrhenius-like, rather than VFT-like. Equation (4) can be used in modeling maximum of ionic conductivity when the interfacial effect is negligible, or assist determining the interfacial effect when the conductivities, morphologies and percolation characteristics are known.

Ionic conductivity of specific charged species $i$ can be linked to the corresponding diffusion coefficients $D_i$ through the Nernst–Einstein equation:

$$\sigma_i = \frac{(zF)^2 c_i D_i}{N_A k_B T} \tag{5}$$

where $F$ is the Faraday constant, $k_B$ is the Bolzmann constant, $c_i$ is the concentration of specific charged ionic species, and $N_A$ is the Avogadro constant. The specific charged species concentration can be estimated from the infrared (IR) or Raman spectroscopy.

Diffusion coefficients are often determined by pulsed-field gradient nuclear magnetic resonance (PFG-NMR). This technique, however, processes the average Fickian diffusion of all species containing the specific probed nucleus, charged and uncharged. The mismatch between the tracer diffusion coefficient measured by PFG-NMR, $D_{\text{tracer}}$, and the diffusion coefficient calculated from the ionic conductivity when total ionic dissociation is assumed, $D_\sigma$ (Equation 3), the so-called Haven ratio, has been used extensively in the field of solid state ionic conductors:

$$H_R = \frac{D_{\text{tracer}}}{D_\sigma} \tag{6}$$

In highly associated electrolytes, Haven ration is higher than unity. In soft mater ionic conductors, in particular in ILs, an inverse term "ionicity" has been proposed for the ratio of molar ionic conductivities ($\Lambda = \sigma / c_{salt}$) measured by EIS and PFG-NMR.

Walden rule approach

$$\Lambda\eta = \text{const} \tag{7}$$

is another useful tool to estimate the temperature dependence of ion pairing in soft matter electrolytes. For the ideal case, in which the temperature influences the ionic conductivity through viscosity to which the ionic motion is fully coupled, the plot of log $\Lambda$ vs log $1/\eta$ should be a straight line of a unit slope. If the ion speciation changes with temperature and/or decoupling of conductivity from viscosity occurs, different slopes and positions will be observed.

Cell concentration gradients are strongly related to the ion transport number or transference number, defined as a fraction of the total current (or ionic conductivity) carried by the particular ionic species $i$:

$$t_i = \frac{I_i}{\sum I_i} = \frac{\sigma_i}{\sum \sigma_i} \tag{8}$$

This definition requires $t_i > 0$ and $\sum t_i = 1$.

A number of different techniques for determination of lithium (sodium) transference numbers have been proposed, including moving boundary method, low-frequency EIS measurement, and most widespread potentiostatic polarization method.[10–13] The moving boundary method was developed for liquid electrolytes and is based on the electrolysis cell with several compartments. The experimental issues arising from this method are the accuracy of analysis due to the negligible concentration changes, control of mixing between different compartments, and precise temperature control.

It is often neglected that low-frequency EIS and potentiostatic polarization method employ partially blocking lithium (or sodium) electrodes and thus take into account lithium (sodium) transport through ion pairs when they are present.[14–17] Indeed, lithium transference numbers, $t_{Li}$, measured by these methods are typically higher than the ones related to the lithium transport only through solvated lithium species, $t_+$, but are practically more relevant for battery application.

The low-frequency impedance spectroscopy relies on the possibility of observation of the low-frequency diffusion resistance with diameter $Z_d$ (Figure 1a). Lithium transference number can subsequently be calculated independently from the SEI formation as:

$$t_{Li} = \frac{1}{1 + \frac{Z_d(0)}{R_b}} \tag{9}$$

where $R_b$ being the bulk electrolyte resistance. The precision of the method is reduced in the cases of large SEI resistance and small low-frequency diffusion resistance.

Unlike the low frequency EIS method, potentiostatic dc polarization (Figure 1b) is strongly influenced by the SEI resistance and its possible growth since:

$$t_{Li} = \frac{I_\infty(\Delta V - I_0 R_{SEI,0})}{I_0(\Delta V - I_\infty R_{SEI,\infty})} \tag{10}$$

where $I_\infty$ is the steady-state current upon applying voltage $\Delta V$, $I_0$ is the initial current and $R_{SEI,0}$ and $R_{SEI,\infty}$ are the resistances of SEI at the beginning and the end of the potentiostatic polarization.

Typical experimental issues are the high reactivity of the electrolyte and thus shift of the SEI semicircle toward higher frequencies ($R_{b,\infty} > R_{b,0}$) and/or erroneous determination of the $I_0$ without taking into account Ohm's law ($I_0 = \Delta V / R_{tot,0}$). Galvanostatic polarization has been recently suggested as a possible improvement of the $I_0$ determination.[18]

## 3 SOFT MATTER LITHIUM AND SODIUM ELECTROLYTES

### 3.1 Salt-in-Solvent and Ionic-Liquid Based Electrolytes

In salt-in-solvent electrolytes, the ingredients are chosen according to their bulk physicochemical properties and compatibilities. Typically, aprotic polar solvents with low viscosity and low vapor pressure are coupled with highly acidic lithium (sodium) salts of good thermal stability to achieve dissociated electrolytes with high ionic conductivity (≈mS cm$^{-1}$ around 1 M salt concentrations), high chemical and electrochemical stability, and wide temperature operation range (−30 to 40 °C). Common solvents include cyclic carbonates (ethylene carbonate, EC, and its fluorinated version, FEC), linear carbonates (dimethylcarbonate, DMC), linear carboxylates (ethyl acetate, EA), alkoxy-nitriles (acetonitrile, AN), siloxanes and silanes, ethers (dimethyl ether, DME; methyl nanofluorobutyl ether, MFE), and organo-sulfur compounds (dimethyl sulfoxide, DMSO). Typical anions include perchlorate ($[ClO_4]^-$) hexafluorophosphate ($[PF_6]^-$), trifluoromethyl sulfonate ($[SO_3CF_3]^-$, Tf), bis(oxalate)

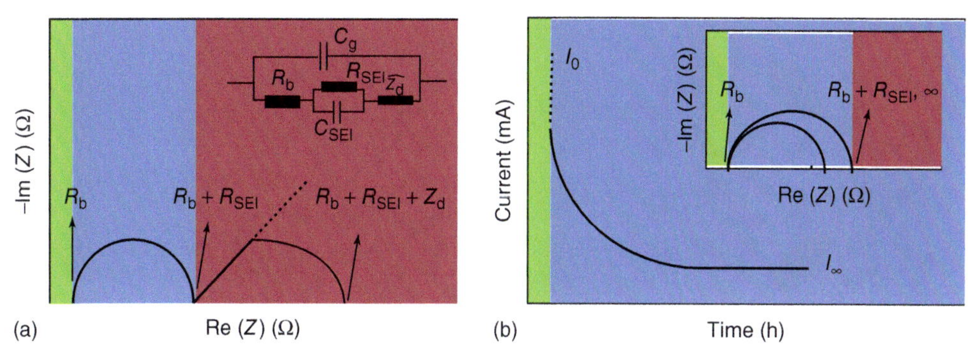


**Figure 1** Methods for determination of lithium transference numbers in Li/electrolyte/Li cells: (a) low-frequency electrochemical impedance spectroscopy (EIS) method (inset shows the corresponding equivalent circuit), (b) galvanostatic polarization linked with EIS. The dotted line in (b) shows a possible error in determination of the initial current. Different gray colors in (a) and (b) indicate regimes of the EIS/polarization experiments

borate ($[B(C_2O_4)_2]^-$), bis(trifluoromethylsulfonyl) imide $[(CF_3SO_2)_2N]^-$, TFSI) and bis(fluorosulfonyl) imide ($[(F_2SO_2)_2N]^-$, FSI).[19] Improved electrochemical performance is expected for bulky, less-mobile anions where ion-pairing is even sterically hindered.[20–23] Recent theoretical study suggests strong energetic, inter-ionic distance, and solubility differences in Li and Na-salt-in-solvent electrolytes.[24] For improved control of the electrode/electrolyte interfaces, additives such as vinylene carbonate ($C_3H_2O_3$, VC) or other salts have been proven effective.[25–27]

Electrolytes in which ILs are used as a solvent are a newer class of materials, mostly investigated due to the higher thermal stability, somewhat broader electrochemical window (up to 5 V vs $Li/Li^+$), and superior long-term performance. The major pitfall of such electrolytes is the complexity introduced by a number of different ionic species (at least three if the anions of the salt and IL match), high viscosity and thus lower ionic conductivity compared to the conventional salt-in-solvent electrolytes. In addition, they exhibit very low salt dissociation as a consequence of strong Coulomb interactions, negligible lithium (or sodium) transference numbers ($t_{Li(Na)} \approx 0.1$), and high price of production. Thoroughly investigated systems involve imidazolium, pyrrolidinium, phosphonium, and ammonium-based ILs, containing salts with anions such as FSI and TFSI.[28–31] It is important to note that the chemical purity and water content (even in the ppm range) of IL-based lithium (sodium) are of immense importance for their electrochemical properties as they can change the predominant transport mechanisms and the interfacial behavior.

Organic–ionic plastic crystals are room temperature solids consisting of cation–anion pairs, analogous to ILs.[32,33] In such materials, the ions have orientational or rotational disorder but remain in long-range order keeping the final crystalline structure. The ion transport mechanism is complex, yet not fully understood and dependent on disorder modes in matrix ions, formation of secondary phases, as well as on the presence of structural defects (vacancies, slip planes). The highly conductive plastic phase (so-called Phase I) in pyrrolidium, imidazolium, and ammonium-based organic ionic plastic crystals typically appears just below the melting point, making the majority of organic–plastic crystal materials interesting only for the high-temperature applications. Reduction of temperature at which Phase I appears to room temperature is possible with FSI, TFSI, fluorohydrogenate ($[(FH)_2F]^-$), or cyanate ($[OCN]^-$) anions.[34,35]

In the recent years, highly concentrated salt-in-solvent electrolytes (>4 M) are gaining importance, in particular, in combination with graphitic and metal anodes as they enable cosolvent intercalation and stable SEIs.[36–42] The physicochemical bulk properties of highly concentrated electrolytes are similar to the ones observed in ILs; glyme-based concentrated electrolytes are even called "solvate ILs".[43] Advantages of highly concentrated salt-in-solvent electrolytes include increased oxidative/reductive stability, high carrier density, and low volatility. The solution structure consisting predominantly of ion pairs and aggregates (Figure 2) is governed by anion's charge distribution and structure, whereas the liquid range is ruled by the thermodynamic stability of the clusters and the solvent structure. So far, the compositions studied in detail include LiFSA in DME or EC/DMC and LiTFSA in triglyme or water. In order to obtain liquid state at high salt concentrations without salt or salt–solvent complex crystallization, phase diagrams of salt–solvents are of great importance. The pitfalls of highly concentrated salt-in-solvent electrolytes remain relatively low room temperature ionic conductivity, low lithium transference numbers and high material costs.

### 3.2 Polymer-Based Electrolytes

The term polymer-based electrolyte is used to describe systems comprising a salt dissolved in high molecular weight polar polymer matrix (salt-in-polymer),

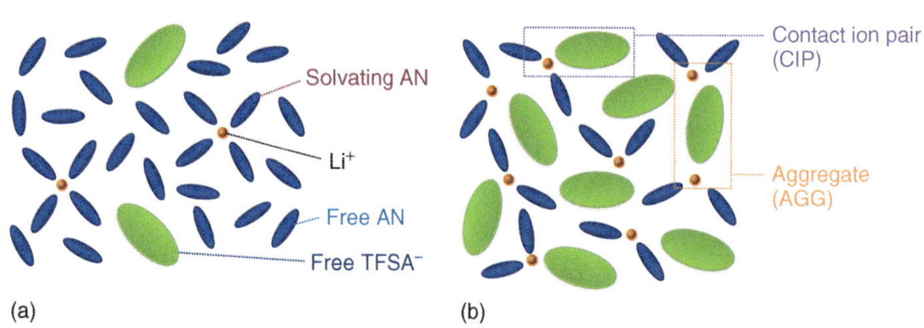


**Figure 2** Solution structure in dilute (a) and highly concentrated (b) salt-in-solvent LiTFSA/AN electrolytes. In highly concentrated electrolytes, a network of mutually interacting $TFSA^-$ anions and $Li^+$ cations in the presence of two AN molecules solvating each $Li^+$ forms. [Reprinted with permission from Zhao, C.-Z., Zhang, X.-Q., Cheng, X.-B., Zhang, R., Xu, R., Chen, P.-Y., Peng, H.-J., Huang, J.-Q., Zhang, Q. (2017) Proc. Natl. Acad. Sci. U.S.A., 114 (42) 11069. © 2014 American Chemical Society]

mixtures of salt/IL/polymer or salt/solvent/polymer (gel polymer electrolytes), and polymerized ILs.

First, salt-in-polymer electrolytes were based on polyethers, polyesters, polyimines, and polythioethers containing a Lewis base that can solvate alkali ion in crown ether like complexes. Of all the accessible chemistries, poly(ethylene oxide) (PEO) proved to be most chemically and electrochemically stable, a consequence of strong unstrained C—O, C—C, and C—H bonds. Although versatile and with very good mechanical properties, lithium salt containing PEO is predominantly an anionic conductor ($t_{\mathrm{Li(Na)}} < 0.4$) with low room temperature ionic conductivity ($\approx 10^{-6}\,\mathrm{S\,cm^{-1}}$), and high ionic association limiting its application to elevated temperatures (above 70 °C) and conventional electrodes. Contrary to the pure Fickian diffusion in salt-in-solvent electrolytes, the complex polymer chain motion additionally assists the ionic motion in salt-in-polymer electrolytes (Figure 3).

With the mechanism of lithium conduction in mind, more flexible polymer backbone and low glass transition temperature, $T_g$, are beneficial. Early research focused mainly on the investigation of highly conductive amorphous phases above $T_g$. However, high lithium conductivity is achievable in crystalline phases of PEO as well; PEO chains may form ordered cylindrical tunnels where lithium ions freely move.[45] Additional way of boosting the ionic conductivity is by decoupling of structural and conductivity relaxation, in particular in highly concentrated polymer-in-salt electrolytes, or by ordering of conductive phases through mechanical deformation (e.g., stretching), magnetic or electric fields.[46–48] In highly concentrated polymer-in-salt electrolytes, the suggested conduction pathway is by ion hopping between different percolating salt clusters. It is important to note that the ion transport mechanisms in salt-in-polymer electrolytes are complexly dependent on the polymer structure, type, phases present, and the molecular weight.

Nowadays, research is mainly focused on finding purely lithium ion polymeric conductors with high ionic conductivities (goal above $10^{-4}\,\mathrm{S\,cm^{-1}}$ at room temperature) by linking the anions to the polymer backbone or by addition of the anion trapping agents. Typical anion receptor is a Lewis base able to accept an electron pair, such as cyclic boron-based compounds, calix[4]arene, or calix[*n*]pyrroles. Linking of the anions to the polymer backbone is successful in block copolymers, where A is the pure lithium conducting block and B is providing the necessary mechanical strength as in poly(styrene trifluoromethanesulphonylimide of lithium), P(STFSILi), linked with PEO in P(STFSILi)–PEO–P(STFSILi).[49] Depending on the compatibility of the blocks, temperature and salt concentration, block copolymers might offer beneficial self-assembled morphologies in the <100 nm range.

In gel polymer electrolytes, a number of different polymer matrices such as PEO, polyvinyl chloride (PVC), polyacrylonitrile (PAN), poly(methyl methacrylate) (PMMA), polyvinylidene fluoride (PVDF), and poly(propylene carbonate maleate) (PPCMA) are extensively used.[50,51] If the final mixture with the salt-in-solvent electrolytes is heterogeneous (typically prepared in a two-step procedure—preparation of the porous membrane and infiltration), the liquid is the ionically conductive phase. Gel polymer electrolytes have electrochemical properties similar to the conventional salt-in-solvent electrolytes, slightly better mechanical properties, and comparable pitfalls.

Polymerized ILs with controlled chain length and morphology are easily prepared either by free radical polymerization of the IL monomers or through post-polymerization processing.[52–54] The ionic conductivity

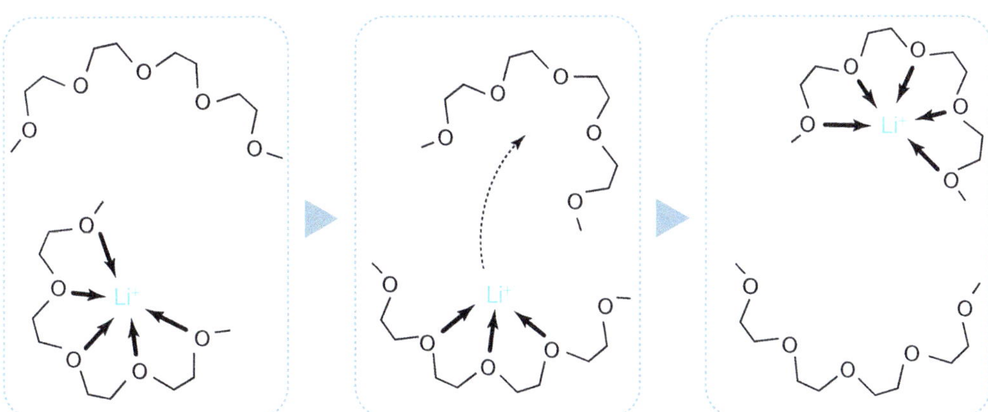

**Figure 3** Lithium conduction mechanism in lithium salt containing PEO polymer electrolyte. The polymer chain motion assists the motion of the $Li^+$ ion solvated in crown ether like complexes. [Reproduced with permission from Sorensen, P. S., Jacobsen, T. (1982) Electrochem. Acta, 27, 1671. © The Royal Society of Chemistry, 1982]

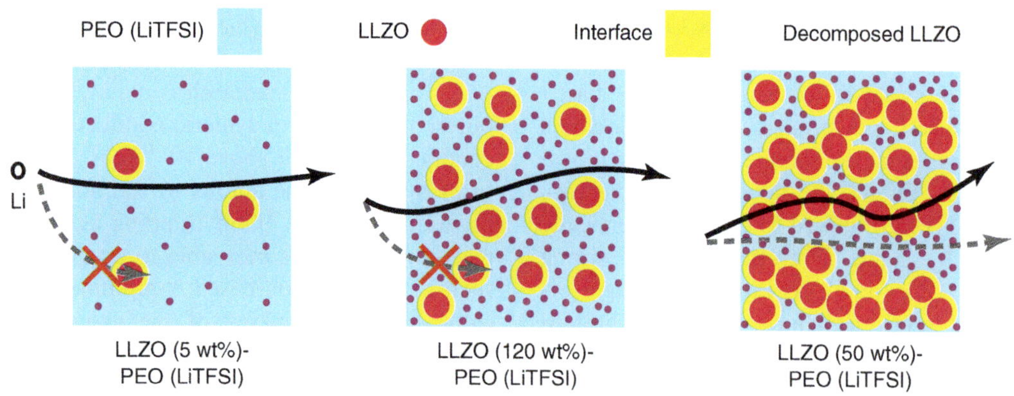


**Figure 4** Lithium ion pathways in $Li_7La_3Zr_2O_{12}$ (LLZO)/(LiTFSI + PEO) composite material at different volume fractions of LLZO. In this case, inorganic and organic phase have similar ionic conductivity. At a critical LLZO wt%, upon the beneficial particle network formation, the predominant $Li^+$ conduction pathway is through the inorganic phase. [Reprinted with permission from Wang, H., Bai, Y., Luo, X., Wu, C., Wu, F., Lu, J., Amine, K. (2015) ACS App. Mater. Interfaces, 7 (1), 80. © 2018 American Chemical Society]

of polymerized ILs is reduced compared to the monomers that constitute them, a consequence of the reduced ionic mobility. Thus, polymerized ILs reach satisfactory performance only in their gel form, when swollen with salt-in-solvent electrolytes.

### 3.3 Inorganic/Organic Composite Electrolytes

Inorganic/organic (hybrid) composite electrolytes are a versatile class of materials including ceramic-in-polymer electrolytes, "soggy sand" electrolytes (oxide particles dispersed in salt-in-solvent electrolyte), and liquid/solid composite electrolytes. Although structurally similar, in the first case, beneficial electrochemical properties are achieved by creation of highly conductive amorphous regions inside the polymer matrix, while in the other cases, a beneficial interfacial effect through anionic adsorption on the particle surface, and possibly a highly lithium conductive electrical double layer or a space charge zone are aimed at. If the fillers are particles, it has generally been observed that the smaller sizes are favorable, as they offer higher surface area and thus more ion adsorption sites; they form networks that are more suitable and do not sediment. Inorganic/organic electrolytes offer liquid-like wettability, flexibility, make the separator use dispensable, and give high possibility of blocking dendrite penetration in alkali metal batteries.

Some of the studied ceramic-in-polymer electrolytes include addition of up to 10 wt% of insulating fillers such as $Al_2O_3$, $SiO_2$, $TiO_2$ or $BaTiO_3$ in PEO/$LiClO_4$, or PEO/$LiSO_3CF_3$, or even higher amounts of the ionically conductive fillers such as $Li_3N$, $LiAlO_2$, $Li_{0.35}La_{0.55}TiO_3$, $Li_{1-x}Al_xTi_{1-x}(PO_4)_3$, $Li_{1.4}Al_{0.4}Ge_{1.6}(PO_4)_3$ or $Li_{6.4}La_3Zr_{1.4}Ta_{0.6}O_{12}$ in well-known salt-in-polymer electrolytes or even nonconductive polymers.[44,55–61] The ordering of the phases is a key challenge in ceramic-in-polymer electrolytes, and bicontinuous 3D networks of filler particles are necessary (Figure 4). In addition, reports on the improvement of the stability vs lithium electrode in ceramic-in-polymer electrolytes are not rare. This is most probably a result of trapping of the residual impurities (most dominantly water) on the surface of the filler particles.

"Soggy sand" effects resulting in high ionic conductivities ($\approx$mS cm$^{-1}$) and satisfactory lithium transference numbers ($t_{Li} \approx 0.6$) are observed in up to 5 vol% $SiO_2$ glyme electrolytes since they contain low dielectric constant solvents (thus large number of ion pairs), and the particle networks formed are percolating.[63–65] Interestingly, lithium ions can additionally be transported inside the pores, when mesoporous fillers such as MSU-H silica are used.[66] Pure lithium conductivity is expected to be achieved at high particle volume fractions when only the electrochemical double layer contributes to the ionic conduction mechanism. This is however, experimentally hardly achievable as the mixtures become solid above 10 vol% of particles. As in other electrolytes with fillers, percolating network formation is crucial, hard to tune and effects such as lack of reproducibility, coarsening/sedimentation are to be expected.

Liquid/solid electrolytes can be considered an extension of the "soggy sand" idea to the monolithic materials infiltrated with a salt-in-solvent electrolyte. The advantage lies in achieving high surface area without the particle network formation issue, possibility of preparing materials with higher volume fraction of solid, and the electrolyte rigidity. If the radius of the pores is proportional to the Debye length of the salt-in-solvent electrolyte, lithium transference number can even approach unity at lower solid volume fraction.[67] Materials with high electrochemical performance were obtained by infiltration of both porous $SiO_2$ and $Al_2O_3$.[18,68–70]

## 4 STABILITY OF THE ELECTROLYTE/ELECTRODE INTERFACES

Understanding of the electrolyte/electrode interfaces is crucial for developing safe batteries with good cycling performances. Since the introduction of the SEI concept in the end of 1970s, excessive amount of work has been done in identifying the morphology and chemical composition of its structure forming between electrodes and soft matter electrolytes, mainly by using combination of techniques such as EIS, X-ray photoelectron spectroscopy (XPS), time of flight secondary ion mass spectrometry (TOF-SIMS), IR or Raman spectroscopy, scanning electron microscopy (SEM) or transmission electron microscopy (TEM), and atomic force microscopy (AFM). Most important limitations of the aforementioned techniques are low vibrational signals if the SEI is too thin (IR, Raman), similar functional groups found in many SEI components (IR, Raman), risk of radiation damage (XPS, SEM, TEM), nontrivial data analysis (IR, XPS), difficulties with sample preparations and handling. Numerous recent studies focus on in situ/in operando determination of the SEI evolution, some of them involving auger electron spectroscopy, sum frequency generation vibrational spectroscopy, X-ray reflectivity, and neutron reflectometry.[71–74]

One of the prevailing models of SEI in salt-in-solvent lithium electrolytes is the "mosaic" structure or polyhetero microphase, in which SEI consists of a compact inorganic layer in the vicinity of the electrode made of components such as $Li_2O$, $Li_2CO_3$, and LiF, and more organic compounds (semicarbonates, dicarbonates, polyolefins) in the porous layers closer to the electrolyte (Figure 5).[75,76] The chemical composition of the SEI is highly dependent on the electrolyte constituents (including additives) and at

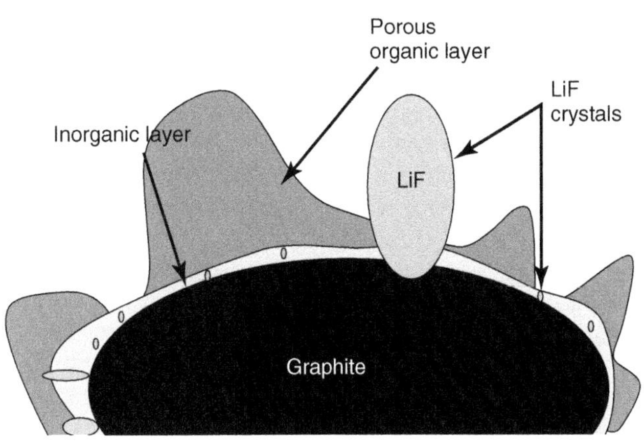


**Figure 5** Solid electrolyte interphase on graphite electrode in salt-in-solvent electrolytes. [Reprinted with permission from Edström, K., Herstedt, M., Abraham, D. P. (2006) J. Power Sources, 153 (2), 380. © Elsevier, 2006]

salt concentrations above 1 M probably more dependent on the anion than solvent chemistry.

The goal is controlled formation of thin, homogeneous, dense, flexible, electronically insulating and ionically conducting SEI containing no metastable phases. The lithium transport inside the SEI is complexly dependent on the ionic conductivity and the spatial distribution of its phases. The stability and rigidity of the SEI is on the other hand linked to the potential electronic conductivity (current leakage) and its mechanical properties. The chemical structure of SEI in polymer electrolytes is similar to the one occurring in salt-in-solvent electrolytes; however, the electrode/electrolyte contacts are poorer and growing voids can be created, in particular on lithium electrodes.

## 5 SPECIFIC ELECTROLYTE ISSUES IN Li–S AND Li–$O_2$ BATTERIES

Lithium–sulfur batteries offer very high theoretical specific capacities (~2600 W h kg$^{-1}$), but their performance is often severely limited by the electrolyte degradation. The reduction of sulfur on the cathode side produces a number of polysulfide species ($L_2S_n$), soluble in most of the common salt-in-solvent electrolytes. The dissolved species such as $S_n^{2-}$, $n = 3–9$, can subsequently migrate and react irreversibly with the metallic lithium consuming the electrolyte and causing increased resistance, or reversibly, inducing the so-called shuttle-effect, lowering the Coulombic efficiency and triggering a continuous loss of the active cathode material. Another important issue in Li–S cells is gas evolution (mainly $H_2$ and $CH_4$) as a result of decomposition of typically used DME, dioxolane, and glyme-based electrolytes and the lithium salts (Figure 6).[78] A fair amount of research has been performed in order to find a lithium anode polysulfide protection agent including $LiNO_3$ or LI additives, $Al_2O_3$ deposition, grafting of lithium with tween polymer or its treatment with elastomers.[78–80] On the other hand, large body of research has been also focused on electrolyte formulation in which polysulfide species are insoluble such as ILs, solvent-in-salt, triflamide-based, polymer, and hybrid electrolytes.[81–85]

Reaction mechanism in Li–$O_2$ batteries is complex and at discharge involves oxygen reduction in an oxygen radical, $O_2^-$, formation of metastable $LiO_2$, subsequently resulting in $Li_2O_2$. The oxygen radical is a highly reactive species that often consumes the electrolyte in a number of parasitic reactions through a nucleophilic attack of the $O_2^-$ on the carbon atom of the solvent molecule or on the hydrogen atom via H-abstraction (Figure 7).[81] Low stability of the electrolyte towards the nucleophilic attack is expected when the formation energy of the [solvent-$O_2^-$]$^-$ complex and its reaction barrier are low, for example, in carbonate-based salt-in-solvent electrolytes. Here, instead

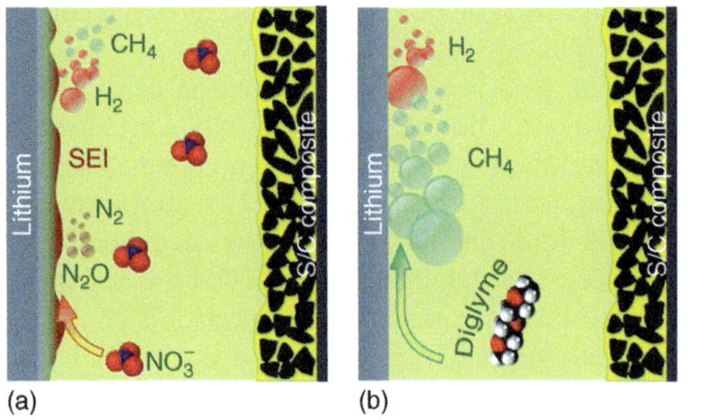


**Figure 6** Sketch of reaction pathways for gas evolution in Li–S batteries containing diglyme electrolyte with (a) and without (b) $LiNO_3$ additive. When $LiNO_3$ is present, formed $LiNO_2$ reacts with sulfur to produce $N_2O$ gas. $N_2O$ may subsequently react with lithium to form $N_2$ gas. In pristine diglyme electrolyte, methyl radicals may react with the electrolyte to form $CH_4$. [Reprinted with permission from Jozwiuk, A. Berkes, B. B., Weiß, T., Sommer, H., Janek, J., Brezesinski, T (2016) Energy Environ. Sci., 9, 2603. © Royal Society of Chemistry, 2016]

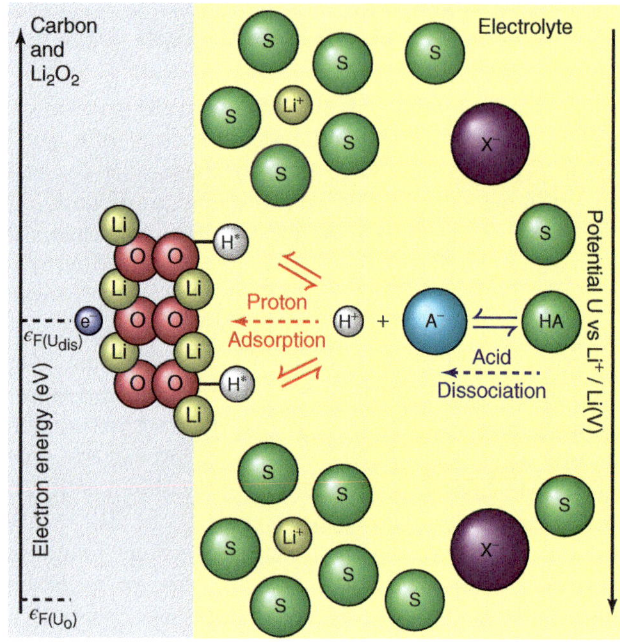


**Figure 7** H-abstraction from nonaqueous solvents in Li–$O_2$ batteries induced by H adsorption at the $Li_2O_2$/carbon cathode interface. The solvent molecule, S, is denoted as a weak acid, HA, which is in equilibrium with its conjugate base, $A^-$, and proton, $H^+$. The protons are in equilibrium with the $Li_2O_2$/carbon cathode interface where they can be adsorbed at multiple sites. [Reprinted with permission from Khetan, A., Luntz, A., Viswanathan, V. (2015) J. Phys. Chem. Lett., 6, 1254. © 2015 American Chemical Society]

of pure $Li_2O_2$, byproducts such as $Li_2CO_3$, LiOH, and $HCO_2Li$ are formed. As a consequence, ether, sulfone, and amide-based electrolytes have been more successfully employed in Li–$O_2$ cells.[87–91] The H-abstraction can be suppressed by keeping the initial $O_2^-$ concentration low in a solvent with high acidity and low donor/acceptor number.[86,92]

Recently, it has been shown that singlet oxygen, $^1O_2$, forming in the cathode during discharge and from the onset charge, rather than oxygen radical, might be responsible for the majority of the byproducts found in Li–$O_2$ batteries.[93] Contrary to what has been observed in other battery electrolytes, addition of small amounts of water enhances the performance in Li–$O_2$ batteries possibly by enhancing the initial oxygen reduction reaction. Polymer electrolytes based on PEO are a poor choice in Li–$O_2$ cells since they undergo auto-oxidation. A well-performing class of materials in Li–$O_2$ cells are gel electrolytes.[94,95]

## 6 ELECTROLYTES FOR MULTIVALENT BATTERIES

Multivalent batteries based on $Mg^{2+}$, $Ca^{2+}$, and $Al^{3+}$ ions give promise of higher energy densities, lower cost, and enhanced safety. However, the challenges in finding suitable electrolyte chemistries are higher in multivalent batteries owing to the phenomena such as low solubility of multivalent salts, low ion mobility in the electrolyte bulk, larger desolvation energy at the electrode interface, and ion-blocking surface films on metallic anodes.

In magnesium batteries, the main request is an electrolyte stable against reduction with high-enough operational voltage, and good compatibility with possible cathode materials. Some of the initially successful mixtures include organohaloaluminate, magnesium and/or aluminum dichloro complex, and phenyl complex Grignard-based electrolytes.[96,97] However, these materials still suffer from low oxidative stability, reproducibility issues (related with complex cycling history of Mg electrode), and potential corrosion of the metallic battery components. Although it has been long believed that standard salt-in-solvent electrolytes are not suitable in magnesium batteries, it was recently demonstrated that $Mg(TFSI)_2$ in glyme electrolyte does undergo magnesium deposition and dissolution if the metal electrode is mechanically and chemically pre-treated.[98] Other salts showing good promise are borohydrates ($Mg(BH_4)_2$) and dodecaborates ($MgB_{12}H_{12}$).[99,100]

The electrochemical performance of calcium batteries is largely dependent on the surface films present on the calcium electrode. In most solvents, $Ca^{2+}$ mobility is very poor in such surface films and calcium dissolution requires high overvoltages and a breakdown of such an SEI. Presently, very few calcium salts are commercially available with high purity (e.g., $Ca(ClO_4)$, $Ca(TFSI)_2$, $CaTf_2$, $Ca(NO_3)_2$), and the synthesis of new salts still remains an important topic. Reversible stripping/platting behavior has been observed in $Ca(BF_4)_2$ in EC/PC-based electrolytes at 100 °C, and $Ca(BH_4)_2$ in tetrahydrofuran at room temperature.[101,102] Importantly, it has been observed that the calcium (and magnesium) electrodes show large open circuit potential drifts as a consequence of high impedance of the electrodes, needed to be taken into account when interpreting the electrochemical behavior.

Major advantage of aluminum vs other multivalent battery systems is its ease of handling under ambient conditions enhancing the safety, coupled with very large volumetric/gravimetric capacity. First electrolytes used for secondary aluminum cells were based on different chloroaluminate melts and molten salt chemistries. Recently, IL electrolytes containing $AlCl_3$ have gained importance as they offer rapid solvation/desolvation of negatively charged aluminum species involved in intercalation, for example, $AlCl_4^-$, $Al_2Cl_7^-$.[62,103,104] Similar species enabling reversible stripping/plating of aluminum have been found in the solutions of $AlCl_3$ in glymes and dialkylsulfones. An alternative electrolyte based on solvated $Al^{3+}$ species is $AlTf_3$ in a mixture of *N*-methylacetamide and urea.

## 7 CONCLUSIONS AND OUTLOOK

The ionic conductivity of soft matter electrolytes is directly connected to the mechanism of ionic conduction dependent on the materials' class (Figure 8). For good battery performance, materials with ionic conductivity ≈ S $cm^{-1}$, predominant lithium (sodium) ion conductivity ($t_{Li} > 0.6$), wide electrochemical window (up to 5 V) and stable SEI (in particular on metal anodes) are aimed at. Soft matter electrolytes are currently still more suitable for battery application than their inorganic counterparts often exhibiting high grain boundary and charge transfer resistances, as well as poor electrode wetting properties.

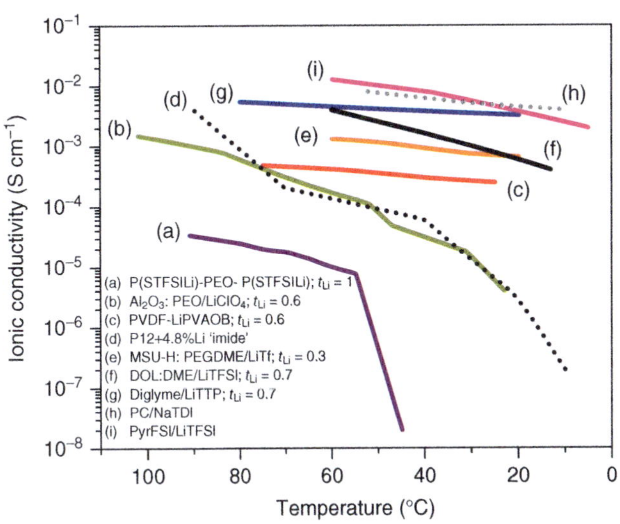


**Figure 8** Comparison of temperature-dependent ionic conductivities of some soft matter lithium electrolytes. The values measured by electrochemical impedance spectroscopy are taken from Refs. 23, 30, 32, 39, 49, 51, 56, 66, 105. Room temperature lithium transference numbers are given when available

With the development of new battery technologies (Li–S, Li–$O_2$, or multivalent systems), soft matter electrolytes gained even more importance as they are the most crucial component determining the future of a certain cell type. Forthcoming studies on soft matter electrolyte chemistries should thus involve deep understanding of the species relevant for the ionic transport as well as kinetics of the interfacial reactions determining the electrolyte/electrode interphases. Increase in the complexity of chemistries used also raises the need for more sophisticated in situ/operando techniques for determination of concentration gradients, lithium (sodium) ion local environments and ionic conductivity mechanisms directly at the interfaces.

## 8 ABBREVIATIONS AND ACRONYMS

AFM = atomic force microscopy; AN = acetonitrile; DMC = dimethylcarbonate; DME = dimethyl ether; DMSO = dimethyl sulfoxide; EA = ethyl acetate; EIS = electrochemical impedance spectroscopy; IL = ionic liquid; IR = infrared; MFE = methyl nanofluorobutyl ether; P(STFSILi) = poly(styrene trifluoromethanesulphonylimide of lithium); PAN = polyacrylonitrile; PEO = poly(ethylene oxide); PFG-NMR = pulsed-field gradient nuclear magnetic resonance; PMMA = poly(methyl methacrylate); PPCMA = poly(propylene carbonate maleate); PVC = polyvinyl chloride; PVDF = polyvinylidene fluoride; SEI = solid electrolyte interphase; SEM = scanning electron microscopy; TEM = transmission electron microscopy; TOF-SIMS = time of flight secondary ion mass spectrometry; VC = vinylene carbonate; XPS = X-ray photoelectron spectroscopy

## 9 REFERENCES


1. K. Xu, *Chem. Rev.*, 2014, **114**(23), 11503.
2. M. Watanabe, M. L. Thomas, S. Zhang, K. Ueno, T. Yasuda and K. Dokko, *Chem. Rev.*, 2017, **117**, 7190.
3. Q. Yang, Z. Zhang, X.-G. Sun, Y.-S. Hu, H. Xing and S. Dai, *Chem. Soc. Rev.*, 2018, **47**, 2020.

4. M. Keller, A. Varzi and S. Passerini, *J. Power Sources*, 2018, **392**, 206.

5. A. Varzi, R. Raccichini, S. Passerini and B. Scrosati, *J. Mater. Chem. A*, 2016, **4**, 17251.

6. D. Golodnitsky, E. Strauss, E. Peled and S. Greenbaum, *J. Electrochem. Soc.*, 2015, **162**(14), A2551.

7. J. Mindemark, M. J. Lacey, T. Bowden and D. Brandell, *Prog. Polym. Sci.*, 2018, **81**, 114.

8. H. Zhang, C. Li, M. Piszcz, E. Coya, T. Rojo, L. M. Rodriguez-Martinez, M. Armand and Z. Zhou, *Chem. Soc. Rev.*, 2017, **46**, 797.

9. A. Ponrouch, D. Monti, A. Boschin, B. Steen, P. Johansson and M. R. Palacin, *J. Mater. Chem. A*, 2015, **3**, 22.

10. D. A. MacInnes and L. G. Longsworth, *Chem. Rev.*, 1932, **11**(2), 171.

11. P. S. Sorensen and T. Jacobsen, *Electrochem. Acta*, 1982, **27**, 1671.

12. J. Evans, C. A. Vincent and P. G. Bruce, *Polymer*, 1987, **28**(13), 2324.

13. M. Watanabe, S. Nagano, K. Sanui and N. Ogata, *Solid State Ion.*, 1988, **28–30**, 911.

14. P. G. Bruce and C. A. Vincent, *Faraday Discuss. Chem. Soc.*, 1989, **88**, 43.

15. J. Maier, *Adv. Funct. Mater.*, 2011, **21**(8), 1448.

16. J. Maier, *Electrochim. Acta*, 2014, **129**, 21.

17. J. Popovic, C. Pfaffenhuber, J. P. Melchior and J. Maier, *Electrochem. Commun.*, 2015, **60**, 195.

18. J. Popovic, G. Hasegawa, I. Moudrakovski and J. Maier, *J. Mater. Chem. A*, 2016, **4**(19), 7135.

19. R. Younesi, G. M. Veith, P. Johansson, K. Edström and T. Vegge, *Energy Environ. Sci.*, 2015, **8**, 1905.

20. C. Liao, K. S. Han, L. Baggetto, D. A. Hillesheim, R. Custelcean, E.-S. Lee, B. Guo, Z. Bi, D. Jiang, G. M. Veith, E. W. Hagaman, G. B. Brown, C. Bridges, M. P. Paranthaman, A. Manthiram, S. Dai and X.-G. Sun, *Adv. Energy Mater.*, 2014, **4**(6), 1301368.

21. L. Qia, Z. Cui, B. Chen, G. Xu, Z. Zhang, J. Ma, H. Du, X. Liu, S. Huang, K. Tang, S. Dong, X. Zhou and G. Cui, *Chem. Sci.*, 2018, **9**, 3451.

22. J. Popovic, D. Höfler, J. P. Melchior, B. List and J. Maier, *J. Phys. Chem. Lett.*, 2018, **9**, 5116.

23. S. Chen, J. Ishii, S. Horiuchi, M. Yoshizawa-Fujita and E. I. Izgorodina, *Phys. Chem. Chem. Phys.*, 2017, **19**, 17366.

24. D. Aurbach, K. Gamolsky, B. Markovsky, Y. Gofer, M. Schmidt and U. Heider, *Electrochem. Acta*, 2002, **47**(9), 1423.

25. K. Sato, L. Zhao, S. Okada and J. Yamaki, *J. Power Sources*, 2011, **196**, 5617.

26. M. C. Smart, B. L. Lucht, S. Dalavi, F. C. Krause and B. V. Ratnakumar, *J. Electrochem. Soc.*, 2012, **159**(6), 739.

27. A. Fernicola, F. Croce, B. Scrosati, T. Watanabe and H. Ohno, *J. Power Sources*, 2007, **174**(1), 342.

28. J.-K. Kim, A. Matic, J.-H. Ahn and P. Jacobsson, *J. Power Sources*, 2010, **195**(22), 7639.

29. G. A. Elia, U. Ulissi, S. Jeong, S. Passerini and J. Hassoun, *Energy Environ. Sci.*, 2016, **9**, 3210.

30. N. Salem, S. Zavorine, D. Nucciarone, K. Whitbread, M. Moser and Y. Abu-Lebdeh, *J. Electrochem. Soc.*, 2017, **164**(8), H5202.

31. D. R. MacFarlane, J. H. Huang and M. Forsyth, *Nature*, 1999, **402**, 792.

32. D. R. MacFarlane and M. Forsyth, *Adv. Mater.*, 2001, **13**(12–13), 957.

33. M. Forsyth, T. Chimdi, A. Seeber, D. Gunzelmann and P. C. Howlett, *J. Mater. Chem. A*, 2014, **2**, 3993.

34. L. Jin, P. C. Howlett, J. M. Pringle, J. Janikowski, M. Armand, D. R. MacFarlane and M. Forsyth, *Energy Environ. Sci.*, 2014, **7**, 3352.

35. Y. Yamada, K. Furukawa, K. Sodeyama, K. Kikuchi, M. Yaegashi, Y. Tateyama and A. Yamada, *J. Am. Chem. Soc.*, 2014, **136**(13), 5039.

36. L. Suo, O. Borodin, T. Gao, M. Olguin, J. Ho, X. Fan, C. Luo, C. Wang and K. Xu, *Science*, 2015, **350**(6263), 938.

37. J. Wang, Y. Yamada, K. Sodeyama, C. H. Chiang, Y. Tateyama and A. Yamada, *Nat. Commun.*, 2016, **7**, 12032.

38. L. Suo, Y.-S. Hu, H. Li, M. Armand and L. Chen, *Nat. Commun.*, 2013, **4**, 1481.

39. L. Suo, O. Borodin, Y. S. Wang, X. H. Rong, W. Sun, X. L. Fan, S. Y. Xu, M. A. Schroeder, A. V. Cresce, F. Wang, C. Y. Yang, Y.-S. Hu, K. Xu and C. S. Wang, *Adv. Energy Mater.*, 2017, **7**(21), 1701189.

40. T. Shiga, C. Okuda, Y. Kato and H. Kondo, *J. Phys. Chem. C*, 2018, **122**, 9738.

41. S. R. Chen, J. M. Zheng, D. H. Mei, K. S. Han, M. H. Engelhard, W. G. Zhao, W. Xu, J. Liu and J.-G. Zhang, *Adv. Mater.*, 2018, **30**(21), 1706102.

42. K. Ueno, K. Yoshida, M. Tuschuya, N. Tachikawa, K. Dokko and M. Watanabe, *J. Phys. Chem. B*, 2012, **116**(32), 11323.

43. Z. Gadjourova, Y. G. Andreev, D. P. Tunstall and P. G. Bruce, *Nature*, 2001, **412**, 520.

44. C.-Z. Zhao, X.-Q. Zhang, X.-B. Cheng, R. Zhang, R. Xu, P.-Y. Chen, H.-J. Peng, J.-Q. Huang and Q. Zhang, *Proc. Natl. Acad. Sci. USA*, 2017, **114**(42), 11069.

45. Z. Stoeva, I. Martin-Litas, E. Staunton, Y. G. Andreev and P. G. Bruce, *J. Am. Chem. Soc.*, 2003, **125**, 4619.

46. C. A. Angell, C. Liu and E. Sanchez, *Nature*, 1993, **365**, 137.

47. E. Livshits, R. Kovarsky, N. Lavie, Y. Hayashi, D. Golodnitsky and E. Peled, *Electrochem. Acta*, 2005, **50**(19), 3805.

48. R. Bouchet, S. Maria, A. Aboulaich, L. Lienafa, J.-P. Bonnet, T. N. T. Phan, D. Bertin, D. Gigmes, D. Devaux, R. Denoyel and M. Armand, *Nat. Mater.*, 2013, **12**, 452.
49. Y.-S. Lee, J. H. Lee, J.-A. Choi, W. Y. Yoon and D.-W. Kim, *Adv. Funct. Mater.*, 2013, **23**(8), 1019.
50. Y. Zhu, S. Xiao, Y. Yang, Y. Hou and Y. Wu, *Adv. Energy Mater.*, 2014, **4**(1), 1300647.
51. W. Huang, Z. Zhu, L. Wang, S. Wang, H. Li, Z. Tao, J. Shi, L. Guan and J. Chen, *Angew. Chem.*, 2013, **52**(35), 9162.
52. M. A. B. H. Susan, T. Kaneko, A. Noda and M. Watanabe, *J. Am. Chem. Soc.*, 2005, **127**(13), 4976.
53. A.-L. Pont, R. Marcilla, I. De Meatza, H. Grande and D. Mecerreyes, *J. Power Sources*, 2009, **188**(2), 558.
54. J. E. Weston and B. C. H. Steele, *Solid State Ion.*, 1982, **7**(1), 75.
55. F. Croce, G. B. Appetecchi, L. Persi and B. Scrosati, *Nature*, 1998, **394**, 456.
56. Z. Tu, Y. Kambe, Y. Lu and L. A. Archer, *Adv. Energy Mater.*, 2013, **4**(2), 1300654.
57. J. Bae, Y. Li, J. Zhang, X. Zhou, F. Zhao, Y. Shi, J. B. Goodenough and G. Yu, *Angew. Chem.*, 2018, **130**(8), 2118.
58. S. Zekoll, C. Marriner-Edwards, A. K. Ola Hekselman, J. Kasemchainan, C. Kuss, D. E. J. Armstrong, D. Cai, R. J. Wallace, F. H. Richter, J. H. J. Thijssen and P. G. Bruce, *Energy Environ. Sci.*, 2018, **11**, 185.
59. D. Lin, P. Y. Yuen, Y. Liu, W. Liu, N. Liu, R. H. Dauskardt and Y. Cui, *Adv. Mater.*, 2018, **30**(32), 1802661.
60. J. Zheng and Y.-Y. Hu, *ACS Appl. Mater. Interfaces*, 2018, **10**(4), 4113.
61. A. Bhattacharyya and J. Maier, *Adv. Mater.*, 2004, **16**(9–10), 811.
62. H. Wang, Y. Bai, X. Luo, C. Wu, F. Wu, J. Lu and K. Amine, *ACS Appl. Mater. Interfaces*, 2015, **7**(1), 80.
63. A. Jarosik, C. Pfaffenhuber, A. Bunde and J. Maier, *Adv. Funct. Mater.*, 2011, **21**, 3961.
64. A. Bhattacharyya, *J. Phys. Chem. Lett.*, 2012, **3**(6), 744.
65. C. Pfaffenhuber, M. Göbel, J. Popovic and J. Maier, *Phys. Chem. Chem. Phys.*, 2013, **15**, 18318.
66. C. Pfaffenhuber, F. Hoffmann, M. Fröba, J. Popovic and J. Maier, *J. Mater. Chem. A*, 2013, **1**, 12560.
67. J. Jorne, *Nano Lett.*, 2006, **6**(12), 2973.
68. Y. Lu, Z. Tu and L. A. Archer, *Nat. Mater.*, 2014, **13**, 961.
69. J. Zhang, Y. Bai, X.-G. Sun, Y. Li, B. Guo, J. Chen, G. M. Veith, D. K. Hensley, M. P. Paranthaman, J. B. Goodenough and S. Dai, *Nano Lett.*, 2015, **15**(5), 3398.
70. Z. Tu, M. J. Zachman, S. Choudhury, S. Wei, L. Ma, Y. Yang, L. F. Kourkoutis and L. A. Archer, *Adv. Energy Mater.*, 2017, **7**(8), 1602367.
71. M. Steinhauer, M. Stich, M. Kurniawan, B.-K. Seidlhofer, M. Trapp, A. Bund, N. Wagner and K. A. Friedrich, *ACS Appl. Mater. Interfaces*, 2017, **9**(41), 35794.
72. C.-Y. Tang, Y. Ma, R. T. Haasch, J. H. Ouyang and S. J. Dillon, *J. Phys. Chem. Lett.*, 2017, **8**(24), 6226.
73. J. Z. Olson, P. K. Johansson, D. G. Castner and C. W. Chlenker, *Chem. Mater.*, 2018, **20**(4), 1239.
74. Y. Horowitz, H.-G. Steinrück, H.-L. Han, C. Cao, I. I. Abate, Y. Tsao, M. F. Toney and G. A. Samorjai, *Nano Lett.*, 2018, **18**(3), 2150.
75. K. Edström, M. Herstedt and D. P. Abraham, *J. Power Sources*, 2006, **153**(2), 380.
76. E. Peled and S. Menkin, *J. Electrochem. Soc.*, 2017, **164**(7), A1703.
77. A. Jozwiuk, B. B. Berkes, T. Weiß, H. Sommer, J. Janek and T. Brezesinski, *Energy Environ. Sci.*, 2016, **9**, 2603.
78. J. Liu, D. P. Lu, J. M. Zheng, P. F. Yan, B. Q. Wang, X. L. Sun, Y. Y. Shao, C. M. Wang, J. Xiao, J. G. Zhang and J. Liu, *ACS Appl. Mater. Interfaces*, 2018, **10**(26), 21916.
79. F. Wu, J.-T. Lee, J. Nitta, H. Kim, O. Borodin and G. Yushin, *Adv. Mater.*, 2015, **27**, 101.
80. J. Liu, I. Qian, M. F. Wang, J. Q. Zhou, N. Xu and C. L. Yan, *Nano Lett.*, 2018, **18**(7), 4598.
81. A. Shyamsunder, W. Beichel, P. Klose, Q. Pang, H. Scherer, A. Hoffmann, G. K. Murphy, I. Krossing and L. F. Nazar, *Angew. Chem.*, 2017, **56**(22), 6192.
82. J. Hassoun and B. Scrosati, *Angew. Chem.*, 2010, **49**(13), 2371.
83. J.-W. Park, K. Yamauchi, E. Takashima, N. Tachikawa, K. Ueno, K. Dokko and M. Watanabe, *J. Phys. Chem. C*, 2013, **117**(9), 4431.
84. L. N. Wang, J. Y. Liu, S. Y. Yuan, Y. G. Wang and Y. Y. Xia, *Energy Environ. Sci.*, 2016, **9**(1), 224.
85. C. B. Bucur, M. Jones, M. Kopylov, J. Spear and J. Muldoon, *Energy Environ. Sci.*, 2017, **10**(4), 905.
86. A. Khetan, A. Luntz and V. Viswanathan, *J. Phys. Chem. Lett.*, 2015, **6**, 1254.
87. S. Freundberger, Y. Chen, N. E. Drewett, L. J. Hardwick, F. Barde and P. G. Bruce, *Angew. Chem.*, 2011, **50**, 8609.
88. F. Barde, Y. Chen, L. Johnson, S. Schaltin, J. Fransaer and P. G. Bruce, *J. Phys. Chem. C*, 2014, **118**(33), 18892.
89. B. D. Adams, R. Black, Z. Williams, R. Fernandes, M. Cuisinier, E. J. Berg, P. Novak, G. K. Murphy and L. F. Nazar, *Adv. Energy Mater.*, 2015, **5**(1), 1400867.
90. D. Sharon, P. Sharon, D. Hirshberg, M. Salama, M. Afri, L. J. W. Shimon, W.-J. Kwak, Y.-K. Sun, A. A. Frimer and D. Aurbach, *J. Am. Chem. Soc.*, 2017, **139**(34), 11690.
91. X. Gao, Y. Chen, L. Johnson and P. G. Bruce, *Nat. Mater.*, 2016, **15**, 882.
92. A. Khetan, H. Pitsch and V. Viswanathan, *J. Phys. Chem. Lett.*, 2014, **5**, 2419.
93. N. Mahnke, B. Schafzahl, C. Leypold, M. Leypold, S. Grumm, A. Leitgeb, G. A. Strohmeier, M. Wilkening, O. Fontaine, D. Kramer, C. Slugovc, S. M. Borisov and S. A. Freundberger, *Nat. Energy*, 2017, **2**, 17036.

94. C. V. Amanchukwu, H.-H. Chang, M. Gauthier, S. Feng, T. P. Batcho and P. T. Hammond, *Chem. Mater.*, 2016, **28**, 7167.

95. Q.-C. Liu, T. Liu, D.-P. Liu, Z.-J. Li, X.-B. Zhang and Y. Zhang, *Adv. Mater.*, 2016, **28**(38), 8413.

96. D. Aurbach, Z. Lu, A. Schechter, Y. Gofer, H. Gizbar, R. Turgeman, Y. Cohen, M. Moshkovich and E. Levi, *Nature*, 2000, **407**, 724.

97. D. Aurbach, G. S. Suresh, E. Levi, A. Mitelman, O. Mizrahi, O. Chusud and M. Brunelli, *Adv. Mater.*, 2007, **19**, 4260.

98. S.-Y. Ha, Y.-W. Lee, S. W. Woo, B. Koo, J.-S. Kim, J. Cho, K. T. Lee and N.-S. Choi, *ACS Appl. Mater. Interfaces*, 2014, **6**, 4063.

99. R. Mohtadi, M. Matsui, T. S. Arthur and S.-J. Hwang, *Angew. Chem. Int. Ed.*, 2012, **51**, 9780.

100. T. J. Carter, R. Mohtadi, T. S. Arthur, F. Mizuno, R. Zhang, S. Shirai and J. W. Kampf, *Angew. Chem. Int. Ed.*, 2014, **53**, 3173.

101. A. Ponrouch, C. Frontera, F. Barde and M. R. Palacin, *Nat. Mater.*, 2016, **15**, 169.

102. D. Wang, X. Gao, Y. Chen, L. Jin, C. Kuss and P. G. Bruce, *Nat. Mater.*, 2018, **17**, 16.

103. M.-C. Lin, M. Gong, B. Lu, Y. Wu, D.-Y. Wang, M. Guan, M. Angell, C. Chen, J. Yang, B.-J. Hwang and H. Hai, *Nature*, 2015, **520**, 324.

104. K. K. Kravchyk, S. Wang, L. Pivetau and M. V. Kovalenko, *Chem. Mater.*, 2017, **29**(10), 4484.

105. A. Plewa-Marczewska, T. Trzeciak, A. Bitner, L. Niedzicki, M. Dranka, G. Z. Zukovska, M. Marcinek and W. Wieczorek, *Chem. Mater.*, 2014, **26**, 4908.

# Modeling Solid State Batteries

**Ting Hei Wan, Ziheng Lu and Francesco Ciucci**

*Hong Kong University of Science and Technology, Hong Kong, China*



## 1 INTRODUCTION

Since the development of the first prototype in the 1990s, solid-state batteries (SSBs) have gained significant interests from both the industrial and the scientific communities as one of the most promising energy storage devices. SSBs utilize an ionic conducting solid as the electrolyte. This choice prevents the safety hazards imposed by the use of flammable organic electrolyte typical of conventional lithium-ion batteries (LIBs). Moreover, the use of a solid electrolyte (SE) allows the utilization of Li metal anodes that can lead to the enhancement of power density. However, many challenges, which hinder the development of SSBs, remain. In particular, SEs that possess a high ionic conductivity and broad electrochemical stability against high potential difference are still in need to reduce the potential and capacity loss during operation.[1] Another major obstacle is the interfacial resistance at the electrode|electrolyte interface due to space charge layer formation, phase and mechanical instability.[2]

To overcome the challenges mentioned above, various fast ions conducting material, including NASICON materials, garnet oxides, amorphous sulfides, and alkali ion rich anti-perovskites, have been developed. Moreover, efforts have been made to reduce the interfacial resistance and enhance the interfacial stability of SSBs, for example, by introducing an additional interlayer at the electrode|electrolyte interface. Nonetheless, a comprehensive understanding of the physics and electrochemistry of the SSBs is necessary to guide further investigations and rational design. In this regard, drawing conclusions about the physics and chemistry of various phenomena from experiments are often not trivial. In particular, one may need to repeat the same test for many materials in order to identify the suitable combinations.[3] More importantly, determining the physical characteristics as diverse as the interphase composition, the Li diffusion coefficient, or the extent of the space charge layer requires complicated or extensive experimentation.[4,5] Although experimental methods, such as the impedance spectroscopy and the cyclic voltammetry, have been developed to quantify the bulk and interfacial properties of SSBs, identifying the performance limiting factors from these experiments without theoretical support is extremely difficult.[6]

In this regard, both atomistic-scale and continuum-scale methods can help elucidate the various physical phenomena taking place in SSBs. In particular, atomistic simulations are useful for assessing many material properties, including ionic conductivity, doping energies, reaction energies with other materials, and activation energies of ionic diffusion. For example, the information obtained from atomistic simulations can help experimentalist to search for electrolyte materials that have high ionic conductivity and that are compatible with the electrode materials. Moreover, continuum-level simulation can also provide valuable insight, especially at the meso- and macro-scale. Continuum-level simulations are useful because they can link physical phenomena, such as bulk conductivities, thermodynamic potentials, and interfacial charge transfer rates, with experimental data, including charge and discharge responses and impedance spectra. Using continuum-scale method thus helps explain the impact of the battery architecture and the material combinations on the overall SSB performance.

In this chapter, we review the recent development of SSB modeling from both atomistic and continuum scales. In particular, this chapter is divided into two. In the first part, we will introduce the atomistic modeling

*Inorganic Battery Materials*. Edited by Hailiang Wang and Boniface P.T. Fokwa
 ISBN 978-1-119-43199-2

of SSBs. We will first review the theory of two categories of atomistic models, namely, the forcefield-based method and the electronic structure-based method. After that, we will illustrate the application of these two methods for the Li diffusion and the interfacial stability of SEs. Finally, we will discuss possible future research directions in this area. In the second part of this chapter, we will review the continuum scale modeling of SSBs. We will introduce the theoretical framework underlying continuum SSB modeling. After that, we will show how to apply continuum-scale simulations to study the bulk transportation of charged species, the space charge layer formation, and the mechanical stresses arising in batteries. Using both atomistic and continuum simulations helps optimize SSBs. The utilization of modeling also facilitates new SSB designs with higher performance and better reliability.

## 2 ATOMISTIC MODELING FOR SOLID-STATE BATTERIES

By definition, atomistic models use computational methods to simulate material properties and chemical processes on the atomic/molecular level. Unlike continuum modeling techniques, such methods are particularly useful when one wants to obtain information on the fundamental properties of materials such as formation enthalpies, elastic constants, band structures, and ionic conductivity. Atomistic modeling was usually used to complement experiments and to provide atomic-level mechanistic insights. In recent years, due to the advances in the predictive power of atomistic modeling techniques and the computing speed of hardware, tremendous progress has been made in computer-assisted materials design. Such a design process usually involves the identification of material property of interest, selection of proper computational methods, generation of candidate structures, computational prediction, and, eventually, experimental realization.[7] A schematic for such design process is shown in Figure 1. For SSBs, the computer-assisted design is used to develop new electrolyte, electrode, and coating materials. A list of critical properties and related computational techniques are listed in Table 1.

Other than approximating specific material properties, atomistic modeling is also useful in simulating microscopic processes, such as chemical reactions at interfaces. In batteries, such simulations are usually used to understand the detailed reaction mechanisms at the electrode|electrolyte interfaces. In conventional LIBs, new phases form at such interfaces, which are known as solid electrolyte interphases (SEIs). Owing to the complexity of the composition, current experimental techniques are

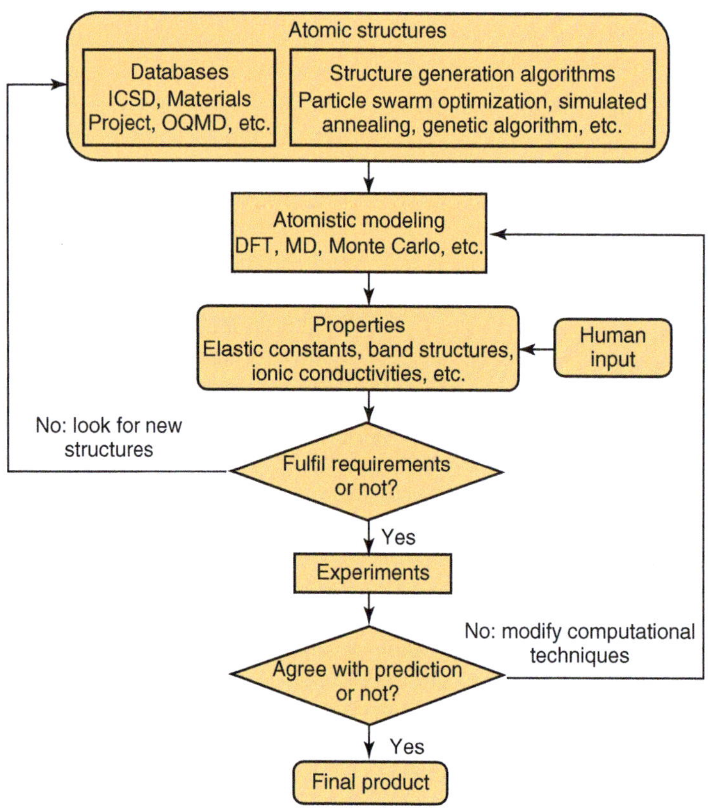


**Figure 1** Flow diagram for computer-assisted materials design

**Table 1** A list of critical parameters that can be predicted using atomistic modeling and the corresponding computational methods

| Components of SSBs | Properties | Atomistic modeling techniques | Underlying theory |
|---|---|---|---|
| Electrodes | Phase stability | Phase diagram construction, phonon calculation | DFT |
| | Voltage | (Grand potential) phase diagram construction | DFT |
| | Thermal stability | Phonon calculation, MD | DFT, forcefield |
| | Ionic mobility | MD, NEB | DFT, forcefield |
| | Volume expansion during cycling | Structural relaxation, MD | DFT, forcefield |
| | Reactivity with SEs | Phase diagram construction, MD | DFT, forcefield |
| Solid electrolytes (SE) | Phase stability | Phase diagram construction, phonon calculation | DFT |
| | Electrochemical window | Grand potential phase diagram construction, band structure | DFT |
| | Thermal stability | MD, phonon calculation | DFT, forcefield |
| | Ionic mobility | MD, NEB | DFT, forcefield |
| | Electron conductivity | Band structure, nonequilibrium green-function | DFT |
| | Reactivity | Phase diagram construction, MD | DFT, forcefield |
| Coating materials | Phase stability | Phase diagram construction | DFT |
| | Ionic mobility | MD, NEB | DFT, forcefield |
| | Mechanical properties | Structural relaxation, MD | DFT, forcefield |
| | Reactivity with other components | Phase diagram construction, MD | DFT, forcefield |

limited when one wants to detect the local structures of an SEI. Atomistic simulations are powerful tools to predict the local atomic arrangements and corresponding properties of SEI when experimental characterizations are not available.[8] Similar to LIBs, the reactions at the interfaces in SSBs can also be simulated using atomistic methods.[9]

Atomistic modeling can, in general, be categorized based on the underlying theory into two broad categories, i.e., semi-empirical interatomic-potential-based (or forcefield-based) methods, and first-principle electronic structure-based methods. In the following section, we will first review these two approaches and introduce their typical scope of use. In Section 2.2, we will illustrate with examples how such methods can be employed to extract insight that is useful for SSBs. In particular, we will introduce the study of lithium diffusion in SEs in Section 2.2.1. In Section 2.2.2, we will illustrate how atomistic modeling can be used to predict the stability of materials and interfaces in SSBs. In Section 2.3, we will discuss the limitations of current atomistic models and propose the future directions for modeling SSBs.

## 2.1 Overview of Computational Methods

### 2.1.1 *First-Principles Electronic Structure Methods*

The fundamentals of modern electronic structure theory are built upon the quantum theory of electrons. In principle, all chemical properties of a material can be predicted if the many-body Schrodinger equation can be solved. However, due to the high complexity of this equation, simplifications have to be made. The two most common theoretical frameworks developed are the wave function-based post-Hartree–Fock (post-HF) theory and the electron density-based density functional theory (DFT).[10,11] For the modeling of SSBs, DFT is the most widely used one to obtain the electronic ground state because of its comparatively low computational cost. Following the Kohn–Sham formalism,[12] the electronic structure can be solved by

$$\left(-\frac{1}{2}\nabla^2 + V_{\mathrm{H}}\left[\rho(r)\right] + V_{\mathrm{N}}(r) + V_{\mathrm{xc}}[\rho(r)]\right)\psi_i = \varepsilon_i\psi_i \quad (1)$$

where $-\frac{1}{2}\nabla^2$ is the electronic kinetic energy, $V_{\mathrm{H}}$ is the Hartree term representing the electrostatic energy of an electron subject to the mean field due to other electrons, $V_{\mathrm{N}}$ is the contribution from the nuclei, and $V_{\mathrm{xc}}$ is the exchange-correlation energy. Unlike the other terms, the exact form of $V_{\mathrm{xc}}$ is unknown. Thus, some approximations have developed and proved to give reasonable results including the local density approximation (LDA) and the generalized gradient approximation (GGA). In principle, GGA can give a better description of the energies of metal and semiconductors and is more commonly used in the modeling of battery materials.[13] However, due to the electron self-interaction errors GGA and LDA are not able to give good results on strongly correlated systems such as transition metal oxides, which is the major family of cathode materials for batteries. To solve this problem, the DFT + U method and the hybrid functionals have been developed.[11,14] These functionals can mitigate the self-interaction problem and give satisfactory results of specific systems.

**Table 2** Interatomic potentials for different components in SSBs

| Components of SSBs | Systems | Typical elements | Forcefields |
|---|---|---|---|
| Electrodes | Metals | Li, Na, In, Sn, Al | UFF, EAM |
| | Inorganic nonmetals | Si, P, Ge, S | UFF, EAM, ReaxFF |
| | Carbon/hydrocarbons | C, H, O | UFF, EAM, REBO, AIREBO, Tersoff, ReaxFF |
| | Ceramics | Li, Co, Mn, Ni, O | UFF, Qeq-MEAM, Buckingham Coulombic |
| Solid electrolytes | Ceramics | Li, La, Ta, Ti, Zr, Ge, P, O, S | Buckingham Coulombic, softBV |
| | Polymers | C, H, O, N, P, S | COMPASS, CVFF, DREIDING, many-body polarizable force fields, AMBER |
| Interfaces | Reactive systems | C, H, O, N, Si, Li | ReaxFF |

The DFT computes the ground state electron density. Therefore all calculated results correspond to 0 K. In order to obtain properties at finite temperatures, two methods are widely used. One is based on the lattice dynamics, or phonon, where the entropic contribution of free energy by the lattice vibrations is considered.[15] Owing to the high computational cost of phonon dispersion using first-principles methods, the finite temperature properties such as formation free energies are often not calculated, and the 0 K formation energies are usually used as an approximation. Such approximation is valid for most SSB modeling cases because the phases involved are most likely solids thus the entropic contribution is relatively insignificant compared with the 0 K formation energy.[13] Another method often used is molecular dynamics (MD) where the nuclei are assumed to follow Newton's law, and the motion of the nuclei is mimicked by solving Newton's equation:

$$\mathbf{F} = m\mathbf{a} \tag{2}$$

where **a** is the acceleration of nuclei, $m$ is the mass of the atom, and **F** is the force acting on the nuclei which can be calculated based on the ground state electronic structure using the Hellmann–Feynman theorem.[16,17] MD simulations based on first-principles electronic structure theory is known as ab-initio molecular dynamics (AIMD) simulations.[16]

### *2.1.2 Interatomic Potential Forcefield Methods*

Abovementioned first-principles electronic-structure-based methods can be used to obtain energies, forces, and electronic structures of a system. However, the computational cost of such methods is usually high. Also, in some cases, only the potential energy surface (PES), which is the energy of the system as a function of the nuclei coordinates, is needed and it is not necessary to acquire the detailed electronic structure of a system. For example, to simulate the dynamics of atoms in a system, one only needs to know the forces, i.e., the derivatives of energy with respect to the coordinates (or the gradient of PES). One can calculate such energy function using first-principles electronic structure methods such as DFT, a more natural way is to fit the PES explicitly against the DFT results. This fit can be done by assigning the interatomic potential or forcefield. Models which express the total energy of a system as a function of its nuclear coordinates are called the interatomic potential methods. For materials with different chemical nature, the forms of interatomic potential are different. Table 2 gives a list of useful forcefields for SSB modeling.

In SSBs, the interatomic potential methods are often used to perform MD simulations because of the lower computational cost compared with AIMD. Such simulations can predict the ionic conductivities and provide detailed information on the Li transport mechanisms in solids, which will be discussed in the next section. It is worth noting that such interatomic-potential-based MD simulations can now be regularly run for timescales on the order of nanoseconds whereas for AIMD it is usually tens to hundreds of picoseconds.[18,19] Also, potential-based MD simulations can handle systems with a more substantial number of atoms, from thousands to millions, whereas for AIMD, the number of atoms is usually not greater than 500. Such a feature enables the simulation of systems with a low concentration of defects.

## 2.2 Application of Atomistic Modeling in Solid-State Batteries

### *2.2.1 Lithium Diffusion in Solid-State Electrolytes*

Compared with conventional LIBs, SSBs are usually characterized by a lower rate capability. One of the reasons for such a reduced rate performance is the sizeable Ohmic resistance originating from the low ionic conductivity of SEs. Therefore, tremendous effort has been devoted to the search of novel SEs with high ionic conductivities.

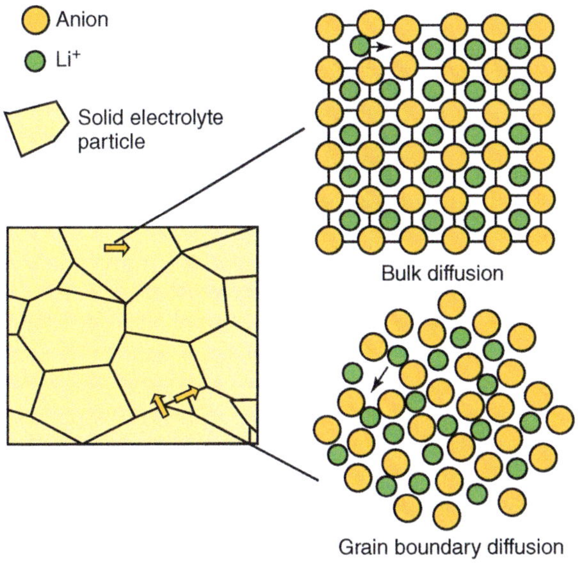


**Figure 2** Illustration of Li-ion diffusion mechanisms in SEs

Atomistic models play an important role in understanding and developing such materials because the vital feature concerning the performance of a SE, i.e., the Li ionic conductivity, can be calculated directly using various techniques such as MD simulations.[18] Further, the motion of Li ions can be visualized using the trajectories obtained from such simulations. The analysis of these trajectories can provide significant insight into the diffusion mechanism.[20,21]

As shown in Figure 2, the Li-ion conductivity can be attributed to two distinct regions of an SE, namely, the bulk and the grain boundary. Both conductivity values can be obtained experimentally using electrochemical impedance spectroscopy (EIS). Computationally, the bulk conductivity is usually computed using MD simulations. In general, the self-diffusion coefficient $D^*$ of $Li^+$ ions can be extracted from the MD trajectories using the Green–Kubo relation[22]:

$$D^* = \frac{1}{3N}\int_0^{\infty}\sum_{i=1}^{N}\langle v_i(t_0)\cdot v_i(t+t_0)\rangle_{t_0}\mathrm{d}t \tag{3}$$

where $v_i(t)$ is the velocity of ion $i$ at time $t$ and $\langle\cdot\rangle_{t_0}$ denotes the ensemble average over time $v_0$. $N$ is the number of mobile ions in the system. Another way to calculate the diffusivity is to utilize the Einstein relation:

$$D^* = \frac{1}{6}\lim_{t\to\infty}\frac{\partial}{\partial t}\frac{1}{N}\left\langle\sum_{i=1}^{N}\left[r_i\left(t+t_0\right)-r_i(t_0)\right]^2\right\rangle_{t_0} \tag{4}$$

where $r_i(t_0)$ is the position of ion $i$ at time $t$. $\frac{1}{N}\left\langle\sum_{i=1}^{N}\left[r_i\left(t+t_0\right)-r_i(t_0)\right]^2\right\rangle_{t_0}$ is also called the mean square displacement (MSD) of the ions where the bracket denotes the ensemble average over time. Theoretically, the two methods are strictly equivalent. However, considering the short simulation time and relatively sparse hopping events, Einstein relation usually gives more reliable results.[18] The ionic conductivity $\sigma$ can further be calculated using the self-diffusion coefficient with the Nernst–Einstein equation:

$$\sigma = \frac{\rho\,(zF)^2}{RT}D^* \tag{5}$$

where $\rho$, $z$, $F$, $R$, $T$ denotes the molar density, the unit charge of the ion, the Faraday constant, the gas constant, and the temperature (in K), respectively. Such relation assumes no correlation between the mobile ions. The discrepancy from the real situation, i.e., charge diffusivity $D_\sigma$, which is more or less correlated can be measured using the Haven ratio[23]:

$$H_R = \frac{D^*}{D_\sigma} \tag{6}$$

where $H_R = 1$ indicates uncorrelated diffusion. Usually, value of $H_R$ is between 0.3 and 0.6 for fast ionic conductors. At low room-temperature, the lithium hopping events are infrequent giving the small simulation box (small sample size) and short simulation time (short sampling time). Thus, the conductivity at room temperature is usually extrapolated using the Arrhenius relationship:

$$D = D_0\exp\left(-\frac{E_a}{k_B\mathrm{T}}\right) \tag{7}$$

where $E_a$ is the activation energy and $k_B$ is the Boltzmann constant. Since $D_0$ does not change with $T$, one can combine this equation with Equation 5 and derive a more commonly used form for experimentalists:

$$\sigma T = \sigma_0\exp\left(-\frac{E_a}{k_B T}\right) \tag{8}$$

Using this computational framework, one can predict the ionic conductivity of materials with the input being the structure. Figure 3(a) shows the conductivity of $Li_7P_3S_{11}$ calculated using AIMD in comparison with the experimental values. One can see that such computational predicts the ionic conductivities. Besides the Li conductivity, MD simulations also provide information on the transport mechanism in SEs. Figure 3(b) shows the simulated Li probability density in $Li_7P_3S_{11}$. Such distribution indicates that Li ions tend to diffuse through the interconnected $S_4$ tetrahedra sites. Accordingly, the hopping between tetrahedral sites in a bcc-like anionic framework has been proposed as the low-barrier mechanism.[25]

In principle, using MD/AIMD simulations, one can screen a wide range of materials based on existing

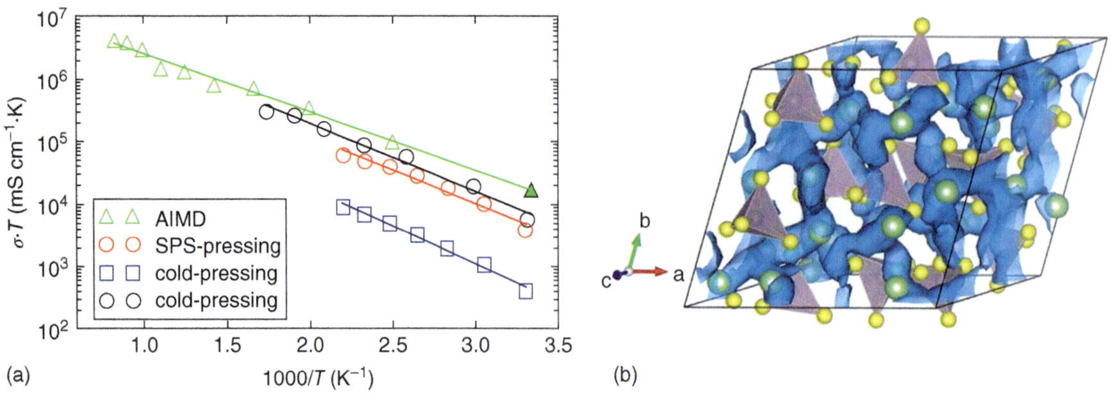


**Figure 3** (a) Ionic conductivity of LGPS from and experiments; (b) Li probability density distribution of $Li_7P_3S_{11}$ from AIMD simulations.[24] [Reprinted with permission from I.-H. Chu, H. Nguyen, S. Hy, Y.-C. Lin, Z. Wang, Z. Xu, Z. Deng, Y.S. Meng, S.P. Ong, Insights into the Performance Limits of the Li7P3S11 Superionic Conductor: A Combined First-Principles and Experimental Study, ACS Applied Materials & Interfaces, 8 (2016) 7843–7853. © American Chemical Society 2016]

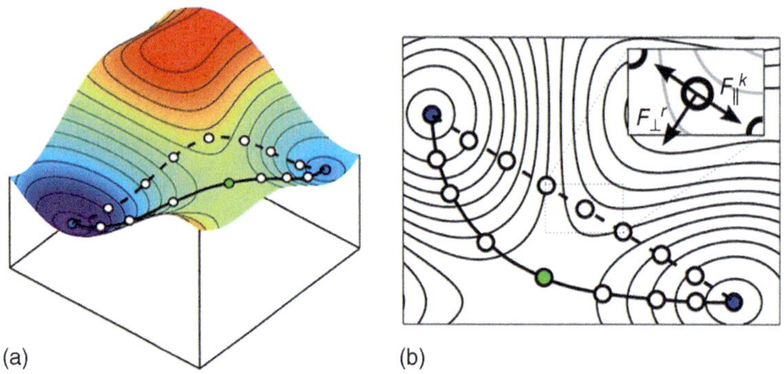


**Figure 4** The principle underlying the NEB method. The dashed line shows the path corresponding to the initial guess for the images. After applying the NEB method, the configurations lie along the minimum energy path (continuous line). The green dot corresponds to the transition state (saddle point).[29] [Reproduced by permission of A. Kraych, http://umet.univ-lille1.fr/Projets/RheoMan/en/to-learn-more-about/nudged-elastic-band.php. © Antoine Kraych, 2018]

databases. However, there are constraints to this method. For MD simulations based on interatomic potentials, the transferability of the parameters of the semi-empirical potential may lead to low credibility and requires tuning based on specific systems. For AIMD simulations, the computational cost is expensive.[18] Thus, MD and AIMD are better suited for mechanistic studies and tailoring the materials' properties after the structure and potentials are known rather than large-scale screening. To identify good ionic conductors, one can compute other parameters that are less computationally demanding. For example, one can compute the cationic hopping barrier using transition-state theory via the formula[26]:

$$K(T) = v^*(T)\exp\left[-\frac{\Delta G^{\ddagger}(T)}{k_B T}\right] \tag{9}$$

where $K$ is the hopping rate, $v^*$ is the attempt frequency, and $\Delta G^{\ddagger}(T)$ is the Gibbs free energy difference between the transition and initial states. One should note that in principle the value of $\Delta G^{\ddagger}(T)$ is temperature dependent.[27,28] However, for systems that follow Arrhenius relation, $\Delta G^{\ddagger}$ can be assumed to be temperature independent and the 0 K energy barrier is a good approximation. To estimate the diffusivity $D$, one only needs to know the distance between two Li sites $a$ and the result can be calculated using $D(T) = a^2(T)K(T)$.[28] The remaining question is how to compute the 0 K energy barrier $\Delta G^{\ddagger}$. As illustrated in Figure 4, the transition state is defined as the first-order saddle points on the PES of chemical species of interest. In principle, one should calculate the entire PES to obtain such states. However, this is computationally too intensive even for systems with only a few atoms. Several methods have been developed to capture the transition-state without knowing the entire PES. The nudged elastic band (NEB) method is most commonly used for searching transition states of cationic diffusion.[30] In this method, several intermediate structures are created by linear interpolation

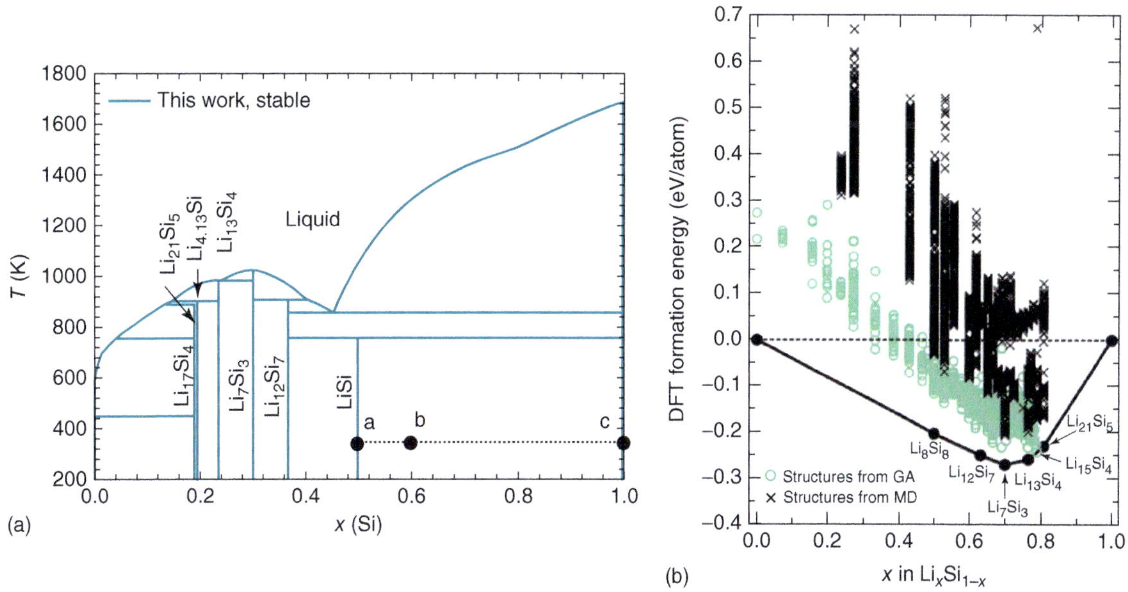


**Figure 5** (a) Experimental[31] and (b) computed phase diagrams of the Li–Si system.[32] [(a) Reproduced with permission from M.H. Braga, A. Dębski, W. Gąsior, Li–Si Phase Diagram: Enthalpy of Mixing, Thermodynamic Stability, and Coherent Assessment, Journal of Alloys and Compounds, 616 (2014) 581–593. © Elsevier 2014; (b) Reprinted from N. Artrith, A. Urban, G. Ceder, Constructing First-Principles Phase Diagrams of Amorphous $Li_xSi$ using Machine-Learning-Assisted Sampling with an Evolutionary Algorithm, The Journal of Chemical Physics, 148 (2018) 241711, with the permission of AIP Publishing]

of the atomic coordinates between the initial and final states. These intermediate structures are also called images. They are "connected" by an elastic band which constitutes harmonic springs between images, as shown in Figure 4(a). Such a band is able to drag the images to their lowest energy states without collapsing them all to the energy minima as illustrated in Figure 4(b). In other words, the minimum energy path (MEP) can be obtained by minimizing the spring forces on each image. The transition state is the image with the maximum energy on the MEP, as highlighted with the green dot in Figure 4.

### 2.2.2 *Stability of Materials and Interfaces*

Apart from the conductivity of SEs, the stability of materials and interfaces is also crucial to the design of an SSB. For example, the formation energy of a predicted SE or electrode material gives insights on whether this material can be synthesized. The reactivity between electrodes and SEs at the interfaces determines whether the SSB can be cycled stably. Atomistic modeling is particularly useful in predicting the thermodynamic stability by constructing phase diagrams. We first introduce how to compute phase diagrams to predict the stability of a particular phase (material) and then discuss how to obtain information on the interfaces.

To predict whether a phase is stable under certain conditions, one usually turns to the phase diagram which represents the thermodynamic phase equilibria of a multicomponent system. For example, Figure 5(a) shows the phase diagram of the Li–Si system.[31] If one tries to determine the composition of a compound with the nominal composition $Li_2Si_3$ (denoted as b in Figure 5(a)) the room temperature stable phases are a mixture of pure Si (phase c in Figure 5(a)) and LiSi (phase b in Figure 5(a)) and the percentage can be determined using lever rule:

$$x_a = \frac{l_{bc}}{l_{ac}} \tag{10}$$

where $x_a$ is the fraction of phase a in $Li_2Si_3$, $l_{bc}$ is the length of the segment between point b and point c, $l_{ac}$ is the length of the segment between point a and point c. However, to construct such phase diagrams, a significant amount of expensive and time-consuming experiments are needed. One solution to this is to construct these phase diagrams by calculating the energy of known compounds in the given system using DFT. Figure 5(b) shows the computed binary phase diagram of Li–Si system, where the crosses and the open circles represent the formation energy of each computed structure.[32] The black line represents the edge of the convex hull of all the computed phases, which connect the stable phases as denoted by black dots. Only the phases at the vertices of the convex hull are stable, and all the other nominal compositions are a mixture of the stable phases. This idea can also be extended to ternary, quaternary systems as well as systems with higher dimensionality.

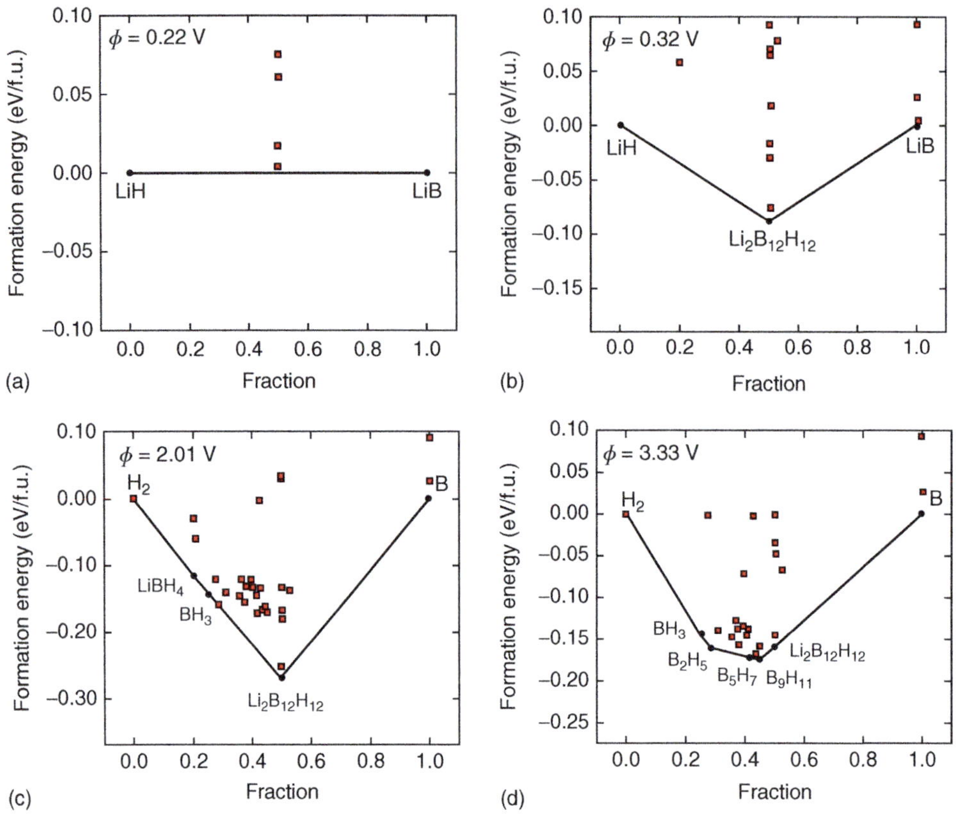


**Figure 6** Grand potential phase diagrams of the Li–B–H system at different potential Φ.[36] [Reprinted with permission from Z. Lu, F. Ciucci, Metal Borohydrides as Electrolytes for Solid-State Li, Na, Mg, and Ca Batteries: A First-Principles Study, Chemistry of Materials, 29 (2017) 9308–9319. © American Chemical Society 2017]

One thing to note is that such phase diagrams are based on 0 K and 0 atm DFT calculations and thus only represents the ground states. In reality, above 0 K, entropic contributions may affect the Gibbs free energy and affect the phase stability. Therefore, phases that are not so far from the convex hull at 0 K may still be synthesized at sufficiently high temperatures and possibly be brought back to room temperature. These phases are called meta-stable phases. A well-known example in SSB is the sulfide electrolyte $Li_{10}GeP_2S_{12}$ which is predicted to decompose at 0 K.[33] However, such phase has been successfully synthesized and is known to be one of the best ionic conductors.[34] Kinetics may also play a role in determining the phase stability. For example, the previously mentioned $Li_xSi$ compounds are usually in amorphous forms instead of their ground states, i.e., a mixture of pure crystalline phases. Such behavior can be attributed to the slow kinetics for crystallization.[35] The electrochemical window of an SE, i.e., the voltage range vs Li/Li$^+$ where the SE is stable, can also be predicted by constructing phase diagrams. Under these conditions, the Li chemical potential ($\mu_{Li}$ in eV) is a variable of electrode potential $\phi$:

$$\mu_{Li}(\phi) = \mu^0_{Li} - e\phi \tag{11}$$

where $\mu^0_{Li}$ is the chemical potential of Li metal. Such phase diagram is called the grand potential phase diagram. An example of Li–B–H system is shown in Figure 6 where the phase stability of Li, B, H-containing phases at different potential $\phi$ vs Li/Li$^+$ is presented.[36] One can see that $Li_2B_{12}H_{12}$, a recently found Li-ion conductor,[37] is stable at voltages as low as 0.32 V indicting good stability under electrochemical reduction. At voltages above 3.33 V, this material becomes unstable (does not appear on the convex hull) and tends to decompose into $B_xH_y$ phases.

Interfaces are crucial and inevitable components of SSBs. In principle, the stability of the interfaces determines whether an SSB can operate. One needs to consider both the chemical and the electrochemical stability in this case. The chemical stability, which tells us whether the electrode and the electrolyte will react when put in contact, can be accounted for by constructing a compositional phase diagram as explained above. The electrochemical stability, where the effect of voltage is also considered, can instead be computed using grand potential phase diagrams.[38] We must remark that the SEs often react with the electrodes. For example, SE lithium phosphorous oxynitride (LiPON) reacts with Li to form $Li_3P$, $LiPN_2$, and $Li_2O$. However, this does not imply the SSB using LiPON as SE, and Li metal as anode cannot operate.[38] In fact quite the opposite, such SSB has been commercialized by the company STMicroelectronics.[39] To understand why LiPON works and more broadly to predict what types interface enables the stable cycling of an SSB, we must consider what the reaction products are and their properties. In general, three situations are particularly representative of battery interphases, i.e., the reaction product can either be (i) Li insulating, (ii) Li conductive and electron conductive, or (iii) Li conductive and electron insulating.

Case (i) usually results in the sudden expiry of the SSB. Since the interphase is Li-insulating, the impedance of the SSB ramps up spontaneous, and the battery cannot be cycled. The second instance (ii) leads to a rapid capacity fade during cycling. Owing to the Li and electron conductivity, there is a constant supply of both mass and charge during charge and discharge. Therefore, the interphase builds up upon cycling, and the internal resistance increases with time. For example, Ti and Ge containing electrolytes such as $Li_{10}GeP_2S_{12}$ (LGPS), $Li_{1.5}Al_{0.5}Ge_{1.5}(PO_4)_3$ (LAGP), and $La_{0.57}Li_{0.29}TiO_3$ (LLTO) tend to be reduced by Li metal and the reaction products are electron conductive.[38,40] The batteries using these electrolytes and Li metal anode cannot be stably cycled. The third case (c) allows the stable operation of the SSBs. One example is the previously mentioned lithium phosphorous oxynitride (LiPON) material. The reaction products of LiPON and Li metal are $Li_3P$, $LiPN_2$, and $Li_2O$. All of the products are electron insulating while $Li_3P$ and $LiPN_2$ are lithium conductive. This prohibits the reaction from further happening but allows Li to pass.[38]

## 2.3 Future Direction for Atomistic Modeling of Solid-State Batteries

The field of atomistic modeling of SSBs has progressed significantly over the last decades and has become a vital research tool for battery researchers. However, atomistic models of SSBs still suffer from several limitations and need improvements. In this section, we discuss the drawbacks and the future directions for this area.

First, all computationally manageable methods still can not reach chemical accuracy (4 kJ mol$^{-1}$ or 0.04 eV atom$^{-1}$).[41] In SSB research, the most popular approach is DFT with Perdew–Burke–Ernzerhof GGA because of its relatively small computational cost.[13,27] However, such a method leads to errors in the formation enthalpies of the order of ~0.2 eV atom$^{-1}$. These errors are still not satisfactory, especially for chemically dissimilar systems that are commonly seen at interfaces in SSBs. Improvements are needed to find more accurate methods for electronic structure calculations. For example, one approach might be to construct new nonempirical functionals that satisfy the constraints applicable

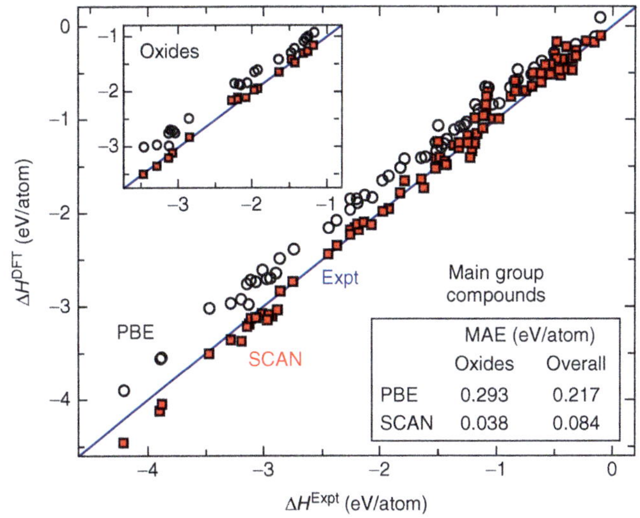


**Figure 7** The formation enthalpy of main group compounds, and oxides in particular. [Source: Zhang (41), https://www.nature.com/articles/s41524-018-0065-z#rightslink. Licensed under CC BY 4.0]

to the exchange-correlation functionals. Recently, Sun *et al.* developed a strongly constrained and appropriately normed (SCAN) meta-GGA functional.[42] DFT based on SCAN gives a significantly lower error in formation enthalpy (~0.120 eV atom$^{-1}$) compared with the PBE-GGA scheme, see Figure 7.[41] Although we are still far from chemical accuracy, this method gives a direction for DFT development.

Second, parameter-dependent corrections are necessary for achieving meaningful predictions from DFT calculations, when the parameters are system dependent. As mentioned previously, the DFT-based methods are generally not able to achieve chemical accuracy. However, if certain corrections are used one could still get meaningful predictions from DFT calculations. Often, these corrections are parameter- and system-dependent, most notably the U parameters in the DFT + U models for highly correlated systems and the dispersion coefficients in the DFT-vdw corrections.[43] The choices of such parameters significantly affect the results such as formation enthalpies, and limit the predictive power of the calculations. Therefore, to make useful predictions for SSBs, one needs to have expertise in both electronic structure theory and electrochemistry. One possible solution is to develop a theory of DFT for parameter-free schemes that are universally applicable.[41] Recent advances in quantum computing may also pave the way for increasing accuracy because it bears the possibility to solve the Schrodinger equations directly.[44]

Third, the atomistic models used in the current simulations are still far from being realistic, especially those for interfaces and systems with feature lengths greater than 100 nm. Current computational models can calculate the properties of small or highly symmetrical systems, for example, crystalline materials thanks to the high translational symmetry which allows the system to be described using tens of atoms with periodic boundary conditions. However, for complex systems such as interfaces in SSBs, the current methods such as DFT and MD can hardly give predictive results. Although one can in principle use MD simulations to capture to the evolution of the interface, the timescale is usually far beyond computationally manageable. Thus, it is necessary to develop alternative approaches to predict the properties of complex systems. One example is using the thermodynamics to get insights into the kinetics of interfaces as discussed in Section 3.2. Other solutions lie in the development of novel multistate computational methods that could accommodate significantly larger models and longer simulation timescales.[45]

Last, the computational data obtained from DFT, AIMD, and MD methods is mostly unused. Novel data science techniques can be applied to obtain more insights from the simulation results. MD simulation could serve as an example. In most of the MD simulations on SEs, the only parameter extracted is the diffusion coefficients of lithium. However, the computational output usually consists gigabytes of data for trajectories and terabytes of information on the electronic structures. Nearly all these information are lost during data post-processing. Recently, significant attention has been drawn to this field, where machine learning and data mining are used to extract patterns of Li diffusion in fast ion conductors.[20] However, the use of electronic structure information is still limited despite the influence of electronic structure on the ion diffusion has been known for over two decades.[46] Other option is to design novel electrolytes using the structures of previously known fast ionic conductors. For example, Sendek *et al.* combined large-scale computational screening with data mining and found 21 structures that show promise as electrolytes.[3]

## 3 CONTINUUM MODELING FOR SOLID STATE BATTERY

Continuum level modeling has been widely used in the past century for the investigation of LIBs. Models of this kind are based on the assumption that the materials properties, such as diffusion coefficients, conductivities, and permittivities, are locally homogeneous in the system under study. Concomitantly, as one divides the system into elements whose size tends to zero, its physical properties (e.g., concentrations, electrochemical, and electrostatic potentials) converge to a local thermal equilibrium limit. Continuum level modeling is particularly useful for investigating LIB devices, systems, and electrochemical experiments because it provides information regarding various physical phenomena, including but not limited to the bulk transportation rates, charge transfer kinetics as well as the space charge layer evolution.

The theoretical framework of the continuum level modeling for SSBs is mostly inspired by prior work on conventional liquid electrolyte based LIBs. Therefore, in the following section, we will first review the theoretical framework used in the continuum level modeling of traditional LIBs. Based on this conceptual framework, researchers have developed models with different degree of complexity to study various aspects of SSBs. Therefore, in Section 3.2, we will illustrate with examples the continuum level investigation of SSB under multiple assumptions. In particular, we will introduce the electroneutral models in Section 3.2.1. In Section 3.2.2, we will discuss how these electroneutral models can be relaxed, enabling the study of space charge layers present at electrode|electrolyte interfaces. In Section 3.2.3, we will describe the emerging mechanical coupling field in the SSB from the continuum level point of view.

### 3.1 Theoretical Framework for Conventional LIBs

#### *3.1.1 Poisson–Nernst–Planck (PNP) Equations*

The Poisson–Nernst–Planck (PNP) equations are the primary governing equations for solving the transportation of the charged species and the electrical potential inside the LIBs under the continuum hypothesis. In particular, the PNP set of equations is written as follows[47]:

$$\frac{\partial \rho_i}{\partial t} = -\nabla \cdot \boldsymbol{J}_i \tag{12a}$$

$$-\nabla \cdot (\varepsilon \nabla \phi) = \sum_i \rho_i + \rho_B \tag{12b}$$

where the index $i$ denotes the $i$th charged species within the electrochemical system. Equations 12a and 12b are known as the Nernst–Planck equation and the Poisson's equation, respectively. In particular, Equation 12a corresponds to charge conservation and equates the divergence of the current relative to the species $i$, $\boldsymbol{J}_i$, with the partial derivative of charge density $\rho_i$ with respect to time. Here, $\rho_i = z_i e c_i$, where $c_i$, and $z_i e$ are the concentration and the elementary charge of charged species $i$, respectively. Also, Equation 12b links the electrical potential $\phi$ with the total charge density, which, in turn, is the sum of the charge of all species and the background charge density $\rho_B$. $\boldsymbol{J}_i$ can be assumed to be the product of the gradient of the reduced electrochemical potential $\widetilde{\mu}_i^*$ and the charged species conductivity $\sigma_i$,[47–49] i.e.,

$$\boldsymbol{J}_i = -\sigma_i \nabla \widetilde{\mu}_i^* \tag{13}$$

where $\widetilde{\mu}_i^* = \widetilde{\mu}_i / z_i e$, with $\widetilde{\mu}_i$ being the electrochemical potential, defined as the sum of the chemical potential $\mu_i$ and the electrical energy $z_i e \phi$:

$$\widetilde{\mu}_i(c_i, T) = \mu_i + z_i e \phi = \mu_i^0 + k_B T \ln\left(\gamma\left(c_i\right)\frac{c_i}{c_i^0}\right) + z_i e \phi \tag{14}$$

Here, $\mu_i$ is the sum of the standard chemical potential $\mu_i^0$ and a concentration depending term, $k_B T \ln\left(\gamma\left(c_i\right)\frac{c_i}{c_i^0}\right)$. In the latter, $k_B$ is the Boltzmann constant, $\gamma(c_i)$ is the activity coefficient, and $c_i^0$ is the standard concentration of the charged species. The Nernst–Einstein relation connects $\sigma_i$ with the species concentration $c_i$ and diffusion coefficient $D_i$ as follows:

$$\sigma_i = \frac{D_i c_i (z_i e)^2}{k_B T} \tag{15}$$

Typically, the electroneutrality approximation is utilized for modeling conventional LIBs. Such an approximation consists in discarding (12b) by assuming zero local charge within the system, i.e., $\sum_i \rho_i + \rho_B = 0$ at all points and times. This approximation can be obtained by singular expansion away from interfaces under the hypothesis that the Debye length, defined as:

$$\lambda_D = \sqrt{\frac{\varepsilon k_B T}{\sum_i z_i^2 e^2 c_i^0}} \tag{16}$$

is far smaller than the characteristic length scale $L$ of the system, i.e., $\lambda_D \ll L$. As a result of the singular expansion, the electroneutrality condition holds only in the bulk but not at the interfaces.

### *3.1.2 Charge Transfer Kinetics*

In addition to the PNP equations, one needs to consider the charge transfer kinetics associated with the insertion/extraction of Li-ion over the electrode|electrolyte interfaces in order to estimate the charge transportation and electrical potential. In particular, one takes into account the charge kinetics of the interfacial reaction, $Li^+$ (elyte) + $e^-$ (electrode) $\rightleftharpoons$ Li (electrode), where (elyte) and (electrode) denote the electrolyte and the electrode domain, respectively. During Li insertion from the electrolyte to the electrode, Li ions in the electrolyte need to overcome an energy barrier, which is characterized by the thermal state of the interface and its local electric potential (Figure 8). The electronic current density of the charge transfer process is often modeled by the well-known Butler–Volmer equation[51]:

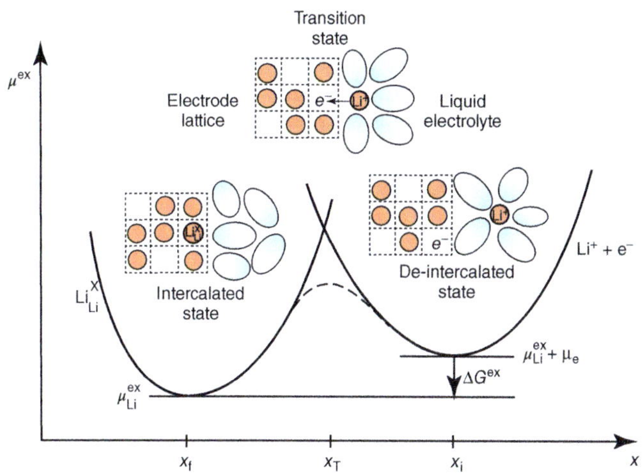


**Figure 8** The kinetics landscape for the Li interaction reaction, $Li^+ + e^- \rightarrow Li$, occurring at the electrode|electrolyte interface. The left and right parabolas show the excess chemical potential for the reduced state and oxidized state respectively. Li interaction occurs when the system overcomes the activation barrier at the intersection of the two parabolas, or just below, as indicated by the dashed curve.[50] [Reprinted with permission from M.Z. Bazant, Theory of Chemical Kinetics and Charge Transfer based on Nonequilibrium Thermodynamics, Accounts of chemical research, 46 (2013) 1144–1160. © American Chemical Society 2013]

$$j_e = j_0\left(\exp\left(\alpha_a \frac{e\eta}{k_B T}\right) - \exp\left(-\alpha_c \frac{e\eta}{k_B T}\right)\right) \tag{17}$$

$j_0$ is the exchange current density, $\alpha_a$ and $\alpha_c$ are the anodic and cathodic symmetry factors, such that $\alpha_a + \alpha_c = 1$. $\eta = \phi(\text{electrode}) - \phi(\text{elyte}) - E_{eq}$ is the overpotential, where $E_{eq}$ is the equilibrium potential difference, $\phi$(electrode) and $\phi$(elyte) are the electric potentials at the electrode surface (either cathode or anode), and that at the electrolyte in the vicinity of the electrode, respectively. By solving the governing equation as described in section 3.1.1, one can evaluate the local values of $\phi$(electrode) and $\phi$(elyte). $\eta < 0$ corresponds to the situation where electrons are produced and collected by the current collector. Li ions are deintercalated from the electrode, giving an anodic current. On the other hand, $\eta > 0$ implies a cathodic current where electrons are consumed upon intercalating Li ions into the electrode. Microscopically, the Butler–Volmer equation has been justified by the Marcus theory,[51] which models the charge transfer current as[50]

$$j_e = j_0 \exp\left(-\alpha_c \frac{\left(z_e e\eta\right)^2}{4 k_B T \lambda}\right)\left(\exp\left(\alpha_a \frac{z_e e\eta}{k_B T}\right) - \exp\left(-\alpha_c \frac{z_e e\eta}{k_B T}\right)\right) \tag{18}$$

where $\lambda = (k/2)(x_O - x_R)^2$ is known as the reorganization energy, $k$ is a proportionality constant (e.g., the force constant for stretching the bond length during transition state), $x_O$ and $x_R$ are the reaction coordinate at the oxidized state and reduced state, respectively (Figure 8). Then, the kinetic barriers for the cathodic reaction $\Delta G_c^{ex}$ and anodic reaction $\Delta G_a^{ex}$ are:

$$\Delta G_c^{ex} = \frac{\lambda}{4}\left(1 + \frac{\Delta G^{ex}}{\lambda}\right)^2 \tag{19}$$

$$\Delta G_a^{ex} = \frac{\lambda}{4}\left(1 - \frac{\Delta G^{ex}}{\lambda}\right)^2 \tag{20}$$

where $\Delta G^{ex}$ refers to the standard free energy of the charge transfer reaction, as illustrated in Figure 8. While the Butler–Volmer equation is frequently used for evaluating the charge transfer kinetics at the macroscopic scale, it is limited by the fact that several assumptions have been taken during its derivation. For instance, the Butler–Volmer equation is derived under the dilute solution approximation. In this regard, Bazant[50] has applied nonequilibrium thermodynamics to correct the exchange current $j_0$ in order to generalize both the Butler–Volmer equation and Marcus theory for concentrated solutions. In addition to the dilute solution approximation, the Butler–Volmer equation assumes a single rate-limiting step for the charge transfer. This is not sufficient if competitive side reactions occur between the electrode and its surrounding phases. To address this limitation, Kee and coworkers have developed an alternative model based on elementary reaction kinetics,[52] and related the current density to the net reaction rate of $n$ parallel elementary reactions, as follows:

$$j_e = e\sum_{m=0}^{n} \nu_{e^-,m}\dot{q}_m \tag{21}$$

where $\nu_{e^-,m}$ is the electron stoichiometric coefficient for the $m$th reaction and $\dot{q}_m$ is the corresponding net production rate (with units of No. particles/m$^2$s). By utilizing this formula for the charge-transfer boundary condition, multiple parallel pathways can be incorporated. One can, for example, apply such a boundary condition for the study of the growth of SEI in conventional LIBs.[52]

## 3.2 Application to Continuum Level Modeling of SSB

### 3.2.1 *Electroneutral Model*

As mentioned in Section 3.1, one of the most common approximations used in continuum scale simulation for conventional LIBs is to assume electroneutrality. Starting from the electroneutral approximation, several SSB models have been developed. The one dimensional mathematical model of Li/LiPON/$LiCoO_2$ developed by Danilov *et al.*[53] is among the pioneering work. In this framework, Danilov and coworkers have identified three overpotentials components contributing to the SSB voltage drop, namely the overpotentials due to (i) the charge transfer at interfaces, (ii) mass transfer over in the electrolyte and (iii) Li diffusion in the cathode. To enforce electroneutrality in the electrolyte, these authors have assumed that Li-ion transport in the SE is activated by the ionization of lattice Li into mobile $Li^+$ ions and the simultaneous presence of static negative charges $n^-$. The model was then fitted against the experimental discharge data at different C-rates to obtain the Li diffusion coefficient, the charge-transfer reaction constant, and the lattice Li ionization rate constant in the electrolyte. The same authors also applied the framework to the thermal management of SSBs.[54]

The work of Danilov *et al.*[53] has later been extended to higher dimensions. For example, Bates *et al.* have studied the time-dependent electrochemical reaction, distribution of charged species, and the heat transfer of a 2D model.[55] Talin and coworkers[56] have studied the impact of a conical microcolumns structure to the charge transportation of the SSB (Figure 9). In fact, the structural inhomogeneity of such SSB may cause a highly nonuniform internal current density distribution, leading to poor cathode utilization, and hence inferior power performance in comparison to the 1D planar case.[56]

To overcome the discrepancy of some of the estimated parameters against experimental literature, Fabre *et al.*[57] have constructed a somewhat simple model, which features a cathode governed by the diffusion equation and an electrolyte treated merely as an Ohmic resistor. These authors were able to reproduce both the discharge potential and the impedance spectra and retrieve physical parameters, such as the interfacial reaction constant and the diffusion coefficient. Importantly, the estimated $LiCoO_2$ diffusion coefficient is in reasonable agreement with that obtained from the Galvanostatic Intermittent Titration Technique (GITT) experiments.[57]

Following the works of Danilov *et al.*[53] and Fabre *et al.*,[57] Finsterbusch and collaborators[58] have recently reported the simulation of an SSB with 3D microstructure. In particular, they simulated an SSB using its 3D reconstructed composite positive electrode, composed of the mixture of the $LiCoO_2$ cathode and Ta-substituted $Li_7La_3Zr_2O_{12}$ (LLZTO) electrolyte, as shown in Figure 10. Although these simulations could not adequately predict the performance loss at room temperature, the simulated electrochemical behavior is in good agreement with the discharge potentials measured at elevated temperature (100 °C). Parametrically studying the model has shown that the interface phenomena, such as the charge transfer resistance at the interface and the poor contact between the active cathode particles and the electrolyte, may probably

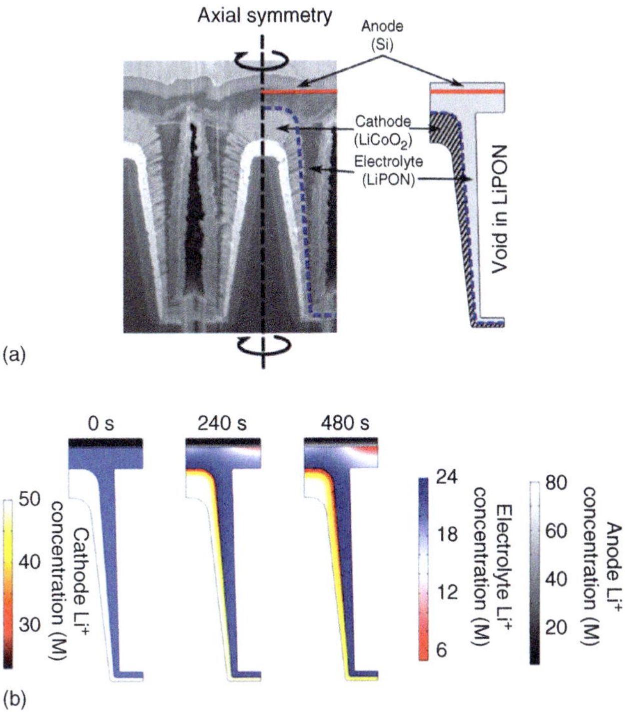


**Figure 9** The 3D SSB with the conical micro-columnar structure as studied by Talin *et al.*[56] Panel (a) shows the SEM image of the SSB and the corresponding simplified geometry upon considering the axial symmetry. Panel (b) shows the time-dependent Li-ion distribution within the SSB.[56] [Reprinted with permission from A.A. Talin, D. Ruzmetov, A. Kolmakov, K. McKelvey, N. Ware, F. El Gabaly, B. Dunn, H.S. White, Fabrication, Testing, and Simulation of All-Solid-State Three-Dimensional Li-Ion Batteries, ACS Applied Materials & Interfaces, 8 (2016) 32385–32391. © American Chemical Society 2016]

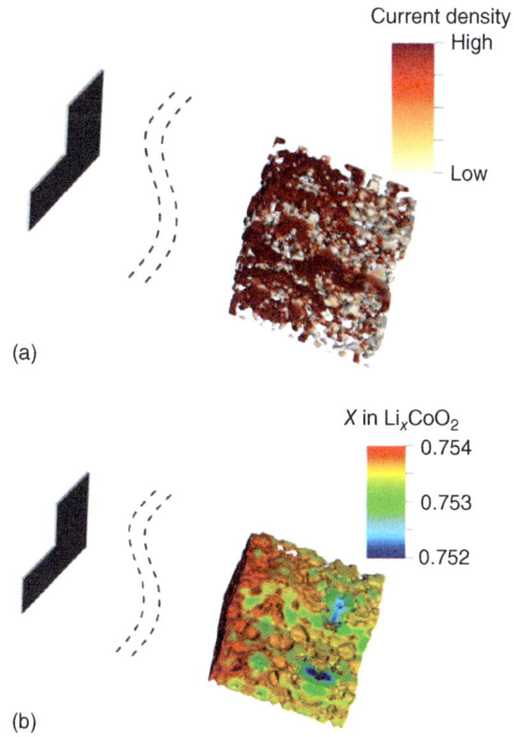


**Figure 10** The continuum scale simulation carried out by Finsterbusch *et al.*[58] on a 3D-microstructure resolved SSB. The cathode considered is a composite electrode composed of $LiCoO_2$ and LLZTO electrolyte. The microstructure is obtained via 3D FIB-SEM reconstruction Panel (a) shows the local current density inside the SE phase. Panel (b) shows the Li content within the active phase (i.e., $LiCoO_2$).[58] [Reprinted with permission from M. Finsterbusch, T. Danner, C.-L. Tsai, S. Uhlenbruck, A. Latz, O. Guillon, High Capacity Garnet-Based All-Solid-State Lithium Batteries: Fabrication and 3D-Microstructure Resolved Modeling, ACS Applied Materials & Interfaces, 10 (2018) 22329–22339. © American Chemical Society 2018]

be the major cause that limits the capacity of the SSB at room temperature.[58] This provides a direction for further optimizing similar SSBs.

### *3.2.2 Space Charge Layer Model*

The models discussed in the section above are based on the electroneutrality approximation, which assumes no space charge layer forms at the electrode|electrolyte interfaces. However, this is not the case as the interfaces are intrinsically charged. In particular, as a nonuniform electrical field is established at the electrode|electrolyte interface, charges in the electrolyte and the electrode rearrange. The resulting accumulation/depletion of charged species at the interface is known as the double layer (or space charge layer), whose thickness is characterized by the Debye length $\lambda_D$ (Equation 16). Generally, $\lambda_D$ in aqueous electrolytes is on the order of 1–100 nm, whereas in the case of SE, it is often comparable to the lattice parameter.[59] In the latter case, $\lambda_D$ can be smaller than any other length scale typical for continuum simulation, which implies that the bulk electrolyte remains quasi-neutral and that all double layers are extremely thin.[60]

One of the earliest developed double layer models is the one proposed by Gouy, Chapman, and Stern, as derived from the Poisson–Boltzmann equation.[61–63] This model suggests that the overall double layer capacitance is the sum of the capacitances due to Helmholtz (also known as the Stern Layer) and diffuse-layers.[51] Since the early work of Gouy, Chapman, and Stern model, substantial efforts have been put into improving the double layer model. For instance, a phenomenological theory has been developed to determine whether the overscreening or crowding type of double layer occurs in dense ionic liquids electrolytes.[64] The impact of chemical reactions can also be included. For example, Bazant *et al.*[65] have presented the analytical solution of current–voltage response in an electrochemical thin film. Such analytical solution is derived based on Poisson–Nernst–Planck equations and subsequent asymptotic expansions in the limit of thin double layers. Their analysis illustrates that whether reaction-limited or diffusion-limited current would occur across the thin film is determined by its dimension, Stern-layer capacitance as well as the interfacial reaction rate.

Moreover, Dreyer *et al.*[66] noticed the thermodynamic inconsistencies exist in the conventional PNP equation framework due to the negligence of the mechanical interactions between constituents in the Nernst–Planck model.[67] In fact, as a nonuniform electric field is established, a nonnegligible body force arises modifying the mechanical stresses at the interfaces.[68] Dreyer and co-workers have generalized the nonequilibrium thermodynamics version of the Nernst–Planck model.[66] In particular, they included an additional momentum balance equation based on the incompressible mixture assumption and coupled this equation with the drift-diffusion equations, linking pressure within the electrolyte to the features of the interfacial charge profile.[66] This also results in a thermodynamically consistent model for the double layer.[69]

Based on the theory developed for the double layer in liquid electrolytes, space charge layer theories have been extended to general SEs. One of the first such models dates back to the 1990s, as Horrocks and Armstrong[59,70] extended the Gouy–Chapman–Stern solution to the metal–SE interfaces. Since $\lambda_D$ for the SE is comparable with the lattice spacing, these authors assumed that the charges in the electrolyte are at quantized positions, i.e., on the lattice plane. The recent interests to the SSBs have facilitated the development of the space charge layer model under the context of SSBs. Indeed, Luntz and coworkers[60] have suggested that the space charge layer may be one of the key challenges that limit the performance of SSBs. This is because, for conventional liquid electrolyte batteries, screening of electric potential at each electrode|electrolyte interfaces is done by both the rearrangement of mobile electrolyte ions and the polarization of liquid solvent particles. In contrast, the charge screening in SSBs is achieved only by the accumulation/depletion of a single mobile charged species at interfaces. The accumulation/depletion of charge species may be so extreme that it may reduce the exchange current density and increase the charge transfer resistance.

To investigate the space charge layer properties within an SSB, Braun *et al.*[71] derived a semi-analytical solution to the SE with blocking electrode using the nonequilibrium thermodynamics framework proposed by Dreyer *et al.*[66] To study the behavior of the space charge layer, Braun and coauthors have carried out a parametric study on the framework concerning the applied potentials and the dielectric susceptibilities. Also, Landstorfer *et al.*[72] included the effect of a constant Li discharge flux to the study of the space charge layer. The same authors also considered a Stern layer type of boundary condition and applied the generalized Frumkin–Butler–Volmer equation described earlier, see Ref. 65 for the kinetics of intercalation/deintercalation reaction. This framework allows one to study the interrelation between the evolution of the space charge layer and the bulk transportation in the electrode.

### *3.2.3 Electro-Chemo-Mechanical Coupling*

In addition to the charged species transportation and the space charge layer properties, continuum scale modeling can also elucidate the mechanical stresses arising in SSBs. In fact, such a topic has widely been studied in conventional LIBs because some electrode materials undergo significant volumetric changes upon Li insertion and de-insertion.[73] For example, traditional graphite-based anodes expand by around 10% upon lithiation.[74] This problem is even more severe for Si, whose chemical expansion can be as high as 400% during lithiation.[75] As the volumetric changes of electrode material are often constrained by the surroundings, significant mechanical stresses can result. The stresses arising from the chemical expansion may cause mechanical failure of the electrode and the surrounding phases. In addition, they may even affect the diffusional properties and the reactivity of the electrode.[76] The coupling between the electrochemical performance and the mechanics is known as electro-chemo-mechanical coupling.

Numerous investigations have addressed the chemo-mechanical coupling occurring in conventional LIBs. For examples, scientists have studied single particles, thin films, phase change material, and porous electrodes. Since an SSB composes entirely by solid materials, the mechanical constraint and the associated stress are expected to be more significant in an SSB than that in a conventional organic-electrolyte-based LIB. Therefore, some SSB electro-chemo-mechanical models have emerged. In particular, Bucci *et al.*[77] have constructed a comprehensive chemo-mechanical model for the stress associated with species diffusion based on a thermodynamically consistent rationale albeit with the electroneutrality assumption. This model couples the nonlinear elastic deformation with the electrochemical potentials and allows for the investigation of the stress associated with ambipolar diffusion and kinetic demixing in ionic conducting material. Bucci and coworkers later applied this model to a 2D system with randomly distributed silicon electrode particles surrounded by a rigid SE matrix,[78,79] see Figure 11. The corresponding model quantified the compressive hydrostatic stress exerted onto the Si electrode during lithiation.[78] Such compressive hydrostatic stress limits the access of Li-ion into the electrode, and hence, leads to capacity fading of the electrode. On the other hand, the model predicts the existence of tensile stresses inside the SE matrix. These tensile stresses can lead to the formation of microcracks during cycling[79] as shown in Figure 11.

Moreover, it has been suggested in some literature that Li dendrites may grow through the SE,

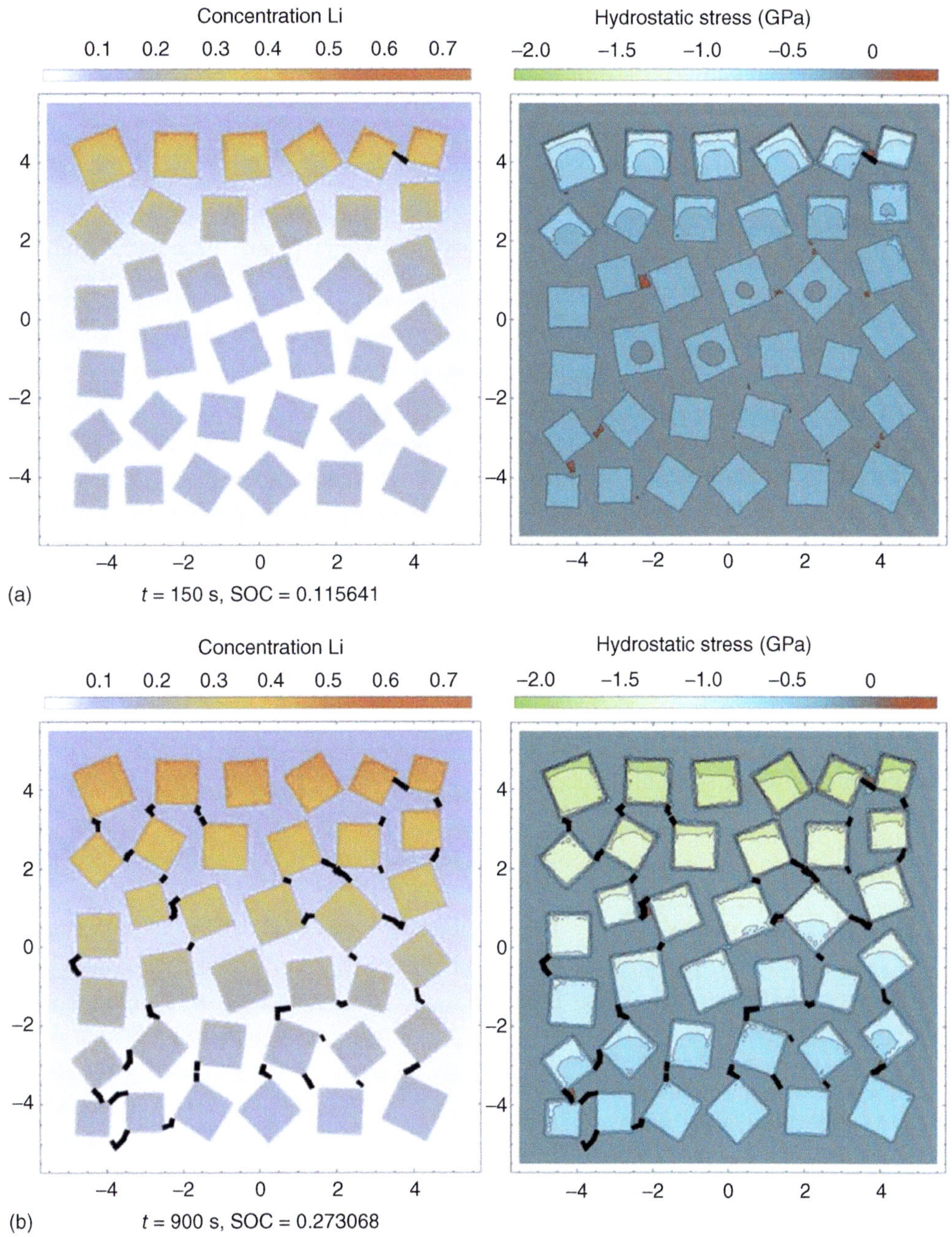


**Figure 11** The chemo-mechanical simulation carried out by Bucci *et al.*[79] on the evolution of stress and micro-crack at following states of charge. The system studied is a negative electrode with randomly distributed Si particles (represented by the squares) surrounded by an SE matrix. Upon Li insertion, the active Si particle undergoes chemical expansion. This leads to stress accumulated inside the microstructure. The resulting tensile stress leads to crack initiation (as marked by the black lines) in the SE matrix.[79] [Republished with permission of The Royal Society of Chemistry from G. Bucci, T. Swamy, Y.-M. Chiang, W.C. Carter, Modeling of Internal Mechanical Failure of All-Solid-State Batteries during Electrochemical Cycling, and Implications for Battery Design, Journal of Materials Chemistry A, 5 (2017) 19422–19430. Permission conveyed through Copyright Clearance Center, Inc]

which can short-circuit the SSEs. Regarding such issue, the continuum level modeling to the electro-chemo-mechanical coupled phenomena can also provide critical physical insight. As an example, Natsiavas *et al.*[80] developed an electroneutral model for the interfacial instability of electrode|electrolyte interfaces in a LiPON based SSB under the presence of prestress. The model was then used to assess the effect of prestress on the delamination and stability at the interface. The result shows that the formation of deleterious Li dendrites can be suppressed by applying compressive stress to the electrode|electrolyte interface.

### 3.2.4 *Short Summary*

In summary, we have reviewed the critical theoretical framework and some of the significant advances in the field of continuum modeling of SSBs. We have illustrated that the framework for continuum level study of the SSBs is highly related with the investigation for conventional LIBs. Continuum level research can enhance our understanding of various macro-level phenomena taking place in SSBs, including the bulk transportation, material properties, space charge layer evolution, and stress propagation. Moreover, we need to point out that continuum level modeling may provide insight to other phenomena in SSBs, like the interfacial reaction and cation diffusion. This can help us determine the reaction rates and the rate-limiting steps of various electrochemical processes taking place at the electrode|electrolyte interface. In addition, models that link the physical phenomena in SSBs with major diagnostic tools in the field, e.g., impedance spectroscopy and cyclic voltammetry experiment, are yet to be developed. Such a model will help experimentalists to extract critical physical parameters, such as the diffusion coefficient, the interfacial reaction constant, and the characteristic relaxation times of various physical processes. As a whole, the understanding gain from continuum level modeling will lead to the improvement to the SSBs design at the assembly level point of view.

## 4 ABBREVIATIONS AND ACRONYMS

AIMD = ab-initio molecular dynamics; DFT = density functional theory; EIS = electrochemical impedance spectroscopy; GGA = generalized gradient approximation; GITT = Galvanostatic Intermittent Titration Technique; LDA = local density approximation; LIBs = lithium-ion batteries; MD = molecular dynamics; MEP = minimum energy path; MSD = mean square displacement; PES = potential energy surface; PNP = Poisson–Nernst–Planck; post-HF = post-Hartree–Fock; SCAN = strongly constrained and appropriately normed; SE = solid electrolyte; SEIs = solid electrolyte interphases; SSBs = solid-state batteries

## 5 REFERENCES


1. J. C. Bachman, S. Muy, A. Grimaud, H.-H. Chang, N. Pour, S. F. Lux, O. Paschos, F. Maglia, S. Lupart, P. Lamp, L. Giordano and Y. Shao-Horn, *Chem. Rev.*, 2016, **116**, 140.
2. J. M. Tarascon and M. Armand, *Nature*, 2001, **414**, 359.
3. A. D. Sendek, Q. Yang, E. D. Cubuk, K.-A. N. Duerloo, Y. Cui and E. J. Reed, *Energy Environ. Sci.*, 2017, **10**, 306.
4. K. Yamamoto, Y. Iriyama, T. Asaka, T. Hirayama, H. Fujita, C. A. J. Fisher, K. Nonaka, Y. Sugita and Z. Ogumi, *Angew. Chem. Int. Ed.*, 2010, **49**, 4414.
5. K. Yamamoto, Y. Iriyama, T. Asaka, T. Hirayama, H. Fujita, K. Nonaka, K. Miyahara, Y. Sugita and Z. Ogumi, *Electrochem. Commun.*, 2012, **20**, 113.
6. E. Barsoukov and J. R. Macdonald, 'Impedance Spectroscopy: Theory, Experiment, and Applications', John Wiley & Sons, Hoboken, NJ, 2005.
7. Z. W. Seh, J. Kibsgaard, C. F. Dickens, I. Chorkendorff, J. K. Nørskov and T. F. Jaramillo, *Science*, 2017, **355**.
8. S.-P. Kim, A. C. Van Duin and V. B. Shenoy, *J. Power Sources*, 2011, **196**, 8590.
9. T. Cheng, B. V. Merinov, S. Morozov and W. A. Goddard III, *ACS Energ. Lett.*, 2017, **2**, 1454.
10. C. J. Cramer, 'Essentials of Computational Chemistry: Theories and Models', John Wiley & Sons, West Sussex, 2013.
11. E. K. Gross and R. M. Dreizler, 'Density Functional Theory', Springer Science & Business Media, New York, 2013.
12. W. Kohn and L. J. Sham, *Phys. Rev.*, 1965, **140**, A1133.
13. A. Jain, S. P. Ong, G. Hautier, W. Chen, W. D. Richards, S. Dacek, S. Cholia, D. Gunter, D. Skinner and G. Ceder, *APL Mater.*, 2013, **1**, 011002.
14. A. Liechtenstein, V. Anisimov and J. Zaanen, *Phys. Rev. B*, 1995, **52**, R5467.
15. K. Parlinski, Z. Li and Y. Kawazoe, *Phys. Rev. Lett.*, 1997, **78**, 4063.
16. G. Kresse and J. Hafner, *Phys. Rev. B*, 1993, **47**, 558.
17. G. Kresse and J. Hafner, *Phys. Rev. B*, 1994, **49**, 14251.
18. Z. Zhu, Z. Deng, I.-H. Chu, B. Radhakrishnan and S. P. Ong, *Comput. Mater. Syst. Des.*, 2018, 147.
19. T. Wyttenbach, N. A. Pierson, D. E. Clemmer and M. T. Bowers, *Annu. Rev. Phys. Chem.*, 2014, **65**, 175.
20. C. Chen, Z. Lu and F. Ciucci, *Sci. Rep.*, 2017, **7**, 40769.
21. Z. Lu and F. Ciucci, *J. Mater. Chem. A*, 2016, **4**, 17740.
22. M. S. Green, *J. Chem. Phys.*, 1952, **20**, 1281.
23. G. Murch, *Solid State Ion.*, 1982, **7**, 177.
24. I.-H. Chu, H. Nguyen, S. Hy, Y.-C. Lin, Z. Wang, Z. Xu, Z. Deng, Y. S. Meng and S. P. Ong, *ACS Appl. Mater. Interf.*, 2016, **8**, 7843.
25. Y. Wang, W. D. Richards, S. P. Ong, L. J. Miara, J. C. Kim, Y. Mo and G. Ceder, *Nat. Mater.*, 2015, **14**, 1026.
26. G. H. Vineyard, *J. Phys. Chem. Solids*, 1957, **3**, 121.
27. A. Urban, D.-H. Seo and G. Ceder, *NPJ Comput. Mater.*, 2016, **2**, 16002.
28. D. Morgan, A. Van der Ven and G. Ceder, *Electrochem. Solid-State Lett.*, 2004, **7**, A30.
29. A. Kraych (2014) http://umet.univ-lille1.fr/Projets/RheoMan/en/to-learn-more-about/nudged-elastic-band.php (accessed 8 December 2018).

30. H. Jónsson, G. Mills and K. W. Jacobsen, in 'Classical and Quantum Dynamics in Condensed Phase Simulations', World Scientific, 1998, p. 385.

31. M. H. Braga, A. Dębski and W. Gąsior, *J. Alloys Compd.*, 2014, **616**, 581.

32. N. Artrith, A. Urban and G. Ceder, *J. Chem. Phys.*, 2018, **148**, 241711.

33. Y. Mo, S. P. Ong and G. Ceder, *Chem. Mater.*, 2011, **24**, 15.

34. N. Kamaya, K. Homma, Y. Yamakawa, M. Hirayama, R. Kanno, M. Yonemura, T. Kamiyama, Y. Kato, S. Hama and K. Kawamoto, *Nat. Mater.*, 2011, **10**, 682.

35. L.-F. Cui, R. Ruffo, C. K. Chan, H. Peng and Y. Cui, *Nano Lett.*, 2008, **9**, 491.

36. Z. Lu and F. Ciucci, *Chem. Mater.*, 2017, **29**, 9308.

37. L. He, H.-W. Li, H. Nakajima, N. Tumanov, Y. Filinchuk, S.-J. Hwang, M. Sharma, H. Hagemann and E. Akiba, *Chem. Mater.*, 2015, **27**, 5483.

38. Y. Zhu, X. He and Y. Mo, *ACS Appl. Mater. Interf.*, 2015, **7**, 23685.

39. S. Larfaillou, D. Guy-Bouyssou, F. Le Cras and S. Franger, *J. Power Sources*, 2016, **319**, 139.

40. F. Han, Y. Zhu, X. He, Y. Mo and C. Wang, *Adv. Energy Mater.*, 2016, **6**.

41. Y. Zhang, D. A. Kitchaev, J. Yang, T. Chen, S. T. Dacek, R. A. Sarmiento-Pérez, M. A. Marques, H. Peng, G. Ceder and J. P. Perdew, *NPJ Comput. Mater.*, 2018, **4**, 9.

42. J. Sun, A. Ruzsinszky and J. P. Perdew, *Phys. Rev. Lett.*, 2015, **115**, 036402.

43. S. Grimme, A. Hansen, J. G. Brandenburg and C. Bannwarth, *Chem. Rev.*, 2016, **116**, 5105.

44. A. Ho, J. McClean and S. P. Ong, *Joule*, 2018, **2**, 810.

45. E. Weinan and B. Engquist, *Commun. Math. Sci.*, 2003, **1**, 87.

46. M. J. Castiglione, M. Wilson and P. A. Madden, *J. Phys.: Condens. Mat.*, 1999, **11**, 9009.

47. W. Lai and F. Ciucci, *Electrochim. Acta*, 2010, **56**, 531.

48. F. Ciucci, W. C. Chueh, D. G. Goodwin and S. M. Haile, *Phys. Chem. Chem. Phys.*, 2011, **13**, 2121.

49. W. Lai, *J. Power Sources*, 2011, **196**, 6534.

50. M. Z. Bazant, *Acc. Chem. Res.*, 2013, **46**, 1144.

51. A. J. Bard and L. R. Faulkner, 'Electrochemical Methods: Fundamentals and Applications', Wiley, New York, 1980.

52. A. M. Colclasure and R. J. Kee, *Electrochim. Acta*, 2010, **55**, 8960.

53. D. Danilov, R. A. H. Niessen and P. H. L. Notten, *J. Electrochem. Soc.*, 2011, **158**, A215.

54. M. Shadman Rad, D. L. Danilov, M. Baghalha, M. Kazemeini and P. H. L. Notten, *Electrochim. Acta*, 2013, **102**, 183.

55. A. Bates, S. Mukherjee, N. Schuppert, B. Son, J. G. Kim and S. Park, *Int. J. Energy Res.*, 2015, **39**, 1505.

56. A. A. Talin, D. Ruzmetov, A. Kolmakov, K. McKelvey, N. Ware, F. El Gabaly, B. Dunn and H. S. White, *ACS Appl. Mater. Interf.*, 2016, **8**, 32385.

57. S. D. Fabre, D. Guy-Bouyssou, P. Bouillon, F. Le Cras and C. Delacourt, *J. Electrochem. Soc.*, 2011, **159**, A104.

58. M. Finsterbusch, T. Danner, C.-L. Tsai, S. Uhlenbruck, A. Latz and O. Guillon, *ACS Appl. Mater. Interf.*, 2018, **10**, 22329.

59. R. D. Armstrong and B. R. Horrocks, *Solid State Ion.*, 1997, **94**, 181.

60. A. C. Luntz, J. Voss and K. Reuter, *J. Phys. Chem. Lett.*, 2015, **6**, 4599.

61. G. Gouy, *J. Phys.*, 1910, **9**, 457.

62. D. L. Chapman, *Philos. Mag. J. Sci.*, 1913, **25**, 475.

63. O. Stern, *Z. Elektrochem.*, 1924, **30**, 1014.

64. M. Z. Bazant, B. D. Storey and A. A. Kornyshev, *Phys. Rev. Lett.*, 2011, **106**, 046102.

65. M. Bazant, K. Chu and B. Bayly, *SIAM J. Appl. Math.*, 2005, **65**, 1463.

66. W. Dreyer, C. Guhlke and R. Müller, *Phys. Chem. Chem. Phys.*, 2013, **15**, 7075.

67. S. R. De Groot and P. Mazur, 'Non-Equilibrium Thermodynamics', Courier Corporation, North Chelmsford, 2013.

68. I. Müller and T. Ruggeri, 'Rational Extended Thermodynamics', Springer Science & Business Media, Berlin, 2013.

69. W. Dreyer, C. Guhlke and R. Muller, *Phys. Chem. Chem. Phys.*, 2015, **17**, 27176.

70. B. R. Horrocks and R. D. Armstrong, *J. Phys. Chem. B*, 1999, **103**, 11332.

71. S. Braun, C. Yada and A. Latz, *J. Phys. Chem. C*, 2015, **119**, 22281.

72. M. Landstorfer, S. Funken and T. Jacob, *Phys. Chem. Chem. Phys.*, 2011, **13**, 12817.

73. S. Bishop, D. Marrocchelli, C. Chatzichristodoulou, N. Perry, M. B. Mogensen, H. Tuller and E. Wachsman, *Annu. Rev. Mater. Res.*, 2014, **44**, 205.

74. J. Dahn, *Phys. Rev. B*, 1991, **44**, 9170.

75. P. G. Bruce, B. Scrosati and J. M. Tarascon, *Angew. Chem. Int. Ed.*, 2008, **47**, 2930.

76. Y. Gao, M. Cho and M. Zhou, *J. Mech. Sci. Technol.*, 2013, **27**, 1205.

77. G. Bucci, Y.-M. Chiang and W. C. Carter, *Acta Mater.*, 2016, **104**, 33.

78. G. Bucci, T. Swamy, S. Bishop, B. W. Sheldon, Y.-M. Chiang and W. C. Carter, *J. Electrochem. Soc.*, 2017, **164**, A645.

79. G. Bucci, T. Swamy, Y.-M. Chiang and W. C. Carter, *J. Mater. Chem. A*, 2017, **5**, 19422.

80. P. P. Natsiavas, K. Weinberg, D. Rosato and M. Ortiz, *J. Mech. Phys. Solids*, 2016, **95**, 92.

# PART 5

# Advanced Characterizations of Inorganic Battery Materials

# TEM Studies on Electrode Materials for Secondary Ion Batteries

**Sooyeon Hwang and Dong Su**

*Center for Functional Nanomaterials, Brookhaven National Laboratory, Upton, NY, USA*



## 1 INTRODUCTION

Transmission electron microscopy (TEM) is a powerful characterization tool for exploring physical, chemical, and biological sciences.[1–5] It may realize three different functions of imaging, diffraction, and spectroscopy with proper detectors. In operation of a TEM, we normally use two modes: TEM and scanning-transmission electron microscopy (STEM). At TEM mode, parallel electron beam illuminates the specimen and the interfered image/diffraction patterns are formed at the screen. At STEM mode, the electron beam is focused to a small probe and being scanned while post-sample detectors collect either transmitted or scattered electrons to form images. If we use a high-angle annular dark-field (HAADF) detector to collect the dark-field signal in STEM mode, the contrast of the image is approximately dependent on the atomic number $Z^{1.7}$ (called $Z$-contrast imaging). At present, one can achieve sub-Angstrom resolution in both TEM and STEM modes. While the contrast of HAADF imaging can be directly interpreted from the mass-thickness, the contrast of a high-resolution TEM image in TEM mode is mainly related to the interference between the electrons (phase contrast), which needs the image simulation in order to be correctly interpreted. Diffraction techniques, for example, selected area electron diffraction (SAED) can give the structural information about the reciprocal space of the sample. To identify the phases of nanocrystals, nano-beam electron diffraction (NBED), and convergent beam electron diffraction (CBED) have been widely used. With spectroscopy techniques, we can also retrieve chemical information of the specimen from the inelastically scattered electrons. When the high-energy electron beam goes through a thin specimen, it interacts with both the nuclei and the electrons surrounding the nuclei via Coulombic interaction. An inelastic scattering event may happen between the fast electrons and the electron cloud. The electrons in the inner shell of the atom may absorb certain amount of energy and excite to the conduction band. A hole will be left in the shell and consequently, an outer shell electron will transfer to the inner shell with electromagnetic radiation in the form of an X-ray photon. The energy dispersive X-ray spectroscopy (EDS) records spectrum of the X-ray emitted by the outer electrons. In the meantime, the fast electrons which lost energy to the core electrons will be attributed to the electron energy-loss spectroscopy (EELS). Furthermore, due to the fruitful interaction between fast electron and specimen, we can collect a number of signals, such as secondary electrons, cathodoluminescence signal, Auger electrons, etc. using specific detectors for each signal. Depending on the applications, we have the advantages of TEM to combine these techniques to retrieve multiple information we need as shown in Figure 1. For example, combining scanning transmission electron microscopy–high angular annular dark field (STEM–HAADF) imaging

*Inorganic Battery Materials*. Edited by Hailiang Wang and Boniface P.T. Fokwa
 ISBN 978-1-119-43199-2

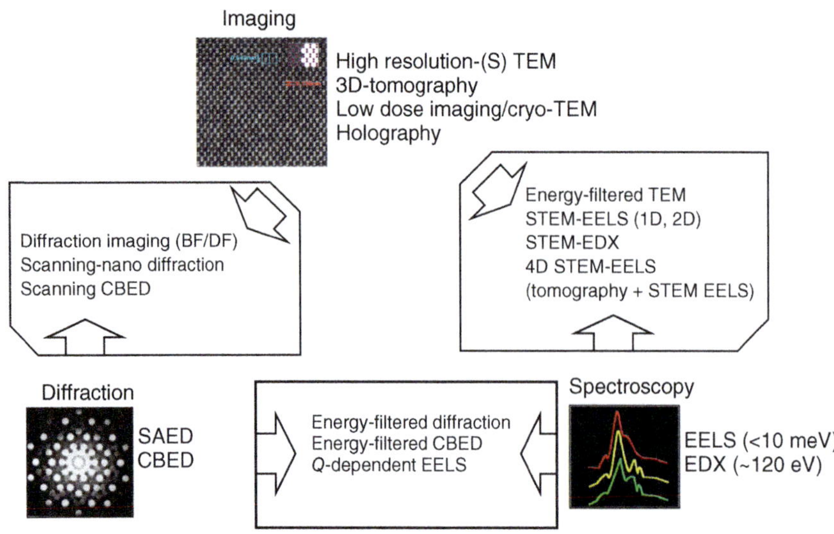


**Figure 1** Schematic of main TEM techniques: imaging, diffraction, and spectroscopy

and EELS or EDS, we can achieve scanning transmission electron microscopy–electron energy loss spectroscopy (STEM–EELS) or scanning transmission electron microscopy–energy dispersive X-ray spectroscopy (STEM–EDS) at STEM mode, which can acquire both structural and chemical information at atomic resolution simultaneously.[6–9]

The development of aberration correction (mainly spherical aberration (Cs)-correction) system renders modern TEMs to achieve much enhanced spatial resolution. For example, a spatial resolution of 0.04 nm has been achieved in both TEM and STEM imaging.[10] Furthermore, the energy resolution of a monochromatized TEM can be better than 5 meV, which is comparable to the resolution of synchrotron soft X-ray absorption spectroscopy.[11] This high-energy resolution will be beneficial to the investigations of low-loss EELS region, especially to the excitations due to electron–phonon interactions.

Comparing to the X-ray-based techniques, TEM has a few of advantages for the characterization of battery materials: (i) high spatial resolution in real space, (ii) high sensitivity to both elastic and inelastic interactions, (iii) easy combination of the techniques in different disciplines, and (iv) capability of studying defect structures. Whereas, there are also disadvantages of TEM: (i) the difficulty of sample preparation: the TEM specimen has to be very thin (<100 nm), and sample preparation is complicated and original sample is easy to be damaged during the sample preparation; (ii) the difficulty in quantitative analysis of the results; and (iii) the issue of electron beam-induced artifacts. Therefore, combined study of X-ray-based techniques and TEM can provide a full and in-depth understanding of the nature of materials at comparable length scales.

TEM techniques have been widely used in the field of batteries for visualizing the morphologies and investigating the structures of battery components (mostly cathode and anode materials). In order to examine the electrode materials at certain conditions, samples are generally prepared after electrochemical tests in coin cells. After reaching designated conditions (cutoff voltage, state of charge, etc.), coin cells are immediately disassembled and electrodes are cleaned with dimethyl carbonate (DMC), which is one of solvent in conventional electrolyte thus DMC should not make any side products on the sample. TEM samples are acquired by abrading electrodes, then drop-casted on TEM grids in DMC solution. These procedures are generally performed in a controlled environment (dry room or glove box) to minimize the exposure of air and moisture. Samples for TEM analysis should be thin enough (<100 nm) for transmission of electron beams. Considering that size of active cathode particle is generally over micron meter, we either only study the verge of particles or use some complicated sample-preparation techniques (microtome, ion-mill, focused ion beam (FIB)) to prepare suitable TEM sample. However, it is important to have the samples under certain controlled environment and being cooled to liquid $N_2$ temperature to reduce the $Ar/Ga^+$ beam-induced damage artifacts.[12]

The high-energy electron beam used in a TEM can cause temporary or permanent changes in the surface or bulk structure of a specimen.[13] For metallic materials, the charge induced by incident electrons is delocalized and there is no charge accumulation process. The elastic scattering effect plays the dominant role in the beam-induced damage of conductive materials. Assuming that $E_d$ is the energy transfer required for atomic displacement, only if the energy transferred from the incident electron to the atom is larger than $E_d$, the atom in the specimen can be displaced (damaged). The elastic process (or knock-on effect) involves only certain transfer of energy and momentum from fast electrons to atoms through an elastic collision (not all kinetic energy of fast electrons can be transferred to the atom). Other damage mechanisms—including radiolysis (for ionization), thermal effects (for phonon excitation), and electric-field-driven migration—are also applicable to battery materials.[14,15] In general, the total damage rate depends on the acceleration voltage, the atomic number $Z$, the electron beam dose, and the composition/geometry of the sample. In practice, lowering the electron beam intensity and shortening illumination time (both dose rate and dose) are the first two choices to avoid the electron-beam damage. Recently, the cryo-EM approach has been applied in the characterization of battery materials which will be introduced in later text.

In this chapter, we will introduce the general research principles of TEM and the recent research progresses on the electrode materials for secondary ion batteries. The first three sections are about ex situ, postmortem studies on electrode materials (cathode and anode) and in situ/operando TEM studies. We will emphasize on the TEM studies on lithium-ion batteries (LIBs) but also cover the representative work on other kinds of secondary ion batteries. In the last section, we will outlook the ongoing TEM approaches, which could be beneficial to the battery community. We assume readers already have had the basic knowledge about TEM techniques. We also provide the references for further reading. Last, within limited space, this chapter could not be exhaustive in covering all TEM-related works, but instead, put an emphasis on those published in the last several years and are of representing value.

## 2 CATHODE MATERIALS

Cathode materials for LIBs mostly work with insertion mechanism—lithium ions are inserted into empty space and extracted without significant structural changes. The crystallographic structure of cathode materials is a key factor to govern the electrochemical performances. X-ray-based diagnostic characterizations (such as XRD, X-ray absorption spectroscopy, pair distribution function (PDF), etc.) have been extensively performed to investigate structural evolutions induced by electrochemical reactions. As these techniques provide averaged bulk structural information, complementary tools are additionally needed to study the nascent and localized structural changes for complete understanding of materials. In this section, detailed structural analysis of cathode materials performed by TEM will be introduced.

### 2.1 $LiTMO_2$, Lithium Transition Metal Oxides

Layered lithium transition metal oxides ($LiTMO_2$ where TM: transition metals, e.g., Ni, Co, Mn) have been representative cathode materials for decades since Sony commercialized Li-ion batteries with a cathode of $LiCoO_2$ in 1991. While Li ions are extracted from (charge)/inserted into (discharge) $LiTMO_2$, TMs are oxidized/reduced in order to fulfill the charge neutrality in cathode materials. The redox reaction of TMs plays a crucial role in determining the behavior of cathode materials and the electrochemical performances of $LiTMO_2$ can be tuned by controlling the type and composition of TMs in layered compounds, for example, $LiNi_{0.8}Co_{0.15}Al_{0.05}O_2$ (NCA), $LiNi_xMn_yCo_zO_2$ (NMC, where $x+y+z=1$).

Most $LiTMO_2$ materials adopt an $\alpha$-$NaFeO_2$ layered structure with a space group of $R\bar{3}m$—closed packed oxygen framework with alternative Li and TM layers at octahedral sites, which is a deformed rock-salt structure (Figure 2). The layered structure of $LiTMO_2$ provides two-dimensional diffusion path of Li ions. In order to keep its original layered structure during reliable operation, the amount of extractable Li ions is restricted, for instance, only 50% of Li can be removed from $LiCoO_2$. In contrast, as the Ni $e_g$ band of $LiNiO_2$ does not overlap with oxygen 2p band, more lithium ions can be removed without disturbing oxygen framework. Thus, Ni-rich $LiTMO_2$ materials have drawn a lot of attention due to their possible high capacities. However, higher degree of charging may lead to instability of the crystal structure, which results in phase transformations to disordered spinel or rock-salt phase. Layered, spinel, and rock-salt phases share the same oxygen framework, thereby, phase transformations would happen with TM ion movements within an identical

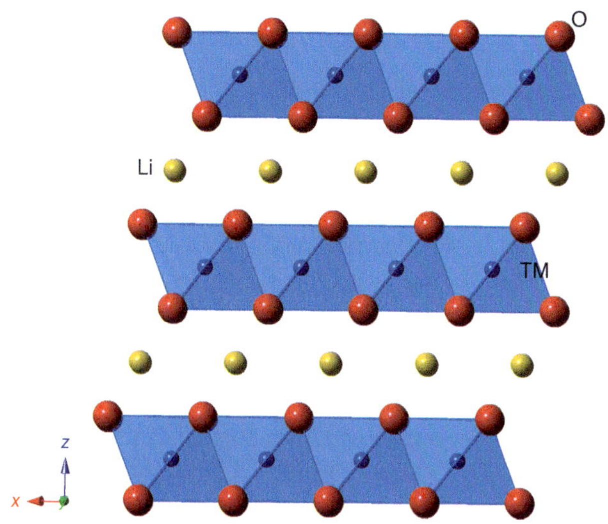


**Figure 2** Atomic model of the layered oxide $LiTMO_2$

oxygen arrangement. Upon Li-ion extraction, TMs of layered oxide compounds may migrate into the octahedral sites in Li slab and this cation mixing results in structural evolution of spinel or rock-salt phase, which are both thermodynamically more stable than layered structure[16] but unfortunately electrochemically inactive. Therefore, to take full advantage of lithium ions in $LiTMO_2$, it is of importance to understand incipient structural modifications and their propagation during repetitive Li movements.

Hwang *et al.* investigated the nascent phase transformations of $LiNi_{0.8}Co_{0.15}Al_{0.05}O_2$ (NCA) at different states of charge.[17] Using SAED and oxygen K-edge EELS, the evolution of crystallographic and electronic structures were simultaneously examined: localized phase transformations to disordered spinel and rock-salt structures occur with reduction of TM at the surface of NCA even after initial charge (Figure 3a and b). Zhang *et al.* explicated detailed steps of atomic migration in NCA during phase transformation from layered structure to spinel structure using high-resolution TEM imaging.[19] Zheng *et al.* demonstrated development of layers with random distribution of Li and TMs at grain boundaries of NCA after one electrochemical cycle (2.5–4.2 V) using high angle annular dark field scanning transmission electron microscopy (HAADF-STEM) imaging technique.[18] From a *Z*-contrast HAADF-STEM image, TM migration into Li layer can be intuitively identified (Figure 3c and d). While HAADF-STEM imaging is more favorable to discern heavy atoms, lighter atoms (e.g., Li and O) are also be identified using an annular bright field (ABF)-STEM technique. By comparing HAADF- and ABF-STEM images (Figure 4), Zhou *et al.* demonstrated intermediate phases of NMC materials after fast charging (30C), which have Li ions at pseudotetrahedral sites.[20] Effect of high-rate reaction on structural changes was also systematically investigated in NCA[21] and NMC materials[22] during fast charge/discharge.

EELS, which reflects the density of states above Fermi level,[23] has provided considerable insight about the chemical evolution of $LiTMO_2$ cathode materials upon Li-ion extraction/uptake via oxygen K-edge or 3d TM white line ($L_{2,3}$ edge) EELS. O K-edge is composed of two prominent peaks: a pre-edge (<530 eV), which is attributed to a transition of electrons from the O 1s state to empty 2p-TM 3d hybridized states and a main peak (~540 eV), which is originated from a transition of electrons from 1s to unoccupied O 2p-TM 4sp hybridized states.[24] Oxygen pre-edge peak is sensitive to valence of binding TM: it shifts to higher energy loss with reduction of TM, and vice versa. Transition of electrons from TM 2p states to highly localized TM 3d states near Fermi level forms $L_{2,3}$ edges of TM. The splitting of $L_{2,3}$ edges results from degenerate 2p states being split into $2p_{1/2}$ and $2p_{3/2}$ levels because of spin–orbit coupling.[25] Both onset energy of $L_{2,3}$ edge and $L_3/L_2$ intensity ratio are dependent on the oxidation states of TMs: onset energy loss becomes lower and $L_3/L_2$ intensity ratio becomes higher when TM is reduced, and vice versa. Distinctive changes occur at O K, and TM $L_{2,3}$ EELS with reduction or oxidation of TM, which allows us to examine the chemical evolution in $LiTMO_2$ upon electrochemical reactions.

Graetz *et al.* elucidated charge compensation mechanism in $LiNi_{0.8}Co_{0.2}O_2$ during extracting Li ions. Li K, O K, and Ni and Co white lines were examined over a wide range of lithium concentration, suggesting a participation of Ni and O ions in charge compensation during Li-ion removal.[26] STEM–EELS analysis of NCA cathode materials after 500 cycles demonstrated the Ni reduction near surfaces and grain boundaries, which causes capacity fading (Figure 5).[27] Surface TM reduction was also found in NMC materials and the reduced TM becomes different with original compositions of NMC cathode materials. Hwang *et al.* compared surface degradation in two compositions of NMC materials ($LiNi_{0.4}Mn_{0.3}Co_{0.3}O_2$ and $LiNi_{0.8}Mn_{0.1}Co_{0.1}O_2$) using STEM–EELS line scan, demonstrating Mn reduction in Mn-rich composition ($LiNi_{0.4}Mn_{0.3}Co_{0.3}O_2$) and Ni reduction in Ni-rich composition ($LiNi_{0.8}Mn_{0.1}Co_{0.1}O_2$).[28]

## 2.2 Li-Rich Lithium Transition Metal Oxides

Li-rich lithium transition metal oxides have drawn extensive attention as promising cathode materials since they can deliver capacities higher than the theoretical capacities of $LiTMO_2$ based on the redox reaction of TMs. The over-stoichiometry of Li ions brings about ambiguities in the crystal structure and there has been an open debate on it. One opinion is that Li-rich lithium transition metal oxides are a composite of $LiTMO_2$ and $Li_2MnO_3$. Structure of $Li_2MnO_3$ (Figure 6a) is analogous to that of $LiTMO_2$, but a fraction of Li-ions replaces TM sites. Lithium ions in TM layer are surrounded by 6 TM ions (Figure 6b). ABC stacking of $LiMn_2$ layer results in monoclinic *C2/m* structure, reducing six-fold rotation symmetry to two-fold.[29] Symmetry reduction brings about two different cross-sectional views of the $LiMn_2$ stacking sequence along <100> or <110> direction (Figure 6c and d, respectively). These characteristic atomic arrangements of monoclinic phase are favorable to HAADF-STEM analysis due to *Z*-contrast. Yu *et al.* determined Li and Mn-rich $Li_{1.2}Mn_{0.567}Ni_{0.166}Co_{0.067}O_2$ is composed of nano-domains of both rhombohedral $LiTMO_2$ (Figure 7a and b) and monoclinic $Li_2MnO_3$-like structures (Figure 7c and d) using electron diffraction and ABF-/HAADF-STEM imaging.[30] On the other hand, Shukla *et al.* claimed that the bulk of $Li_{1.2}N_{0.13}Mn_{0.54}Co_{0.13}O_2$ particle is aperiodic crystal composed of nanodomains which correspond to three different variants of monoclinic phase[31] (Figure 7e–k). The domains usually flip between

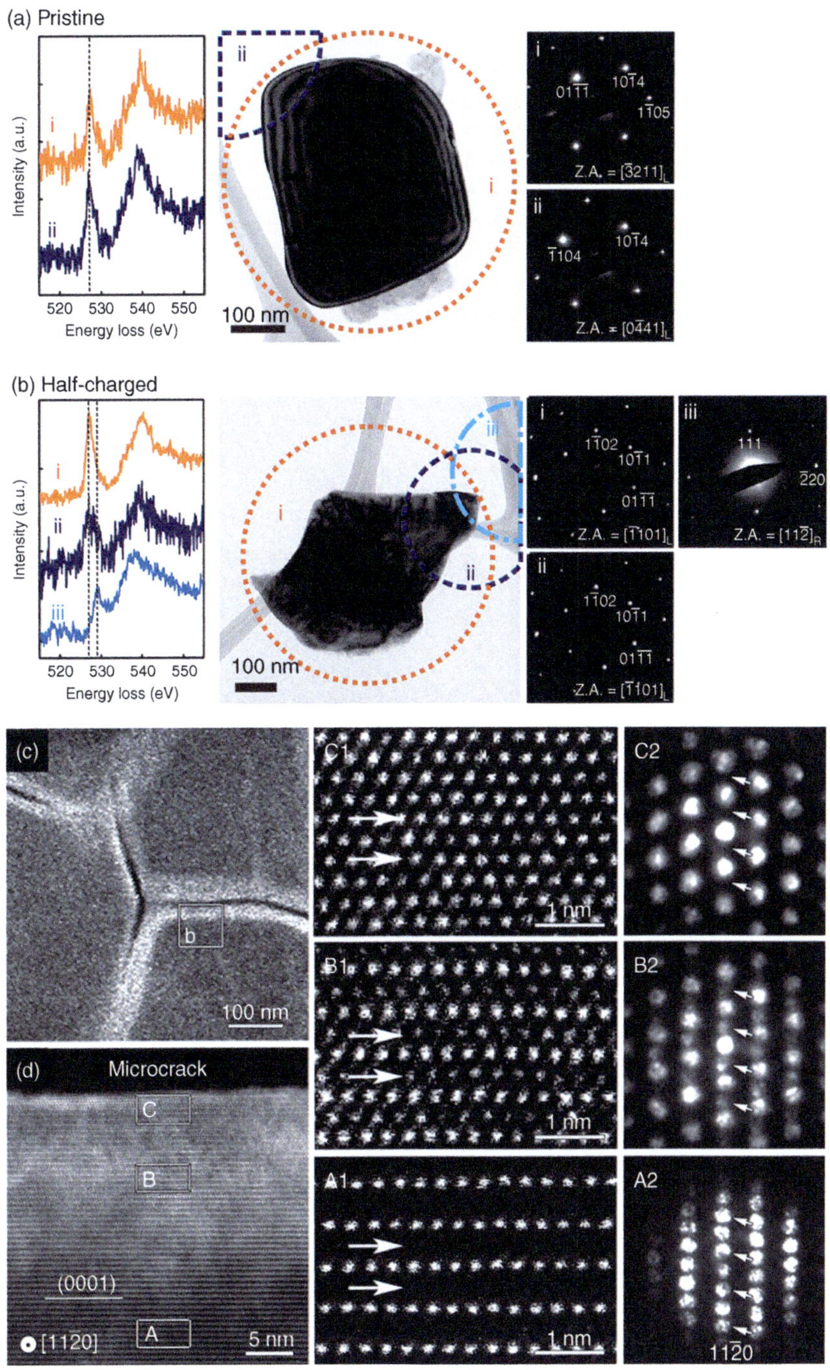


**Figure 3** BF images, SAEDs, and EEL spectra of the oxygen K-edge from (a) pristine, (b) half charged NCA, respectively. (c) STEM image showing typical grain boundaries in the single cycled NCA sample (d) a high magnification STEM corresponding to the framed region in (c); (A1)–(C1) STEM images acquired from regions A–C in (d), respectively, in which two Li layers are marked with arrows; (A2)–(C2) microdiffraction patterns corresponding to STEM images (A1)–(C1), respectively. [(a, b) Reprinted with permission from Ref. 17. © 2014 American Chemical Society, (c, d) Reproduced with permisssion from Ref. 18. © The Electrochemical Society, 2011]

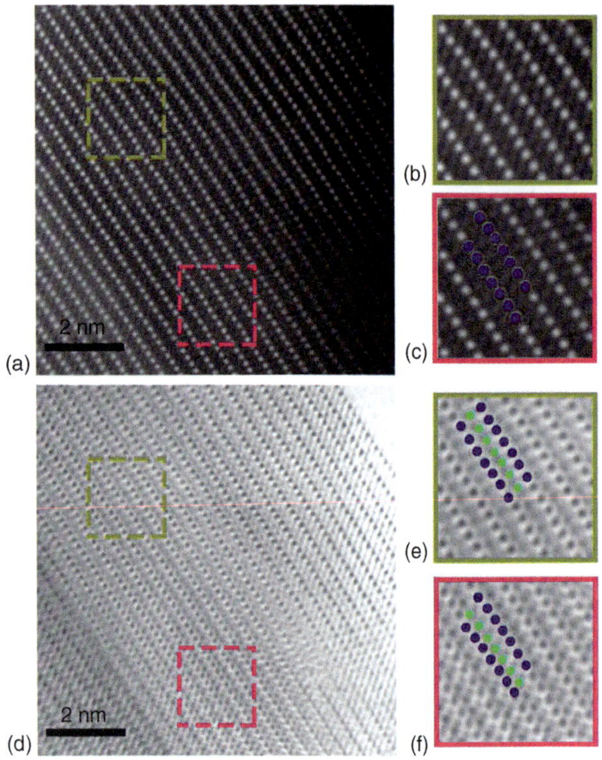


**Figure 4** Structure of half charged NMC at 30C rate. (a) Typical HAADF image taken along the [110] zone axis of the NMC electrode after 55 s charging at the current rate of 30C. (b, c) The zoom-in image of the areas marked with orange and pink squares, respectively. (d) Corresponding ABF images of NMC electrode after 55 s charging. (e, f) The zoom-in image of the areas marked with orange and pink squares, respectively. The blue and green dots indicate the TM ions and Li ions, respectively. [Reprinted with permission from Ref. 20, © WILEY-VCH Verlag GmbH & Co. 2016]

$[100]_M$ and $[110]_M$ or $[1\bar{1}0]_M$ where M stands for the monoclinic unit cell. It is noteworthy that monoclinic phase is hard to differentiate from stoichiometric $LiTMO_2$ depending on the projection direction (such as [010], [001]). Therefore, analysis at different zone axis should be carried out to unravel the crystal structure of material.

Besides of crystallographic uncertainty and variation of chemical distribution in Li-rich compounds have been reported using STEM–EELS or STEM–EDS. Gu *et al.* took advantage of EDS mapping and tomography to elucidate uneven three-dimensional Ni distribution in $Li_{1.2}Ni_{0.2}Mn_{0.6}O_2$ (LMR), which may influence the kinetics of Li transport.[32] Yan *et al.* found pristine Ni segregation layer in a LMR cathode material along $(20\bar{2})$ planes.[33] In case of $Li_{1.2}Ni_{0.13}Co_{0.13}Mn_{0.54}O_2$, Ni segregation was found at $(200)_M$ facet while Co segregation layers were developed on $(20\bar{2})_M$ plane.[34] Plane selectivity of Ni and Co segregation layers depends on the TM compositions and synthesis methods.

While the capacities of stoichiometric $LiTMO_2$ solely depends on the redox capability of TM ions, it is known that the high capacities of Li-rich compounds are attributed to cumulative cationic ($TM^{n+} \leftrightarrow TM^{(n+1)+}$) and anionic ($O^{2-} \leftrightarrow O_2^{2-}$) redox because of M(d)–O(sp) hybridization.[35,36] During the first charge (Li-ion extraction), voltage profiles are manifested as a staircase with a high voltage plateau but the voltage curves become to S-like shape while Li-ion insertion and following cycles, which may indicate different redox mechanisms. Large first-cycle irreversible capacity loss occurs due to oxygen liberation from cathode materials and partial electrolyte oxidation.[37,38] Despite high capacities, Li-rich compounds face great challenges including voltage and capacity fading over cycles and slow charge/discharge rates. TEM has been employed to address reaction mechanisms as well as degradation mechanisms of Li-rich compounds.

McCalla *et al.* visualized the formation of peroxo-like species in delithiated $Li_{0.5}IrO_3$ sample, which are responsible for the capacity gain in Li-rich layered electrode materials.[39] Figure 8(a) and (b) show HAADF- and ABF-STEM images, demonstrating formation of shortened black dumbbells in Figure 8b and lengthened projected O–O separations. The shortened projected O–O represents the peroxo-like dimers. Ito *et al.* showed movement of TM ions to Li layers in $Li[Ni_{0.17}Li_{0.2}Co_{0.07}Mn_{0.56}]O_2$, lead to the phase transformation to a spinel structure. Yan *et al.* proposed a sequential phase transitions of $C2/m \rightarrow I41 \rightarrow$ Spinel at LMR after cycles (Figure 8c–h), which may be responsible for voltage and capacity fading.[33] While pristine LMR shows a homogeneous structure from surface to bulk (Figure 8c), newly evolved phases were found in the vicinity of the surface after electrochemical cycles (Figure 8d–f). In order to decide the phase of surface reconstruction layer (SRL), experimental and simulation results of diffraction patterns, HRTEM and HAADF-STEM images were systematically compared. Zheng *et al.* suggested that structural evolution during cycling is highly dependent on the initial structure of cathode material and preparation conditions. LMR cathode prepared by hydrothermal assisted method showed phase transformation from $C2/m$ to $R\bar{3}m$ after activation, then defect spinel, and disordered rock-salt structures after long-term cycling.[40]

## 2.3 Other Intercalation-Type Cathode Materials

The olivine-type lithium iron phosphate, $LiFePO_4$, has been a strong candidate of cathode materials for LIBs as well as the layered lithium transition metal oxides. Phosphate materials are stable in contact with organic electrolyte. In addition, Fe-based materials are relatively inexpensive and environmentally benign. In spite of advantages of $LiFePO_4$, poor electronic and ionic conductivity have been serious challenges to be overcome.

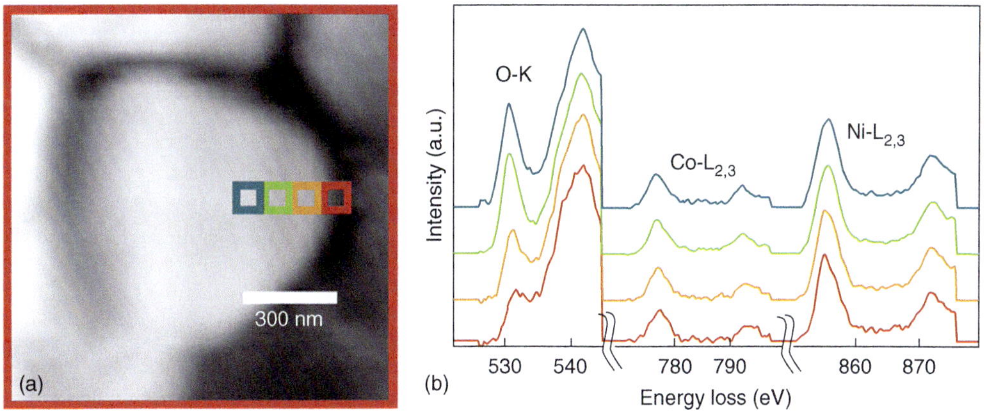


**Figure 5** (a) HAADF image and (b) O–K, Co-$L_{2,3}$, and Ni-$L_{2,3}$ ELNES spectra of NCA cathode materials after 500 test cycles at 80 °C. Spectra at (b) aligned from top to bottom were acquired the areas in (a) from surface to bulk. [Reproduced with permisssion from Ref. 27. © The Electrochemical Society, 2009]

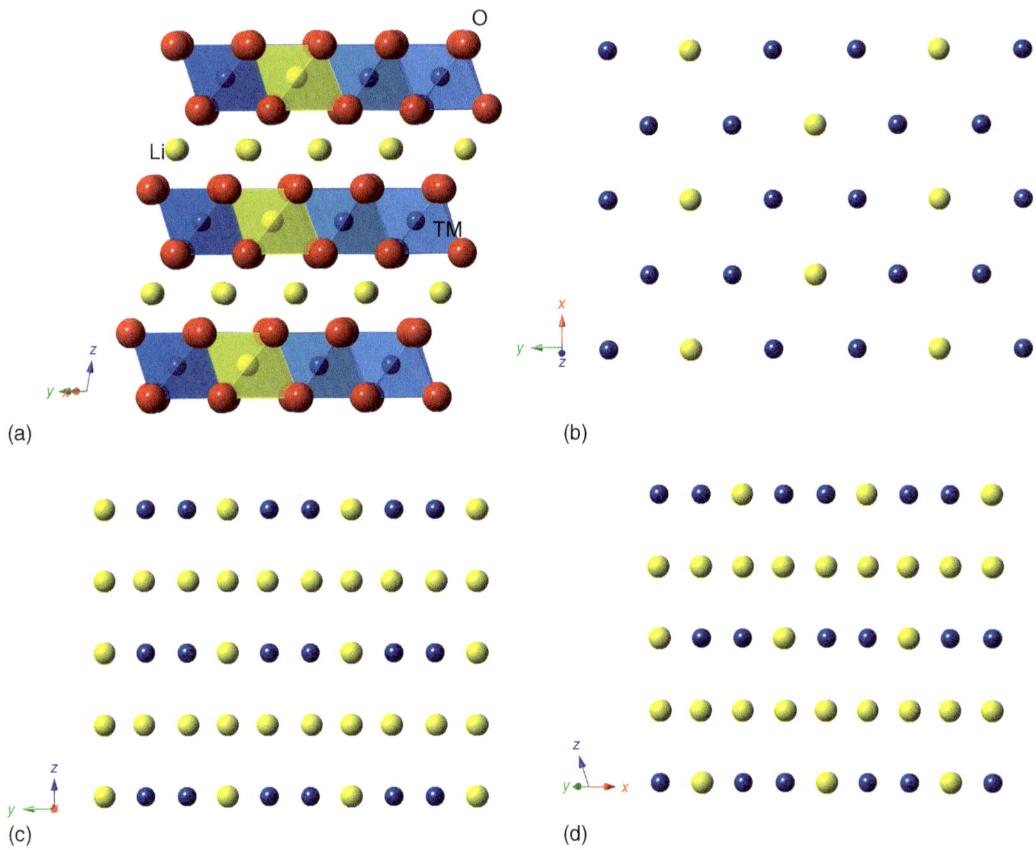


**Figure 6** (a) Atomic model of Li-rich lithium transition metal oxides. Atomic arrangement (b) at transition metal layer (c) along 100 [100] and (d) along 110 [110] direction, respectively

The olivine structure is composed of $PO_4^{3-}$ tetrahedra with $TM^{2+}$ ions at corner-shared octahedral positions and Li ions in chains of edge-sharing octahedra. $LiFePO_4$ and delithiated form of $FePO_4$ have same olivine structure with different occupancy of Li ions. The mechanism of isostructural change from $LiFePO_4$ to $FePO_4$ has been addressed by a variety of analytic tools including TEM. Chen *et al.* examined this phase transition after chemical delithiation,[41] suggesting that phase transformation proceeds along *a*-axis while lithium ions were extracted

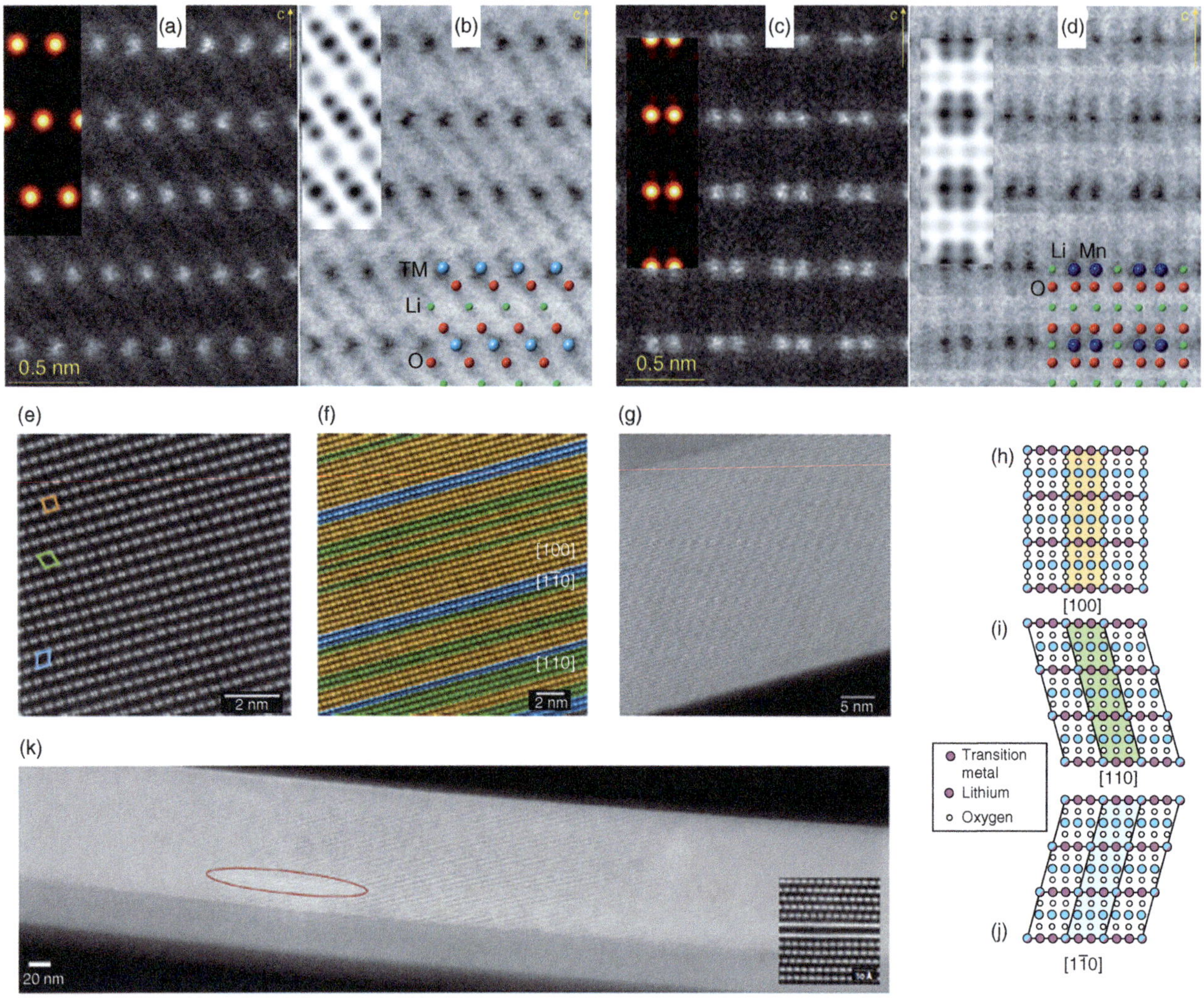


**Figure 7** (a) and (b) HAADF and ABF STEM images of the rhombohedral $LiTMO_2$ structure projected along the $[100]_{rh}$ direction. (c) and (d) HAADF and ABF STEM images of the monoclinic $Li_2MnO_3$-like structure projected along the $[100]_{mon}$ direction. Insets: (top left) simulated HAADF and ABF STEM images of the rhombohedral $LiTMO_2$ and monoclinic $Li_2MnO_3$-like structures based on fast-Fourier transform multi-slice algorithm. Insets: (bottom right) of (b) and (d) are the crystal-structure image of the rhombohedral $LiTMO_2$ and monoclinic $Li_2MnO_3$-like structures, respectively. (e) HAADF image of Li-rich Mn-rich transition metal oxides sample. (f) Color-coded HAADF image showing the variants of the monoclinic phase. (g) HAADF image with a larger field of view showing that domains of a single-phase, monoclinic structure are present throughout the primary particle. (h–j) Atomic model showing the monoclinic structure along [110], [110], and [110] directions, respectively. (k) HAADF image showing that the structure is observed over the entire primary particle. Occasionally, transition metal-rich defects are observed as shown in the inset, which corresponds to the defect shown in the red oval. [(a–d) Reprinted with permission from Ref. 30, © WILEY-VCH Verlag GmbH & Co. KGaA, Weinheim, 2013, (e–j) Reprinted with permission from Ref. 31, © Springer Nature, 2015]

along *b*-axis. In spite of the existence of disordered transition zone, newly evolved $FePO_4$ domains have high crystallinity. Laffont *et al.* took advantage of a combination of STEM and EELS to investigate the chemical evolution of $LiFePO_4/FePO_4$ two-phase system.[42] Line scan results of O K-edge and Fe $L_{2,3}$ edge EELS indicate that partially delithiated $Li_xFePO_4$ is composed of separate domains of $FePO_4$ and $LiFePO_4$. Between two phases, interface layer exists, where is a linear combination of $LiFePO_4$ and $FePO_4$. Fracture of $LiFePO_4$ particle was observed after electrochemical and chemical delithiation: fracture surface was parallel to (100) planes after chemical delithiation while that was parallel to either (100) or (010) plane after electrochemical charge.[43] Gu *et al.* suggested staging mechanism of delithiation, which is inconsistent with prementioned two-phase mechanism.[44] ABF-STEM images directly showed that Li ions exist at every other row in $Li_{0.5}FePO_4$ when image was acquired from [010] direction after electrochemical charging.

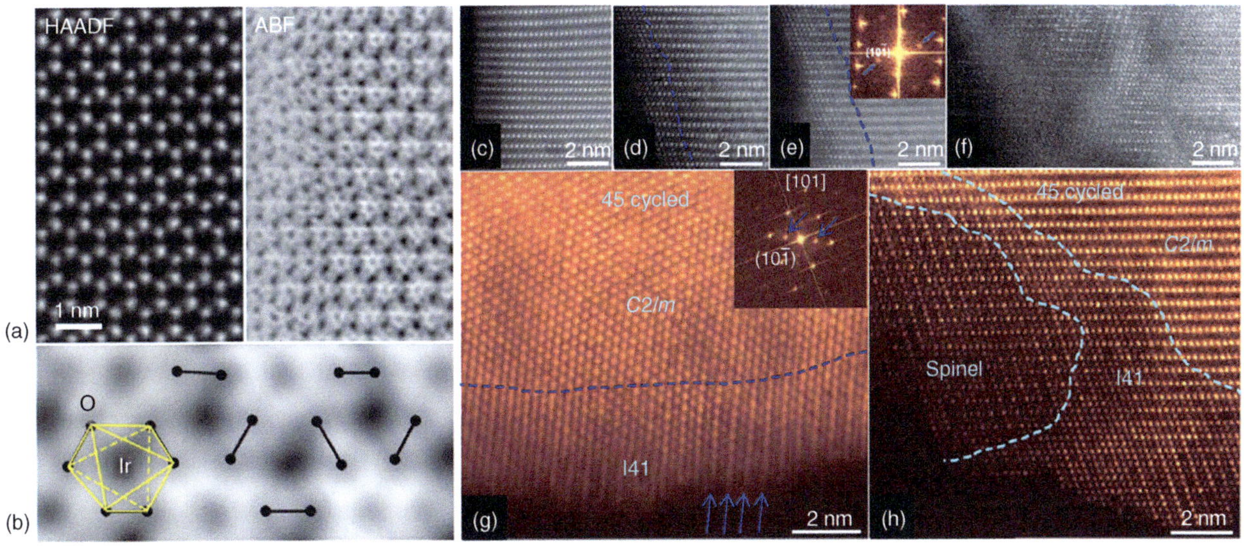


**Figure 8** (a) [001] HAADF-STEM and ABF-STEM images of $Li_{0.5}IrO_3$. The noise in the ABF-STEM image is suppressed by applying a low-pass Fourier filter). (b) Enlarged ABF-STEM image. O–O pairs with short projected distances are marked with dumbbells. The O–O pairs arise from twisting the opposite triangular faces of the $IrO_6$ octahedra (shown in yellow) (c–f) High-resolution STEM-HAADF images to show the cycling induced structure change on $Li_{1.2}Ni_{0.2}Mn_{0.6}O_2$ particle surfaces. Pristine samples (c) shows homogeneous structure from surface to bulk. Dashed lines in (d, e) highlight the thickness of the SRL. In (f), the whole areas were transformed. (g) STEM-HAADF image along [101] zone axis and its FFT image. Blue arrows indicate the ordered features of $(20\bar{2})$ planes and extra diffraction spots. (h) [010] zone axis STEM-HAADF image to show spinel structure and I41 structure in a 45 cycled sample. [(a, b) Reproduced with permission from Ref. 39, © American Association for the Advancement of Science 2015; (c–h) Reprinted with permission from Yan, P., Nie, A., Zheng, J., Zhou, Y., Lu, D., Zhang, X., Xu, R., Belharouak, I., Zu, X., Xiao, J., Amine, K., Liu, J., Gao, F., Shahbazian-Yassar, R., Zhang, J.-G., Wang, C.-M. (2015) Evolution of Lattice Structure and Chemical Composition of the Surface Reconstruction Layer in Li1.2Ni0.2Mn0.6O2 Cathode Material for Lithium Ion Batteries. Nano Lett., 15 (1), 514–522, © American Chemical Society 2015]

## 2.4 Conversion-Type Cathode Materials

Intercalation reaction requires vacant spaces for lithium ions to be inserted, thus, the number of Li ions which can participate in electrochemical reaction is highly limited by crystal structure. Some of oxide, sulfide, fluoride materials can store Li (or Na) ions via conversion reaction—which is expressed as $M_aX_b + (b{\cdot}n)\ Li^+$ (or $Na^+$) $+ (b{\cdot}n)\ e^- \leftrightarrow aM + bLi_nX$ (or $bNa_nX$) where M represents metal and X = O, S, F, etc. Multiple Li ions and electrons are associated during conversion reaction, delivering much higher capacity than intercalation-type electrode materials. Metal fluorides possess relatively high electrochemical potential, thereby considered as promising cathode materials of high capacity.

Conversion mechanism of $FeF_2$ and $CuF_2$ was examined after electrochemical discharge.[45] Li ion's insertion into $FeF_2$ results in the formation of ultrafine metallic Fe nanoparticles (less than 5 nm) surrounded by insulating LiF matrix. Fe particles are connected each other, providing local electron pathways. Evolution of metallic Cu and LiF was also observed but Cu particles were relatively large (5–12 nm in diameter) and separated by LiF, which may partially attribute to poor reversibility of $CuF_2$ electrode. Li *et al.* investigated the structural changes of $FeF_3$ during conversion/deconversion.[46] High-resolution TEM images and corresponding fast Fourier transformation (FFT) results suggested that interconnected Fe networks with LiF matrix were formed after the first discharge. Lithium extraction induces reverse conversion reaction, but the product was defective trirutile $FeF_3$ rather than original rhombohedral $FeF_3$, which may be responsible for capacity fading. In order to enhance cyclability of metal fluorides, oxygen substitution has been attempted. Kim *et al.* explored reaction nature of $FeO_{0.7}F_{1.3}$ to elucidate the effect of mixed-anion on electrochemical properties and structural evolutions.[47] The structure of $FeO_{0.7}F_{1.3}$ is rutile, which is same as that of $FeF_2$ but chemical distribution was not homogeneous: surface is O-rich while particle core is F-rich, evidenced by STEM–EELS elemental mapping. Conversion reaction preferentially occurred at the core, results in the formation of metallic Fe while Li–Fe–O–F rock-salt phase were evolved at the surface. Between $Fe^0$ and rock-salt phase, high lattice coherency was observed, which may be beneficial for ionic and electronic transport. Lithium extraction leads to the formation of core/shell like inhomogeneous structure—F-rich rutile (core) and O-rich rock-salt (shell). O-rich shell may protect active core, providing enhanced mechanical and chemical stability of $FeO_xF_{2-x}$ cathode materials.

## 3 ANODE MATERIALS

Currently, the standard anode materials for commercial LIBs are carbonaceous materials including graphite.[48] A few of other intercalation-type anode materials also have been developed, such as $TiO_2$, $Li_4Ti_5O_{12}$, etc. However, these intercalation compounds can only accommodate a limited amount of Li ions in their structure (low capacity). To achieve a higher energy density, new compounds have been explored for the utilization of anode.[49,50] These materials have mainly two types of reaction chemistries: (i) materials with alloying reaction, such as Si, Ge, and $SiO_2$, (ii) materials with conversion reaction, such as rock-salt oxides ((MO; M = Mn, Fe, Co, Ni, or Cu) and spinel oxide compounds (($AB_2O_4$, A/B = Mn, Fe, Co, Ni, or Cu). Some materials store Li ions via multiple reactions such as the intercalation–conversion reaction or conversion–alloying reaction. What determines which reaction mechanism works and how these different types of reactions are engaged each other are still not well known.

Owing to the large volumetric change during the lithiation, the alloying and conversion compounds exhibit much worse cycling stability than the intercalation compounds. In practice, researchers have applied different nanoengineering techniques to improve the performances of these two kinds of compounds, such as nanosized structure, surface coating, integrated with carbon nanotube (CNT)/graphene or 3D network/porous materials.[50–53] The electrochemical reactions associated with the real electrode are complicated and take place in highly nonequilibrium states which are not fully understood yet. Thus, when carrying out a TEM work, it demands us to be very careful to consider all possible factors in sample preparation and reactions. In this section, we will introduce the structural characterization works on intercalation, conversion, and alloy types of anode materials. This section will emphasize the ex situ TEM work. The in situ TEM work on nonequilibrium reaction processes will be discussed in the next section.

### 3.1 Intercalation-Type Anode

The most widely used anode materials with intercalation reaction are graphite, $TiO_2$, and $Li_4Ti_5O_{12}$ (LTO). Graphite undergoes an intercalation reaction from graphite (space group: $P6_3mc$) to $LiC_6$ (space group: $P6/mmm$) and delivers a theoretical capacity of 372 mAh $g^{-1}$.[54] Intensive studies have been done to understand the intercalation process of lithium. Despite the similar chemistry of $Li^+$ and $Na^+$, it is known that graphite is not favorable to accommodate sodium ions. However, recently, expanded graphite (EG) was successfully synthesized as an anode material for sodium-ion batteries. EG is a graphite-derived material formed by a two-step oxidation–reduction process and it retains the long-range ordered, layered structure of graphite. The structure of pristine graphite (PG), graphite oxide (GO), EG-1 h and EG-5 h are shown in the HRTEM images of Figure 9.[55] EG shows a much better conductivity than GO (EG-0 h) because of the removal of oxygen during thermal reduction. The interlayer spacing of EG can be experimentally controlled: 0.43 nm for EG-1 sample while

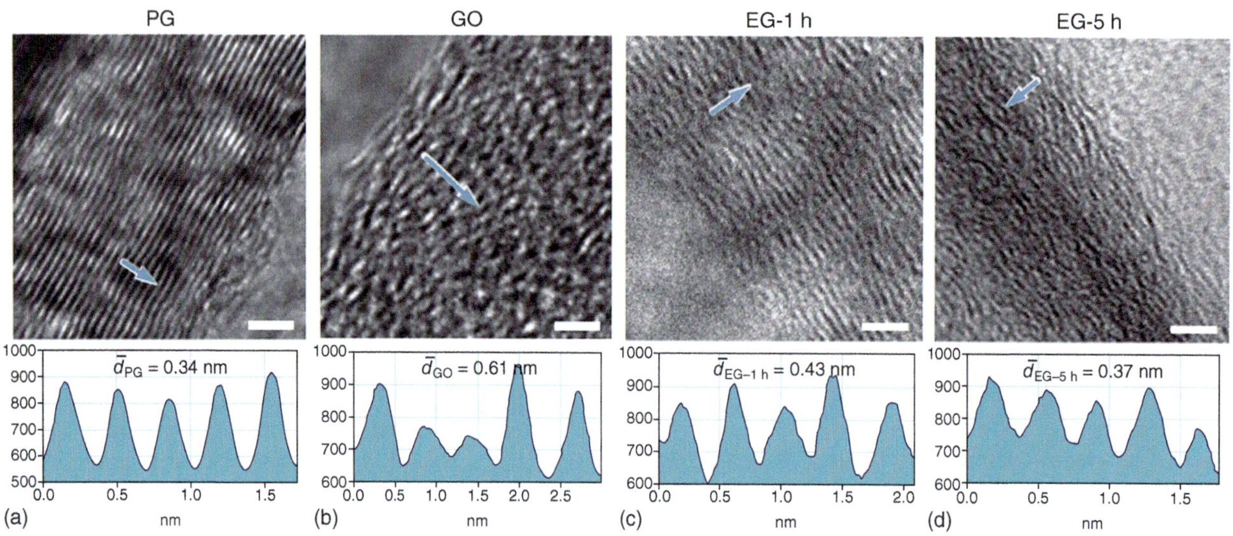


**Figure 9** HR TEM images showing cross-sectional layered structures for (a) pristine graphite (PG), (b) graphite oxide (GO), (c) expended graphite (EG)-1 h and (d) EG-5 h. Scale bars, 2 nm. Contrast profiles along the arrows indicate interlayer spacings of corresponding samples.[55] [Reprinted with permission from Wen, Y., He, K., Zhu, Y., Han, F., Xu, Y., Matsuda, I., Ishii, Y., Cumings, J. Wang, C. (2014). Expanded graphite as superior anode for sodium-ion batteries. Nature communications, 5, 4033. © Springer Nature, 2014]

0.33 nm for graphite. It is found that $Na^+$ ions indeed can be reversibly inserted into or extracted from the EG-1.

Nanoengineered titanium oxides having anatase, rutile, and $TiO_2$(B) structures have been investigated as anode materials for secondary ion batteries. Researchers have tried to mix $TiO_2$ with highly conductive materials such as CNTs and graphene to boost the electronic conductivity, and hence to improve the electrochemical performance of $TiO_2$.[56,57] Su *et al.* compared the electrochemical performances of hollow nanospheres of amorphous $TiO_2$, mixed anatase/rutile $TiO_2$, and pure anatase $TiO_2$.[58] The morphology of three $TiO_2$ samples is shown in Figure 10. Although these nanostructures have similar size and shape, pure anatase shows the best rate-performance due to the lowest energy barriers for the insertion of $Na^+$.

Spinel LTO exhibits an excellent cycling reversibility and a high thermodynamic stability due to its high voltage plateau of 1.55 V versus $Li/Li^+$, and almost negligible volumetric change during the charge and discharge processes. In its structure, the 32e positions are occupied by oxygen atoms, 5/6 of the 16d positions are taken by Ti atoms and the rest of the 16d positions are taken by Li atoms. The tetrahedral (8a) sites are occupied by Li atoms while the octahedral (16c) sites are empty. This structure

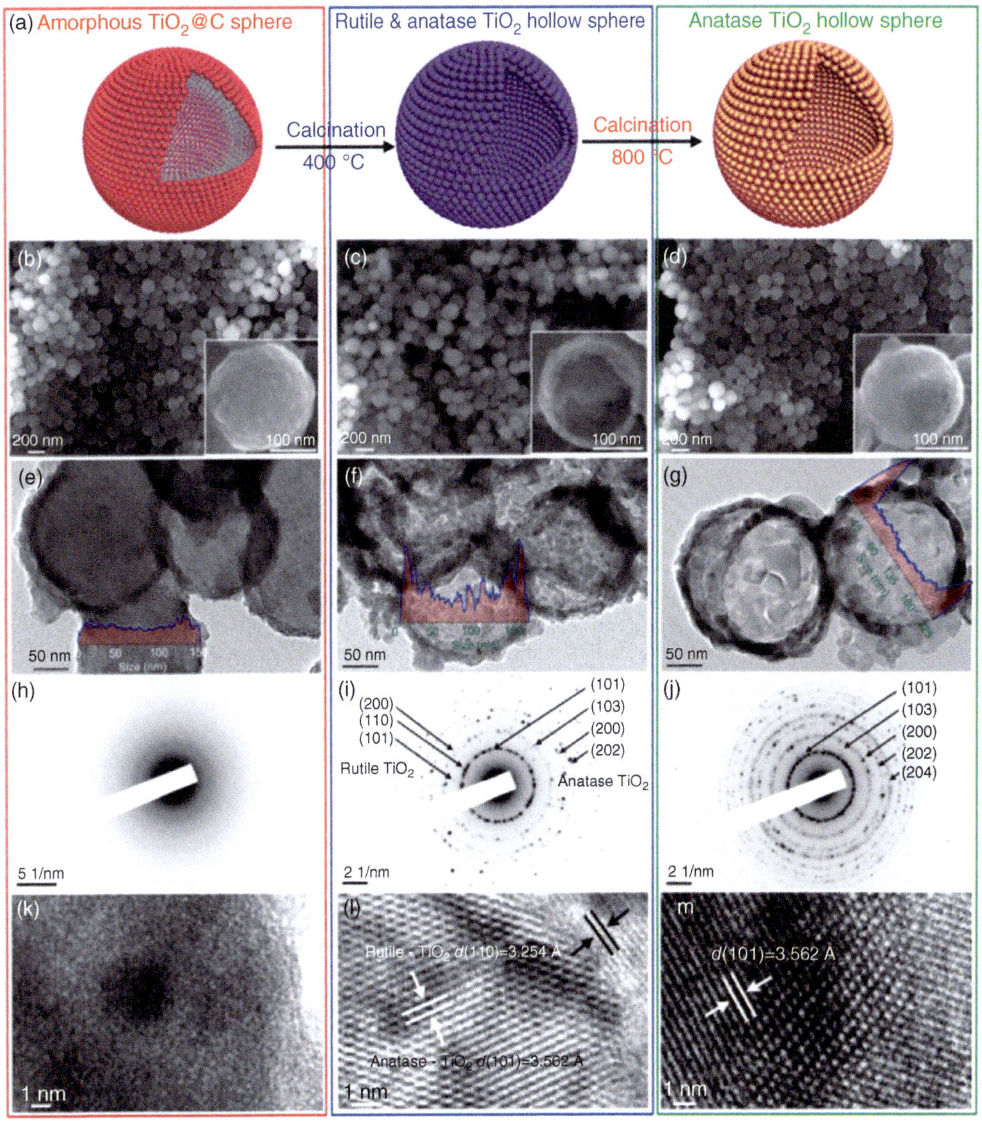


**Figure 10** (a) Schematic diagram of the transformation from amorphous $TiO_2$@C nanospheres to the rutile and anatase $TiO_2$ hollow nanosphere composite, and then to pure anatase $TiO_2$ hollow nanospheres. (b–d) FE-SEM images of these three samples, with the insets showing enlargements of single nanospheres. (e–g) TEM images of these three samples. Insets show the electron signals on the cross section of a $TiO_2$ nanosphere. (h–j) SAED patterns and (k–m) Lattice resolution HRTEM images of these three samples. [Reprinted with permission from Su, D., Dou, S., Wang, G. (2015) Anatase $TiO_2$: better anode material than amorphous and rutile phases of $TiO_2$ for Na-ion batteries. Chemistry of Materials, 27 (17), 6022–6029. © 2015 American Chemical Society]

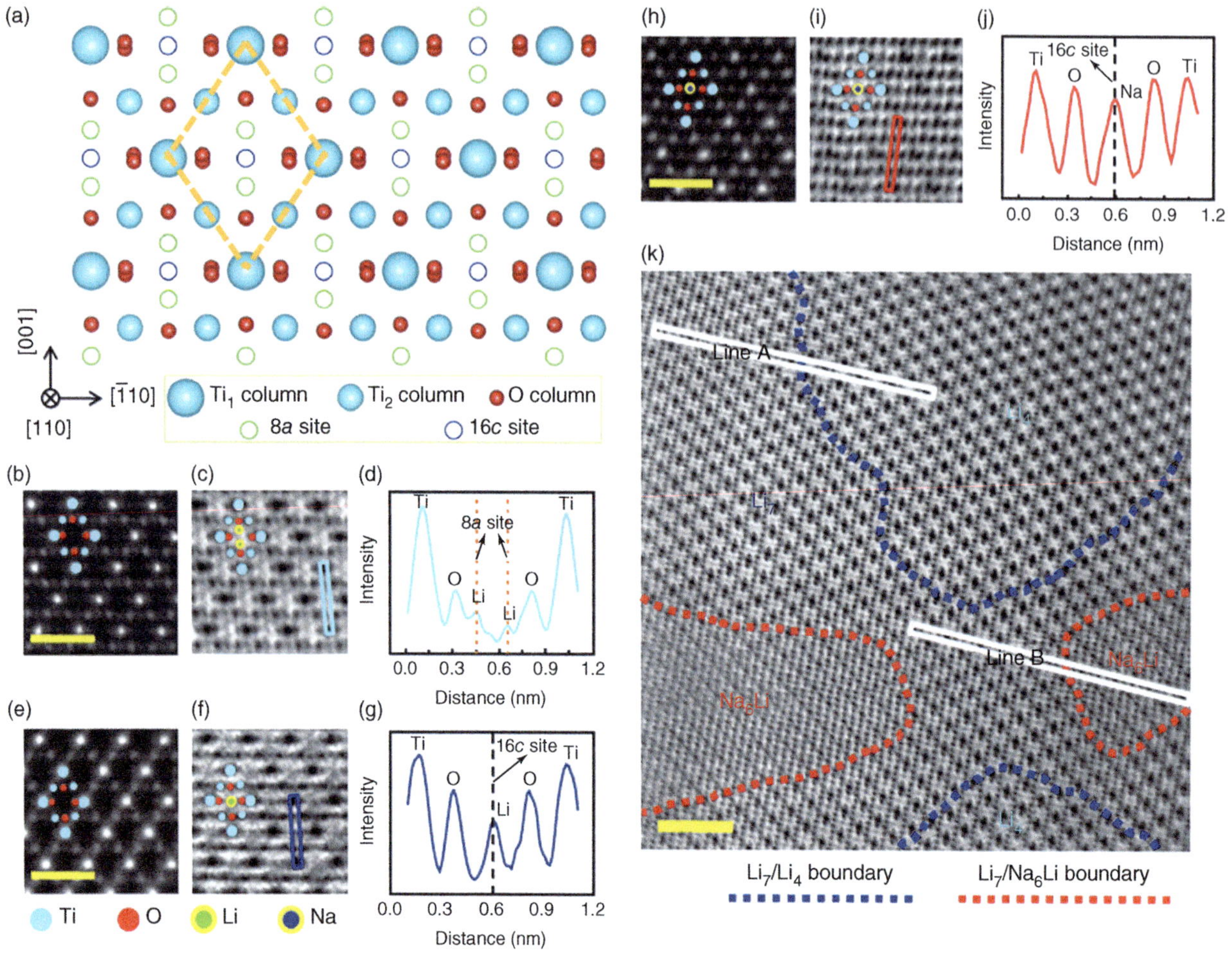


**Figure 11** (a) Crystal structure of spinel $Li_4Ti_5O_{12}$ viewed from the [110] crystallographic direction showing separated Ti and O columns. Ti1 and Ti2 columns with different atom densities are represented by balls of different size. HAADF (b), ABF (c) images and ABF line profile (d) of $Li_4Ti_5O_{12}$ ($Li_4$) phase. Scale bar, 1 nm. HAADF (e), ABF (f) images and ABF line profile (g) of $Li_7Ti_5O_{12}$ ($Li_7$) phase. Scale bar, 1 nm. HAADF (h), ABF (i) images and ABF line profile (j) of $Na_6LiTi_5O_{12}$ ($Na_6Li$) phase. Scale bar, 1 nm. (k) ABF image in the half electrochemically sodiated $Li_4Ti_5O_{12}$ nano-particle. In ABF line profile, the contrast is inverted for a convenient visualization. Scale bar, 2 nm. [Reprinted with permission from Sun, Y., Zhao, L., Pan, H., Lu, X., Gu, L., Hu, Y.S., Li, H., Armand, M., Ikuhara, Y., Chen, L., Huang, X. (2013) Direct atomic-scale confirmation of three-phase storage mechanism in $Li_4Ti_5O_{12}$ anodes or room-temperature sodium-ion batteries. Nature communications, 4, 1870. © Springer Nature 2013]

can be denoted as $[Li]_{8a}[Li_{1/3}Ti_{5/3}]_{16d}O_4$. Upon lithiation, three Li can be inserted into the 16c sites accompanied with movement of three Li ions from the 8a sites to the 16c sites. Used for the anode of sodium ion battery, the sodiation of LTO may induce the formation of two different phases with $Li^+$ or $Na^+$ at 16c sites, respectively. Figure 11 shows the atomic scale characterization of these two phases using STEM-ABF imaging technique.[59] This technique allowing to directly observe the light elements can be very helpful to identify the atomic structure of alkali oxide compounds. Using the same technique, edge structure of LTO nanosheets were directly revealed as terminated with rutile-$TiO_2$.[60] It is found that rutile-$TiO_2$ surface layer can work as a kind of carbon-free coating, which improves the kinetics of LTO toward fast lithium insertion/extraction.

### 3.2 Alloying-Type Anode

Si has drawn a huge attention as the next-generation anode material due to its high capacity (4200 mAh $g^{-1}$ when lithiated to $Li_{4.4}Si$). However, the huge volume expansion (300%) during the reaction is regarded as the main obstructive problem for its widespread applications. To accommodate the volumetric changes, Cui's group has modified the nanostructures of Si materials, for example, Si nanowires, Si–carbon yolk–shell structure, and hollow Si.[52,61–63] They also investigated the shape and volume changes of crystalline Si nanopillars with different orientations upon first lithiation, as shown in Figure 12. While uptake of Li ions, the original circular cross sections of nanopillars with ⟨100⟩, ⟨110⟩, and ⟨111⟩ axial orientations expand into cross, ellipse, and hexagonal

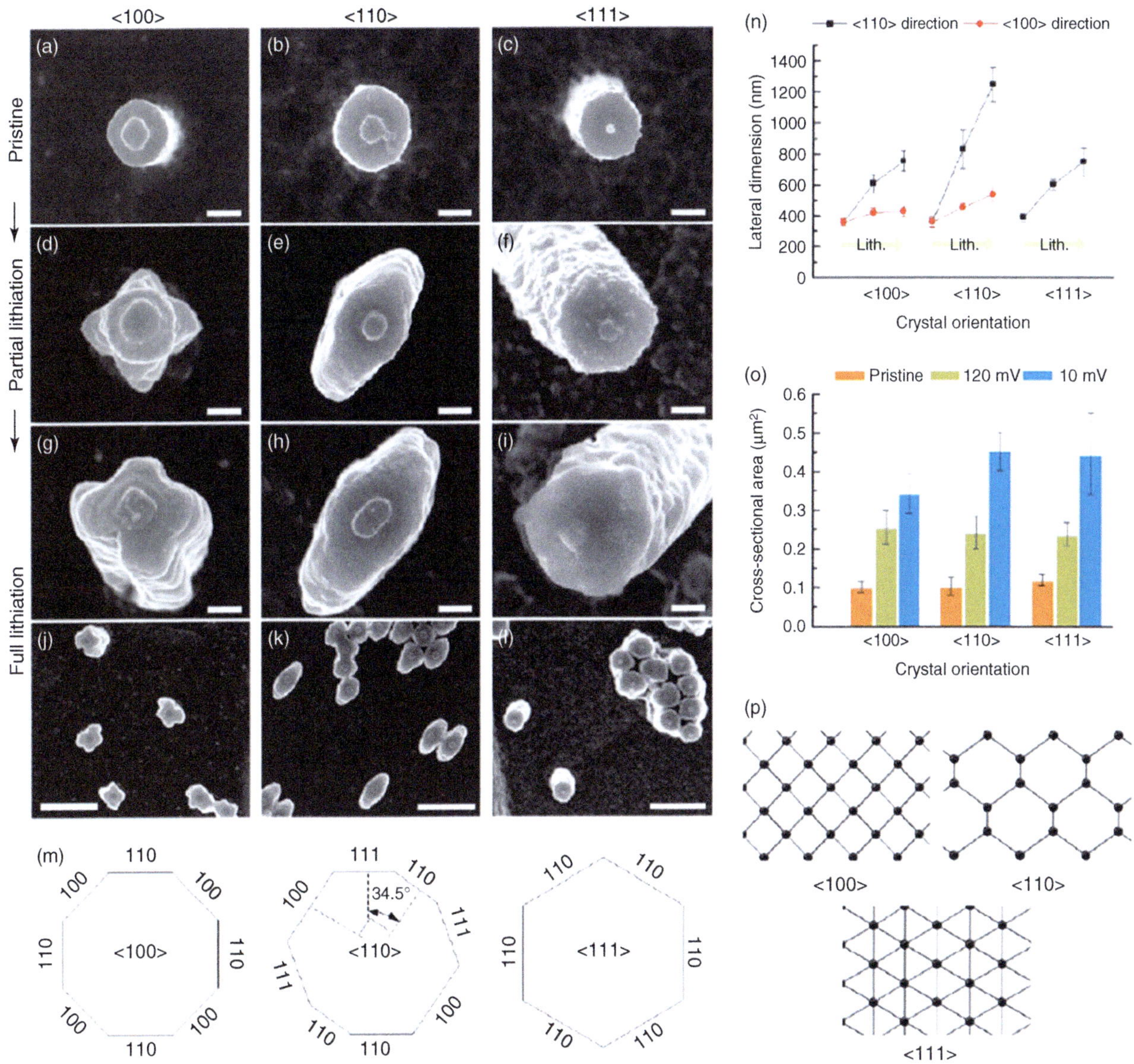


**Figure 12** Anisotropic lateral expansion of crystalline Si nanopillars with three different axial orientations (<100>, <110>, and <111>) upon lithiation. (a–l) Top-view SEM images of Si nanopillars of each crystal orientation and each lithiation state. The <100> axially oriented pillars are shown in the left column, <110> pillars are shown in the middle column, and <111> pillars are shown in the right column. The top row shows pristine pillars, the second row shows partially lithiated pillars held at 120 mV vs $Li/Li^+$, and the third row shows fully lithiated pillars held at 10 mV vs $Li/Li^+$. The images in the fourth row show low-magnification views of fully lithiated pillars of each axial orientation. Scale bars from (a–i) are 200 nm and from (j–l) are 2 μm. (m) Schematic diagram of the crystallographic orientation of the facets on the sidewalls of each of the pillars. (n) Statistical data of the changes in cross-sectional dimensions for the three types of nanopillars. (o) Statistical data of the changes in cross-sectional area for the three types of nanopillars. (p) View along three different directions of the diamond cubic lattice.[64] [Reprinted with permission from Lee, S. W., McDowell, M.T., Choi, J.W., Cui, Y. (2011) Anomalous shape changes of silicon nanopillars by electrochemical lithiation. Nano letters, 11 (7), 3034–3039. © American Chemical Society 2011]

shapes, respectively.[64] The in situ TEM study revealed new features of the phase evolution from Si to $Li_{4.4}Si$, which will be discussed later.

$SnO_2$ is another promising candidate for anode. It undergoes a two-step reaction: $SnO_2 \rightarrow Sn + Li_2O \rightarrow Li_xSn$, which are conversion and alloying.[65] Similar to the case of Si, large volume change induces a severe capacity fading over cycles. Thus, hybrid structure of a nanosized $SnO_2$ binding with graphene or other materials was attempted to achieve long cycle life. In case of alloying anode materials, ex situ analytical TEM may help to identify the microscopic changes (morphology, phase, composition, etc.) before and after cycling.

### 3.3 Conversion-Type Anode

Nanosized metal oxides have been explored as conversion-type anode materials since 2000.[66] After conversion reaction, a composite of $Li_2O$ nanoparticles and reduced metal nanoparticles is formed as the product. It was proposed that this nanocomposite nature may help to improve the cycling performance. However, similar to the alloying reaction, the significant volume change is believed to be troublesome for the long cycle life. There are many kinds of oxides for anode materials. Two typical anode compounds are rock-salt and spinel oxides. For rock-salt oxide, the reaction (with $Li^+$ and $Na^+$ ions) is a simple: direct conversion from metal oxide to metallic nanoparticles. However, due to various sources of heterogeneous reaction (overpotential, stain, etc.), there can be different reaction modes. He *et al.* used analytical TEM to characterize the lithiation process of NiO nanosheet and they found two reaction modes: shrinking-core mode and finger mode, which controls reaction speed when the redox reaction propagates from the surface to internal area of sample.[67] Using STEM-HAADF imaging and STEM 3D tomography technique, He *et al.* further investigated the reaction interface during a conversion reaction with $Na^+$ ions ($NiO + 2Na^+ = Na_2O + Ni$).[68] High-resolution STEM image, FFT and modeling results identified a lattice coherency between reduced Ni nanoparticles and NiO nanosheet as shown in Figure 13(a)–(c). 3D electron tomography in Figure 13(d) and the STEM–EELS mapping results identify the 3D morphology of this reaction front on the surface of NiO. Comparing its different reaction nature with $Li^+$ ions and $Na^+$ ions may help us to understand the intrinsic mechanism of the redox reactions with different ion species.

Spinel oxides, such as $Fe_3O_4$ (inverse spinel), $Co_3O_4$, and $Mn_3O_4$, experience two-step's reaction during lithiation. The initial insertion of $Li^+$ pushes the TM ions at the 8a site into the adjacent octahedral 16c site to form a rock-salt structure. This reaction step is very similar to the lithiation reaction occurring at $Li_4Ti_5O_{12}$. Further lithiation may cause the decomposition of the rock-salt structure into a nanocomposite of $Li_2O$ and metallic nanoparticles. An extrusion reaction, the insertion of Li ions pushes TM ions out of the framework, was suggested as a possible intermediate process.[69] Using HRTEM and FFT techniques, as shown in Figure 14, Li *et al.* identify the intermediate rock-salt phase and the two phases of the composite structure.[70] Spinel oxide–graphene hybrid/composite anodes have been synthesized, which show good performance with large reversible capacities.[71–73] Figure 15 shows the morphology of the composite of graphene nanosheet with $Fe_3O_4$ nanoparticles. It is believed that graphene layers play a "flexible confinement" function by enwrapping $Fe_3O_4$ particles, which can compensate for the volume change of $Fe_3O_4$ and prevent their detachment and agglomeration.[72]

## 4 IN SITU TEM

As we have discussed in previous sections, in situ TEM can provide more information than static analysis: one can observe dynamic structural and chemical changes occurring at samples in real time under the external stimuli such as mechanical, electrical, magnetic, etc. This allows one to directly link the structure–processing–property relationships,[74–76] which significantly improves our understandings in materials' properties.

### 4.1 In Situ Heating

Temperature inside of batteries can increase during operation for a number of reasons: failure of battery management system, mechanical stress, overcharge, high-temperature environment. At high temperatures, oxygen evolution from cathode materials is one of critical safety threats since oxygen gas may react with flammable electrolyte inside LIBs, resulting in either thermal runaway or battery failure. If LIBs are being incorporated into larger applications such as electric vehicles, safety concerns should be more carefully assessed as the impact of catastrophic reaction would be much greater in larger batteries. Thus, thermal stability of electrode materials, particularly cathode materials, has been investigated with in situ heating holders, which are able to increase temperatures around the sample inside TEM.

It has been reported that layered cathode materials, $LiTMO_2$, undergo phase transformations to disordered spinel and rock-salt structures at high temperature, which accompanies liberation of oxygen gas.[77] Hwang *et al.* demonstrated that surface of charged $Li_xNi_{0.8}Co_{0.15}Al_{0.05}O_2$ (NCA) cathode materials becomes porous at high temperatures: development of porosity occurs at lower temperature with more lithium extraction (Figure 16).[78] Oxygen K-edge EELS was used for tracking the structural changes of several particles at same state of charge with increasing temperature. There are evident particle-to-particle variations in thermal decomposition, showing that some particles can undergo severe structural changes even below 100 °C, which is reachable temperature during normal operation. This suggests that thermal runaways could be initiated at those particles at much lower temperatures than usually expected. Sharifi-Asl *et al.* demonstrated that phase transformations and oxygen evolution of $LiCoO_2$ cathode initiate from the side-facets of particles like {012} and {014} facets using a combination of TEM analysis (dark-field TEM imaging and atomic resolution images) and ab initio molecular dynamics simulations.[79] Hwang *et al.* examined the thermal stability of $Li_xNi_yMn_zCo_{1-y-z}O_2$ (NMC) cathode materials with different compositions of Ni, Mn, Co.[80] White line ratios of Ni, Mn, Co EELS demonstrated that reduction of Ni starts at the lowest temperature, then reduction of Co

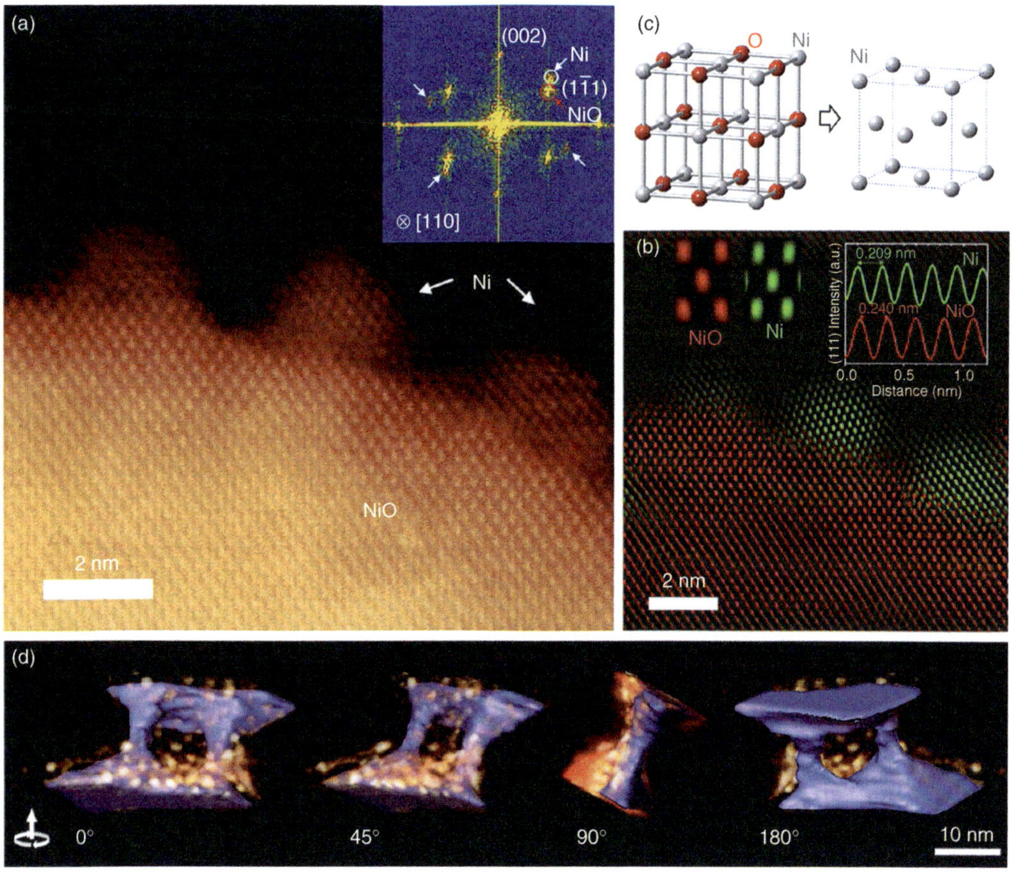


**Figure 13** Atomic structures after sodiation. (a) STEM-ADF image showing the atomically resolved crystal structures of the sodiated electrode materials in a cross-sectional view. The bulk material is an unreacted NiO single-crystal with reduced Ni single-crystal nanoparticles on the surface, both on the zone axis of [110] direction. The inset FFT showing two distinct sets of diffraction spots, attributed to NiO and Ni (indicated by arrows) as labeled. (b) Reconstructed atomic structures of (a) after separating the NiO (red) and Ni (green) reflections by imposing inverse-FFT filters. The inset images of crystal unit cells (top-left) indicate that the lattice parameter decreases as NiO (red) transforms into Ni (green), consistent with the decrease of the lattice spacing between (111) planes (top-right). (c) Models of crystal structures for NiO and Ni. (d) Reconstructed 3D tomograms at different viewing angles showing that the reduced Ni nanoparticles (in orange) are attached on the surface of an incompletely reacted NiO slab (in purple). No sodiation occurred at the back side of NiO contacting the carbon film (not shown in reconstruction). [Reprinted with permission from He, K., Lin, F., Zhu, Y., Yu, X., Li, J., Lin, R., Nordlund, D., Weng, T.-C., Richards, R. M., Yang, X.-Q., Doeff, M. M., Stach, E. A., Mo, Y., Xin, H. L., Su, D. (2015). Sodiation kinetics of metal oxide conversion electrodes: a comparative study with lithiation. Nano letters, 15 (9), 5755–5763. (c) 2015 American Chemical Society]

and Mn follows at higher temperature regardless of NMC composition. Among the NMC compositions in this work, $LiNi_{0.6}Mn_{0.2}Co_{0.2}O_2$ has well-balanced properties of high capacity and good safety, suggesting that controlling TM composition can be an effective way to optimize the properties of cathode materials. Yan *et al.* clearly showed the crack formation and propagation in delithiated $LiNi_{0.6}Mn_{0.2}Co_{0.2}O_2$, suggesting the coupled effect among the electrochemical, thermal, mechanical processes.[81]

## 4.2 In Situ Lithiation: Dry Cell

As secondary ion batteries work via movement of working ions (e.g., $Li^+$ and $Na^+$) between anode and cathode, understanding the reaction mechanisms of electrode materials with those ions are of great importance in order not only to improve electrochemical properties but also to mitigate current challenges of electrode materials. In 2010, Huang *et al.* reported a construction of nanowire battery device, which enables to watch electrochemical reactions at nanoscale inside TEM.[82] Figure 17(a) shows a schematic of the experimental setup. A scanning tunneling microscopy (STM)-TEM holder was utilized. At the one end, target electrode materials (in this case, $SnO_2$) were attached to a gold rod, which can be controlled with a piezo manipulator. On the other side, Li source (here, $LiCoO_2$) were loaded and ionic liquid were prepared on the surface of $LiCoO_2$ as electrolyte. The ionic liquid is a molten

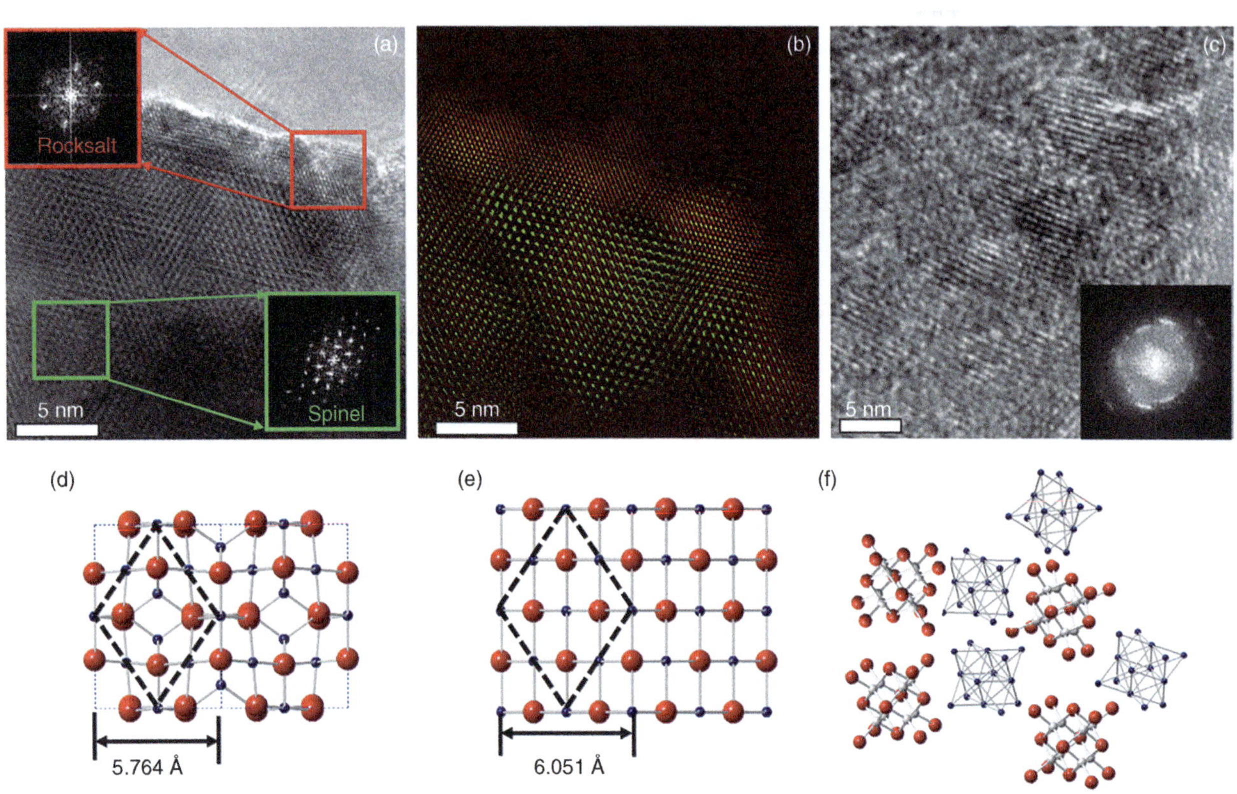


**Figure 14** (a) HRTEM image of a $Co_3O_4$ nanoparticle indicating the coexistence of both spinel and rock-salt structure. Insets show the FFT of the spinel structure (green) and rock-salt structure (red) along the [011] zone axis. (b) Filtered images of (a) show the corresponding phase distribution using two sets of spinel and rock-salt FFTs. (c) HRTEM images of the composite electrode after the conversion reaction. (d–f) Atomic structures of pristine $Co_3O_4$ with spinel structure, rock-salt ($Li_xCo_3O_4$), and composite of $Li_2O$ plus metallic Co, respectively. [Reprinted with permission from Li, J., He, K., Meng, Q., Li, X., Zhu, Y., Hwang, S., Sun, K., Gan, H., Zhu, Y., Mo, Y., Stach, E. A., Su, D. (2016) Kinetic Phase Evolution of Spinel Cobalt Oxide during Lithiation. ACS Nano, 10 (10), 9577–9585. (c) 2016 American Chemical Society]

organic salt having very low vapor pressure, thereby it is usable even in high vacuum of TEM column. This pioneer work has inspired many researchers to address various issues on electrode materials using similar experimental setups. In recent years, Li metal and $Li_2O$ layer, automatically formed on Li surface while transfer the TEM-STM holder into microscope, have been employed as lithium supply and solid electrolyte, respectively. By applying voltage to the sample, electrochemical reaction may be triggered.

### 4.2.1 *Alloying Type Electrode Materials*

Since alloying reaction normally accompanies huge volume expansion, pulverization of original material and contact failure from current collector or other electrode components (binder, conductive carbon) have been huge obstacles for commercialization of alloying type materials. Real-time observation of alloying/dealloying reactions allows one to understand the reaction nature and furthermore, provides ideas to resolve problems.

With in situ dry cell approach, silicon anode materials have been actively investigated because there is much room for improvements to take full advantage of their high capacity. Size-dependence on fracture behavior of lithiated Si nanoparticle was discovered via real-time observation of single Si nanoparticle during lithiation.[84] Critical Si particle size was determined as about 150 nm: bigger (>150 nm) particles experience the cracking while smaller (<150 nm) particles can accommodate volume expansion. McDowell *et al.* claimed that both crystal[85] and amorphous[86] Si nanoparticles undergo a two-phase reaction during the first lithiation, which has clear phase boundaries between pristine and lithiated areas. Even though considerable stress may build around the interface because of volume expansion of lithiated area, amorphous Si is more resistant to fracture: the critical size of amorphous Si is around 870 nm. Liu *et al.* observed anisotropic volume expansion of Si nanowire during in situ lithiation: preferential swelling along <110> over along <111>.[87] These anisotropic volumetric changes result in an intermediate dumbbell-shaped cross section, which

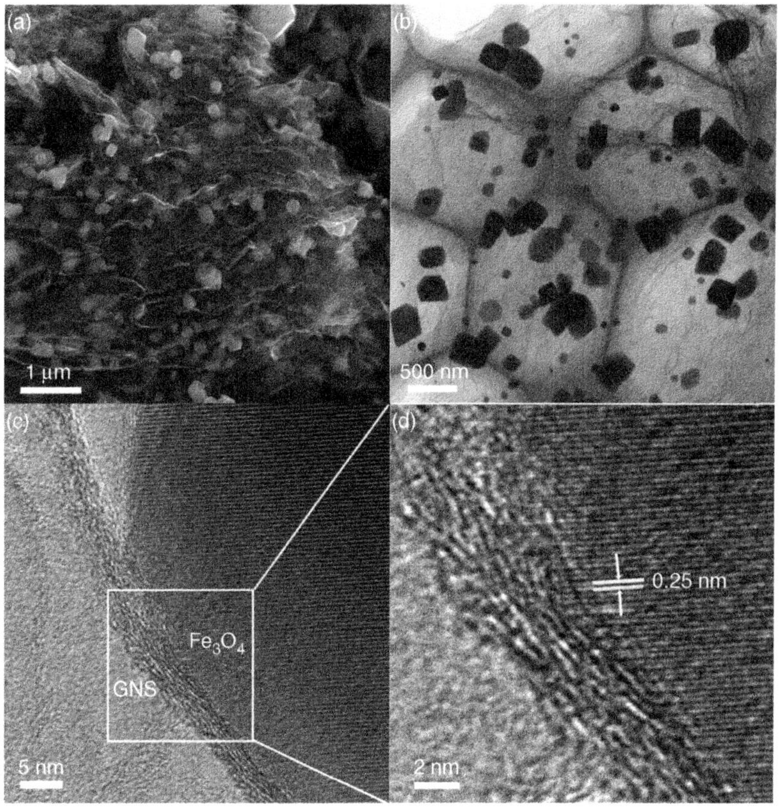


**Figure 15** (a) SEM image of the cross-section of GNS/$Fe_3O_4$ composite, (b) TEM and (c) high-resolution TEM images of GNS/$Fe_3O_4$ composite, and (d) detailed interface structure of the square area in (c). [Reprinted with permission from Zhou, G., Wang, D.-W., Li, F., Zhang, L., Li, N., Wu, Z.-S., Wen, L., Lu, G. Q., Cheng. H.-M. (2010) Graphene-wrapped $Fe_3O_4$ anode material with improved reversible capacity and cyclic stability for lithium ion batteries. Chemistry of Materials. 22 (18), 5306–5313. Copyright (2010) American Chemical Society]

builds tensile stress inside nanowire and eventually leads to fracture. Atomistic mechanism of lithium insertion into Si was elucidated with nanowire with growth direction of <111>.[83] Figure 17(b)–(e) show bright field images and corresponding selected area electron diffraction patterns of Si nanowire before and after partial lithiation, respectively. Li–Si alloy phase was formed at the expense of crystalline Si with considerable volume expansion. There is a clear phase boundary between crystalline Si and amorphous $Li_xSi$. High-resolution TEM images of interface area (Figure 17f–h) demonstrated that lithiation occurs with migration of the interface: lateral movement of ledges on close-packed {111} planes. $Li_xSi$ amorphous phase is developed as a result of layer-by-layer peeling of {111} facets, suggesting orientation-dependent reaction kinetics.

By replacing lithium source in Figure 17(a) with sodium or potassium metal, the real-time observation of sodiation or potassition is also possible. For example, Sn–Na alloying reaction was reported, suggesting two-step sodiation: formation of amorphous $NaSn_2$ in the first step and amorphous $Na_9Sn_4$, $Na_3Sn$ and crystalline $Na_{15}Sn_4$ phase in the second step.[88]

### *4.2.2 Conversion Type Electrode Materials*

During conversion reaction, crystal structure of original electrode material (MX) is destroyed and a composite of metal (M) nanoparticles and LiX (or NaX) matrix are produced, indicating dramatic structural changes occur as a result of lithium (sodium) insertion. Identifying reaction mechanisms and kinetics are critical for understanding and practical engineering of this type of electrode materials.

The experimental setup of a dry-cell was modified to track the electrochemical reactions in $FeF_2$ nanoparticles.[89] Instead of the samples of nanorod/nanowire (Figure 17a), conventional carbon coated TEM half-grid was loaded into TEM-STM holder (Figure 18a). Here, carbon layer can successfully hold nanoparticle samples and act as conducting agent as well. Direct conversion from $FeF_2$ to a mixture of Fe and LiF results in noticeable morphological change (Figure 18b)—formation of ultrafine nanoparticles in matrix. During sodiation of $FeF_2$, morphological evolution seems similar to that of lithiation; however, careful analysis

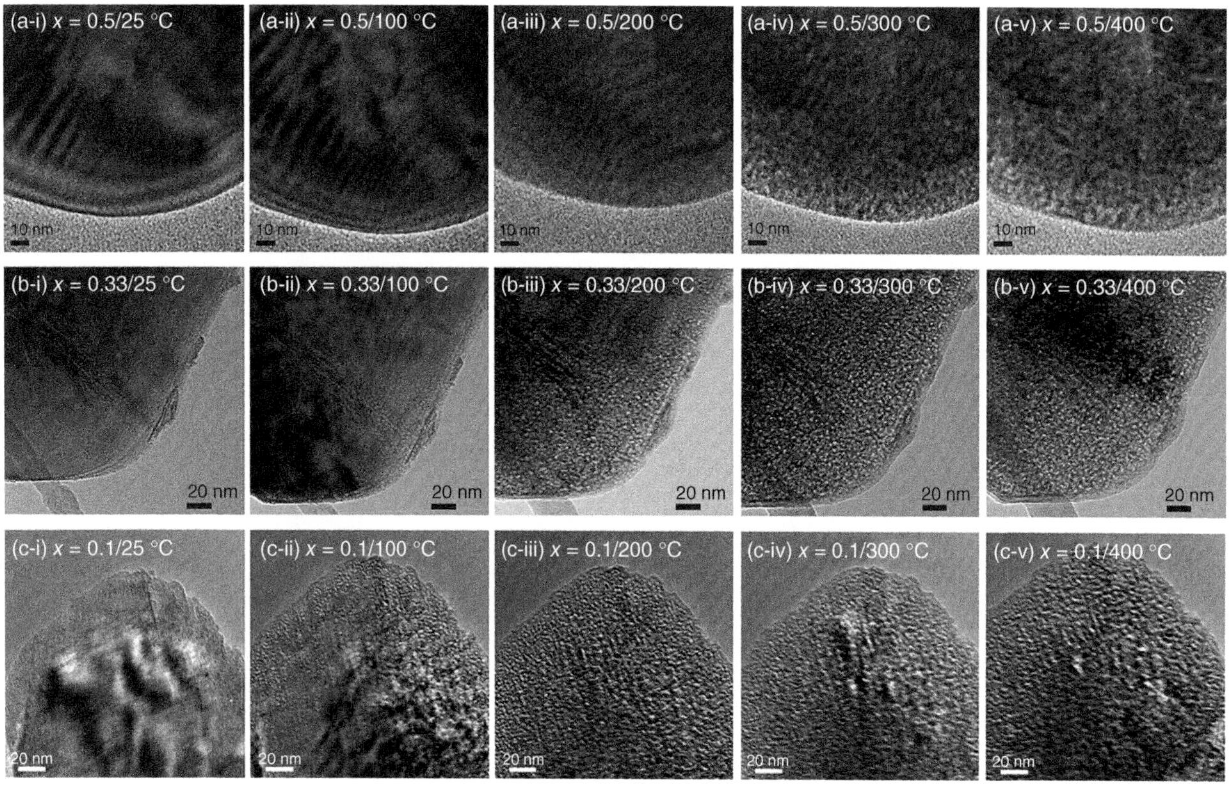


**Figure 16** Real-time BF images of the surface of (row a) $Li_{0.5}NCA$, (row b) $Li_{0.33}NCA$, and (row c) $Li_{0.1}NCA$ during heating. [Reprinted with permission from Hwang, S., Kim, S. M., Bak, S.-M., Cho, B.-W., Chung, K. Y., Lee, J. Y., Chang, W., Stach, E. A. (2014) Investigating the Local Degradation and Thermal Stability of Charged Ni-Based Cathode Materials through Real-Time Electron Microscopy. ACS Appl. Mater. Interfaces, 6 (17), 15140–15147. Copyright (2014) American Chemical Society]

based on in situ and ex situ TEM techniques demonstrated that $FeF_2$ experiences heterogeneous reactions: direct conversion on the surface and disproportionation in the core of the particle, forming ultrafine Fe nanoparticles and a core–shell $Na_3FeF_6$/NaF mixture phase.[91]

Oxide materials having spinel structure usually take more complicated reaction path. In spinel structure, cations only occupy half of octahedral sites and one-eighth of tetrahedral sites. Less packed spinel structure can provide more degree of freedom in structural evolution when foreign ions are being inserted. Atomic structural changes in $Co_3O_4$ during lithium insertion and extraction were elucidated through in situ TEM, suggesting the formation of intermediate phases of lithiated $Co_3O_4$ and nanosized Li–Co–O cluster before conversion reaction happens at the first discharge. When Li ions were extracted, CoO nanoparticles were found as a product of reverse conversion reaction.[92] He *et al.* investigated the intercalation-conversion two-step reaction of $Fe_3O_4$ lithiation using strain-sensitive bright-field (BF) STEM technique.[90] During intercalation, there is minimal change in crystal structure and morphology of materials; thus, observing intercalation reaction is challenging. BF-STEM images are strain sensitive because images are generated from the electrons transmitted or scattered to lower angles, allowing one to visualize small changes during intercalation. A time series of BF-STEM images of $Fe_3O_4$ (Figure 18c) shows the movement of phase boundary between $Fe_3O_4$ and $Li_xFe_3O_4$ before subsequence conversion reaction occurs. Low angle annular dark field (LAADF)-STEM imaging technique was also employed to discern intercalation-conversion two-step reaction of $Co_3O_4$ since LAADF-STEM is also strain sensitive.[70]

In copper sulfides, lithiation reaction happens different way from a viewpoint of crystal structure even though the chemical formula interpreting reaction is similar to that of conversion reaction. When lithium ions are introduced into $Cu_2S$,[93] metallic Cu and $Li_2S$ are formed meanwhile sulfur framework is maintained. Li ions replace Cu ions in $Cu_2S$, evidenced with in situ HRTEM. Considering that conversion reaction induces a complete decomposition of original structure, displacement of Li and Cu in $Cu_2S$ is the distinct lithiation nature from conversion.

Metal sulfides having two-dimensional layered structure exhibit multiple steps of reactions. Li ions can be

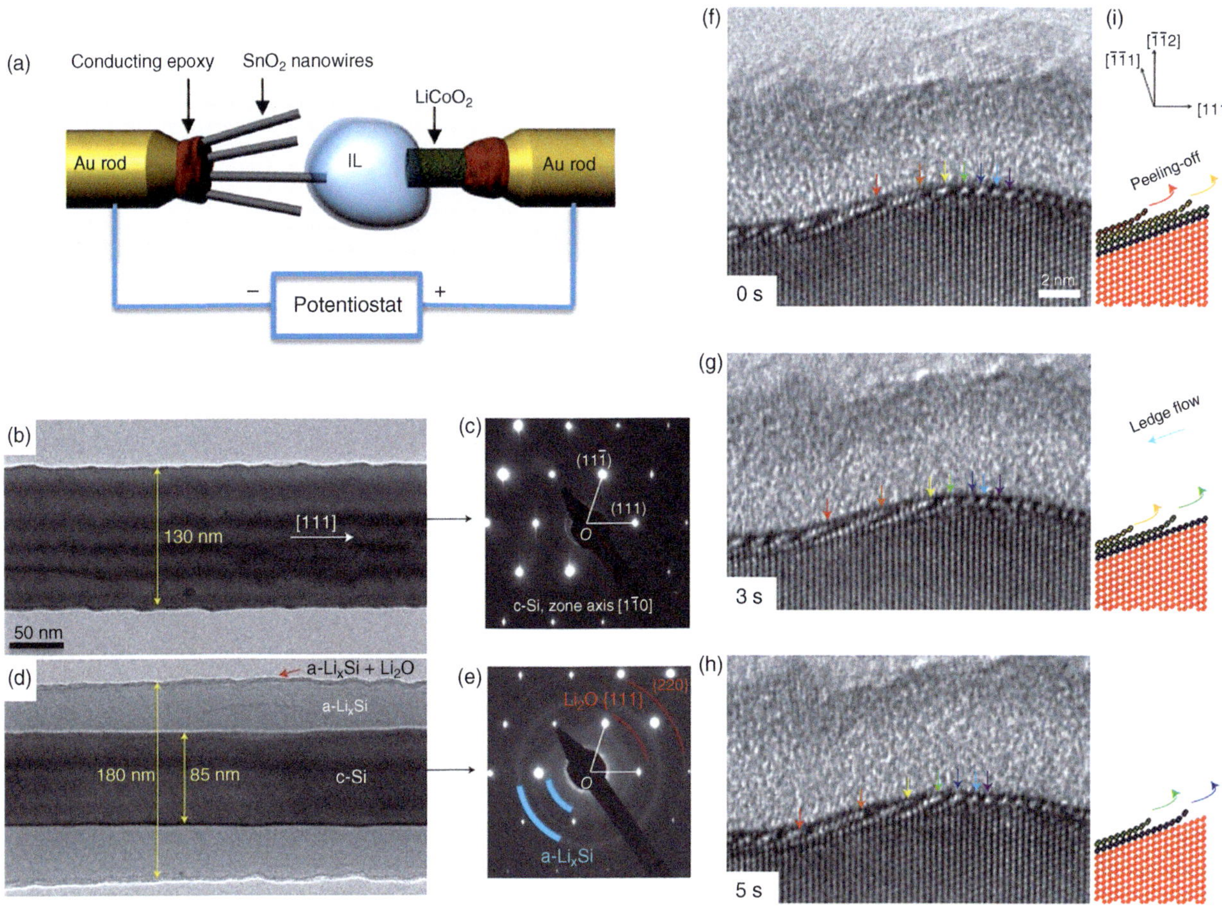


**Figure 17** (a) Schematic of the experimental setup of dry cell approach (b, c) bright field image and electron diffraction pattern of (b, c) pristine silicon nanowire and (d, e) partially lithiated Si nanowire with $Li_xSi$ layer surrounding the crystalline Si core. (f–h) High-resolution TEM image sequences showing the lithiation process by lateral ledge flow (marked by colored arrows) in the amorphous/crystalline interface (i) (d) schematic showing the peeling-off of the $(\overline{1}\overline{1}1)$ atomic layers by continuous ledge flow in the planes during lithiation. [(a) Reprinted with permission from Huang, J. Y., Zhong, L., Wang, C.-M., Sullivan, J. P., Xu, W., Zhang, L. Q., Mao, S. X., Hudak, N. S., Liu, X. H., Subramanian, A., Fan, H., Qi, L., Kushima, A., Li, J., (2010) In Situ Observation of the Electrochemical Lithiation of a Single $SnO_2$ Nanowire Electrode. Science, 330 (6010), 1515–1520. Copyright (2010) American Association for the Advancement of Science; (b–h) Reprinted with permission from Liu, X. H., Wang, J. W., Huang, S., Fan, F., Huang, X., Liu, Y., Krylyuk, S., Yoo, J., Dayeh, S. A., Davydov, A. V., Mao, S. X., Picraux, S. T., Zhang, S., Li, J., Zhu, T., Huang, J. Y. (2012) In situ atomic-scale imaging of electrochemical lithiation in silicon. Nat. Nanotechnol., 7, 749–756. Copyright (2012) Springer Nature]

intercalated between S–S layers then Cu extrusion reaction follows in case of CuS.[94] Intercalation and conversion reactions happen at $MoS_2$.[95] Tin sulfides experience four steps of lithiation: intercalation, disordering, conversion, and alloying.[96] Figure 19 presents time sequence ADF-STEM images of $SnS_2$ with false colors, which represent each lithiation step and corresponding atomic model. During lithiation of $SnS_2$, after Li interaction reaction ($Li_xSnS_2$), Li and Sn disordering reaction occurs, resulting in the formation of rock-salt-like phase ($Li_ySnS_2$). Sn nanoparticles and $Li_2S$ matrix are formed as a result of conversion reaction and Sn–Li alloying reaction follows. Unexpected cation disordering was clearly demonstrated, which is an undoubtable merit of in situ TEM analysis.

### *4.2.3 Intercalation Type Electrode Materials*

As Li ions are being inserted into empty space of electrode materials, only subtle structural modification occurs during intercalation. Thus, extra care should be exerted in order to examine the small changes in intercalation type electrode materials.

Expansion of interlayer of carbonaceous materials was observed during lithiation,[97] sodiation, and potassiation.[98] High-resolution lattice images and SAED patterns were employed to detect lattice expansion. Cesium and potassium intercalation into single-wall nanotube were examined in real time after deposition of Cs or K on nanotube bundles.[99] Layered structure of $MoS_2$ also allows Li ions to be intercalated between S–S layers.[95] However,

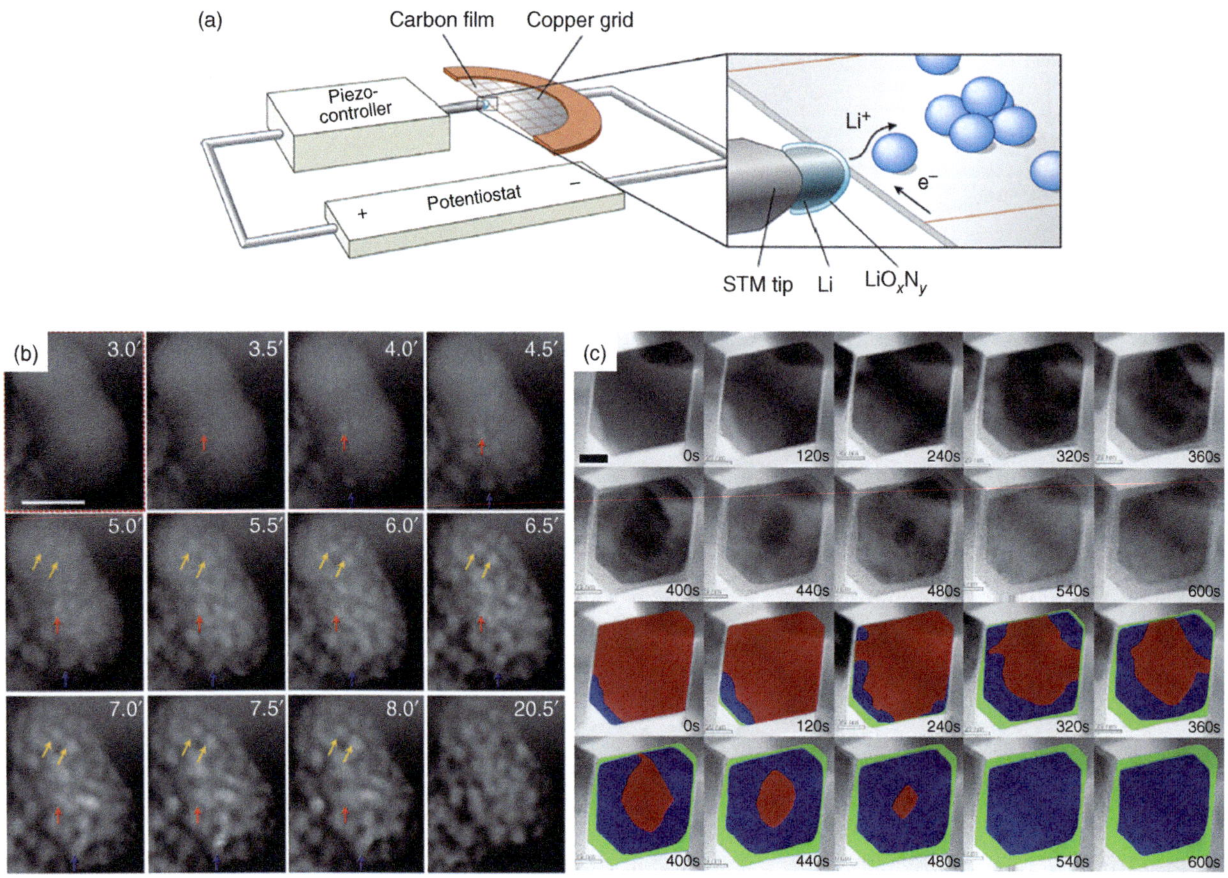


**Figure 18** (a) Dry cell approach for observing nanoparticle samples supported by half TEM grid (b) morphological evolution of a single $FeF_2$ nanoparticle, revealing the rapid formation of ultrafine (sub-nm) Fe particles on the surface, followed by gradual formation of larger ones (1–3 nm) within the domain of the original $FeF_2$ particle (arrows used to identify specific Fe particles). Scale bar, 10 nm. (c) BF-STEM image series showing phase evolution during lithiation. The overlaid false colours indicate different phases: pristine $Fe_3O_4$ (red), Li-inserted $Li_xFe_3O_4$ (blue) and Fe·$Li_2O$ composite after conversion (green). [(a, b) Reprinted with permission from Wang, F., Yu, H.-C., Chen, M.-H., Wu, L., Pereira, N., Thornton, K., Van der Ven, A., Zhu, Y., Amatucchi, G. G., Graetz, J. (2012) Tracking lithium transport and electrochemical reactions in nanoparticles. Nat. Commun., 3, 1201. Copyright (2012) Springer Nature; (c) Reprinted with permission from He, K., Zhang, S., Li, J., Yu, X., Meng, Q., Zhu, Y., Hu, E., Sun, K., Yun, H., Yang, X.-Q., Zhu, Y., Gan, H., Mo, Y., Stach, E. A., Murray, C. B., Su, D. (2016) Visualizing non-equilibrium lithiation of spinel oxide via in situ transmission electron microscopy. Nat. Commun., 7, 11441. Copyright (2016) Springer Nature]

noticeable changes in crystal symmetry were observed as a result of the phase transformation from 2H-$MoS_2$ to 1T-$Li_xMoS_2$. Sliding of slabs in $MoS_2$ with lithium insertion gradually results in the formation of 1T-$MoS_2$.

Phase boundary migration and anisotropic lithium insertion in $FePO_4$ particles were observed via in situ HRTEM and corresponding FFTs. Interface of $FePO_4$/$LiFePO_4$ was determined to align along (010) plane and to move towards [010] direction, which confirms theoretical expectation of anisotropic lithiation in $FePO_4$.[100] Nanobattery was constructed with $LiMn_2O_4$ nanowires, ionic liquid electrolyte, and LTO anode material and observed in real time to investigate the structural changes of $LiMn_2O_4$ during operation.[101] SAED patterns were obtained at different potentials. During discharge, cubic–orthorhombic–tetragonal phase transformations occur sequentially and the formation of tetragonal phase is understood by local lithium accumulation. Nie *et al.* reported that effect of twin boundaries on lithium-ion transport in single crystalline $SnO_2$.[102] In situ TEM results show that lithium ions are preferentially intercalated around twin boundaries. Strain and Li composition maps also confirmed higher concentration of strain and Li concentration vicinity of twin defects. This work suggests novel approach of materials design—smart design of defects to facilitate lithium transport.

### 4.3 In Situ TEM with Liquid Cell

Mimicking working environment of device inside TEM can provide tremendous insights in structural and

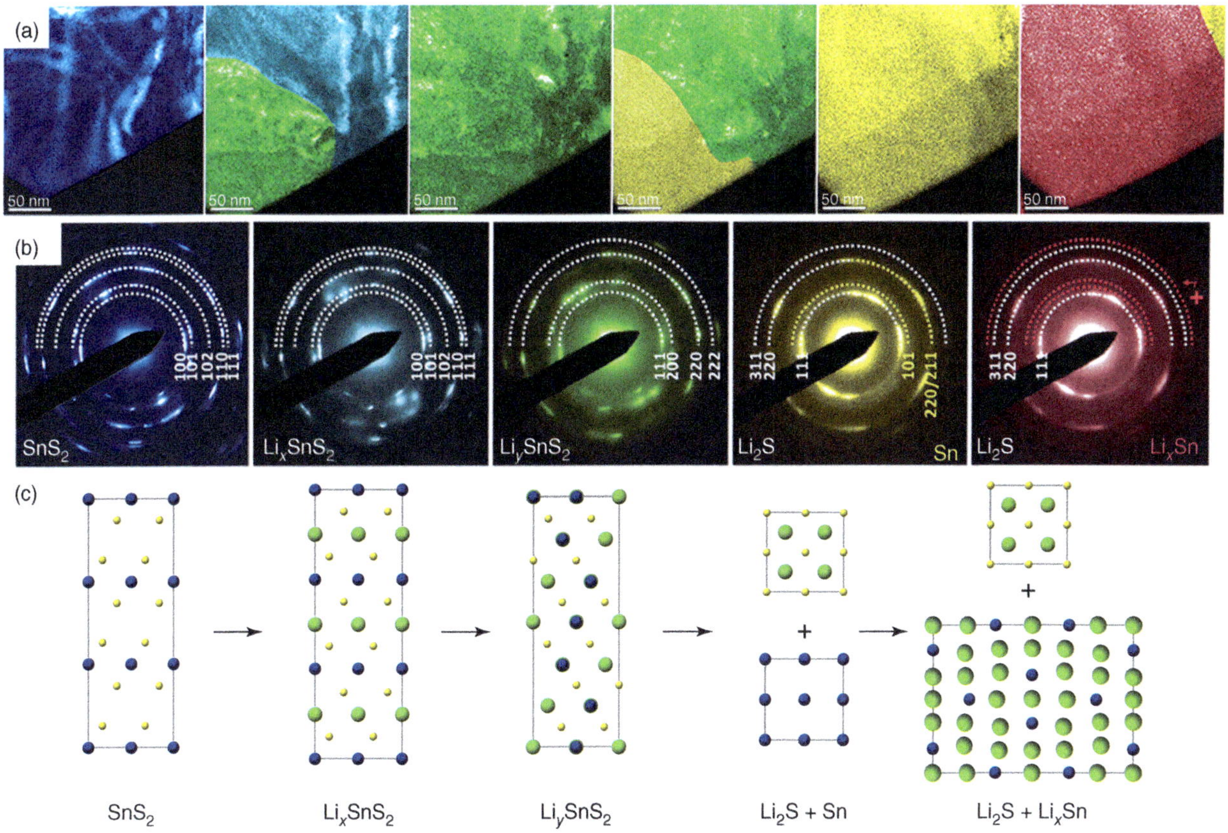


**Figure 19** (a) Time-sequence ADF-STEM images of $SnS_2$ during in situ lithiation. False color represents each reaction step of intercalation (blue), disordering (green), conversion (yellow), and alloying (red) (b) electron diffraction pattern and (c) atomic model of each step. [Reprinted with permission from Hwang, S., Yao, Z., Zhang, L., Fu, M., He, K., Mai, L., Wolverton, C., Su, D. (2018) Multistep Lithiation of Tin Sulfide: An Investigation Using in SituElectron Microscopy. ACS Nano, 12 (4), 3638–3645. Copyright (2018) American Chemical Society]

chemical evolutions occurring materials during operation. In case of batteries, realization of similar working condition inside the microscope is a grant challenge because of liquid electrolyte. Even though the structural evolution of active electrode materials was successfully examined at atomic scale using dry-cell TEM, the dynamic information of electrode–electrolyte interface has been still missing due to lack of organic electrolyte. Fortunately, confining liquid for TEM analysis have been achieved with in-house MEMS-based liquid cells,[103] graphene liquid cells,[104] or commercial TEM holders,[105] which may allow us to build more realistic battery systems in TEM.

Abellan *et al.* examined degradation of a number of electrolyte solutions for Li-ion batteries under electron beam.[106] Two-windowed cells and commercial fluid stage were used for this work. Electron beam itself may induce electrochemical reaction and degradation of electrolyte. As a result of electrolyte breakdown, particle growth was observed, which, however, could be an artifact while observing in situ electrochemical reactions.

During charge/discharge cycles, various electrochemical reactions can happen at the interface of electrolyte and electrode, such as initiation of lithium insertion into electrode, formation of solid–electrolyte interphase (SEI), Li dendritic growth. While dry cell approach, we discussed before, is able to investigate the lithium insertion processes, SEI formation and Li dendrite growth are only observable with liquid electrolyte. Therefore, to fully understand the interfacial reactions, real-time observation of electrode/common liquid electrolyte interface should be performed.

Electrodeposited Li metal is an ideal anode material for lithium-ion batteries; thus, uniform microstructural evolution during Li deposition/stripping is a key factor in deciding battery performance and safety. Zeng *et al.* visualized electrochemical lithiation and delithiation of Au electrode with commercial electrolyte of $LiPF_6$ dissolved in ethylene carbonate (EC) and diethyl carbonate (DEC).[107] Home-made electrochemical liquid cells[108] (Figure 20a) were employed for in situ TEM experiments.

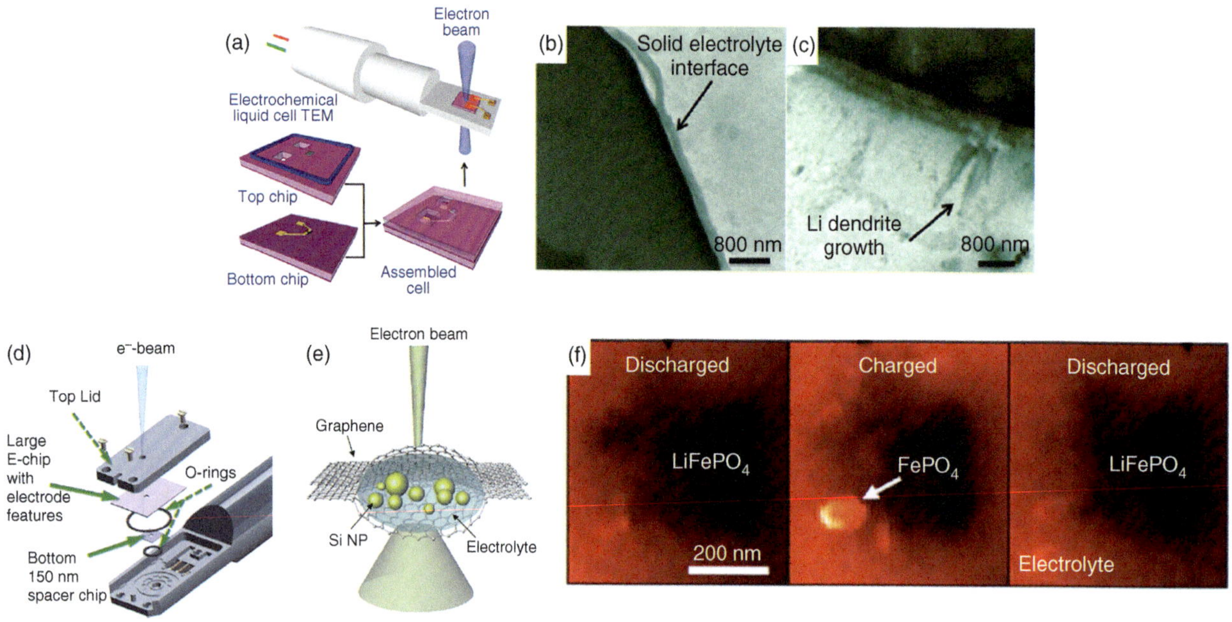


**Figure 20** (a) Home-made electrochemical liquid cell. (b, c) Interface reactions of Au anode in conventional liquid electrolyte. [(a–c) Reprinted with permission from Zeng, Z., Liang, W.-I, Liao, H.-G., Xin, H. L., Chu, Y.-H., Zheng, H. (2014) Visualization of Electrode-Electrolyte Interfaces in $LiPF_6$/EC/DEC Electrolyte for Lithium Ion Batteries via in Situ TEM. Nano Lett., 14 (4), 1745–1750. Copyright (2014) American Chemical Society]; (d) Commercially available liquid cell holder. [Reprinted with permission from Mehdi, B. L., Qian, J., Nasybulin, E., Park, C., Welch, D. A., Faller, R., Mehta, H., Henderson, W. A., Xu, W., Wang, C. M., Evans, J. E., Liu, J., Zhang, J.-G., Mueller, K. T., Browning, N. D. (2015) Observation and Quantification of Nanoscale Processes in Lithium Batteries by Operando Electrochemical (S)TEM. Nano Lett., 15 (3), 2168–2173. Copyright (2015) American Chemical Society]; (e) graphene liquid cell. [Reprinted with permission from Yuk, J. M., Seo, H. K., Choi, J. W., Lee, J. Y. (2014) Anisotropic Lithiation Onset in Silicon Nanoparticle Anode Revealed by in SituGraphene Liquid Cell Electron Microscopy. ACS Nano, 8 (7), 7478–7485. Copyright (2014) American Chemical Society]; (f) 5 eV spectroscopic EFTEM of $LiFePO_4$ upon electrochemical reactions. [Reprinted with permission from Holtz, M. E., Yu, Y., Gunceler, D., Gao, J., Sundararaman, R., Schwarz, K. A., Arias, T. A., Abruña, H. D., Muller, D. A. (2014) Nanoscale Imaging of Lithium Ion Distribution During In Situ Operation of Battery Electrode and Electrolyte. Nano Lett., 14 (3), 1453–1459. Copyright (2014) American Chemical Society]

Inhomogeneous lithiation of Au electrode, growth of lithium dendrite, formation of gas bubbles from electrolyte decomposition, and SEI formation were observed during electrochemical reactions (Figure 20b and c). Leenheer *et al.* constructed customized liquid cell to demonstrate the dynamics of lithium electrodeposition with conventional $LiPF_6$ in EC and DMC.[112] Li deposition and stripping were observed on Ti electrode. Since current density is a determining factor of Li dendrite growth, current densities from 1 to 25 mA cm$^{-2}$ were applied to observe any variations in structural evolution. At high current density of 25 mA cm$^{-2}$, needle-like Li crystals were generated while no dendrite was observed under a current density of 1 mA cm$^{-2}$. Electron beam induced surface film formation, which may alter Li's microstructures. Mehdi *et al.* exploited commercially available liquid cell TEM holder (Figure 20d) and suggested the contrast reversal in HAADF-STEM images during Li deposition and dissolution.[109] Contrast of newly formed phase was darker than background in HAADF-STEM images, indicating that the average atomic number of deposited particles is lower than average atomic number of the electrolyte. Only Li metal is less dense than the electrolyte. In addition, since thickness of sample also determines the contrast in STEM images, the amount of deposited Li metal was quantified from the images.

The electrolyte/electrode interfacial reactions also have been actively investigated with liquid electrolytes. Yuk *et al.* examined the chemical lithiation of Si nanoparticles in 1 M $LiPF_6$ in EC, DMC, and DEC solution.[110] Using graphene liquid cell (Figure 20e), Si nanoparticles in liquid electrolyte were encapsulated for further analysis. It was claimed that Li atoms were formed as a result of electron beam induced electrolyte decomposition. Diffusion of Li atoms into Si nanoparticles takes place preferentially along <110> but once the degree of lithiation reaches certain level, lithium diffusion occurs isotropically. Lithiation dynamics of Si nanowires were compared with experimental setups: dry cell approach and liquid cell methods.[113] In the liquid cell, nanowires were immersed into liquid electrolyte, global response of whole single nanowire can be observed. On the other hand, only a small portion of Si nanowires contacts with solid electrolyte of dry cell ($Li_2O$);

thus, it is beneficial to investigate the intrinsic reaction nature along controlled lithium diffusion path. General lithiation/delithiation behavior of Si was consistent regardless of experimental setup. Reaction dynamics of $LiFePO_4$ was investigated using liquid cell in situ TEM coupled with energy-filtered TEM (EFTEM).[111] Detecting Li by core-loss EELS is challenging since EELS signal is degraded by multiple scattering in thick liquid samples and lithium K-edge may be superimposed by bulk plasmon peak of thick liquid layers. Here, valence EELS was used for tracking lithiation and delithiation reactions because lithium-ion movements are associated with filling and emptying valence bands, resulting in the changes of electronic structure. Only delithiated $FePO_4$ has a prominent peak at 5 eV and using this peak, EFTEM images (at 5 eV with a 5-eV wide energy window) were acquired at different of states charge to visualize lithiation and delithiation reactions (Figure 20f). Zheng *et al.* reported real-time observation of electrochemical reaction of $MoS_2$.[114] During lithium insertion, irreversible breakdown of $MoS_2$ nanosheets occurs, resulting in fragmentation into nanoparticles. Nanobeam diffraction at STEM mode, which acquires a series of diffraction patterns at each pixel within scanned area, was exploited to identify the phase of fragmented particles. Diffraction patterns, BF and dark-field images can be reconstructed from nanobeam diffraction. Inhomogeneous reaction nature was observed: some $MoS_2$ nanosheets undergo swelling or deformation without fragmentation.

### 4.4 In Situ TEM with Solid Electrolytes

All-solid-state batteries (ASSB) are highly promising energy storage devices for next-generation. As flammable liquid electrolyte in conventional battery is replaced with solid-state electrolyte, safety of whole battery system can be much enhanced. Li-metal anode could be incorporated into ASSB as mechanical strength of solid electrolyte may inhibit Li dendrite growth, which enables to achieve high energy density. However, large interfacial resistance between solid electrolyte and electrode should be overcome for their widespread application. Grain boundaries of each component also may degrade the performance of ASSB. Therefore, it is of importance to understand the structural and chemical evolution of ASSB during operation, particularly at various kinds of interfaces and solid electrolyte itself, which has not been addressed before.

Ruzmetov *et al.* fabricated solid-state nanowire core–multishell LIBs—Si nanowire template and Ti/Pt/Ti (current collector), $LiCoO_2$ (cathode), nitrogen-doped $Li_3PO_4$ (LiPON, solid electrolyte), and amorphous Si (anode).[115] Thickness of solid electrolyte is critical factor for battery performance: thick electrolyte may delay charge-transport kinetics but too thin electrolyte can experience electronic current flow because of space-charge-limited electronic conduction. Here, stability of LiPON was investigated with a function of thickness. In case of 110 nm thick of LiPON, self-discharge and void formation at electrode/electrolyte interface were observed during operation, suggesting electrical and chemical failure. By increasing thickness of LiPON, electrochemical performances were much improved. Lithium-ion transport and corresponding charge transfer were elucidated by STEM–EELS.[116] $LiCoO_2$/LiPON/Si thin film battery were prepared as a TEM sample using FIB method and galvanostatically biased inside of TEM. STEM–EELS spectrum imaging demonstrated that disordered interfacial layer was formed between $LiCoO_2$ cathode and LiPON after a charge to 4.2 V, claimed as disordered $LiCoO_2$ where Co ions are relatively oxidized along with lithium oxide and lithium peroxide. These results suggest that changes in chemistry of electrode/electrolyte interface may increase the interfacial impedance. Gong *et al.* reported structural changes of $LiCoO_2$ in working ASSB with Y and Ta doped LLZO solid electrolyte ($Li_{6.75}La_{2.84}Y_{0.16}Zr_{1.75}Ta_{0.25}O_{12}$).[117] HAADF- and ABF-STEM images were acquired to investigate heavy and light elements simultaneously. Single crystal of $LiCoO_2$ became polycrystals with sizes of 5–10 nm, facing each other with coherent twin and antiphase domain boundaries after charging up to 5.1 V vs Li/$Li^+$. It is noteworthy that with liquid electrolyte, phase transformations of $LiCoO_2$ occurs at high voltage but here, other crystal phases were not observed, indicating that materials may behave differently upon working environments.

## 5 OUTLOOK

TEM is a dynamic field and being developed quickly now. As seen at the report of US DOE Basic Energy Sciences workshop on "Electron Scattering and Diffraction" in 2014,[118] multi-dimensional atomic-resolution electron microscopy, ultrafast electron microscopy, and environmental TEM are being identified as the key directions of developing TEM. Lots of instrumental improvements can be beneficial to investigate the electrode materials for batteries. For our point of view, there are three kinds of techniques which might be beneficial to the battery society:

1. Low-dose/cryogenic electron microscopy. These approaches can help to reduce the beam damage which is one of the most challenge issues for the TEM investigations on battery materials. Recently, Li *et al.* have applied the cryo-TEM methods to atomically resolve individual lithium metal atoms and their interface with SEI with cooling the sample down to liquid $N_2$ temperature, as shown in Figure 21.[119] Zachman *et al.* have prepared TEM samples using a cryo-FIB

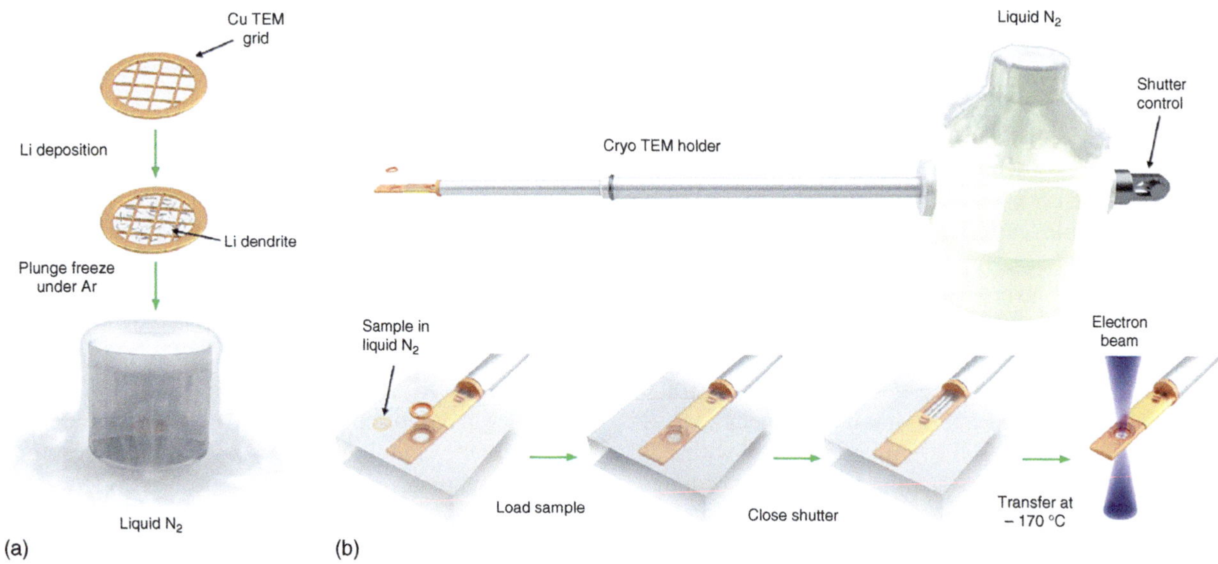


**Figure 21** Procedure of cryo-TEM used to study SEI and lithium metal at −170 °C. [Reprinted with permission from Li, Y., Li, Y., Pei, A., Yan, K., Sun, Y., Wu, C. -L., Joubert, L.-M., Chin, R., Koh, A. L., Yu, Y., Perrino, J., Butz, B., Chu, S., Cui, Y. (2017). Atomic structure of sensitive battery materials and interfaces revealed by cryo-electron microscopy. Science, 358 (6362), 506–510. © 2017 American Association for the Advancement of Science]

technique and studied the dendrite growth using cryo-STEM imaging and cryo-EELS.[120] Cryo-TEM is becoming a standard method to investigate the solid electrolyte interphase (SEI) as well as other beam sensitive alkali electrode materials.

2. Operando electron microscopy. As discussed above, in situ TEM has been widely used for battery research. However, operando TEM technique, indicating the simultaneous TEM observations and the performance measurement, is still challenging at present.[121] The design of special sample holder, the setup of performance testing, and dedicated sample preparation are all very important. Likely, one has to realize all these requirements using all possible techniques including MEMS, FIB and microfabrication. How to measure the performance from a tiny sample would be also challenging issue. Among lots of work being under development, operando TEM approach on solid-state-batteries would be very promising to make breakthrough in near future.[117]
3. Ultrafast electron microscopy. Recent developments on high-brightness pulse electron source, high-efficient electron detector, and high-speed shutter enable an ultrafast electron microscopy. Ultrafast diffraction and ultrafast imaging may help us to probe the physical and chemical changes happened in battery materials within time range from picoseconds to nanoseconds.[122,123] The improvement in temporal resolution may help us to understand the rapid changes and intermediate phases during the reaction which can open another dimension in understanding the reaction dynamics.

## 6 RELATED ARTICLE

Solid State and Materials Chemistry for Sodium-Ion Batteries

## 7 ABBREVIATIONS AND ACRONYMS

ABF = annular bright field; ASSB = all-solid-state batteries; BF = bright-field; CBED = convergent beam electron diffraction; CNT = carbon nanotube; DEC = diethyl carbonate; DMC = dimethyl carbonate; EC = ethylene carbonate; EDS = energy dispersive X-ray spectroscopy; EELS = electron energy-loss spectroscopy; EG = expanded graphite; FFT = fast Fourier transformation; FIB = focused ion beam; GO = graphite oxide; HAADF = high-angle annular dark-field; HAADF-STEM = high angle annular dark field scanning transmission electron microscopy; LAADF = Low angle annular dark field; LIBs = lithium-ion batteries; NBED = nano-beam electron diffraction; NCA = $LiNi_{0.8}Co_{0.15}Al_{0.05}O_2$; PDF = pair distribution function; PG = pristine graphite; SAED = selected area electron diffraction; SEI = solid–electrolyte interphase; SEI = solid electrolyte interphase; STEM = scanning-transmission electron microscopy; STEM–EDS = scanning transmission electron microscopy–energy dispersive X-ray spectroscopy; STEM–EELS = scanning transmission electron microscopy–electron energy loss spectroscopy; STEM–HAADF = scanning transmission electron microscopy–high angular annular dark field; STM = scanning tunneling microscopy; TEM = Transmission electron microscopy; TM = transition metal

## 8 REFERENCES


1. D. B. Williams and C. B. Carter, 'Transmission Electron Microscopy: A Textbook for Materials Science', 2nd edition, Springer, New York, 2009.
2. S. J. Pennycook and P. D. Nellist eds., 'Scanning Transmission Electron Microscopy: Image and Analysis', Springer, New York, 2011.
3. P. Hawkes and J. C. H. Spence eds., 'Science of Microscopy', Springer, New York, 2007.
4. R. F. Egerton, 'Electron Energy-Loss Spectroscopy in the Electron Microscope', 3rd edition, Springer, New York, 2011.
5. J. C. H. Spence and J. M. Zuo, 'Electron Microdiffraction', Plenum Press, New York/London, 1992.
6. D. A. Muller, L. F. Kourkoutis, M. Murfitt, J. H. Song, H. Y. Hwang, J. Silcox, N. Delby and O. L. Krivanek, *Science*, 2008, **319**(5866), 1073.
7. O. L. Krivanek, M. F. Chisholm, V. Nicolosi, T. J. Pennycook, G. J. Corbin, N. Dellby, M. F. Murfitt, C. S. Own, Z. S. Szilagyi, M. P. Oxley, S. T. Pantelides and S. J. Pennycook, *Nature*, 2010, **464**(7288), 571.
8. F. Lin, I. M. Markus, D. Nordlund, T.-S. Weng, M. D. Asta, H. L. Xin and M. M. Doeff, *Nat. Commun.*, 2014, **5**, 3529.
9. B. Xu, C. R. Fell, M. Chi and Y. S. Meng, *Energy Environ. Sci.*, 2011, **4**(6), 2223.
10. S. Morishita, R. Ishikawa, Y. Kohno, H. Sawada, N. Shibata and Y. Ikuhara, *Microscopy*, 2017, **67**(1), 46.
11. O. L. Krivanek, T. C. Lovejoy, N. Dellby, T. Aoki, R. W. Carpenter, P. Rez, E. Soignard, J. Zhu, P. E. Batson, M. J. Lagos and R. F. Egerton, *Nature*, 2014, **514**(7521), 209.
12. L. A. Giannuzzi and F. A. Stevie, *Micron*, 1999, **30**, 197.
13. D. Su, F. Wang, C. Ma and N. Jiang, *Nano Energy*, 2013, **2**(3), 343.
14. R. F. Egerton, P. Li and M. Malac, *Micron*, 2004, **35**(6), 399.
15. N. Jiang, *Rep. Prog. Phys.*, 2015, **79**(1), 016501.
16. L. Wang, T. Maxisch and G. Ceder, *Chem. Mater.*, 2007, **19**(3), 543.
17. S. Hwang, W. Chang, S. M. Kim, D. Su, D. H. Kim, J. Y. Lee, K. Y. Chung and E. A. Stach, *Chem. Mater.*, 2014, **26**(2), 1084.
18. S. Zheng, R. Huang, Y. Makimura, Y. Ukyo, C. A. J. Fisher, T. Hirayama and Y. Ikuhara, *J. Electrochem. Soc.*, 2011, **158**(4), A357.
19. H. Zhang, K. Karki, Y. Huang, M. S. Whittingham, E. A. Stach and G. Zhou, *J. Phys. Chem. C*, 2017, **121**(3), 1421.
20. Y.-N. Zhou, J.-L. Yue, E. Hu, H. Li, L. Gu, K.-W. Nam, S.-M. Bak, X. Yu, J. Liu, J. Bai, E. Doryhee, Z.-W. Fu and X.-Q. Yang, *Adv. Energy Mater.*, 2016, **6**(21), 1600597.
21. E. Jo, S. Hwang, S. M. Kim and W. Chang, *Chem. Mater.*, 2017, **29**(7), 2708.
22. S. Hwang, E. Jo, K. Y. Chung, K. S. Hwang, S. M. Kim and W. Chang, *J. Phys. Chem. Lett.*, 2017, **8**(23), 5758.
23. V. J. Keast, A. J. Scott, R. Brydson, D. B. Williams and J. Bruley, *J. Microsc.*, 2001, **203**(2), 135.
24. Y. Koyama, T. Mizoguchi, H. Ikeno and I. Tanaka, *J. Phys. Chem. B*, 2005, **109**(21), 10749.
25. J. Graetz, C. C. Ahn, H. Ouyang, P. Rez and B. Fultz, *Phys. Rev. B*, 2004, **69**(23), 235103.
26. J. Graetz, C. C. Ahn, R. Yazami and B. Fultz, *J. Phys. Chem. B*, 2003, **107**(13), 2887.
27. S. Muto, Y. Sasano, K. Tatsumi, T. Sasaki, K. Horibuchi, Y. Takeuchi and Y. Ukyo, *J. Electrochem. Soc.*, 2009, **156**(5), A371.
28. S. Hwang, S. Y. Kim, K. Y. Chung, E. A. Stach, S. M. Kim and W. Chang, *APL Mater.*, 2016, **4**(9), 096105.
29. J. Bareño, C. H. Lei, J. G. Wen, S.-H. Kang, I. Petrov and D. P. Abraham, *Adv. Mater.*, 2010, **22**(10), 1122.
30. H. Yu, R. Ishikawa, Y.-G. So, N. Shibata, T. Kudo, H. Zhou and Y. Ikuhara, *Angew. Chem. Int. Ed.*, 2013, **52**(23), 5969.
31. A. K. Shukla, Q. M. Ramasse, C. Ophus, H. Duncan, F. Hage and G. Chen, *Nat. Commun.*, 2015, **6**, 8711.
32. M. Gu, I. Belharouak, A. Genc, Z. Wang, D. Wang, K. Amine, F. Gao, G. Zhou, S. Thevuthasan, D. R. Baer, J.-G. Zhang, N. D. Browning, J. Liu and C.-M. Wang, *Nano Lett.*, 2012, **12**(10), 5186.
33. P. Yan, A. Nie, J. Zheng, Y. Zhou, D. Lu, X. Zhang, R. Xu, I. Belharouak, X. Zu, J. Xiao, K. Amine, J. Liu, F. Gao, R. Shahbazian-Yassar, J.-G. Zhang and C.-M. Wang, *Nano Lett.*, 2015, **15**(1), 514.
34. P. Yan, J. Zheng, J. Zheng, Z. Wang, G. Teng, S. Kuppan, J. Xiao, G. Chen, F. Pan, J.-G. Zhang and C.-M. Wang, *Adv. Energy Mater.*, 2016, **11**(9), 1502455.
35. M. Sathiya, G. Rousse, K. Ramesha, C. P. Laisa, H. Vezin, M. T. Sougrati, M.-L. Doublet, D. Foix, D. Gonbeau, W. Walker, A. S. Prakash, M. B. Hassine, L. Dupong and J.-M. Tarascon, *Nat. Mater.*, 2013, **12**(9), 827.
36. K. Luo, M. R. Roberts, R. Hao, N. Guerrini, D. M. Pickup, Y.-S. Liu, K. Edström, J. Guo, A. V. Chadwick, L. C. Duda and P. G. Bruce, *Nat. Chem.*, 2016, **8**(7), 684.
37. A. R. Armstrong, M. Holzapfel, P. Novák, C. S. Johnson, S.-H. Kang, M. M. Thackeray and P. G. Bruce, *J. Am. Chem. Soc.*, 2006, **128**(26), 8694.
38. N. Yabuuchi, K. Yoshii, S.-T. Myung, I. Nakai and S. Komaba, *J. Am. Chem. Soc.*, 2011, **133**(12), 4404.

39. E. McCalla, A. M. Abakumov, M. Saubanère, D. Foix, E. J. Berg, G. Rousse, M.-L. Doublet, D. Gonbeau, P. Novák, G. V. Tendeloo, R. Dominko and J.-M. Tarascon, *Science*, 2015, **350**(6267), 1516.
40. J. Zheng, P. Xu, M. Gu, J. Xiao, N. D. Browning, P. Yan, C. Wang and J.-G. Zhang, *Chem. Mater.*, 2015, **27**(4), 1381.
41. G. Chen, X. Song and T. J. Richardson, *Electrochem. Solid-State Lett.*, 2006, **9**(6), A295.
42. L. Laffont, C. Delacourt, P. Gibot, M. Y. Wu, P. Kooyman, C. Masquelier and J. M. Tarascon, *Chem. Mater.*, 2006, **18**(23), 5520.
43. H. Gabrisch, J. Wilcox and M. M. Doeff, *Electrochem. Solid-State Lett.*, 2008, **11**(3), A25.
44. L. Gu, C. Zhu, H. Li, Y. Yu, C. Li, S. Tsukimoto, J. Maier and Y. Ikuhara, *J. Am. Chem. Soc.*, 2011, **133**(13), 4661.
45. F. Wang, R. Robert, N. A. Chernova, N. Pereira, F. Omenya, F. Badway, X. Hua, M. Ruotolo, R. Zhang, L. Wu, V. Volkov, D. Su, B. Key, M. S. Whittingham, C. P. Grey, G. G. Amatucci, Y. Zhu and J. Graetz, *J. Am. Chem. Soc.*, 2011, **133**(46), 18828.
46. L. Li, F. Meng and S. Jin, *Nano Lett.*, 2012, **12**(11), 6030.
47. S.-W. Kim, N. Pereira, N. A. Chernova, F. Omenya, P. Gao, M. S. Whittingham, G. G. Amatucci, D. Su and F. Wang, *ACS Nano*, 2015, **9**(10), 10076.
48. J. R. Dahn, T. Zheng, Y. Liu and J. S. Xue, *Science*, 1995, **270**(5236), 590.
49. M. V. Reddy, G. V. Subba Rao and B. V. R. Chowdari, *Chem. Rev.*, 2013, **113**(7), 5364.
50. R. Yazami, 'Nanomaterials for Lithium-Ion Batteries: Fundamentals and Applications', Pan Stanford, New York, 2013.
51. K. T. Lee and J. Cho, *Nano Today*, 2011, **6**(1), 28.
52. C. K. Chan, H. Peng, G. Liu, K. Mcilwrath, X. F. Zhang, R. A. Huggins and Y. Cui, *Nat. Nanotechnol.*, 2008, **3**(1), 31.
53. J. Sun, H.-W. Lee, M. Pasta, H. Yuan, G. Zheng, Y. Sun, Y. Li and Y. Cui, *Nat. Nanotechnol.*, 2015, **10**(11), 980.
54. M. S. Dresselhaus and G. Dresselhaus, *Adv. Phys.*, 2002, **51**(1), 139.
55. Y. Wen, K. He, Y. Zhu, F. Han, Y. Xu, I. Matsuda, Y. Ishii, J. Cumings and C. Wang, *Nat. Commun.*, 2014, **5**, 4033.
56. H. Liu, W. Li, D. Shen, D. Zhao and G. Wang, *J. Am. Chem. Soc.*, 2015, **137**(40), 13161.
57. W. Donghai, D. Choi, J. Li, Z. Yang, Z. Nie, R. Kou, D. Hu, C. Wang, L. V. Saraf, J. Zhang, I. A. Aksay and J. Liu, *ACS Nano*, 2009, **3**(4), 907.
58. D. Su, S. Dou and G. Wang, *Chem. Mater.*, 2015, **27**(17), 6022.
59. Y. Sun, L. Zhao, H. Pan, X. Lu, L. Gu, Y. S. Hu, H. Li, M. Armand, Y. Ikuhara, L. Chen and X. Huang, *Nat. Commun.*, 2013, **4**, 1870.
60. Y. Q. Wang, L. Gu, Y. G. Guo, H. Li, X. Q. He, S. Tsukimoto, Y. Ikuhara and L. J. Wan, *J. Am. Chem. Soc.*, 2012, **134**(18), 7874.
61. Y. Yao, M. T. McDowell, I. Ryu, H. Wu, N. Liu, L. Hu, W. D. Nix and Y. Cui, *Nano Lett.*, 2011, **11**(7), 2949.
62. N. Liu, H. Wu, M. T. McDowell, Y. Yao, C. Wang and Y. Cui, *Nano Lett.*, 2012, **12**(6), 3315.
63. H. Wu and Y. Cui, *Nano Today*, 2012, **7**(5), 414.
64. S. W. Lee, M. T. McDowell, J. W. Choi and Y. Cui, *Nano Lett.*, 2011, **11**(7), 3034.
65. X. Zhou, L. J. Wan and Y. G. Guo, *Adv. Mater.*, 2013, **25**(15), 2152.
66. P. Poizot, S. Laruelle, S. Grugeon, L. Dupont and J.-M. Tarascon, *Nature*, 2000, **407**(6803), 496.
67. K. He, H. L. Xin, K. Zhao, X. Yu, D. Nordlund, T.-C. Weng, J. Li, Y. Jiang, C. A. Cadigan, R. M. Richards, M. M. Doeff, X.-Q. Yang, E. A. Stach, J. Li, F. Lin and D. Su, *Nano Lett.*, 2015, **15**(2), 1437.
68. K. He, F. Lin, Y. Zhu, X. Yu, J. Li, R. Lin, D. Nordlund, T.-C. Weng, R. M. Richards, X.-Q. Yang, M. M. Doeff, E. A. Stach, Y. Mo, H. L. Xin and D. Su, *Nano Lett.*, 2015, **15**(9), 5755.
69. M. M. Thackeray, S. D. Baker, K. T. Adendorff and J. B. Goodenough, *Solid State Ion.*, 1985, **17**(2), 175.
70. J. Li, K. He, Q. Meng, X. Li, Y. Zhu, S. Hwang, K. Sun, H. Gan, Y. Zhu, Y. Mo, E. A. Stach and D. Su, *ACS Nano*, 2016, **10**(10), 9577.
71. H. Wang, L.-F. Cui, Y. Yang, H. S. Casalongue, J. T. Robinson, Y. Ling, Y. Cui and H. Dai, *J. Am. Chem. Soc.*, 2010, **132**(40), 13978.
72. G. Zhou, D.-W. Wang, F. Li, L. Zhang, N. Li, Z.-S. Wu, L. Wen, G. Q. Lu and H.-M. Cheng, *Chem. Mater.*, 2010, **22**(18), 5306.
73. Z. S. Wu, W. Ren, L. Wen, L. Gao, J. Zhao, Z. Chen, G. Zhou, F. Li and H. M. Cheng, *ACS Nano*, 2010, **4**(6), 3187.
74. E. A. Stach, *Mater. Today*, 2008, **11**(S), 50.
75. Y. Yuan, K. Amine, J. Lu and R. Shahbazian-Yassar, *Nat. Commun.*, 2017, **8**, 15806.
76. T. Shang, Y. Wen, D. Xiao, L. Gu, Y. S. Hu and H. Li, *Adv. Energy Mater.*, 2017, **7**(23), 1700709.
77. S.-M. Bak, K.-W. Nam, W. Chang, X. Yu, E. Hu, S. Hwang, E. A. Stach, K.-B. Kim, K. Y. Chung and X.-Q. Yang, *Chem. Mater.*, 2013, **25**(3), 337.
78. S. Hwang, S. M. Kim, S.-M. Bak, B.-W. Cho, K. Y. Chung, J. Y. Lee, W. Chang and E. A. Stach, *ACS Appl. Mater. Interfaces*, 2014, **6**(17), 15140.
79. S. Sharifi-Asl, F. A. Soto, A. Nie, Y. Yuan, H. Asayesh-Ardakani, T. Forrozan, V. Yurkiv, B. Song, F. Mashayek, R. F. Klie, K. Amine, J. Lu, P. B. Balbuena and R. Shahbazian-Yassar, *Nano Lett.*, 2017, **17**(4), 2165.

80. S. Hwang, S. M. Kim, S.-M. Bak, S. Y. Kim, B.-W. Cho, K. Y. Chung, J. Y. Lee, E. A. Stach and W. Chang, *Chem. Mater.*, 2015, **27**(11), 3927.

81. P. Yan, J. Zheng, T. Chen, L. Luo, Y. Jiang, K. Wang, M. Sui, J.-G. Zhang, S. Zhang and C. Wang, *Nat. Commun.*, 2018, **9**, 2437.

82. J. Y. Huang, L. Zhong, C.-M. Wang, J. P. Sullivan, W. Xu, L. Q. Zhang, S. X. Mao, N. S. Hudak, X. H. Liu, A. Subramanian, H. Fan, L. Qi, A. Kushima and J. Li, *Science*, 2010, **330**(6010), 1515.

83. X. H. Liu, J. W. Wang, S. Huang, F. Fan, X. Huang, Y. Liu, S. Krylyuk, J. Yoo, S. A. Dayeh, A. V. Davydov, S. X. Mao, S. T. Picraux, S. Zhang, J. Li, T. Zhu and J. Y. Huang, *Nat. Nanotechnol.*, 2012, **7**, 749.

84. X. H. Liu, L. Zhong, S. Huang, S. X. Mao, T. Zhu and J. Y. Huang, *ACS Nano*, 2012, **6**, 1522.

85. M. T. McDowell, I. Ryu, S. W. Lee, C. Wang, W. D. Nix and Y. Cui, *Adv. Mater.*, 2012, **24**(45), 6034.

86. M. T. McDowell, S. W. Lee, J. T. Harris, B. A. Korgel, C. Wang, W. D. Nix and Y. Cui, *Nano Lett.*, 2013, **13**(2), 758.

87. X. H. Liu, H. Zheng, L. Zhong, S. Huang, K. Karki, L. Q. Zhang, Y. Liu, A. Kushima, W. T. Liang, J. W. Wang, J.-H. Cho, E. Epstein, S. A. Dayeh, S. T. Picraux, T. Zhu, J. Li, J. P. Sullivan, J. Cumings, C. Wang, S. X. Mao, Z. Z. Ye, S. Zhang and J. Y. Huang, *Nano Lett.*, 2011, **11**(8), 3312.

88. J. W. Wang, X. H. Liu, S. X. Mao and J. Y. Huang, *Nano Lett.*, 2012, **12**(11), 5897.

89. F. Wang, H.-C. Yu, M.-H. Chen, L. Wu, N. Pereira, K. Thornton, A. Van der Ven, Y. Zhu, G. G. Amatucchi and J. Graetz, *Nat. Commun.*, 2012, **3**, 1201.

90. K. He, S. Zhang, J. Li, X. Yu, Q. Meng, Y. Zhu, E. Hu, K. Sun, H. Yun, X.-Q. Yang, Y. Zhu, H. Gan, Y. Mo, E. A. Stach, C. B. Murray and D. Su, *Nat. Commun.*, 2016, **7**, 11441.

91. K. He, Y. Zhou, P. Gao, L. Wang, N. Pereira, G. G. Amatucci, K.-W. Nam, X.-Q. Yang, Y. Zhu, F. Wang and D. Su, *ACS Nano*, 2014, **8**(7), 7251.

92. L. Luo, J. Wu, J. Xu and V. P. Dravid, *ACS Nano*, 2014, **8**(11), 11560.

93. M. T. McDowell, Z. Lu, K. J. Koski, J. H. Yu, G. Zheng and Y. Cui, *Nano Lett.*, 2015, **15**(2), 1264.

94. K. He, Z. Yao, S. Hwang, N. Li, K. Sun, H. Gan, Y. Du, H. Zhang, C. Wolverton and D. Su, *Nano Lett.*, 2017, **17**(9), 5726.

95. L. Wang, Z. Xu, W. Wang and X. Bai, *J. Am. Chem. Soc.*, 2014, **136**(18), 6693.

96. S. Hwang, Z. Yao, L. Zhang, M. Fu, K. He, L. Mai, C. Wolverton and D. Su, *ACS Nano*, 2018, **12**(4), 3638.

97. X. H. Liu, J. W. Wang, Y. Liu, H. Zheng, A. Kushima, S. Huang, T. Zhu, S. X. Mao, J. Li, S. Zhang, W. Lu, J. M. Tour and J. Y. Huang, *Carbon*, 2012, **50**(10), 3836.

98. Y. Liu, F. Fan, J. Wang, Y. Liu, H. Chen, K. L. Jungjohann, Y. Xu, Y. Zhu, D. Bigio, T. Zhu and C. Wang, *Nano Lett.*, 2014, **14**(6), 3445.

99. S. Suzuki, C. Bower and O. Zhou, *Chem. Phys. Lett.*, 1998, **285**(3–4), 230.

100. Y. Zhu, J. W. Wang, Y. Liu, X. Liu, A. Kushima, Y. Liu, Y. Xu, S. X. Mao, J. Li, C. Wang and J. Y. Huang, *Adv. Mater.*, 2013, **25**(38), 5461.

101. S. Lee, Y. Oshima, E. Hosono, H. Zhou, K. Kim, H. M. Chang, R. Kanno and K. Takayanagi, *J. Phys. Chem. C*, 2013, **117**(46), 24236.

102. A. Nie, L.-Y. Gan, Y. Cheng, Q. Li, Y. Yuan, F. Mashayek, H. Wang, R. Klie, U. Schwingenschlogl and R. Shahbazian-Yassar, *Nano Lett.*, 2015, **15**(1), 610.

103. M. J. Williamson, R. M. Tromp, P. M. Vereecken and F. M. Ross, *Nat. Mater.*, 2003, **2**(8), 532.

104. J. M. Yuk, J. Park, P. Ercius, K. Kim, D. J. Hellebusch, M. F. Crommie, J. Y. Lee, A. Zettl and A. P. Alivisatos, *Science*, 2012, **336**(6077), 61.

105. J. E. Evans, K. L. Jungjohann, N. D. Browning and I. Arslan, *Nano Lett.*, 2011, **11**(7), 2809.

106. P. Abellan, B. L. Mehdi, L. R. Parent, M. Gu, C. Park, W. Xu, Y. Zhang, I. Arslan, J.-G. Zhang, C.-M. Wang, J. E. Evans and N. D. Browning, *Nano Lett.*, 2014, **14**(3), 1293.

107. Z. Zeng, W.-I. Liang, H.-G. Liao, H. L. Xin, Y.-H. Chu and H. Zheng, *Nano Lett.*, 2014, **14**(4), 1745.

108. M. Sun, H.-G. Liao, K. Niu and H. Zheng, *Sci. Rep.*, 2013, **3**(1), 3227.

109. B. L. Mehdi, J. Qian, E. Nasybulin, C. Park, D. A. Welch, R. Faller, H. Mehta, W. A. Henderson, W. Xu, C. M. Wang, J. E. Evans, J. Liu, J.-G. Zhang, K. T. Mueller and N. D. Browning, *Nano Lett.*, 2015, **15**(3), 2168.

110. J. M. Yuk, H. K. Seo, J. W. Choi and J. Y. Lee, *ACS Nano*, 2014, **8**(7), 7478.

111. M. E. Holtz, Y. Yu, D. Gunceler, J. Gao, R. Sundararaman, K. A. Schwarz, T. A. Arias, H. D. Abruña and D. A. Muller, *Nano Lett.*, 2014, **14**(3), 1453.

112. A. J. Leenheer, K. L. Jungjohann, K. R. Zavadil, J. P. Sullivan and C. T. Harris, *ACS Nano*, 2015, **9**(4), 4379.

113. M. Gu, L. R. Parent, B. L. Mehdi, R. R. Unocic, M. T. McDowell, R. L. Sacci, W. Xu, J. G. Connell, P. Xu, P. Abellan, X. Chen, Y. Zhang, D. E. Perea, J. E. Evans, L. J. Lauhon, J.-G. Zhang, J. Liu, N. D. Browining, Y. Cui, I. Arslan and C.-M. Wang, *Nano Lett.*, 2013, **13**(12), 6106.

114. Z. Zeng, X. Zhang, K. Bustillo, K. Niu, C. Gammer, J. Xu and H. Zheng, *Nano Lett.*, 2015, **15**(8), 5214.

115. D. Ruzmetov, V. P. Oleshko, P. M. Haney, H. J. Lezec, K. Karki, K. H. Baloch, A. K. Agawal, A. V. Davydov, S. Krylyuk, Y. Liu, J. Y. Huang, M. Tanase, J. Cumings and A. A. Talin, *Nano Lett.*, 2012, **12**(1), 505.

116. Z. Wang, D. Santhanagopalan, W. Zhang, F. Wang, H. L. Xin, K. He, J. Li, N. Dudney and Y. S. Meng, *Nano Lett.*, 2016, **16**(6), 3760.

117. Y. Gong, J. Zhang, L. Jiang, J.-A. Shi, Q. Zhang, Z. Yang, D. Zou, J. Wang, X. Yu, R. Xiao, Y.-S. Hu, L. Gu, H. Li and L. Chen, *J. Am. Chem. Soc.*, 2017, **139**(12), 4274.

118. Report of the Basic Energy Sciences Workshop on "Electron Scattering and Diffraction" (2014); https://science.energy.gov/~/media/bes/pdf/reports/files/Future_of_Electron_Scattering.pdf.

119. Y. Li, Y. Li, A. Pei, K. Yan, Y. Sun, C.-L. Wu, L.-M. Joubert, R. Chin, A. L. Koh, Y. Yu, J. Perrino, B. Butz, S. Chu and Y. Cui, *Science*, 2017, **358**(6362), 506.

120. M. J. Zachman, Z. Tu, S. Choudhury, L. A. Archer and L. F. Kourkoutis, *Nature*, 2018, **560**, 345.

121. H. Zheng, Y. S. Meng and Y. Zhu, *MRS Bull.*, 2015, **40**(1), 12.

122. R. J. D. Miller, *Science*, 2014, **343**(2014), 1108.

123. A. H. Zewail, *Annu. Rev. Phys. Chem.*, 2006, **57**, 65.

# Synchrotron-Based Soft X-Ray Spectroscopy for Battery Material Studies

**Wanli Yang**

*Advanced Light Source, E O Lawrence Berkeley National Laboratory, Berkeley, CA, USA*



## 1 INTRODUCTION

Today's sustainable energy applications such as electric vehicles (EVs) and electric grid require high-efficiency low-cost electric energy storage solution that is hard to manage through existing battery systems. Nonetheless, rechargeable battery systems through reversible electrochemical storage/release of ions and electrons remain one of the most promising candidates. While rechargeable Li-ion batteries (LIBs) provide so far the best energy density for EVs, the rechargeable Na-ion battery (NIB) holds the promise of low-cost for grid-scale storage. In the past decades, tremendous research and development efforts have been made to achieve both a systematic understanding and technological optimization of batteries by researchers in the academy, government, and industry. Despite all these efforts, technological breakthrough remains formidable to overcome the bottleneck problems related to the battery stability, safety, and cost issues, which are impeding further developments of the battery technology to meet the requirements of large-scale EV and grid applications. In the meantime, modern material characterization tools have empowered the researchers in the field with multimodal approaches for both the fundamental understanding and practical optimization of the battery systems. The optimism lies in the hope that systematic and detailed understanding of the evolving material properties and chemistry in batteries could provide the rationality and guideline for technological breakthroughs.

Among the many technical challenges for realizing a high-performance battery system, almost all components in battery devices, from electrodes to electrolyte, from surfaces/interfaces to bulk, from current collectors to packaging, need to be optimized to meet the requirements for a safe and stable operation of high-energy batteries. Most anode materials are carbon based, with Si material starts to be introduced into commercial products at this time. Anodes for LIBs require a stable interphase layer on the carbon anode surface, now called solid-electrolyte-interphase (SEI), which is often responsible for the safety

*Inorganic Battery Materials*. Edited by Hailiang Wang and Boniface P.T. Fokwa
 ISBN 978-1-119-43199-2

and lifetime issues of LIBs. Anodes for NIBs are still under active debates in both fundamental mechanism and practical developments. Compared with anodes that are relatively less of a concern on storage capacity, current cathode component in a commercial battery system is the bottleneck of the storage capacity. In addition to the capacity issue, conventional wisdom also considers the electronic state configuration and interplay between transition metal (TM) 3d states and ligand anion, typically oxygen, p states fundamentally determines the intrinsic voltage limit and safety of a battery.[1] For both electrodes, most practical electrode materials are of the so-called intercalation type, which allows the Li-ion ($Li^+$) insertion and extraction through a more or less open and stable lattice structure. Such a process triggers a number of structural, physical, and chemical changes in the electrode material upon electrochemical cycling. The complicated nature of battery operation is indicated by the lengthy developments of almost all LIB components. The optimism now is to develop new strategies for speedy innovation of battery technologies based on better fundamental understandings of the mechanism and guidelines for the optimization.

In general, there are two critical topics on the central components of electrode and electrolyte that need to be clarified. One is the surface of the electrode particles and their interfaces/interphases with the electrolyte. Unfortunately, the chemical environment at the interface is in the approximation to the bulk electrode/electrolyte materials. Other than the structural probes through microscopy, direct measurements of the chemical and physical properties at the interface require incisive tools with superior chemical sensitivity, so that the subtle but crucial effect of the interface could be resolved in experiments. The other is the structural and chemical evolution of the active electrode material itself. Relatively, structural studies of battery materials have become standard characterization together with electrochemical profiles. However, as for any electronic device, information on electronic state evolution is an indispensable part of the battery operation mechanism but remains relatively limited partially due to the unavailability of suitable techniques for direct measurements. Therefore, synchrotron-based techniques have attracted more and more attention recently for studying battery materials,[2] especially for revealing the chemical reaction mechanism with elemental and chemical sensitivities under both ex situ and in situ conditions.

In addition, new concepts and demonstrations for improving energy and power density of battery electrodes emerge and challenge the conventional spectroscopic approaches. For example, novel redox couples, especially oxygen redox in oxide electrodes, have recently revitalized the field.[3] Other unusual chemical states of TMs could also be produced through the electrochemical process in batteries, especially atypically high valence states, for example, $Fe^{4+}$, which also hold the promise to achieve improved capacity in cathodes. For such novel redox couples, a reliable detection is a nontrivial issue, which is partially the reason why such topics are often under active debate in both practical applications and fundamental understandings. For example, the anionic redox in oxide cathodes has been proposed to originate from peroxo-like oxygen dimers, oxygen holes that are localized, or enabled by some specific structural and chemical configurations.[4] Although conventional core-level spectroscopy, especially X-ray photoelectron spectroscopy (XPS) and soft X-ray absorption spectroscopy (sXAS) have been extensively employed, their reliability for detecting the intrinsic bulk oxygen chemistry is questionable due to the limitations on the probe depth and strong effects from other contributions.[5] Therefore, incisive tools beyond conventional spectroscopy are keenly required for a reliable characterization to clarify both the interface and bulk reactions involved in the battery material performance.

This chapter covers synchrotron-based soft X-ray spectroscopic (SXS) studies of battery materials on both the interfaces/interphases and bulk electrodes, with the focus on core-level photon-in-phonon-out (PIPO) SXS in the methodology, analysis, and technical innovations. We provide a summary of conventional sXAS on both interface and electrode studies but concentrate more on the recent developments and demonstrations, especially recent technical developments through the high-efficiency mapping of resonant inelastic X-ray scattering (mRIXS).

Overall, this chapter is organized as following: We first introduce the fundamentals of SXS through simplified atomic models, with emphasis on the different detection channels and the technical comparison between sXAS and mRIXS. We then showcase a general example of utilizing sXAS and soft X-ray emission spectroscopy (sXES) for probing the critical unoccupied/occupied states of the charged/discharged electrodes, respectively. After these general introduction and demonstrations, our focused discussions are separated into five parts: (i) sXAS studies of low-Z elements, including C and O of the particle surface binder and interphases; (ii) TM cationic redox couples that could be quantified through straightforward linear combination fitting of the sXAS spectra; (iii) limitations of sXAS on studying solvation shell and demonstrations of the superior chemical sensitivity of RIXS; (iv) limitations of sXAS and demonstrations of mRIXS on detecting an unusual Mn(I/II) redox state in anode materials; (v) detection of oxygen redox reactions in battery cathodes and the clarifications of the popular misunderstanding of oxygen sXAS. At the end, we provide our perspectives on utilizing the advanced SXS for battery materials researches, especially the in situ/operando SXS experiments, which provides unique opportunities for studying both the interfaces and the anionic/cationic redox couples in batteries.

## 2 SYNCHROTRON BASED SXS TECHNIQUES

### 2.1 Continuously Tunable X-Rays from Synchrotron Light Source

The techniques discussed in this chapter rely on X-ray light sources that could deliver photons with continuously tunable energies (wavelength). For X-ray spectroscopy, the typical unit of electron volt (eV) is used here, instead of wavelength. The reason synchrotron radiation could provide the tunable X-ray photons at a wide range stems is based on an important theorem of classical electrodynamics: if a charged particle undergoes a change of direction at a relativistic speed, electromagnetic radiation will be emitted. In the simplest case, a single dipole (bending) magnet cause the electrons to move along a curved trajectory following the right-hand rule, corresponding to an (centrifugal) acceleration that emits a wide fan of electromagnetic radiation that covers a broad energy range, as schematically shown in Figure 1(a). More intense X-ray can be emitted in a periodic magnetic structure, so-called undulator, by adding up the radiation coherently, as indicated in Figure 1(b). A complete review of the physics of synchrotron science is beyond the scope of this chapter. Interested readers are referred to existing reviews and books on synchrotron radiation.[6] Note that high-intensity X-ray source is critical for some SXS techniques, such as RIXS that becomes more and more useful for studying battery materials.

Elements involved in practical battery materials are typically low-Z elements such as C, O, N, F, etc. and 3d TMs. It is known that the chemical activities of low-Z and TM elements are dominated by the 2p and 3d electron states. This makes the SXS tool-of-choice as the most direct probes of these states because, as elaborated in the next section, the dipole selection rule allows the photon excitations directly into the interested p and 3d states in the soft X-ray energy range. Figure 2 shows the electron excitations that are accessible through soft X-rays with excitation photon energy of 50 eV to 2 keV. Except for the M- and N-edges of the heavy elements that are not plotted here, SXS covers the energy range of the 1s–2p excitations of the low Z elements up to Mg and the 2p–3d transitions of the 3d TM elements. Therefore, the valence electronic states (2p states for low Z elements and 3d states for 3d TMs) can be directly accessed through SXS.

### 2.2 Excitations and Decay Channels in SXS

Figure 3 displays a general atomic schematic of typical SXS experiments with data collected in reflection, that is, non-transmission, modes.[8] The spectroscopic process starts with the absorption of the continuously tunable synchrotron photons by a core electron that is excited to an unoccupied state (Figure 3a). XAS thus provides the information of the unoccupied electronic structure at the excited states with the existence of core holes. Note "sXAS" is used in this work to differentiate soft X-ray XAS from hard X-ray XAS, which is significantly different in many aspects.[2] In principle, the intrinsic absorption coefficient corresponds to the attenuation of incident X-rays; however, the short penetration depth of soft X-rays limits the direct transmission-rate measurement of the attenuation to only ultra-thin samples. For battery studies, except for scanning transmission X-ray microscopy (STXM), the XAS is mostly collected through the intensity of decay products of the core holes. Two broadly practiced channels are the decay through the Auger electron yield (AEY), or total electron yield (TEY) (Figure 3b), and the total yield of fluorescence photons (TFY) (Figure 3c). The short escape length of electrons makes the AEY and TEY very surface sensitive with probing depth of only several nanometers. The TFY mode is relatively bulk sensitive with a probe depth of hundreds of nanometers depending on the photon energy. At this time, almost all soft XAS system allows the simultaneous data collection through both the TEY and TFY channels, which often reveal the contrast between surface and bulk states of battery samples. However, caution needs to be taken with any quantitative analysis based on TFY, because serious lineshape distortion from the so-called self-absorption effect may occur, which is due to the abrupt change of the penetration depth of the X-rays when the incident photon energy scans through an absorption edge.

For battery material studies, an important advance on XAS technique is the recently developed inverse partial fluorescence yield (iPFY), which could potentially solve the technical problem of the self-absorption distortion in fluorescence yield signals. In contrast to TEY and TFY, which measure the excited core-hole states by scanning the excitation photon energy through the absorption edge of interest, iPFY is effectively a measure of the X-ray attenuation length, akin to transmission measurement. For battery material studies, iPFY provides the opportunity of a quantitative analysis of the TM states in the bulk.[9]

Other than collecting the total number of released particles (electrons or photons) through the decay process in XAS experiments, the decay process indicated by Figure 3(c) leads to various features in experimental data: (i) If the excited state decays back to the same ground state by releasing the same amount of energy as the incident X-ray, it will generate an elastic (no change on energy) feature (Figure 4a, left). (ii) If the decay corresponds to filling up the core hole by a valence-band electron, the emitted photon energy will be pinned by the energy difference between the valence electron and the core hole, independent of the incident X-ray. This is called normal X-ray emission spectroscopy (XES) or normal fluorescence (Figure 4a, middle). Therefore, XAS signals corresponds to the occupied valence states. Experimentally,

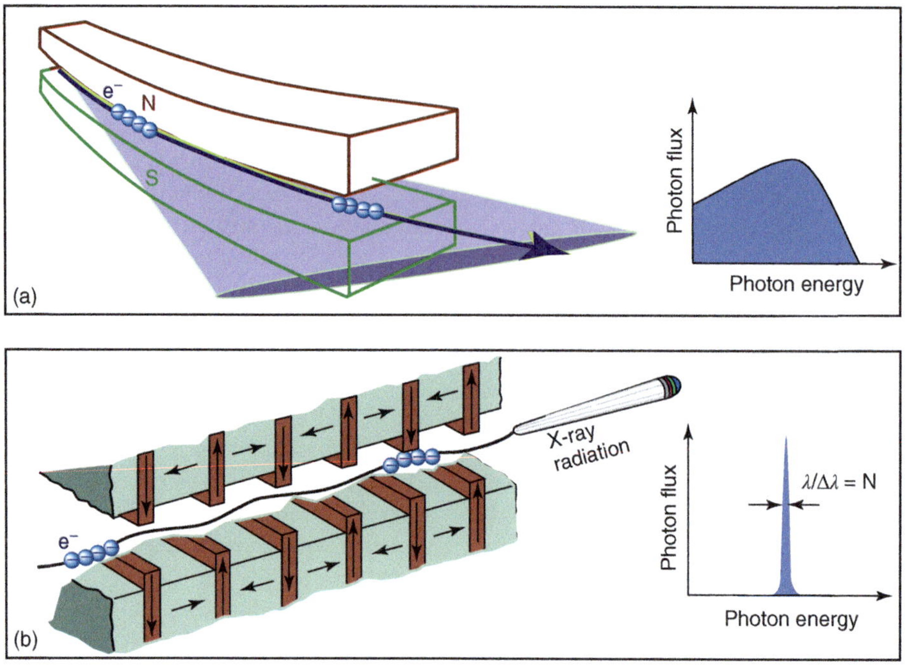


**Figure 1** Tunable X-ray source based on synchrotron radiation by injecting electron bunches into (a) one group of bending magnets, or (b) an undulator with a parodic magnetic structure. The photon spectra on the right show the broad energy distribution of the radiation based on bending magnets (a), compared with the narrow energy range with a small angular divergence of the beam based on undulator (b).[2] [Reprinted with permission from F. Lin, Y. Liu, X. Yu, L. Cheng, A. Singer, O. G. Shpyrko, H. L. Xin, N. Tamura, C. Tian, T. C. Weng, X. Q. Yang, Y. S. Meng, D. Nordlund, W. Yang and M. M. Doeff, Chem. Rev., 2017, 117, 13123–13186. © American Chemical Society 2017]

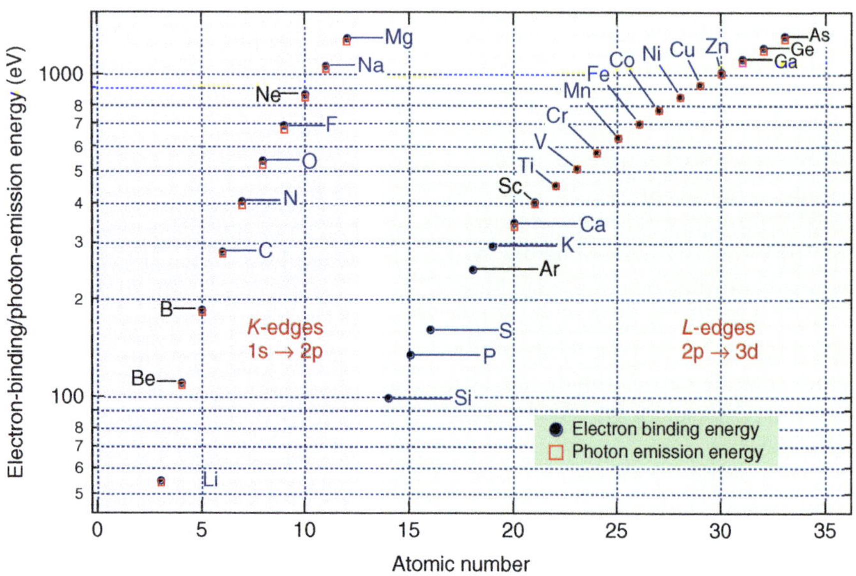


**Figure 2** Core-level binding energies (solid dots) and associated X-ray emission energies (hollow squares) that are accessible through SXS. This figure focuses on the practical materials for batteries and the M- and N-edges of the heavy elements are not plotted here.[7] [Reproduced with permission from Q. Li, R. M. Qiao, L. A. Wray, J. Chen, Z. Zhuo, Y. Chen, S. Yan, F. Pan, Z. Hussain and W. Yang, J. Phys. D. Appl. Phys., 2016, 49, 413003. © Institute of Physics and the Physical Society 2016]

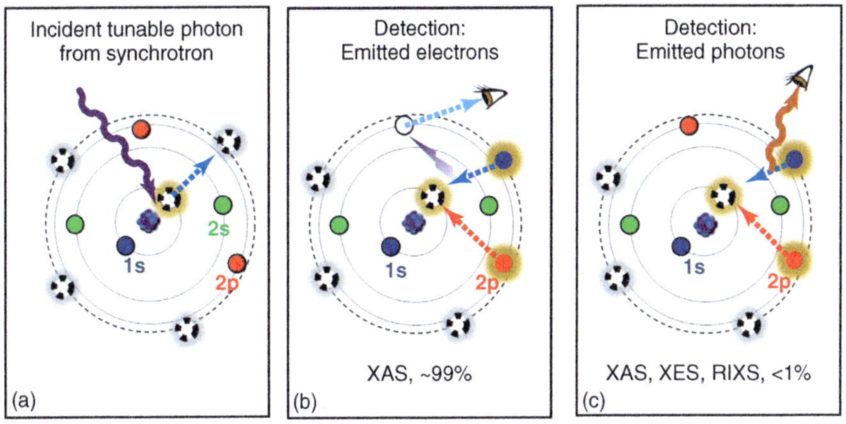


**Figure 3** Atomic schematic of the general excitations and decay process involved in core-level XAS, XES, and RIXS. (a) Core-electron excitation, (b) photon-in-electron-out XAS, and (c) photon-in-photon-out XAS, XES, and RIXS.[8] [Reproduced with permission from W. Yang, X. Liu, R. Qiao, P. Olalde-Velasco, J. D. Spear, L. Roseguo, J. X. Pepper, Y.-d. Chuang, J. D. Denlinger and Z. Hussain, J. Electron. Spectrosc. Relat. Phenom., 2013, 190, 64–74. © Elsevier 2013]

XES is typically collected by exciting the core electron to the high-energy continuum well above the absorption threshold. (iii) In order to maintain the excited state after the absorption process, the core hole of the intermediate state exerts a strong potential to the outer shell electrons to screen itself. Such an effect could lead to various other excitations at much lower energy scale, such as lattice vibrations (phonons), spin flips (spinons or magnons), and charge transfer excitations. When the core hole is eventually filled, these excitations are left in the system, leading to an energy loss of the emitted photon energy compared with the incident X-ray. Therefore, through the excitation (absorption) and decay process, the system may reach three different types of final states, as summarized in Figure 4.

## 3 GENERAL SXS DEMONSTRATIONS OF BATTERIES

### 3.1 Critical Unoccupied and Occupied States in $FePO_4$ Electrodes

Various information of the outer-shell electron states of a material could be obtained by detecting the emitted photons. The total count of the emitted photons is actually the fluorescence yield signal in sXAS experiments, which naively corresponds to the unoccupied conduction-band states with the core-hole existence. XES, on the other hand, corresponds to the occupied valence-band states because it mainly involves the decay from valence-band electrons to the core hole. Therefore, the combination of sXAS and XES probe both the unoccupied and occupied states in the vicinity of the Fermi level.

The applications of sXAS and XES for studying battery interfaces and materials have been reviewed in more details with examples before.[2,8] Here, we only point out that the occupied/unoccupied states close to Fermi level in the discharged/charged states in electrodes are the most relevant electron states to batteries, because electrons are removed/added from/to these states during the electrochemical operations. Often the concern of the safety in LIBs is on the oxygen stability because oxygen occupied state may stay too close to Fermi level and get oxidized during the deintercalation process. As an example, Figure 5 shows the XES of $LiFePO_4$ displays an interesting reshuffling of the Fe-3d occupied states due to specific spin configurations and el–el Coulomb interactions, resulting in an occupied Fe-3d state close to the Fermi level (Figure 5). This guarantees that the key electron states involved in the electrochemical charging are dominated by Fe-3d states in $LiFePO_4$, which naturally provides the fundamental rationality on "Why $LiFePO_4$ is a safe battery electrode."[10]

### 3.2 mRIXS and sXAS Experiments of $LiNi_{1/3}Mn_{1/3}Co_{1/3}O_2$

As explained above, RIXS results contain the information of three types of decay processes and their associated excitations. In fact, RIXS has been used as a powerful technique for probing low-energy excitations in fundamental physics. Because RIXS does not directly represent a static configuration of electron density of states, understanding RIXS often requires complicated theoretical approaches. However, through a direct comparison of RIXS results with conventional sXAS data (Figure 2), one could easily understand the technical meaning of RIXS, and why it is an elemental and orbital sensitive tool with superior chemical sensitivity than conventional XAS.

Figure 6 presents a direct comparison between mRIXS and various sXAS spectra of a

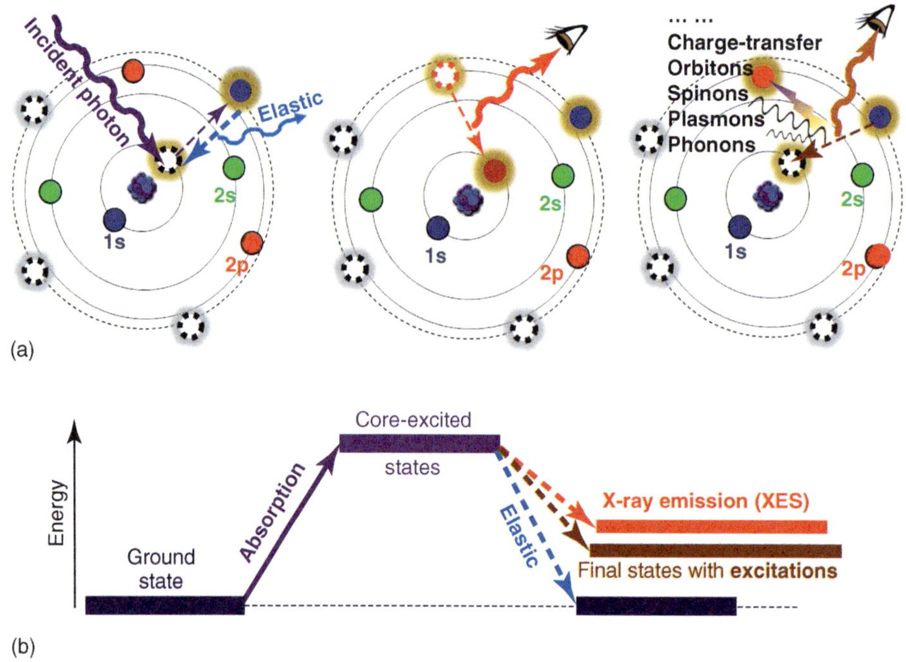


**Figure 4** Schematic of PIPO SXS. (a) Simplified atomic models of sXAS (left), XES (middle), and RIXS (right). (b) Displays three different types of signals through the RIXS experiments.[5] [Reproduced with permission from W. Yang and T. P. Devereaux, J. Power Sources, 2018, 389, 188–197. © Elsevier 2018]

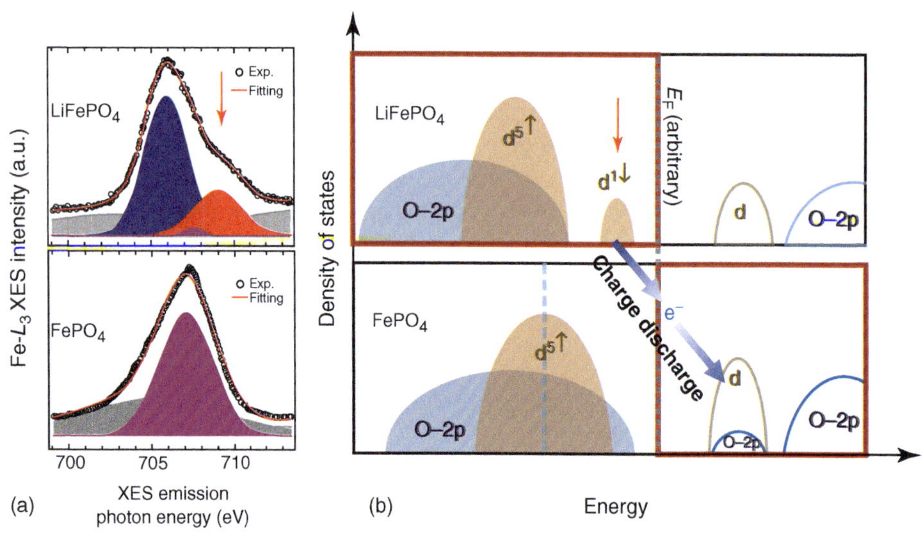


**Figure 5** (a) Fe-$L_3$ XES collected on $LiFePO_4$, contrasting that of fully delithiated (charged) $FePO_4$. The extra Fe-3d occupied state in the fully lithiated $LiFePO_4$ (red arrow) is pushed up in energy toward Fermi level, which naturally becomes the active state with almost pure Fe-3d character during charging. (b) Schematic of the critical occupied and unoccupied electron states that are involved in a battery operation. These states are both of Fe-3d characters in $LiFePO_4$, as experimentally identified through XES and sXAS, respectively. [Reproduced with permission from Ref. 10. © PCCP Owner Societies 2015]

$LiNi_{1/3}Mn_{1/3}Co_{1/3}O_2$ electrodes with incident X-ray energy (vertical axis) at the Mn-*L* range. At each incident energy, the emitted photons from the sample are differentiated by a RIXS spectrometer upon their emission energies (horizontal axis). Therefore, if one integrates all the RIXS intensity with the same excitation energy (horizontal integration), it will generate a single spectrum of Mn-*L* sXAS. Depending on the range of emission energy

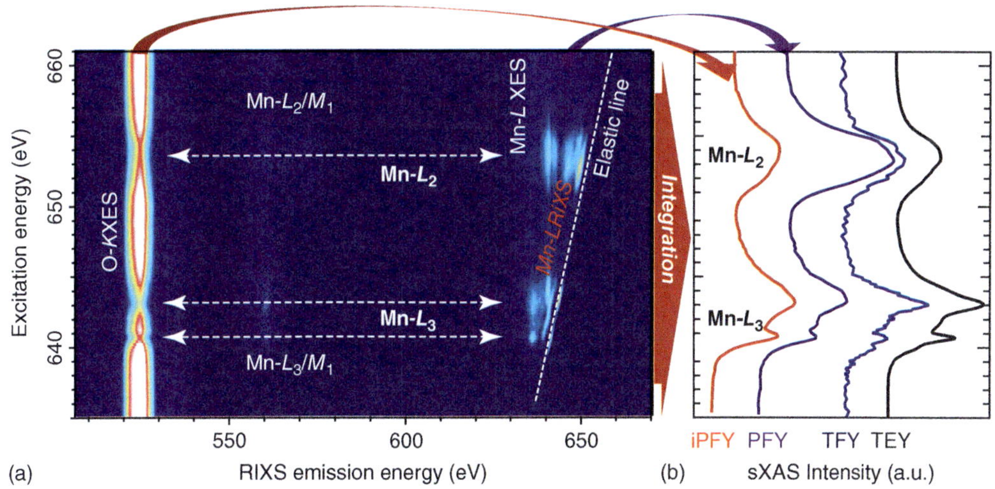


**Figure 6** (a) mRIXS collected from $LiNi_{1/3}Mn_{1/3}Co_{1/3}O_2$ with incident X-ray excitation energy at the Mn-*L* edge range (vertical axis). Horizontal axis is the energy of the emitted photons detected by a RIXS spectrometer. Color indicates the intensity distribution of the emitted photons, with light color represents high intensity. (b) Integration of mRIXS intensity along horizontal direction leads to different types of sXAS spectra. The conventional sXAS from total electron yield (TEY) and total fluorescence yield (TFY) channels are plotted for comparison. TFY spectrum is severely distorted due to the competing contributions from both O and Mn (white arrows in (a)). Depending on which section the mRIXS intensity is integrated, partial fluorescence yield (PFY) and inverse partial fluorescence yield (iPFY) could be obtained by integrating the intensity in the Mn-*L* and O-*K* emission range, respectively. Note iPFY is a bulk-probe PIPO channel without any lineshape distortion found in TFY sXAS. [Reproduced with permission from Ref. 11. © AIP Publishing LLC 2017]

for such an integration, different channels of sXAS signals could be obtained simultaneously: Integration in the Mn-*L* emission energy range from 630 to 650 eV generates the partial fluorescence yield (PFY) sXAS. The inverse of the integrated intensity in the O-*K* range, 520–530 eV, generates the iPFY sXAS. The integration of all the RIXS intensity is the total fluorescence yield (TFY) of sXAS.

Several important messages emerge through the comparison in Figure 6. First, with the same excitation energy, RIXS further resolves the single data point in sXAS into an energy distribution curve. RIXS, therefore, provides a new dimension of information along emission energy that is completely missing in sXAS. As shown throughout this chapter, this provides extra sensitivity to the chemical states in battery materials. Second, it is clear that the "elemental sensitive" of the sXAS does not necessarily mean all signals are from the same element in the material. Here, Mn-*L* sXAS contains significant contribution of emissions from oxygen. Such a mixing of emission signals in sXAS is the direct reason of the severe distortion of the TFY sXAS spectral lineshape because the intensity change of the O and Mn contributions are exactly opposite (white arrows in Figure 2a) due to the sudden change of the attenuation length at an absorption edge. Third, it is worth noting that, as shown by the seminal work of Achkar *et al.* with silicon-drift detector, iPFY is a bulk probe of sXAS without any lineshape distortion.[12] This makes iPFY particularly useful for quantitative analysis of the bulk TM states, and mRIXS enables iPFY of all edges through its much improved emission energy resolution. Fourth, resolving the emission energy in mRIXS enables a new dimension for differentiating critical RIXS features within a particular emission energy range, allowing a more detailed study of a particular chemical state that could be completely buried in sXAS. If such an emission energy range is very small, for example, in one or two eVs, the generated spectrum, dubbed super-partial fluorescence yield (sPFY), becomes intrinsically different from the conventional sXAS. While sPFY is actually RIXS signals plotted upon the same sXAS axis, it provides unique technical capabilities to quantify specific RIXS features for chemistry and material studies. Fifth, other unpopular decay channels could also be detected in such high-efficiency mRIXS. For example, decays from Mn 3s to 2p states (dipole allowed) are shown in the mRIXS around 560 eV emission energy (Figure 2a). Such a core–core partial fluorescence yield (ccPFY) provides another channel for obtaining non-distorted sXAS bulk signal, especially when iPFY cannot be implemented due to the lack or low-energy edges.

RIXS is a photon-hungry technique, due to the low branching ratio of the particular core hole decay channel involved. This is the reason why XES, discovered in 1934 since the early days of quantum mechanics, was largely replaced by XPS until the third generation synchrotron light source was developed. RIXS has made remarkable progress and provided invaluable contributions to the fields of highly correlated physics in the last two decades.

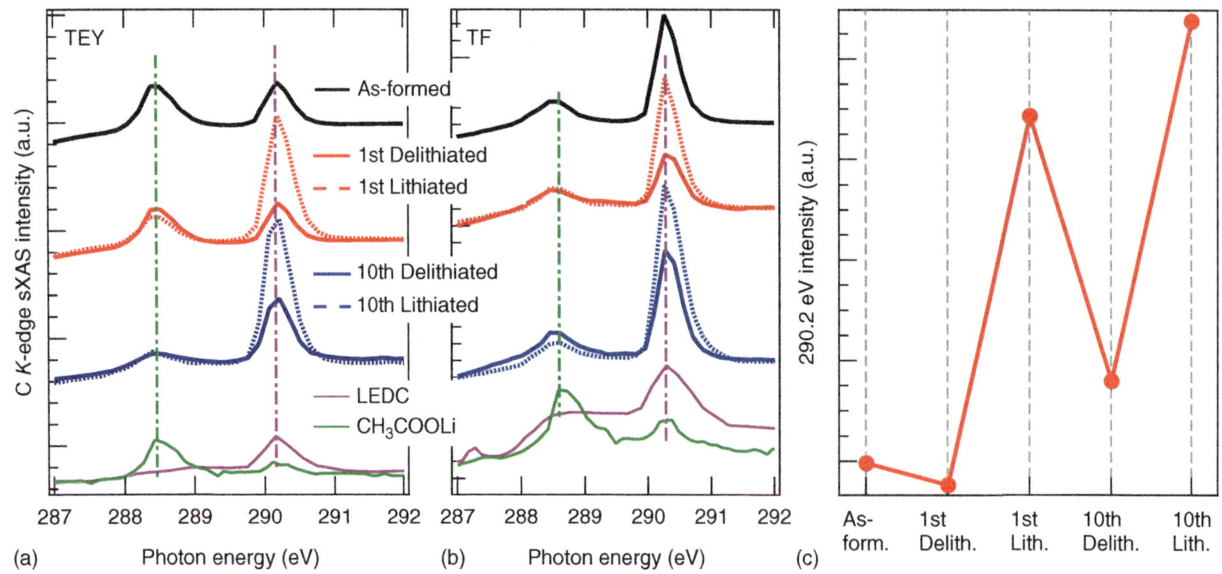


**Figure 7** (a) The C *K*-edge TEY spectra of the as-formed SEI (black), and the SEIs at charged (solid lines) and discharged (dotted lines) states at the 1st (red) and 10th (blue) cycles. (b) The C *K*-edge TFY spectra of the same series of SEIs. (c) The intensity evolution of the carbonate C-K TEY feature at 290.2 eV plotted upon different electrochemical state.[15] [Reproduced with permission from Z. Zhuo, P. Lu, C. Delacourt, R. Qiao, K. Xu, F. Pan, S. J. Harris and W. Yang, Chem. Commun. (Camb.), 2018, 54, 814–817. © Royal Society of Chemistry 2018]

Today's mRIXS was made possible through the recent developments of ultra-high efficiency spectrographs in the soft X-ray range.[11] Other than the superior chemical sensitivity of mRIXS, the many new sXAS channels beyond conventional TEY, TFY, and PFY signals, for example, iPFY, sPFY, ccPFY as explained above, are simply byproducts of mRIXS, and they open up new opportunities for studying both the surface and bulk chemistry of energy materials through soft X-rays.

## 4 *K*-EDGE SXAS OF LOW-Z ELEMENTS IN BATTERIES

### 4.1 Surfaces and Interfaces Probed by sXAS

Interfaces play critical roles in all the battery operation. Decades of fundamental studies and practical tests have been performed to understand and optimize the battery interfaces in both academia and industry. In conventional batteries with liquid electrolyte, the SEI on the anode surface is formed through electrolyte decomposition during the early cycles. The SEI behaves as a passivation layer to stabilize the battery operation by preventing electron transportation but allowing efficient ion diffusion and depletion.[13] For the cathode, especially high voltage compounds, the surface reconstruction, and surface layer formation have also attracted much research attention. Surface treatments and modifications have been implemented for most commercial cathodes, but the control and optimization of the electrode surface for a stable high-voltage operation remain a major technical challenge.

Owing to the shallow escape depth of electrons, the TEY channel of sXAS is a surface sensitive probe with probe depth of about 10 nm. The contrast between TEY and TFY (probe depth of about 100 nm) often provides valuable information on the surface and interface chemistry in batteries.[8] Such experiments are based on the sensitivity of sXAS to the relevant chemical species and chemical bonds. Particularly benefiting from the shallow probe depth, sXAS also allows the investigation of the chemical contrast of the different electrode surfaces facing electrolyte and current collector at high voltages, and specific molecular configurations on electrode surfaces.

For the complex SEI systems, sXAS could directly reveals the contrast on chemical composition of the SEIs. For example, SEIs formed on Sn anodes with different surface orientations display distinct sXAS lineshapes, indicating carbonates and fluorides on the Sn (100) and (001) surfaces from different decomposition reactions of the solvent and salt, respectively.[14] Figure 7 shows an intriguing electrochemical activity of the SEI formed on model Cu electrodes.[15] The sXAS data show that there is a strong and systematic variation of the SEI signal upon electrochemical charge and discharge process. In particular, the peak that corresponds to carbonate species in SEI gets much enhanced in the lithiated states, and reversibly decreases at each delithiation cycle. Such sXAS observations indicate that the formation process of SEI could be much more complex than a continuous deposition of the electrolyte decomposition products. The SEI could evolve upon electrochemical potentials during the formation process and displays an oscillating growth pattern in both the chemical compositions and thickness.[15]

sXAS in TEY and TFY channels have also been used to study the surface behavior of solid electrolyte particles used for solid-state batteries (SSBs). For example, an counterintuitive air-exposure effect was found on Al-substituted $Li_7La_3Zr_2O_{12}$ (LLZO) particle surface through the sXAS studies by Cheng *et al.* By evaluating the contribution ratio of the O-*K* features from the LLZO and carbonates, sXAS suggests that much less carbonates are formed on the small-grain sample than on the large-grain samples. This is interesting because small-size particles with large specific surface area are typically more prone to the surface reaction upon air exposure than large particles. Further experimental and theoretical analysis reveals that the surface of the small LLZO particles contains less lithium and more aluminum than that of the large particles, leading to a less air-sensitive surface. In addition, considering the 10 nm probe depth of the TEY channel of sXAS, the thickness of the carbonate coating could be roughly estimated to be couple nm for the LLZO solid electrolyte particles.[16]

### 4.2 Key Binder State Revealed by sXAS

As elaborated above, soft XAS corresponds to the unoccupied conduction band or lowest unoccupied molecular orbital (LUMO) derived states. Therefore, the low energy leading edge represents the upper edge of the band gap in a non-metallic material, which gives information about electric conductivity. Figure 8 shows the C-*K* soft XAS collected on a series of polyfluorene (PF) type battery binders with different functional groups. Binder is an important component in batteries to maintain both the mechanical and electrical integrity. However, a high volume of conductive additive reduces both the volumetric and gravimetric lithium-ion storage capacity. In addition, for high-capacity active materials, electrical disconnection takes place due to the severe volume expansion during electrochemical operation, for example, silicon anode is known to suffer the volume expansion for almost four times at the lithiated state. An ideal binder system should integrate mechanical stability and electrical conductivity, so both factors could be maintained through the electrochemical operation.

As shown in Figure 8, soft XAS reveals a LUMO state sitting at a very low energy in samples with carbonyl groups. This carbonyl induced $\pi^*_{C=O}$ transition is even lower in energy compared with that of the typically low-energy $\pi^*_{C=C}$ states. This soft XAS finding is also consistent with ab-initio density functional theory (DFT) calculations. The calculation further predicts that charge transfer to this low-energy state, which improves the overall electric conductivity, will take place if such materials are used under reducing environment in batteries. Such information provides unambiguous understanding of the improved electric conductivity in such a battery binder system. Electrodes made by mixing commercial Si particles with the Polyfluorene with carbonyl and methylbenzoic-ester (PFFOMB) polymer binder and without any other conductive additives show improved cycling performance

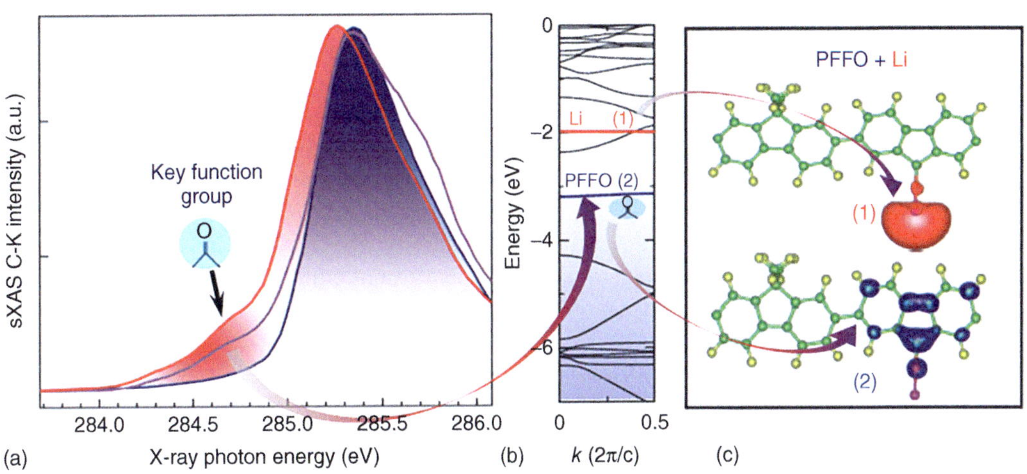


**Figure 8** (a) C-K XAS spectra of a series of conductive polymers as Li-ion battery anode binder materials. Blue: polyfluorene (PF); purple: PF with carbonyl C=O group (PFFO); red: PF with both carbonyl and methylbenzoicester groups (PFFOMB). The carbonyl group in the PF type conductive polymer introduces a low-energy LUMO state at 284.7 eV. (b) The calculated band structure of PFFO. The low energy electron state is from the carbonyl group, and sits below the Li level, leading to electron charge transfer from Li to the polymer backbone, as indicated by the isosurface plots (c).[17] [Reproduced with permission from G. Liu, S. Xun, N. Vukmirovic, X. Song, P. Olalde-Velasco, H. Zheng, V. S. Battaglia, L. Wang and W. Yang, Adv. Mater., 2011, 23, 4679. © Wiley 2011]

compared to those made with conventional formulations.[17] These data advance development of a functioning silicon electrode and provide fundamental understanding of the performance contrasts between different conductive polymer materials.

## 5 sXAS FOR QUANTIFYING TM CATIONIC REDOX REACTIONS

Compared with hard X-ray XAS, one of the inherent advantages of TM *L*-edge SXS is the strong dipole-allowed TM 2p–3d excitations, leading to arguably the most direct probe of the oxidation state of 3d TMs in battery compounds.[2] The TM *L*-edge spectral lineshapes are defined largely by atomic multiplet effects and therefore sharp and well-defined features can be observed via sXAS.[18] The high sensitivity and well-defined features of TM sXAS enable various kinds of in situ/operando and microscopic studies of the TM oxidation states.[19] In addition, lineshape analysis of the sharp multiplet features in TM-*L* sXAS provides so far the most reliable quantification of the 3d TM oxidation state distributions in battery materials at different electrochemical stages.[7]

Figure 9 is a summary of three selected examples of quantitative soft sXAS studies of Mn, Fe, and Ni cations in battery cathodes. A perfect quantitative fitting of all the high-resolution sXAS experimental features could be achieved through a simple linear combination of multiple reference spectra (for Mn, Figure 9a), two end-members for two-phase systems (for Fe, Figure 9b), or theoretical calculations of the TFY spectra (For Ni TFY, Figure 9c).[7] Even for the more complicated Co system that is of more solid-solution type, quasi-quantitative analysis could be achieved by the main peak position. Such quantitative analysis could reveal some key TM species on the surface of the electrodes, and define unambiguously the TM redox involved in electrode cycling, as specifically reviewed before for most 3d TMs in battery materials.[7]

However, technical challenges remain for a reliable detection of the bulk TM redox reactions due to the inherent lineshape distortion in the bulk-sensitive sXAS TFY channel, as explained above. As directly shown in Figure 6, the distortion is particular severe for TMs with other edges close by, for example, Mn,[12] and developments of iPFY and ccPFY become critical for a non-distorted bulk sXAS detection channel. We note again that, with the high resolution of the emission energy, mRIXS naturally includes both iPFY and ccPFY signals of all TM-*L* sXAS. The quantitative sXAS analysis demonstrated here could thus be extended to bulk TM analysis in exactly the same way but based on iPFY and/or ccPFY experimental data from mRIXS.

## 6 RIXS FOR DETECTING THE SUBTLE CHEMICAL CONTRAST

While sXAS remains a sensitive tool for detecting the surface chemistry of many systems, we would like to emphasize that, often the chemistry at the interface of two chemical species only shows subtle but critical difference compared with the bulk materials involved. It could be technically challenging to detect such weak contrast through conventional sXAS. One of the extreme case is to detect the ion-solvation shell in liquid electrolyte, where the molecules in the solvation shell (the "interface" of ions and liquid) are essentially the same as in the bulk electrolyte. The ion-solvation shell is a critical configuration in conventional battery systems and is directly involved in the charge transfer process throughout the electrochemical operations.[13] Unfortunately, experimental detection of the solvation shell remains one of the grand challenges. Figure 10 displays the sXAS and RIXS spectra of a model aqueous system with different KCl salt concentrations.[20] As expected, the sXAS spectral lineshape does not change much for different salt concentrations and shows characteristic pre- (~534 eV), main- (~536 eV), and post-edge (~540 eV) features that are typical for water. This is expected because the O-*K* sXAS measured here are from the solvent molecule, water, which is the same for pure water and the water in the ion solvation shells. In contrast, a significant change could be observed in the RIXS spectra collected at the 534.2 eV excitation energy (one horizontal cut in mRIXS as explained in Figure 6). The strong evolution of the RIXS spectral weight toward 534.2 eV emission energy (elastic peak) suggests that low-energy excitations of vibrational modes (phonons) in RIXS are greatly suppressed in the ion-solvation shell, and higher resolution analysis of the phonon excitations in RIXS could shed light on the details of the molecular configuration in the solvation shell.[20] Such strong spectral changes, however, are completely missing in the sXAS data because, technically, the integrated area of the RIXS spectra are roughly the same after normalization. This example is thus a typical demonstration of the superior chemical sensitivity of RIXS for interface studies due to the new dimension of emission energy distributions in RIXS.

## 7 mRIXS FOR DETECTING NOVEL mn REDOX STATES

Compared with sXAS, the new information by introducing the emission energy into mRIXS naturally provides another dimension to resolve the entangled states in sXAS. This becomes particularly important for determining the novel redox couples of both TMs and oxygen. Such an advantage of mRIXS is directly shown in Figure 11,

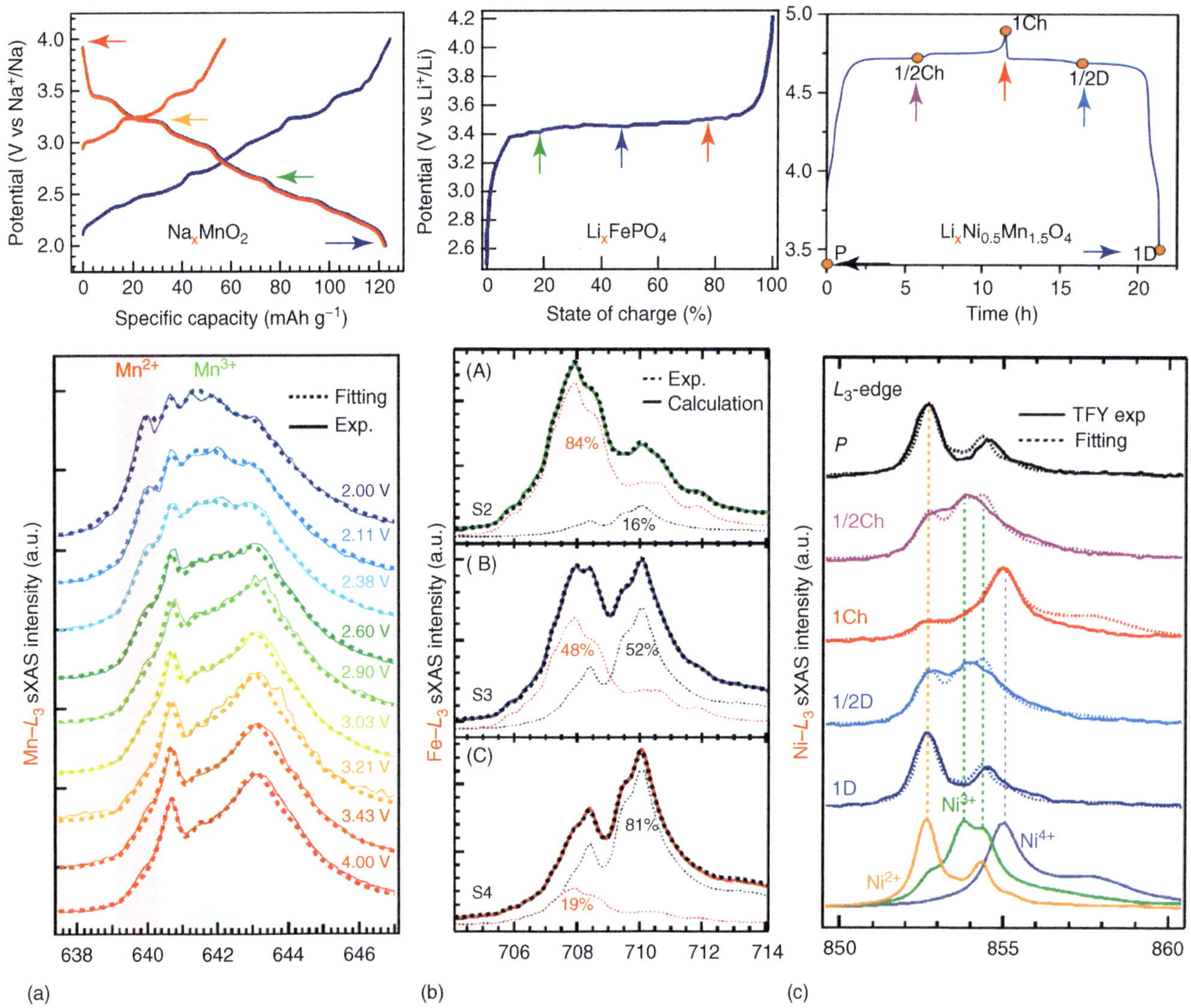


**Figure 9** Quantitative definition of transition-metal valence distributions through sXAS of Mn, Fe, and Ni in the electrodes of Na-ion and LIBs. Dotted spectra are simulation results. (a) A quantitative analysis of Mn L-edge soft XAS on a series of $Na_{0.44}MnO_2$ electrode cycled to different electrochemical states. (b) A quantitative analysis of Fe states in $LiFePO_4$. Lithiation level could thus be precisely determined. (c) A quantitative fitting of the Ni-$L$ sXAS of $LiNi_{0.5}Mn_{1.5}O_4$ cathodes collected through the TFY channel, based on the calculated $Ni^{2+}$, $Ni^{3+}$, and $Ni^{4+}$ TFY spectra at different electrochemical states.[2] [Reprinted with permission from F. Lin, Y. Liu, X. Yu, L. Cheng, A. Singer, O. G. Shpyrko, H. L. Xin, N. Tamura, C. Tian, T. C. Weng, X. Q. Yang, Y. S. Meng, D. Nordlund, W. Yang and M. M. Doeff, Chem. Rev., 2017, 117, 13123–13186. © American Chemical Society 2017]

where a 90-year speculation of $Mn^{1+}$ state in coordinated compound is clearly revealed in a charged battery electrode through mRIXS, but not sXAS.[21]

The sXAS and mRIXS collected on a charged and discharged anode with a Prussian blue analog material, $Na_{1.24}Mn[Mn(CN)_6]\cdot 2.1H_2O$, are shown in Figure 11. An intriguing low-voltage plateau is found through the Na-ion intercalation with an excellent rate performance, which suggests that this material could be used as a high-rate NIB anode with very low production cost. Further reduction of the pristine $Mn^{2+}$ to $Mn^{1+}$ through the ion intercalation is very unusual because Mn typically takes only zero or bivalence states. In fact, $Mn^{1+}$ was proposed in such coordinated compounds since 1928 but was not verified experimentally through conventional spectroscopy including XPS and sXAS. Because the spin states of the TMs in such material could be well defined depending on their coordination sites, such material displays distinct signatures of the typical TM states in sXAS. However, for novel chemical state, the contrast on the sXAS is very weak for the cycled electrodes (Figure 11). Only a hint of difference around 643.5 eV could be seen, although the lineshape change does correspond to the $Mn^{2+/1+}$ states expected by multiplet theory. In sharp contrast, the

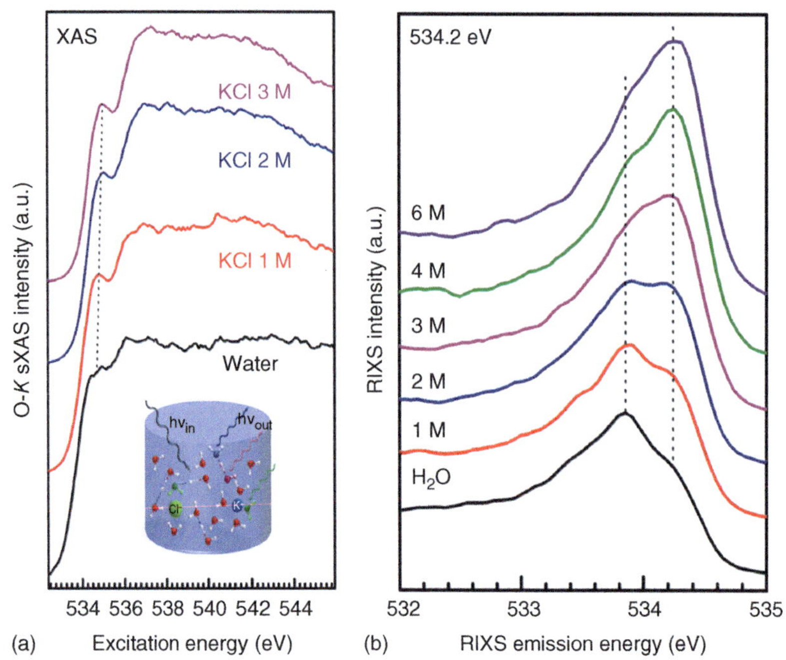


**Figure 10** In situ sXAS and RIXS of a model aqueous salt solution with different KCl salt concentrations. (a) The sXAS displays more or less the same lineshape with different salt concentrations. (b) The RIXS cuts at 534.2 eV excitation energy (dotted line in sXAS) show a dramatic lineshape change along the emission energy due to the ion-solvation effect.[20] [Reprinted with permission from Y. L. Jeyachandran, F. Meyer, S. Nagarajan, A. Benkert, M. Bär, M. Blum, W. Yang, F. Reinert, C. Heske, L. Weinhardt and M. Zharnikov, J. Phys. Chem. Lett., 2014, 5, 4143–4148. © 2014 American Chemical Society]

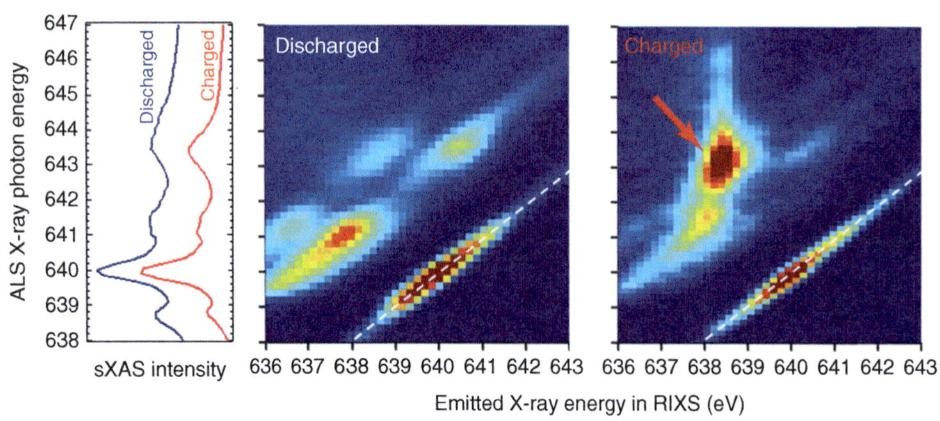


**Figure 11** The SXS contrast between a discharged and charged anode based on a Prussian Blue Analogue material, $Na_{1.24}Mn[Mn(CN)_6]$. A small contrast in sXAS could be seen around 643.5 eV photon energy (vertical axis), but mRIXS shows a significant intensity redistribution, indicating a novel $Mn^{1+/2+}$ redox in this anode material. [Source: Firouzi *et al.* 21, https://www.nature.com/articles/s41467-018-03257-1. Licensed under CC BY 4.0]

mRIXS reveals distinct features of the charged electrode that corresponds to the d–d excitations (excitation of an occupied TM 3d electron to an unoccupied state within the 3d orbitals) of a low-spin $Mn^{1+}$ system, thus providing the ultimate experimental verification of the $Mn^{1+/2+}$ redox in the anode.[21] This demonstration not only showcase the power of mRIXS for revealing novel chemistry, more importantly, it directly confirms that chemical states in battery environment could be well beyond what we would expect in conventional systems.

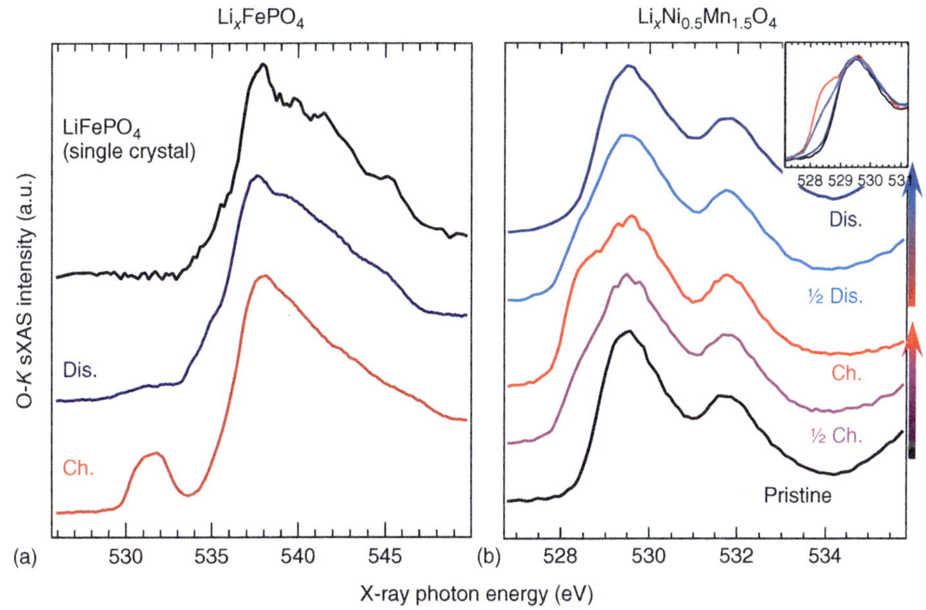


**Figure 12** O-K sXAS spectra collected from charged (Ch.) and discharged (Dis.) electrodes of (a) olivine $LiFePO_4$, and (b) spinel $LiNi_{0.5}Mn_{1.5}O_4$. Strong variation in the area intensity and lineshape could be observed for the pre-edge features below 535 eV without involving oxygen redox. [(a) Reproduced with permission from Ref. 10. © PCCP Owner Societies, 2015; (b) Reprinted with permission from R. Qiao, L. A. Wray, J.-H. Kim, N. P. W. Pieczonka, S. J. Harris and W. Yang, J. Phys. Chem. C, 2015, 119, 27228–27233. © 2015 American Chemical Society]

## 8 mRIXS FINGERPRINTS OXYGEN REDOX STATES

Anionic redox, that is, oxygen redox in oxide cathodes, holds a promise of delivering much higher capacity and power compared with the conventional TM redox-based batteries,[3] provided the oxygen redox could be maintained to be stable and reversible. In general, the evolution of oxygen has long been a critical topic in battery field ever since the first-generation Li-ion battery with $LiCoO_2$ cathode, however, has typically been considered detrimental to the stability of battery materials. Reversibility is the most critical parameter on whether the intensive effort on studying oxygen redox in batteries is meaningful at all. A crucial question associated with this concern is whether the oxygen redox mechanism naturally triggers the unfavorable oxygen evolution and gas release in batteries. Therefore, detecting and understanding the fundamental mechanism of the oxygen redox involved in battery electrodes have recently become one of the most pursued topics in the battery field.

### 8.1 Limitations of O-*K* sXAS and X-Ray Transmission Microscopy

Due to the inherent elemental sensitivity, O-*K* spectroscopy including XPS and sXAS have quickly become popular tools to tackle oxygen redox issues. Discussions on oxygen redox-based on O-*K* sXAS have focused on either the intensity or the lineshape of the so-called "pre-edge" features below 535 eV. Firstly, the seminal 1989 work by Frank de Groot *et al.* concluded that these features are "related to the metal 3d states". These features are from TM-3d states hybridized to O-2p states, which is then involved in the O-*K* sXAS. The intensity (area) of the pre-edge features is dominated by the strength of the hybridization, which is naturally enhanced in charged states because the high valence state of the TMs results in stronger hybridizations with O. This is most clearly shown by the contrast of charged and discharged $LiFePO_4$, where obvious pre-edge features including two peaks is shown in O-*K* sXAS of charged electrodes, in sharp contrast with the negligible pre-edge intensity of the discharged state (Figure 12a).[10] Such a pre-edge area enhancement due to strengthened hybridization during the charging process has been found almost universally for electrodes with or without oxygen redox, for example, olivine, spinel, and non-Li-rich layered compounds. The area analysis of these features shows that the intensity increases even before the oxygen redox voltage is reached, also indicating it is merely from the enhanced hybridization from the increasing TM valence, not a signature of oxygen redox.

Secondly, the lineshape change of the O-*K* pre-edge features upon cycling are due to two factors. One is the overall broadening of the spectral features due to the enhanced hybridization and covalency in charged states as

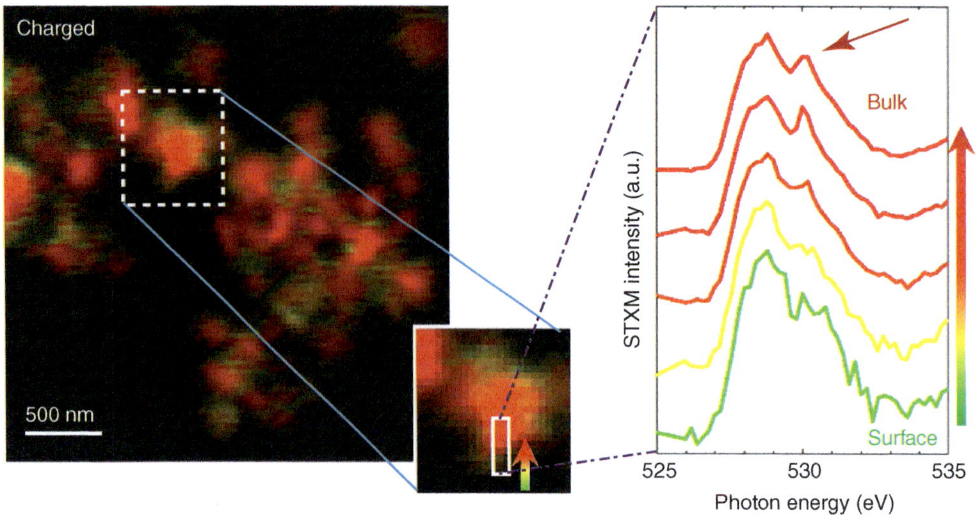


**Figure 13** Spatially resolved STXM spectra collected on a fully charged $Li_{1.17}Ni_{0.21}Co_{0.08}Mn_{0.54}O_2$ particle from the surface (bottom) to the bulk (top). The large image shows the overall STXM image. Spectra were collected by moving the probe along the white rectangular area indicated in the middle image. A 531 eV peak shows up when the probe is shifted to the bulk particle area (red arrow), but not at the surface regime. [Source: Gent *et al.* 23, https://www.nature.com/articles/s41467-017-02041-x. Licensed under CC BY 4.0]

described above. The other is that the changed TM valence upon cycling also changes the O-*K* pre-edge peak positions, leading to different lineshapes. A typical example of such lineshape changes could be seen in Figure 6(b) collected on spinel $LiNi_{0.5}Mn_{1.5}O_4$ electrodes. Besides the overall intensity enhancement, an extra feature emerges at the low energy leading edge in charged state due to the highly oxidized $Ni^{4+}$ (Figure 12b).[22] Therefore, the overall broadening and lineshape changes are also predominantly from TM changes, not signatures of oxygen redox.

It should be clarified that, although neither the intensity nor the lineshape of the O-*K* pre-edge features is a reliable probe of oxygen redox, the sXAS feature of oxidized oxygen in reference compounds, for example, $Li_2O_2$ and $O_2$, was indeed found to be around 530–532 eV within the same energy range as the pre-edge features. In fact, arguments based on possible $Li_2O_2$ feature contribution have been used for some meaningful discussions on oxygen redox. Unfortunately, the possible peroxide contribution cannot be conclusive due to the dominating effects on both the peak broadness and position changes, and it cannot differentiate the specific oxidized oxygen state either. In another word, the oxygen redox feature is indeed concealed in the O-*K* pre-edge regime, but unfortunately cannot be distinguished through sXAS due to the dominating effects from TM states in both the intensity and lineshape.

Another technical issue with conventional O-*K* sXAS experiments for studying oxygen redox reactions stems from the limitation on probe depth and lineshape distortion. As explained above, while sXAS TEY signals are non-distorted, they cannot probe the bulk oxygen states. sXAS TFY has a much deeper probe depth, however, suffers lineshape distortion, thus cannot be reliable on quantitative analysis. In order to directly address the surface to bulk contrast on the oxygen states in the oxygen-redox systems, in Figure 13, we present the spatially resolved STXM results with the probe area moving from the surface to the bulk part of charged $Li_{1.17}Ni_{0.21}Co_{0.08}Mn_{0.54}O_2$ electrode particles.[23] STXM shows that a 531 eV feature emerges when the probe switches from the surface to the bulk area of a charged particle (Figure 13). We note that such feature is completely buried in the conventional sXAS data in both TFY and TEY mode due to the technical issues discussed above. In addition, the energy of the feature falls in the wide range of oxidized oxygen, for example, peroxide and $O_2$. Therefore, for some electrode systems, STXM from the bulk particle through transmission mode measurements could reveal the signature of oxygen redox states. Unfortunately, due to the short penetration length, STXM samples are limited to nanoparticles. Also, the oxygen redox feature still sits on the strong background from the TM-O hybridization signals, leading to complication in many other systems. Therefore, the critical challenge remains for differentiating the intrinsic oxygen redox feature from the strong hybridization features in O-*K* spectra.

## 8.2 mRIXS Signature of the Oxygen Redox States

As elaborated by Figure 6, mRIXS is a technique that further deciphers the single data spot of a particle counting number in sXAS into a full spectrum upon a new dimension of the emission energy. This provides the superior chemical sensitivity to distinguish specific

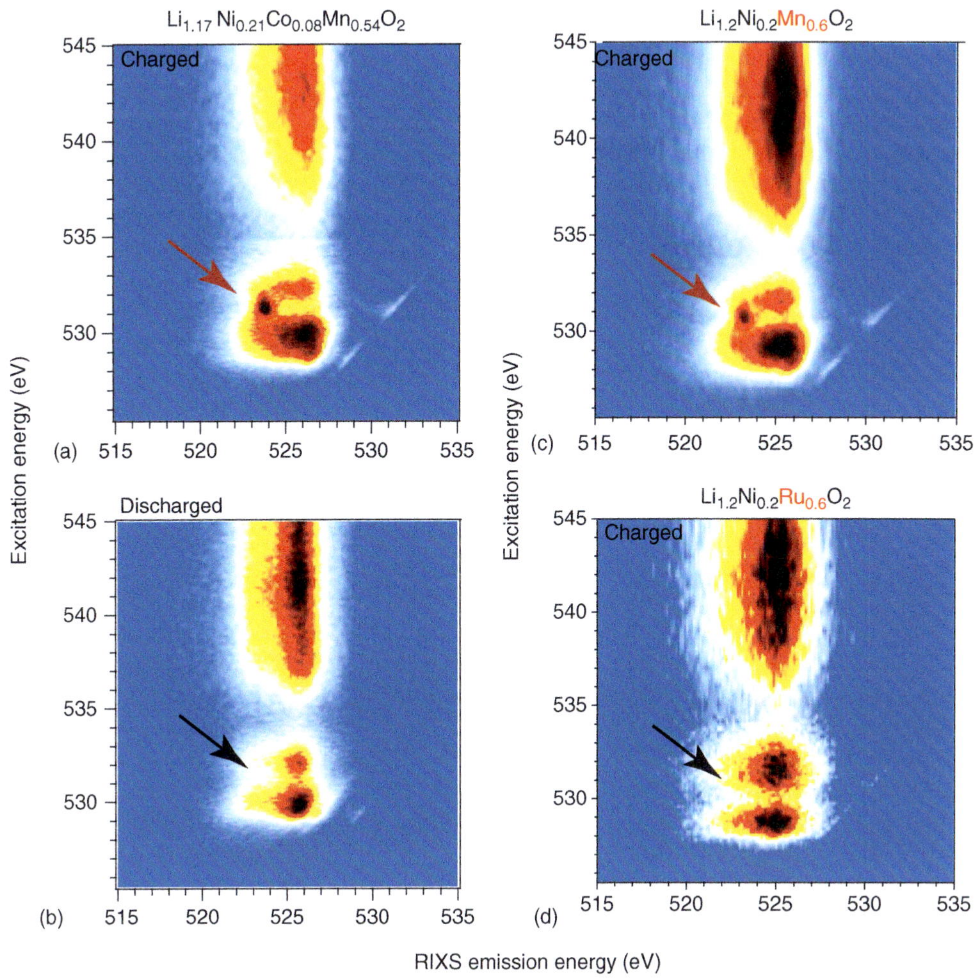


**Figure 14** (a) and (b) shows the contrast in mRIXS results of the charged and discharged $Li_{1.17}Ni_{0.21}Co_{0.08}Mn_{0.54}O_2$ electrodes, respectively. A sharp feature at 531 eV excitation and 523.8 eV emission energy appears in the charged electrode, corresponding to the bulk oxygen redox state, that is, oxidized non-divalent oxygen. (c) and (d) shows the mRIXS of charged $Li_{1.2}Ni_{0.2}Mn_{0.6}O_2$ and $Li_{1.2}Ni_{0.2}Ru_{0.6}O_2$ electrodes, one with and one without the specific oxygen redox feature, respectively. Arrows indicate the disappearance (black) and appearance (red) of the oxygen redox mRIXS feature. [Source: Wang 24, https://www.nature.com/articles/s41467-018-03403-9 Licensed under CC BY 4.0]

chemical states that cannot be clearly defined in sXAS. Such an improvement is exactly what we need to tackle the entangled oxygen redox and TM-O hybridization features in sXAS pre-edge regime. In addition, mRIXS is a PIPO technique with probe depth of about 100 nm in the O-*K* range. Compared with PIPO sXAS, the bulk sensitivity of mRIXS could be further enhanced if the bulk signal is differentiated from surface signal on the emission energy, which is, fortunately, the case for oxygen redox features, as discussed below. Therefore, mRIXS is potentially an ideal tool to reveal the intrinsic bulk (lattice) oxygen redox states in battery electrodes.

Figure 14 displays the mRIXS results of several electrode systems at different electrochemical states. Figure 14(a) and (b) shows the mRIXS of charged and discharged $Li_{1.17}Ni_{0.21}Co_{0.08}Mn_{0.54}O_2$ electrodes, respectively. It could be clearly seen that a sharp feature at 531 eV excitation energy and 523.8 eV emission energy emerges when the electrode is charged (red arrow in Figure 14a). As a matter of fact, this feature appears only if the charge potential reaches the high-voltage plateau of this Li-rich electrode, a voltage range that oxygen redox reactions take place.[23] The comparison between this oxygen-redox mRIXS feature with the STXM results (Figure 13) suggests that, although RIXS has nominally the same probe depth as TFY sXAS, the specific oxygen redox feature observed in mRIXS corresponds to only the bulk oxygen redox signal. The new dimension of information on the emission in mRIXS successfully distinguishes the lattice oxygen redox feature (523.8 eV emission energy) from the dominating hybridization features (broad features around 525 eV emission energy) of the standard $O^{2-}$ species.

Note that, although the mRIXS observations are consistent with the single-energy cut of RIXS at 531 eV excitation energy, mRIXS provides the full profile of the oxygen redox feature along both excitation and emission energy, which is critical for its theoretical interpretation to reveal the fundamental mechanism of the oxygen redox. Compared with STXM, RIXS experiments could be performed directly on any battery electrodes at different electrochemical stages, instead of the TEM-style samples for STXM.

Striking results have been reported based on the signatures of oxygen redox through mRIXS experiments. It has been found that, regardless of the types of TMs (3d or 4d) that are involved in the oxide electrodes, for example, $Li_{1.2}Ni_{0.2}Mn_{0.6}O_2$ and $Li_2RuO_3$, once the oxygen redox is activated in the electrode, the different systems display the same mRIXS feature. This strongly indicates a universal mechanism that fundamentally drives the oxygen redox in all the electrode systems. On the other hand, the activation of oxygen redox strongly depends on the TM compositions. However, this TM dependence does not simply reflect a TM-O hybridization model. For example, changing Mn to Ru in $Li_{1.2}Ni_{0.2}M_{0.6}O_2$ (M=Mn, Ru) completely suppresses the oxygen redox feature as revealed by mRIXS (Figure 14c, d). mRIXS confirms that $Li_{1.2}Ni_{0.2}Mn_{0.6}O_2$ is oxygen redox active; however, Ru is the active cationic redox in the $Li_{1.2}Ni_{0.2}Ru_{0.6}O_2$ without oxygen redox.[24] The independence of TMs on the oxygen-redox feature and the dependence of TMs on the activation of oxygen redox imply that the mechanism is more complex than what has been proposed so far. A fundamental foundation of a strategically architectural mechanism is the key to take the mystery out of the oxygen redox reactions, and mRIXS provides clear experimental evidences to guide the theoretical efforts. A direct simulation of the mRIXS oxygen-redox feature holds the promise to reveal the true fundamental nature of oxygen redox, which will lead to the ultimate understanding of oxygen redox by further incorporating other factors like structure and specific chemical configurations.

## 9 PERSPECTIVES ON IN SITU SXS OF BATTERIES

SXS techniques introduced in this chapter are based on the principle that dipole-selected soft X-ray excitations of the relevant elements in batteries directly reach the critical valence states in the materials, which fundamentally controls the chemical reactions. However, because of the inherent requirement of ultra-high vacuum for most soft X-ray detection channels, SXS used to be limited to solid-state samples that are exposed directly in vacuum. Today's SXS techniques have greatly advanced, and experiments of liquid and gaseous phases could be measured with sophisticated in situ cells, as extensively reported and reviewed previously. [19,25] Most of these in situ/operando setups are based on very thin (typically 100 nm thickness) membrane windows that separate the liquid/gas samples from vacuum chambers. Different types of in situ cells for SXS PIPO techniques have been designed and fabricated at synchrotron facilities. Figure 15(a) shows a static cell design that Guo successfully utilized for studying the hydrogen bonding in liquid water and organic solvent. In addition to the static cells for SXS studies, flow-through gas and liquid cells have also been designed and manufactured with a continuous flow of gas/liquid samples during the soft X-ray spectroscopic experiments. Figure 15(b) displays the images of a liquid flow cell that were reported by Blum *et al.*[26] Recent developments of flow cells have realized further controls of the sample temperature ranging from liquid-nitrogen temperature to several hundred degrees, and the pressure limit up to 1 atm. These soft X-ray in situ cells and their many derivatives have utilized more and more for studying battery materials.

In the meantime, solid-state electrolyte has recently been paid much attention in the battery field, due to its promise of safe operations with high energy density. With solid-state electrolyte, complications involving soft X-ray membrane could be avoided, and SXS experiments could be performed directly on a real-world battery by designing the top layer as in the schematic of Figure 15(c). For example, controlled micrometer-sized detection windows could be fabricated on a typical metal foil current collector for measuring the electrode materials.[27] A thin film top-layer of metal, for example, Al, or Au, or a metallic mesh, could also serve as the current collector while allowing most soft X-ray photons passing through. In particular, if the electrode deposition is also controlled to achieve an electrode thickness of less than 100 nm, in situ/operando PIPO SXS could reach the electrode–electrolyte interface directly. With the superior chemical sensitivity of mRIXS, we expect unprecedented information will be uncovered on both the bulk electrodes and interfaces in solid-state battery systems.

## 10 SUMMARY AND CONCLUSIONS

This chapter provides a comprehensive summary in both the techniques and analysis of the state-of-art SXS for revealing the critical chemistry in batteries. We provide an atomic overview on all the relevant excitations involved in SXS process. Through a direct comparison with sXAS, mRIXS is introduced for battery researches and is explicated as further resolved sXAS signals along a new energy dimension, the emission energy. In addition to the

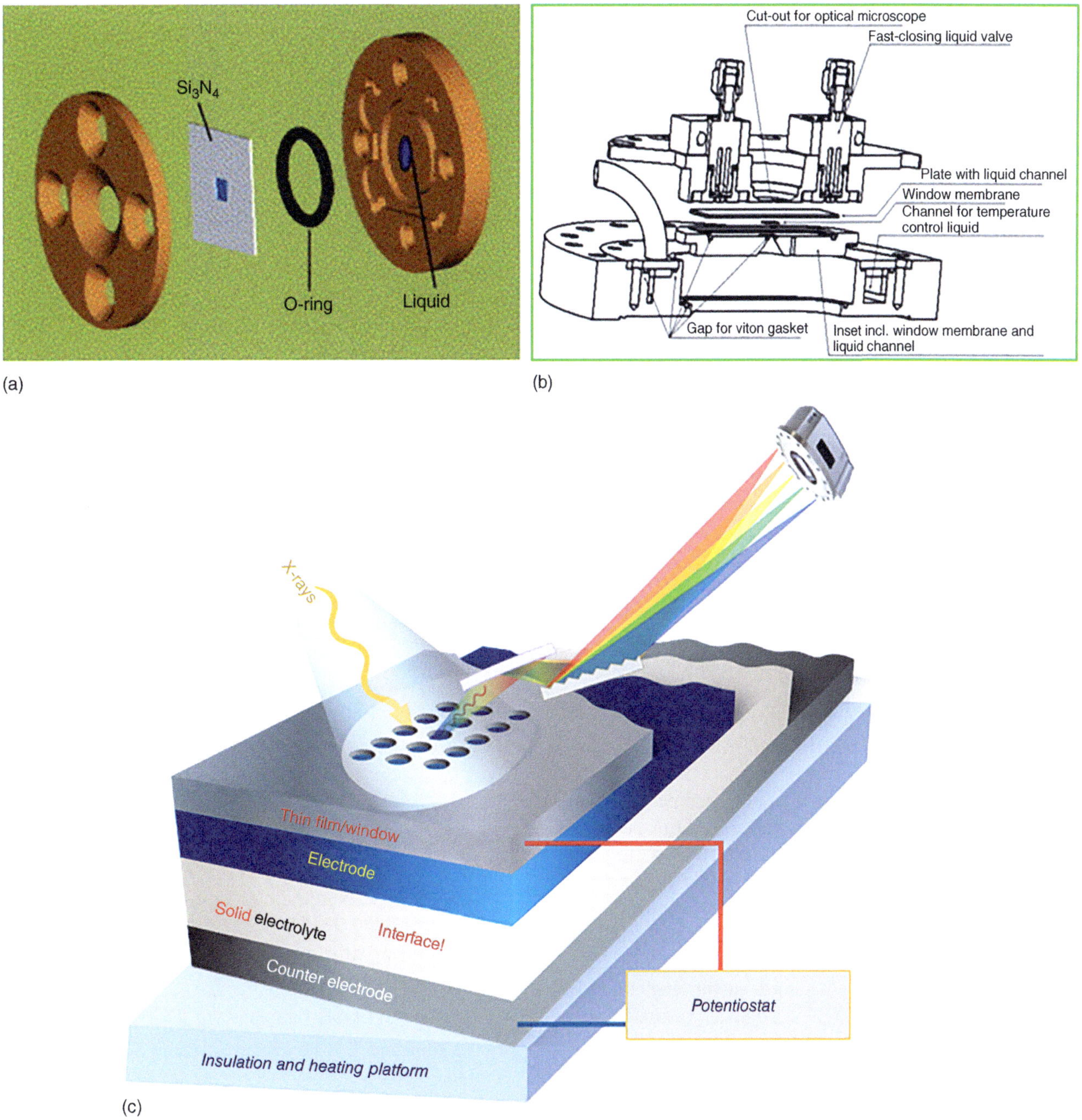


**Figure 15** Examples of static (a) and flow-through (b) cells designed for SXS experiments of gaseous and liquid samples.[25,26] [(a) Reproduced with permission from J. Guo, J. Electron. Spectrosc. Relat. Phenom., 2013, 188, 71–78. © Elsevier 2013; (b) Reproduced with permission from M. Blum, L. Weinhardt, O. Fuchs, M. Bar, Y. Zhang, M. Weigand, S. Krause, S. Pookpanratana, T. Hofmann, W. Yang, J. D. Denlinger, E. Umbach and C. Heske, Rev. Sci. Instrum., 2009, 80, 123102–123106. © AIP Publishing LLC 2009] Soft X-ray membranes are employed to separate the real-world sample environment from the soft X-ray vacuum chamber in these designs. (c) Schematic of a typical in situ/operando SXS experiment based on solid-state electrolyte. Top layer could be conventional current collector with detection windows, mesh current collector, or a thin film of Al or Au. Controlled thickness of the electrode will allow the probe depth to reach the electrode–electrolyte interface.[5] [Reproduced with permission from W. Yang and T. P. Devereaux, J. Power Sources, 2018, 389, 188–197. © Elsevier 2018]

conventional TEY and TFY channels of sXAS, mRIXS simultaneously contains four more types of PIPO signals, that is, PFY, iPFY, sPFY, and ccPFY. The combination of these signal channels provides complete and powerful characterizations of both the surface/interface and the bulk properties of battery materials.

We then discuss SXS studies of battery materials in different aspects. First, we introduced the combination of sXAS and sXES for probing the critical unoccupied and occupied states of battery electrodes. These states in the vicinity of the Fermi level defines the states directly involved in the redox reactions during electrochemical operations. Second, we show *K*-edge sXAS of low-Z elements is a chemical sensitive probe to detect the surface and interfaces in batteries. Studies on SEI and binder coating are presented. Third, on the cationic TM redox reactions, we show that, due to the sharp multiplet features in TM *L*-edge sXAS, lineshapes of TM-*L* spectra could quantitatively analyzed to define the TM states through a straightforward linear combination fittings. Fourth, RIXS is first introduced as a powerful tool to distinguish even the subtle chemical contrast between the ion-solvation shell and bulk electrolyte. Fifth, we further showcase the power of mRIXS for revealing a long-standing speculation of novel $Mn^{1+}$ state found in a NIB electrode. Sixth, on the anionic oxygen redox reactions, we first clarify explicitly that neither the intensity nor the lineshape of the O-*K* sXAS pre-edge features should be used as evidence of oxygen redox. A reliable spectroscopic probe of oxygen redox reactions is demonstrated through a specific feature in mRIXS. We discuss some important observations through mRIXS on understanding the oxygen redox.

At the end, we provide our perspectives on SXS studies of battery materials under in situ/operando conditions. We note that, while this chapter focuses on core-level SXS techniques with mostly chemical information, the family of SXS has kept growing and evolving. In addition to the recently developed mRIXS that is emphasized in this chapter, other modern SXS techniques are also emerging and advancing quickly into energy materials researches, such as resonant soft X-ray scattering with the combined structural and chemical information. We foresee that applications of advanced SXS techniques will lead to more and more significant contributions to chemical characterizations, fundamental understandings and material explorations for a broad range of topics in advanced battery technologies.

## 11 ACKNOWLEDGMENT

This research used resources of the Advanced Light Source, which is a DOE Office of Science User Facility under Contract No. DE-AC02-05CH11231.

## 12 RELATED ARTICLES

Li–$CO_2$ Batteries; Solid State and Materials Chemistry for Sodium-Ion Batteries

## 13 ABBREVIATIONS AND ACRONYMS

AEY = Auger electron yield; ccPFY = core–core partial fluorescence yield; DFT = density functional theory; eV = electron volt; EV = electric vehicles; iPFY = inverse partial fluorescence yield; LIB = Li-ion batteries; LLZO = $Li_7La_3Zr_2O_{12}$; LUMO = lowest unoccupied molecular orbital; mRIXS = mapping of resonant inelastic X-ray scattering; NIB = Na-ion battery; PF = polyfluorene; PFY = partial fluorescence yield; PIPO = photon-in-phonon-out; SEI = solid-electrolyte-interphase; sPFY = super-partial fluorescence yield; SSBs = solid-state batteries; STXM = scanning transmission X-ray microscopy; sXAS = soft X-ray absorption spectroscopy; sXES = soft X-ray emission spectroscopy; SXS = soft X-ray spectroscopic; TEY = total electron yield; TFY = total fluorescence yield; TFY = total yield of fluorescence photons; TM = transition metal; XES = X-ray emission spectroscopy; XPS = X-ray photoelectron spectroscopy

## 14 REFERENCES

1. J. B. Goodenough and Y. Kim, *Chem. Mater.*, 2010, **22**, 587.
2. F. Lin, Y. Liu, X. Yu, L. Cheng, A. Singer, O. G. Shpyrko, H. L. Xin, N. Tamura, C. Tian, T. C. Weng, X. Q. Yang, Y. S. Meng, D. Nordlund, W. Yang and M. M. Doeff, *Chem. Rev.*, 2017, **117**, 13123.
3. A. Grimaud, W. T. Hong, Y. Shao-Horn and J. M. Tarascon, *Nat. Mater.*, 2016, **15**, 121.
4. G. Assat and J.-M. Tarascon, *Nat. Energy*, 2018, **3**, 373.
5. W. Yang and T. P. Devereaux, *J. Power Sources*, 2018, **389**, 188.
6. D. Attwood, 'Soft X-rays and Extreme Ultraviolet Radiation', Cambridge University Press, Cambridge, UK, 2000.
7. Q. Li, R. M. Qiao, L. A. Wray, J. Chen, Z. Zhuo, Y. Chen, S. Yan, F. Pan, Z. Hussain and W. Yang, *J. Phys. D. Appl. Phys.*, 2016, **49**, 413003.
8. W. Yang, X. Liu, R. Qiao, P. Olalde-Velasco, J. D. Spear, L. Roseguo, J. X. Pepper, Y.-d. Chuang, J. D. Denlinger and Z. Hussain, *J. Electron Spectrosc. Relat. Phenom.*, 2013, **190**, 64.
9. A. J. Achkar, T. Z. Regier, E. J. Monkman, K. M. Shen and D. G. Hawthorn, *Sci. Rep.*, 2011, **1**, 182.

10. X. Liu, Y. J. Wang, B. Barbiellini, H. Hafiz, S. Basak, J. Liu, T. Richardson, G. Shu, F. Chou, T. C. Weng, D. Nordlund, D. Sokaras, B. Moritz, T. P. Devereaux, R. Qiao, Y. D. Chuang, A. Bansil, Z. Hussain and W. Yang, *Phys. Chem. Chem. Phys.*, 2015, **17**, 26369.

11. R. Qiao, Q. Li, Z. Zhuo, S. Sallis, O. Fuchs, M. Blum, L. Weinhardt, C. Heske, J. Pepper, M. Jones, A. Brown, A. Spucces, K. Chow, B. Smith, P.-A. Glans, Y. Chen, S. Yan, F. Pan, L. F. J. Piper, J. Denlinger, J. Guo, Z. Hussain, Y.-D. Chuang and W. Yang, *Rev. Sci. Instrum.*, 2017, **88**, 033106.

12. A. J. Achkar, T. Z. Regier, H. Wadati, Y. J. Kim, H. Zhang and D. G. Hawthorn, *Phys. Rev. B*, 2011, **83**, 081106.

13. K. Xu, *Chem. Rev.*, 2004, **104**, 4303.

14. R. Qiao, I. T. Lucas, A. Karim, J. Syzdek, X. Liu, W. Chen, K. Persson, R. Kostecki and W. Yang, *Adv. Mater. Interfaces*, 2014, **1**, 1300115.

15. Z. Zhuo, P. Lu, C. Delacourt, R. Qiao, K. Xu, F. Pan, S. J. Harris and W. Yang, *Chem. Commun. (Camb.)*, 2018, **54**, 814.

16. L. Cheng, E. J. Crumlin, W. Chen, R. Qiao, H. Hou, S. Franz Lux, V. Zorba, R. Russo, R. Kostecki, Z. Liu, K. Persson, W. Yang, J. Cabana, T. Richardson, G. Chen and M. Doeff, *Phys. Chem. Chem. Phys.*, 2014, **16**, 18294.

17. G. Liu, S. Xun, N. Vukmirovic, X. Song, P. Olalde-Velasco, H. Zheng, V. S. Battaglia, L. Wang and W. Yang, *Adv. Mater.*, 2011, **23**, 4679.

18. F. M. Frank de Groot, *Coord. Chem. Rev.*, 2005, **249**, 31.

19. X. Liu, W. Yang and Z. Liu, *Adv. Mater.*, 2014, **26**, 7710.

20. Y. L. Jeyachandran, F. Meyer, S. Nagarajan, A. Benkert, M. Bär, M. Blum, W. Yang, F. Reinert, C. Heske, L. Weinhardt and M. Zharnikov, *J. Phys. Chem. Lett.*, 2014, **5**, 4143.

21. A. Firouzi, R. Qiao, S. Motallebi, C. W. Valencia, H. S. Israel, M. Fujimoto, L. A. Wray, Y. D. Chuang, W. Yang and C. D. Wessells, *Nat. Commun.*, 2018, **9**, 861.

22. R. Qiao, L. A. Wray, J.-H. Kim, N. P. W. Pieczonka, S. J. Harris and W. Yang, *J. Phys. Chem. C*, 2015, **119**, 27228.

23. W. E. Gent, K. Lim, Y. Liang, Q. Li, T. Barnes, S. J. Ahn, K. H. Stone, M. McIntire, J. Hong, J. H. Song, Y. Li, A. Mehta, S. Ermon, T. Tyliszczak, D. Kilcoyne, D. Vine, J. H. Park, S. K. Doo, M. F. Toney, W. Yang, D. Prendergast and W. C. Chueh, *Nat. Commun.*, 2017, **8**, 2091.

24. J. Xu, M. Sun, R. Qiao, S. E. Renfrew, L. Ma, T. Wu, S. Hwang, D. Nordlund, D. Su, K. Amine, J. Lu, B. D. McCloskey, W. Yang and W. Tong, *Nat. Commun.*, 2018, **9**, 947.

25. J. Guo, *J. Electron Spectrosc. Relat. Phenom.*, 2013, **188**, 71.

26. M. Blum, L. Weinhardt, O. Fuchs, M. Bar, Y. Zhang, M. Weigand, S. Krause, S. Pookpanratana, T. Hofmann, W. Yang, J. D. Denlinger, E. Umbach and C. Heske, *Rev. Sci. Instrum.*, 2009, **80**, 123102.

27. X. Liu, D. Wang, G. Liu, V. Srinivasan, Z. Liu, Z. Hussain and W. Yang, *Nat. Commun.*, 2013, **4**, 2568.

# Solid Electrolyte Interphase in Lithium-Based Batteries

**Feifei Shi**

*Stanford University, Stanford, CA, USA*

and

**Philip N. Ross**

*Lawrence Berkeley National Laboratory, Berkeley, CA, USA*



## 1 INTRODUCTION

Energy storage is increasingly important in modern times, with intense and widespread demand among most industries, spanning from large-scale electrical grid storage to electric vehicles and consumer electronics. The high energy density of lithium-based batteries has revolutionized "wireless" electronic devices and become the dominant technology on the market. After decades of intensive research on each component of Li-ion batteries, researchers have achieved the theoretical capacity limit and excellent initial performance. However, lithium batteries usually experience degradation over prolonged cycling, which typically leads to undesired capacity loss. In a battery system, the electrode/electrolyte interface is where electronic charge transfer, ion migration, and solvent molecular adsorption/desorption occur. Therefore, it is reasonable to believe that the poor battery cyclability is related to the reversibility of the electrode/electrolyte interface.[1]

Due to the large output voltage of lithium batteries, the electrolyte is not thermodynamically stable, that is, the electrolyte is reduced and oxidized on the anode and cathode surfaces, respectively.[1] The decomposed electrolyte forms a film on the electrode surfaces, which passivates the electrodes and inhibits further parasitic reactions, while allowing the conduction of the $Li^+$ ions. Because it acts as an interphase with the properties of a solid electrolyte, this film is termed solid electrolyte interphase (SEI). This passivation-layer model was first proposed by Peled *et al.* in 1979[2] and is applicable to alkali metals (like lithium metals) in nonaqueous battery systems (Figure 1a). Since then, the "Mosaic" model has been generally accepted and simply applied to the SEI generated on the other anode materials. However, the structure and components of the SEI are largely influenced by the anode materials and their surface conditions.[3] For example, graphite anodes are highly anisotropic and contain edge and basal plane surfaces.[4,5] The basal plane surface contains hardly any surface functional groups, whereas the edge surface is covered by oxygen-containing species and is the main location of ion transport and chemical reactions. As shown in Figure 1(b), the anisotropy and oxygen-containing functional groups

*Inorganic Battery Materials*. Edited by Hailiang Wang and Boniface P.T. Fokwa
 ISBN 978-1-119-43199-2

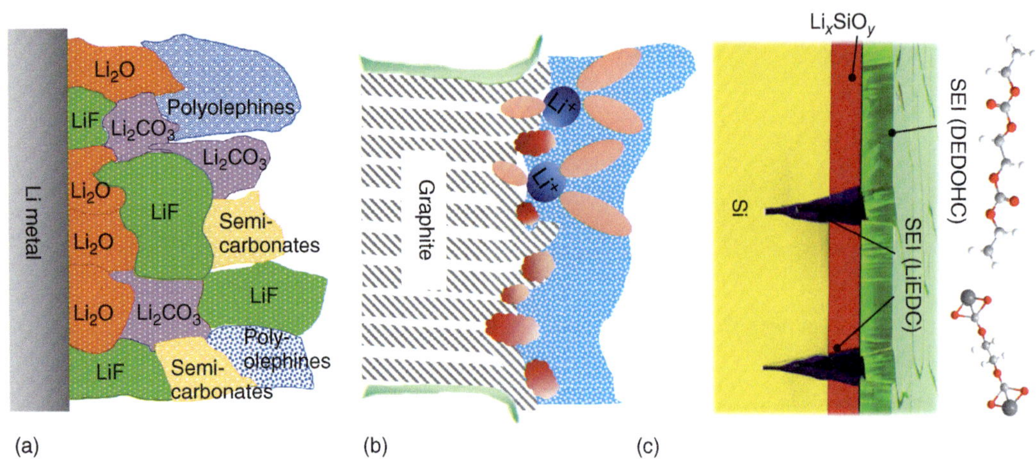


**Figure 1** Solid electrolyte interphase (SEI) schematic on (a) lithium metal, (b) graphite, (c) silicon electrode

of the anode material induce the nonuniformity of the SEI. The thickness of the SEI on the edge is greater than that on the basal plane,[4] and the chemical components of the SEI also vary according to the oxygen-containing functional groups.[6] For alloying anode materials such as Si and Sn, their SEI structures are largely influenced by the large volume expansion/shrink and accompanied pulverization. The freshly exposed lithiated surface causes a different reduction path compared to that on the native-oxide-covered surface (Figure 1c).[7] Hence, a case-by-case model should be proposed for different interface systems.

An overview of the fundamental concepts and principles of the SEI has been published in books such as "Nonaqueous Electrochemistry",[8] "Lithium Ion Batteries: Solid-Electrolyte Interphase",[9] and several review chapters.[3,10,11] The scope of this chapter is to review the studies of interfacial chemistry in lithium-based batteries, with emphasis on utilizing a surface science approach to study the passive layer on lithium. The catalytic effect, mechanical effect, and properties of the SEI, which have received little attention, will be discussed. Furthermore, potential causes of controversy and discrepancy in surface characterization results are revealed. At last, we make several recommendations on future research of the SEI. Our main objective is to illustrate the revolution that is taking place in this field and thus provide inspiration for future developments in battery chemistry.

## 2 THE PASSIVE LAYER ON LITHIUM

### 2.1 Reaction with Air

Lithium has been recognized as a promising anode material for decades in industry. Meanwhile, lithium metal batteries are widely criticized because of their poor cycle life and safety concerns. One of the fundamental reasons for these critiques is the uncontrollable electrodeposition of lithium, which leads to needle-like lithium crystals. Upon extended cycles, these large surface area particles consume more and more electrolyte to form the SEI and dry out the cell. Pulverization of lithium electrodes leads to severe safety concerns, due to their high reactivity. After the 1990s, the commercialization of lithium-ion batteries shifted the market's attention from lithium metal to intercalation materials. Since then, lithium-ion batteries are the leading technology in the battery industry and battery research community. However, Li-ion batteries cannot keep up with the constantly increasing demands from products such as electric vehicles. This has revived the field of high energy density batteries with a lithium metal anode.[12] Due to its naturally low chemical potential, the SEI growth on lithium shares a lot of similarity as formed on the other lithiated anode materials. Hence, we first discuss the formation of a passivation layer on lithium.

Active materials such as lithium are inevitably covered with surface films, which are formed by the spontaneous reaction with atmospheric species. The reactivity of lithium with air ($O_2$, $N_2$, $H_2O$ and $CO_2$) has been thoroughly investigated by Auger electron spectroscopy (AES), in situ ellipsometry, and X-ray photoelectrons spectroscopy (XPS) in ultra-high vacuum (UHV) systems (e.g., $<10^{-11}$ torr).[13–15] The reaction with $O_2$ in dry air is extremely fast and continuous. The porous structure of the $Li_2O$ film makes the delivery of $O_2$ to fresh lithium easier, which is the main reason why the passivation reaction cannot be kinetically stopped. The reaction rate of lithium with $N_2$ is orders of magnitude lower than that of $O_2$.[16] Therefore, without the freshly deposited lithium in the UHV set up, the normal operated Li foil in glovebox

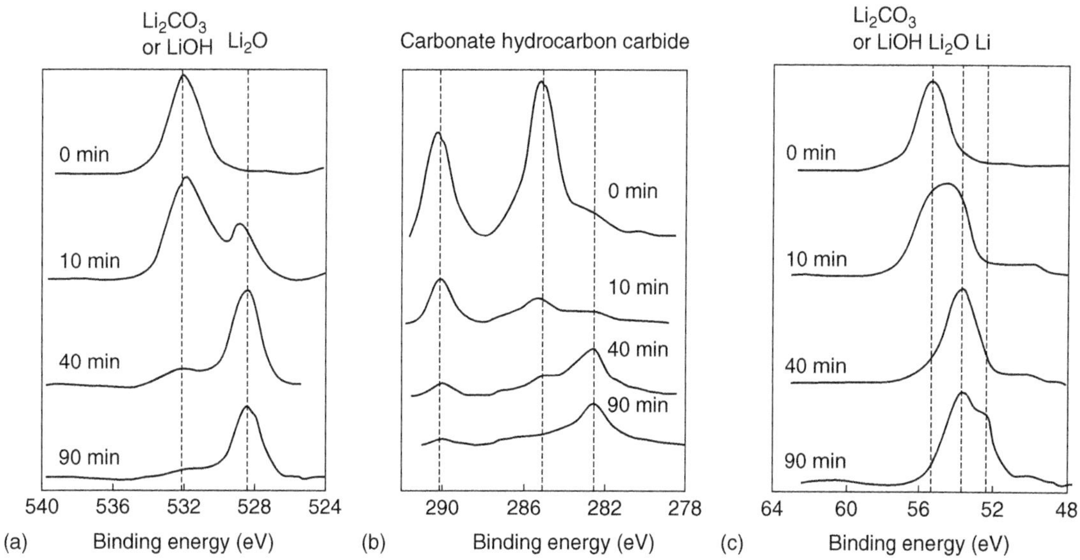


**Figure 2** XPS spectra (a) O 1s, (b) C 1s, and (c) Li 1s of the native film on lithium.[17] [Reprinted with permission from K. Kanamura, H. Tamura, S. Shiraishi, Z. i. Takehara, J. Electrochem. Soc. 1995, 142, 340–347. © The Electrochemical Society 1995]

always has $Li_2O$, LiOH, and $Li_2CO_3$ covering the surface. Figure 2 shows the XPS spectra of lithium with these native oxide species.[17]

### 2.2 Reaction with Electrolytes

Upon contact with electrolyte, fresh lithium immediately reacts with the organic solvent molecules due to their low chemical potential (−3.04 V vs NHE). The UHV system provides the conditions for ideal experiments to evaluate the reactivity of lithium. The Ross group applied UHV surface preparation/characterization techniques, which let liquid solvents condense onto the freshly evaporated lithium at liquid nitrogen temperatures and react upon slow warming.[18–20] Many of the common lithium battery solvents were studied, including tetrahydrofuran (THF), dioxolane (DOL), propylene carbonate (PC), and dimethyl carbonate (DMC). Figure 3 shows the XPS spectra of fresh lithium exposed to gaseous DMC molecules at temperatures of 130–620 K. The reaction between DMC and lithium is initiated at 180 K, a temperature much lower than the bulk melting temperature of DMC (277 K). The low temperature (180–270 K) product is lithium methyl carbonate and methyllithium; while at elevated temperatures, the product starts to decompose to $Li_2O$, elemental carbon, and alkyl lithium. The surface oxides and hydroxides are highly nucleophilic, while many polar aprotic solvents are highly electrophilic. DMC can be easily cleaved to methoxy radicals, with the assistance of oxide species on lithium.[21]

The electrolyte immersion experiments produce more of the intermediate state of lithium radicals than the UHV experiments. The standard reduction potential of lithium is lower than that of the solvated-electron system (like highly purified ammonia, amines, and ethers). It resulted in the formation of "blue solutions" of solvated electrons. The solvated electron may take part in the early stages of SEI formation and in the break-repair processes during lithium plating and stripping. The solvent molecules can be reduced through both inner-sphere electron transfer (via electron transfer to adsorption species) and outer-sphere electron transfer (via solvated electrons transfer to species in the bulk electrolyte), which partially explains why, as an insulator layer, the thickness of SEI layer can exceed the electron tunneling distance (<2 nm).[22] The solute molecule may react with metallic lithium as well as the solvent molecule, so the final composition of the SEI is a balance of several competing reduction processes of solvent and salt anions.[8] For salts such as $LiClO_4$ and $LiBF_4$, whose reaction rate constants with solvated electron are smaller than $10^6\ M^{-1}\ s^{-1}$, the reduction products (LiCl and boron ($B^0$)) are rarely found in the SEI formed in highly reactive solvents (e.g., carbonate and ester). On the contrary, when electrolytes containing salts with reactive $SO_2CF_3$ groups ($LiSO_3CF_3$, $LiN(SO_2CF_3)_2$ and $LiC(SO_2CF_3)_3$) are used, the SEI is mainly composed of anion reduction products. This is particularly obvious when the solvent is kinetically stable ether (like THF), whose reaction rate constant is lower than $10^7\ M^{-1}\ s^{-1}$.[23,24] The surface formation process is very fast, which leads to rapid changes of the film properties. Hence, it is almost impossible to study the surface film formation by standard electrochemical techniques.[25] The use of microelectrodes or ultramicroelectrodes (<25 μm diameter) makes it

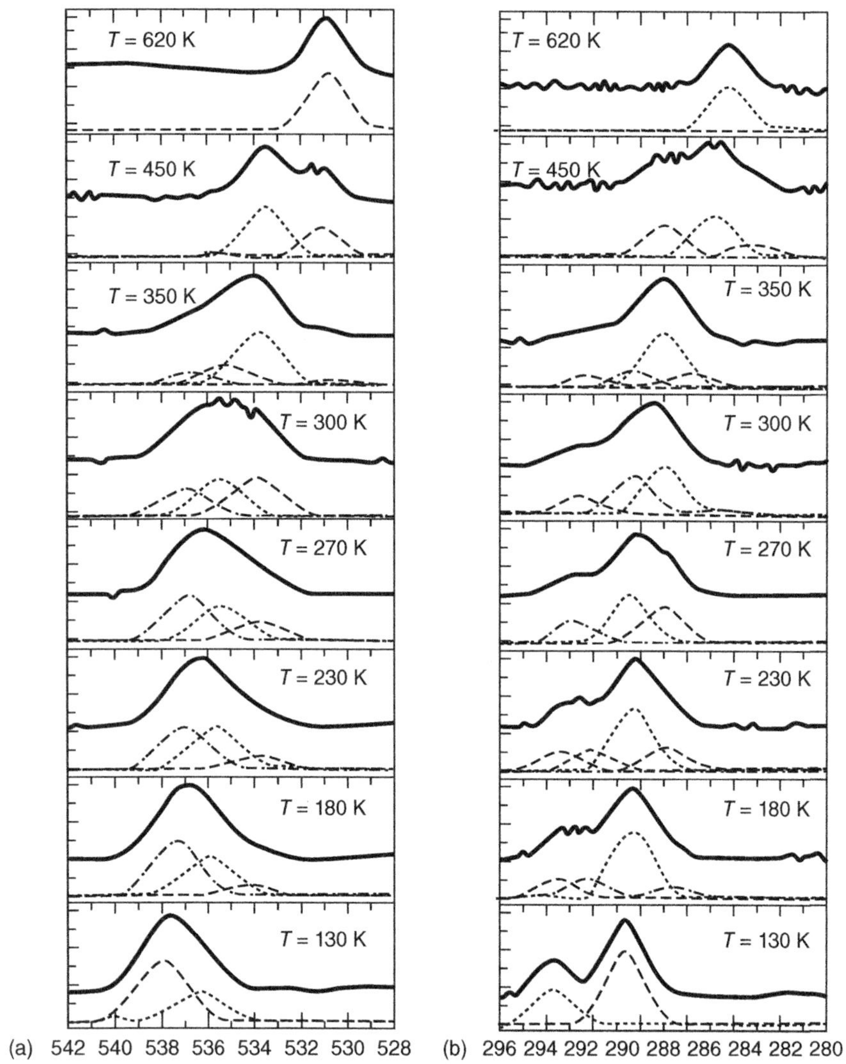


**Figure 3** XPS spectra of (a) O 1s and (b) C 1s acquired as a function of temperature for an evaporated Li surface dosed with DMC at 130–620 K.[18] [Reprinted with permission from G. Zhuang, Y. Chen, P. N. Ross, Langmuir 1999, 15, 1470–1479. © American Chemical Society 1999]

possible to study the fast Li deposition dissolution processes at high scan rates (100–1000 $mV\ s^{-1}$) without polarization of counter electrode. Fast potentiodynamics could then be measured on Li electrodes before the corrosion of the Li deposits.[26–32]

## 3 CATALYSIS EFFECT

### 3.1 Surface Dependence of Electrolyte Reduction Potential

Previous studies mainly focused on the effect of electrolyte and additive on SEI formation mechanism and composition.[10,33] The solvation shell mechanism is one of the widely accepted theories. It suggests that high dielectric constant solvents (e.g., cyclic carbonate) dominate the primary Li ion solvation sheath. As a result, the components of the SEI are dominated by the reduction of cyclic carbonate. However, recent studies have shown that the reduction of electrolyte and composition of the SEI is surface dependent. As shown in Figure 4, the reduction peak in 1.2 M $LiPF_6$ ethylene carbonate (EC)/ethyl methyl carbonate (EMC) (3:7 w/w) electrolyte is 1.5–1.8 V vs $Li/Li^+$ for Ni electrodes and 0.7–1 V vs $Li/Li^+$ for graphite electrodes;[34,35] the reduction peak in 1 M $LiPF_6$ EC/diethyl carbonate (DEC) (1:2 v/v) electrolyte is 0.5–0.6 V vs $Li/Li^+$ for Au electrodes and 1.2–1.5 V vs $Li/Li^+$ for Sn electrodes.[36] The potential (on the thermodynamic scale) for adiabatic electron transfer, that is, an outer-sphere noncatalytic electron transfer,

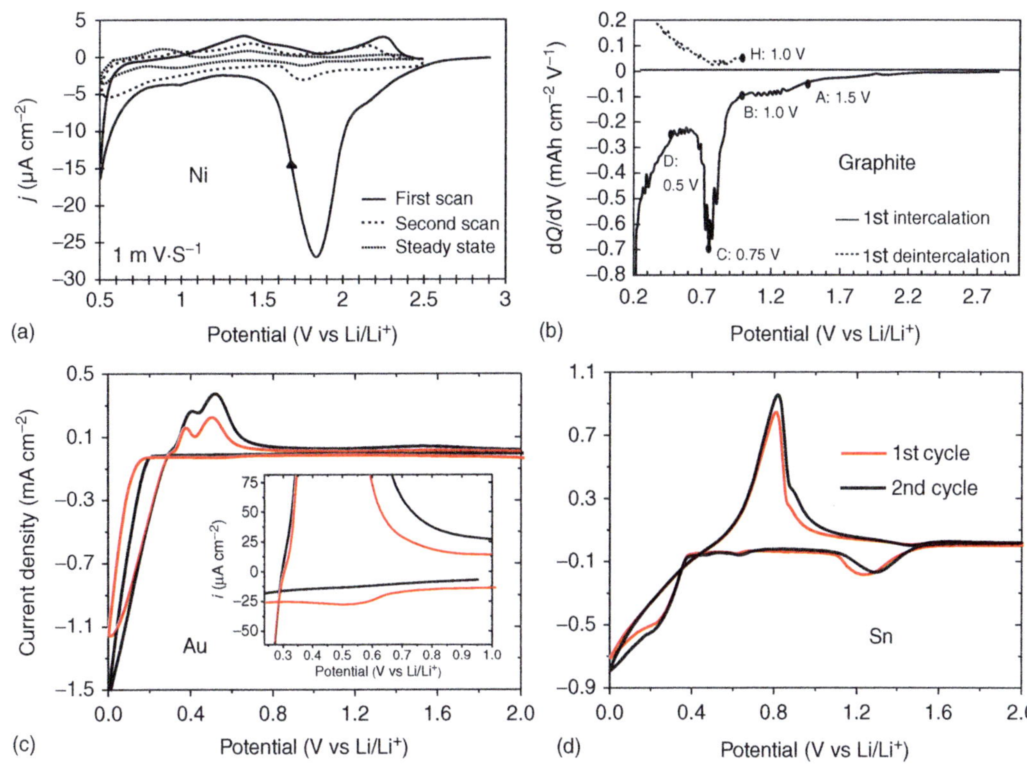


**Figure 4** Surface dependence of electrolyte reduction. (a) Cyclic voltammograms of Ni electrode in 1.2 M $LiPF_6$/EC:EMC (3:7 w/w) electrolyte. The sweep rate was 1 mV $s^{-1}$.[34] [Reproduced with permission from G. V. Zhuang, K. Xu, H. Yang, T. R. Jow, P. N. Ross, J. Phys. Chem. B 2005, 109, 17567–17573. © 2005 American Chemical Society]; (b) d*Q*/d*V* vs potential of the graphite anode in the first complete charge/discharge cycle at current density of 50 μA $cm^{-2}$.[35] [Reproduced with permission from Ref. 35. © The Electrochemical Society, 2005]; (c) Au and (d) Sn electrodes in 1 M $LiPF_6$ electrolyte (EC:DEC = 1:2 v/v) for 1 mV $s^{-1}$ scan rate. The inset in (c) is a magnified plot of the CV region revealing the formation of SEI.[36] [(c, d) Reprinted with permission from F. Shi, P. N. Ross, H. Zhao, G. Liu, G. A. Somorjai, K. Komvopoulos, J. Am. Chem. Soc. 2015, 137, 3181–3184. © American Chemical Society 2015]

to EC molecules solvating a lithium ion is found to be ca. 0.9 V versus $Li/Li^+$. The electron transfer produces a ring-opened transition state that is the precursor to lithium ethyl dicarbonate (LiEDC) (by dimerization of the radical anion).[37] These results suggest that reductions at potentials above 1.0 V are likely electrocatalytic and may produce other products.[38] Therefore, a thorough understanding of how the primary Li ion solvation structure and electrode material influence the electrolyte decomposition process is essential.

### 3.2 Selectivity of Electrolyte Reduction

An in situ attenuated total reflection-Fourier transform infrared (ATR-FTIR) spectroscopy approach was developed to study the electrolyte reduction process.[36] Considering the complexity of surface functional groups on real battery electrodes, two model electrode materials were chosen for this study, namely, Au as a metal electrode without any functional groups and Sn as an electrode material with oxygen functional groups that can be controlled and/or identified.[39–42] As a noble metal, the Au surface is relatively inert and barely interacts with aprotic electrolyte solvent, whereas Sn, as a typical active metal, forms a native oxide layer upon exposure to the atmosphere. When electrodes with active functional groups (nucleophilic sites) are introduced into polar aprotic electrolyte solutions, their oxygen species firstly react with the electrolyte solvent upon electroreduction.

In Figure 5(a) five new peaks (2920, 2851, 1576, 1557, and 1115 $cm^{-1}$) appear in the spectrum of the Au electrode at 0.6 V potential. These peaks are assigned to LiEDC (1115 $cm^{-1}$) and lithium propionate (2920, 2851, and 1565 $cm^{-1}$). The peaks of 2920, 2851, 1576, and 1557 $cm^{-1}$ in Figure 5(b) at the potential of 1.4 V are assigned to lithium propionate, whereas the peak at 1744 $cm^{-1}$ is assigned to diethyl 2,5-dioahexane dicarboxylate (DEDOHC). The intensity of the peak at 1197 $cm^{-1}$ (assigned to the –C–O– stretch of Li-solvated EC) decreases with the potential concurrently with the formation of DEDOHC. In brief, lithium propionate is identified as a common reduction product on both Au and Sn electrodes, whereas LiEDC and DEDOHC are detected only

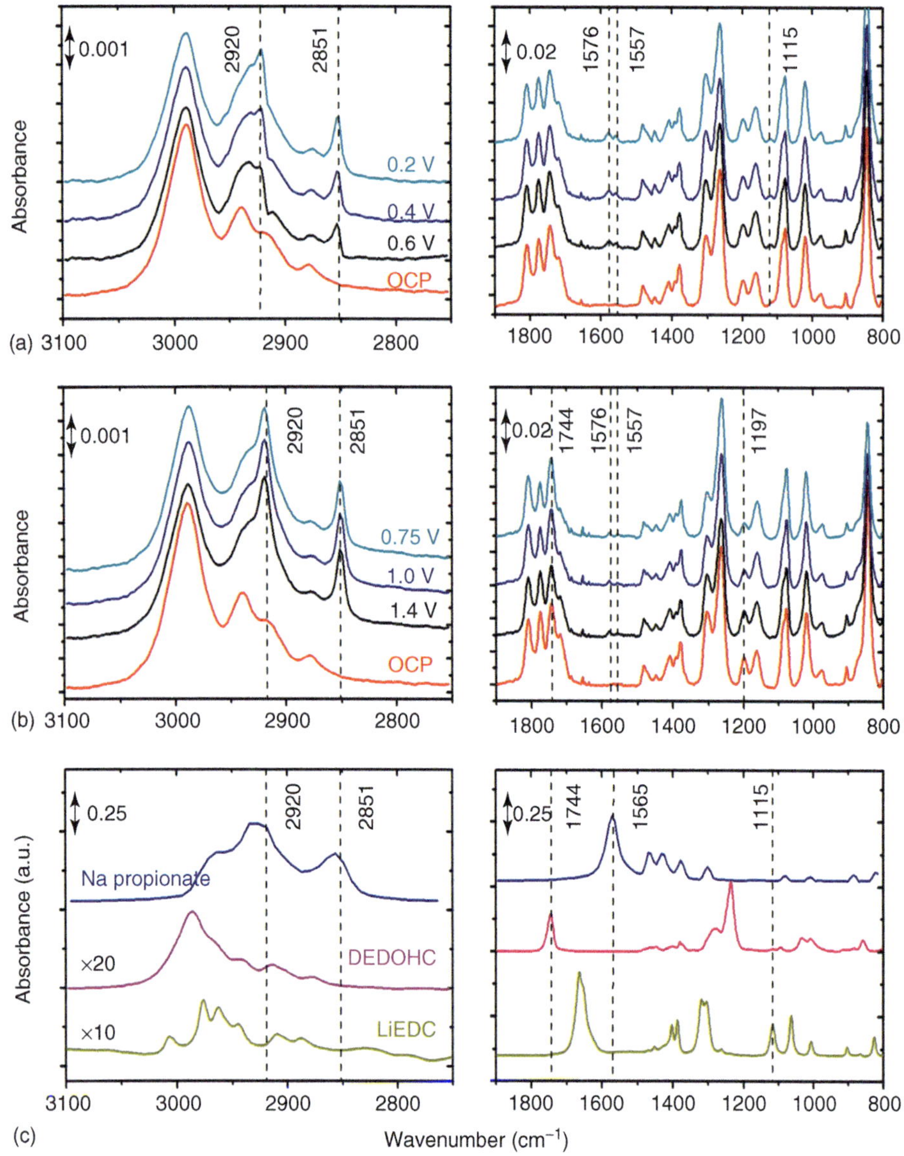


**Figure 5** In situ ATR-FTIR spectra of (a) Au electrode obtained after applying a potential of open circuit potential (OCP), 0.6, 0.4, and 0.2 V and (b) Sn electrode obtained after applying a potential of OCP, 1.4, 1.0, and 0.75 V (both electrodes are in EC/DEC 1 M $LiPF_6$ electrolyte). (c) ATR-FTIR spectra of sodium propionate (reference spectrum for lithium propionate), DEDOHC, and LiEDC.[36] [Reprinted with permission from F. Shi, P. N. Ross, H. Zhao, G. Liu, G. A. Somorjai, K. Komvopoulos, J. Am. Chem. Soc. 2015, 137, 3181–3184. © American Chemical Society 2015]

on the Au and Sn electrode surfaces, respectively. The presence of different chemical species is direct evidence that the surface of electrode materials can exhibit catalytic behavior, that is, lower reaction overpotential and different product selectivity.

The noncatalytic and catalytic reduction paths on Au and Sn are summarized in Scheme 1(a) and 1(b), respectively. As shown in Scheme 1(a), in the absence of strong interactions between the Au electrode and the electrolyte solvent, the solvated molecule is reduced by electron transfer, yielding the thermodynamically favorable product LiEDC. On the active Sn electrode, which is covered by oxygen spices, the main reduction product is DEDOHC. As a transesterification compound of the electrolyte, the formation mechanism of DEDOHC has been systematically investigated.[6,43–45] Ethyl oxide radical is an important catalytic intermediate during its formation. As shown in Scheme 1(b), the strong interaction between DEC and Sn oxide produces ethyl oxide, which triggers the decomposition of EC and the formation of DEDOHC. Thus, the catalytic effect is reflected in both the reduction potential and selectivity of solvent reduction.

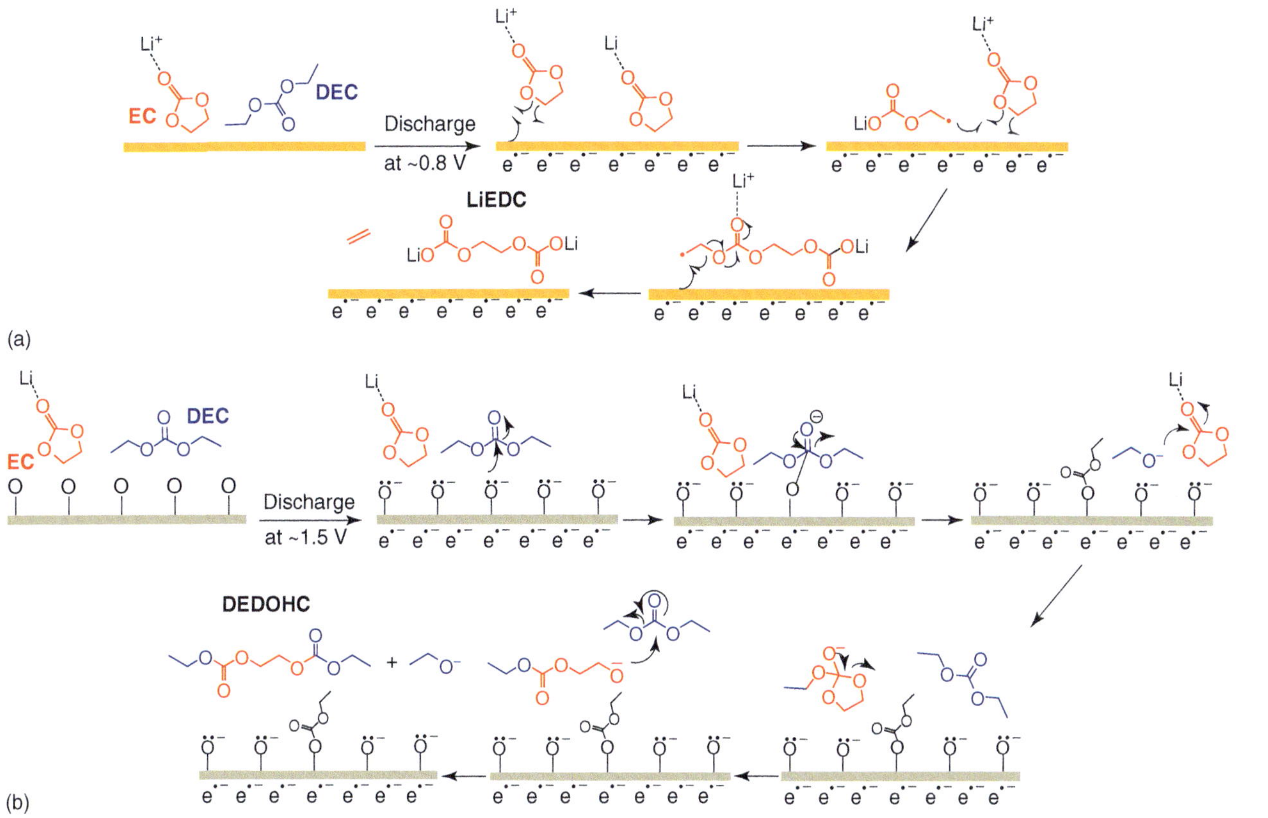


**Scheme 1** Proposed formation mechanisms of reduction products. (a) A noncatalytic path leads to the formation of LiEDC on the Au electrode surface. The inactive electrode primarily serves as a current transfer medium; thus, the state of ion solvation is the dominant factor in the electrolyte decomposition process. (b) A catalytic path leads to the formation of DEDOHC on the Sn electrode surface.[36] [Reprinted with permission from F. Shi, P. N. Ross, H. Zhao, G. Liu, G. A. Somorjai, K. Komvopoulos, J. Am. Chem. Soc. 2015, 137, 3181–3184. © American Chemical Society, 2015]

## 4 MECHANICAL EFFECT

### 4.1 Chemical and Mechanical Coupling in Batteries

Similar to other traditional rechargeable batteries, LIBs have two electrodes, which reversibly host the Li ion insertion/extraction. To achieve larger capacity density, unit electrode materials have to host more Li ions, which causes larger volume expansion and irreversible electrode deformation and fracture. Thus, electrode fracture is one of the most important reasons for the capacity decay. Figure 6 shows a typical fracture of high capacity electrodes, which are made of amorphous silicon thin films, crystalline silicon particles, and SnO particles. Silicon is one of the most promising next-generation anode materials, due to its high capacity (3579 mAh $g^{-1}$ compared to 372 mAh $g^{-1}$ of graphite), but it is prone to fracture due to excessive volume expansion and contraction (270% expansion of $Li_{15}Si_4$ compared to 10% of $LiC_6$). Consequently, the mechanical degradation of the silicon electrode results in severe capacity and power fade, thereby greatly limiting its implementation in commercial batteries. Thus, in this section we focus on single-crystal silicon electrodes as model system and explore the mechanical effect on its SEI.

In order to mimic the extended cycle effect on the fracture of silicon electrode, Shi *et al.* cycled a single crystal of silicon (100) for 30 cycles, as shown in Figure 7(a)–(c).[49] The onset of the reduction current is observed in the first cycle at ~0.1 V, corresponding to the initial alloying of the crystalline silicon with lithium; the two anodic peaks at ~0.3 and ~0.5 V are associated with Li dissolution.[50–52] The magnitudes of the oxidation and reduction peaks increase with the number of cycles in Figure 7(b), which is attributed to the continuous increase in electrochemically active silicon in each cycle. After 30 cycles, the cyclic voltammogram stabilizes as shown in Figure 7(c). With continued cycling of the silicon electrode, the redox current area begins to decrease, indicating the loss of active silicon material.

Figure 7(d)–(f) shows the evolution of the surface morphology of the silicon electrode after 3, 8, and 50 cycles, and Figure 7(g)–(i) shows magnified scanning electron microscope (SEM) images of the electrode surface obtained after 30 cycles. The dominant feature is the

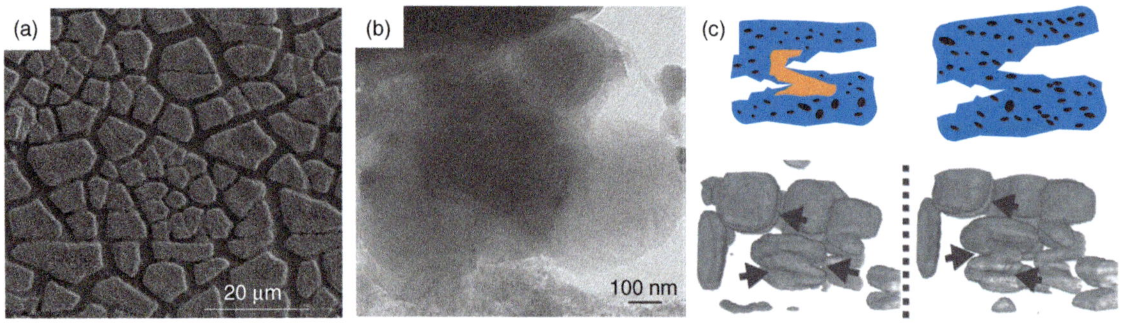


**Figure 6** Electrode materials fracture in Li-ion battery (a) amorphous silicon thin film of 500 nm is cracked after the 10 cycles of charge.[46] [Reprinted with permission from J. Li, A. K. Dozier, Y. Li, F. Yang, Y.-T. Chenga, J. Electrochem. Soc. 2011, 158, A689-A694. © The Electrochemical Society 2011]; (b) Nano-fracture in a 540 nm crystalline silicon particle after 138 s of lithiation.[47] [Reproduced with permission from X. H. Liu, J. Y. Huang, Energy Environ. Sci. 2011, 4, 3844–3860. © Royal Society of Chemistry 2011]; (c) 20 μm SnO particles are cracked into parts after lithiation.[48] [Reproduced with permission from M. Ebner, F. Marone, M. Stampanoni, V. Wood, Science 2013, 342, 716–720. The American Society for the Advancement of Science © 2013]

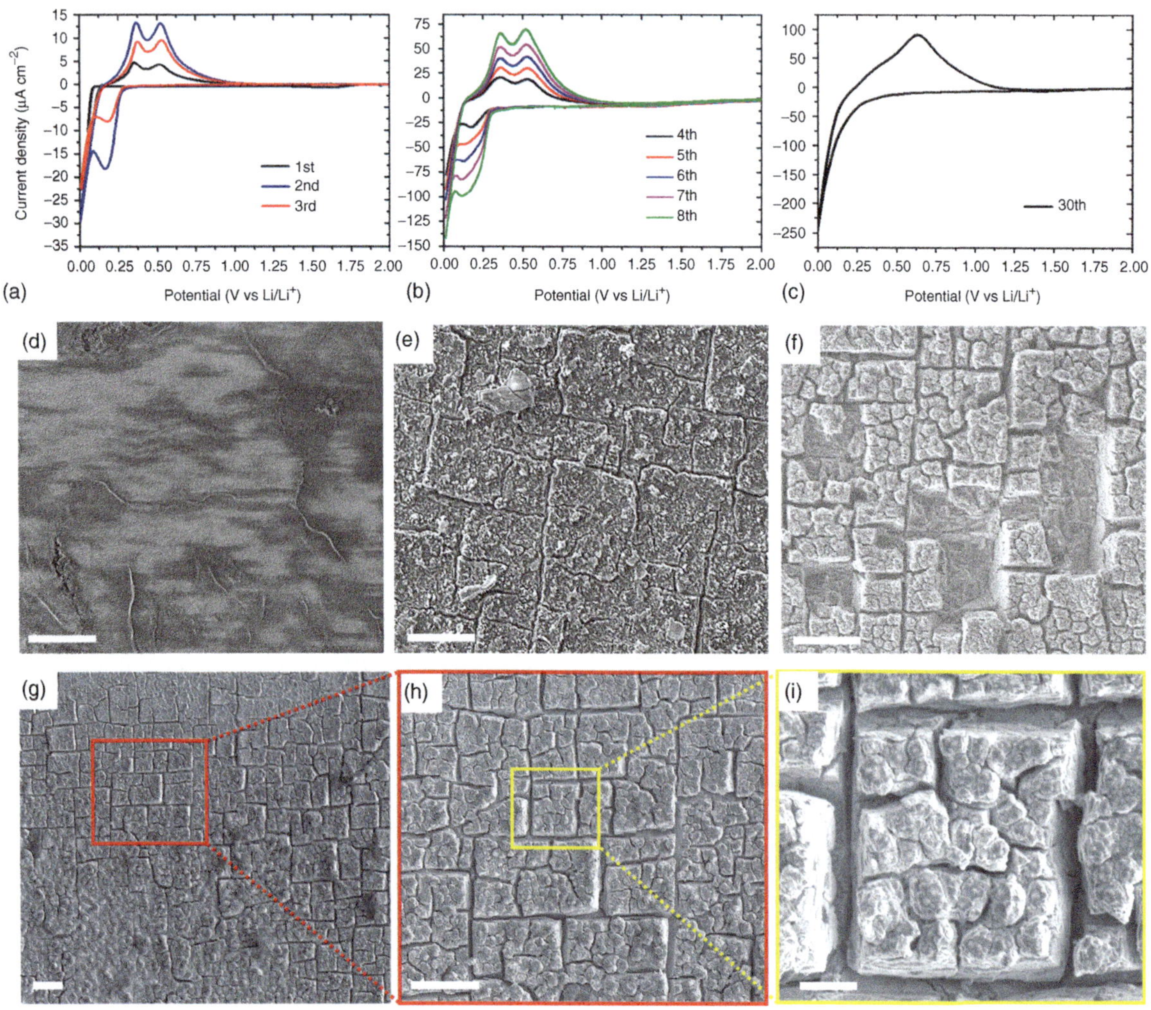


**Figure 7** Electrode cyclic voltammetry response and surface morphology. Current–voltage curves of a Si (100) electrode cycled between 2.0 and 0.01 V at a scan rate of 0.1 mV $s^{-1}$ for (a) 1–3, (b) 4–8, and (c) 30 successive lithiation/delithiation cycles. Top-view scanning electron microscope images of electrode surface morphology obtained after (d) 3, (e) 8, (f) 50, and (g–i) 30 cycles. Scale bars, 20 μm in (d–h) and 3 μm in (i)[49]

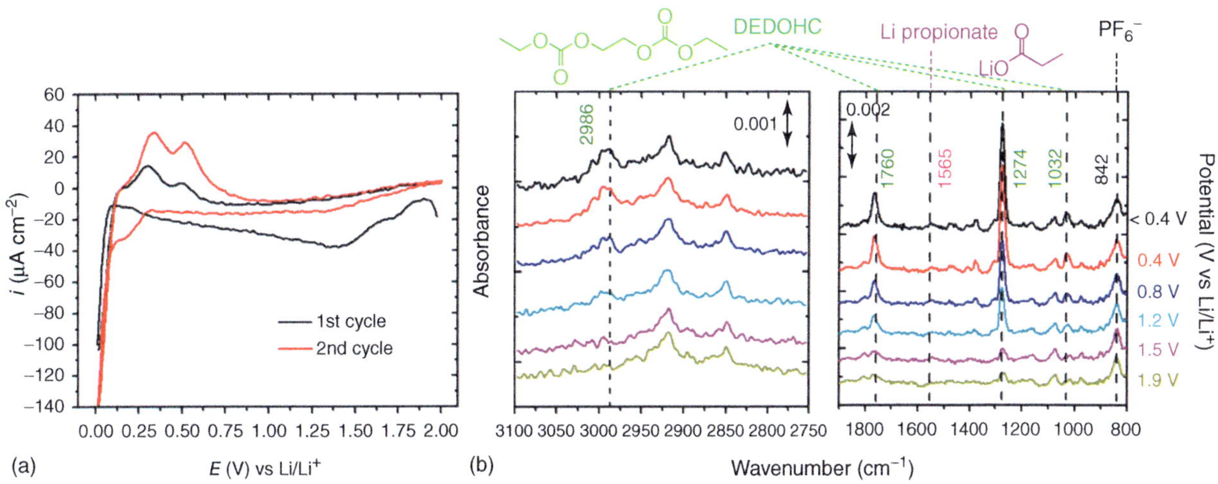


**Figure 8** Typical cyclic voltammetry response of silicon electrodes and in situ ATR-FTIR spectra. (a) Cyclic voltammograms of a Si(100) electrode in 1 M $LiPF_6$/EC:DEC (1:2 v/v) electrolyte for a scan rate of 1 mV s$^{-1}$ and (b) corresponding spectra for a potential in the range of 1.9–<0.4 V and 65° incidence angle.[7] [(b) Reprinted with permission from F. Shi, P. N. Ross, G. A. Somorjai, K. Komvopoulos, J. Phys. Chem. C 2017, 121, 14476–14483. © American Chemical Society, 2017]

development of surface cracks in two orthogonal directions, which results in isolated small squares that form a regular repetitive pattern on the electrode surface. These crack patterns further account for the formation of square cavities, shown in Figure 7(f). Apparently, the damage to the electrode surface is not a rapid and catastrophic process but occurs in a gradual and cumulative fashion. Although damage of the electrode surface is insignificant during the first few cycles, shallow surface cracks initiate after a critical number of cycles and propagate with further cycling, eventually causing delamination at the electrode surface.

## 4.2 SEI Formation with/without Electrode Fracture

Shi *et al.* further used in situ ATR-FTIR to study the SEI formation on a silicon single crystal (100).[7] Figure 8(a) shows the voltammograms for the first two cycles from 2 to 0.005 V at a scan rate of 1 mV s$^{-1}$. In the first cycle, the first electrolyte reduction peak starts at ~1.8 V, while the maximum current occurs at ~1.3 V, which indicates SEI formation. Figure 8(b) shows ATR-FTIR spectra obtained by applying a potential sweep from 1.9 to <0.4 V, which is the potential region before lithiation. Based on the spectra of the synthesized reference compounds,[36,53] the new peaks at 2986, 1760, 1274, and 1032 cm$^{-1}$ are assigned to the DEDOHC components. Both C=O (1760 cm$^{-1}$) and C–O (1274 cm$^{-1}$) peaks are slightly shifted compared with the bulk DEDOHC spectra (1744 and 1238 cm$^{-1}$, respectively), presumably due to the adsorption of DEDOHC at the surface. The C=O feature at 1032 cm$^{-1}$ is assigned to Li propionate compound, which is the co-product of DEDOHC.

Further lithiation of silicon will cause excessive volume change, electrode fracture, and exposure of fresh lithiated silicon to the electrolyte.[49] To understand the formation of SEI on lithiated silicon, a constant potential of 5 mV was applied in the chronoamperometry (CA) experiment, which led to the formation of a metastable phase ($Li_{15}Si_4$), without Li metal plating.[54] The incidence angle of the infrared beam was varied to obtain the composition depth profile of the SEI. Figure 9 shows that the peak at 2986 cm$^{-1}$ disappears when incidence angle is above 45°, which corresponds to the C–H signal of bulk electrolyte. The C=O peaks at 1805, 1770, 1736, and 1713 cm$^{-1}$ vanish with the increase of incidence angle above 60°, and new peaks appear at 3008, 2975, 2961, 2943, 1664, and 1310 cm$^{-1}$, which are assigned to LiEDC.[34] Clearly, reactions under a potential step of 5 mV bypass the catalytic effect of the native silicon oxide for DEC reduction, instead favor the lithiation of silicon oxide, which accelerates the fluorination of silicon oxide compared with long-term aging. As shown by the SEM images presented in Figure 10, the lithiation of silicon and outward expansion produces surface cracks in the silicon electrode, which facilitates the reduction of EC molecules on the lithiated silicon and results in the formation of LiEDC.[7,36]

Further evidence of the reduction of LiEDC due to the reaction of EC with exposed lithiated silicon can be obtained by observing the evolution of the spectra with cycling.[7] Figure 10(a) shows in situ ATR-FTIR spectra of a lithiated silicon electrode at incidence angle of 65° after 0.5, 1, 5, and 10 lithiation/delithiation cycles. During the first half-cycle, the electrochemical reduction product is mainly DEDOHC (peaks at 1760 and 1274 cm$^{-1}$), which passivates the electrode surface, preventing further electrolyte reduction. In the following 5–10 cycles, LiEDC emerges (peaks at 3008, 2975, 2961, 2943, 1664, and 1310 cm$^{-1}$), while DEDOHC peaks disappear.

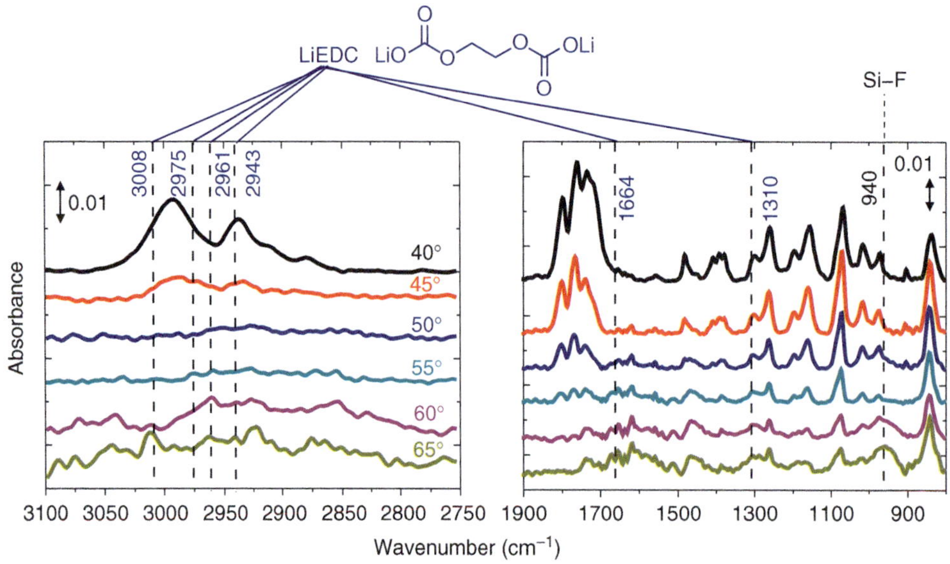


**Figure 9** In situ ATR-FTIR spectra of lithiated silicon electrodes. Lithiation of a Si(100) electrode in 1 M $LiPF_6$/EC:DEC (1:2 v/v) electrolyte by applying a potential of 5 mV. The potential was applied for 1 h before acquiring the spectra, while the incidence angle was varied between 40° (bulk electrode) and 65° (near-surface electrode region).[7] [Reprinted with permission from F. Shi, P. N. Ross, G. A. Somorjai, K. Komvopoulos, J. Phys. Chem. C 2017, 121, 14476–14483. © American Chemical Society, 2017]

Figure 10(b)–(e) shows top-view SEM images of the electrode's surface morphology after 1, 5, 10, and 30 lithiation/delithiation cycles, whereas Figure 10(f) and (g) shows top-view and cross-sectional high-magnification SEM images of the electrode after 30 cycles, respectively. Before imaging, the electrode was rinsed with DMC to remove the residual electrolyte. After the first cycle, the silicon surface is intact; however, there is evidence of microscopic fissure formation in the silicon-oxide film as shown in Figure 10(b). Further cycling causes cracks to initiate at the electrode surface and then grow in the depth direction shown in Figure 10(c)–(g). Figure 10(c) shows flaking of the lithium silicate layer due to the volume expansion mismatch between the silicon-oxide film and bulk silicon. As the number of cycles accumulates, the oxide flakes start to detach as shown in Figure 10(f) and cracks start to grow into the electrode bulk (Figure 10g).

Scheme 2 summarizes the formation mechanism of the SEI on silicon electrodes. Sweeping from the open circuit potential (OCP) to the lithiation potential drives the selective reduction of DEC, which leads to the formation of DEDOHC on the native silicon-oxide film. Passivation of the electrode is triggered when the potential is lower than ~1.3 V and is characterized by the accumulation of DEDOHC. After the first cycle, the SEI mainly consists of DEDOHC on $Li_xSiO_y$. The DEDOHC has been reported to increase the battery impedance, causing the charge/discharge performance to degrade.[55] The formation of DEDOHC consumes extra electrons, whereas the co-product—lithium propionate—traps the Li ions,[36] contributing to the low Coulombic efficiency of silicon-based batteries in the first cycle. Applying a potential of 5 mV results in the simultaneous lithiation of the silicon-oxide film and bulk silicon and bypasses the formation of DEDOHC. Under this condition, solvated EC molecules are directly reduced to LiEDC on lithiated silicon, which is exposed through the cracks driven by mechanical stress during lithiation. Extended cycling leads to electrode cracking, producing LiEDC on the freshly cracked lithiated silicon surface, while the original native silicon-oxide film remains covered with DEDOHC. After prolonged cycling, the continuous low-rate electrolyte consumption and electrical isolation of active silicon electrode materials are the major factors contributing to capacity fading.

## 5 PROPERTIES OF SEI

### 5.1 Ionic Conductivity of SEI

The kinetics of ion transport through the SEI usually includes three stages: desolvation of $Li^+$ near the electrolyte/SEI interface, $Li^+$ migration through the SEI, and then charge transfer on the SEI/anode interface. In Li-metal batteries, where the SEI layer is thick and the charge transfer is fast on the anode, the $Li^+$ migration becomes the rate-determining step. Thus, the ionic conductivity of the SEI layer greatly determines the morphology of lithium deposits and the cyclability of Li anodes. The ionic conduction is also vital to the rate capability of the Li-ion battery, which may induce large

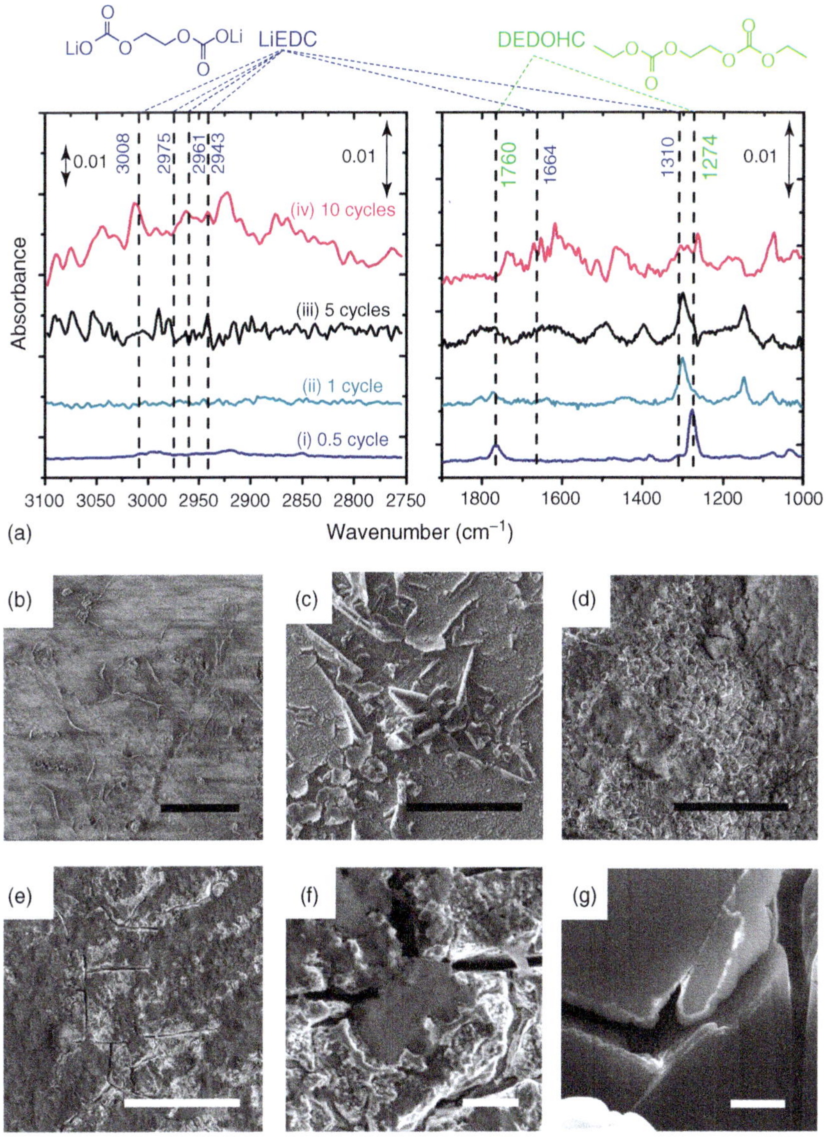


**Figure 10** In situ ATR-FTIR spectra and surface morphology of lithiated silicon electrodes. (a) Spectra of a Si(100) electrode in 1 M $LiPF_6$/EC:DEC (1:2 v/v) electrolyte cycled between 5 mV and 2 V at a scan rate of 1 mV $s^{-1}$ for 0.5–10 cycles for 65° incidence angle. Top-view SEM images of the electrode surface morphology obtained after (b) 1, (c) 5, (d) 10, and (e) 30 cycles. High-magnification (f) top-view and (g) cross-sectional SEM images of the electrode surface morphology after 30 cycles. The scale bar is (b–e) 30 μm, (f) 2.5 μm, and (g) 1 μm.[7] [Reprinted with permission from F. Shi, P. N. Ross, G. A. Somorjai, K. Komvopoulos, J. Phys. Chem. C 2017, 121, 14476–14483. © American Chemical Society, 2017]

electrode polarization under fast charge. Here, we firstly discuss the ion conduction properties of the SEI.

The transport of ions through inorganic crystals takes place through Schottky or Frenkel defects. Transport is much more complex when considering a composite layer like the SEI, since the contribution of the grain boundaries, polymerized, and amorphous parts must be taken into consideration. According to the multilayer model, the SEI is comprised of two or more separate layers of different composition and properties.[56] The most interior layer is thin and compact, while the exterior layer is a more porous or structurally open layer, which suppresses the mass transport of ions in the electrolyte filling its pores. Despite its obvious importance to lithium battery operation, the mechanism of Li ion transport through the SEI has rarely been investigated. By performing time-of-flight secondary ion mass spectroscopy (TOF-SIMS) on isotopically labeled samples, it was confirmed that the exterior layer of SEI is porous, and Li ion transport through the inner layer happens by ion exchange.[57] It was further determined that the dominant diffusion carrier in inorganic components ($Li_2CO_3$) below the voltage range of SEI formation is

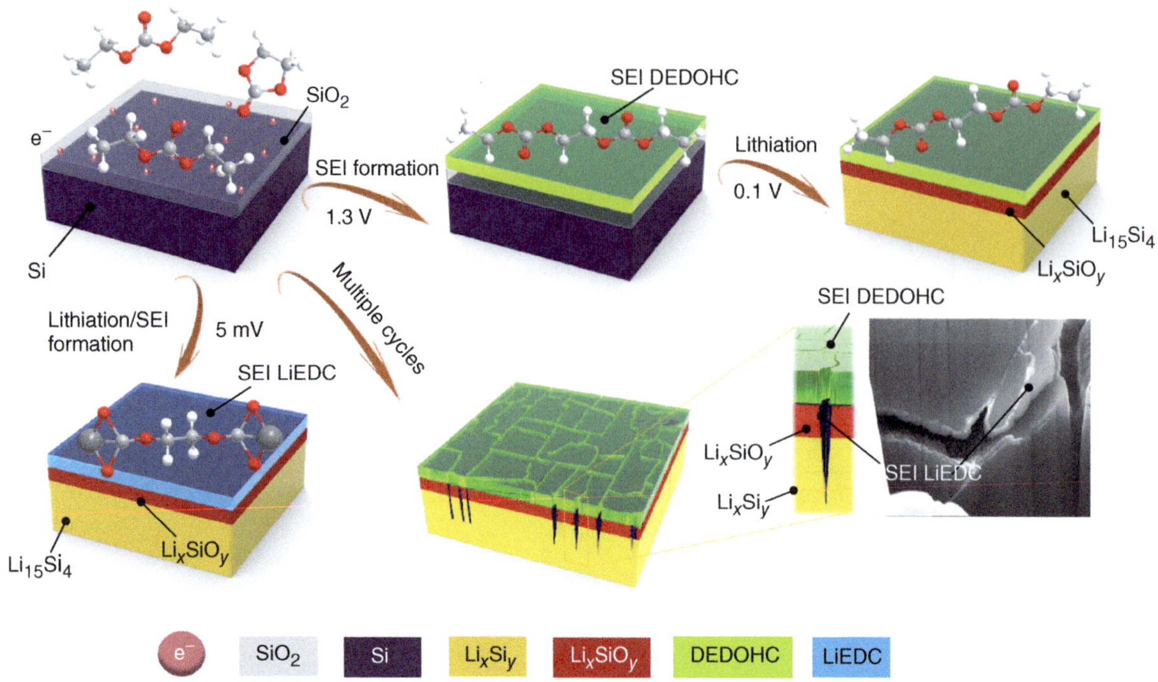


**Scheme 2** Potential and cycle number dependence of EC/DEC electrolyte reduction on silicon electrodes. DEDOHC forms on the native oxide film covering the silicon electrode, whereas LiEDC forms on the lithiated silicon electrode. After the first cycle, SEI mainly consists of DEDOHC on $Li_xSiO_y$. After multiple cycles, electrode fracture emerges and the SEI gradually changes to LiEDC on freshly exposed $Li_xSi_y$ inside the forming cracks. The schematics are not to scale, that is, the $SiO_2$ and $Li_xSiO_y$ are much thinner than the SEI and $Li_xSi_y$.[7] [Reprinted with permission from F. Shi, P. N. Ross, G. A. Somorjai, K. Komvopoulos, J. Phys. Chem. C 2017, 121, 14476–14483. © American Chemical Society, 2017]

excess interstitial Li ion, which diffuses via a knock-off mechanism to maintain higher O-coordination, rather than direct-hopping through empty spaces in the $Li_2CO_3$ lattice. Mesoscale diffusion equations were then formulated upon a new two-layer/two-mechanism model: pore diffusion in the outer layer and knock-off diffusion in the inner layer (Figure 11).

Due to the compact nature, migration of Li ion through the inorganic layer of the SEI is usually the rate-determining step. Table 1 summarizes the ionic conductivity of the major components of SEI ($Li_2CO_3$, LiEDC, LiF) and recently reported artificial components (LiF, $Li_3N$). The measured ionic conductivity largely relies on the crystallinity, grain size, impurity content, interphase structure, and electronic contact of the measured sample, which can lead to a variation of reported ionic conductivity of the same materials. It should be noted that the ionic conductivity is also potential dependent (state of charge dependent),[62,63] which complicates the comparison between different components. The conductivities derived from simulations are usually lower than the experimental data, which may be attributed to the inaccurate knowledge of impurities, defects, and vacancies on specific samples.[64] Due to the heterogeneity between electrode, SEI, and electrolyte, there are ion diffusion junctions among them. With the conversion of Equation (1), we can compare the ionic conductivity of SEI to the diffusion coefficient of Li ions in anode and cathode electrodes.

$$D(T) = \frac{kT}{Ne^2}\sigma(T) \tag{1}$$

The diffusivity of lithium atoms ($^7$Li) in solid lithium metal ($^7$Li) at 25 °C is $5.5\times10^{-11}\,cm^2\,s^{-1}$.[65] The typical diffusion coefficient of lithium in graphite anodes is $\sim1.12\times10^{-11}\,cm^2\,s^{-1}$ and $\sim10^{-12}\,cm^2\,s^{-1}$ in nano-silicon anodes.[66,67] The diffusivity of lithium in cathode materials is about the same order, for example, $2.48\times10^{-15}\,cm^2\,s^{-1}$ for carbon coated $LiFePO_4$ and $\sim10^{-10}\,cm^2\,s^{-1}$ for $Li_xCoO_2$.[68,69] A matching diffusivity between SEI and electrode could lower the polarization effect (Li ion accumulation or depletion) under high charging rate.

## 5.2 Solubility of SEI

The SEI layer forms instantaneously upon contact with the electrolyte, including both insoluble and slightly soluble species. Due to the porous structure of the native oxide layer, the solvent molecules can also diffuse to the

**Table 1** The ionic conductivity ($S\,cm^{-1}$) and diffusion coefficient ($cm^2\,s^{-1}$) of different SEI species at room temperature under open circuit potential

| | $Li_2CO_3$ (Ref. 58) | LiEDC (Ref. 59) | LiF (Ref. 60) | $\alpha$-$Li_3N$ (Ref. 61) |
|---|---|---|---|---|
| Ionic conductivity | $1.373 \times 10^{-13}$ | $10^{-9}$ | $<10^{-11}$ | $5.767 \times 10^{-4}$ |
| Diffusion coefficient | $1.1 \times 10^{-7}$ (knock-off)<br>$8.4 \times 10^{-12}$ (direct-hopping) | $1.373 \times 10^{-13}$ | $<1.373 \times 10^{-15}$ | $7.91807 \times 10^{-8}$ |

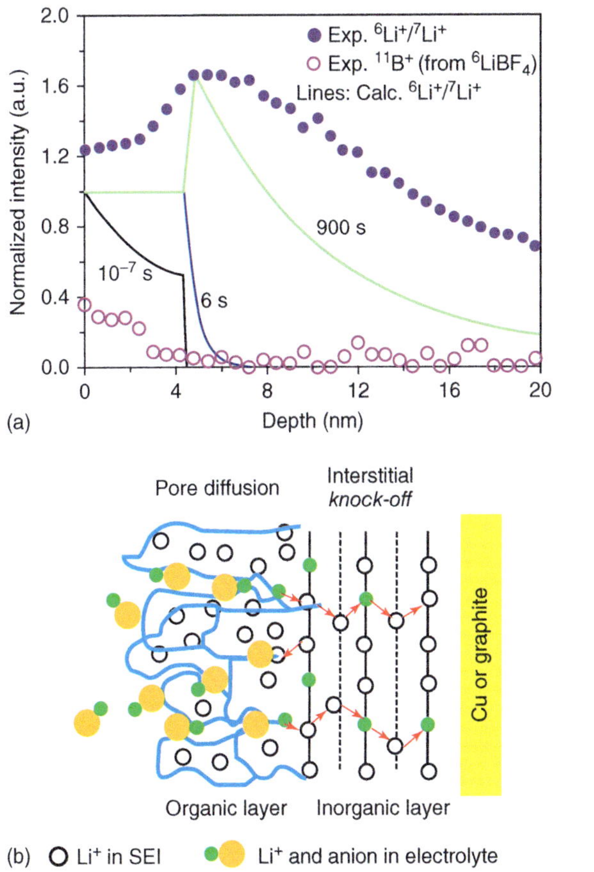


**Figure 11** (a) TOF-SIMS measured (by the isotope exchange experiment) depth profiles of $^6Li^+/^7Li^+$ and $^{11}B^+$ (symbols) for the SEI growing on a Cu substrate after 900 s soaking and calculated depth profiles of $^6Li^+/^7Li^+$ (solid lines) after $10^{-7}$, 6, and 900 s soaking. (b) Schematic drawing of pore diffusion in the porous organic layer of SEI and knock-off diffusion in the dense inorganic layer of SEI.[58] [(b) Reprinted with permission from S. Shi, P. Lu, Z. Liu, Y. Qi, L. G. Hector, H. Li, S. J. Harris, J. Am. Chem. Soc. 2012, 134, 15476–15487. © American Chemical Society, 2012]

lithium interface. Thus, the first interaction between the native passivation layer and the electrolyte includes both physical and chemical dissolution.[17,70–72] Because of the similar physicochemical properties between the SEI and cosolvents in the electrolyte, this dissolution of the SEI is likely universal, which would undoubtedly reduce the performance of lithium batteries. During long-term cycling, the partially dissolved SEI is repaired constantly by the same electrochemical reactions that occurred in the initial formation process. As a result, the finite lithium ion source in the cell is consumed and the impedance continually increases at the electrode/electrolyte interface.

Tasaki *et al.* evaluated the solubilities of a series of lithium salts in both EC and DMC using atomistic simulations. These salts included oxides, alkoxides, carbonates, semicarbonates, fluorides, and oxalates and represented almost all possible ingredients of an interphase formed on graphite anodes.[73] They found that the order in which the salts are likely to dissolve is:

$$LiOCO_2CH_3 > LiOCO_2C_2H_5 > LiOH > LiF > (LiCO_2)_2 > Li_2CO_3$$

The organic salts were more likely to dissolve than the inorganic salts in DMC. We would like to point out that low solubility is not the only requirement for stable passivation during the aging or long-term cycling. Due to their intrinsic nucleophilic properties, some components (e.g., LiOH or $Li_2O$) will cause electrolyte degradation chemically, thus increasing the impedance of the cell.

### 5.3 Mechanical Properties of the SEI

As discussed in previous sections, during the cycling, the electrodes experience volume change. As an adhesion layer on the anode, the SEI is also exposed to mechanical stresses. Therefore, the mechanical properties of the SEI will largely determine whether this passivation layer is prone to breakdown upon cycling. Young's modulus of the SEI on silicon thin films has been extracted by scanning force spectroscopy.[74] The 0.5 GPa modulus sits at the boundary of inorganic and organic components, implying that the material is a hybrid of these two categories. It has been reported that high modulus coating, or interface, can effectively suppress lithium dendrite growth or silicon fracture.[75,76] Young's moduli of lithium and silicon are 4.9 and 47 GPa, respectively. Compared to the SEI generated in normal electrolyte, an SEI layer tuned by additives is needed to prevent the electrode underneath from fracturing.

Electrolyte additives, which are typically used to modify the chemical properties of the SEI,[22,23] can influence the mechanical properties of the SEI and, consequently, electrode degradation. A Fourier transform infrared spectroscopy investigation has shown that the

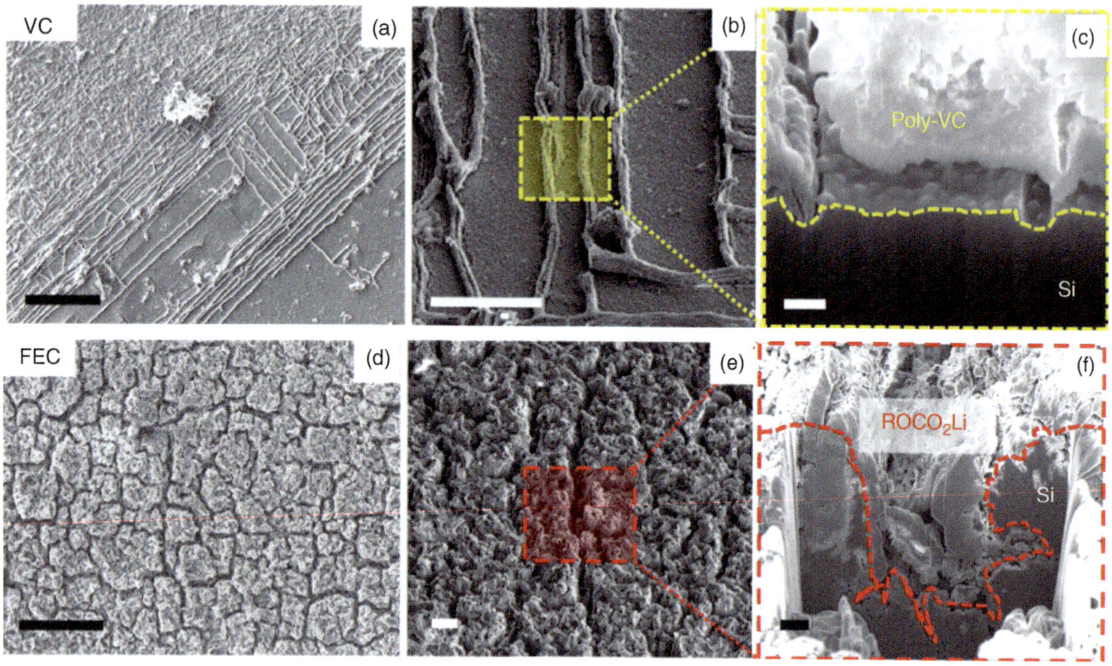


**Figure 12** Electrode crack arrest by the reduction products of electrolyte additives. Top-view and cross-sectional scanning electron microscope images of a Si(100) electrode cycled between 2.0 and 0.01 V for 30 cycles at a scan rate of 0.01 mV $s^{-1}$ in EC/DEC/$LiPF_6$ electrolyte with (a)–(c) 5% VC additive and (d)–(f) 5% FEC additive. The dashed rectangles shown in (b) and (e) indicate the focused-ion-beam milling area. The reduction products of the VC and FEI additives (poly-VC and $ROCO_2Li$, respectively) modify the mechanical properties of the solid electrolyte interphase (SEI) film and inhibit crack growth. (Scale bar: (a, d) 100 µm, (b, e) 20 µm, (c) 300 nm, and (f) 3 µm.)[49] [Reprinted with permission from F. Shi, Z. Song, P. N. Ross, G. A. Somorjai, R. O. Ritchie, K. Komvopoulos, Nat. Commun. 2016, 7, 11886. © Springer Nature Limited, 2016]

reduction product of vinylene carbonate (VC) additive is poly-VC, whereas that of the fluoroethylene carbonate (FEC) additive is alkyl carbonate $ROCO_2Li$ salt.[26–32,34–77] Both reaction products can chemically passivate the electrode surface before the decomposition of the electrolyte solvent. The poly-VC aggregates fill the vertical surface cracks of silicon (Figure 12a and b) and promote crack-face bridging (Figure 12c). As a result, the fracture toughness is increased, and further crack propagation is inhibited. The crack depth after 30 cycles is ~100 nm, which is significantly smaller than the 5–8 µm crack depths observed with additive-free electrolyte. The $ROCO_2Li$ salt uniformly precipitates on the electrode surface (Figure 12d and e), as evidenced by the roughening of the electrode surface. These $ROCO_2Li$ salts fill and passivate the surface cracks, thereby significantly decelerating crack growth.

## 6 POTENTIAL CAUSES OF CONTROVERSY DURING SURFACE CHARACTERIZATION

### 6.1 Ex Situ Sample Preparation

Beyond any doubt, the electrode/electrolyte interface constitutes the foundation for the state-of-the-art lithium ion chemistry and has become the most active research topic during the past decade. However, the characterization of many key attributes of the corresponding surface chemistries has proven very difficult, and the results can be rather controversial. The elusive nature of these interfaces is believed to arise from the sensitivity of the SEI components. This controversy can be quite significant when ex situ measurements are employed. Vacuum environments must be used for most SEM images or electron spectroscopy experiments. The low vapor pressure of the co-solvent (like DMC, DEC, DOL, 1,2-dimethoxyethane (DME)) in the electrolyte will be pumped out before measurement, which inevitably affects the morphology and thickness of the SEI. Furthermore, inorganic salts in the electrolytes usually precipitate on top of the dried sample and dominate the major feature/signal of results. For ex situ vibration spectroscopy, dried-out samples are also applied in an inert atmosphere. In order to reduce or eliminate the contribution from residual electrolyte, Zhuang *et al.* firstly reported a controlled DMC process to enhance the SEI signal among the residual electrolyte.[35] Figure 13 shows a gradual increase of SEI signal with increasing DMC washing time. Thus, rinsing process could profoundly affect the spectroscopy or imaging results, which makes quantitative assessment very inaccurate. We also want to point out that over-rinsing may

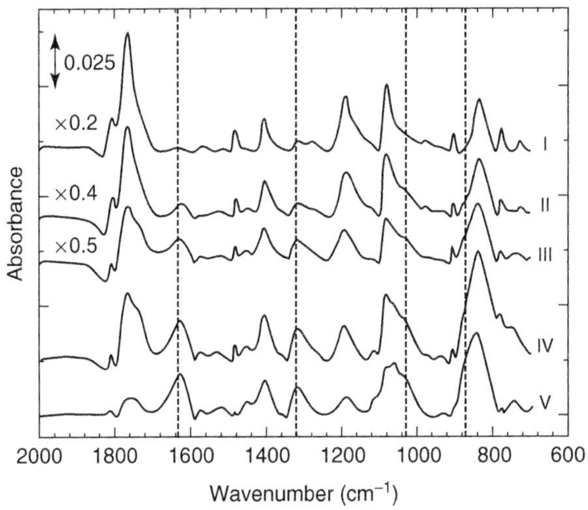


**Figure 13** Evolution of FTIR spectra on Mag-10 graphite anode after first cycle and stopped at 1 V vs Li/Li$^+$. The electrolyte used is 1.2 M $LiPF_6$/ethylene carbonate (EC): ethyl methyl carbonate (EMC) (3:7 w/w). Ex situ sample has been subject to (I) 0 s; (II) 20 s; (III) 30 s; (IV) 40 s, and (V) 60 s DMC washing.[35] [Reproduced with permission from G. V. Zhuang, H. Yang, B. Blizanac, P. N. Ross, Electrochem. Solid-State Lett. 2005, 8, A441-A445. © The Electrochemical Society, 2005]

also cause chemical reactions between the lithium/lithiated-state anode and solvent, which consequently generates "artificial features". Hence, caution should always be used when interpreting ex situ results. The audiences will have a better understanding of the data, if researchers could provide more details and discussion about the ex situ sample preparation in their papers.

### 6.2 Beam Damage and Ion-Sputtering Effect

X-ray absorption spectroscopy (XAS) and XPS have been widely used for the characterization in lithium metal batteries due to their high sensitivity to chemical, bonding information and compatibility with in situ cell designs. The effect of beam damage has been broadly observed on the organic compounds under X-ray-based characterizations, which is a major concern for sensitive materials like the SEI.[78,79] For the elements inside the SEI (e.g., Li, C, O, F, …), soft X-rays are specifically used during the XAS measurement. Due to the intrinsic penetration depth and high cross-section for interaction with electrons, more serious irradiation damages from soft X-rays are often observed than hard X-rays. Specifically, the secondary electrons generated from soft X-ray photon-excitation process could strongly interact with the molecules bonding on surface, leading to undesired species. Qiao *et al.* showed that even the inorganic components such as $Li_2O_2$ and $Li_2CO_3$ turn to $Li_2O$ under the soft X-ray irradiation.[80] XPS also employs a high-energy X-ray source (~1200–1500 eV for Mg and Al K$\alpha$). The highly energetic beam spot may lead to degradation of the SEI components and may alter their chemical nature.

Due to the surface sensitivity of some techniques, depth profiles are largely obtained on SEI layers with destructive techniques, such as sputter-assisted XPS, AES, and TOF-SIMS.[81–84] Although these methods provide valuable information about the distribution of organic and inorganic chemical species, they inevitably damage and contaminate the sample and increase the surface roughness, thus complicating the interface definition.[85] For example, during the destructive depth analysis of inhomogeneous films, the sputtering rate is not the same for hard and soft components. The inorganic materials need longer sputtering time to reach a certain depth compared to soft organic materials. Sputtering may also lead to roughening of the surface and side reactions of the active species to form undesirable contaminants.

## 7 FUTURE PROSPECTS

Energy storage is an emerging field that is full of challenges and opportunities. Comprehensive understanding of the processes at critical surfaces and interfaces is crucial to successfully develop next-generation batteries. Beyond the work mentioned in this chapter, the following topics are also interesting and important to explore in future work.

### 7.1 Exploring New Diagnostic Tools for In Situ Measurements

The passivation layer makes the electrode surface kinetically stable and is responsible for the reversibility of Li$^+$-intercalation/alloy; therefore, it is the most crucial component in the batteries.[10,11] Characterization of the SEI is difficult because of its small (~10 nm) thickness, chemical sensitivity, and elusive pathway of formation. Moreover, the selective sensitivity to certain chemical species and simultaneous blindness to others often prevent most spectroscopy methods to provide a comprehensive understanding about the interphase. Although significant advances have been made during the past decade with the development of various new models and tools, we still lack in situ data, which would eliminate artificial effects and bring forth new understandings.

### 7.2 Interfacial Problems in All-Solid-State Batteries

Much of the research and development effort in lithium batteries today is directed toward all-solid-state architecture by employing solid-state electrolytes in direct contact with lithium metal.[86] The nature of the interface between lithium and the solid electrolyte is the focus of much research but involves fundamentally different chemistries from what we have discussed above for liquid electrolytes. The most fundamental difference and the one we shall focus on here is the mobility of the components of the electrolytes. In single ion conductors, lithium ion is the only mobile species (at near room temperature), reactions between the metallic lithium and the anions of the electrolyte, for example, oxide or phosphate, would be inherently limited by the mobility. The methods for diagnostic studies of the interfacial reactions are even more challenging than is the case with liquid electrolytes, since probes must be more penetrating even when employed ex situ, for example, there is no "rinsing" process of the bulk electrolyte.

## 8 ACKNOWLEDGMENTS

The authors acknowledge the Office of Science, Basic Energy Sciences, Chemical Sciences Division and Assistant Secretary for Energy Efficiency and Renewable Energy, Office of Vehicle Technologies, of the US Department of Energy, for supporting our research for many years.

## 9 RELATED ARTICLE

$Li-CO_2$ Batteries

## 10 ABBREVIATIONS AND ACRONYMS

AES = Auger electron spectroscopy; ATR-FTIR = attenuated total reflection-Fourier transform infrared; CA = chronoamperometry; DEC = diethyl carbonate; DEDOHC = diethyl 2,5-dioahexane dicarboxylate; DMC = dimethyl carbonate; DME = 1,2-dimethoxyethane; DOL = dioxolane; EC = ethylene carbonate; EMC = ethyl methyl carbonate; FEC = fluoroethylene carbonate; LiEDC = lithium ethyl dicarbonate; OCP = open circuit potential; SEI = solid electrolyte interphase; SEM = scanning electron microscopy; THF = tetrahydrofuran; TOF-SIMS = time of flight-secondary ion mass spectroscopy; UHV = ultra-high vaccum; VC = vinylene carbonate; XAS = X-ray absorption spectroscopy; XPS = X-ray photon spectroscopy

## 11 REFERENCES


1. J. B. Goodenough and Y. Kim, *Chem. Mater.*, 2010, **22**, 587.
2. E. Peled, *J. Electrochem. Soc.*, 1979, **126**, 2047.
3. M. Winter, *Z. Phys. Chem.*, 2009, **223**, 1395.
4. D. Bar-Tow, E. Peled and L. Burstein, *J. Electrochem. Soc.*, 1999, **146**, 824.
5. J. P. Olivier and M. Winter, *J. Power Sources*, 2001, **97–98**, 151.
6. T. Sasaki, T. Abe, Y. Iriyama, M. Inaba and Z. Ogumi, *J. Electrochem. Soc.*, 2005, **152**, A2046.
7. F. Shi, P. N. Ross, G. A. Somorjai and K. Komvopoulos, *J. Phys. Chem. C*, 2017, **121**, 14476.
8. D. Aurbach, 'Nonaqueous Electrochemistry', Marcel Dekker, Inc., New York, 1999.
9. Y. Wang, 'Lithium-Ion Batteries: Solid–Electrolyte Interphase', Imperial College Press, London, 2004.
10. K. Xu, *Chem. Rev.*, 2004, **104**, 4303.
11. K. Xu, *Chem. Rev.*, 2014, **114**, 11503.
12. D. Lin, Y. Liu and Y. Cui, *Nat. Nanotechnol.*, 2017, **12**, 194.
13. K. Wang, P. N. Ross, F. Kong and F. McLarnon, *J. Electrochem. Soc.*, 1996, **143**, 422.
14. G. Zhuang, P. N. Ross, F. p. Kong and F. McLarnon, *J. Electrochem. Soc.*, 1998, **145**, 159.
15. G. Zhuang, Y. Chen and P. N. Ross, *Surf. Sci.*, 1998, **418**, 139.
16. E. F. McFarlane and F. C. Tompkins, *Trans. Faraday Soc.*, 1962, **58**, 997.
17. K. Kanamura, H. Tamura, S. Shiraishi and Z. i. Takehara, *J. Electrochem. Soc.*, 1995, **142**, 340.
18. G. Zhuang, Y. Chen and P. N. Ross, *Langmuir*, 1999, **15**, 1470.
19. C. Lee, B. Mun and P. N. Ross, *J. Electrochem. Soc.*, 2002, **149**, A1286.
20. G. R. Zhuang, K. Wang, Y. Chen and P. N. Ross, *J. Vac. Sci. Technol. A*, 1998, **16**, 3041.
21. P. Tundo and M. Selva, *Acc. Chem. Res.*, 2002, **35**, 706.
22. T. R. Jow, K. Xu, O. Borodin and M. Ue, 'Electrolytes for Lithium and Lithium-Ion Batteries', Springer, New York, 2014.
23. D. Aurbach, O. Chusid, I. Weissman and P. Dan, *Electrochim. Acta*, 1996, **41**, 747.
24. E. Peled and S. Menkin, *J. Electrochem. Soc.*, 2017, **164**, A1703.
25. M. Odziemkowski and D. E. Irish, *J. Electrochem. Soc.*, 1992, **139**, 3063.
26. K. S. Aojula, J. D. Genders, A. D. Holding and D. Pletcher, *Electrochim. Acta*, 1989, **34**, 1535.
27. M. W. Verbrugge and B. J. Koch, *J. Electroanal. Chem.*, 1994, **367**, 123.

28. D. Pletcher, J. F. Rohan and A. G. Ritchie, *Electrochim. Acta*, 1994, **39**, 1369.
29. D. Pletcher, J. F. Rohan and A. G. Ritchie, *Electrochim. Acta*, 1994, **39**, 2015.
30. J. Xu and G. C. Farrington, *Solid State Ion.*, 1994, **74**, 125.
31. F. Shi, A. Pei, A. Vailionis, J. Xie, B. Liu, J. Zhao, Y. Gong and Y. Cui, *Proc. Natl. Acad. Sci.*, 2017, **114**, 12138.
32. F. Shi, A. Pei, D. T. Boyle, J. Xie, X. Yu, X. Zhang and Y. Cui, *Proc. Natl. Acad. Sci.*, 2018, **115**, 8529.
33. K. Xu and A. V. Cresce, *J. Mater. Chem.*, 2011, **21**, 9849.
34. G. V. Zhuang, K. Xu, H. Yang, T. R. Jow and P. N. Ross, *J. Phys. Chem. B*, 2005, **109**, 17567.
35. G. V. Zhuang, H. Yang, B. Blizanac and P. N. Ross, *Electrochem. Solid-State Lett.*, 2005, **8**, A441.
36. F. Shi, P. N. Ross, H. Zhao, G. Liu, G. A. Somorjai and K. Komvopoulos, *J. Am. Chem. Soc.*, 2015, **137**, 3181.
37. Y. Wang, S. Nakamura, M. Ue and P. B. Balbuena, *J. Am. Chem. Soc.*, 2001, **123**, 11708.
38. P. N. Ross, *Catal. Lett.*, 2014, **144**, 1370.
39. Y. Chen and M. W. Kanan, *J. Am. Chem. Soc.*, 2012, **134**, 1986.
40. T. Hatchard, S. D. Beattie, A. Bonakdarpour, K. C. Hewitt and J. R. Dahn, *J. Electrochem. Soc.*, 2003, **150**, A701.
41. I. A. Courtney and J. R. Dahn, *J. Electrochem. Soc.*, 1997, **144**, 2405.
42. R. Qiao, I. T. Lucas, A. Karim, J. Syzdek, X. Liu, W. Chen, K. Persson, R. Kostecki and W. Yang, *Adv. Mater. Interfaces*, 2014, **1**, 1300115.
43. H. Yoshida, T. Fukunaga, T. Hazama, M. Terasaki, M. Mizutani and M. Yamachi, *J. Power Sources*, 1997, **68**, 311.
44. T. Sasaki, T. Abe, Y. Iriyama, M. Inaba and Z. Ogumi, *J. Power Sources*, 2005, **150**, 208.
45. G. Gachota, S. Grugeona, M. Armanda, S. Pilardb, P. Guenotc, J.-M. Tarascona and S. Laruelle, *J. Power Sources*, 2008, **178**, 409.
46. J. Li, A. K. Dozier, Y. Li, F. Yang and Y.-T. Chenga, *J. Electrochem. Soc.*, 2011, **158**, A689.
47. X. H. Liu and J. Y. Huang, *Energy Environ. Sci.*, 2011, **4**, 3844.
48. M. Ebner, F. Marone, M. Stampanoni and V. Wood, *Science*, 2013, **342**, 716.
49. F. Shi, Z. Song, P. N. Ross, G. A. Somorjai, R. O. Ritchie and K. Komvopoulos, *Nat. Commun.*, 2016, **7**, 11886.
50. M. Green, E. Fielder, B. Scrosati, M. Wachtler and J. S. Moreno, *Electrochem. Solid-State Lett.*, 2003, **6**, A75.
51. B. R. Long, M. K. Y. Chan, J. P. Greeley and A. A. Gewirth, *J. Phys. Chem. C*, 2011, **115**, 18916.
52. K. W. Schroder, H. Celio, L. J. Webb and K. J. Stevenson, *J. Phys. Chem. C*, 2012, **116**, 19737.
53. F. Shi, H. Zhao, G. Liu, P. N. Ross, G. A. Somorjai and K. Komvopoulos, *J. Phys. Chem. C*, 2014, **118**, 14732.
54. J. Li and J. R. Dahn, *J. Electrochem. Soc.*, 2007, **154**, A156.
55. T. Sasaki, S.-K. Jeong, T. Abe, Y. Iriyama, M. Inaba and Z. Ogumi, *J. Electrochem. Soc.*, 2005, **152**, A1963.
56. E. Peled, D. Golodnitsky and G. Ardel, *J. Electrochem. Soc.*, 1997, **144**, L208.
57. P. Lu and S. J. Harris, *Electrochem. Commun.*, 2011, **13**, 1035.
58. S. Shi, P. Lu, Z. Liu, Y. Qi, L. G. Hector, H. Li and S. J. Harris, *J. Am. Chem. Soc.*, 2012, **134**, 15476.
59. O. Borodin, G. V. Zhuang, P. N. Ross and K. Xu, *J. Phys. Chem. C*, 2013, **117**, 7433.
60. T. Oi, *Mater. Res. Bull.*, 1984, **19**, 451.
61. W. Li, G. Wu, C. M. Araújo, R. H. Scheicher, A. Blomqvist, R. Ahuja, Z. Xiong, Y. Feng and P. Chen, *Energy Environ. Sci.*, 2010, **3**, 1524.
62. K. Persson, V. A. Sethuraman, L. J. Hardwick, Y. Hinuma, Y. S. Meng, A. van der Ven, V. Srinivasan, R. Kostecki and G. Ceder, *J. Phys. Chem. Lett.*, 2010, **1**, 1176.
63. H. Xia, L. Lu and G. Ceder, *J. Power Sources*, 2006, **159**, 1422.
64. L. Benitez and J. M. Seminario, *J. Electrochem. Soc.*, 2017, **164**, E3159.
65. A. Lodding, J. N. Mundy and A. Ott, *Phys. Status Solidi*, 1970, **38**, 559.
66. P. Yu, B. N. Popov, J. A. Ritter and R. E. White, *J. Electrochem. Soc.*, 1999, **146**, 8.
67. N. Ding, J. Xu, Y. X. Yao, G. Wegner, X. Fang, C. H. Chen and I. Lieberwirth, *Solid State Ion.*, 2009, **180**, 222.
68. C. K. Park, S. B. Park, S. H. Oh, H. Jang and W. I. Cho, *Bull. Korean Chem. Soc.*, 2011, **32**, 836.
69. K. A. Striebel, C. Z. Deng, S. J. Wen and E. J. Cairns, *J. Electrochem. Soc.*, 1996, **143**, 1821.
70. K. Kanamura, H. Takezawa, S. Shiraishi and Z. i. Takehara, *J. Electrochem. Soc.*, 1997, **144**, 1900.
71. D. Aurbach, *J. Electrochem. Soc.*, 1989, **136**, 1606.
72. D. Aurbach, *J. Electrochem. Soc.*, 1989, **136**, 1611.
73. K. Tasaki, A. Goldberg, J.-J. Lian, M. Walker, A. Timmons and S. J. Harris, *J. Electrochem. Soc.*, 2009, **156**, A1019.
74. J. Zheng, H. Zheng, R. Wang, L. Ben, W. Lu, L. Chen, L. Chen and H. Li, *Phys. Chem. Chem. Phys.*, 2014, **16**, 13229.
75. C. Monroe and J. Newman, *J. Electrochem. Soc.*, 2005, **152**, A396.
76. C. Wang, H. Wu, Z. Chen, M. T. McDowell, Y. Cui and Z. Bao, *Nat. Chem.*, 2013, **5**, 1042.
77. F. Shi, H. Zhao, K. Komvopoulos, G. Somorjai, and P. N. Ross, 227th Electrochemical Society Meeting, 2015.
78. P. S. Johnson, P. L. Cook, X. Liu, W. Yang, Y. Bai, N. L. Abbott and F. J. Himpsel, *J. Chem. Phys.*, 2011, **135**, 044702.

79. J. Cazaux, *J. Microsc.*, 2003, **188**, 106.

80. R. Qiao, Y.-D. Chuang, S. Yan and W. Yang, *PLOS ONE*, 2012, **7**, e49182.

81. Y. M. Lee, J. Y. Lee, H.-T. Shim, J. K. Lee and J.-K. Park, *J. Electrochem. Soc.*, 2007, **154**, A515.

82. C. Pereira-Nabais, J. Swiatowska, A. Chagnes, F. Ozanam, A. Gohier, P. Tran-Van, C.-S. Cojocaru, M. Cassir and P. Marcus, *Appl. Surf. Sci.*, 2013, **266**, 5.

83. K. W. Schroder, A. G. Dylla, S. J. Harris, L. J. Webb and K. J. Stevenson, *ACS Appl. Mater. Interfaces*, 2014, **6**, 21510.

84. E. Radvanyi, E. D. Vito, W. Porcher and S. J. S. Larbi, *J. Anal. Atom. Spectrom.*, 2014, **29**, 1120.

85. J. F. Watts and J. Wolstenholme, 'An Introduction to Surface Analysis by XPS and AES', John Wiley & Sons Ltd, West Sussex, 2003.

86. A. Manthiram, X. Yu and S. Wang, *Nat. Rev. Mater.*, 2017, **2**, 16103.

# Application of In Situ Electrochemical-Cell Transmission Electron Microscopy for the Study of Rechargeable Batteries

**Wentao Yao**

*Michigan Technological University, Houghton, MI, USA*

and

**Reza Shahbazian-Yassar**

*University of Illinois at Chicago, Chicago, IL, USA*



## 1 INTRODUCTION

Ever since its invention in the 1930s, transmission electron microscopy (TEM) has found its crucial role as a major analytical technique in the field of materials science, physics, chemistry, and biological sciences. Compared with other analytical tools, such as, scanning electron microscopy (SEM) and atomic force microscopy (AFM), TEM has several advantages (i) ability to identify the crystalline phase, atomic order and defects, and chemical signature of materials. Unlike SEM, which uses scattered electrons for surface imaging, TEM utilizes the electrons that pass through thin samples to construct images. The transmitted electrons contain not only morphology information of the samples but also local information about atomic orders, the identity of atoms, crystal orientations, and internal defects. (ii) Ultra-high spatial and analytical resolutions for imaging and spectroscopy. Recent development in aberration correction techniques has successfully brought the resolution of scanning transmission electron microscopy (STEM) down to 0.5 Å.[1] This ultra-high resolution enables direct imaging of atoms in crystal structures. The advancement of electron energy loss spectroscopy (EELS) allows for chemical analysis with energy resolutions better than 0.1 eV in monochromated STEMs.[2]

Although traditional TEM can provide extensive characterization of spatial and chemical analysis for different samples, it still suffers from the time delay between reactions/processes and the characterization because of the complicated sample preparation/loading steps and the different environment between reactions/processes and TEM vacuum chamber.[3] This limitation calls for the development of in situ TEM techniques. This technique enables real-time tracking of structural or chemical changes inside the samples during reactions/processes, such as physical/mechanical evolutions in nanostructured materials,[4,5] grain boundaries migration,[6–8] and electrochemical reactions in battery electrode materials.[9–11] Standard open-cell design of in situ TEM has its own limitations, especially for the study of electrochemical reactions in a liquid environment. The point-contact open-cell configuration deviated itself from the real liquid environment, where the electrode is fully immersed and reactions take place at the same time along with all the solid–liquid interfaces. The open-cell design also limited the use of liquid electrolytes

*Inorganic Battery Materials*. Edited by Hailiang Wang and Boniface P.T. Fokwa
 ISBN 978-1-119-43199-2

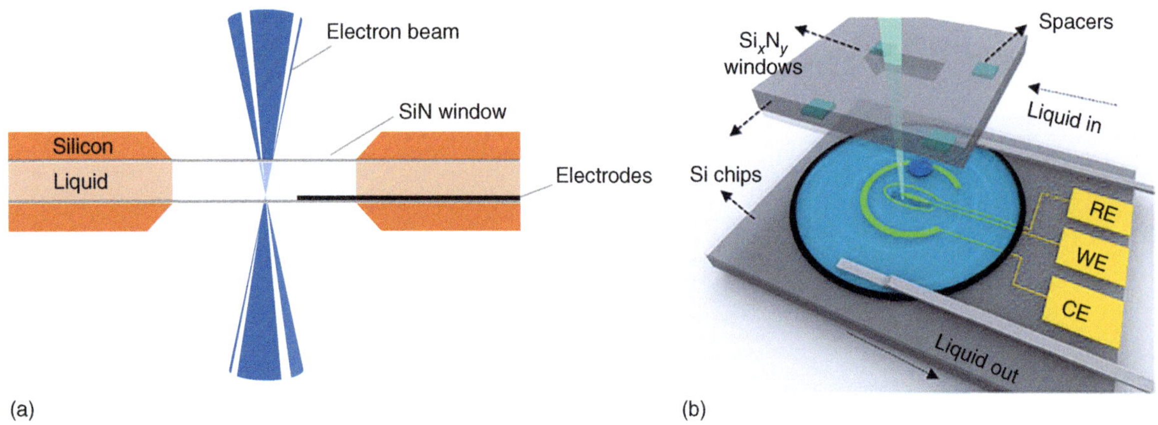


**Figure 1** (a) Schematic image showing the configuration of liquid cell design for (S)TEM imaging. (b) In situ liquid flow-cell design with three electrodes for the study of lithium-air batteries. [(b) Reproduced with permission from K. He *et al.*, Operando liquid cell electron microscopy of discharge and charge kinetics in lithium-oxygen batteries. Nano Energy 49, 338–345 (2018). © Elsevier 2018]

to solid-state metal oxides or ionic liquid because of the vacuum condition in the specimen chamber.

To overcome these limitations, the liquid-cell design was developed (Figure 1a),[13] thanks to the microfabrication techniques and utilizing electron-transparent SiN membranes (20–100 nm). Carbon film and graphene sheets were also used to fabricate the liquid cells for TEM characterizations.[14–16] However, compared with graphene liquid cells, SiN membranes are more stable, easy to fabricate with controllable thicknesses, and capable of introducing flowing liquid environment (Figure 1b).

The in situ liquid-cell (operando) TEM has found wide applications in the study of nanocrystal growth and sintering,[17–20] biomaterial structures,[21,22] electrochemical deposition processes,[23,24] and most recently the study of energy storage devices,[12,25] for example, lithium-ion batteries, metal–air batteries, etc. In this chapter, we will summarize the recent progress specifically in the study of rechargeable batteries using in situ electrochemical-cell TEM technique. Compared with traditional point-contact in situ TEM design, liquid cell design enables not only the monitoring of lithiation/sodiation behaviors of electrode materials, but also the formation of solid electrolyte interphase (SEI), lithium dendrite growth, and the degradation of the battery electrolyte itself under the irritation of electron beams.

## 2 APPLICATION IN LITHIUM-ION BATTERIES

### 2.1 Lithiation Behavior of the Electrode Materials—Liquid-Cell vs. Open-Cell

The lithiation behavior of electrode materials in the liquid-cell and open-cell design depends on the electrode itself. The in situ liquid-cell TEM study for battery materials was first demonstrated by Gu *et al.* in 2013, where the lithiation/delithiation behavior of silicon nanowire was investigated in a real battery electrolyte.[26] In that case, lithium perchlorate ($LiClO_4$) in ethylene carbonate (EC) and dimethyl carbonate (DMC) was used instead of ionic liquid or solid $Li_2O$ used in open-cell in situ TEM design. Lithium metal was used as the counter electrode. The lithiation behavior of the silicon nanowire inside the liquid cell showed similar core–shell mode compared with that in the open-cell design (Figure 2). The core–shell lithiation process was related to fast diffusion of lithium ions in the surface of the silicon nanowires.[26] The lithiation behaviors of crystalline silicon nanoparticles between the liquid-cell design and point-contact design also showed similar lithiation behaviors.[16,27] Anisotropic lithiation of silicon particles along <110> directions was observed inside a graphene liquid cell filled with the standard battery electrolyte (1 M $LiPF_6$ in EC/DMC/DEC (diethyl carbonates), 1:1:1 volumetric ratio).[16] Open-cell lithiation of silicon nanoparticles also showed fast swelling along <110> direction.[27]

However, different lithiation behaviors for $SnO_2$ nanowires were reported by Zhong *et al.* in an open-cell setup versus a flooding geometry setup.[11,28] In the partially flooding geometry of $SnO_2$ nanowires in an ionic liquid electrolyte, multiple-stripe lithiation behavior was observed along the (020) planes with the generation of high-density dislocations and enlarged interplanar spacing.[28] This mechanism is different from the open-cell observations reported by the same group.[11] In the open-cell design, the ionic liquid electrolyte was in point-contact with the $SnO_2$ nanowire. A reaction front containing high-density dislocations was observed instead of multiple-stripe along certain crystal planes. Although different lithiation behaviors were observed in the early lithiation stage, at the end of lithiation, the same crystalline-to-amorphous phase

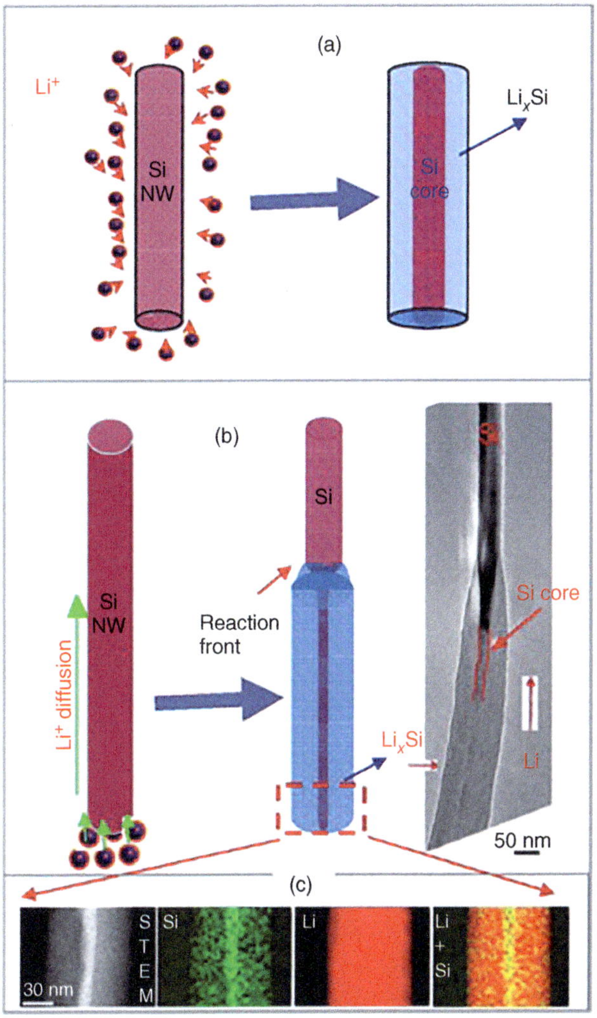


**Figure 2** Comparison between the lithiation behavior of Si nanowires in (a) the liquid cell design and (b) the open-cell design. (c) STEM, Si, Li, and overlaid Si and Li EELS maps showing the distribution of Si and lithium during the lithiation. [Reprinted with permission from Ref. 26. © 2013 American Chemical Society]

transition was observed. The differences between the lithiation behaviors might be related to the preferred lithiation direction along the (020) planes. Under the point-contact mode, this preferred lithium-ion diffusion was limited because of the limited contact areas between the lithium source and the electrode surfaces. Although the flooding geometry used in this case is not entirely liquid-cell design and only the ionic liquid electrolyte were used, the different findings between two designs proved that a conformal contact between the electrode and the battery electrolyte can reveal new details for the lithiation behaviors. These findings can be adopted for the fabrication of electrode materials with certain orientations to further improve their electrochemical performance. Wang *et al.* also studied the in situ lithiation behavior of the $SnO_2$ nanowire soaked in an ionic liquid electrolyte and found severe deformation during the initial phase transition which is similar to that observed in the point-contact lithiation mode.[29]

The different observations of the lithiation behaviors in open-cell and liquid-cell designs may also relate to the different resolutions of the TEM under these two situations. The spatial resolution for liquid-cell TEM is generally lower than open-cell TEM due to the scattering of the electron beam by the silicon nitride window and the thin liquid layer.[30,31] Atomic-scale resolution can be easily reached in the open-cell design. For example, Wang *et al.* reported a trigonal prismatic (2H)-octahedral (1T) phase transition during the lithiation of $MoS_2$ nanosheets using in situ high-resolution TEM.[32] A liquid-cell study of the lithiation behavior of $MoS_2$ nanosheets, however, showed different results.[33] Apart from structural expansion and deformation, the lithiation of $MoS_2$ in a $LiPF_6$/EC/DEC commercial electrolyte decomposed the $MoS_2$ nanosheets into nanoparticles. The 2H to 1T phase transition was hard to demonstrate due to the resolution of the liquid-cell TEM (Figure 3b).

The morphology change of other electrode materials during the lithium alloying was also investigated using in situ liquid-cell TEM, such as Sn,[34] Au,[35] etc. Using an enclosed TEM liquid cell, Leenheer *et al.* compared the lithium alloying behaviors with amorphous Si, crystalline Al, and crystalline Au thin films; clear phase boundaries were found to form and propagate during the lithiation of the crystalline thin film electrodes, which further lead to highly concentrated stresses within the electrode.[36]

Further studies of different electrode materials inside liquid-cell configurations will definitely reveal lithiation behaviors that are close to the real conditions inside batteries, especially for crystalline materials with preferred lithiation directions. On the other hand, computational efforts might be beneficial for the predictions of these lithiation behaviors in aid of the in situ liquid-cell TEM studies.

### 2.2 Lithium Dendrite Growth

Lithium metal anode showed the highest theoretical capacity (3860 mAh $g^{-1}$) and lowest electrochemical potential, which is highly important for the development of high-energy-density Li-ion, Li–S, and Li–air batteries.[37] The biggest challenge for lithium metal anode is known to be its dendritic growth behavior, which can lead to thermal runaway and cell short-circuit. Understanding the lithium dendrite growth behavior will stimulate new designs that can reach ultimately dendrite-free lithium deposition upon battery cycling.

In situ liquid-cell TEM offers the chance to directly visualize real-time lithium dendrite growth.[38–41] Immersing Au electrodes in a commercial $LiPF_6$/EC/DEC electrolyte, Zeng *et al.* observed lithium dendrite growth and

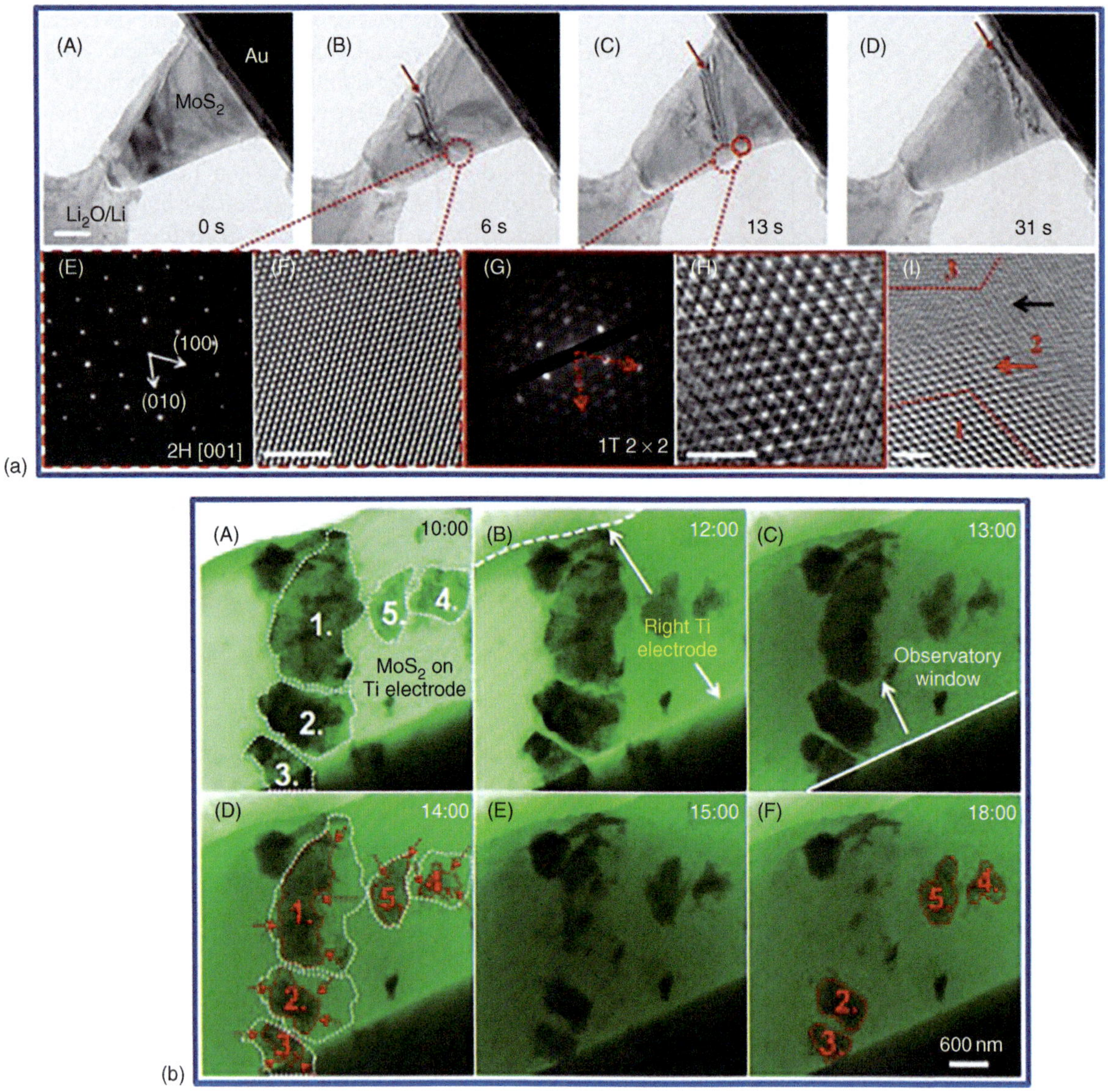


**Figure 3** (a) In situ TEM observation of the lithiation behavior of $MoS_2$ nanosheets, showing a 2H to a 1T phase transition. [Reprinted with permission from Ref. 32. © 2014 American Chemical Society]; (b) In situ liquid-cell TEM study of the lithiation behavior of $MoS_2$ nanosheets in a commercial $LiPF_6$/EC/DEC electrolyte, showing the dissolution of the electrode materials. [Reprinted with permission from Ref. 33. © 2015 American Chemical Society]

stripping in an electrochemical liquid TEM cell upon electrochemical cycling.[38] The lithium dendrite was observed to grow from the base point of the passivation layer (Li–Au alloy) upon plating and dissolved from the tip and kink points upon stripping.[38] Using Au working electrodes and $LiPF_6$ EC/DMC electrolyte, Sacci *et al.* demonstrated the dendritic SEI was formed before lithium plating in a liquid TEM flow cell, suggesting the composition of the electrolyte can affect lithium deposition.[39]

One challenge for imaging lithium dendrite is that lithium is a light element. Pure lithium metal is even lighter than the electrolyte, which makes it hard to image using traditional TEM. Taking the advantage of STEM high-angle annular dark field (HAADF) image and an image contrast reversal, Mehdi *et al.* successfully distinguished the lithium metal dendrite from the surrounding electrolyte, SEI layer, and electrode surfaces (Figure 4); the mass of lithium dendrite and the evolution of the SEI layer were quantified directly from the image.[41] Using bright field (BF) STEM images, Leenheer *et al.* studied lithium deposition/dissolution on a Ti electrode in a $LiPF_6$ EC/DMC electrolyte under various beam current densities.[40] Higher current density was found to facilitate the formation of more needle-like lithium dendrite over a higher surface area. Moreover, the TEM electron beam can induce the surface film and change the lithium structure.[40] Sacci

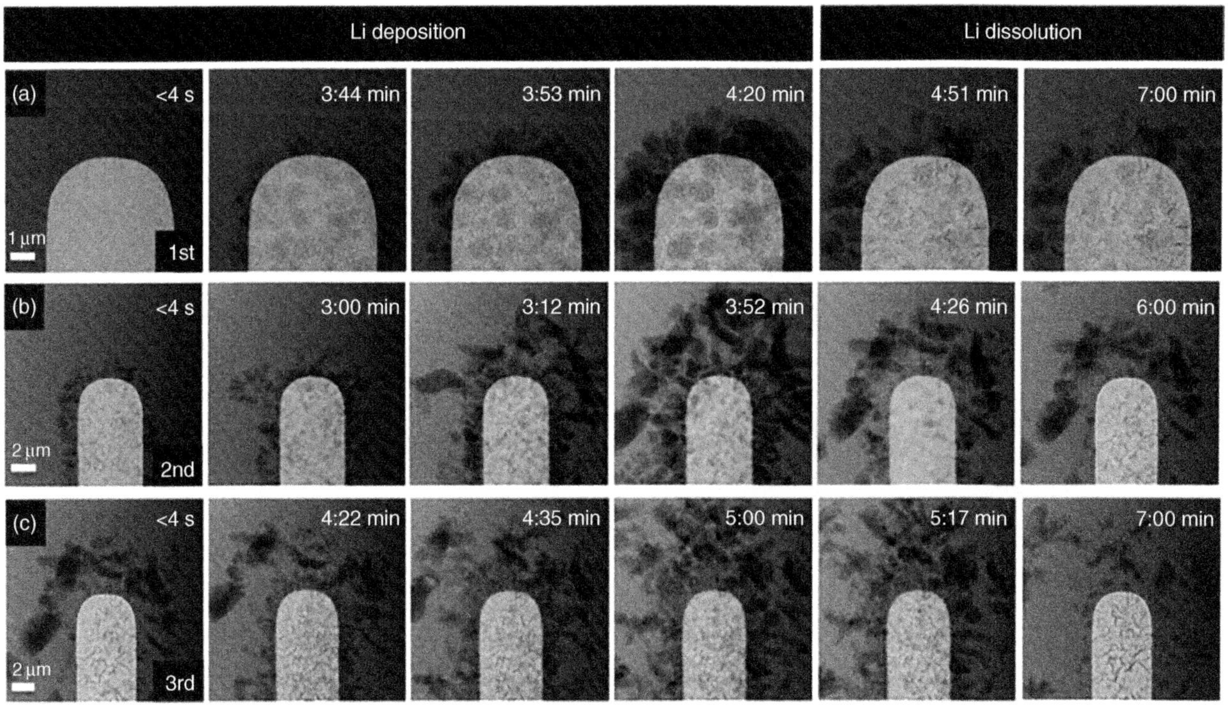


**Figure 4** High-angle annular dark field (HAADF) images showing the Li deposition and dissolution at the interface between the Pt working electrode and the $LiPF_6$/PC electrolyte during the (a) first, (b) second, and (c) third charge/discharge cycles of the operando cell. [Reprinted with permission from Ref. 41. © 2015 American Chemical Society]

*et al.* further coupling quantitative electrochemical measurements with STEM imaging (BF/ADF) to image and analyze the structural and chemical changes of the SEI and lithium deposition; the lithium metal was found to nucleate and grow on the surface and edge of the glassy carbon electrode based on the ADF STEM image intensity.[42]

Recently, other in situ liquid-cell techniques were reported to investigate the lithium dendrite growth and dissolution.[43,44] Cheng *et al.* reported the use of operando transmission X-ray microscopy (XTM) to study lithium dendrite growth; dendritic and mossy lithium was found to grow under lower and higher current densities, respectively; and dendritic lithium contributed more than the mossy one to form "dead lithium".[43] Rong *et al.* reported the use of liquid-phase electrochemical scanning electron microscopy (EC-SEM) to study lithium stripping/plating in a liquid electrolyte (LiTFSI in DOL/DME) and investigated the effect of electrolyte additives in regulating the growth of lithium dendrites.[44]

### 2.3 Electrode–Electrolyte Interphase

Another important application of in situ liquid-cell TEM is the study of lithium-ion distribution, SEI formation, and evolution along the electrode–electrolyte interphase.[38,39,42,45–48] Combing the electrochemical liquid-cell with valence energy-filtered transmission electron microscopy (EFTEM), Holtz *et al.* reported a nanoscale imaging of lithium-ion distribution along the $LiFePO_4$ cathode particles and the surrounding electrolyte during charging and discharging.[48] As shown in Figure 5, during charging, the particles show more bright regions, which correspond to delithiated $FePO_4$; during discharging, the bright regions turned back to dark $LiFePO_4$.[48]

The formation and evolution of the SEI layer have been investigated by several groups using in situ liquid-cell TEM. Noh and Dillon reported the formation and partial dissolution of the SEI layer during the lithiation and delithiation of a Sn electrode in a commercial liquid electrolyte ($LiPF_6$ in EC/DMC), which would contribute to capacity fading during the battery cycling.[34] Using Au electrodes and a self-designed electrochemical liquid cell, Zeng *et al.* reported real-time formation and evolution of the SEI layer on the Au electrode surrounded by a battery electrolyte 1 M $LiPF_6$ in EC–DEC (Figure 6).[38] Combing electrochemical measurement and STEM imaging, Sacci *et al.* realized a quantitative analysis of the SEI layer formed on a Pt electrode in a standard organic electrolyte ($LiPF_6$ in EC/DMC); the SEI layer was estimated to be twice as dense as the electrolyte based on the STEM images and electron scattering theory.[42] Cheong *et al.* used the graphene liquid cell to study the SEI formation on $SnO_2$ nanotubes in a $LiPF_6$ in EC/DEC electrolyte.[49] The

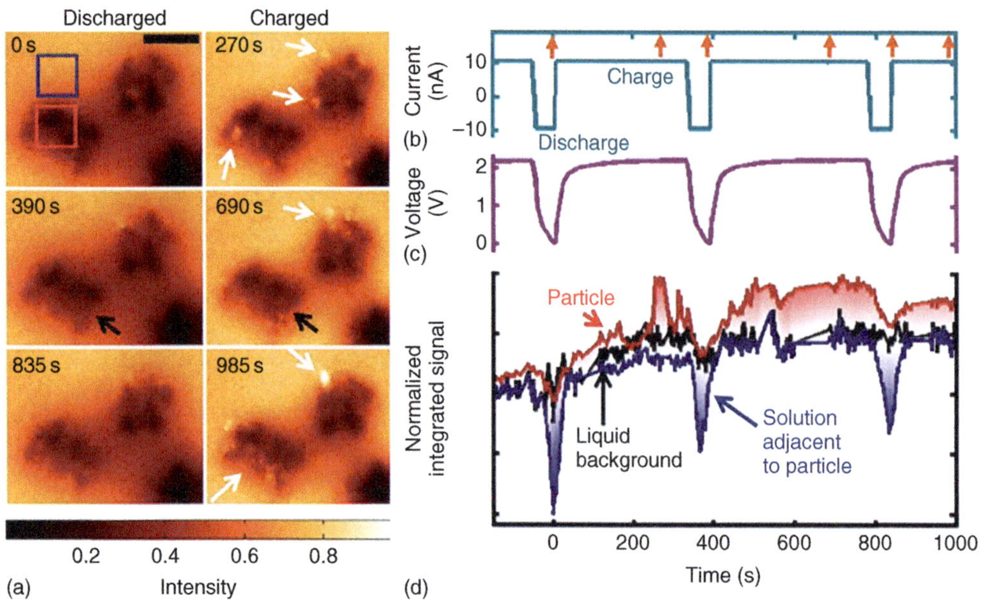


**Figure 5** (a) Spectroscopic EFTEM images showing the charging and discharging $LiFePO_4$ particles in a $Li_2SO_4$ electrolyte. The corresponding current (b), voltage (c), and integrated signal intensities (d) upon cycling. [Reprinted with permission from Ref. 48. © 2014 American Chemical Society]

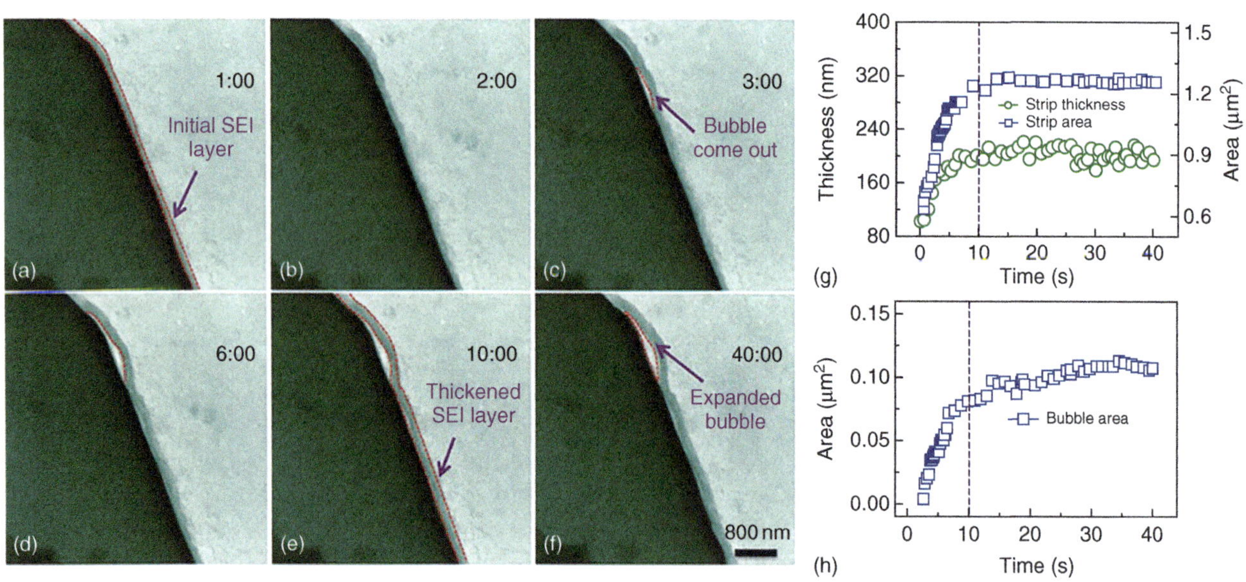


**Figure 6** (a–f) Formation and evolution of the SEI film on the Au electrode in a commercial $LiPF_6$/EC/DEC electrolyte, and the corresponding statistical analysis of the SEI thickness (g) and area (h). [Reproduced with permission from Ref. 38. © ACS 2014 Publications]

lithiation and electrolyte decomposition were induced by the TEM electron beam; SEI layers were formed based on the agglomerates of the decomposed electrolyte.[49]

Recently, in situ liquid secondary ion mass spectroscopy (SIMS) was also reported to study the SEI formation in a $LiPF_6$/PC electrolyte combing in situ liquid cell TEM; the deposition of lithium was found to induce the condensation of electrolyte molecules around the electrode, which further affect the initial formation of SEI layer.[47]

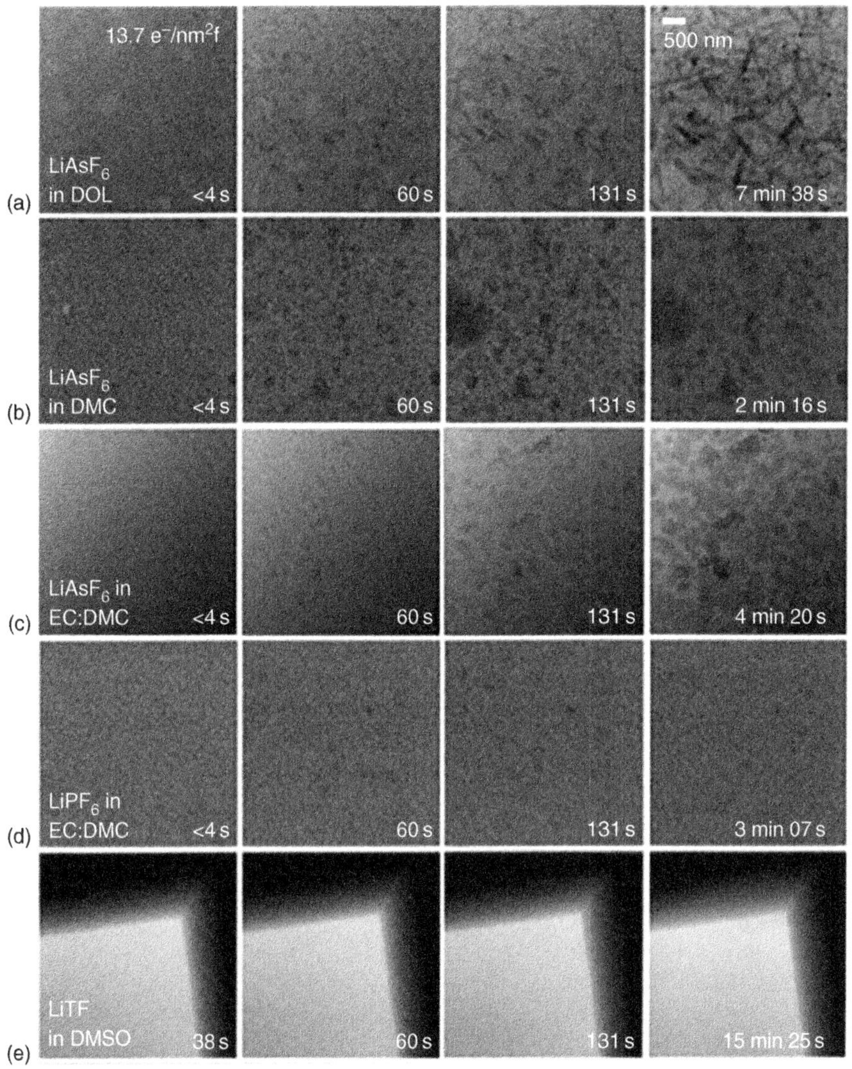


**Figure 7** Bright field STEM images showing the stability of various electrolyte solution under the same dose values per image with different exposure times. [Reprinted with permission from Ref. 50. © 2014 American Chemical Society]

## 2.4 Electrolyte Degradation

Abellan *et al.* reported the study of degradation mechanism for a range of inorganic/salt electrolyte using in situ liquid-cell TEM; the electrolyte breakdown was induced by the TEM electron beam.[50] The in situ stability results (Figure 7) are consistent with previous electrochemical stability tests: The LiTf in DMSO showed the highest stability; the $LiAsF_6$ in DOL or DMC showed fast degradation upon electron beam irradiation with the formation of LiF particles.[50]

# 3 APPLICATION IN METAL–AIR BATTERIES

In situ (operando) liquid-cell TEM recently found application in studying the formation and decomposition of discharge products in the cathode of metal–air batteries.[12,51–53] He *et al.* adopted operando liquid-cell to study the nucleation and decomposition of $Li_2O_2$ particles in a Li–air battery (Figure 8); the nucleation of $Li_2O_2$ was observed both in the solution and the electrode surfaces (Figure 8a); the decomposition of $Li_2O_2$ in the solution followed a side-to-side mode while the dissolution of $Li_2O_2$ on the electrode surface started at the electrode/$Li_2O_2$ interface (Figure 8b).[12] Similar work was reported by Liu *et al.*, where the relationship between discharge/charge overpotentials and the nucleation, growth, and decomposition of $Li_2O_2$ under a constant current density was thoroughly investigated.[52] Hou *et al.* studied the catalytic effect of $RuO_2$ on the formation and decomposition of $Li_2O_2$ using the liquid TEM cell; $RuO_2$ was found to act as the preferred sites for the formation of $Li_2O_2$; the decomposition of $Li_2O_2$ preferred to

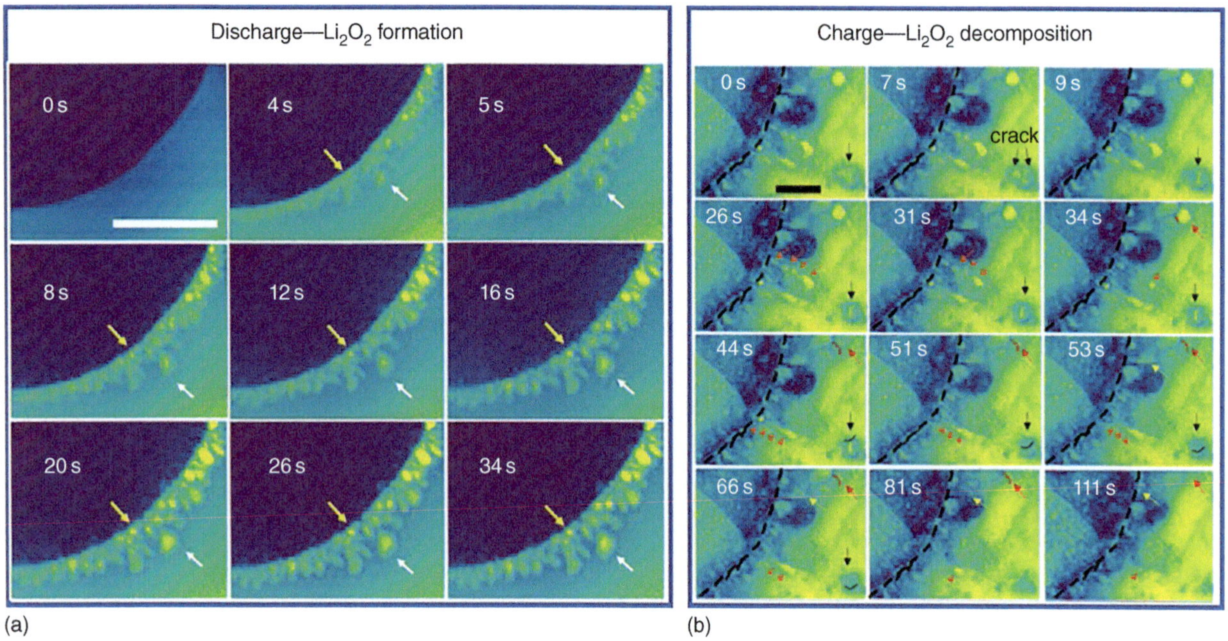


**Figure 8** The $Li_2O_2$ (a) formation and (b) decomposition observed on a glassy carbon electrode in 1 M $LiCF_3SO_2$ in $O_2$-rich TEGDME electrolyte. [Reprinted with permission from Ref. 12. © 2018 American Chemical Society]

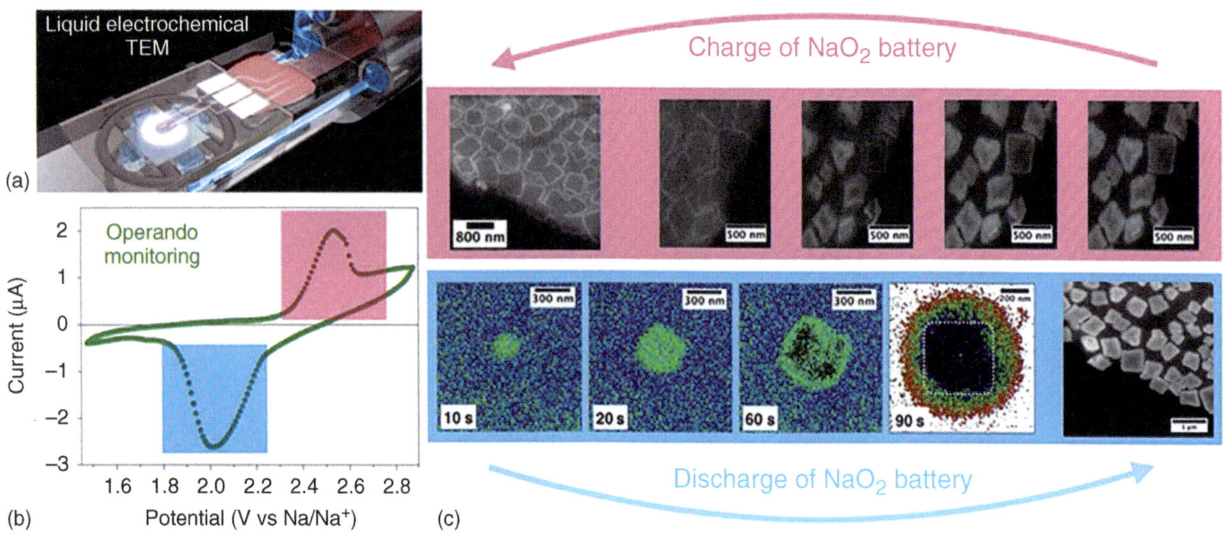


**Figure 9** (a) Schematic image showing the in situ liquid electrochemical TEM holder. (b) The electrochemical cyclic voltammetry curve of the Na–$O_2$ battery. (c) The formation and decomposition of $NaO_2$ during the discharge and charge of the Na–$O_2$ battery. [Reprinted with permission from Ref. 53. © 2018 American Chemical Society]

take place at the electrolyte-$RuO_2$-$Li_2O_2$ triple-phase interfaces.[51]

Recently, Lutz *et al.* reported the use of in situ liquid-cell TEM to investigate the charge and discharge processes in a Na–air battery.[53] The Na–air liquid-cell battery was fabricated with Pt as the reference and counter electrodes, and glassy carbon as the working electrode; the electrolyte was 0.5 M oxygen-saturated $NaPF_6$ dissolved into dimethoxyethane (DME) (Figure 9a). During discharge, the $NaO_2$ was formed by a solution-mediated nucleation process; moreover, a parasitic shell was formed on the $NaO_2$ cubes, which was related to the reaction between the $NaO_2$ and the battery electrolyte and this parasitic shell was believed to affect the cyclability of the Na–$O_2$ batteries.

During charging, the $NaO_2$ particles were oxidized based on a solution mechanism.[53] The observed mechanism here will provide important guidance for the selection of battery electrolyte to further improve the performance of $Na-O_2$ batteries.

## 4 CONCLUSIONS AND OUTLOOK

The application of in situ liquid-cell TEM in the study of rechargeable batteries is reviewed based on the research progresses in the field of lithium-ion batteries and metal–air batteries. Specifically for lithium-ion batteries, in situ liquid-cell TEM showed prominent advantages in the real-time imaging of lithium dendrite growth, SEI formation, and evolution, as well as the electrolyte degradation. As for metal–air batteries, in situ liquid-cell TEM directly visualized the formation and decomposition processes of $Li_2O_2$ and $NaO_2$.

Despite these advantages, further improvements are still needed for the liquid-cell configuration, such as improving the spatial resolution of the liquid cell by reducing the liquid thickness or using high-atomic-number electrodes.[54] Another disadvantage for the liquid-cell TEM is its limitation in atomic-scale structural and chemical analysis. For EELS analysis using the in situ liquid-cell, the inelastic core-loss signal tends to be covered by the multiple scattering through the liquid and SiN windows.[48] In addition, energy dispersive spectroscopy (EDS) analysis in liquid-cell TEM may face some challenges because the in situ TEM holder and silicon chips can block the X-rays signals generated from the cell.[55] New holder designs have shown successful EDS signal collection from samples immersed in microfluidic devices.[56,57] Recently, Unocic *et al.* reported the use of Dual EELS to acquire low-loss and core-loss EELS spectra at the same time and use zero-loss peak in the low-loss spectra to correct the plural scattering effect in core-loss spectra, which could be adopted for future EELS analysis in liquid TEM cells.[55] Other techniques, such as the reported EFTEM,[48] could also be used for future chemical analysis in liquid TEM cells.

The application of in situ liquid-cell TEM for the study of metal–air batteries will continue to expand, especially at the cathode side, for example, the role of different cathode catalysts in the formation and decomposition of the discharge products; the interactions among discharge products, battery electrolyte, and cathode catalysts.

## 5 ACKNOWLEDGMENTS

This work was financially supported by NSF-DMR Award No. 1620901. W.T.Y. acknowledges the support from Multi-Scale Technologies Institute at Michigan Technological University.

## 6 RELATED ARTICLE

Solid State and Materials Chemistry for Sodium-Ion Batteries

## 7 ABBREVIATIONS AND ACRONYMS

AFM = atomic force microscopy; BF = bright field; DEC = diethyl carbonate; DMC = dimethyl carbonate; DME = dimethoxyethane; EC = ethylene carbonate; EC-SEM = electrochemical scanning electron microscopy; EELS = electron energy loss spectroscopy; EFTEM = energy-filtered transmission electron microscopy; HAADF = high-angle annular dark field; SEI = solid electrolyte interphase; SEM = scanning electron microscopy; SIMS = secondary ion mass spectroscopy; STEM = scanning transmission electron microscopy; TEM = transmission electron microscopy; XTM = X-ray microscopy

## 8 REFERENCES

1. C. Kisielowski, B. Freitag, M. Bischoff, H. van Lin, S. Lazar, G. Knippels, P. Tiemeijer, M. van der Stam, S. von Harrach, M. Stekelenburg, M. Haider, S. Uhlemann, H. Müller, P. Hartel, B. Kabius, D. Miller, I. Petrov, E. A. Olson, T. Donchev, E. A. Kenik, A. R. Lupini, J. Bentley, S. J. Pennycook, I. M. Anderson, A. M. Minor, A. K. Schmid, T. Duden, V. Radmilovic, Q. M. Ramasse, M. Watanabe, R. Erni, E. A. Stach, P. Denes and U. Dahmen, *Microsc. Microanal.*, 2008, **14**, 469.
2. H. H. Rose, *Sci. Technol. Adv. Mater.*, 2008, **9**, 014107.
3. Y. Wu and N. Liu, *Chem. Rev.*, 2018, **4**, 438.
4. Z. Wang, P. Poncharal and W. De Heer, *J. Phys. Chem. Solids*, 2000, **61**, 1025.
5. D. Qian, E. C. Dickey, R. Andrews and T. Rantell, *Appl. Phys. Lett.*, 2000, **76**, 2868.
6. F. Mompiou, D. Caillard and M. Legros, *Acta Mater.*, 2009, **57**, 2198.
7. M. Legros, D. S. Gianola and K. J. Hemker, *Acta Mater.*, 2008, **56**, 3380.
8. Y. Zhong, F. Xiao, J. Zhang, Y. Shan, W. Wang and K. Yang, *Acta Mater.*, 2006, **54**, 435.
9. X. H. Liu, Y. Liu, A. Kushima, S. Zhang, T. Zhu, J. Li and J. Y. Huang, *Adv. Energy Mater.*, 2012, **2**, 722.
10. X. H. Liu and J. Y. Huang, *Energy Environ. Sci.*, 2011, **4**, 3844.
11. J. Y. Huang, L. Zhong, C. M. Wang, J. P. Sullivan, W. Xu, L. Q. Zhang, S. X. Mao, N. S. Hudak, X. H. Liu and A. Subramanian, *Science*, 2010, **330**, 1515.

12. K. He, X. Bi, Y. Yuan, T. Foroozan, B. Song, K. Amine, J. Lu and R. Shahbazian-Yassar, *Nano Energy*, 2018, **49**, 338.
13. M. Williamson, R. Tromp, P. Vereecken, R. Hull and F. Ross, *Nat. Mater.*, 2003, **2**, 532.
14. K. Nishijima, J. Yamasaki, H. Orihara and N. Tanaka, *Nanotechnology*, 2004, **15**, S329.
15. N. Mohanty, M. Fahrenholtz, A. Nagaraja, D. Boyle and V. Berry, *Nano Lett.*, 2011, **11**, 1270.
16. J. M. Yuk, H. K. Seo, J. W. Choi and J. Y. Lee, *ACS Nano*, 2014, **8**, 7478.
17. E. A. Sutter and P. W. Sutter, *J. Am. Chem. Soc.*, 2014, **136**, 16865.
18. N. áde Jonge, *Chem. Commun.*, 2015, **51**, 16393.
19. J. Yang, C. M. Andrei, G. A. Botton and L. Soleymani, *J. Phys. Chem. C*, 2017, **121**, 7435.
20. B. Song, K. He, Y. Yuan, S. Sharifi-Asl, M. Cheng, J. Lu, W. A. Saidi and R. Shahbazian Yassar, *Nanoscale*, 2018, **10**(33), 15809.
21. D. B. Peckys and N. de Jonge, *Nano Lett.*, 2011, **11**, 1733.
22. D. B. Peckys, P. Mazur, K. L. Gould and N. de Jonge, *Biophys. J.*, 2011, **100**, 2522.
23. M. Sun, H.-G. Liao, K. Niu and H. Zheng, *Sci. Rep.*, 2013, **3**, 3227.
24. A. J. Leenheer, J. P. Sullivan, M. J. Shaw and C. T. Harris, *J. Microelectromech. Syst.*, 2015, **24**, 1061.
25. N. D. Browning, B. L. Mehdi, A. Stevens, W. Xu, W. A. Henderson, J.-G. Zhang, K. T. Mueller, H. Mehta, L. Kovarik and A. Liyu, *Microsc. Microanal.*, 2017, **23**, 1970.
26. M. Gu, L. R. Parent, B. L. Mehdi, R. R. Unocic, M. T. McDowell, R. L. Sacci, W. Xu, J. G. Connell, P. Xu and P. Abellan, *Nano Lett.*, 2013, **13**, 6106.
27. X. H. Liu, L. Zhong, S. Huang, S. X. Mao, T. Zhu and J. Y. Huang, *ACS Nano*, 2012, **6**, 1522.
28. L. Zhong, X. H. Liu, G. F. Wang, S. X. Mao and J. Y. Huang, *Phys. Rev. Lett.*, 2011, **106**, 248302.
29. C.-M. Wang, W. Xu, J. Liu, J.-G. Zhang, L. V. Saraf, B. W. Arey, D. Choi, Z.-G. Yang, J. Xiao, S. Thevuthasan and D. R. Baer, *Nano Lett.*, 2011, **11**, 1874.
30. F. M. Ross, *Science*, 2015, **350**.
31. N. De Jonge and F. M. Ross, *Nat. Nanotechnol.*, 2011, **6**, 695.
32. L. Wang, Z. Xu, W. Wang and X. Bai, *J. Am. Chem. Soc.*, 2014, **136**, 6693.
33. Z. Zeng, X. Zhang, K. Bustillo, K. Niu, C. Gammer, J. Xu and H. Zheng, *Nano Lett.*, 2015, **15**, 5214.
34. K. W. Noh and S. J. Dillon, *Scr. Mater.*, 2013, **69**, 658.
35. Z. Zeng, W.-I. Liang, Y.-H. Chu and H. Zheng, *Faraday Discuss.*, 2015, **176**, 95.
36. A. J. Leenheer, K. L. Jungjohann, K. R. Zavadil and C. T. Harris, *ACS Nano*, 2016, **10**, 5670.
37. D. Lin, Y. Liu and Y. Cui, *Nat. Nanotechnol.*, 2017, **12**, 194.
38. Z. Zeng, W.-I. Liang, H.-G. Liao, H. L. Xin, Y.-H. Chu and H. Zheng, *Nano Lett.*, 2014, **14**, 1745.
39. R. L. Sacci, N. J. Dudney, K. L. More, L. R. Parent, I. Arslan, N. D. Browning and R. R. Unocic, *Chem. Commun.*, 2014, **50**, 2104.
40. A. J. Leenheer, K. L. Jungjohann, K. R. Zavadil, J. P. Sullivan and C. T. Harris, *ACS Nano*, 2015, **9**, 4379.
41. B. L. Mehdi, J. Qian, E. Nasybulin, C. Park, D. A. Welch, R. Faller, H. Mehta, W. A. Henderson, W. Xu, C. M. Wang, J. E. Evans, J. Liu, J. G. Zhang, K. T. Mueller and N. D. Browning, *Nano Lett.*, 2015, **15**, 2168.
42. R. L. Sacci, J. M. Black, N. Balke, N. J. Dudney, K. L. More and R. R. Unocic, *Nano Lett.*, 2015, **15**, 2011.
43. J.-H. Cheng, A. A. Assegie, C.-J. Huang, M.-H. Lin, A. M. Tripathi, C.-C. Wang, M.-T. Tang, Y.-F. Song, W.-N. Su and B. J. Hwang, *J. Phys. Chem. C*, 2017, **121**, 7761.
44. G. Rong, X. Zhang, W. Zhao, Y. Qiu, M. Liu, F. Ye, Y. Xu, J. Chen, Y. Hou and W. Li, *Adv. Mater.*, 2017, **29**, 1606187.
45. R. R. Unocic, X.-G. Sun, R. L. Sacci, L. A. Adamczyk, D. H. Alsem, S. Dai, N. J. Dudney and K. L. More, *Microsc. Microanal.*, 2014, **20**, 1029.
46. R. L. Sacci, J. M. Black, N. Balke, N. J. Dudney, K. L. More and R. R. Unocic, *Microsc. Microanal.*, 2015, **21**, 2437.
47. Z. Zhu, Y. Zhou, P. Yan, R. S. Vemuri, W. Xu, R. Zhao, X. Wang, S. Thevuthasan, D. R. Baer and C.-M. Wang, *Nano Lett.*, 2015, **15**, 6170.
48. M. E. Holtz, Y. Yu, D. Gunceler, J. Gao, R. Sundararaman, K. A. Schwarz, T. A. Arias, H. D. Abruña and D. A. Muller, *Nano Lett.*, 2014, **14**, 1453.
49. J. Y. Cheong, J. H. Chang, H. K. Seo, J. M. Yuk, J. W. Shin, J. Y. Lee and I.-D. Kim, *Nano Energy*, 2016, **25**, 154.
50. P. Abellan, B. L. Mehdi, L. R. Parent, M. Gu, C. Park, W. Xu, Y. Zhang, I. Arslan, J.-G. Zhang, C.-M. Wang, J. E. Evans and N. D. Browning, *Nano Lett.*, 2014, **14**, 1293.
51. C. Hou, J. Han, P. Liu, C. Yang, G. Huang, T. Fujita, A. Hirata and M. Chen, *Nano Energy*, 2018, **47**, 427.
52. P. Liu, J. Han, X. Guo, Y. Ito, C. Yang, S. Ning, T. Fujita, A. Hirata and M. Chen, *Sci. Rep.*, 2018, **8**, 3134.
53. L. Lutz, W. Dachraoui, A. Demortière, L. R. Johnson, P. G. Bruce, A. Grimaud and J.-M. Tarascon, *Nano Lett.*, 2018, **18**, 1280.
54. D. Qian, C. Ma, K. L. More, Y. S. Meng and M. Chi, *NPG Asia Mater.*, 2015, **7**, e193.
55. R. R. Unocic, L. Baggetto, G. M. Veith, J. A. Aguiar, K. A. Unocic, R. L. Sacci, N. J. Dudney and K. L. More, *Chem. Commun.*, 2015, **51**, 16377.
56. B. R. Van Devener, A. H. Vollmer and N. N. Youssef, *Microsc. Microanal.*, 2018, **24**, 1204.
57. M. Ge, M. Lu, Y. Chu and H. Xin, *Sci. Rep.*, 2017, **7**, 16420.

# Index

Printed and bound by CPI Group (UK) Ltd, Croydon, CR0 4YY